Network Traffic Engineering

Network Traffic Engineering

Stochastic Models and Applications

Andrea Baiocchi
University of Roma "La Sapienza"
Via Eudossiana 18
Rome, Italy

Registered Offices
John Wiley & Sons, Inc., 111 River Street, Hoboken, NJ 07030, USA
John Wiley & Sons Ltd., The Atrium, Southern Gate, Chichester, West Sussex, PO19 8SQ, UK

Editorial Office
111 River Street, Hoboken, NJ 07030, USA

For details of our global editorial offices, customer services, and more information about Wiley products visit us at www.wiley.com.

Wiley also publishes its books in a variety of electronic formats and by print-on-demand. Some content that appears in standard print versions of this book may not be available in other formats.

Library of Congress Cataloging-in-Publication Data

Names: Baiocchi, Andrea, 1962- author.
Title: Network traffic engineering : stochastic models and applications /
   Andrea Baiocchi, University of Roma, Rome, IT.
Description: First edition. | Hoboken, NJ : John Wiley & Sons, Inc., 2020.
   | Includes bibliographical references and index.
Identifiers: LCCN 2020008307 (print) | LCCN 2020008308 (ebook) | ISBN
   9781119632436 (cloth) | ISBN 9781119632504 (adobe pdf) | ISBN
   9781119632511 (epub)
Subjects: LCSH: Computer networks–Mathematical models. | Queuing theory.
Classification: LCC TK5105.5 .B326 2020 (print) | LCC TK5105.5 (ebook) |
   DDC 004.601/51982–dc23
LC record available at https://lccn.loc.gov/2020008307
LC ebook record available at https://lccn.loc.gov/2020008308

Cover Design: Wiley
Cover Image: © ktsdesign/Shutterstock

Set in 9.5/12.5pt STIXTwoText by SPi Global, Chennai, India

10 9 8 7 6 5 4 3 2 1

MIX
Paper from
responsible sources
FSC® C013604

*To Laura, Alessandro,
Claudia, and Cloppa*

Contents

Preface

This book is an outgrowth of lecture notes, conceived originally as a support for a course of Network Traffic Engineering for graduate students at University of Roma "La Sapienza." The scope has broadened as the text took shape.

Communication networks, pervasive systems for smart environments, cloud computing, big data management and analysis, Intelligent Transportation Systems, smart energy grid, industrial automation, logistics, and inventory management are but few examples that highlight the fast-paced evolution of networked systems. The complexity of those systems and of their inter-relationships, and the need to avoid wasting limited resources, call for adequate mathematical modeling and quantitative assessment.

In this fast-moving and rich context, I think that a gap is developing between the highly skilled technological knowledge of many students, researchers, and practitioners and their ability to abstract from the details of a specific system and capture its essential features, defining a manageable quantitative model and assessing its performance trade-offs. Sophisticated models are often presented without a deep awareness of their relevance to the real-life problem that should motivate them in the first place. Conversely, lack of careful quantitative modeling and evaluation often drives toward design choices that turn out to be suboptimal, sometimes even unsatisfactory, when implemented.

Many years of research work with colleagues, master's and doctorate students, and of lecturing both at undergraduate and graduate levels, have convinced me that successful performance evaluation has two main sides: (i) mastering a wide array of well-established modeling tools; and (ii) understanding the relevance of a given model in capturing real system characteristics and in gaining insight into its trade-offs. Step (i) is necessary, but not sufficient. Step (ii) is the target, but it is unfeasible if step (i) is not accomplished accurately. Confining our work to mathematical aspects—a must (and possibly a pleasure) for researchers advancing the state-of-the-art of queueing and probability theory—is not an appealing target for those in need of gaining a sound understanding of models and how they apply to real-life problems, in nontrivial, often innovative ways. On the other hand, complex applications and the need to extend the range of applicability to more sophisticated systems, make often not fully satisfactory the material provided by manuals and books focusing on technology, rather than methodology. The desire to find a good balance of these two requirements is one key motivation of this book.

Crafting models is more an art than a science. Besides systematic learning of mathematical tools, examples and successful models applied to significant fields of networked systems can provide a guide to build this ability. The book aims at covering the first step in that direction, offering the reader a substantial knowledge of fairly advanced queueing and traffic theory, and showing how to apply them to networked systems evaluation and design.

The topics covered in the book are selected to give a consistent and reasonably self-contained coverage of modeling tools and applications of network traffic engineering. Specific topics are chosen for their relevance, so as to provide a broad view of most useful models. Examples are drawn from real technical problems or engineering applications of interest to current and foreseeable future systems (e.g., LTE, Wi-Fi, ad-hoc networks, automated vehicles, reliability). Examples are not merely numerical application of models. They show how insight on real problems can be gained by means of quantitative modeling. As application examples in Part III of the book I have selected three major topics: multiple access, congestion control, and quality of service. These topics have been selected for their relevance in networked service systems, for their applicability range and also for the beauty of the results of relevant theories. I feel that some simple issues also deserve to be dealt with along with more theoretical topics to provide a book that is not only for highly specialized scientists.

I hope this book can attract interest on the exciting and relevant field of performance evaluation and system modeling. Quantitative reasoning is a characteristic trait of science and engineering. Moreover, in a limited resource universe, it is becoming of increasing importance to guarantee a widespread consciousness of the need of careful trade-off evaluation and of the ability to size the amount of resource in an optimal way.

A book is the result of interactions with a lot of people: colleagues, students and friends. I am grateful to those with whom I have shared scientific work and fruitful discussions, and from whom I have learned interesting ideas—too many over my career to cite them all. A warm thank you goes to the Wiley editorial staff, who has been generously supportive in the journey from manuscript to the final book product. Finally, my family has provided continuous and precious help throughout my work. Their love and patience were a great encouragement when my mind so often strayed into a maze while pondering the ideas developed in this book.

June, 2020   *Andrea Baiocchi*
Roma, Italy

Acronyms

ACK	Acknowledgment
ADSL	Asymmetric Digital Subscriber Line
AIMD	Additive Increase, Multiplicative Decrease
AP	Access Point
AQM	Active Queue Management
ARP	Access Reservation Procedure
ARQ	Automatic Repeat reQuest
ASTA	Arrivals See Time Averages
ATM	Asynchronous Transfer Mode
AWGN	Additive White Gaussian Noise
BA	Basic Access
BBR	Bottleneck Bandwidth and RTT
BDP	Bandwidth Delay Product
BEB	Binary Exponential Back-off
BM	Brownian Motion
BMAP	Batch Markovian Arrival Process
BS	Base Station
CBF	Contention-Based Forwarding
CBFQ	Credit-Based Fair Queueing
CBR	Constant Bit Rate
CCDF	Complementary Cumulative Distribution Function
CDF	Cumulative Distribution Function
CDMA	Code-Division Multiple Access
CDT	Carrier Detect Threshold
CLT	Central Limit Theorem
COV	Coefficient Of Variation
CSMA	Carrier-Sense Multiple Access
CSMA/CA	Carrier-Sense Multiple Access/Collision Avoidance
CTMC	Continuous-Time Markov Chain

CTS	Clear To Send
CWV	Congestion Window Validation
DBMAP	Discrete Batch Markovian Arrival Process
DCA	Dynamic Channel Allocation
DCF	Distributed Coordination Function
DCTCP	Data Center TCP
DIFS	DCF Inter-Frame Space
DMAP	Discrete Markovian Arrival Process
DRR	Deficit Round Robin
DSRC	Dedicated Short Range Communications
DT	Defer Threshold
DTMC	Discrete Time Markov Chain
DWDM	Dense Wavelength Division Multiplexing
ECN	Explicit Congestion Notification
EIFS	Extended Inter-Frame Space
EMC	Embedded Markov Chain
eNB	evolved Node B
ETSI	European Telecommunications Standards Institute
FB	Foreground-Background
FBM	Fractional Brownian Motion
FCA	Fixed Channel Allocation
FCLT	Functional Central Limit Theorem
FCFS	First-Come, First Served
FIFO	First-In, First-Out
GbE	Gigabit Ethernet
GPS	Generalized Processor Sharing
HOL	Head-Of-Line
HTTP	HyperText Transfer Protocol
ICT	Information and Communications Technology
i.i.d.	independent and identically distributed
IETF	Internet Engineering Task Force
IFS	Inter-Frame Space
IoT	Internet of Things
IP	Internet Protocol
ISP	Internet Service Provider
JBT	Join Below Threshold
JIQ	Join Idle Queue
JSQ	Join Shortest Queue
LA	Location Area
LAN	Local Area Network
LAS	Least Attained Service

LCFS	Last-Come, First-Served
LIFO	Last-In, First-Out
LMGF	Log Moment-Generating Function
LTE	Long-Term Evolution
LOS	Line-Of-Sight
MAC	Medium-Access Control
MAP	Markovian Arrival Process
MCS	Modulation and Coding Set
MMPP	Markov Modulated Poisson Process
MGF	Moment-Generating Function
MPDU	MAC Protocol Data Unit
MPEG	Moving Picture Experts Group
MPLS	Multi-Protocol Label Switching
MPR	Multi-Packet Reception
MRTG	Multi Router Traffic Grapher
MSS	Maximum Segment Size
MTU	Maximum Transfer Unit
MVA	Mean Value Analysis
NAV	Network Allocation Vector
NLOS	Non-Line-Of-Sight
NUM	Network Utility Maximization
OBU	On-Board Unit
ODE	Ordinary Differential Equation
OFDM	Orthogonal Frequency-Division Multiplexing
PASTA	Poisson Arrivals See Time Averages
PDF	Probability Density Function
PDU	Protocol Data Unit
PER	Packet Error Ratio
PHY	Physical (layer)
PLCP	Physical Layer Convergence Protocol
PPDU	Physical Protocol Data Unit
PS	Processor Sharing
QAM	Quadrature Amplitude Modulation
QBD	Quasi Birth-Death
QNA	Queueing Network Analyzer
QoS	Quality of Service
QPSK	Quadrature Phase Shift Keying
RAN	Radio Access Network
RB	Resource Block
RBM	Reflected Brownian Motion
RED	Random Early Detection

RO	Random Order
RTO	Retransmission Time-Out
RTS	Request To Send
RTT	Round-Trip Time
SCOV	Squared Coefficient Of Variation
SDH	Synchronous Digital Hierarchy
SDN	Software-Defined Network
SIFS	Short Inter-Frame Space
SIM	Subscriber Identity Module
SJF	Shortest Job First
SMP	Semi-Markov Process
SNIR	Signal-to-Noise plus Interference Ratio
SNR	Signal-to-Noise Ratio
SRPT	Shortest Remaining Processing Time
SSL	Secure Sockets Layer
TCP	Transmission Control Protocol
ToS	Type of Service
UDP	User Datagram Protocol
UE	User Equipment
VBR	Variable Bit Rate
VoIP	Voice over Internet Protocol
VM	Virtual Machine
WFQ	Weighted Fair Queueing
WiFi	Wireless Fidelity

Part I

Models for Service Systems

1

Introduction

Quelli che s'innamoran di pratica sanza scienza son come il nocchiere, ch'entra in navilio sanza timone o bussola, che mai ha certezza dove si vada.[1]

Leonardo da Vinci

1.1 Network Traffic Engineering: What, Why, How

Engineering is the application of scientific principles and results to the design and optimization of machine, processes, systems. The typical approach of engineers consists of understanding objectives and requirements, abstracting a model of the system to be designed, defining a solution approach and testing for its suitability, i.e., checking if relevant performance metrics meet the prescribed requirements.

A key point is the ability of deriving a simplified model from a description of the system or function to be designed. The model should be simple enough to lend itself to analysis and provide understanding of performance trade-offs, yet it should not miss any feature having significant impact on the relevant performance indicators.

Optimization of the model is a second key step. This can be often stated as a constrained optimization problem, where constraints come from performance requirements, costs, physical limits of the system.

The entire modeling and design process can be conceived as a double loop (see Figure 1.1). First, comparison with simulations or experimental measurements leads to the refinement of the model, so that it can reliably match the relevant dynamics of the system to be modeled. Once the model is assessed, it is used to

---

1 "Those who are fond of practice without science are like the helmsman of a ship without rudder or compass, so that he's never sure of where he's heading."

Network Traffic Engineering: Stochastic Models and Applications, First Edition. Andrea Baiocchi.
© 2020 John Wiley & Sons, Inc. Published 2020 by John Wiley & Sons, Inc.

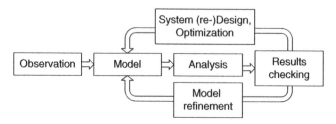

Figure 1.1 Scheme of system modeling and design process: from system observation and description, to model definition and refinement, based on comparison with simulations or experimental measurements (lower loop), then model usage for system dimensioning and optimization, according to an iterative refinement process based on performance results checking (upper loop).

refine the system design and to pursue its optimization, according to the results of the analysis, leading to new (hopefully better) performance results.

The very concise sketch of the engineering approach to problem solving is a general one. Traffic engineering refers to the design and optimization of a class of systems and processes: *networked service systems*.

Let us examine the keywords one by one.

Service system is an abstraction of any physical or logical function under Quality of Service (QoS) constraints. This is where the essence of *service* is.

Networked refers to the fact that multiple interconnected systems carry out the assigned task(s). For that purpose, "traffic" moves from one service system to another one, according to the topology of the interconnection and subject to the capacity of the network. We use terms in an informal way in this introductory section. So, by *network capacity* we mean the capability of the network to transfer resources (e.g., information, goods, vehicles) depending on the kind of service, hence network, we are considering (e.g., communication network, logistic network, transportation network) to provide service to users' demand.

Traffic can be defined informally as the stochastic process describing the users' service demand, as regards both time of demand submission to the system (arrival) and duration of service. Users of the service system (e.g., applications, persons, machines) require the service system to carry out its tasks to meet their service demand. Times when service demand is submitted to the system as well as the amount of work required to meet the specific demand can be characterized as random variables. Hence, traffic engineering is intimately connected with probability and stochastic processes theory and its applications, a prominent position being reserved to *queueing theory*. That is the preferential "language" of traffic engineering, even if also other mathematical tools are often used (e.g., fluid approximation theory, optimization theory, game theory, to mention a few).

Performance evaluation is at the heart of traffic engineering. A service system encompasses three major aspects: (i) users' traffic demand; (ii) serving capability and resources provided by the system; and (iii) QoS constraints. The aim of traffic engineering is the design of the service system to meet the expected users' demand under the prescribed quality constraints. Minimization of cost, both capital expenditure to set up the system and operational costs, is of paramount importance to the system provider. This is usually in conflict with meeting an assigned level of QoS, which is instead of primary relevance to the system users. Trading off costs for QoS, given the users' demand, is the core "business' of traffic engineering. Conversely, estimating the admissible demand for the desired level of QoS, given the available resources and the way the system is designed, is another key task of traffic engineering, leading to the definition of algorithms and procedures to rule the access of users to the system resources and to manage those resources (priority, scheduling, flow control, congestion control, multiple access).

The reason why such a discipline has been developed and has grown as a recognized field is that no 'free" resource is given in any service system. Hence, rational design of what resources to use, how much of them, and how to use them, still providing a "useful" service (i.e., meeting a specified QoS level) is key to making design of service systems viable from a technical-economic point of view. This is why the design of service systems calls for suitable quantitative methods, able to provide predictions of key performance indicators.

Since traffic engineering is based on system modeling and abstractions, it has long been recognized that many different technical fields give rise to networked service systems that lend themselves to common models, independent of details of the specific technology or application field. Mathematical tools have been developed that can be used across many different application areas to a very large spectrum of systems. To mention some of them, communication networks, computing systems, transportation networks, logistic networks, power grid networks, production processes, all can be cast into the service system abstraction and therefore be designed by resorting to network traffic engineering tools. Each example application domain is itself a highly structured and complex system, encompassing a huge variety of physical resources and processing logic (we could refer to them as "hardware" and "software," borrowing a classic terminology of information technologies).

Networked service systems can be modeled and analyzed by using different approaches. More in-depth, analysis, dimensioning, and optimization of service systems can be faced along three main lines:

1. Analytical models
2. Simulations
3. Experiments

Analytical models provide a mathematical description of the system that yields to tractable analysis (closed formulas) or, more often, to numerical investigation. This is the most powerful approach for a quick and nontrivial understanding of the performance trade-offs, to gauge stability margins of the system, to assess the impact of key system parameters on performance, to provide a setting for stating optimization problems. While producing an effective analytical model requires hard study and a bit of talent to strike the best balance between simplified assumptions and a representative model, the time and computational effort required to use an analytical model make it the least costly among the three approaches listed above. The real difficulty of an analytical model is not really in solving the model once stated (books are there just to provide a guide for that purpose). It is rather the ability "to make things simpler, but not easier," to say it with Albert Einstein's words. The art of modeling consists in making all sort of assumptions leading to the simplest model that still captures the aspects that are decisive to give a sensible answer to questions on the system. A fluid model that disregards completely the discrete nature of packet traffic in a communication network, such as the Internet, can be perfectly acceptable when we set out to study algorithms for congestion control, whereas it is definitely inadequate if we are interested in characterizing the delay jitter of a packet voice multiplexer or the collision probability of a random access protocol.

Among analytical models, a major role is played by stochastic process theory and queueing theory. The most useful class of stochastic process for service system analysis relates to Markov chains. The success of stochastic process theory and queueing theory as tools for network traffic engineering motivates the space devoted to them in this book.

Analytical models provide answers to basic questions in a quick and lightweight way. We can gain valuable insight on the system performance cheaply. When it comes to assessing second-order effects or we need to relax assumptions on the system model in a way that does not yield to analytical tractability any more, computer simulation is often a valid approach.

Computer simulation for network traffic engineering purposes amounts to defining a detailed operational model of the system and reproducing all processes involved in this detailed model by means of a computer program. This is obviously not a real-life system; it is, rather, a virtual image of how a simplified version of the real-life system would work. The limits of this approach reside only in the limits of the coding language and of the available computational resources. Extremely complex models can be simulated. As a matter of example, simulating a vehicular traffic management applications implies modeling these features:

- The road map. This includes meta-data describing road lanes, directions, signals, and surrounding environment (buildings, trees, tunnels).

- Vehicle mobility of different types of vehicles (cars, trucks, buses, motorcycles, bicycles, etc.), possibly also pedestrians. This includes modeling every useful detail of the mobility mechanics (end-to-end flow vehicles, routes, motion laws, overtaking, reaction times).
- Communication equipment on board vehicles, e.g., cellular transponders or other equipment for local communications (WiFi, vehicle-to-vehicle communication devices). This means modeling all communication architecture layers, from the physical layer to the application layer, with all relevant details of protocols of evert layer, modeling radio propagation among devices, modeling telecommunications traffic generation processes.
- Incidents that change the mobility environment.
- The logic of the application for vehicular traffic management, including the relevant message exchange among vehicles and the fixed network infrastructure.
- Feedback on vehicle mobility due to the information acquired through the traffic management application, according to the logic of the application.

Clearly, the software implementing all of these aspects must be highly complicated. A possible simulation framework able to support this kind of modeling is provided by VEINS (`https://veins.car2x.org`), a software package that integrates a module for the simulation of urban mobility (SUMO), a module to simulate communication network protocol stack (OMNET++), and all the logic required to develop simulation software of customized application logic, to import roadmaps and any other meta-data required to parametrize the simulation experiments.

Even this brief example suggests how powerful simulation can be. On the down side, setting up a simulation software requires a relatively high software coding skill level and possibly extensive training on specialized software packages (dedicated simulation software packages have been developed in many application fields of science and engineering, e.g, communication networks, transportation systems, power distribution). Mastering one of those specialized softwares typically requires several weeks up to months. Another drawback of the simulation-based approach is the limited flexibility of the model (making modifications can be very costly in terms of person-time effort). The computational burden of simulation could also be a problem, often making difficult to obtain a quick answer to what-if questions. The availability of extremely large processing resources in the public cloud and the ever-decreasing cost of computing power relax the boundary of feasibility continuously, but the bottom line is that developing a simulation model and running simulation experiments is worth it when one already has a quite firm vision of how the system could be designed and needs more stringent answers to a number of detailed issues before delving into the actual realization of the system.

The last word on system performance is real-life experiments on a possibly scaled-down prototype. Here we need not make any assumptions or simplifications on reality. However, development time and required skill level, as well as material cost, can grow to a significantly higher level than with simulations. At the same time, an experimental setting is even less flexible and easy to use than simulations. This is by far the less desirable alternative when it comes to characterizing the performance of a system, providing input for its design, and optimizing algorithms. It is however, the only way to give evidence that: (i) the proposed service system, or at least key parts of it, can be actually realized in a viable way; and (ii) key assumptions made in the development of the analytical or simulation model are backed up by experimental data. Both points are an unavoidable step in the (long) process leading from an idea to a successful product, whether it be a physical system or a process. Many technology success stories start with a theoretical idea, first proved by means of analytical models that point to possibly large gains of exciting breakthroughs, *provided that* some assumptions hold. Simulations and ultimately experiments give evidence of whether the theory is well grounded and promising, given the feasible technology context and the application opportunities.

1.2 The Art of Modeling

Analysis and design of service systems go through an abstraction process that leads from real-life systems or processes to a simplified formal description that yields to mathematical description. This abstraction process is more of an art than a science, since there is no single way of doing it, nor is the resulting model unique. The decision on what part of the original system is good to simplify, i.e., the list of assumptions, and the choice of the mathematical tool to be used are to a large extent a subjective decision-making process, highly dependent on the background, competence, and experience of the person(s) doing the job. A second crucial point is the application context and the very purpose we set out to develop a model, i.e., the questions for which we are seeking an answer. In a sense, a good model is one that leads as straightforwardly as possible (i.e., with the least effort) to a satisfactory answer (i.e., within the degree of accuracy we need) to the specific questions that matter for the problem at hand. There are obviously general techniques (e.g., Markov chain theory, queueing theory), established results, best practices. Yet, it is not infrequent that a useful model needs "customization," depending on the application context, the purpose of the modeling, the amount of time, computational resources, and skill available to those that have to provide results.

Modeling is the approach taken to answer questions arising in some application context and involving ultimately the investment of workforce effort and the use of physical or logical valuable resources. The practical problems that trigger traffic engineering modeling are the essential motivation for it and push continuously the development of new tools and theories. As for many other applied science branches, sometimes traffic engineering theories are studied for themselves, i.e., extensions and generalization are investigated even beyond the specific questions that promoted initially the development of the model. Nevertheless, traffic engineering cannot be merely a theoretical investigation of mathematical tools. The art of developing a useful model from a description of a system, to answer questions on its performance and design, is a fundamental ingredient of traffic engineering. Besides a solid understanding of models, it is therefore highly recommended to be confronted with several application examples. The art of modeling is probably best grasped through a learn-by-experience approach. Let us then give two simplified yet meaningful examples of the kind of issues arising in this process.

Consider a store selling some specific kind of goods, e.g., vegetables and fruit. It can be thought of as a service system, where "users" are people coming to buy products sold at the store and the service provided by the system is the possibility of finding a selection of products with some specified quality (variety of products, freshness, packaging standards, bio-compatible production chain, special characteristics, like gluten-free). A number of questions could be posed on the system, such as in which part of the town should the shop be located, which kind of installation should be employed, how many people should be hired to run it, with which kind of skill and tasks, what is the best shop opening schedule, where and at what prices products sold at the store are best procured, and what selling prices should be applied. The kind of model to be developed depends in a crucial way on what the questions we seek an answer for are.

As a matter of example, let us ask how many employees we should hire for the store. That depends primarily on the volume of demand (how many people come to the store during a regular working day, how much vegetables and fruit they buy). It also depends on the store installation, whether it is a big store, with a reserved parking lot, a large building for goods exhibition, and large warehouses; a street shop, with a relatively small warehouse and limited space for exposing goods; or even a market stall. The economic viability of the proposed solution is the target of the modeling, under constraints on customer satisfaction, e.g., the average amount of time that a customer has to wait before being served during peak hours. The amount of people coming to the shop could vary during the day, on weekends with respect to working days, or on a seasonal basis. It depends on the local density of residential population and on how many competitors are located nearby. It is clear that many of the variables describing customers and their habits can be characterized only as random variables, with parameters that can only be predicted or measured within some level of accuracy. The complexity of the model should therefore be tuned to the accuracy of the knowledge of customer demand (there is no benefit in adopting a sophisticated customer arrival model requiring several parameters to be tuned, if we have only partial or inaccurate data to fit the model: better to use a very simple parsimonious model, with as few parameters as possible). The specific question we pose (how many employees to hire) could be answered by defining a queueing model of the store and applying known results to size the number of "servers" to meet the waiting time requirement. We could also try to state an optimization problem, e.g., what is the optimal number of employees, given a model of customer impatience (if we hire a lot of people, customer satisfaction will be excellent, we will attract a lot of customers, but the serving basin is anyway limited, so expenses for personnel would eventually exceed the growth of income; conversely, limiting the number of employees makes us lose customers and potential revenue, and that could possibly bring the store to shrink its customer basis to a level too small to survive).

As a second example, let us consider a road crossing. "Service" here consists of vehicles switching from one road to another one through the crossing. "Customers" are vehicles. The "server" is the crossing area, with all its features (e.g., traffic lights, number of lanes, roundabout or cross-shaped intersection). Depending on the way the crossing is used, it could be modeled as a single server or a set of multiple servers (e.g., in case multiple vehicles could engage the crossing simultaneously). The quality of service can be measured by the time a vehicle has to wait before being able to access the crossing and by the probability that a collision (accident) occurs at the crossing. A trade-off exists between the two. That is, if we introduce traffic lights at the crossing, we expect to reduce the probability of accidents, but also to increase the time that a vehicle has to wait before it can engage the intersection. Assume we have to optimize the green times of the crossing traffic lights to minimize the average vehicle waiting time. We could consider a first-order approximation of the system as a single server system with an infinite waiting line. This model applies to a single lane on a single road arriving at the crossing. It disregards interaction of vehicles engaging the crossing, i.e., inter-dependencies among the serving capacities of the servers representing the roads converging to the crossing. The simple model does not account either for vehicle mechanics (acceleration, speed, vehicle size and length) and for human reaction times, traffic light overhead times (dead times when switching from red to green and vice versa). A refined model could be made up of a network of queues, one for each road (or maybe, each lane) arriving at the crossing. A refined model, accounting for those details and including sophisticated statistical modeling of vehicle arrivals could be set up and analyzed by means of simulations. This is still a model, simulations being virtual processes that mimic real-life situation, still with a number of simplifications. The last word would be to construct a possibly scaled-down, real system, where experiments with real vehicles are run. This last approach is extremely costly and time-consuming. Moreover, it does not lend itself to stress the system at high traffic levels, since high traffic entails involving a large number of vehicles, making experiments unfeasible. This is why analytical modeling or simulation are extremely useful tools to understand performance trade-offs, design algorithms and optimize real-life service systems.

In the following we discuss in some detail an example. Starting from the description of a real-life issue, we derive a model and develop mathematical analysis and simulations. The assumptions required to derive the model are discussed as well as the lesson learned from performance results.

Before delving into the detailed development of a model, it is worth noting the relationship between modeling and machine learning. Machine learning encompasses a broad field of theories, algorithms, and applications that has been growing over the last several decades, leveraging on the progress of information and communication technologies. The last decade has witnessed an impressive growth of the application range of machine learning, boosted by the ever-increasing availability of computational power, of big data, and by breakthroughs in algorithm implementation (e.g., deep learning networks[2] [28,67]). Since this book is entirely devoted to modeling tools and examples of their applications to networked service systems, it is useful to take a quick look at an approach that might be alternative to modeling (as intended here).

A basic view of supervised machine learning may be stated as a problem of function identification (see Figure 1.2). As a matter of example, let us consider a set of "objects." An object is associated with a label y and is described through a set of features x. We can think of the features x as a vector of $\mathbb{R}^n$ (generalizations to qualitative features, belonging to non metric spaces, are possible). We assume that a functional relationship exists between x and y, i.e., it is $y = f(x)$ for a suitable (unknown) function $f(\cdot)$.

We aim at identifying the function $f(\cdot)$ (at least a good approximation of it) with a data-driven approach. We assume therefore that we are assigned a set of

Figure 1.2 Illustration of a basic concept of supervised machine learning.

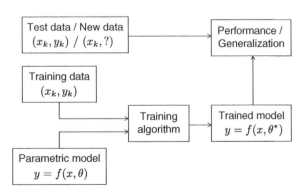

---

2 Recent works point out at apparently fundamental limits of current algorithms, e.g., see [75].

couples $(x_k, y_k), k \in \mathcal{I}$ (the so-called "ground truth"). We choose a parametric set of models for these data, say $f(x, \theta), \theta \in S$. We split the available data into two sets: the training set $\mathcal{I}_{\text{train}}$, used to select one model within the chosen family, and a test set $\mathcal{I}_{\text{test}}$, used to evaluate the performance of the selected model.

Using the training data $(x_k, y_k), k \in \mathcal{I}_{\text{train}}$, and a suitable *learning algorithm* (training algorithm), we synthesize the "best" possible model, say $f(x, \theta^*)$ (typically, one that minimizes the error with respect to the ground truth in the training set).

Using the test data $(x_k, y_k), k \in \mathcal{I}_{\text{test}}$, we assess the performance of the synthesized function $f(x, \theta^*)$.

A key point is the *generalization* capability of the model $f(x, \theta^*)$, i.e., the ability of the selected model to yield the right value y when fed with a previously unseen input x (i.e., the ability to reconstruct a *new* couple, not belonging to the training set).

Even this very brief description of the machine learning approach highlights its key aspects: data-driven modeling, generalization capability. In a sense, this is a black-box approach. We do not start from a functional description of the system producing the objects, from which we try to find a model able to predict the output y for any given input x. Rather, we collect a (possibly large) set of *examples* by "running" the system (input-output couples), then approximate the functional relationship that we assume to exist between the input and the output data. We are therefore not required to *understand* the laws governing the internal working of the system that produces the objects. We are giving up to insight into the system. On the other hand, we are able to define a "working" model, that provides us (hopefully) useful answers, even for unmanageably complex systems, for which it is too hard to derive useful models in the "traditional" sense[3] .

Both approaches, modeling and machine learning, have their strong and weak points, both have their use cases. If we are able to state a model of the relationship between system input and output, say $y = \hat{f}(x)$, and to define an efficient algorithm to evaluate $\hat{f}(x)$ for any interesting x, there is no need to resort to machine learning. If we cannot collect enough data, in the form of couples (x_k, y_k), machine learning is not applicable, either. So what are use cases for machine learning?

- When we are not able to state a model.
- When we manage to state a model, but it is unfeasible to "solve" it, that is, to use it for deriving predictions on the studied system.
- When we have a model and feasible algorithms, but we can achieve a significant computational complexity reduction resorting to machine learning algorithms.
- When we can collect data in the form of couples (x_k, y_k) at a reasonable cost.

---

3 Note however that the machine learning approach cannot forget about a good grasp on the system to be modeled, even if modeling is data-driven. In fact, assuming that there exists a functional relationship between the input x and the output y is already calling for some field-expert knowledge on the specific system.

The first three items point at cases where machine learning offers a viable or preferable alternative, whereas "traditional" modeling could be at a deadlock or too hard. The last point is a precondition for machine learning to be a practical alternative.

We do not advocate either approach. Both have merits and should be considered when confronted with a real problem. "Integrated" solutions are also possible, that use both approaches to build a composite model[4] .

In the rest of this book we present tools and examples oriented to developing modeling skills. It is important to bear in mind that other approaches, besides "traditional" modeling, exist for designing and optimizing service systems. Which way to go depends on available skills and time, on design objectives, on technological opportunities.

1.3 An Example: Delay Equalization

Let us consider a streaming application in the Internet. Streaming is used for audio/video retrieval and play-out. The content is usually available in servers located in data centers (the "cloud"), possibly replicated in temporary cache memories, close to potential users (content delivery networks). The user has a play-out application (often embedded into a web browser), essentially consisting of a decoder and a graphical user interface. The encoded audio/video data is downloaded from the server to feed the decoder. Smooth play-out requires feeding the decoder with audio/video data according to exactly the same timing as produced by the encoder. This implies in turn that data delivery delay through the network should be constant and no piece of information should be lost.

In the real Internet occasional packet loss is possible (typically a few percent of packets are lost in moderately congested links). In the application example at hand, packet loss can be concealed by redundant coding, since human perception can be deceived up to a certain amount of missing information in the reconstruction of the audio/video streaming. More importantly however, packets sent through the Internet suffer variable delays: packet belonging to the same end-to-end flow, even if they follow the same network path, encounter different levels of node congestion. The inevitable result is that delays of successive packets are different and there is *no way* to avoid this impairment (unless changing the fundamental principles of data transfer through the Internet).

The contrast between application requirement (constant delay) and network operation, resulting in variable delays, can be reconciled by introducing a *delay equalization buffer* in front of the decoder. Delay equalization is performed by

---

4 To make a simple example, one might say that channel estimation and adaptive equalization in telecommunication receivers is a form of machine learning, embedded into a system that is designed and optimized based on mathematical models of the signal, the channel impairments, and the algorithms applied at the receiver.

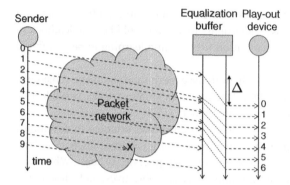

Figure 1.3 Illustration of the delay equalization of a streaming flow sent through a packet network. The 'x' denotes packet loss. The slope of the dashed arrows through the network corresponds to the delay suffered by packets. The initial play-out delay Δ is shown: once play-out starts, frames are read from the buffer at the same rate as they are produced by the sender.

imposing an *additional* delay to each arriving packet, so that the sum of the delay suffered by the packet in the network plus the equalization delay equals a fixed delay, *same* for all packets (hence the name of the algorithm: equalization has the same root as "equal").

Figure 1.3 illustrates the system components for this example: the server sending audio/video data, the intermediate network, the delay equalization buffer, and the decoder. The detailed explanation of the delay equalization algorithm is developed in Section 1.3.1.

In the following we pose performance questions, unveiling issues with the equalization algorithm (starvation, buffer overflow). We derive models that provide a quantitative tool to gain insight into the delay equalization algorithm and to dimension its parameters.

1.3.1 Model Setting

We assume that the encoder produces fixed length data frames at a fixed rate. Let L be the length of a frame and T the time interval between the emission of two consecutive frames. We assume that a frame fits into a single packet, i.e., there is a one-to-one correspondence between application-level data units (frames) and network-level data units (packets). In general, multiple packets are required to carry a single frame. Capturing this feature requires however a significantly more complex model (see Problem 1.1).

The data flow throughput sustained through the network is $\Lambda = L/T$. We assume that a transport capacity at least equal to Λ is available through the network.

Let $t_k = t_0 + kT$, $k \geq 0$ be the time when the k-th frame is released by the encoder and sent into the network. The network introduces a variable delay, say D_k for frame k. Occasionally, a frame can get lost and be never delivered at the destination. Out-of-order delivery is a possible outcome as well. In normal operating conditions, frame loss and out-of-order delivery are actually sporadic events: they could typically affect less than a few percent of the frames. For the time being we neglect frame loss and out-of-order delivery. Those issues are reconsidered in Section 1.3.3.

We also assume that the D_k's are independent, identically distributed (i.i.d.) random variables, admitting a probability density function (PDF) $f_D(x)$, defined for $x > 0$. Let also $G_D(x) = P(D > x)$ be the complementary cumulative distribution function (CCDF) of the random variable D. Assuming that the D_k's have a same probability distribution amounts to require that the stochastic processes that cause delays inside the network are stationary during the data transfer. This can well be the case if the time dynamics of the network traffic (i.e., the time scale over which the average network link loads have a significant variation) is much bigger than the duration of the audio/video data transfer. As for statistical independence among the D_k's, this is a reasonable assumption, if the time dynamics of buffers within routers along the end-to-end path of the frames is smaller than the inter-frame time T. If that is true, the queue states sampled by two consecutive frames are weakly correlated. For a buffer of size B_r and a link of capacity C_r, the time scale of the buffer queue is roughly of the order of B_r/C_r. As a matter of example, with $B_r = 10$ Mbyte and $C_r = 10$ Gbit/s, we have $B_r/C_r = 8 \cdot 10^7$ bit/10^{10} bit/s $= 8$ ms. If $T > 8$ ms, it is plausible that the delays encountered by two consecutive frames be negligibly correlated, hence the independence assumption is reasonable.

Play out at the receiver is started after a delay Δ from the reception of the 0-th frame (see Figure 1.3). After that, frames are consumed, one frame every T seconds, to feed the decoder at the receiver. Variable delays can be compensated by storing the incoming frames in an equalization buffer temporarily. Two issues must be faced then: (i) *starvation*, i.e., not finding the expected frame in the buffer at the time it must be used by the decoder; and (ii) *buffer overflow*, i.e., an arriving frame must be stored until play-out, but there is no more space left in the equalization buffer. We consider these two issues in the rest of this section, by dimensioning properly the initial delay Δ of the play-out and the equalization buffer size B.

1.3.2 Analysis by Equations

To keep the analysis manageable, we resort to approximations, a customary approach in system modeling. We assume the equalization buffer is infinite, i.e., we neglect the overflow issue when dimensioning the initial delay Δ.

If we apply an initial delay Δ, the k-th frame is expected at the decoder for play-out at time $t_{\text{out},k} = t_0 + D_0 + \Delta + kT$, $k \geq 0$. Starvation of the decoder is triggered by either of two possible events: (i) the k-th frame was late; or (ii) the k-th frame arrived in time, but it found a full buffer and was dropped. According to our approximation (infinite buffer size), the second event is ruled out. Hence, starvation is equivalent to the first event, i.e., it occurs if and only if $t_k + D_k > t_{\text{out},k}$, that is $D_0 + \Delta < D_k$. Denoting the probability of starvation with S, we get:

$$S = P(D_k > \Delta + D_0) = \int_0^\infty f_D(x) G_D(x + \Delta) dx \tag{1.1}$$

For example, if $f_D(x) = \mu \exp(-\mu x)$ for $x > 0$, i.e., the delay through the network has negative exponential probability distribution with mean $E[D] = 1/\mu$, we get

$$S = \frac{1}{2} e^{-\mu\Delta} = \frac{1}{2} e^{-\Delta/E[D]} \tag{1.2}$$

If we require that $S < \epsilon_S$, then it must be $\Delta > E[D]| \log(2\epsilon_S)|$.

Let us now relax the assumption of an infinite buffer and consider a buffer of size B. We can upper bound the probability of overflow of the buffer of size B with the probability that the occupancy level of the infinite buffer exceeds B.

Let $\tilde{N}(t)$ and $N(t)$ be the number of frames stored in the finite buffer and in the infinite buffer at time t, respectively. Let $\tilde{P}$ denote the overflow probability of a buffer of size B, i.e., the joint probability of the events $\{\tilde{N}(t) = B\}$ and $\mathcal{A}(t) =$ {a frame arrives at time t}. Since it is[5] $\tilde{N}(t) \leq N(t)$ for all t, the event $\{\tilde{N}(t) = B\} \& \mathcal{A}(t)$ implies the event $N(t) > B$. Hence $\tilde{P} = P(\text{buffer overflow}) \leq P(N(t) > B)$, i.e., the probability that the buffer content of the infinite buffer exceeds the threshold B is an upper bound of the overflow probability of the finite buffer.

Let N_k be the number of frames stored in the buffer at the time immediately preceding play-out of the k-th frame, i.e., $N_k \equiv N(t_{\text{out},k-})$. Let $N = B/L$ (we assume B is an integer multiple of the fixed frame size). We have $\tilde{P} \leq P(N_k > N) \equiv P$.

In the following we evaluate the probability distribution of N_k and use its tail to dimension the equalization buffer size. This is an example of how bounds and approximations help deriving performance results.

It is $N_k > n$ if and only if the arrival time of the $(k + n)$-th frame is less then the time $t_{\text{out},k}$, that is $t_{k+n} + D_{k+n} < t_{\text{out},k} = t_0 + D_0 + \Delta + kT$. Therefore, by letting $Q(n) = P(N_k > n)$, we have

$$Q(n) = P(t_0 + (k + n)T + D_{k+n} < t_0 + D_0 + \Delta + kT) = P(D_{k+n} < D_0 + \Delta - nT) \tag{1.3}$$

---

5 This is also true if we account for packet loss event in the network.

Then

$$Q(n) = \int_0^\infty f_D(x) P(D_{k+n} < x + \Delta - nT) dx$$

$$= \int_{\max\{0,nT-\Delta\}}^\infty f_D(x)[1 - G_D(x + \Delta - nT)] dx$$

$$= G_D(\max\{0, nT - \Delta\}) - \int_{\max\{0,nT-\Delta\}}^\infty f_D(x) G_D(x + \Delta - nT) dx \quad (1.4)$$

For example, with negative exponential network delays, we obtain

$$Q(n) = \begin{cases} \frac{1}{2} e^{-\mu(nT-\Delta)} & nT \geq \Delta, \\ 1 - \frac{1}{2} e^{\mu(nT-\Delta)} & nT \leq \Delta. \end{cases} \quad (1.5)$$

The upper bound P of the overflow probability we are looking for is then $P = Q(N)|_{N=B/L}$. The requirement $\tilde{P} \leq \epsilon_B$ on the overflow probability of the equalization buffer is guaranteed by imposing that $P \leq \epsilon_B$.

Let us assume that $NT \geq \Delta$. From eq. (1.5) we have $Q(N) = e^{-\mu(NT-\Delta)}/2$, with $\mu = 1/E[D]$. Imposing $Q(N)|_{N=B/L} \leq \epsilon_B$, we find

$$B \geq L \frac{\Delta + E[D]|\log(2\epsilon_B)|}{T} \quad (1.6)$$

Substituting the expression found for the initial delay Δ and recalling that the end-to-end throughput of the audio/video packet flow is $\Lambda = L/T$, we get

$$B \geq \Lambda E[D](|\log(2\epsilon_S) + \log(2\epsilon_B)|) \quad \Rightarrow \quad \frac{B}{\Lambda E[D]} \geq |\log(4\epsilon_B\epsilon_S)| \quad (1.7)$$

which gives the minimum required buffer size B as a function of the quantity $\Lambda \cdot E[D]$, the so called bandwidth-delay product (BDP). The BDP is a key parameter in many networking problems. The dimensioning criterion of the buffer size B in eq. (1.7) exemplifies in a clear way the role of the quality of service constraints (the parameters ϵ_B and ϵ_S) and of key system parameters (the BDP in this case).

For $\epsilon_B = \epsilon_S = 10^{-3}$, we have $B \approx 12.43 \cdot (\Lambda \cdot E[D])$. With a buffer of 256 kbytes we can face a mean network delay of about 41.2 ms for a throughput of 4 Mbit/s.

The expression in eqs. (1.2) and (1.7) hold for a negative exponential distribution of network delays. The analytical model developed above can in fact be used with a general distribution of network delays, to dimension the initial delay Δ and the buffer size B. The analytical formulas found for a general network delay distribution (eqs. (1.1) and (1.4)) can be used, at least numerically, to evaluate the starvation probability S and the upper bound P of the overflow probability for given values of the model parameters.

In the following we show the numerical results obtained by estimating the probability distribution of the network delay from a sample of measured round trip times (RTTs). The RTT trace has been collected between a host in a WiFi access

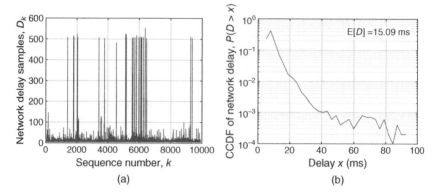

Figure 1.4 Left plot: Sequence of RTT values measured between a host in a WiFi access network and a server on public Internet. The sequence has been collected using `ping` with an interval of 0.1 *s* between consecutive message sending times. Right plot: empirical CCDF of the network delay *D* based on the RTT sequence shown in the left plot.

network and a server on the public internet, both located in Italy. The sequence of collected samples is shown in Figure 1.4(a). Figure 1.4(a) suggests that the network crossed by the packets exhibits a relatively large "random" variability of the delays, on top of which there are occasional delay spikes, according to an apparently random pattern. This hints to occasional heavy congestion phenomena, or to some recurring high-priority task, carried out by the target server used to collect RTT values, that causes a large delay of the `echo_reply` message.

Figure 1.4(b) illustrates the empirical CCDF of network delay. The estimated mean delay is 15.1 ms, while the estimated standard deviation is 43.2 ms. The bulk of probability is around the mean, yet there is a rather long tail that can hardly be estimated, given the available number of samples. The variability of network delay is evident also from the high value of the ratio of the standard deviation to the mean[6] .

The resulting numerical values of the starvation probability and of the overflow probability are shown in Figure 1.5(a) as a function of Δ and in Figure 1.5(b) as a function of *B*, respectively. We have assumed $L = 1400$ bytes and $T = 33$ *ms* (30 video frames per second).

The values of the initial delay and of the buffer size that meet the performance requirements $\varepsilon_S = \varepsilon_B = 10^{-2}$ are $\Delta_{\text{req}} = 220.7$ ms and $B = 19.6$ kbytes, respectively. The effect of the slow decay of the CCDF of network delays appears in the slow decay part of the curve of the starvation probability *S* as a function of Δ.

---

6 We have selected an extreme case, exhibiting a rarely seen large variability, to make numerical results more interesting in the chosen model setting. For the same reason, we have directly used the sequence of measured RTTs as representative of end-to-end network delays, without halving them.

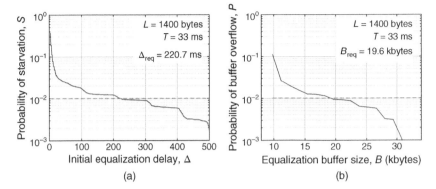

Figure 1.5 Left plot: starvation probability as a function of the initial delay Δ, assuming an infinite size equalization buffer. The minimum initial delay that meets the starvation probability requirement (dashed line) is $\Delta_{req} = 220.7$ ms. Right plot: buffer overflow probability as a function of the equalization buffer size B. The minimum buffer size that meets the packet loss probability requirement (dashed line) is $B_{req} = 19.6$ kbytes.

Even if numerical results are obtained by estimating the probability density function of the network delays from real data, still there are a number of assumptions underlying the model. Packet delays are assumed to be i.i.d. random variables, the equalization buffer is assumed to be infinite, out-of-order packet delivery has been neglected. To the cost of setting up a detailed simulation code, we can remove all of these assumptions and check results, which we do in the next section.

1.3.3 Analysis by Simulation

We can investigate the performance metrics of the delay equalization buffer by means of simulations. At the cost of developing the simulation model, coding it, and bearing the computational cost of running simulations (typically much more expensive than evaluating the analytical model), we gain the possibility to relax the assumptions we have made in the derivation of the analytical results of the previous section. Specifically, we can evaluate the frame loss probability (FLP) with the given buffer size B and initial equalization delay Δ, without the need of assuming an infinite buffer or resorting to upper bounds. As a consequence, starvation at time $t_{out,k}$ occurs at the output of the buffer for two possible causes: (i) late arrival, i.e., frame k has not arrived yet; (ii) loss, i.e., frame k had already arrived at the input of the buffer, but it was dropped because of a full buffer at the time it arrived. The FLP is defined as the probability of starvation, whichever the cause.

Given the values of T and L, the simulation of the equalization buffer depends only on the parameters Δ and B, i.e., the initial delay and the buffer size.

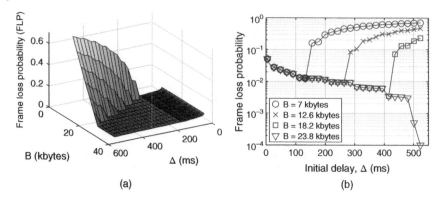

Figure 1.6 Simulations of the equalization buffer. Left plot: frame loss probability as a function of the initial delay Δ and the buffer size B. Right plot: frame loss probability as a function of the initial delay Δ for four different values of the buffer size B.

We account for the effect of the network still using the RTT experimental trace. The buffer size is converted into the maximum number of packets it can contain, $N_{max} = \lfloor B/L \rfloor$.

Figure 1.6(a) shows the 3-D plot of the FLP, as a function of Δ and B. As expected, the FLP decreases, eventually going to 0 as the buffer size is increased[7] . It might appear counterintuitive that the FLP increases sharply with Δ for a given value of B. This is especially evident for the smaller sizes of the buffer.

The reason for this behavior is that FLP is the sum of two components. Some frames get lost because they arrive at the buffer *after* due time for play-out; other frames are lost because they are dropped due to buffer overflow. The first component is dominant at low Δ values (up to the order of a few standard deviations of the network delay) then fades away quickly as Δ grows. The second component of FLP is dominant when the buffer is small compared to the amount of frames that can arrive during the initial delay Δ. Since frames arrive at an average rate of $1/T$, on the average $\Delta \cdot L/T$ bytes arrive, before play out can start. If B is less than this quantity, frame loss is massive.

A clearer picture of the phenomenon can be appreciated by plotting FLP as a function of Δ for some values of B, as done in Figure 1.6(b). All curves have an initial common behavior, independent of B, that corresponds to the operation region dominated by late arrivals. This is the behavior correctly predicted by the analytical model. Then, as Δ grows up further, the subsequent behavior breaks up into

---

7 The FLP can actually hit 0, since this is a data-driven simulation, where packet delays are taken from a file of 10000 measured delays, hence there is a maximum delay and correspondingly a maximum initial delay beyond which no frame loss occurs, for a large enough buffer size. There are exactly 0 lost frames as soon as $\Delta > \Delta_{max} = \max_k \{D_k\} - D_1 \approx 533$ ms and for $B > \Delta_{max} L/T \approx 22.6$ kbytes.

different branches as a function of *B*. The FLP grows sharply with Δ, when the frame loss due to buffer overflows becomes dominant. The bigger *B*, the wider is the favorable dimensioning interval, where we achieve a low FLP value.

The comparison between Figure1.5(b) and Figure 1.6(b) gives a striking visual evidence of the gap between analytical model predictions and simulation results. Due to assumptions required to make the model tractable, the analytical curve of Figure 1.5(b) captures correctly the lower branch of the FLP curve resulting from simulations, but it misses completely the sudden increase of FLP when Δ grows beyond a threshold depending on *B*.

1.3.4 Takeaways

Consider a typical network traffic engineering problem, delay equalization. The highlights are as follows:

1. Defining an analytical model entails major simplifications and assumptions, the stronger the simpler the obtained results and potentially more insightful.
2. Analytical models may lose relevant effects or hold only for limited range of system parameters. Care must be taken when drawing conclusions on the basis of analytical models. They provide invaluable help in guiding the setup of more detailed evaluation tools (e.g., simulations or measurements), but they could miss some phenomena.
3. Simulations can be powerful, since they allow detailed modeling and enable us to relax a lot of assumptions. Still, it must be considered that simulation is based on models, Moreover, understanding the system dynamics by brute force simulation can turn into searching for a needle in a haystack.
4. The design and dimensioning tasks become much more effective when one has a good intuition and solid expectations on the system behavior.
5. Making illustrative graphs of the performance results, as well as visualizing the data, can help a lot. Hence, it is worth spending a significant fraction of the time allowance for performance evaluation on this task.

1.4 Outline of the Book

In this section we first give a concise account of the content of the next chapters. Then we discuss possible uses of this book for courses and self-learning. Finally, we introduce definitions and notation of general use throughout the book.

1.4.1 Plan

The book comprises ten chapters besides this one, plus an Appendix.

Chapter 2 is devoted to service system definition, the role of queueing models, and general properties of queues in equilibrium. A service system is an abstraction of a physical or a logical element providing a function (the service) to users according to a given level of quality, as measured by relevant performance metrics. The service system is deemed to be out of service if the quality of service constraints are violated. In network traffic engineering, examples of service systems can be a router, a web server, a virtual machine, a protocol, a (sub-)network, or a data center. The structural elements of a service system are defined in Section 2.1, and service demand is characterized in Section 2.2. Section 2.6 is devoted to the formal definition of the traffic process. General properties of service systems in statistical equilibrium are addressed. First, the notion of stationarity of a random process is discussed. Then, Little's law is stated and proved. The probability distributions of the state seen at different event epochs of a service system are characterized (Palm's probability distributions). Finally, the most common performance key indicators or metrics are introduced in Section 2.7.

The material in Chapter 3 is functional to applications to the modeling of networked service systems. The Poisson process is first introduced and characterized. Generalizations of the Poisson process are considered, namely nonhomogeneous and spatial Poisson processes as well as the Markov modulated Poisson process. Then, renewal processes are treated. Some operations on renewal process are analyzed in detail, namely excess variables and superposition. Finally, two special classes of processes with many applications in network and service system analysis are introduced: birth-death and branching processes.

Chapter 4 aims at a comprehensive account of the analysis of single server queues. The $M/G/1$ queueing system is analyzed in depth, with specific attention devoted to numerical evaluation of the probability distribution and to finite waiting line systems. An asymptotic approximation of the loss probability of the $M/G/1/K$ queue is presented as the queue size K grows. An account is given also of the $G/M/1$ model and of extensions to matrix-geometric models of single server queues, specifically the queues described by a quasi-birth-death (QBD) process. The general result known as Reich's formula, holding for any single server queue, closes the chapter. More results for the general $G/G/1$ queue can be obtained only via approximations and are dealt with in Chapter 8.

Chapter 5 focuses on queueing models with multiple parallel servers. The general $G/G/m$ model does not yield to closed-form analysis. Then, we address special, yet relevant, cases, mostly based on Poisson arrivals. We consider both loss- and wait-oriented systems. The first category is represented by the Erlang model, i.e., the $M/G/m/0$ queue, where there is no wait. This is of primary importance in many practical applications, e.g., in modeling cellular networks. We give several examples thereof. Then, we consider the $M/M/m$ queue, that is completely tractable and allows a good insight into the working of multi-server queues. We use this model to discuss a classic problem of service system, namely the comparison between separate versus shared queues. Finally, we analyze infinite server models. In spite of their seemingly only theoretical relevance, they have highly useful applications. We discuss the application of an infinite server model to the analysis of message propagation in a line network.

Differentiated treatment of traffic flows and sharing of a communication or processing resource are key issues in telecommunication networks as well as in many other networked service systems (transportation, computing, energy distribution, to mention few of them). We leverage on results of single server queueing, specifically on the $M/G/1$ queue, to derive models of priority queueing systems in Chapter 6. We address head-of-line, shortest job first, shortest remaining processing time, and preemptive policies. We use those results to understand the basic trade-offs of service policy differentiation. Ultimately, from a traffic engineering perspective, we aim at characterizing the impact of introducing prioritized service classes on performance perceived by different customers. The second part of the chapter is devoted to scheduling. Major examples of scheduling are introduced and analyzed, namely processor sharing, weighted fair queueing, credit-based fair queueing, least attained service. A specific attention is payed to weighted fair queueing, that has laid the conceptual ground on which one of the major attempts of providing quality of service in the Internet has been founded. Finally, we review optimal queueing disciplines for different classes of queueing system, where the classification is based on what information can be exploited by the service policy.

Chapter 7 is intended to provide a solid introduction to the vast topic of queueing networks. There is a large body of literature on queueing networks, given both the fundamental theoretical interest of the model by itself and its numerous applications to communications networking, cloud computing, transportation systems, manufacturing, inventory and storage management. We address first the Jackson-type model of a queueing network, considering both open and closed queueing networks. The general theory is discussed in detail. Optimization problems are defined, as well as extensive examples of use of those models applied mostly to communication and transportation networks. The famous Braess paradox is discussed as a highly instructive warning on how even apparently simple queueing network models can turn out to be deceptive to intuition. In addition, we introduce loss networks, since they are a completely different model with respect to most other queueing networks, so that they deserve an ad hoc treatment. Here too we devote significant space to application examples of the model. Finally, we discuss the stability of queueing networks, a topic of growing interest.

In Chapter 8 we review basic results and approaches for obtaining approximate results with more general models than those for which exact solutions are available. First, we consider G/G queues, both single server and multi-server.

We derive bounds and approximations for the mean system time. We also introduce an asymptotic bound for the waiting time probability distribution and the Gaussian approximation based on the Brownian motion process. Approximate analysis of the mean system time is extended to network of G/G queues. We then cover the fluid approximation, both as an asymptotic description of a properly scaled process and as a continuous state approximation of discrete systems. Within this framework, we address also stochastic fluid models. As an application example, we apply the fluid model to the performance evaluation of a packet multiplexer loaded with intermittent (On-Off) traffic sources.

The last three chapters form the third part of the book, devoted to application of modeling and performance evaluation tools to three broad fields of networked service systems: multiple access, congestion control and quality of service guarantees.

Models for Slotted ALOHA and carrier-sense multiple access (CSMA) are introduced in Chapter 9 as an application of the tools defined in the previous parts. Selected topics are discussed out of the huge existing literature. The main target is twofold: (i) grasp how the general analysis tools of the previous chapters can be applied to a specific technical context; (ii) give concrete examples of how the working of a system can be understood and performance trade-offs characterized by means of a model. Under this respect, multiple access systems are one of the major examples of the potential of Markov chains. We first look at Slotted ALOHA, devoting special attention to the stabilization of the protocol. Pure ALOHA is considered as well, specifically in the general case of variable length packets, which leads to a new, nonclassic analysis. CSMA is then examined in detail. We consider models able to describe a general multi-packet reception setting. Stabilization is investigated as well. The remaining part of the chapter is devoted to the famous WiFi MAC protocol, the CSMA/CA. We derive the saturation throughput, access delay performance and give a thorough discussion of the drawbacks of the binary exponential back-off mechanism, advocating an alternative back-off adaptation algorithm, the so-called idle sense. Finally, the fairness issue of WiFi is discussed and evaluated.

Congestion control is among the most important topics in network traffic engineering. In Chapter 10, we address specifically congestion control in the Internet, even if the considered models can be applied to other contexts, abstracting from technology details. We address closed-loop congestion control as realized by the Transmission Control Protocol (TCP). First, general ideas and definitions are laid out. Then, several variants of the TCP congestion control algorithms are reviewed (classic TCP, CUBIC, Vegas, DCTCP, BBR). As for the models, the fluid approximation is used to gain insight into the dynamics of a TCP connection. First, a simple constant capacity single bottleneck scenario is considered. Then, a variable capacity model is introduced, thus showing the usefulness of the fluid approach and at the same time, identifying a resonance phenomenon of TCP congestion control with the time scale of the bottleneck time-varying capacity. We consider then fluid models of multiple TCP connections sharing a same fixed capacity bottleneck link. We review models for classic TCP (with a drop-tail buffer and with a buffer running an Active Queue Management algorithm) and models for DCTCP. The fairness concept is explored and a general framework is introduced, based on Network Utility Maximization (NUM). Besides giving a general approach to the definition of fairness, NUM allows revisiting TCP congestion control, interpreting the classic TCP operations as a distributed, iterative algorithm for the solution of a global network optimization problem, namely the maximization of a social utility function under link capacity constraints. Finally, we review the main traffic engineering issues that TCP faces in current networking practice, highlighting state-of-the-art approaches to solve them and open problems.

Chapter 11 is devoted to models of traffic sources, sharing a common network resource, under strict quality of service (QoS) requirements. This framework is apt for so called inelastic or inflexible traffic sources, that require their throughput and delay to lie in suitable ranges to provide an effective service. We consider first the deterministic traffic theory. It is based on nontrivial *deterministic* bounds that describe the traffic source behavior and the service provided by network elements. The main result it provides is a kind of "system theory" that allows us to give worst-case end-to-end performance bounds for networked service systems and to dimension network elements that guarantee a prescribed level of quality of service. The down side is that performance bounds can sometimes be quite loose. Moreover, the stochastic nature of traffic is not canceled altogether, since we need to analyze and dimension devices (the "traffic shapers") that enforce deterministic bounds on the stochastic traffic flows offered to the network. We move then to stochastic models of the multiplexing of inelastic traffic sources. Here we introduce the concept of effective bandwidth and give some major results on the relevant theory. We show how effective bandwidth can be used to analyze and dimension a network of service elements, exploiting the stochastic variability of the offered traffic to reap the so-called multiplexing gain.

Finally, a primer on probability, random variables, and stochastic processes is presented in the Appendix. It gives essential definitions and properties to ease the reader of this book that needs a quick reference to refresh its background of probability and Markov chains.

1.4.2 Use

This book is meant as an advanced textbook, suitable for senior undergraduate, graduate, and PhD students. It can be consulted also by those who need a solid introduction to performance evaluation in an applied context. It aims to provide a self-contained source text to cover traffic theory, queueing theory, and their application to networked service systems.

The objective of the book is to provide a comprehensive guide to these topics:

- What traffic is, and how it is characterized and applied to networked systems.
- The performance evaluation tools from queueing theory, used to model, analyze, and dimension service systems to which traffic is offered.
- Applications of performance evaluation tools to major aspects of networked systems, selected for their interest both from theoretical and application points of view: multiple access, congestion control, quality of service.

The book can be a primary reference for classes in engineering, computer science, data science, and statistics, for students desiring to gain understanding of fundamentals of performance evaluation as well as aiming at consolidating a basic knowledge with more advanced material. The book assumes a basic knowledge of probability (a concise refresher is provided in the Appendix, tailored to topics required in the book) and programming (any language will do; what matters is having firmly understood the logic of programming). To appreciate fully several examples of the book, it is useful to know the basics of TCP/IP networking and of communication systems (especially cellular and wireless ones). Most undergraduate students of computer, electrical, telecommunications, and industrial engineering and of computer science take classes on communication and networking at an introductory level, which is more than enough to understand application examples of this textbook. Occasional readers of this textbook could be found among graduate students of data science, transportation engineering, physics, mathematics, statistics as far as they need to manage, design and optimize service systems within their work. As a matter of example, transportation is a prolific field for application of queueing and traffic theory. It is not by chance that some examples presented in this textbook are drawn from transportation systems. Moreover, many scientists make extensive use of networking and computing facilities, which they often need to tailor to their special needs.

A set of exercises is proposed at the end of each chapter. They provide a self-test to assess the comprehension level of the subject of each chapter.

The book can be used modularly, given the dependencies among chapters, as depicted in Figure 1.7.

A directed arc between two chapters represents a major dependence. An arc labeled by section number means that the section of the source chapter is relevant to the target chapter. A label with two section numbers, connected by an arrow, means that the source chapter section is relevant for the target chapter section in the label. If the lecturer wishes to focus on the mathematical tools (with possibly some application examples), Parts I and II can be used. This way, a solid introduction to traffic and queueing theory is provided, moving from first principles

Figure 1.7 Dependencies among chapters.

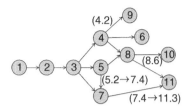

and basic definition to rather advanced topics. If a short course is to be set up, besides the whole of Part I, Chapters 4 and 5 provide an introductory cornerstone to queueing theory (possibly skipping some more advanced topic, e.g., Sections 4.3, 4.7, 4.8, 5.5). According to available time, class interest and skills, the queueing theory core could be extended by considering advanced topics in Chapters 4 and 5, covering also priority and scheduling in Chapter 6 or generalizing stand-alone queueing systems to network of queues, dealt with in Chapter 7 or choosing topics from Chapter 8, to introduce bounds and approximations that go beyond queueing models yielding to mathematical analysis in closed form.

To offer a course leaning toward applications, Part I plus selected chapters from Part II can be sampled and then Part III can be covered. The minimum set of material to be covered in Part II in this case comprises Section 4.2 (the concept of embedded Markov chain and its application to a single server queue) and Section 8.6 (the fluid approximation). A richer selection of Part II chapters can be organized according to the guidelines mentioned above.

1.4.3 Notation

Scalar variables are denoted usually with plain letters (e.g., x), while vectors are denoted with small-capital boldface letters (e.g., $\mathbf{x}$) and matrices with capital bold-face letters (e.g., $\mathbf{X}$). The identity matrix is denoted with $\mathbf{I}$, while $\mathbf{e}$ stands for a column vector of 1's.

Time is usually denoted with t. Hence, $Q(t)$ denotes a function of time. Space variables are usually denoted with x, y, and z.

The usual mathematical notation is used for the sets of integer and real numbers, $\mathbb{Z}$ and $\mathbb{R}$ respectively. The notation $\mathbb{R}^+$ ($\mathbb{Z}^+$) indicates the set of non-negative real (resp., integer) numbers.

We use sometimes the $o(\cdot)$ and $O(\cdot)$ notation. Writing $g(x) \sim o(f(x))$ for $x \to x_0$ means that $\lim_{x \to x_0} g(x)/f(x) = 0$. Instead $g(x) \sim O(f(x))$ for $x \to x_0$ means that the ratio $|g(x)/f(x)|$ remains bounded as $x \to x_0$.

A $-$ (+) subscript on a variable x corresponds to approaching x from the left (right). As an example, $f(x_-)$ stands for $\lim_{\epsilon \to 0} f(x - |\epsilon|)$. Analogously $f(x_+) = \lim_{\epsilon \to 0} f(x + |\epsilon|)$. For a continuous function, it is $f(x_+) = f(x_-)$, whereas a function for which the two limits are finite, but $f(x_+) \neq f(x_-)$, is said to have a jump at x.

The probability of event E is denoted with $P(E)$. $I(E)$ denotes the indicator function of the event E: it is equal to 1 if and only if event E occurs.

Capital letters denote random variables, while sample values are usually denoted with a small capital letter, e.g., the random variable V that takes a specific value x is written as $V = x$.

The cumulative distribution function (CDF) of a random variable V is denoted with $F_V(x) \equiv P(V \leq x)$, the complementary CDF (CCDF), sometimes also referred to as *survivor function*, with $G_V(x) = P(V > x) = 1 - F_V(x)$. When existing, also the PDF is used and it is given by $f_V(x) = F'(x) = -G'(x)$.

The expectation operator associated with the random variable X is denoted with $E_X[\cdot]$. The subscript X is dropped unless it is necessary to avoid ambiguity. Given a function $g : D \mapsto \mathbb{R}$, where D is the domain of the random variable X, it is $E[g(X)] = \int_D g(x)dF_X(x)$. Specifically, the mean value of X is denoted with $E[X]$. The variance is denoted with $\sigma_X^2 = E[X^2] - (E[X])^2$. Sometimes the notation $Var(X)$ is used for the variance of the random variable X. The coefficient of variation (COV) of the random variable X is defined as $C_X = \sigma_X/E[X]$. Also the squared COV (SCOV) C_X^2 is used.

The Laplace transform of the PDF of a non-negative random variable V is denoted with $\varphi_V(s)$, and it can be calculated from $\varphi_V(s) = E[e^{-sV}] = \int_{0_-}^{\infty} f_V(x)e^{-sx} \, dx$.

For random variables defined over the entire real axis, we define the moment generating function (MGF) $\phi_V(\theta) = E[e^{\theta V}] = \int_{-\infty}^{\infty} f_V(x)e^{x\theta} \, dx$.

As for discrete random variables, the CDF, CCDF, and probability distribution of a discrete random variable N are denoted with $F_N(k) = P(N \leq k)$, $G_N(k) = P(N > k)$ and $p_N(k) = P(N = k)$ for $k \in \mathbb{Z}$. Note that in the discrete case the equality sign in the definition of the CDF is important.

The MGF of a non-negative discrete random variable N is denoted with $\phi_N(z)$. It can be calculated from $\phi_N(z) = E[z^N] = \sum_{k=0}^{\infty} p_N(k)z^k$.

The notation $A \sim B$ means that the random variables A and B have the same probability distribution. The notation $Z \sim \mathcal{N}(a, b)$ means that Z is a Gaussian random variable with mean a and variance b, while $Z \sim \mathcal{U}(a, b)$ denotes a random variable Z uniformly distributed in the interval $[a, b]$. A random variable Z with negative exponential probability distribution and mean a is denoted with $Z \sim Exp(a)$.

Measure units follow the scientific International System, i.e., meters, seconds and multiples thereof. The symbols and values of multiples and sub-multiples of measure units used in this text are listed in Table 1.1 for reader's ease.

A list of acronyms can be found at the beginning of the book.

Table 1.1 Symbols of multipliers and sub-multipliers of measure units.

Symbol	Name	Value	Symbol	Name	Value
k	kilo	10^3	m	milli	10^{-3}
M	Mega	10^6	μ	micro	10^{-6}
G	Giga	10^9	n	nano	10^{-9}
T	Tera	10^{12}	p	pico	10^{-12}
P	Peta	10^{15}	f	femto	10^{-15}

1.5 Further Readings

Many books can be consulted to integrate the material of this book or to provide in-depth follow-ups.

A first group of references addresses applied probability and queueing theory [54, 86, 94, 121, 130, 131, 196]. This list contains textbooks biased toward or definitely devoted to queueing theory. While being excellent sources for learning on queues, they are not concerned with any specific application. [94, 130, 131] are classic introductory textbooks on queueing theory. [86] is a more recent textbook on queueing theory and stochastic networks, with some emphasis on fluid approximations. [54, 121] are mainly monographs on stochastic and queueing networks. The comprehensive textbook [196] provides an excellent and thorough introduction that covers everything from probability, to Markov chains, queues, and simulation.

Another group of books, [30, 58, 98, 122, 137, 139, 193], leans more toward applications to networked service systems, often in the realm of information and communications technologies. [30] is a classic textbook on communication network performance evaluation. It represents one of the first examples of a perfect mix of technological aspects coupled with rigorous modeling, analysis and dimensioning approaches, based on queueing theory and Markov chains. A much more recent attempt to introduce modeling and performance analysis approaches starting from real-life problems in offered by [58], with specific reference to social networks and communication networks. It uses a wide range of mathematical tools, mostly at an introductory level, and does not give any account of queueing theory, except of elementary notions. The ponderous book of Kumar et al. [137] overviews models and performance results for all aspects of communication networks (multiplexing, switching, routing). It assumes that the reader is already familiar with the required basic theory. It focuses only on performance models. [98] gives a full introductory account of queueing theory, constantly coupling theory and application to computer networks. The book is mostly at an introductory level. [122] is a very nice and concise book on rather advanced modeling, mostly applied to communication and networking problems. This deep book is a neat example of a smart balance between presenting methodological tools and applying them to technical systems, even if examples are rather at high level and do not touch many technical details of real systems. [139] is mainly an introductory level queueing theory book, the last chapter overviewing applications to communication networks. The recent book [193] applies optimization, game and control theories to modeling, analysis and dimensioning of communication networks.

Finally, to expand fundamental theories on probability and Markov chains, the reader might refer to the following classic textbooks, being advised that many other excellent books can be found: for probability theory [76, 90]; for Markov chains and stochastic processes [172, 117, 60, 40]. A concise and rigorous introduction to statistics can be found in Part II of [87], while statistics applied to performance evaluation is presented in [145].

Since networks and networked system have been mentioned several times, it is worth spending a word on a terminology clarification. Here "network" is meant to be a technological network, i.e., the interconnection of service systems (either physical or logical), set up to support a class of applications, e.g., telecommunications, computing, transportation, energy distribution, logistics and inventory, industrial production. The word *network* is also meant sometimes to address graph-based models of interconnected entities. This is more precisely referred to as *network science*. Two reference textbooks on the subject are those by Albert László Barabási [24] and by Mark E. J. Newman [169]. While graphs are a widely used model for any network, the focus of network science is on understanding the properties of graphs and what they mean to the specific "environment" being modeled.

The focus of this book is to apply rigorous mathematical methodology to the performance evaluation, design, and optimization of technological networks. This is essentially the difference between science and engineering (or, more broadly, between fundamental and applied science). The former is concerned with modeling reality to understand it; the second aims at understanding in order to act on reality.

Problems

1.1 Let an audiovisual (AV) frame have a constant length F bigger than the maximum packet payload L, so that each AV frame requires $m = \lceil F/L \rceil$ packet to be conveyed to the destination. Play-out of an AV frame requires all packets carrying the frame information to have been received in due time. Generalize the analysis of the delay equalization buffer of this chapter to the multi-packet per frame case.

1.2 Generalize further the model of Problem 1.1 by letting the number M of packets (each having a fixed length L) per AV frame be a random variable with probability distribution $q_m \equiv \mathcal{P}(M = m)$, $m \geq 1$. Give an expression of the starvation probability for an assigned PDF of the network delay $f_D(x)$, under the same assumptions as in Section 1.3.1.

2

Service Systems and Queues

Things may come to those who wait, but only the things left by those who hustle.
Abraham Lincoln

2.1   Service System Structure

A service system is defined by three elements:

1. *Structure.* What parts the service system is composed of. Specifically, structure relates to the resources the system can use to serve the demands posed by the users.
2. *Policies.* These address the rules according to which the service system operates on the demands of the users, exploiting its resources.
3. *Capacity.* This is maximum (upper bound) amount of work that the service system can provide per unit time, given its structure and policies, to meet the user demand as it grows unboundedly.

For example, let us consider the output line of a router. According to packet switching principle (store and forward), packets destined to that output link are handled by the network level entity of the output port. If the output line is idle (no packet transmission is under way), the packet is immediately sent out onto the link. Otherwise, packets arriving at a busy link are stored in the buffer of the output port, waiting for their opportunity to be sent out.

The service system structure is the network entity of the link, equipped with buffer space B and capacity C provided by the lower-layer entity (data link layer entity). The former is measured in terms of the maximum number of bytes or packets that can be stored in the buffer. The second one is measured by the maximum number of bits that the lower layer entity accepts (to be delivered onto the link)

Network Traffic Engineering: Stochastic Models and Applications, First Edition. Andrea Baiocchi.

per unit time. These two elements are the key components of the structure of the service system at hand.

The policies specify the rules according to which a new packet is selected out of the buffer for service (delivery onto the output link) as soon as the link is available. An example of queueing discipline is FIFO (first-in, first-out), the well-known policy that prescribes that packets be served in the same order as they have arrived. This discipline is quite simple to realize, since it can be implemented by using two circular pointers (addresses) and a sequential memory. One address points at the first byte of the oldest packet in the buffer (the first to go, as soon as the link is available), the other address points at the tail of the queue, that is at the first empty byte of the buffer space.

Other disciplines are: (i) LIFO (last-in, first-out), the one used by a CPU serving procedure calls; (ii) HOL (head-of-line) priority queueing, where each incoming packet carries a label that assigns a priority to the packet. With HOL priority, packets with higher priority are served first. A packet with priority j is served only if there is no packet with priority $i < j$ waiting for service. Packets with a same priority are served in FIFO order. Priorities can also be dependent on the time it takes to serve a packet or they can be time-varying, e.g., a packet priority can grow up with its queueing time proportionally to a constant b_i for packets belonging to class i.

Yet another example of discipline is the fair queueing (FQ) or its limiting case PS (processor sharing), where the server gives slices of service to each active "task" according to a round robin schedule (possibly weighted), until its service is completed. More complicated policies than round robin are sometimes referred to under the hat of FQ, where competing active "tasks" are served according to some credit they accumulate. Each time a service slice is granted, some credit is spent. Then, the service system picks the task with the highest credit for the next slice.

Examples of other queueing disciplines can depend on the amount of work demanded by a task, e.g., shortest job first (SJF) or least attained service (LAS). With the former one, the task requiring the least amount of work is chosen as the next one to serve. In the latter one, the task that requires the least amount of time to complete its demand is selected for the assignment of the service slice.

Finally, the capacity of the service system in our example (the output link of a router) is the link capacity C, offered by the lower layer to carry packets.

When referring to a generic service system, we name the service requests as *customers*. The facilities of the service system that are assigned to a service request are called *servers*. In each given context, specific terms are used to replace the generic terminology *customer* and *server* (e.g., packet and link in the previous example).

Besides structure, policy and capacity, a service system description encompasses the characterization of the service demand, which we address in the next section.

2.2 Arrival and Service Processes

Service demand is made up of two elements:

1. Service request submission to the system, or *customer arrivals*
2. Amount of service needed by each service request, or *service time* of a customer

The first element is given by the *arrival* process, a (generally statistical) description of the pattern of the service requests on the time axis. Let T_k denote the time elapsing between the $(k-1)$-th and the k-th request arrivals. That quantity is called the k-th *inter-arrival* time. It is positive and can be characterized as a random variable with the cumulative distribution function (CDF) $F_{T_k}(t) = P(T_k \leq t)$ or the complementary cumulative distribution function (CCDF) $G_{T_k}(t) = P(T_k > t) = 1 - F_{T_k}(t), t \geq 0$, sometimes also referred to as the *survivor function*. We define also the cumulative arrival time, from the initial time until the k-th arrival, $S_k = T_1 + T_2 + \cdots + T_k$, for $k \geq 1$.

For a stationary arrival process, the probability distribution of T_k does not depend on k. Hence we simplify notation and denote the random variable sharing the common probability distribution of all inter-arrival times with T. The corresponding survivor function and the probability density function (PDF)[1] are, respectively, $G_T(t)$ and $f_T(t) = -G'_T(t)$. The mean arrival rate λ is by definition the reciprocal of the mean inter-arrival time $E[T]$, i.e., $\lambda = 1/E[T]$, with

$$E[T] = \int_0^\infty t f_T(t) \, dt = \int_0^\infty G_T(t) \, dt \tag{2.1}$$

It is possible that, at each arrival, more than a single service request is posed to the service system (*batch arrivals*). Then, we have to add another discrete, positive random variable, G_k, that gives the size of the k-th batch. It is characterized by the probability distribution $p_{G_k}(n) = P(G_k = n)$ for $n \geq 1$. For a stationary batch process, the probability distribution of G_k does not depend on k. In a batch arrival process, inter-arrival times refer to batch arrivals.

An alternative way to describe a customer arrival process is by means of a *counting process*. Let $A(0, t)$ be a non-negative integer representing the number

---

1 In principle, the PDF, when it exists, is equivalent to the CDF as a description of the statistical properties of a random variable. In practice, there are a number of reasons why it is often preferable to use the CDF (or the CCDF), especially in numerical computations: (i) CDF and CCDF are probabilities; as such they are always comprised between 0 and 1, whereas a PDF can take any non-negative value; (ii) CDF and CCDF have monotonicity properties, the PDF has none in general; (iii) CDF and CCDF are nondimensional, whereas a PDF has dimension of the reciprocal of the dimension of the random variable; e.g., the PDF of an inter-arrival time is measured as a frequency, the PDF of a distance is measured as the inverse of a length; and (iv) in case of probability masses concentrated at given points, the CDF and the CCDF have jumps, whereas the PDF requires using the Dirac delta function.

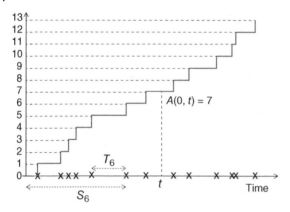

Figure 2.1 Example of counting process: the inter-arrival time T_6, between arrival events 5 and 6, is highlighted along with the cumulative time S_6, from time 0 up to arrival 6.

of arrivals in the interval $(0, t]$. This quantity can be characterized as a random process, whose realizations are step functions, that are almost always constant and jump by one (or by a random integer amount, in case of batch arrivals) at the times immediately after a customer arrival. An example of arrival counting process is depicted in Figure 2.1.

A fundamental relationship can be written between the cumulative arrival time and the counting process of the arrivals. It is easy to verify that the event $S_k > t$ is equivalent to the event $A(0, t) < k$. Therefore,

$$P(A(0, t) < k) = P(S_k > t) \tag{2.2}$$

or $P(A(0, t) < k) = 1 - F_{S_k}(t)$; then

$$P(A(0, t) = k) = F_{S_k}(t) - F_{S_{k+1}}(t), \qquad k \geq 0 \tag{2.3}$$

with the convention that $F_{S_0}(t) \equiv 1$ for $t \geq 0$. As a special case, we find $P(A(0, t) = 0) = 1 - F_{T_1}(t) = P(T_1 > t)$. If the mean of $A(0, t)$ exists, eq. (2.3) yields

$$E[A(0, t)] = \sum_{k=1}^{\infty} F_{S_k}(t) \tag{2.4}$$

As for the service times, each customer arrival brings with it an amount of service demand, that can be measured by how many (and also what type of) resources it needs to accomplish its service and how long it will keep those resources engaged. For example, a packet arriving at the output link of a router requires being sent out. It engages the output link for a time equal to its transmission time. For a constant capacity output link, the service time of the k-th packet is simply L_k/C, where L_k is the length of the k-th packet. In general, we denote with X_k the service time of the k-th customer. It can be characterized as a random variable with survivor function $G_{X_k}(x) = P(X_k > x), x \geq 0$.

Note that the amount of the demand that customers pose to a service system depends *both* on arrival and service processes. If we just say that there is one packet arriving per ms on the average, we are not giving a complete picture of the load on the output link. The amount of time that the link is busy transmitting (=serving) packets is in fact proportional to the packet lengths, so it makes a big difference whether incoming packets are 1500 bytes long or only 52 bytes long. If one observes the link out of a local router (i.e., a router at the access of the network), when downloading a file from a remote site, it will probably be the case that the input bit rate (also referred to as downlink bit rate in the ADSL context) is something like 30–40 times the output bit rate (uplink bit rate). The reason is that incoming packets carry TCP data segments (e.g., for http page download or file transfer), while outgoing packets just carry TCP ACKs, with no payload. Their service times being proportional to packet lengths, it is apparent that the load on the uplink is much lighter than the load on the downlink.

Any description of the demand on a service system must encompass both arrivals and service times. By putting them together, we can define the *unfinished work* or *workload* $U(t)$ of a service system. It is the amount of work that servers have to carry out to empty the system at time t, provided no more customers arrive at the system. This is also known as the *virtual waiting time*, since $U(t)$ would be the amount of time an hypothetical (hence virtual) customer arriving at time t should wait for service, if FIFO discipline is adopted. In general $U(t)$ is a stochastic process[2], since it is driven by random customer arrivals and service demands. A sketch of a sample path of $U(t)$ for a work-conserving[3], *single-server* system is shown in Figure 2.2.

The time elapsing between an arrival that finds the system idle and the subsequent departure that leaves the system idle is called *busyperiod*. The time interval during which the system is continuously idle is called *idle time* (see Figure 2.2).

The slope of the decay of $U(t)$ between two consecutive arrivals is -1, i.e., 1 sec of work for sec of elapsed time, until it hits the x-axis; then it stays at 0, until a new arrival occurs. At each new arrival, the function $U(t)$ jumps up by the amount of the service time brought into the system by the arriving customer, provided it can enter the system.

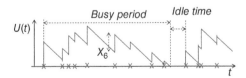

Figure 2.2 Example of workload process of a work-conserving, single-server system: a sample service time, X_6, is highlighted along with samples of a busy period and an idle time.

---

2 See the Appendix at the end of the book for a concise review of main probability theory definitions and some results useful for the topics dealt with in this book.

3 See Section 2.3 for a formal definition.

Given the sequence of inter-arrival times $\{T_k\}_{k\geq 1}$ and service times $\{X_k\}_{k\geq 1}$, it is easy to write a recursion giving the value of the unfinished work seen by arriving customers. Let $U_k \equiv U(t_{k-})$ be the amount of unfinished work seen by the customer arriving at time t_k. We have

$$U_{k+1} = \max\{0, U_k + X_k - T_{k+1}\}, \qquad k \geq 1, \tag{2.5}$$

with the initial condition $U_1 = 0$. The recursion holds for a work-conserving, single-server system that accepts all arriving customer (no rejection).

If the mean duration of the busy period is finite, we can write a simple relationship between the mean busy period, the mean idle time, and the utilization coefficient of the server, i.e., the mean fraction of time that the server is busy. It is

$$\rho = \frac{E[Y]}{E[Y] + E[I]} \tag{2.6}$$

where Y and I denote the random variables associated with the busy period and the idle time.

Example 2.1 A simple variant of (2.5) is obtained if we assume that arriving customers do not join the queue if the system time exceeds a threshold. Let us assume that the k-th arriving customer has a time deadline D_k, and that it can evaluate the amount of work to be done before it can get its own service done (response time). With a FIFO discipline the response time of the k-th customer is $U_k + X_k$ (including customer's service time). Then, the k-th customer joins the queue if $U_k + X_k \leq D_k$, it leaves immediately if the opposite is true. The recursion becomes:

$$U_{k+1} = \begin{cases} \max\{0, U_k + X_k - T_{k+1}\} & U_k + X_k \leq D_k, \\ \max\{0, U_k - T_{k+1}\} & U_k + X_k > D_k. \end{cases} \tag{2.7}$$

Example 2.2 Yet another variant is obtained if we define a utility the customer gains for each second of received service, say u, and a penalty it suffers for each second of waiting time before service, say c. The customer joins the system if its net utility is positive, i.e., only if $uX_k - cU_k > 0$. The parameters u and c could possibly be customer specific, i.e., they could depend on k.

2.3 The Queue as a Service System Model

A queue is a mathematical model of a service system. It is defined by

1. the arrival process;
2. the service process;
3. the structural elements.

The last ones are the waiting line size, the queue discipline (service policy), and the number of servers. It is assumed that servers are interchangeable and fully accessible, unless otherwise specified. Moreover, we consistently assume that considered service systems are work-conserving, single arrival, single service.

Work-conserving means that no work is created or destroyed in the system, i.e., a customer entering the system will eventually complete its service with probability 1 and, conversely, any server that becomes idle will immediately pick another customer from the waiting line, if there is at least one.

Single arrival and single service means that the event that two or more customers arrive at the same time or terminate their respective services at the same time is ruled out. As opposed to this assumption, bulk arrival (bulk service) means that groups of customers arriving at the same time are possible (groups of customer served in a bundle all together are possible). Bulk arrivals (services) are sometimes referred to as batch arrivals (services).

The queue evolution is described by stochastic processes that represent the amount of unfinished work in the queue at a given time, $U(t)$, or the number of customers in the system, $Q(t)$, the waiting time of an arriving customer accepted into the system at time t, $W(t)$, the number of servers engaged into service, $M(t)$, and similar others.

A notation to denote the main characteristics of a queueing system has been defined by David G. Kendall. It consists of a five-tuple $A/B/m/K/N$.

The letters A and B stand for acronyms, denoting the arrival and service process respectively. Used acronyms are: (i) M for Poisson process, i.e., negative exponential inter-arrival or service times, identically distributed and independent of one another (see Section 3.2); (ii) D stands for deterministic; and (iii) G (sometimes noted GI) is a general renewal process (see Section 3.4); iv) $MMPP$ is the Markov Modulated Poisson Process, i.e., a Poisson process whose mean arrival rate is a function of the state of a finite, discrete-state Markov process (see Section 3.3), and so on and so forth.

As for the last three fields of the notation, m is an integer number ≥ 1 that denotes the number of servers, while K is the size of the waiting line, i.e., the maximum number of customers that can be accommodated in the system waiting for service. Finally, N denotes the size of the population of customers that offer service requests to the queue. Either of these three integer numbers can be infinite. If they are infinite, the parameters K and N are omitted in the notation. For example, $M/M/1$ denotes a single server, infinite size queueing system whose arrivals follow a Poisson process and service times are exponentially i.i.d. random variables. Such a model can only delay customer service requests, since no request gets rejected for lack of room in the queue.

In the following we will use consistently the notation T and X to refer to customer inter-arrival and service times, respectively. We will also use the notation λ

and μ to denote the mean customer arrival rate and server service completion rate, i.e., $\lambda = 1/E[T]$ and $\mu = 1/E[X]$.

2.4 Queues in Equilibrium

2.4.1 Queues and Stationary Processes

A queueing model of a service system is described by random processes that give, e.g., the number of customers in the queue, $Q(t)$, the unfinished work, $U(t)$, the waiting time of the n-th customer, W_n. As pointed out by even these few examples, processes of interest include integer and real valued quantities, indexed either by discrete or by continuous time.

A stochastic process usually describes the behavior of a dynamical system. Starting from an initial state, the evolution of a dynamical system goes through a *transient* phase, after which it eventually settles on a steady-state equilibrium. For stochastic systems, the steady-state equilibrium corresponds to fluctuations of the system variables that are governed by probability laws *invariant with respect to a time shift*. System variables are *not* constant at steady-state. Rather, their probability distribution does not depend on time. While transient is unavoidable when starting off a given initial state, steady-state can possibly be never attained., e.g., unstable system never settle on a steady-state statistical equilibrium. The concepts of transient and stationary phases applies to stochastic processes as well. The reason why we are interested in stationary processes is that performance metrics evaluated on a stationary stochastic system do not depend on the time when we observe the system, so that they are informative of a general property of the system itself. That description of the system holds as long as we can assume or guarantee that stationarity of the queue processes holds.

For example, let us consider a 10 Gbit/s link connecting the campus network of University of Rome La Sapienza to the GARR backbone, that is to say the backbone network of academic and research institutions in Italy (this backbone is, in turn, interconnected to the Internet)[4]. The so-called Multi Router Traffic Grapher (MRTG) diagrams of the link are plotted in Figure 2.3. Each graph shows the input and output profile of the traffic in Gbit/s versus time. The time scale changes across the four diagrams shown in Figure 2.3.

The top plot runs over the day time scale. It shows the plots of the input/output traffic for two consecutive days, Friday and Saturday. The day cycle of traffic, with a fast rise in the morning, fluctuations over the working day (with a notch around lunch time) and a fall going toward the evening is typical of links where

---

4 The data displayed in Figure 2.3 is available at https://gins.garr.it, under the heading "Statistics."

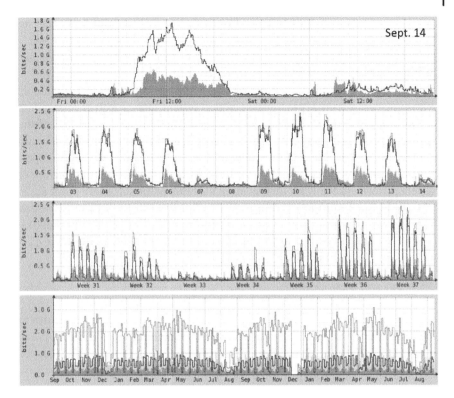

Figure 2.3 MRTG diagrams of a 10 Gbit/s link connecting the campus network of University of Rome La Sapienza to the Italian academic backbone and to the Internet. The diagrams refer to different time scales. From top to bottom, the time scale goes respectively over two consecutive days, weeks, months, years. Shaded profile curve: input traffic. White profile curve: output traffic.

human-related traffic is dominant and we are in a working context. Traffic profile for residential connections may have a traffic peak in the evening and relatively low traffic intensity during the late morning/early afternoon.

The second plot corresponds to a week time scale. Apparently working days present a similar traffic profile (the traffic process can be described as cyclo-stationary), except of weekend days, where traffic is remarkably lower. Again, this is typical of working environments, where a major component of the traffic is generated by human activity.

The third graph is over the month time scale. Here we can appreciate a seasonal effect. Since the measurements have been run in mid-September, the month time scale encompasses weeks from the months of August and September. Mid-August is clearly a slack time for this link, as expected, since most human activities are paused during the central part of August in Italy, where the link is.

The fourth diagram has a year time scale going from mid-September to mid-September over two years. Apart from notches around mid-August and a hole at the end of the second instance of December (problably due to maintenance of the system), the traffic profile appears to be quite regular.

In general, stationarity is a property of a mathematical model that can apply with more or less accuracy to a given practical system. Only a deep understanding of the main driving forces that generate demand and of the mechanisms of the service system being modeled can assist the designer to judge whether stationary models can be assumed and for which time scale. Conversely, equilibrium performance predictions obtained from the model can provide a faithful picture of system performance only for time spans in the order of the stationarity time scale.

A queue is a mathematical model of a service system (see Section 2.3). Such a system can be described by a state, whose evolution is driven by external processes. The input-state-output representation of systems apply to queues, e.g., if we consider the number of customers in the queue, $Q(t)$, we can write

$$Q(t + u) = Q(t) + A(t, t + u) - D(t, t + u) \tag{2.8}$$

where $A(a, b)$ and $D(a, b)$ are the number of customer arrivals and departures in the time interval $(a, b]$, respectively. The arrival counting process $A(a, b)$ is the external "force" driving the system, while the departure process is a consequence of the interaction of the arrivals with the system structure and working rules (serving policy).

As usual with dynamical system description, we can identify a transient regime and a (possibly nonexistent) stationary regime. When started in an arbitrary state, the system evolves and eventually it settles on a stationary regime, if attainable. For a stochastic system, a stationary regime is characterized by invariance with respect to time shifts of the statistical description of the system processes. Obviously, $Q(t)$ is always variable with time, and fluctuations of $Q(t)$ can well be expected. What becomes invariant with time is the (limiting) probability distribution of the random variable $Q(t)$ and hence its moments. In other words, the random variables sampled by the process realizations at different times share the same probability distribution. By following this approach, we are led to define the steady-state probability distribution of a queue as the limit of the probability distribution of the processes that describe the queue, and specifically of the number of customers in the queue, if the limit exists.

Formally, let $p_k(t) = P(Q(t) = k)$ for $k \in \{0, \dots, m + K\}$, m and K being the number of servers and the waiting line size, respectively. Given an initial probability distribution $p_k(0) = P(Q(0) = k) = q_k$, we say $Q(t)$ admits a steady-state, if the limit

$$\lim_{t \to \infty} p_k(t) = p_k, \qquad 0 \le k \le m + K \tag{2.9}$$

exists, and the limiting constants form a probability distribution, i.e., $p_k \in [0, 1]$ and $\sum_{k=0}^{m+K} p_k = 1$. The limiting probability distribution characterizes what we call the *statistical equilibrium* of the random process $Q(t)$. The equilibrium probability distribution (when it exists) describes the long-term behavior of the system.

We define also the *stationary* probability distribution of a system: The probability distribution $\{s_k\}_{0 \le k \le m+K}$ is said to be stationary, if letting $p_k(0) = s_k, k = 0, \ldots, m + K$ implies that $p_k(t) = s_k, k = 0, \ldots, m + K$ for any $t > 0$. The stationary probability distribution does not necessarily exist, nor must it be unique or coincide necessarily with the limiting probability distribution. In most applications, stochastic modeling of the service system leads to a Markov process model (see the Appendix at the end of the book for a concise review). Often enough, for some values of the involved parameters, such models yield a unique stationary probability distribution, a unique limiting one and the two coincide. A thorough discussion of these issues for Markov chains is given in [196, Ch. 9].

In general, the limiting probability distribution, if it exists, depends on the chosen initial distribution q_k. For example, consider a service system with servers that double their serving capacity when the queue content exceeds a threshold θ and back it off to normal capacity as soon as the queue length falls below θ. Assume that the customer population arrival rate increases, if the quality of service of the system is good while it decreases, if a poor quality of service is provided. If the system is started at time 0 from a state $k > \theta$, customer demand reacts to the good quality (doubled service capacity) by a fast increase and the system tends to settle on high throughput levels, with possibly a large average occupancy. If on the contrary the initial state is $k < \theta$, the poor perceived quality of service will drive customers away, so that the system can settle on a low level equilibrium, consistently below the threshold level θ. If an observer samples the system at equilibrium (steady-state), it could find either a high or low average queue size level and the observation outcome depends on the initial state of the system. As another example, the distance traveled by a message forwarded in a vehicular network by means of multi-hopping will depend essentially on the initial vehicle spatial distribution, e.g., if it is clustered with large gaps in between, hence possible disconnections, or uniform along the road lanes.

We are very interested in service systems that admit a statistical equilibrium as $t \to \infty$ and such that the limiting state distribution does not depend on the initial conditions. Those systems have *typical* equilibria, that is to say, when they settle on their statistical equilibrium, their dynamics is the same no matter the way they were started off and what the transient path to equilibrium has been.

We say that a queue is *ergodic* if the limit in eq. (2.9) exists and it is independent of the initial probability distribution q_k. The "physical" intuition of an ergodic system is one that touches all possible states with some given probability, when it is in

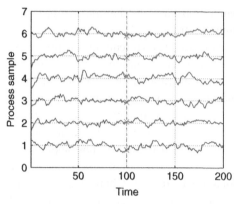

Figure 2.4 Illustration of time average and ensemble average for a stochastic process.

equilibrium, given that evolution rolls out for a time interval long enough (mathematically: as the time horizon tends to infinity). Then, time averages on a single (sufficiently long) realization yield the same result as ensemble averages, conceptually done on all realizations at a fixed time of the equilibrium, as if they were running in infinitely many parallel universes.

An illustrative example of this concept is given in Figure 2.4. Here six realizations of a correlated Gaussian process $X(t)$ are shown. The average value of the process can be found by averaging a *single* realization over time (any one of the horizontal curves), or by selecting a generic time point and averaging the values of the six process realizations at that time (e.g., the time 100 in the figure, where the values assumed by the realizations are marked with a cross sign). The two procedures lead to the same result for an ergodic process.

When we write $E[X(t_0)]$, we refer to the second approach, i.e., we refer to the limit (with probability 1) of $\frac{1}{N} \sum_{j=1}^{N} x_j(t_0)$ as $N \to \infty$. Here $x_j(t)$ is the j-th realization of the stochastic process $X(t)$. If the stochastic process $X(t)$ is ergodic, then $E[X(t_0)]$ can be obtained as well by the limit of $\frac{1}{T} \int_{t_0-T/2}^{t_0+T/2} x(t)\, dt$ as $T \to \infty$. Here $x(t)$ denotes a generic realization of the stochastic process $X(t)$. Note that in both definitions we assume that the resulting value does not depend on t_0 (stationary or statistical equilibrium).

Ergodicity is a key property to reconcile the performance predictions based on modeling with estimates worked out from system realizations (either by means of computer simulations or measurements). In the latter case, we deal with a single realization of the stochastic process observed for some time interval. Therefore, performance metrics are obtained as time averages. On the other hand, analytic modeling yields expectations, i.e., ensemble averages, obtained by averaging over different realizations of the stochastic process. Only if the quantities obtained according to these two approaches lead to same values can we rely on analytical performance predictions, validate a model with measurements, or interpret simulation results as estimates of the performance metrics of the system.

2.4.2 Little's Law

Let us consider a service system that is offered a flow of customers arriving according to a stationary stochastic process. The system provides each customer with a service that lasts an amount of time drawn from a stationary probability distribution. Customers entering the system receive service with probability 1, and they get out of the system within a finite time with probability 1. We assume a single-arrival, single-service system, i.e., bulk arrivals or service are ruled out.

Let us fix an arbitrary time origin and consider the time interval $(0, T]$. Let $A(t)$ be the number of arrivals in $(0, t]$ for the chosen time origin and any positive t. We assume that the arrival process is nondegenerate, i.e., that for any positive k we have $P(A(t) \geq k) \to 1$ as $t \to \infty$. Note that $\lim_{T \to \infty} \frac{A(T)}{T} > 0$ for a nondegenrate arrival process. Let $0 < t_{a,1} < t_{a,2} < \cdots < t_{a,A(T)}$ be the arrival times in the interval $(0, T]$, if $A(T) > 0$. The arrival process being stationary and nondegenerate, a first arrival must appear in a finite time with probability 1. Since we are interested in the limit for $T \to \infty$, we can safely assume that $A(T) > 0$ hereinafter.

Let $D(t)$ be the number of departures from the system in the interval $(0, t]$ for any positive t. Let also $0 < t_{d,1} < t_{d,2} < \cdots < t_{d,D(T)}$ be the departure times in the interval $(0, T]$, if $D(T) > 0$. Again, since service times are assumed to be finite with probability 1, $D(T)$ becomes positive with probability 1 as time T grows. Note that the indexing of the departure times follows the temporal ordering. Let t_{d,j_k} be the departing time of the customer arrived at time $t_{a,k}$, $k = 1, \ldots, A(T)$.

We consider the function

$$N(t) = N(0) + A(t) - D(t) \tag{2.10}$$

It represents the number of customers in the system at time t, by the very definition of $A(t)$ and $D(t)$. The functions $A(t)$ and $D(t)$ are step-wise, monotonously nondecreasing functions, with unit steps at arrival and departures times, respectively. Also, it must be $A(t) \geq D(t), \forall t$. The area between the two functions is simply the area of rectangles with unit height and length equal to $t_{d,k} - t_{a,k}$ for the k-th one. Figure 2.5 shows a sample path of the arrival counting process and of the departure counting process of a queue.

Figure 2.5 Example of arrival and departure processes of a queue. The number of customers at time t, $N(t)$, can be found as the difference between the upper step-wise curve and the lower step-wise curve at t. Crosses mark arrival epochs. Departures occur at jumps of the lower step-wise curve.

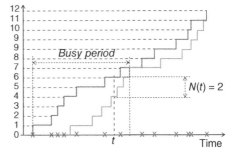

The time average of $N(t)$ is

$$\langle N(t)\rangle_T \equiv \frac{1}{T}\int_0^T N(t)dt = \frac{1}{T}\sum_{k=-N(0)+1}^{A(T)}(\min\{T,t_{d,k}\} - \max\{0,t_{a,k}\})$$

$$= \frac{1}{T}\sum_{k=-N(0)+1}^{0}\min\{T,t_{d,k}\} + \frac{1}{T}\sum_{k=1}^{A(T)}(\min\{T,t_{d,k}\} - t_{a,k})$$

$$= \frac{1}{T}\sum_{k=-N(0)+1}^{0}\min\{T,t_{d,k}\} + \frac{1}{T}\sum_{k=1}^{D(T)}(t_{d,k} - t_{a,k}) + \frac{1}{T}\sum_{k=D(T)+1}^{A(T)}(T - t_{a,k})$$

$$= \frac{1}{T}\sum_{k=-N(0)+1}^{0}\min\{T,t_{d,k}\} + \frac{1}{T}\sum_{k=1}^{A(T)}(t_{d,k} - t_{a,k}) - \frac{1}{T}\sum_{k=D(T)+1}^{A(T)}(T - t_{d,k})$$

$$(2.11)$$

where arrivals with nonpositive index (from $-N(0)+1$ to 0) refer to the arrival times of those users that at time 0 are found inside the system. In eq. (2.11) a sum is intended as equal to 0 if the lower index is greater than the upper index.

The first sum appearing in eq. (2.11), apart from the factor $1/T$, is upper bounded by the amount of unfinished work found into the system at the initial time 0, i.e., the amount of work required to clear all the backlog in the system at time 0. The last sum is nothing but the unfinished work of the queue at time T, i.e., the amount of service time required to complete service of all customers in the queue at time T. Let us denote the unfinished work at time t as $U(t)$: then the first term in eq. (2.11) is bounded by $U(0)/T$ and the last term is $U(T)/T$. Therefore

$$\langle N(t)\rangle_T = \frac{1}{T}\sum_{k=-N(0)+1}^{0}\min\{T,t_{d,k}\} + \frac{1}{T}\sum_{k=1}^{A(T)}(t_{d,k} - t_{a,k}) - \frac{U(T)}{T} \qquad (2.12)$$

Since the system is at equilibrium, $U(0)$ must be finite with probability 1. Moreover, assuming that the queue is stable as $T \to \infty$ means that the unfinished work $U(T)$ is bounded above with probability 1, so that $\lim_{T\to\infty} U(0)/T = \lim_{T\to\infty} U(T)/T = 0$ with probability 1.

The key point is that *any* service order corresponds to a permutation of the integer set $\{1, 2, \dots, A(T)\}$. Let the customers be labeled with integers $1, 2, \dots, A(T)$, according to their arrival order. Let j_k denote the index of the customer departing at time $t_{d,k}$. Let $\Theta_k = t_{d,j_k} - t_{a,k}$ be the system time of the k-th customer. Then, we have w.p. 1

$$E[N] = \lim_{T \to \infty} \langle N(t) \rangle_T \qquad \text{[ergodicity]}$$

$$= \lim_{T \to \infty} \frac{1}{T} \sum_{k=1}^{A(T)} (t_{d,k} - t_{a,k}) \qquad \text{[neglect the 1st and 3rd terms in eq.2.11]}$$

$$= \lim_{T \to \infty} \frac{1}{T} \sum_{k=1}^{A(T)} (t_{d,k} - t_{a_{j_k}}) \qquad \text{[sum commutativity]}$$

$$= \lim_{T \to \infty} \frac{1}{T} \sum_{k=1}^{A(T)} \Theta_{j_k} = \lim_{T \to \infty} \frac{1}{T} \sum_{k=1}^{A(T)} \Theta_k \qquad \text{[by definition of } \Theta\text{'s]}$$

$$= \lim_{T \to \infty} \frac{A(T)}{T} \frac{1}{A(T)} \sum_{k=1}^{A(T)} \Theta_k = \lambda \cdot E[\Theta]$$

In deriving the above result we exploit the assumption that $A(T)$ goes to infinity so that the ratio $A(T)/T$ tends to a finite, positive limit λ as $T \to \infty$. This is in essence the notion of "stationarity" we need, i.e., the arrival process cannot die out after a finite time or even slow down arrivals so that they become too sporadic. We also use the fact that permuting the order of summands does not change the sum value. We summarize the result as follows, where we adopt the notation most usually found in the Little's theorem statement [148].

Theorem 2.1 In a work-conserving service system at statistical equilibrium the limiting time-average number of customers in the system is equal to a constant $\overline{L}$ with probability 1; the sample path-average system time per customer is also equal to a constant $\overline{W}$ with probability 1, and the two are tied by $\overline{L} = \lambda \overline{W}$, where λ is the time-average arrival rate of customers admitted into the system, i.e., the reciprocal of their average inter-arrival time.

The obtained result is called *Little's law*. It states that, in a service system (queue) at statistical equilibrium, the mean number of customers inside the system at any given time (since we are at equilibrium, the mean is independent of time!) equals the product of the mean rate of customers entering the system by the mean sojourn time of a customer. Since we are at equilibrium, the mean rate in the system is equal to the mean departure rate of customers leaving the system, after completing their service. Little's law is a very general result and that is why it is so useful. Specifically, Little's law holds independently of

- The probability distribution of the inter-arrival times;
- The probability distribution of the sojourn times into the system;
- The number of serving units (servers) into the system;

- The service discipline, i.e., the order according to which customers are selected for service once inside the system.

Let us see some examples.

Example 2.3 *TCP connection* Let us consider a TCP sender and a TCP receiver engaged in a long-lived TCP connection, so that we can consider the transmission window process $W(t)$ to be stationary and ergodic (e.g., influence of initial slow start phase has died out). This requires that the connection duration is much bigger than the time scale of window variations, namely the round trip time (RTT) of the connection. For this to be a reasonable assumption, the intermediate network between the two TCP connection endpoints must be under a stationary traffic load, i.e., the time scale of significant variations of the link and node loads must be large as compared to the duration of the TCP connection (e.g., the connection lasts seconds, while load variation are relevant over minutes).

Under these conditions, we can consider the service system consisting of the entire network in between sender and receiver. Customers are TCP segments emitted by the sender. They can be injected into the "system" only when allowed by the window. We assume the sender is a saturated (greedy) source, i.e., it always has new packets to send. Then, the mean number of segments in the "system" is equal to the mean value of the window, $E[W]$. The mean sojourn time of a segment into the system is the mean of the time elapsing since when the segment is injected into the connection until when the corresponding ACK is received at the sender, thus informing the sender that it is done with that segment ("service" completed). So, the sojourn time is just the average RTT. By applying Little's law we find that the mean throughput of the connection is $\Lambda = E[W]/E[RTT]$ in segments per second.

Example 2.4 *Vehicular traffic* Let us consider a span of road of length L. Vehicles enter the road with a flow rate ϕ (vehicles/h) and the average speed of vehicles traveling on the road span is V (km/h). Here the service system is the road span, customers are vehicles and service consists of traveling the considered road for a length L, from the initial marking point of the span until the end. The average sojourn time of a vehicle in the system is the travel time L/V. By applying Little's law, we find that the average number of vehicles in the road span at equilibrium is $E[N] = \phi L/V$. This can be rearranged as $\phi = V\delta$, where $\delta = E[N]/L$ is the average density of vehicles in the road span. This is the well-known flow equation of transportation systems, namely for a road segment at equilibrium the mean vehicle flow equals the mean vehicle density multiplied by the mean vehicle speed.

2.5 Palm's Distributions for a Queue

Let us consider a queueing system and let $N(t)$ denote the number of customers queued up at time t, including those that are receiving service. We assume that the arrival process is nondegenerate and that the queue is ergodic and stable, i.e., it admits a proper limiting state probability distribution as $t \to \infty$, independent of the initial state.

Let $p_n = \mathcal{P}(N(t) = n)$ the steady state probability that there are n customers in the queue at a randomly sampled time during the statistical equilibrium regime, $n \geq 0$. This is the limiting probability distribution attained by the system, provided it is ergodic, i.e., it admits a limiting state probability distribution that is independent of the initial state (and coincides with the stationary probability distribution). A picture of the interactions of customers with the queueing system is shown Figure 2.6.

Customers that arrive at the system can be rejected or enter the system. In the latter case, they possibly have to stand in the waiting line before receiving service. After completing service, they leave the system.

Let $\mathcal{A}[t, t + \Delta t]$ denote the event of a single arrival occurring in the interval $[t, t + \Delta t]$. We assume that

$$\mathcal{P}(\mathcal{A}[t, t + \Delta t] | N(t) = k) = \lambda_k \Delta t + o(\Delta t) \tag{2.13}$$

for $k \geq 0$ and $\Delta t \to 0$. The non-negative quantity λ_k is the mean arrival rate of the customers when the system state is k. We let $\Lambda \equiv \sum_{k=0}^{\infty} \lambda_k p_k$ be the mean arrival rate of the customers to the system.

Finally, let $\mathcal{B}$ be the set of blocking states of the queue, i.e., those states where new arrivals are turned away and lost to the queue. Let also $\Omega = \mathbb{Z}^+ \setminus \mathcal{B}$ be the set of states where an arriving customer can actually join the queue[5].

Figure 2.6 Interaction of customers with a queueing system.

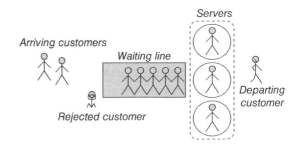

---

5 $\mathbb{Z}^+$ denotes the set of non-negative integers.

We can define the limiting state probability distributions seen by an arrival at the queue, $\{q_{a,n}\}_{n \in \mathbb{Z}^+}$, as follows:

$$q_{a,n} \equiv \lim_{\Delta t \to 0} P(N(t) = n | A[t, t + \Delta t]) \tag{2.14}$$

By using eq. (2.13) and applying Bayes' rule, we get:

$$q_{a,n} = \lim_{\Delta t \to 0} \frac{P(A[t, t + \Delta t]|N(t) = n)P(N(t) = n)}{\sum_{n=0}^{\infty} P(A[t, t + \Delta t]|N(t) = n)P(N(t) = n)} = \frac{\lambda_n p_n}{\Lambda} \tag{2.15}$$

for $n \in \mathbb{Z}^+ = \Omega \bigcup B$ and $\Lambda = \sum_{n \in \Omega \cup B} \lambda_n p_n$.

We define the probability distribution $\{q_{s,n}\}_{n \in \Omega}$ of the customers in the system seen by an arrival that *joins the system*. With a reasoning that parallels the one above, it can be derived that

$$q_{s,n} = \frac{\lambda_n p_n}{\Lambda_s}, \qquad n \in \Omega \tag{2.16}$$

where $\Lambda_s \equiv \sum_{n \in \Omega} \lambda_n p_n$. This is the mean arrival rate of customers that actually join the system. The subscript s stands for "served", since each customer joining the queue will eventually receive the requested service[6].

In the special case when the arrival rates do not depend on the system state, namely $\lambda_n = \lambda, \forall n$, we have $q_{a,n} = p_n, n \in \mathbb{Z}^+$ and $q_{s,n} = p_n/(1 - P_B), n \in \Omega$, with $P_B = \sum_{n \in B} p_n$ being the blocking probability of the system. This is true if arrivals occur according to a Poisson process of mean rate λ (see Chapter 3 for the definition and main properties of the Poisson process). The fact that Poisson arrivals see the same queue probability distribution of the generic steady state time is known as *PASTA property* (PASTA = Poisson Arrivals See Time Averages).

We consider the probability distribution of the number of customers left behind by a departing customer, $\{\pi_n\}_{n \in \Omega}$. Note that a departing customer must leave the system in a nonblocking state, since there is at least the place just left by the departing customer. It can be shown that $\pi_n = q_{s,n}, n \in \Omega$ provided arrivals and departures occur singly (not in batches).

Let $A_n(t)$ denote the number of arrivals in $(0, t)$ that find the queue in state n. Let also $D_n(t)$ denote the number of departures in $(0, t)$ that leave the queue in state n. Since arrivals and departures are single, the number of jumps upward from state n (arrivals) may differ at most by one from the number of jumps downward to the same state (departures), i.e., $|A_n(t) - D_n(t)| \leq 1$. The number in the queue at time t is clearly $N(t) = N(0) + A(t) - D(t)$, where $A(t)$ and $D(t)$ denote the overall number

---

6 We do not consider special systems with "impatient" customers, that may leave the queue before receiving their service.

of arrivals and departures in $(0, t)$, respectively. If the queue admits a steady state independent of the initial state (ergodicity), then we have[7]

$$\frac{A_n(t)}{A(t)} \to^p q_{s,n}, \qquad \frac{D_n(t)}{D(t)} \to^p \pi_n \qquad (t \to \infty) \qquad (2.17)$$

where convergence is in probability. From the simple identity

$$\frac{A_n(t)}{A(t)} = \frac{D_n(t) + A_n(t) - D_n(t)}{D(t) + A(t) - D(t)} = \frac{D_n(t)}{D(t)} \frac{1 + [A_n(t) - D_n(t)]/D_n(t)}{1 + [N(t) - N(0)]/D(t)} \qquad (2.18)$$

it is easily derived that

$$\lim_{t \to \infty} \frac{A_n(t)}{A(t)} \overset{p}{=} \lim_{t \to \infty} \frac{D_n(t)}{D(t)} \qquad (2.19)$$

since both $A_n(t) - D_n(t)$ and $N(t) - N(0)$ stay bounded with probability 1, while $D_n(t)$ and $D(t)$ grow without bound with probability 1 thanks to the fact that the arrival process is nondegenerate and the queue is stable. This shows that $q_{s,n} = \pi_n$, $n \in \Omega$, by virtue of eq. (2.17).

To sum up, for a queue at equilibrium, the probability distribution of the number of customers in the queue at a generic time is $p_n = P(Q(t) = n)$, for $n \in \Omega \bigcup B$; Ω denotes the set of nonblocking states of the queue, B the set of blocking states. We have defined the probability distributions of the number of customers seen by special points of view: they are the so-called *Palm's distributions*. Specifically, we have proved the following result.

Theorem 2.2 Let p_n, $n \in \Omega \bigcup B$ be the limiting probability distribution of the number of customers of a service system at equilibrium, where B is the set of blocking states. Let λ_n be the mean customer arrival rate when there are n customers in the system. Then, the probability distribution of the number of customers

1. seen by an arrival is $q_{a,n} = \lambda_n p_n / \Lambda$, for $n \in \Omega \bigcup B$, with $\Lambda = \sum_{n \in \Omega \bigcup B} \lambda_n p_n$;
2. found in the system by an arriving customer that joins the queue is $q_{s,n} = \lambda_n p_n / \Lambda_s$, for $n \in \Omega$, with $\Lambda_s = \sum_{n \in \Omega} \lambda_n p_n$

Moreover, the probability distribution of the number of customers left behind by a departing customer, π_n, for $n \in \Omega$, is the same as $q_{s,n}$, if the system admits only single customer arrivals and departures.

For single arrival, single service systems, we have $q_{s,n} = \pi_n$ for $n \in \Omega$. For non-blocking systems, we have also $q_{a,n} = q_{s,n}$ for $n \in \mathbb{Z}^+$. In case of Poisson

---

7 Convergence in probability of a sequence of random variables $\{Y_n\}_{n \geq 0}$ to a random variable Y, denoted in the following with $Y_n \to^p Y$, means that $\forall \epsilon > 0$ we have $P(|Y_n - Y| > \epsilon) \to 0$ as $n \to \infty$.

arrivals, we have $q_{a,n} = p_n$, for $n \in \mathbb{Z}^+$ (PASTA property). All of those probability distributions coincide in the case of nonblocking systems with Poisson arrivals.

Let us consider further the special case where the state space of the queue is $\{0, 1, \ldots, K\}$, for some positive integer K. We assume that $\Omega = \{0, 1, \ldots, K-1\}$ and $B = \{K\}$, so that $P_B = p_K$. We assume also that there is a single server, with mean service time $1/\mu$, and that arrivals follow a Poisson process with mean rate λ. Since in the considered case it is $\Lambda_s = \lambda(1 - p_K)$, according to eq. (2.16), we have

$$q_{s,n} = \frac{p_n}{1 - p_K} , \quad n \in \Omega \tag{2.20}$$

By equating the mean flow in and out of the queueing system, we have

$$\lambda(1 - p_K) = \mu(1 - p_0) \quad \Rightarrow \quad 1 - p_K = \frac{1 - p_0}{A} \tag{2.21}$$

with $A = \lambda/\mu$. Putting together these equations, we find

$$\pi_n = q_{s,n} = A \frac{p_n}{1 - p_0} , \quad n = 0, 1, \ldots, K - 1 \tag{2.22}$$

By using this identity for $n = 0$, we get $\pi_0 = A p_0/(1 - p_0)$, whence $p_0 = \pi_0/(\pi_0 + A)$. Plugging this expression of p_0 into eq. (2.21), we have $1 - p_K = 1/(\pi_0 + A)$. From this equality and eq. (2.20) we derive finally:

$$P_n = \begin{cases} \frac{\pi_n}{\pi_0 + A} & n = 0, 1, \ldots, K - 1, \\ 1 - \sum_{n=0}^{K-1} P_n = 1 - \frac{1}{\pi_0 + A} & n = K. \end{cases} \tag{2.23}$$

These relationships tie together the state probability distribution at a generic time with the one seen by a customer leaving the system. They hold for a single server system with Poisson offered traffic.

Summing up, we have proved the following.

Theorem 2.3 Let us consider a single server system with mean serving rate μ, with Poisson customer arrivals of mean rate λ, a single blocking state, $B = \{K\}$. Assume the system has reached statistical equilibrium. Then, we have $q_{s,n} = \pi_n$, $n \in \Omega$, $q_{a,n} = p_n$, $n \in \Omega \bigcup \{K\}$ (PASTA property), and

$$P_n = \frac{\pi_n}{\pi_0 + A} , \quad n \in \Omega \tag{2.24}$$

$$p_K = 1 - \frac{1}{\pi_0 + A} \tag{2.25}$$

where $A \equiv \lambda/\mu$.

Theorem 2.3 tells us that for a single server system with Poisson arrivals, at equilibrium, all Palm's probability distributions can be derived from the knowledge of the probability distribution of the number of customers left behind by a departing customer. We will exploit this result in the analysis of the $M/G/1$ queue.

As a last issue, we compare the probability distributions of the number of customers in two systems with same arrival and service processes, same structure, in terms of number of servers and service discipline, except that the first system has no blocking state ($B = \emptyset$) while the second one has blocking states. It is easy to see that the process $N_\infty(t)$ stochastically dominates $N(t)$, where the former is the queue state in case of no blocking (hence no customer loss). Therefore $P(N_\infty(t) > k) \geq P(N(t) > k)$ for any $k \in \mathbb{Z}^+$. If the blocking state is K, so that $\Omega = \{0, 1, \ldots, K - 1\}$ and $B = \{K\}$, then $P_B = P(N(t) = K) \leq P(N_\infty(t) \geq K)$. Then, the tail probability of the system size in the infinite size queue yields an upper bound of the blocking probability of the finite size corresponding system.

2.6 The Traffic Process

The demand offered to a service system by a flow of user requests defines a *traffic process*. Let us consider an infinite server, work-conserving system. Since there are infinite servers, the system does not reject any request. Let $M(t)$ denote the number of servers busy at time t. That number depends both on the statistics of the arrivals of service requests and on the amount of service time requested by each arrival. $M(t)$ is a continuous-time, discrete-state random process. It is called the *offered traffic process* generated by the given demand in the given service system.

We assume that the offered traffic process is stationary. Its mean value is denoted as $A_o \equiv E[M(t)]$ and named *mean offered traffic*. It is easy to check that the mean offered traffic is equal to the mean arrival rate of service requests Λ_o multiplied by the mean service time, i.e., $A_o = \Lambda_o E[X]$. This is a simple consequence of Little's law applied to the infinite server system. The mean offered load is nondimensional, yet it is usually expressed in so called Erlang units (usually denoted with Erl), in honor of the Danish teletraffic engineer Agner Krarup Erlang (January 1, 1878 to February 3, 1929), who pioneered the use of probability theory in the analysis and dimensioning of traffic engineering problems. His 1917 paper [73] addresses the analysis of a group of circuits that carry telephone traffic offered by a population of users. We can also define the variance of the offered traffic process, namely $Var(M(t)) = E[M^2(t)] - E[M(t)]^2$. We denote it with V_o.

An infinite server model is an abstraction. Practical service systems have a finite service capability. Let us consider a finite server system, and let m be the number of servers. Let us assume that the system *rejects* any service request arriving when the system servers are all busy. We can still consider the process $M(t)$, whose state space is now restricted to $\{0, 1, \ldots, m\}$. Now the evolution of the process $M(t)$ is driven only by the service requests that make it into the system. It is possible that some requests get rejected when offered to the system, because of congestion and of the policy adopted by the system. We refer to the process $M(t)$ in this new situation as *carried traffic*. Its mean value is denoted as A_c; it represents the mean number of busy servers in the system at equilibrium, provided the system ever achieves equilibrium (hence, the offered traffic process must be stationary). As for the offered traffic, we have $A_c = \Lambda_c E[X]$, where Λ_c is the mean rate of service requests accepted by the system. Since $M(t) \leq m$, it is clearly $A_c \equiv E[M(t)] \leq m$. The ratio $\rho = A_c/m$ is named *average utilization factor* and it gives the average amount of carried traffic delivered by each server. If servers are indistinguishable and fully accessible, the utilization factor of any server is equal to a same value that is just the average fraction of carried load per server.

Any traffic measurement carried out in an operational or experimental real system can only estimate the carried traffic statistics.

As an example, let us consider a server that can accept at most $m = 5$ connections, e.g., because it assigns a rate share of $1/m$-th of the overall available link rate (or of its overall processing power) to each incoming connection. If we measure that an average number of 4 connections are up and running at any given time, then $A_c = 4$ and hence $\rho = A_c/m = 4/5 = 0.8$, i.e., servers are busy 80% of the time on average. With measurements on system activity we can only observe the carried traffic. The connections requests that are rejected cannot start, so that they cannot display their duration. Even if we knew the average rate of rejected connections (Λ_l), we could not infer the offered traffic, since we would not know what the service times of the rejected connections were. We can only *assume* (without proof!) that lost connections would have lasted an amount of time drawn from the *same* probability distribution of the connection holding times that we actually observe on the successful connections. Under this assumption, the knowledge of Λ_l allows us to estimate the mean offered traffic. If, for example, connection requests arrive at an average rate of 9 per minute and each connection lasts for 30 s on average, the offered traffic is $A_o = 9/60$ req/s $\cdot$ 30 $s = 4.5$ Erl. Since $A_c = 4$ Erl (measured), it turns out that $A_l = 0.5$ Erl, and therefore $\Lambda_l = 0.5$ Erl/30 s $= 1$ req/min.

Example 2.5 If one measures that the number of packets out of a router connected to a gigabit ethernet (GbE) link is 10^8 in an hour and their average length is 1000 bytes, it can be estimated that $\Lambda_c = 10^8/3600 \approx 2.78 \cdot 10^4$ pkt/s. Since $E[X] = 8 \cdot 1000$ bit/10^9 bit/s $= 8$ μs/pkt, we have $A_c \approx 0.222$. Since the output link is a single server system, we have $\rho = A_c$, hence we can conclude that the average utilization of the output link is about 22%. We can say nothing on the offered traffic, i.e., on the packet flow offered at the ingress of the link under measurement. In fact, that traffic flow is made up of: (i) the packets carried on the link; (ii) the packets that get lost, when offered to the buffer at the link ingress, because they find it full; and (iii) the packets that are stored into the buffer, waiting to be sent on the link. By observing the packets carried on the link, we cannot know anything about packets still stored in the buffer and packets dropped because of buffer overflows. We can only make modeling extrapolations on what it could have been.

An exception to this general rule arises when a service system is operated in a nonblocking regime, i.e., it is so oversized with respect to the current demand that it records no service demand rejection (or, in practice, rejection is marginal, i.e., the fraction of dropped traffic can be neglected). Then, obviously, the mean carried and offered traffic intensities coincide.

Example 2.6 As another example, consider a cellular network base station receiving connection requests from terminals that are roaming in the area covered by the base station. The base station has a number of channels m it can allocate to requesting terminals, for the connection time. If all channels are busy, typically new requests are not put on hold, rather they are rejected.

By measuring the mean number of busy channels at equilibrium (e.g., during the peak hour), we can estimate the mean *carried* traffic intensity. We could even estimate the mean connection holding time, again from measurements taken from the accepted connections.

We can say nothing however on the duration of the connections that have been rejected. Therefore, strictly speaking, we cannot know the offered traffic. We could only make an assumption, e.g., that the average duration of the rejected connection would have been the same as those that have been measured. This is consistent with the assumption that arrival of new connection requests are *independent* of service times (duration of the connection in this example).

The difference between offered and carried traffic for the same input process (demand) is called *lost traffic* $A_l = \Lambda_l E[X]$, where $\Lambda_l = \Lambda_o - \Lambda_c$ is the mean rate of rejected service requests. We denote the probability that a service request gets rejected with P_L, namely the *loss probability*. By definition, it is $\Lambda_l = \Lambda_o P_L$ and then $\Lambda_c = \Lambda_o(1 - P_L)$. A similar equality holds for the mean traffic intensities A_l and A_c. The average utilization factor of the service system is $\rho = A_c/m = A_o(1 - P_L)/m$.

For a single server system it is $m = 1$ and this equality turns into $\rho = A_o(1 - P_L)$. Along with $p_0 = 1 - \rho$, we finally find $P_L = 1 - (1 - p_0)/A_o$, where p_0 denotes the equilibrium probability that an arriving customer finds the system empty.

At least for single-server systems, this result points out that a system working without rest (never empty) is pathological, i.e., if it attains a stationary state, its loss probability equals 1. For a service system to work properly, *it must occasionally remain idle*. This is a consequence of the randomness of arrivals and service times. A corner case arises when purely deterministic arrival and service processes are considered. Under that special settings (no randomness, hence no uncertainty on the timing and amount of demand posed onto the service system) it is possible to have *both* high utilization (probability of the system being empty close to 0) and good quality of service as perceived by the customers.

2.7 Performance Metrics

A quantitative model of a service system relates together the system resources, the service demand and the performance metrics that express the *grade or quality of service* offered by the system. Metrics of a service system ultimately qualify the customer experience and the usage of system resources.

The first category focuses on the performance degradation introduced by the system as the load increases, i.e., the system is driven into congestion. This category is of primary interest to the users. Since the objective of the arriving customer is to get service as soon as possible, degradation can occur in the form of: (i) a rejection of the service request; (ii) delaying of the service request, before actually accessing the server; and (iii) a degraded service, e.g., only part of the requested service time is obtained, or a lower quality service is given. Key performance indicators pertaining this category are: *loss*, *delay*, and *age of information*.

The second category relates to measuring the effectiveness of system resources. This category is especially important to service providers, designers, and planners. Key metrics regard the usage of servers and of system resources in general, the average carried load, and the average rate of served requests. *Throughput* and *utilization* fall into this category.

Three key characteristics are part of any useful definition of a performance indicator. A good performance indicator should be:

1. Meaningful to users (applications, persons) and to designers;
2. Accurately and, possibly, easily measurable;
3. Actionable, i.e., susceptible of modification, by acting on the system structure or parameter configuration.

Balancing these often-contrasting criteria makes defining useful performance indicators by no means a trivial task.

2.7.1 Throughput

Throughput is a measure of the average amount of work that a service system does per unit time in serving customer demand. It is therefore synonymous with the mean carried traffic, even if throughput is typically not normalized and hence expressed in dimensional units, e.g., number of data units delivered per unit time.

The upper limit of the throughput achievable by a service system is the *capacity* of the system. Formally, $C = \lim_{\Lambda_o \to \infty} \Lambda_c$, where Λ_c denotes the throughput and Λ_o the mean offered traffic rate.

For example, the mean amount of bytes delivered over a wireless link between a WiFi access point and a client node of the wireless LAN is the throughput at MAC level achieved by the access point while serving the traffic flow destined to the client node. As another example, the mean number of searches completed by Google per unit time for a given population of users measures the throughput of Google search service (realized by means of its data centers and the hundreds of thousands of servers hosted in those data centers). The mean number of packets transferred through an unreliable link by an ARQ protocol, the mean number of packets processed by a firewall per unit time, and the mean number of vehicles flowing through a road section per unit time provide more examples of throughput measures.

The throughput Λ_c achieved by a service system cannot be more than its capacity C. There are two sources of limitations of the throughput measured on a real system:

1. The customer demand level;
2. The service system efficiency.

To know which of the two situations we are in, we should compare the obtained throughput with the intrinsic capacity of the system. This can be obtained by assuming that the offered traffic has "infinite" intensity (saturated traffic) and observing what the carried traffic (throughput) is in that case. The achieved throughput in steady-state (if steady-state is ever reached) represents the capacity of the system.

For example, let us assume we measure the mean traffic carried over a WiFi network. The measurements record a throughput of 20 Mbit/s. It is interesting to understand why, in the specific scenario where the measurement has been carried out, the resulting throughput has that value. The interest comes from the fact that a service provider would like to know whether it could be worth enhancing the system (investing money) so that it can carry more traffic. This option makes sense only if the limitation of the throughput comes from insufficient resource in the service system.

If we know that the WiFi is using a IEEE 802.11g standard, hence the maximum physical bit rate is 54 Mbit/s, and we analyze carefully all the overhead implied by all protocol layers, then we can find out that the maximum expected throughput we can achieve at application level is around 24 Mbit/s. Therefore, we conclude that the observed throughput is quite "high" (20 Mbit/s out of 24), and little margin is left to more traffic, unless we change the system (e.g., the MAC protocol).

Improving the system capacity makes sense, if we deem that substantially more traffic would be carried, i.e., there is a *latent* demand.

As another example, let us consider a wired local area network. Measurements of the throughput realized on a GbE link could yield a disappointingly small fraction of the link capacity, when averaged over time periods long enough with respect to individual user activities. This is by no means a symptom of a bad design of the GbE system; it is instead related to the typically very limited *long-term average* demand that users pose on such kind of links. Data transfers on a GbE are quite sporadic, yet when a user starts downloading a 1 Gbyte archive, it expects that the operation will be completed as soon as possible, whence the push toward high bit rate on this kind of links. The opportunity of attaining high bit rates is offered by the technological developments that have made fully affordable for local area networks to build electronics and cables cheap enough to be appealing for a mass market, yet capable of bit rates in the multi-Gbit/s range[8].

As a last example, ARQ protocols introduce redundancy check and sequence number fields in each data unit sent over an unreliable channel. Moreover, they introduce acknowledgements, time-outs and retransmissions to provide a reliable transfer of the data of the upper layer entities. All of those provisions consume channel capacity, hence the throughput offered to the upper layer is strictly less than the unreliable channel throughput. The point is that the unreliable channel throughput has a smaller "value" (i.e., it contains errors) with respect of the *added value* throughput offered by the ARQ protocol, which is essentially error-free (i.e., undetected errors have usually negligible probability).

An important observation emerging from the examples above, is that the same "system," measured from different protocol layers, turns out to sustain different throughput levels. The fundamental reason why is that each protocol layer adds functionality at the cost of some overhead, which translates into reduction of the achieved throughput. These considerations carry over to realms other than ICT.

Whenever evaluating a service system, either by means of analytical models, simulations, or measurements, it is important to be able to distinguish between the two root causes of throughput limitation. The first cause has to do with traffic demand forecasting and sizing of the service system consistently with expected demand. The second cause can be acted on by refining or changing the internal working mechanisms of the service system (e.g., by designing a better ARQ or MAC protocol, with reference to the two examples above), or, ultimately, by increasing the amount of resource available in the service system.

Assessing the throughput of a complex service system is by no means a trivial task. Defining what we should mean by throughput in a given situation is a

---

8 The 10 Gbit/s ethernet has been standardized by the early 2000s. The 100 Gbit/s ethernet on copper cables has been standardized around the middle of the 2010s.

deceptively simple task, given the complex architectural layout and layering of most networked systems nowadays. For example, during spring 2007 the author took part in an official technical panel to define quality of service parameters for the ADSL access by the Italian Authority for Telecommunications. Several meetings and intense discussions with people from research institutes and operator companies were spent to agree on a manageable, meaningful, and practical definition of *what* the throughput of an ADSL access system should mean, including a fully detailed operational procedure for the measurement of such a throughput[9].

2.7.2 Utilization

The utilization of a service system offering a service capacity C is measured by the ratio Λ_c/C, that is, the ratio of the throughput to the system capacity. By the definition of throughput, this is a non-negative number, not greater than 1.

From the service provider point of view, it is relevant to know how much servers are actually being used. Let us consider an observation time interval $[t_a, t_b]$ and let $Z(t_a, t_b)$ be the overall amount of time that a tagged server has been working during the observed interval.

The ratio $Z(t_a, t_b)/(t_b - t_a)$ gives an estimate of the average fraction of time that the tagged server is busy. As the time horizon tends to infinity, for a stationary system this ratio converges to the mean fraction of time that the server is busy, ρ. This is referred to as the (mean) server utilization. In an ergodic system, ρ is also the probability of finding the tagged server busy when sampling the system state at random during statistical equilibrium.

If servers are indistinguishable (they can serve any service request indifferently) and fully accessible (any admitted service request can ask for any of the servers), the mean server utilization is the same for all servers.

In a *single server*, work-conserving system we can establish a fundamental relationship between ρ and the probability that a customer arriving at the systems finds it empty, p_0. Thanks to the system being work-conserving, the server can be idle if and only if no customer is in the system, i.e., it is empty. So it must be $p_0 = 1 - \rho$.

2.7.3 Loss

A customer arriving at a service system can be rejected, according to the policies of the system and to the state of engagement of the resources. As a matter of example, if a packet arrives at a router output link, whose buffer is full, the packet gets dropped. If one more connection request arrives at a busy server, where already

---

9 The full blown public project outcome can be seen at https://www.misurainternet.it

established connections saturate the allowed maximum number of parallel flows that can be maintained, the new connection request is rejected.

A typical performance indicator related to service request rejection is the probability of rejection or *loss*, P_L, estimated as the average fraction of service request rejected by the system in a given time interval $[t_a, t_b]$ divided by the number of all requests offered to the system in the same time interval.

The conditions under which a service request is rejected are specific of each service system. In general, there is a subset B of the state space S of the system where system congestion is such that new requests cannot be accepted. States belonging to B are *blocking*, i.e., the system is "blocked" in those states, and it cannot accept new incoming service requests.

For example, under suitable assumptions, the buffered output line of a router can be described by the state variable that gives the number of packets enqueued at the buffer at a given time, $Q(t)$. Then, the entire state space is the set of integers between 0 and Q_{max}, namely $S = \{0, 1, \ldots, Q_{max}\}$, where Q_{max} is the maximum number of packets that can be stored in the buffer. In case of FIFO discipline, the only blocking state is $Q(t) = Q_{max}$, so $B = \{Q_{max}\}$.

In some cases, the service system has multiple servers, but they are not equivalent, or not fully accessible to any service request. Then, different blocking metrics can be defined.

Example 2.7 Let us consider a base station of a cellular network, equipped with m (logical) channels, and working according to a connection oriented, static allocation policy. A new service request is granted a single channel, provided at least one channel is available, until it explicitly releases it. Requests can be classified in two types: (i) new radio sessions, opened by users roaming in the coverage area of the base station; (ii) ongoing radio sessions coming from nearby cells and handed off to the considered base station (briefly: *handoffs*). It makes sense to define two loss probabilities, one for each class, namely a loss probability for new radio session (loss probability, P_L) and a loss probability for incoming handoffs (usually referred to as drop probability, P_D). The reason for having two metrics is that different targets are usually set for these two probabilities, e.g., $P_L \leq 10^{-2}$ and $P_D \leq 10^{-3}$. These targets are achieved, e.g., by setting aside a number m_{ho} of reserved channels, so that the blocking state set for new radio sessions is $B_L = \{m - m_{ho} + 1, \ldots, m\}$ while for handoffs it is $B_D = \{m\}$.

Example 2.8 Consider a network where end-to-end traffic flows are routed on a single shortest path connecting the origin and the destination nodes. Let $\mathcal{F}$ and $\mathcal{L}$ denote the set of active flows and the set of links of the network. Let x_k, $k \in \mathcal{F}$ be the mean rate of flow k and c_h, $h \in \mathcal{L}$ be the capacity of link h. We assume that mean flow rate can take a discrete spectrum of values, $b_1, \ldots, b_m$.

We impose that no link is overloaded, i.e., the sum of the mean rates of flows routed through that link shall not exceed the capacity of the link. Let a_{hk} be a routing coefficient that equals 1 if and only if flow k routing path goes through link h and it is 0 otherwise. The constraint on the number of admissible flows is expressed as:

$$\sum_{k \in F} a_{hk} x_k = \sum_{j=1}^{m} b_j n_{jh} \leq c_h \tag{2.26}$$

where n_{jh} is the number of flows of type j routed through link h. All states such that $\sum_{j=1}^{m} b_j n_{jh} > c_h$ are blocking for link h.

The specification of the "reaction" of the service request is part of the *loss model*. A rejected service request can disappear and bring no effect on future service requests. This is sometimes called the *cleared lost requests* scenario. It is the simplest model, since, after rejection, there is no memory of that event and subsequent requests are offered according to the same demand statistics as any other request. A different model is *repeated requests*. With this model a request that encounters a rejection will be submitted again to the system after a given retrial time. Retrials can go on unconditionally until the request is finally accepted or the request can be eventually canceled and forgot, if it undergoes a maximum number of rejections. The retrial customer model can be generalized to envisage not only that rejected service requests be retried, but that in general a customer that cannot receive service upon arrival can decide to leave the system and wait some time before trying again. During this time the customer is in "orbit." Retried arrivals merge with fresh arrivals to form the complete service request arrival process. The interested reader can find an introduction to retrial queueing models in [94, Ch. 3].

2.7.4 Delay

A customer accepted into a service system can start immediate service, if a server is available, or it can wait until its turn of service comes, if all eligible servers are busy upon arrival and the newly arriving customer does not have a right to overrun on-going server assignments, according to the service policies of the system. We refer to *work-conserving* systems, where customers entering the system will eventually fulfill their full service demand (no customer impatience, i.e., customers leaving the system before receiving complete service) and no server remains idle as long as there is work to be done.

The whole time S spent by the user in the system is usually referred to as *system time* or *response time*. In general, by looking at the system from an input-output point of view, we can mark the events of customer arrival and admission to the service system (time t_a) and of the subsequent customer departure (time t_d). Then, it is $S = t_d - t_a$. If the service time is independent of the system state, then we can write $W = S - X$, since the service requirement X of the user is given at its arrival. This general approach includes cases where service is granted in chunks (e.g., processor sharing disciplines), rather than serving continuously the customer since it starts its service until completion.

The waiting time can be characterized as a non-negative random variable, described by means of its survivor function $G_W(t) = P(W > t)$. Widely used metrics are the mean wait, $E[W]$, and the standard deviation of the waiting time, $\sigma_W = \sqrt{E[(W - E[W])^2]}$. Also waiting time quantiles are often used, e.g., the 90 or 99 quantile. In general, the quantile r is defined as the minimum value w_r such that $P(W > w_r) \leq 1 - r/100$. So, w_{99} is the waiting time level that is exceeded for no more than 1% of the customers.

A last metric on delays is the probability that a service request admitted to the system be served with a positive delay, P_W, i.e., it finds all servers busy and it is forced to wait for service. This metric applies to systems where service is provided in a single shot. It is simply $P_W = P(W > 0) = 1 - P(W = 0)$. In general, this value is non-null for a service system where waiting is allowed. As a consequence, the CDF of the waiting time $F_W(t)$ has a jump at 0, that is to say, $F_W(0_-) = 0$ and $F_W(0_+) = P_W > 0$.

2.7.5 Age of Information

Many applications consist of collecting data, e.g., from sensors, about the status of some device or about an environmental variable and transferring it to a remote processing facility (e.g., a cloud-based application or a control server). Such applications fit the Internet of Things (IoT) paradigm and are spreading rapidly as the IoT becomes pervasive. Major specific examples are telecommunication network equipment monitoring, vehicular traffic monitoring, urban sensing, power grid metering, industrial processes control, and sensor networks. In these cases, typically only the latest piece of information from a given source is relevant. A key requirement of such applications is timeliness of the information provided to the central processing server (along with reliability of the information delivery). Getting the most up-to-date piece of information is the target. The user of the collected information maintains the most up-to-date piece of information it can obtain at any given time. This implies that an out-of-order message (i.e., one arriving later than a subsequent message) will be simply discarded, since the piece of information it carries has been made obsolete by more recent messages that have already arrived.

We can model the general paradigm with an information source, emitting pieces of data within messages repeatedly over time (e.g., in a periodic fashion),

Figure 2.7 Example of age of information plot. A source sends message 1 at time s_1 and that same message arrives at the central processing unit at time a_1.

a communication network that connects the source to the the remote processing facility and the processing facility itself. Let $g(t)$ be the timestamp[10] of the latest message arrived at the processing facility from the tagged source and stored there at time t. Then, the age of the information coming from that source is defined as $a(t) = t - g(t) \geq 0$. As a new message is received, the age drops to a local minimum. Then, it start growing linearly (1 second per second) as the time goes by, until the next piece of information gets to the processing facility. An example of the age of information (AoI) sample path is shown in Figure 2.7.

The times when message 1 is sent by the source (s_1) and when the same message is received by the central processing unit (a_1) are highlighted. The central processing unit learns s_1 by looking at the timestamp carried in the received message. Therefore, it can calculate the age of message 1, namely $a_1 - s_1$.

The interesting point is that minimizing the age of information is *not* the same as minimizing the message transfer delay through the network or maximizing the network throughput. Intuitively, an optimal balance must be struck between a high rate of information refresh from the source, hence a potentially high level of congestion on the network resources, and a low level of congestion of the network, hence low delay.

Summary and Takeaways

Network traffic engineering is all about *service systems*. This is an abstraction for a software function, a subsystem, a piece of hardware or any system that is set up and configured to provide a *service* (i.e., to carry out a function) for a population of customers (persons, machines, applications).

Identifying a service system means focusing on a specific function (e.g., error recovery in a data link protocol, buffer management at the output link of a router, access to disk information, data processing, etc.) and defining how much work per unit time the service system can provide (*capacity*), according to which rules (*policy*), and using which resources (*structure*).

---

10 A message processing device can associate a time mark to a message, e.g., the time that it receives the message or the time when it generates the message. The time mark appended to the message is called timestamp. Here we assume that the source of the message adds its timestamp to the emitted message.

The demand is characterized by two elements: timing of the demand arrival pattern and amount of service required by each demand. Both of them are required to define the *traffic process*. This is but a stochastic process that describes the amount of system resources engaged in serving accepted demand at any given time. As any stochastic process, it is characterized by moments and probability distributions. The mean traffic intensity is nondimensional, yet it is customarily measured in so called Erlang, in honor to the inventor of the teletraffic theory, A. K. Erlang.

The key usage of the service system abstraction is the analysis and dimensioning of resources in the face of a given demand to meet prescribed quality of service objectives. It is therefore fundamental to characterize the performance of a service system. This is done by using metrics such as: throughput, utilization, loss, delay, age of information.

Key features of a good performance metric are: (i) it should be easy to understand and meaningful to customers; (ii) it should be practically measurable; and (iii) it should be influenceable in some way by the service provider, e.g., by a suitable dimensioning of the service system.

The mathematical tool used to describe service systems for the purpose of analytical modeling is queueing theory. A queueing model is denoted with $A/B/m/K/N$, where A indicates the kind of stochastic process describing arrivals, B does the same for the service process, m is the number of serving units available in the queue (*servers*), K specifies the maximum number of customers that can be standing in the queue while waiting for a server, and N gives the number of potential customers.

Stationarity of a stochastic process means invariance of the statistics of the process to time axis translations. Stationarity is akin to statistical equilibrium, i.e., a condition where the process is not constant, yet its variation is around a constant mean. Under additional conditions, a system described by a stochastic process evolves toward a statistical equilibrium that is independent of the initial state (ergodicity). Ergodic processes are especially interesting since we can analyze their equilibrium state and be reassured that the results of the analysis apply to the system described by the stochastic process, except for a limited time after the system is triggered into its initial state (initial transient).

A key result for stationary systems is Little's law [148, 197]. It is a classical statistical equilibrium result, stating that the mean number of customers into a service system equals the mean input flow rate of customers accepted into the system times the mean sojourn time of a customer into the system. The power of this result is its general applicability and its black-box approach.

In general, the probability distribution of the number of customers into a service system depends on the point of view. We restrict to stationary systems that have reached statistical equilibrium. The probability distribution of the number of customers into the system can be sampled at times when customers arrive, are admitted into the systems, leave the system. For all cases the observer is the customer itself, i.e., the sampling of the system is biased by the arrival and/or service processes that govern the evolution of the system itself. The resulting probability distributions are called Palm's distributions. Their importance lies in the fact that relevant events of a service system occur at arrival or service completion epochs (e.g., loss events take place at customer arrivals).

Problems

2.1 The packets arriving at a radio link have the following length (L) probability distribution:

$$
\begin{aligned}
p_1 &= P(L = 160 \text{ bytes}) = 0.45 \\
p_2 &= P(L = 1024 \text{ bytes}) = 0.1 \\
p_3 &= P(L = 1500 \text{ bytes}) = 0.44 \\
p_4 &= P(L = 2240 \text{ bytes}) = 0.005 \\
p_5 &= P(L = 9000 \text{ bytes}) = 0.005
\end{aligned}
\tag{2.27}
$$

The radio link protocol data unit (PDU) has fixed payload size equal to $L_0 = 100$ bytes and header length of $H = 34$ bytes. Packets are adapted to the radio link PDU format by means of fragmentation and padding. How could you describe the arrival process of the radio link PDUs? If packets arrive at an average rate $\lambda = 2000$ pkts/s, what is the average arrival rate of PDUs at the radio link layer? What are the corresponding average bit rates?

2.2 The inter-arrival time T of a stationary arrival process is given by $T = \min\{V_1, V_2\}$, where V_1 and V_2 are two independent random variables with negative exponential PDF, with parameters λ_1 and λ_2, respectively. Find the CCDF of T, $G_T(t) = P(T > t)$.

2.3 In a multi-rate multiplexing system, incoming connections can be granted one out of M possible bit rates $R_j, j = 1, \ldots, M$. Assume all rates are integer multiples of a basic bit rate R_0, i.e., $R_j = n_j R_0, j = 1, \ldots, M$. The overall service capacity is organized in m channels, each of capacity R_0. A request of class j is either assigned n_j channels, if available, or it is rejected immediately. No wait is allowed.

Assume that the state of the system can be described by the array variable $Q(t) = [Q_1(t) \ldots Q_M(t)]$, where $Q_j(t)$ is the number of class j connections active at time t. Find the state space and classify blocking and nonblocking states.

2.4 At a post office there are two clerks on duty. The director of the office notices that they are almost always busy during peak periods. It estimates that the average idle time of each clerk is 1% of the overall time. The director knows that ideally the percent idle time should be 20% to ensure that the clerks stay in good shape and work accurately.

1. If the director hires one more clerk during peak periods, how much idle time would each server have then?
2. When the new clerk adds to the two original ones, the pressure on the clerks relieves, hence they work more carefully and they start devoting a bit more time to each task. As a consequence, service rate reduces by 10%. What now is the percent time each would be idle?

2.5 IP packets arrive at a link with Maximum Transfer Unit (MTU) = 1500 bytes according to a deterministic process. 80% of the packets are 1500 bytes long or less. The remaining 20% can have lengths between 1501 and 2200 bytes with probability 0.9 and length 9000 bytes with probability 0.1. Describe a suitable arrival process model of the stream of packets arriving at the link, by accounting for the effect of the MTU and of the induced IP fragmentation function.

2.6 You are planning to sell your home. After gathering some information, you realize that about 12 houses are on sale at any given time in your neighborhood. New houses go on the market at a rate of 3 new houses per month. How long do you estimate it could take to sell your home? What assumptions your estimate is based on?

2.7 You enter your favorite pizza shop and find a line of 9 persons in front of you. There is unfortunately only one server on duty. Service time has a gamma PDF with mean 3 min and standard deviation 1 min. What is your estimate of the probability that you have to wait more than half an hour? [Note: (i) the sum of N identically distributed, independent gamma random variables, with mean $E[X]$ and variance σ_X^2, is still a gamma random variable, with mean $NE[X]$ and variance $N\sigma_X^2$; (ii) the gamma PDF is given by $f(x) = \alpha \frac{(\alpha x)^{\beta-1}}{\Gamma(\beta)} e^{-\alpha x}$ for $x \geq 0$; $\Gamma(x)$ is the Euler gamma function; and (iii) let

$$\Gamma_{inc}(x, \beta) = \frac{1}{\Gamma(\beta)} \int_x^\infty u^{\beta-1} e^{-u} \, du \qquad (2.28)$$

Then, it is $\Gamma_{inc}(10, 9) \approx 0.3328$.]

2.8 In the university system of Nowhereland a young researcher gets a tenure assistant professor position on the average at 28. On average, a fraction p_1

of the assistant professors gains a promotion to associate professors after 9 years. A fraction p_2 of the associate professors move further to occupy a position as full professors after another 15 years on the average. All tenure positions can be held until retirement age, say 72 years old on average. Those who fail to move to the next step in the career remain in their position until retirement age.

1. Find the ratios of the mean numbers of assistant, associate, and full professors in the system by assuming it is in equilibrium ($p_1 = 0.8$, $p_2 = 0.75$).
2. Find the value of the input rate to the system that guarantees an average faculty staff of 100 persons.
3. Find the promotion rates that guarantee an equal average number of assistant, associate and full professors.

2.9 Table 2.1 lists the number of freshmen students and the overall enrollment of the Schools of Engineering, Computer Science and Statistics of University of Roma "La Sapienza" over eight academic years. Note that the overall numbers in Table 2.1 comprises both undergraduate and graduate level students.

A student that fulfills the undergraduate degree will move to graduate level courses with probability $q = 0.7$. Reports say that students completing both degrees (bachelor's and master's) take 7.5 years on the average.

Table 2.1 Record of first-year students and overall enrollments for the Schools of Engineering, Computer Science and Statistics of University of Roma "La Sapienza."

A.Y.	First-year students	Total enrollment
2018/2019	5309	18052
2017/2018	5112	17382
2016/2017	4895	16793
2015/2016	4523	15913
2014/2015	4300	16500
2013/2014	4079	17011
2012/2013	4209	17788
2011/2012	4298	18643
2010/2011	4537	19652
2009/2010	5037	20316
2008/2009	4817	19850

Table 2.2 Record of start and end times of SSL connections gathered at a WiFi hot spot (times are measured in seconds).

#conn	Start time (s)	End time (s)
1	305.90	660.12
2	306.08	310.65
3	306.25	392.22
4	311.15	312.17
5	311.46	1297.01
6	312.83	314.87
7	314.35	344.47
8	316.32	318.62
9	318.69	320.79
10	320.97	344.54
11	335.31	805.20
12	335.42	465.11
13	335.57	466.13
14	335.68	466.14
15	338.63	3726.45
16	353.70	378.89
17	356.86	681.80
18	415.24	440.41
19	537.04	562.21
20	659.53	660.24

Could you give an estimate of the average time that students take to achieve a bachelor's degree and the average time it takes to obtain a master's degree?

2.10 Table 2.2 reports the start and end times of SSL connections generated by smartphone applications, from a traffic trace collected at a WiFi hot spot. All times are measured in seconds.

By assuming that no other connection arrives, after the first 20 ones, do a manual simulation of the traffic through the WiFi hot spot. Calculate the mean number of connections active at the same time and the duration of the idle time of the WiFi hot spot during the first five minutes of the traffic measurement time (i.e., five minutes starting from time 305.90 sec).

2.11 By collecting large measurements of the queue lengths observed by arriving customers and the queue length left behind by leaving customers at a single-server service facility, you realize that the empirical distribution derived by the two sets of measurements are definitely different. If statistical tests on measurements of the arrival times suggest that arrivals follow a Poisson process, and you know that no batch service is allowed, what can you say about the service facility waiting room? Can you estimate the mean offered traffic of the system, if you know that 25% of customers leave the system empty upon their departures?

2.12 A firewall has a processing capability of μ pkts/s. The time required to analyze a packet can be assumed to be a constant as a first approximation. Packets arrive according to a stationary process from a link with bit rate R. The mean packet length is $\overline{L}$. The average utilization of the link is ρ. The firewall has a buffer that can store up to K packets.
Define a model of the firewall processing unit. Write an expression of the mean packet arrival rate at the firewall.

2.13 From a `ping` probe it is found that the average RTT between the source node A and the destination node B is 149 ms for `ICMP_Echo_Request` messages that are 64 bytes long. When the message length is raised to 1024 bytes, the mean RTT grows up to 173 ms. Is it possible to estimate the capacity of the bottleneck between A and B based on this two probes? Under what assumptions?

2.14 A flow of packets is fed into a path with a mean rate of 1000 packets/s. It is reported that 7% of the packets get lost, while 23% of the packets experience the minimum registered end-to-end delay, i.e., a delay equal to the fixed transmission and propagation times. The average length of the packets is 1000 bytes. How can the bottleneck capacity be estimated, under the hypothesis that the network is in equilibrium and that there is a single bottleneck along the path under test?

2.15 At the university canteen the desk where lunch food can be collected has a maximum throughput of μ students/min. After collecting their meal, students go to the canteen room where 60 tables are available, each provided with four seats. Mean lunch time is 30 min. Shall μ be limited to avoid having many students standings with their trays, waiting for a seat?

2.16 An $M/G/1/K$ queueing system has a loss probability $P_L = 0.05$ and an average queue length $E[Q] = 4$. Knowing that $\lambda = \mu = 3$ customer/sec, find the value of the mean system time $E[S]$, the mean waiting time $E[W]$, the mean carried customer rate Λ_c, the probability that the system be empty p_0 and the utilization coefficient ρ.

3

Stochastic Models for Network Traffic

That which is static and repetitive is boring. That which is dynamic and random is confusing. In between lies art.

John Locke

3.1 Introduction

The purpose of this chapter is to introduce several models largely used to describe traffic processes. The first half of this chapter is concerned with *point processes*. We introduce the most used stochastic process model, the Poisson process, so called in honor of the nineteen-century French mathematician and physicist Siméon-Denis Poisson. Then, we generalize to renewal processes. The second half of the chapter introduces *birth-death* processes and *branching* processes.

A point process is a stochastic process that defines the distribution of points in a metric space. With reference to the unidimensional real line, often interpreted as the time axis, a point process specifies a sequence of instantaneous events that occur on the time axis. In the following we address mainly unidimensional point processes, since those are mostly used in network traffic engineering applications. We devote however a section to spatial (multidimensional) point processes for their increasing interest.

Relevant information for a point process is the counting of the events occurring in a given interval and the probability distribution of the inter-event times. Therefore, we define the following quantities.

$A(t_0, t_0 + t)$ *counting function*: number of events (often referred to as "arrivals") falling in the interval $(t_0, t_0 + t]$, for $t > 0$. Note that, by convention, the counting function is referred to the interval $(t_0, t_0 + t]$, i.e., excluding the left extreme and including the right one.

Network Traffic Engineering: Stochastic Models and Applications, First Edition. Andrea Baiocchi.
© 2020 John Wiley & Sons, Inc. Published 2020 by John Wiley & Sons, Inc.

S_k *event epochs*: the time of occurrence of the k-th event, $k \geq 1$. By convention, we assume that $S_k > 0$ and we let $S_0 = 0$. Here k is an integer.

T_k *inter-arrival times*: the time elapsing since the $(k-1)$-th event and the k-th event, i.e., $T_k = S_k - S_{k-1}$, for $k \geq 1$. T_1 represents the time interval between the time origin and the first event after the time origin.

In the rest of the chapter we review four classes of stochastic processes largely used in network traffic engineering applications: the Poisson process and variants thereof (spatial Poisson processes, modulated Poisson processes); renewal processes; birth-death processes; and branching processes.

3.2 The Poisson Process

Let us introduce the short notation $A(t) \equiv A(0, t)$ to indicate the number of arrivals until time t, from the given time origin. The number of arrival in an interval $(t_1, t_2]$ is given by $A(t_2) - A(t_1)$, by definition of $A(t)$.

Given a positive constant λ, a Poisson process is defined by the following three properties:

1. $\mathcal{P}(A(t + \Delta t) - A(t) = 1) = \lambda \Delta t + o(\Delta t)$, as $\Delta t \to 0$.
2. $\mathcal{P}(A(t + \Delta t) - A(t) > 1) = o(\Delta t)$, as $\Delta t \to 0$.
3. The number of arrivals in nonoverlapping intervals are independent random variables, i.e., the random variables $A(t_2) - A(t_1)$ and $A(t_4) - A(t_3)$ are independent of each other for any $t_1 \leq t_2 \leq t_3 \leq t_4$.

According to properties (1)-(3), we can write

$$P(A(t + \Delta t) = n) = \sum_{k=0}^{n} P(A(t) = k)P(A(t + \Delta t) - A(t) = n - k)|A(t) = k)$$

$$= \sum_{k=0}^{n} P(A(t) = k)P(A(t + \Delta t) - A(t) = n - k) \qquad \text{(Prop. 3)}$$

$$= P(A(t) = n)(1 - \lambda \Delta t) + P(A(t) = n - 1)\lambda \Delta t + o(\Delta t) \qquad (3.1)$$

For a compact notation, let $p_n(t) = P(A(t) = n)$, $n \geq 0, t \geq 0$. Then, from eq. (3.1) we derive

$$\frac{p_n(t + \Delta t) - p_n(t)}{\Delta t} = -\lambda p_n(t) + \lambda p_{n-1}(t) + o(1), \qquad n \geq 1 \qquad (3.2)$$

and taking the limit as $\Delta t \to 0$, we get the ordinary differential equation (ODE) system:

$$\begin{cases} p'_n(t) = -\lambda p_n(t) + \lambda p_{n-1}(t) & n \geq 1 \\ p'_0(t) = -\lambda p_0(t) & n = 0 \end{cases} \qquad (3.3)$$

with initial conditions $p_0(0) = 1$ and $p_n(0) = 0$, $n \geq 1$.

The solution of this system of linear-difference differential equations can be found by resorting to the generating function method. A brief review of generating functions of discrete random variables is provided in the Appendix at the end of the book. Multiplying both sides of the first equation in (3.3) by z^n and summing over n, it is readily found that

$$\sum_{n=1}^{\infty} p'_n(t)z^n = -\lambda \sum_{n=1}^{\infty} p_n(t)z^n + \lambda \sum_{n=1}^{\infty} p_{n-1}(t)z^n = -\lambda[P(z,t) - p_0(t)] + \lambda z P(z,t)$$

(3.4)

where $P(z,t) \equiv \sum_{n=0}^{\infty} z^n p_n(t)$. By using the second equation in (3.3) and interchanging summation and derivation, we find

$$\frac{\partial P(z,t)}{\partial t} = \lambda(z-1)P(z,t)$$

(3.5)

with initial condition $P(z,0) = 1$. It is easy to check that the unique solution is

$$P(z,t) = e^{\lambda t(z-1)} = \sum_{n=0}^{\infty} e^{-\lambda t} \frac{(\lambda t)^n}{n!} z^n$$

(3.6)

whence we derive that the probability distribution of the Poisson process is:

$$p_n(t) = P(A(t) = n) = e^{-\lambda t} \frac{(\lambda t)^n}{n!}, \qquad n \geq 0, \quad t \geq 0.$$

(3.7)

We have just shown the following theorem.

Theorem 3.1 The probability distribution of the counting process $A(t)$ of the stochastic process defined by the properties 1, 2, and 3 above is the Poisson distribution, given in eq. (3.7).

Incidentally, we highlight for future use the result of eq. (3.6), which gives the generating function of the Poisson counting process for a given time t, namely $\phi_A(z) = E[z^{A(t)}] = e^{\lambda t(z-1)}$.

The parameter λ can be given a meaning by evaluating the average of $A(t)$, since $E[A(t)] = \sum_{n=0}^{\infty} n p_n(t) = \lambda t$. Therefore $\lambda = E[A(t)]/t$, i.e., it is the average arrival rate. It can be easily checked that the variance of the number of arrivals in the interval $(0, t]$ is $\mathrm{Var}(A(t)) = \lambda t$.

We can find the explicit form of the probability density function (PDF) of the inter-arrival times of the Poisson process as well.

Theorem 3.2 The inter-arrival times $\{T_k\}_{k \geq 1}$ of the Poisson process are independent, identically distributed (i.i.d.) random variables with a negative exponential PDF with parameter λ, namely, $T_k \sim T$ $\forall k$, with $P(T > t) = e^{-\lambda t}$.

Proof: From the definitions given in Section 3.1, the event $S_1 > t$ is equivalent to the event $A(t) = 0$, i.e., no arrival until time t. Hence,

$$G_T(t) = P(T_1 > t) = P(S_1 > t) = P(A(t) = 0) = e^{-\lambda t} \tag{3.8}$$

where the last equality is a consequence of eq. (3.7) for $n = 0$.

The independence of different inter-arrival times can be shown by considering the cumulative distribution function (CDF) of the sum of the first n arrival times:

$$F_{S_n}(t) = P(T_1 + \cdots + T_n \leq t) = P(A(t) \geq n) = 1 - \sum_{k=0}^{n-1} \frac{(\lambda t)^k}{k!} e^{-\lambda t} \tag{3.9}$$

By deriving both sides, we get the corresponding PDF:

$$f_{S_n}(t) = \lambda \sum_{k=0}^{n-1} \frac{(\lambda t)^k}{k!} e^{-\lambda t} - \lambda \sum_{k=1}^{n-1} \frac{(\lambda t)^{k-1}}{(k-1)!} e^{-\lambda t} = \lambda \frac{(\lambda t)^{n-1}}{(n-1)!} e^{-\lambda t} \tag{3.10}$$

The rightmost hand side is just the gamma PDF with parameters λ and n, which is known to be the convolution of n negative exponential PDFs, all with the same parameter λ. Hence we have shown that the inter-arrival times of the Poisson process form a sequence of i.i.d. random variables with negative exponential PDF given by $\lambda e^{-\lambda t}$.

It follows that $E[T] = 1/\lambda$, which is a pleasingly intuitive result, in view of the meaning of λ as the average arrival rate.

The converse of Theorem 3.2 is true as well, since we can show the following.

Theorem 3.3 A random process with i.i.d. inter-arrival times, having a negative exponential PDF with mean $E[T]$, is a Poisson process with rate $\lambda = 1/E[T]$.

Proof: We have

$$P(A(t) \geq n) = \sum_{k=n}^{\infty} p_k(t) = P(S_n \leq t) \tag{3.11}$$

Taking differences over n, we find:

$$p_n(t) = P(S_n \leq t) - P(S_{n+1} \leq t), \quad n \geq 0, \tag{3.12}$$

where we let $S_0 \equiv 0$ for ease of notation. The random variable $S_n = T_1 + \cdots + T_n$ is by hypothesis the sum of n i.i.d. negative exponential random variables with mean $E[T]$. The Laplace transform of the PDF of S_n is therefore $\varphi_{S_n}(s) = [\varphi_T(s)]^n$, where $\varphi_T(s)$ is the Laplace transform of the inter-arrival time T and $T_k \sim T \ \forall k$. Transforming both sides of eq. (3.12), we get:

$$\varphi_{p_n}(s) = \frac{\varphi_{S_n}(s)}{s} - \frac{\varphi_{S_{n+1}}(s)}{s} = [\varphi_T(s)]^n \frac{1 - \varphi_T(s)}{s} \tag{3.13}$$

Since $\varphi_T(s) = 1/(1 + sE[T])$ by hypothesis, we have:

$$\varphi_{P_n}(s) = \frac{1}{s}\left(\frac{1}{1 + sE[T]}\right)^n\left[1 - \frac{1}{1 + sE[T]}\right] = \frac{E[T]}{(1 + sE[T])^{n+1}} \tag{3.14}$$

By taking the inverse Laplace transform and accounting for the couple

$$\left(\frac{a}{a + s}\right)^n \leftrightarrow \frac{a(at)^{n-1}}{(n-1)!}e^{-at} \tag{3.15}$$

we find

$$p_n(t) = E[T]\frac{(t/E[T])^n}{E[T]n!}e^{-t/E[T]} = \frac{(\lambda t)^n}{n!}e^{-\lambda t}, \qquad n \geq 0, \tag{3.16}$$

where we have let $\lambda \equiv 1/E[T]$.

The Poisson process can be characterized by the "memoryless" property. A point process is said to be memoryless if

$$P(T > t + \tau | T > \tau) = P(T > t), \qquad t, \tau > 0 \tag{3.17}$$

that is to say the probability to wait t more seconds for the next arrival, provided already τ seconds have gone by, is just the same as the probability that the inter-arrival time exceeds t. In other words, the arrival process does not retain any "memory" of the amount of time already elapsed waiting for an arrival.

We show that the Poisson process is the only point process that is memoryless, i.e., this is a unique feature of the Poisson process.

Theorem 3.4 An arrival process with i.i.d. inter-arrival times, distributed according the random variable T, satisfies $P(T > t + \tau | T > \tau) = P(T > t)$ for all $t, \tau > 0$ if and only if it is a Poisson process, i.e., T has a negative exponential probability distribution.

Proof: It is easy to verify that the negative exponential PDF satisfies eq. (3.17). The interesting point is that the converse can be shown as well, namely, if an arrival process is memoryless (with time continuous inter-arrival PDF) then its inter-arrival time PDF must be negative exponential. This implies that the *only* memoryless arrival process is the Poisson one. In fact, the validity of eq. (3.17) entails that

$$G_T(t) = P(T > t) = P(T > t + \tau | T > \tau) = \frac{P(T > t + \tau)}{P(T > \tau)} = \frac{G_T(t + \tau)}{G_T(\tau)} \tag{3.18}$$

where $G_T(t)$ is the CCDF of the inter-arrival time. Then, the identity $G_T(t + \tau) = G_T(t)G_T(\tau)$ holds for $t, \tau \geq 0$. Deriving with respect to τ and letting $\tau = 0$ we get $G'(t) = G'(0)G(t)$, with initial condition $G(0) = 1$. Since $G(t)$ is a monotonously decreasing function of t for a continuous random variable, we can let $a = -G'(0) > 0$. Then, solving the differential equation we obtain $G(t) = e^{-at}$, so that a can be identified as the reciprocal of the mean inter-arrival time.

The properties of a Poisson process are preserved through the application of various operations on the arrival process. For example, let us assume that arriving customers are sampled independently of one another with probability p. Let us consider the interval $(0, t]$ and assume that there are $A(t) = n$ arrivals in that interval. The conditional probability that the sampled process registers $A^{(s)}(t) = k$ events is:

$$P(A^{(s)}(t) = k | A(t) = n) = \binom{n}{k} p^k (1-p)^{n-k}, \quad k = 0, 1, \dots, n. \tag{3.19}$$

Removing the condition, we find:

$$\begin{aligned} P(A^{(s)}(t) = k) &= \sum_{n=k}^{\infty} \frac{(\lambda t)^n}{n!} e^{-\lambda t} \binom{n}{k} p^k (1-p)^{n-k} \\ &= \frac{(p\lambda t)^k}{k!} e^{-\lambda t} \sum_{n=k}^{\infty} \frac{[(1-p)\lambda t]^{n-k}}{(n-k)!} \\ &= \frac{(p\lambda t)^k}{k!} e^{-\lambda t} e^{(1-p)\lambda t} = \frac{(p\lambda t)^k}{k!} e^{-p\lambda t}, \quad k \geq 0. \end{aligned} \tag{3.20}$$

This shows the following result.

Theorem 3.5 The arrival process obtained by sampling independently of one another with probability p the arrivals of a Poisson process with mean rate λ is another Poisson process, with mean rate $p\lambda$.

Let us now consider the superposition of N independent Poisson processes, with mean rate $\lambda_i, i = 1, \dots, N$. The resulting counting function is $A(t) = A_1(t) + \cdots + A_N(t)$, i.e., it is the sum of N independent random variables. The corresponding generating function is obtained as

$$E[z^{A(t)}] = E[z^{A_1(t)+\cdots+A_N(t)}] = \prod_{i=1}^{N} E[z^{A_i(t)}] = \prod_{i=1}^{N} e^{\lambda_i t(z-1)} = e^{\lambda t(z-1)} \tag{3.21}$$

where $\lambda = \sum_{i=1}^{N} \lambda_i$. This shows that the superposition of N independent Poisson processes is still a Poisson process, with mean rate equal to the sum of the N mean rates of the component processes.

Finally, we recall a well-known property of Poisson arrivals, which is particularly useful, e.g., when generating Poisson events in simulations.

Theorem 3.6 Given that n arrivals of a Poisson process occur in the interval $(0, t)$, the PDF of their ordered epochs $t_1 < t_2 < \cdots < t_n$ is $f(t_1, \dots, t_n) = n!/t^n$.

Remark: Note that the joint density of the arrival times given in the theorem is nothing but the order statistics[1] of n independent uniform random variables over the interval $(0, t)$.

Proof: The density $f(t_1, \ldots, t_n)$ can be calculated according to the expression

$$f(t_1, \ldots, t_n) = \frac{\mathcal{P}(\overline{E}_0, E_1, \overline{E}_1, E_2, \overline{E}_2, \ldots, E_n, \overline{E}_n)}{\mathcal{P}(A(t) = n)} \tag{3.22}$$

where we have used the events $E_k = \{A(t_k - dt, t_k) = 1\}$, $k = 1, \ldots, n$ and $\overline{E}_k = \{A(t_k, t_{k+1} - dt) = 0\}$, $k = 1 \ldots, n - 1$, $\overline{E}_0 = \{A(0, t_1 - dt) = 0\}$, $\overline{E}_n = \{A(t_n, t) = 0\}$. Since all events refer to nonoverlapping intervals, they are independent and the probability in the numerator factors out into $2n + 1$ terms:

$$f(t_1, \ldots, t_n) = \frac{e^{-\lambda(t_1 - dt)} \cdot \lambda e^{-\lambda dt} \cdot e^{-\lambda(t_2 - dt - t_1)} \cdot \ldots \cdot \lambda e^{-\lambda dt} \cdot e^{-\lambda(t - t_n)}}{\frac{(\lambda t)^n}{n!} e^{-\lambda t}}$$

$$= \frac{n! \lambda^n e^{-\lambda t}}{\lambda^n t^n e^{-\lambda t}} = \frac{n!}{t^n}, \qquad 0 < t_1 < t_2 < \cdots < t_n < t. \tag{3.23}$$

We close this section by stating an efficient algorithm for generating Poisson random variables on a computer. Let N be a random variable on non-negative integers with PDF given by $\mathcal{P}(N = n) = \frac{a^n}{n!} e^{-a}$, $n \geq 0$. Assume a function `rand()` is available that generates (pseudo-)random samples uniformly distributed over $(0, 1)$. The event $N = n$ is equivalent to $T_1 + \cdots + T_n \leq a$ and $T_1 + \cdots + T_{n+1} > a$, where T_k are i.i.d. random samples of a negative exponential random variable with mean 1, $T_k \sim T \sim Exp(1)$. It is easy to see that $-\log R \sim Exp(1)$ if $R \sim \mathcal{U}(0, 1)$. This follows from $\mathcal{P}(-\log R > x) = \mathcal{P}(R < e^{-x}) = e^{-x}$, holding for any $x \geq 0$.

Checking the events on the times T_k amounts to verifying that $\sum_{j=1}^{n}(-\log R_j) \leq a$ and $\sum_{j=1}^{n+1}(-\log R_j) > a$, that is to say $\prod_{j=1}^{n} R_j \geq e^{-a}$ and $\prod_{j=1}^{n+1} R_j < e^{-a}$. The value of n satisfying both inequalities is the sample of the random variable N. An example code for implementing this algorithm is as follows.

```
function [N] = Poissonrand(a)
% a is the mean of the Poisson r.v.
k = 0;
S = -log(rand);
while S <= a
   k = k+1;
   S = S-log(rand);
end
N = k;
```

---

1 Given n random variables $X_1, X_2, \ldots, X_n$, their order statistics is the joint density of the random vector **Y** obtained by sorting out in ascending order the outcomes of the variables X_k's.

3.2.1 Light versus Heavy Tails

Before overviewing a number of generalizations and extensions of the basic Poisson process, we spend a word on the qualitative behavior of the point process and its relationship with the tail of the PDF of its inter-event times.

We have shown that Poisson events are intimately connected with the negative exponential PDF. The Poisson process inter-event times follow that kind of PDF, with mean equal to the reciprocal of the mean event rate and the CCDF of the inter-event times is $G_{\mathrm{Poi}}(t) = e^{-\lambda t}$.

Figure 3.1 illustrates 1000 samples of Poisson events with unit mean inter-event time. The top plot shows the events over time as vertical bars. The bottom left plot is the sequence of inter-event times. The bottom right plot reports the estimated CCDF of the inter-event times in a semi-log scale. As expected, the last plot exhibits a negative exponential behavior. The marking feature of Poisson events is their "smooth" distribution over time. Inter-event times vary within a relatively narrow range around the mean (the ratio of the standard deviation to the mean of the negative exponential PDF is 1).

It can be expected that Poisson events make a suitable model for events that depart definitely from deterministic, yet do not display any extreme outcome, i.e., no wild variability of the samples is generated by a Poisson process.

Many natural phenomena and artificial processes do not fit into this frame. Instead, a significant number of samples departing even orders of magnitude from the mean are observed, although most values stay below the mean. The marking character of random variables describing this situation is the decaying of the CCDF tail: while Poisson events correspond to an exponential tail, with this highly variable phenomena the tail of the CCDF of the inter-event times decays according to a power law. That is the reason why those probability distributions and the associated phenomena are also known as *heavy tails*. Note that a heavy-tailed random variable has only a limited number of finite moments. If $G(t) \sim 1/t^{\alpha}$, then only moments $E[X^k]$ with $k < \alpha$ do exist. Let us make an example with a simple Pareto random variable T, i.e., $G_{Par}(t) = (\theta/t)^{\alpha}$ for $t \geq \theta$

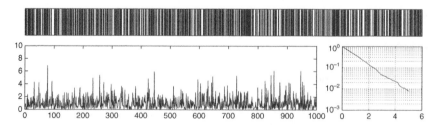

Figure 3.1 Sample of Poisson events (top plot); sequence of inter-event times (bottom, left plot); estimated CCDF of the inter-event times (semi-log scale; bottom, right plot).

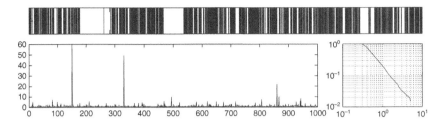

Figure 3.2 Sample of Pareto events (top plot); sequence of inter-event times (bottom, left plot); estimated CCDF of the inter-event times (log-log scale; bottom, right plot).

and $G_{Par}(t) = 1$ for $0 \leq t \leq \theta$. The parameter θ can be chosen to match a given mean, namely $\theta = (1 - 1/\alpha)\mathrm{E}[T]$ for $\alpha > 1$.

Figure 3.2 illustrates 1000 samples of Pareto events with $\alpha = 1.5$ and $\mathrm{E}[T] = 1$. The top plot shows the events over time. The bottom left plot is the sequence of inter-event times. The bottom right plot reports the estimated CCDF of the inter-event times in a log-log scale. As expected, the last one exhibits a power-law behavior, which gives a straight line in the log-log axes. The marking feature of Pareto events is their "widely variable" distribution over time, with time intervals that are crowded and time spans that are void of events. Inter-event times vary on a very wide range, occasionally reaching very high values (up to 50 times the mean in our sample). Note that the chosen Pareto PDF has infinite variance.

Among countless other examples, heavy tails have been found also in network traffic, e.g., lengths of exchanged files, duration of many application sessions, duration of cellular calls, inter-arrivals of packets in LANs. Still the Poisson process (or some generalization thereof, e.g., inhomogeneous Poisson process, spatial Poisson process, Markov modulated Poisson process) is perhaps the most widely used model in network traffic engineering. The main reason is that it yields to analytical treatment. Besides that, Poisson arrivals provide a statistical description that is reasonably accurate for all those phenomena where *multiplexing* plays a key role, i.e., where a large population of sources contribute to the observed event process, each source contributing a "negligible" amount to the overall process.

3.2.2 Inhomogeneous Poisson Process

The Poisson process is stationary, i.e., the statistical characteristics of the random variable $A(t_0, t_0 + t)$ depend only on t, the length of the observed interval, not on the initial time t_0. Equivalently, the PDF of the random variables T_k does not depend on k. In some applications it is useful to generalize this definitions and consider a Poisson process with mean arrival rate λ, which is a function of time.

The properties defining the inhomogeneous Poisson process with mean arrival rate $\lambda(t)$ are

1. $P(A(t + \Delta t) - A(t) = 1) = \lambda(t)\Delta t + o(\Delta t)$, as $\Delta t \to 0$;
2. $P(A(t + \Delta t) - A(t) > 1) = o(\Delta t)$, as $\Delta t \to 0$;
3. The number of arrivals in nonoverlapping intervals are independent CBF random variables, i.e., the random variables $A(t_2) - A(t_1)$ and $A(t_4) - A(t_3)$ are mutually independent for any $t_1 \le t_2 \le t_3 \le t_4$.

Following entirely analogous steps as done in Section 3.2, it can be found that the probability distribution of the counting function of the inhomogeneous Poisson process is

$$P(A(t_1, t_2) = n) = \frac{\Lambda(t_1, t_2)^n}{n!} e^{-\Lambda(t_1, t_2)}, \qquad t_1 \le t_2, n \ge 0, \tag{3.24}$$

where $\Lambda(t_1, t_2) \equiv \int_{t_1}^{t_2} \lambda(t)\, dt$. The nonstationary character of the process is evident, since the PDF of the counting function does depend on both the initial and final time epochs of the observed time interval. Nevertheless, the analytical expression of the distribution is still relatively simple.

Example 3.1 In this example, we show how an inhomogeneous Poisson process arises, when modeling a network communication process. The example refers to a wireless multi-hop relay network, where nodes are spread along a line according to a stochastically *spatial* process. By assuming a unidimensional spatial model along the x-axis, we have to describe the random distance between adjacent nodes. Equivalently, we can give the counting random variable $A(x, y)$, that represents the number of nodes found in the line interval $(x, y]$. We assume that nodes are distributed according to a homogeneous Poisson process with mean density λ (measured in nodes per unit length).

We consider a message dissemination protocol, i.e., a set of rules to deliver a message originating at one node to every other node. A message is launched by an originating node along the positive direction of the x axis. Messages are carried by broadcast packets. At each hop, a node forwarding the message encapsulates the message into a packet and sets the destination address of that packet as "broadcast." A trivial way to distribute the message would be having all nodes forwarding the message as they receive it. In dense scenarios, i.e., in cases where the number of nodes within the coverage radius of the transmitter is much larger than 1, the number of replicas of the message grows fast, eventually unleashing the so-called "broadcast storm." It is clear that some algorithm to realize *selective* forwarding is desirable. We examine a variant of the contention based forwarding (CBF) message dissemination protocol, defined in the GeoNetworking ETSI standard [74].

According to CBF, message dissemination follows two rules:

- *Forwarding rule* : When a node B first receives a message from a source node A, it checks if the source node is within a distance R_{max}; if that is the case, the node starts a timer of value $T_{max}(1 - d_{AB}/R_{max})$, where d_{AB} is the distance between A and B; at timer expiry, the message is forwarded, unless inhibition occurs; the node B does not relay the message, if the distance from the source node A is larger than R_{max}.
- *Inhibition rule* : When a node with a running timer receives further copies of the message scheduled for relaying, it cancels both the copies and the pending message, and it gives up to the forwarding operation.

In either case, the node B will not relay any more copies of a given message, once B has decided whether to forward it or not.

The rationale is to guarantee the dissemination of the message, as long as nodes are within a given range R_{max}, yet avoiding to produce a large number of duplicated receptions of the message. The timer ensures that only a single node will actually relay the message (the one that selects the shortest timer, i.e., the most distant node from the source node within a range R_{max}). At least, this is true so long as all nodes within the distance R_{max} receive the message successfully with probability 1.

The ensuing example of nonhomogenous Poisson process deals with the dimensioning of the parameter R_{max} in the face of a stochastic reception model, i.e., a model that accounts for the event the message reception may fail, depending on the level of the signal-to-noise ratio (SNR) at the receiver. The inhomogeneous Poisson process arises since we need to describe the set of nodes that receive the message successfully. This event depends on nodes' position, hence the inhomogeneity of the spatial process that describes the node density.

Reception is considered to be successful if the SNR at the receiver exceeds a threshold that depends on the link rate. Under a deterministic path loss model, the requirement of a minimum SNR level translates into a *sharp* maximum transmission range R_{th}. Thus, in this case, the probability $P(x)$ of successful reception for a node at distance x from the transmitter is $P(x) = 1, x \in [0, R_{th}]$ and $P(x) = 0, x > R_{th}$. In general, $P(x)$ varies in a more gradual manner so that it is reduced from 1 to 0 as x increases.

To understand the rationale of the R_{max} parameter in the context of timer-based dissemination protocols, where the inhibition rule is used to suppress most message duplicates, consider the following: Let $x = 0$ denote the position of the source node A. When A sends out a message, the nodes in the interval $[0, R_{max}]$ can be divided into two subsets: the set of nodes that have successfully received the message from A, S_A, and the set of nodes that have failed to receive the message, F_A (most probably, those that are located farther away from A.) The nodes in S_A start their forwarding timers; eventually the node with the smallest timer, say B, sends out a new copy of the message, thus inhibiting all nodes in $S_A \cap S_B$. On the contrary, nodes in $(S_A \cap F_B) \bigcup (F_A \cap S_B)$ do have a copy of the message ready to send and are not inhibited, since they have received only a single copy of the message, either from A or from B. The result is that message duplication suppression enforced by the inhibition rule is not fully effective. The ineffectiveness of duplicate suppression can be made negligible as long as R_{max} is such that $P(x) \sim 1$ for $x \in [0, R_{max}]$.

To assess the effect of R_{max} we assume a specific form for $P(x)$, i.e., a Rician fading path loss model, so that the normalized SNR $\hat{\gamma} = \gamma / \bar{\gamma}(x)$ at distance x from the transmitting node is modeled as a random variable with PDF expressed as

$$f_{\hat{\gamma}}(v) = (1 + K)e^{-K}e^{-(1+K)v}I_0(2\sqrt{K(1 + K)v}) , \qquad v \geq 0 \tag{3.25}$$

where $\bar{\gamma}(x)$ is the average SNR value at distance x, K is the ratio between the mean power of the dominating radio propagation path and the mean power of the other paths, and $I_0(\cdot)$ is the modified Bessel function of the first kind of order zero. The average SNR is $\bar{\gamma}(x) = G(x)P_{tx}/P_N$, where P_N is the background noise power and P_{tx} is the transmission power. $G(x)$ depends on the assumed path loss propagation model (e.g., single or dual slope). Here we have assumed a dual slope path loss propagation model with break-point distance $d_b = 120 \ m$ and path loss exponent $\alpha_1 = 2$ for distances up to d_b and $\alpha_2 = 4$ for distances bigger than d_b. Formally, it is

$$G(x) = \begin{cases} \dfrac{\kappa}{x^{\alpha_1}} & x \leq d_b \\ \dfrac{\kappa d_b^{\alpha_2 - \alpha_1}}{x^{\alpha_2}} & d > d_b. \end{cases} \tag{3.26}$$

where $\kappa = 1.637 \cdot 10^{-5}$ and distances are measured in meters.

Let γ_{th} denote the threshold SNR required to sustain the communication link at the desired rate and quality (bit error ratio) level. Then,

$$P(x) = \int_{\gamma_{th}/\bar{\gamma}(x)}^{\infty} f_{\hat{\gamma}}(v) \, dv , \qquad x > 0 \tag{3.27}$$

Figure 3.3(a) shows the behavior of $P(x)$ for three different values of K and for $P_{tx} = 200 \ \text{mW} = 23 \ \text{dBm}$, $P_N = -104 \ \text{dBm}$ and $\gamma_{th} = 10 \ dB$. $K = 0$ corresponds to Rayleigh fading model. A Rician model with low K is more appropriate when the direct, line-of-sight path is absent or has a relatively low strength with respect to reflected and diffracted rays. In the opposite case, a larger value of K must be used. The bigger K, the more dominant the direct radio propagation path.

As results from Figure 3.3(a), by setting $R_{max} = 182 \ \text{m}, 264 \ \text{m}, 409 \ \text{m}$ we guarantee that $P(x) \geq 0.99$ for $x \in [0, R_{max}]$ for $K = 0, 3, 10$ respectively.

Let us consider a tagged node A at $x = 0$. A is the source of a message to be disseminated to nodes located on the semi-axis $x > 0$. Let B be the most distant

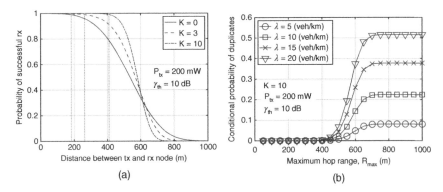

Figure 3.3 Example 3.1: duplicated message delivery under a Rician channel model. Left plot: probability $P(x)$ of correct message detection versus the distance x between transmitter and receiver, for three values of the wireless channel model parameter K. Right plot: probability of duplicated messages P_{dup} versus R_{max}. Transmission power level: $P_{tx} = 200$ mW; SNR threshold: $\gamma_{th} = 10$ dB.

node that successfully receives the message sent by A, among all nodes within distance R_{max} of A. Let Y be the random variable defined as the distance between A and B. Let $N(0, R_{max})$ denote a random variable that represents the number of nodes located in the interval $(0, R_{max})$ that successfully receive the message sent by A. Those nodes form an inhomogeneous Poisson process with mean density $\lambda P(x), x > 0$. The CDF of Y, conditional on there being at least one successful reception in $(0, R_{max}]$, is

$$F_Y(y) = \mathcal{P}(Y \leq y | N(0, R_{max}) > 0) = \frac{P(N(y, R_{max}) = 0 \ \& \ N(0, y) > 0)}{\mathcal{P}(N(0, R_{max}) > 0)} =$$

$$= \frac{\left[1 - e^{-\int_0^y \lambda P(u) \, du}\right] e^{-\int_y^{R_{max}} \lambda P(u) \, du}}{1 - e^{-\int_0^{R_{max}} \lambda P(u) \, du}} = \frac{e^{\int_0^y \lambda P(u) \, du} - 1}{e^{\int_0^{R_{max}} \lambda P(u) \, du} - 1} \quad (3.28)$$

for $0 \leq y \leq R_{max}$.

According to the timer-based dissemination protocol, the most distant node B is the designated message forwarder. Nodes in the interval $(0, R_{max})$ will not forward the message, thanks to the inhibition rule, *provided they receive the message both from A and from B*. Nodes receiving one and only one copy of the message, will not be inhibited, thus causing a *duplicated* message forwarding. We now characterize the probability of such an event.

Let $\mathcal{N}_1(y) = (S_A \cap F_B) \bigcup (F_A \cap S_B)$ denote the set of nodes belonging to $(0, R_{max})$ that receive the message exactly once, after both A's and B's transmissions, conditional on the distance between A and B being $Y = y$. Those nodes form an inhomogeneous Poisson process with mean density $\lambda_1(x|y) = \lambda[P(x)(1 - P(y - x)) + (1 - P(x))P(y - x)]$ for $x \in (0, y)$ and $y \in (0, R_{max})$.

The conditional average number of such nodes is

$$M(y) \equiv \mathrm{E}[\mathcal{N}_1(y)|Y = y] = 2 \int_0^y \lambda P(x) \, dx - 2 \int_0^y \lambda P(x) P(y - x) \, dx \qquad (3.29)$$

Then, the probability that at least one node escapes the inhibition mechanism, i.e., the probability that inhibition is not fully effective and duplicated message forwarding actions take place, is

$$P_{dup} \equiv 1 - \int_0^{R_{max}} e^{-M(y)} \, dF_Y(y) \qquad (3.30)$$

The probability P_{dup} is plotted in Figure 3.3(b) for $P_{tx} = 200 \, \mathrm{mW} = 23 \, \mathrm{dBm}$, $P_N = -104 \, \mathrm{dBm}$, $K = 10$, $\gamma_{th} = 10 \, \mathrm{dB}$ and for the same path loss model as in Figure 3.3(a).

In the case $K = 10$ (a value consistent with a scenario where propagation occurs in essentially open spaces, i.e., fading due to multiple paths has a marginal effect), Figure 3.3(b) highlights that the probability of duplicated messages is negligible if $R_{max} \sim 410 \, m$. For bigger values of R_{max} the message duplication probability grows steeply, up to quite high values, the higher the bigger the mean vehicle density λ.

The conclusion is that the maximum hop range R_{max} should be limited to achieve good performance of the timer-based dissemination protocol. Once a suitable upper bound is selected for R_{max}, so that reception of message is reliable within that range, the random nature of the radio channel impacts the dissemination process in a marginal way.

3.2.3 Poisson Process in Multidimensional Spaces

It is possible to extend the notion of Poisson process to multidimensional spaces. Instead of thinking of customer arrivals scattered along the time axis, one can refer to dots dispersed in a spatial region. Indeed, spatial Poisson process can be used to model the positions of users or of infrastructure equipment over a service area (e.g., cellular base stations, access points).

We extend the notation used in previous sections consistently, by letting $A(B)$ denote the number of points found in a region B of the space. This is nothing but the extension of the counting function from unidimensional lines to multidimensional spaces. We will consider only compact, measurable regions. The measure (area, volume) of the region B is denoted with $|B|$.

We define first a homogeneous or uniform Poisson point process (PPP) in the d dimensional space $\mathbb{R}^d$.

Definition 3.1 Given a real, positive constant λ, a uniform PPP with intensity λ is a point process such that:

- for every compact set $B \subset \mathbb{R}^d$, $A(B)$ is a Poisson random variable with mean $\lambda|B|$;
- for any positive integer m and any collection $B_1, \dots, B_m$ of disjoint bounded sets, $A(B_1), \dots, A(B_m)$ are independent random variables.

Theorem 3.6 carries over to multidimensional spaces in a quite natural way.

Theorem 3.7 Given a homogeneous PPP with intensity λ in $\mathbb{R}^d$ and any region $W \subset \mathbb{R}^d$, with positive and finite measure, then the conditional probability distribution of $A(B)$ for $B \subset W$, given that $A(W) = n$ is binomial with parameters n and $p = |B|/|W|$, i.e.,

$$P(A(B) = k | A(W) = n) = \binom{n}{k} p^k (1 - p)^{n-k}, \qquad k = 0, 1, \dots, n. \qquad (3.31)$$

For the proof as well as for many other results and generalizations, we direct the reader to, e.g., [97].

Theorem 3.7 gives a path to the computer simulation of PPPs. Given a compact region B and a rectangle R covering B, i.e., $B \subset R$, first draw a Poisson distributed integer with mean $\lambda |R|$, say it be n. Then, distribute uniformly at random the n points over R. Ignore points that fall outside B. Those points that fall within B represent a realization of the PPP of intensity λ in B. Note that the only distinction between a binomial process (uniformly random distribution of a *given* number n of points) and a PPP in a given compact set W is that different realizations of the PPP consist of different number of points.

An example of a PPP is illustrated in Figure 3.4(a) for $d = 2$. The considered region is a square of side length 10 (arbitrary units). The intensity of the uniform PPP is $\lambda = 10$ (dots per square length unit).

The definition of PPP can be generalized to the case where the intensity is a measure of $\mathbb{R}^d$, i.e., a function $\Lambda(B)$ that associates a non-negative real number to each measurable set $B \subset \mathbb{R}^d$. Often, it is possible to define a density $\lambda(x), x \in \mathbb{R}^d$, such that

$$\Lambda(B) = \int_B \lambda(x) \, dx \qquad (3.32)$$

The generalized definition is as follows.

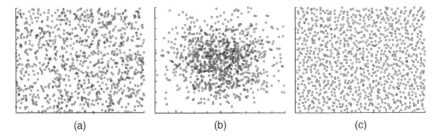

(a)   (b)   (c)

Figure 3.4 Sample realizations of spatial point processes. Left plot: uniform PPP. Middle plot: Gaussian PPP with intensity density as in eq. 3.33 and σ equal to 1/5 of the box side length. Right plot: lattice with edge length equal to 1, perturbed with a random displacement uniformly distributed in $[0, 0.9]$.

Definition 3.2   Given an intensity measure Λ, a general PPP is a point process such that:

- for every compact set $B \subset \mathbb{R}^d$, $A(B)$ is a Poisson random variable with mean $\Lambda(B)$;
- for any positive integer m and any collection $B_1, \ldots, B_m$ of disjoint bounded sets, $A(B_1), \ldots, A(B_m)$ are independent random variables.

For example, Figure 3.4(b) shows the point pattern of a general PPP in $\mathbb{R}^2$, with intensity density

$$\lambda(x) = \frac{1}{2\pi\sigma^2} \exp\left(-\frac{\|x - x_0\|^2}{2\sigma^2}\right), \quad x \in \mathbb{R}^2. \tag{3.33}$$

This is known as a Gaussian Poisson process. In Figure 3.4(b) the central point x_0 is chosen as the center of the square region, i.e., the point $(5,5)$.

Point patterns can follow different statistics, other than Poisson. A simple example of point pattern is the lattice, i.e., a grid of evenly spaced out points. By properly choosing the unit of measurement, we can define the d-dimensional lattice as $\mathbb{Z}^d$. A perturbed lattice can be defined by introducing a family of i.i.d. random variables $X_u, u \in \mathbb{Z}^d$, having a PDF $f(x)$. Then, the perturbed lattice can be defined as $\Phi = \{u \in \mathbb{Z}^d : u + X_u\}$. A simple case is obtained when $X_u \sim \mathcal{U}(0, a)$ with $0 < a < 1$. Figure 3.4(c) depicts a perturbed lattice with $a = 0.9$.

It is useful to make a brief digression to introduce two definitions that are strictly related with point patterns. Given a set of random points $\Phi = \{x_i\}$ in $\mathbb{R}^d$ and a distance, it is possible to define the Voronoi cell associated with x_i as the set of points $x \in \mathbb{R}^d$ that are closest to x_i rather than to any other x_j, $j \neq i$. Namely, we let

$$V(x_i) = \{y \in \mathbb{R}^d : \|y - x_i\| \leq \|y - x_j\|, \forall j \neq i\} \tag{3.34}$$

It is then possible to define a Voronoi tessellation of the space, as the decomposition of $\mathbb{R}^d$ defined by the set of Voronoi cells associated to the points in Φ. Every point in $\mathbb{R}^d$ belongs to a unique Voronoi cell, except boundary points, those that are equidistant from at least two different pattern points of Φ. The Voronoi decomposition of the space induces also a graph relationship among the points of Φ. We connect two points of Φ with an arc if and only if the two Voronoi cells of the considered points share a common boundary. The obtained graph is called a Delaunay triangulation. Notice that both Voronoi decomposition and Delaunay triangulation are completely defined once given the set of points Φ and the distance measure (e.g., the norm induced distance over $\mathbb{R}^d$).

Figure 3.5 gives examples of a Voronoi tessellation and of the corresponding Delaunay triangulation for a set of 100 points uniformly distributed over a bidimensional square region of size 10×10.

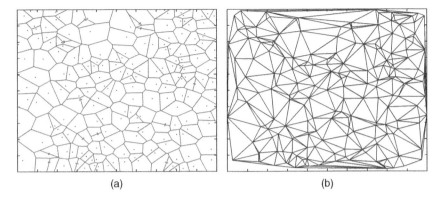

(a)   (b)

Figure 3.5 Sample realization of a Voronoi tessellation with 100 points uniformly distributed at random over a square region (left plot) and the corresponding Delaunay triangulation (right plot).

Example 3.2 As an application we consider the publicly available Cologne dataset (http://sumo.dlr.de/wiki/Data/Scenarios/TAPASCologne). Specifically, we extract the cellular network base station (BS) locations. There are 246 BS locations listed in the dataset, corresponding to the central part of the city of Cologne in Germany.

We assume a simple power-law model for signal path loss, i.e., the power level received at distance d from a transmitting BS is given by $P_{rx}(d) = P_{tx}\min\{1, \kappa/d^\alpha\}$, where P_{tx} is the transmission power level and κ is a constant depending on carrier frequency, antenna gains and other physical parameters. For this numerical example, we let $\alpha = 3$ and $\kappa = -35, 89$ dB with the distance d measured in meters.

If user terminals associate to serving BSs according to a maximum received power criterion, than the set of points forming the serving area of a BS (the so called *cell* of that BS) coincides with the Voronoi cell of that BS.

Figure 3.6(a) illustrates the Voronoi diagram of the Cologne BSs. Areas on the outer border region are unrealistically large due to the spatial truncation of the BS locations available in the dataset.

Real cell shapes are not so sharp due to the complicated electromagnetic field propagation (i.e., path loss does not decay as a simple deterministic power law: that is only a first-approximation model). Moreover, real cells overlap significantly to allow smooth handoff.

The received power level in each point of the whole area is shown in Fig. 3.6(b) as a heat map. The darker the point of the map, the lower the received power level. The transmission power level of BSs, P_{tx}, ranges from 33 dBm to 13 dBm, and it is set so that an SNR level of at least 6 dB is guaranteed to all points falling within the Voronoi cell of the serving BS. Brighter points correspond to higher received power

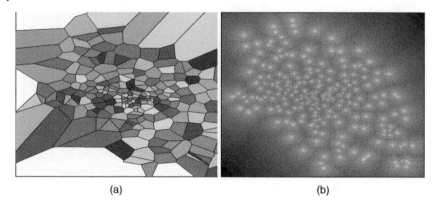

(a)   (b)

Figure 3.6 Example 3.2: Voronoi tessellation of the Cologne cellular base stations (left plot) and heat map of the received power level (right plot).

levels. The brightest points are just BS locations. Increasing the transmitted power of BSs would raise the received power level, at the expense of a larger interference among different cells.

Using a PPP model to generate locations of BSs of a cellular network neglects the fact that BS locations are planned so as to offer a good coverage of potential users with the minimum possible number of BS installations. Therefore, BSs are never too close to each other, whereas a PPP model does not enforce any positive minimum distance between any two points. A hard-core spatial process is more suitable for that task.

Let us now introduce some definitions. Our purpose is to characterize a number of general operations on point processes.

Definition 3.3   Joint probability distribution. The finite dimensional probability distributions of a point process are the joint probability distributions of the random variables $A(B_1), \dots, A(B_m)$ for all integers $m > 0$ and compact sets $B_1, \dots, B_m$.

Definition 3.4   Translation of a point process. Given a point process $\Phi = \{x_1, x_2, \dots \}$, a translation of the process by $x \in \mathbb{R}^d$ is $\Phi = \{x_1 + x, x_2 + x, \dots \}$.

Definition 3.5   Stationarity. A point process is said to be stationary if its finite dimensional distributions are invariant with respect to translations.

Definition 3.6   Isotropy. A point process is said to be isotropic or rotationally invariant if its finite dimensional distributions are invariant with respect to rotations around the origin of the coordinate system in $\mathbb{R}^d$.

Definition 3.7   Motion-invariance. A point process is said to be motion invariant if it is stationary and isotropic.

The class of motion invariant point processes play a role similar to stationary processes over time. An example of more general process (nonstationary) that is analytically tractable is the general PPP. The lattice grid defined above, when perturbed by uniformly distributed random shift, gives rise to an example of stationary, yet anisotropic point process.

Next we define a number of operations on a point process and we give properties of a PPP under these operations.

3.2.3.1   Displacement

Given a point process Φ, its displacement is the point process Φ' defined by

$$\Phi' = \{x + V_x, x \in \Phi\} \tag{3.35}$$

where V_x are independent random variables whose PDF may depend on the location x. Let such a PDF be denoted with $f(x, \cdot)$. Displacement consists of random shifts to the points of the process. The displacements are independent from one point to another. In general we expect the statistics of the modified process to be different from the original one. When this operation is applied to a general PPP Φ with density function $\lambda(x)$ it yields another PPP Φ' with density function

$$\lambda'(x) = \int_{\mathbb{R}^d} \lambda(y) f(y, x - y) \, dy , \qquad x \in \mathbb{R}^d \tag{3.36}$$

Note that, if the original process is a uniform PPP, with $\lambda(x) = \lambda \; \forall x$, then so is the process resulting from random displacement, i.e., a uniform PPP *is stochastically invariant under random, independent, and stationary displacements.* This is true in any dimension d, thus including the basic case of standard Poisson process on a line.

3.2.3.2   Mapping

A point process can be transformed by mapping each point of the process into another point, possibly in a space with different dimensions. Let us restrict our attention to mappings defined by measurable functions $f : \mathbb{R}^d \mapsto \mathbb{R}^c$, such that f does not shrink a nonsingleton compact set to a singleton. Then, it is possible to show the following.

Theorem 3.8   Let Φ be a PPP with intensity measure Λ and density λ. Let f be a mapping of the class defined above with the property that $\Lambda(f^{-1}(y)) = 0, \forall y \in \mathbb{R}^c$. Then $\Phi' = f(\Phi) = \bigcup_{x \in \Phi} \{f(x)\}$ is a PPP with intensity measure $\Lambda'(B') = \Lambda(f^{-1}(B'))$ for all compact $B' \subset \mathbb{R}^c$.

For example, let us consider a uniform PPP in $\mathbb{R}^d$ with density λ. As a mapping we choose $f(x) = ||x||$, hence $f : \mathbb{R}^d \mapsto \mathbb{R}^+$. Given an interval $[0, r]$, the inverse image under f is the ball of center the origin and radius r: let it be denoted with $B(0, r)$. Then

$$\Lambda'([0, r]) = \Lambda(B(0, r)) = \lambda|B(0, r)| = \lambda\beta_d r^d \tag{3.37}$$

where the coefficient β_d is the volume of the d dimensional ball of radius 1. It is

$$\beta_d = \frac{\pi^{d/2}}{\Gamma(d/2 + 1)} \tag{3.38}$$

with $\Gamma(x) \equiv \int_0^\infty u^{x-1}e^{-u}\, du$ (Euler gamma function). In the planar case, $d = 2$ and $\Lambda'([0, r]) = \lambda\pi r^2$. The density is $\lambda' = 2\lambda\pi r$, hence it is clear that the transformed PPP is not uniform.

As another example, let us consider $f(x) = ||x||^d$. Then $[0, r]$ corresponds to the ball $B(0, r^{1/d})$, and hence we get

$$\Lambda'([0, r]) = \Lambda(B(0, r^{1/d})) = \lambda\beta_d r \tag{3.39}$$

and the corresponding density is $\lambda' = \lambda\beta_d$, i.e., it is constant. Therefore, under this second transformation, a uniform PPP in $\mathbb{R}^d$ is turned into another (unidimensional) uniform PPP.

3.2.3.3 Thinning

Thinning consists of removing some points from a point process. Usually removing is governed by probabilistic rules. In general we define a function $p : \mathbb{R}^d \mapsto [0, 1]$ that gives the probability $p(x)$ that a point survives, while a point x is removed with probability $1 - p(x)$. If the decision of removing a point is taken independently for each point of the original point process, we say that we apply independent thinning.

It can be shown that, if we start with a homogeneous PPP with intensity λ and apply independent thinning with probability $1 - p(x)$, then we end up with an inhomogeneous PPP, having density function $\lambda p(x)$. This is clearly a generalization of the sampling result presented in Section 3.2.

Formally, the following theorem can be stated.

Theorem 3.9 Let Φ be a stationary PPP with mean density λ. Removal of points of Φ according to independent thinning with removal probability $1 - p(x)$ leads to an inhomogeneous PPP Ψ with mean density $\lambda p(x)$.

Proof: Let B denote a closed set. By the law of total probability, we can write

$$P(\Psi(B) = k) = \sum_{j=k}^\infty P(\Psi(B) = k|\Phi(B) = j)P(\Phi(B) = j) \tag{3.40}$$

Given that there are j point of Φ in B, their location is uniformly and independently distributed over B for each point. Then, the probability of retaining one such point is $q \equiv \int_B p(x)\, dx / |B|$, where $|B|$ denotes the measure of B. Therefore, since thinning is done independently on each point, we get:

$$P(\Psi(B) = k | \Phi(B) = j) = \binom{j}{k} q^k (1-q)^{j-k}, \quad j \geq k, k \geq 0. \tag{3.41}$$

Substituting into (3.40), we have

$$P(\Psi(B) = k) = \sum_{j=k}^{\infty} \binom{j}{k} q^k (1-q)^{j-k} \frac{(\lambda|B|)^j}{j!} e^{-\lambda|B|}$$

$$= \frac{(\lambda q |B|)^k}{k!} e^{-\lambda|B|} \sum_{j=k}^{\infty} \frac{(\lambda(1-q)|B|)^{j-k}}{(j-k)!}$$

$$= \frac{(\lambda q |B|)^k}{k!} e^{-\lambda|B|} e^{-\lambda(1-q)|B|} = \frac{\left(\int_B \lambda p(x) dx\right)^k}{k!} e^{-\int_B \lambda p(x) dx}$$

that is to say, Ψ is an inhomogeneous PPP, with mean density $\lambda p(x)$.

3.2.3.4 Distances

We consider a uniform PPP Φ with constant density λ. Given a point x of Φ, we can find easily the distance of the point of Φ closest to x, other than x itself. We define the random variable $D = \inf\{||y - x||, y \in \Phi \backslash \{x\}\}$. Given the stationarity and isotropy of Φ, the probability distribution of D does not depend on x. We have

$$P(D > r) = P((\Phi \backslash \{x\}) \cap B(x, r) = \emptyset) = \exp\left(-\int_{B(x,r)} \lambda\, dy\right) = e^{-\lambda \beta_d r^d} \tag{3.42}$$

hence the PDF of D is $f_D(r) = d\lambda \beta_d r^{d-1} e^{-\lambda \beta_d r^d}$, $r \geq 0$. In the special case $d = 2$, we have $P(D > r) = e^{-\lambda \pi r^2}$ and $f_D(r) = 2\lambda \pi r e^{-\lambda \pi r^2}$, $r \geq 0$.

We can even find the distance of n-th closest point, D_n (so $D \equiv D_1$):

$$P(D_n > r) = P(|(\Phi \backslash \{x\}) \cap B(x, r)| < n) = \sum_{k=0}^{n-1} \frac{(\lambda \beta_d r^d)^k}{k!} e^{-\lambda \beta_d r^d} \tag{3.43}$$

For example, a wireless station can associate with an access point (AP) if the level of received power from the AP exceeds a threshold P_{th}. If the path gain at a distance d can be expressed as $G(d) = \min\{1, \kappa/d^\alpha\}$, the received power level is $P_{rx}(d) = P_{AP} G(d)$, where P_{AP} is the transmission power level of the AP. The condition for the association corresponds to $P_{tx}(d) \geq P_{th}$, i.e., $d \leq d_{th} = (\kappa P_{AP}/P_{th})^{1/\alpha}$. If stations are scattered around the AP according to a uniform PPP with mean spatial density λ in three dimensions (e.g., as in office building), then the probability that no more than n stations can associate with the AP is given by $P(D_{n+1} > d_{th})$, with D_n distributed as in eq. (3.43) with $d = 3$.

3.2.3.5 Sums and Products on Point Processes

Given a point process Φ and a function $f(x) : \mathbb{R}^d \mapsto \mathbb{R}$, we can define the sum of f over Φ; formally $\sum_{x \in \Phi} f(x)$. Campbell's theorem is a general result on this kind of sums.

Theorem 3.10 Let Φ be a point process on $\mathbb{R}^d$ with density function $\lambda(x)$ and $f(x) : \mathbb{R}^d \mapsto \mathbb{R}$ a measurable function. Then

$$Y = \sum_{x \in \Phi} f(x) \tag{3.44}$$

is a random variable with mean

$$E[Y] = \int_{\mathbb{R}^d} f(x)\lambda(x) \, dx \tag{3.45}$$

provided the integral is finite.

For example, let us consider a set of transmitting nodes scattered over the plane according to a uniform PPP with density λ. The interference contributed by these transmitters on a probe node located at the origin is

$$I_0 = \sum_{x \in \Phi} P_{tx} G(\|x\|) \tag{3.46}$$

where $G(d)$ is the path gain of the radio channel at distance d and P_{tx} is the transmission power level. Campbell's theorem allows us to calculate the mean interference at the origin, i.e.,

$$E[I_0] = \lambda P_{tx} 2\pi \int_0^\infty G(r) \, dr \tag{3.47}$$

We can also consider products over the point process. Let $v : \mathbb{R}^d \mapsto [0, 1]$ be a measurable function such that $1 - v(x)$ has bounded support. The probability generating functional of point process Φ is defined as

$$G_\Phi(v) = E\left[\prod_{x \in \Phi} v(x) \right] = E\left[\exp\left(\sum_{x \in \Phi} \log v(x) \right) \right] \tag{3.48}$$

The Laplace functional is defined as

$$L_\Phi(u) = G_\Phi(e^{-u}) = E\left[\exp\left(-\sum_{x \in \Phi} u(x) \right) \right] \tag{3.49}$$

For a PPP with density $\lambda(x)$ it can be seen that

$$G_\Phi(v) = \exp\left(-\int_{\mathbb{R}^d} [1 - v(x)]\lambda(x) \, dx \right) \tag{3.50}$$

and

$$L_\Phi(u) = \exp\left(-\int_{\mathbb{R}^d} [1 - e^{-u(x)}]\lambda(x)\, dx\right) \tag{3.51}$$

As an application, we can calculate the outage probability of a link under the interference of a Poisson network, i.e., a set of transmitting nodes scattered on the plane according to a PPP of density $\lambda(x)$. Let us assume that the receiver of the tagged link is located at the origin of $\mathbb{R}^2$. The signal-to-interference-and-noise-ratio (SINR) of the receiver can be written as

$$SINR_0 = \frac{P_{tx}G(r)}{I + P_N} \tag{3.52}$$

where P_N is the background thermal noise power, I is the interference and $G(r)$ is the path gain from the tagged transmitter to the tagged receiver, at distance r. In general, the path gain is modeled as the product of a deterministic component $G_{det}(r)$ (depending mostly on the physical characteristics of the e.m. field and on the geometry of the propagation environment) and a random component, that aims at capturing the effect of obstructions (shadowing) and of the multi-path (fading). The former component is often modeled as a power law with distance, i.e., $G_{det}(r) = \kappa/r^\alpha$, $r \geq r_0$; the latter component is a random variable G_{rnd} with unit mean. In case of Rayleigh fading, G_{rnd} is a negative exponential variable, i.e., $P(G_{rnd} > h) = e^{-h}$.

By definition, we say that the tagged link is in outage if its SINR falls below a threshold θ. With Rayleigh fading, we get

$$P_{out} = \mathcal{P}(SINR \leq \theta) = 1 - \mathcal{P}(SINR > \theta) = 1 - \mathcal{P}\left(G_{rnd} > \theta\frac{I + P_N}{P_{tx}G_{det}(r)}\right) \tag{3.53}$$

Then

$$1 - P_{out} = \mathrm{E}\left[\exp\left(-\theta\frac{I + P_N}{P_{tx}G_{det}(r)}\right)\right] = \exp\left(-\frac{\theta}{SNR_0(r)}\right)\mathrm{E}[e^{-\theta\hat{I}}] \tag{3.54}$$

where $SNR_0(r) \equiv P_{tx}G_{det}(r)/P_N$ is the baseline signal-to-noise ratio of the tagged link and $\hat{I}$ is the interference normalized to the average received power $P_{tx}G_{det}(r)$. Under the PPP assumption, the interference can be written as

$$\hat{I} = \frac{1}{P_{tx}G_{det}(r)}\sum_{x\in\Phi} P_{tx}G_{det}(||x||)G_{rnd,x} = \sum_{x\in\Phi} G_{rnd,x}\frac{r^\alpha}{||x||^\alpha} \tag{3.55}$$

Let us denote the random variable of the Rayleigh fading with H for simplicity, i.e., $G_{rnd,x} \sim H, \forall x \in \Phi$. then

$$\mathrm{E}[e^{-\theta\hat{I}}] = \mathrm{E}\left[\prod_{x\in\Phi} e^{-\theta Hf(x)}\right] = \mathrm{E}_\Phi\left[\prod_{x\in\Phi} \mathrm{E}_H[e^{-\theta Hf(x)}]\right] \tag{3.56}$$

where $f(x) \equiv (r/||x||)^\alpha$ and we have assumed that fading is independent among different interfering nodes. Now, using the definition of the probability generating functional and the result in eq. (3.50), we obtain for $\alpha > 2$:

$$
\begin{aligned}
\mathrm{E}[e^{-\theta \hat{I}}] &= \exp\left(-\int_{\mathbb{R}^2} (1 - \mathrm{E}_H[e^{-\theta H f(x)}])\lambda \, dx\right) \\
&= \exp\left(-\mathrm{E}_H\left[\int_0^\infty (1 - e^{-\theta H(r/y)^\alpha})2\pi\lambda y \, dy\right]\right) \\
&= \exp\left(-\mathrm{E}_H\left[\lambda\pi r^2(\theta H)^{2/\alpha}\int_0^\infty (1 - e^{-z})\frac{2}{\alpha}z^{-2/\alpha-1} \, dz\right]\right) \\
&= \exp\left(-\mathrm{E}_H\left[\lambda\pi r^2(\theta H)^{2/\alpha}\int_0^\infty z^{-2/\alpha}e^{-z} \, dz\right]\right) \\
&= \exp(-\lambda\pi r^2\theta^{2/\alpha}\mathrm{E}[H^{2/\alpha}]\Gamma(1 - 2/\alpha)) \\
&= \exp(-\lambda\pi r^2\theta^{2/\alpha}\Gamma(1 + 2/\alpha)\Gamma(1 - 2/\alpha)) \\
&= \exp\left(-\lambda\pi r^2\theta^{2/\alpha}\frac{2\pi/\alpha}{\sin(2\pi/\alpha)}\right)
\end{aligned}
\tag{3.57}
$$

where the third line is obtained by the variable change $z = \theta H(r/y)^\alpha$, the fourth line is obtained by integrating by parts, the fifth line uses the definition of the Euler gamma function, the sixth line exploits the negative exponential distribution with mean 1 of the Rayleigh fading variable H.

Equations (3.54) and (3.57) provide an explicit expression of the outage probability of a link over distance r, under the interference of a Poisson network (uniform PPP), for power law attenuation and Rayleigh fading.

3.2.3.6 Hard Core Processes

A point process where points cannot be closer than a given distance r is said to be a *hard core* process. It is a useful model for systems where feasibility or opportunity leads to distributing points not too close to each other, e.g., when defining locations of cellular base stations. In that case, planning of cellular coverage leads to designing a geographical distribution of the base stations of an operator where two stations are as separated as possible, i.e., the objective is to minimize the number of base stations required to cover a given service area with an assigned quality of service. Placement of base stations results however in an irregular pattern due to technical and regulatory constraints. Note that a PPP does not guarantee any minimum distance between two nodes.

A classic example of hard core process is the Matern one. It comes in two different brands.

Definition 3.8 Matern process of type I. It starts with a uniform PPP of density λ_p. Each point is marked, if it has a neighbor at a distance less than or equal to r.

Then, all marked points are removed. The resulting point pattern is a realization of the Matern process of type I.

This definition gives a practical way of constructing a Matern process of type I. It leads typically to sparse point patterns, since two points falling at a distance less than r in the original PPP are *both* marked and hence removed. In $\mathbb{R}^d$, the probability that a node has no neighbor at a distance $\leq r$ is $\exp(-\lambda_P \beta_d r^d)$. Hence, the density of the surviving points, after removal of marked ones, is $\lambda_{M1} = \lambda_P e^{-\lambda_P \beta_d r^d}$. For example, in two dimensions, $\lambda_{M1} = \lambda_P e^{-\lambda_P \pi r^2}$.

A more dense process, still fulfilling the requirement on the minimum distance between points, is the Matern process of type II.

Definition 3.9   Matern process of type II. It starts with a uniform PPP of density λ_P. Each point is assigned a random weight in $[0, 1]$, say w_x for point x, independently of all other points. Then, a point is marked, if it has a neighbor within distance r with a weight less than its own, i.e., node y has mark $m_y = 1$ if and only if $\exists\, x \in \Phi : w_x < w_y, \|x - y\| \leq r$. All marked nodes are removed and the resulting point pattern is a realization of the Matern process of type II.

To find the density of the Matern type II process, starting from a uniform PPP with density λ_P, we consider a tagged node and condition on its weight value w. Let us restrict our attention to the points of the original PPP having weight less than or equal to w. Since weights are assigned independently to the points of the PPP, the restricted point process is still a PPP, with density $P(W \leq w)\lambda_P = w\lambda_P$. This is a consequence of the uniform distribution over $[0, 1]$ of the weights. The event that the tagged point is retained, conditional on its weight being w, is realized if and only if the tagged point has no neighbor within distance r and belonging to the restricted point process. The probability q_w of such an event is $q_w = e^{-w\lambda_P \beta_d r^d}$. Removing the conditioning, the unconditional probability q of being retained is

$$q = \int_0^1 q_w\, dw = \int_0^1 e^{-w\lambda_P \beta_d r^d}\, dw = \frac{1 - e^{-\lambda_P \beta_d r^d}}{\lambda_P \beta_d r^d} \tag{3.58}$$

then, the density of the Matern II process is

$$\lambda_{M2} = \lambda_P q = \frac{1 - e^{-\lambda_P \beta_d r^d}}{\beta_d r^d} \tag{3.59}$$

If $d = 2$, we get:

$$\lambda_{M2} = \frac{1 - e^{-\lambda_P \pi r^2}}{\pi r^2} \tag{3.60}$$

As $\lambda_P \to \infty$, eq. (3.60) shows that $\lambda_{M2} \to 1/(\pi r^2)$, which is obviously the highest density compatible with the constraint on the minimum spacing among points.

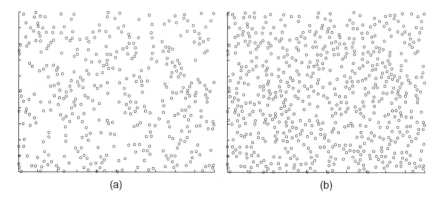

(a)   (b)

Figure 3.7 Matern point processes originating from a uniform PPP with $\lambda_p = 10$ over a square grid 10×10. Left plot: realization of Matern I process with 368 points. Right plot: realization of Matern II process with 608 points.

On the contrary, for $\lambda_p \to 0$, it is $\lambda_{M2} \sim \lambda_p$. This result corresponds to the intuition that the minimum distance constraint has no effect on the point process when it is highly sparse.

Sample outcomes of Matern process are shown in Figure 3.7, starting from a uniform PPP with density 10 points per square unit. Over a region of size 10×10 (hence, the expected number of points of the PPP is 1000), we have obtained 368 points for the example of Matern type I process and 608 points for the type II.

The average density of the PPP and the minimum distance r are set so that $\lambda_p = 1/(\pi r^2)$. This guarantees with high probability that the resulting Matern II process is tightly "packed."

3.2.4 Testing for Poisson

In this section we briefly overview some elementary statistical method to assess whether a collection of time points in a unidimensional space (e.g., the time axis) can be described adequately by the Poisson process. We say that the random mechanism that produces the points is "explained" by the Poisson process or that the Poisson process fits the data. The starting point is a simulation experiment or a measurement yielding a collection of time points, to which we would like to fit a stochastic model. The simplest model to explain a time series is the Poisson process, so we naturally desire to have methods to readily rule out this hypothesis or, on the contrary, to provide evidence that the Poisson model is consistent with the data.

There exist specific tests for the Poisson process, e.g., the Brown-Zhao test [46]. We present here general tests, that could be adapted to other probability distributions, except of the test on the memoryless property, that depends on a special property that characterizes the Poisson process.

Let $\mathcal{T} = \{t_i\}_{0 \leq i \leq n}$ be a set of $n + 1$ points on the time axis, sorted in ascending order. The corresponding inter-arrival times can be calculated as $\tau_i = t_i - t_{i-1}$ for $i = 1, \ldots, n$. The mean inter-arrival time is estimated as $\langle T \rangle = \sum_{i=1}^{n} \tau_i / n$. Then, the estimated average arrival rate is $\langle \lambda \rangle = 1/\langle T \rangle$.

If the time series $\mathcal{T}$ follows a Poisson process, then the CCDF of the inter-arrival times must be negative exponential. This can be readily visualized in a semi-logarithmic plot. If the abscissa is linear and the ordinate is logarithmic, the graph of $G(t) = e^{-\lambda t}$ is a straight line, starting out at $(0, 1)$ and having negative slope proportional to λ. Hence, we estimate $G(t)$ as follows:

$$\tilde{G}(t) = \frac{1}{n} \sum_{i=1}^{n} I(\tau_i > t) = \frac{n_{>t}}{n}, \qquad t \geq 0, \tag{3.61}$$

where $I(E)$ is the indicator function of the event E, i.e., $I(E) = 1$ if and only if E is true, and $n_{>t}$ is the count of the inter-arrival times that fall beyond level t.

We also know that the negative exponential probability distribution is the only one that satisfies the memoryless property in eq. (3.17). We can estimate separately both sides of the identity and compare them. Again, this is neatly visualized in a semi-logarithmic plot. As for the estimates, they are easily computed as follows:

$$P(T > t + x | T > x) \approx \frac{n_{>t+x}}{n_{>x}} \tag{3.62}$$

while $P(T > t)$ is estimated as in eq. (3.61).

Yet another characteristic of the inter-arrival time probability distribution is the QQ-plot. It plots the theoretical quantile values against the estimated ones. If the chosen probability distribution (the negative exponential one with parameter $\langle \lambda \rangle$ in our case) fits the data, the QQ-plot appears as a straight line with slope 1. The q-level quantile of the continuous random variable X with CCDF $G_X(x)$ is the value X_q such that $G(X_q) = 1 - q$. The levels q in the subscript are generally denoted with a percentage, e.g., X_{99} is the quantile at level 0.99, i.e., the value that is exceeded by the random variable X with probability $1 - 0.99 = 0.01$. The quantile of the negative exponential random variable with parameter λ are $X_q = -\log(1 - q)/\lambda$ for $q \in (0, 1)$. As for the estimate, we let $\langle X_q \rangle = \sup\{x : n_{>x}/n \geq 1 - q\}$.

Some numerical examples are provided in the rest of this section. We consider traces of aggregated packet traffic (IP packets), captured on operational links, from which we extract the sequence of inter-arrival times. Specifically, the trace labeled ADSL has been captured on the output link of a domestic ADSL with maximum downlink capacity of 7 Mbit/s. The trace labeled WAN has been obtained from CAIDA (http://www.caida.org): it refers to a high precision measure on a 10 Gbit/s wide area network link from a node in Chicago (IL, USA) to a node in San Jose

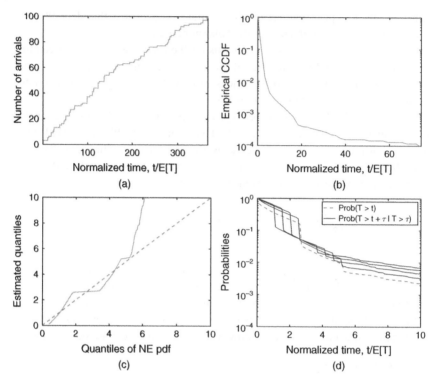

Figure 3.8 Poisson tests for IP packet inter-arrival times on an ADSL link. Sample of the counting function (top left plot). Empirical CCDF (top right plot). QQ-plot (bottom left plot). Assessment of the memoryless property (bottom right plot). The mean inter-arrival time is E[T] = 53.685 ms.

(CA, USA). Several tens of thousands of arrival times have been extracted from those traces, to carry out the statistical tests outlined in this section. In all cases, the sequence of inter-arrival times has been normalized so as to make the mean equal to 1. The actual average inter-arrival time, estimated from the sample data, is reported in the captions.

Figure 3.8 shows the counting function, the empirical CCDF, the QQ-plot, and the test on the memoryless property for the sequence of arrival time of packets in the ADSL trace. The results for the packet arrival process of the WAN measurement are plotted in Figure 3.9.

The test displayed in those figures suggests that the Poisson model fits reasonably with the data from the WAN trace, while it is inadequate to represent the aggregated packet arrival process experienced on the ADSL link. This is somewhat expected and consistent with many observations that lead to the general conclusion that the Poisson model is a good choice to represent a sequence of

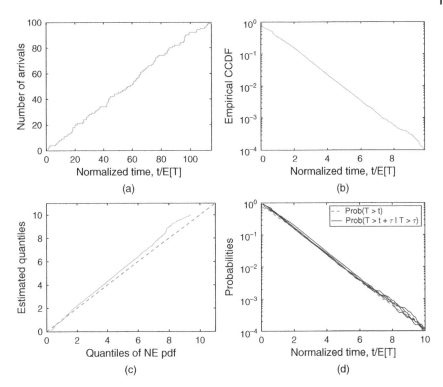

Figure 3.9 Poisson tests for IP packet inter-arrival times on a WAN link. Sample of the counting function (top left plot). Empirical CCDF (top right plot). QQ-plot (bottom left plot). Assessment of the memoryless property (bottom right plot). The mean inter-arrival time is $E[T] = 1.6483\mu s$.

events produced by the superposition of a large number of concurrent flows, each of which gives a marginal contribution to the whole. In networking terms, when there is a high degree of multiplexing of packet flows, there we expect the Poisson model to provide a good fit. Since the ADSL is fed by the traffic of a little number of sources, as typical of a domestic environment, in that case the correlation among inter-arrivals prevails and the Poisson model departs significantly from the actual data.

Finally, we hint at a general test to check the model of the CDF of a set of points, the famous Kolmogorov-Smirnov test. It is particularly interesting since it is backed by a strong theory and it does not require knowledge or estimates of any parameter, except of the points themselves. Given the set of points $\{x_i\}_{1 \leq i \leq n}$, the estimate of the CDF is given by

$$\tilde{F}_n(x) = \frac{1}{n} \sum_{i=1}^{n} I(x_i \leq x) \tag{3.63}$$

The Kolmogorov-Smirnov statistic for a given CDF model $F(x)$ to be tested is

$$D_n = \sup_x |\tilde{F}_n(x) - F(x)| \tag{3.64}$$

Under null hypothesis that the sample comes from the hypothesized distribution $F(x)$, Kolmogorov has shown that, if $F(x)$ is continuous, then the statistics $\sqrt{n}D_n$ converges in distribution to the Kolmogorov random variable K, which has a CDF that does not depend on $F(x)$. Specifically, it can be shown that $\sqrt{n}D_n \sim K$ as $n \to \infty$ with

$$P(K \le x) = 1 - 2\sum_{k=1}^{\infty} (-1)^{k-1} e^{-2k^2 x^2}, \qquad x > 0. \tag{3.65}$$

The goodness-of-fit test or the Kolmogorov-Smirnov (KS) test is constructed by using the critical values of the Kolmogorov distribution (3.65). The null hypothesis is rejected at level α, if $\sqrt{n}D_n > K_\alpha$, where K_α is calculated from $P(K \le K_\alpha) = 1 - \alpha$.

The KS test can be used to check whether the sequence of inter-arrival times collected by the experiment is consistent with the null hypothesis $F(x) = 1 - e^{-x}$, once normalized to the estimated mean value.

3.3 The Markovian Arrival Process

Poisson arrivals lead to relatively simple models that are often amenable to analysis. However, whenever arrivals exhibit non-negligible correlations, the Poisson process is inadequate. A generalization of the Poisson process can be obtained by considering a modulating Markov process. Let $J(t)$ be the state of a continuous time, finite Markov process, over the state space $\{1, \dots, m\}$. Let $\mathbf{D}$ denote the infinitesimal generator of the Markov process.

The Markovian arrival process (MAP) is defined by two matrices, $\mathbf{D}_0$ and $\mathbf{D}_1$, such that $\mathbf{D}_1$ is a non-negative matrix and $\mathbf{D} = \mathbf{D}_0 + \mathbf{D}_1$ is an irreducible infinitesimal generator of the finite Markov process $J(t)$, usually referred to as the *phase* of the MAP. Let $d_r(i,j)$ denote the entry (i,j) of the matrix $\mathbf{D}_r$, $r = 0, 1$. Then, it is $d_1(i,j) \ge 0, \forall i,j = 1, \dots, m$, $d_0(i,j) \ge 0, \forall i \ne j$ and $d_0(i,i) \le 0$, $i = 1, \dots, m$, with $\sum_{j=1}^{m}(d_0(i,j) + d_1(i,j)) = 0$, $i = 1, \dots, m$.

The properties defining the MAP are as follows, for $i \ne j$ and $\Delta t \to 0$:

1. $P(A(t + \Delta t) - A(t) = 0, J(t + \Delta t) = j | J(t) = i) = d_0(i,j)\Delta t + o(\Delta t)$
2. $P(A(t + \Delta t) - A(t) = 1, J(t + \Delta t) = j | J(t) = i) = d_1(i,j)\Delta t + o(\Delta t)$
3. $P(A(t + \Delta t) - A(t) > 1, J(t + \Delta t) = j | J(t) = i) = o(\Delta t)$

Let us define the joint probabilities or arrivals and phase:

$$P_{ij}(n, t) = P(A(t) = n, J(t) = j | A(0) = 0, J(0) = i), \qquad i, j = 1, \dots, m, \tag{3.66}$$

for $n \geq 0$ and $t \geq 0$. Let $\mathbf{P}(n,t)$ be the matrix with entries $P_{ij}(n,t)$ for $i,j = 1, \ldots, m$. For $t = 0$, we have $\mathbf{P}(n,0) = \mathbf{0}$ for $n > 0$ and $\mathbf{P}(0,0) = \mathbf{I}$, where $\mathbf{I}$ is the identity matrix. Let us define the generating function of the time-dependent probabilities::

$$\tilde{P}_{ij}(z,t) \equiv \sum_{n=0}^{\infty} z^n P_{ij}(n,t), \quad t \geq 0; \; i,j = 1, \ldots, m, \tag{3.67}$$

and the corresponding matrix $\tilde{\mathbf{P}}(z,t)$.

Consider the following proof.

Theorem 3.11 For a MAP the generating function of the matrix sequence of arrivals obeys the differential equation $\tilde{\mathbf{P}}'(z,t) = \tilde{\mathbf{P}}(z,t)\mathbf{D}(z)$, with initial condition $\tilde{\mathbf{P}}(z,0) = \mathbf{I}$, where $\mathbf{D}(z) = \mathbf{D}_0 + z\mathbf{D}_1$. The solution is

$$\tilde{\mathbf{P}}(z,t) = \exp(\mathbf{D}(z)t) \tag{3.68}$$

Proof: For $t = 0$, we find $\tilde{\mathbf{P}}(z,0) = \mathbf{I}$. For $n > 0$, we write the probability of transition from state i to state j of the modulating process over a time interval of duration $t + \Delta t$, by breaking it up into first a transition from state i to an intermediate state k in time t and then a further transition from k to j in time Δt. For this last transition, we use the properties of the MAP stated above, in the limit as $\Delta t \to 0$:

$$P_{ij}(n,t+\Delta t) = \sum_{k \neq j} P_{ik}(n,t)d_0(k,j)\Delta t + \sum_{k} P_{ik}(n-1,t)d_1(k,j)\Delta t +$$

$$+ P_{ij}(n,t)\left(1 - \sum_{k \neq j}[d_0(j,k) + d_1(j,k)]\Delta t\right) + o(\Delta t) \tag{3.69}$$

By exploiting the identity $d_0(j,j) = -\sum_{k \neq j}[d_0(j,k) + d_1(j,k)]$, we can rewrite the evolution of the elements $P_{ij}(n,t+\Delta t)$ in matrix form as follows:

$$\mathbf{P}(n,t+\Delta t) = \mathbf{P}(n,t) + \mathbf{P}(n,t)\mathbf{D}_0\Delta t + \mathbf{P}(n-1,t)\mathbf{D}_1\Delta t + o(\Delta t), \quad n > 0. \tag{3.70}$$

This can be rearranged as

$$\frac{\mathbf{P}(n,t+\Delta t) - \mathbf{P}(n,t)}{\Delta t} = \mathbf{P}(n,t)\mathbf{D}_0 + \mathbf{P}(n-1,t)\mathbf{D}_1 + o(1), \quad n > 0. \tag{3.71}$$

and taking the limit for $\Delta t \to \infty$, we obtain finally:

$$\mathbf{P}'(n,t) = \mathbf{P}(n,t)\mathbf{D}_0 + \mathbf{P}(n-1,t)\mathbf{D}_1, \quad n > 0. \tag{3.72}$$

Repeating the same steps for $n = 0$, it is easy to derive that

$$\mathbf{P}'(0,t) = \mathbf{P}(0,t)\mathbf{D}_0 \tag{3.73}$$

By multiplying both sides of eq. (3.72) by z^n, summing over $n \geq 1$ and summing further also eq. (3.73), we get

$$\sum_{n=0}^{\infty} \mathbf{P}'(n, t)z^n = \sum_{n=0}^{\infty} \mathbf{P}(n, t)z^n \mathbf{D}_0 + z \sum_{n=1}^{\infty} \mathbf{P}(n-1, t)z^{n-1} \mathbf{D}_1 \qquad (3.74)$$

whence

$$\tilde{\mathbf{P}}'(z, t) = \tilde{\mathbf{P}}(z, t)(\mathbf{D}_0 + z\mathbf{D}_1) = \tilde{\mathbf{P}}(z, t)\mathbf{D}(z) , \qquad t > 0. \qquad (3.75)$$

The solution with initial condition $\tilde{\mathbf{P}}(z, 0) = \mathbf{I}$ is easily found to be $\exp(\mathbf{D}(z)t)$.

If we let $\alpha_i = P(J(0) = i)$ and $\alpha = [\alpha_1, \ldots, \alpha_m]$, the generating function of the number of arrivals of the MAP is:

$$\tilde{A}(z, t) = \sum_{k=0}^{\infty} z^k P(A(t) = k | A(0) = 0) = \alpha \exp(\mathbf{D}(z)t)\mathbf{e} \qquad (3.76)$$

where $\mathbf{e}$ is a column vector of 1's of size m.

The mean number of arrivals over a time interval $(0, t]$ is obtained by taking the derivative of $\tilde{A}(z, t)$ with respect to z and by setting $z = 1$ in the resulting expression. We find

$$\mathrm{E}[A(t)] = \frac{d}{dz}\tilde{A}(z, t)\Big|_{z=1} = t\alpha \mathbf{D}_1 \exp(\mathbf{D}(1)t)\mathbf{e} = t\alpha \mathbf{D}_1 \mathbf{e} \qquad (3.77)$$

since the matrix $\mathbf{D}(1) = \mathbf{D}$ is an infinitesimal generator, hence $\mathbf{De} = \mathbf{0}$. From eq. (3.77) we find the mean arrival rate of the MAP:

$$\lambda_{MAP} = \alpha \mathbf{D}_1 \mathbf{e} \qquad (3.78)$$

The MAP is a highly versatile model for arrival processes, albeit it could be difficult to gather enough experimental data to fit the large number of parameters that define a general MAP, namely two $m \times m$ matrices $\mathbf{D}_0$ and $\mathbf{D}_1$. This requires assigning values to $2m^2 - m$ values, where we account for the constraint that $\mathbf{D}_0 + \mathbf{D}_1$ be an infinitesimal generator.

A special case of MAP is the markov modulated poisson process (MMPP), that corresponds to the case $\mathbf{D}_1 = \mathbf{L}$ and $\mathbf{D}_0 = \mathbf{R} - \mathbf{L}$, where $\mathbf{L}$ is a diagonal matrix with positive diagonal entries $\lambda_i, i = 1, \ldots, m$ and $\mathbf{R}$ is an infinitesimal generator. The resulting process corresponds to a Poisson process with a time-varying mean arrival rate according to the state of the modulating Markov process $J(t)$, i.e., $\lambda(t) = \lambda_{J(t)}$.

The MAP can be generalized to the batch MAP (BMAP), where bulk arrivals are possible. This is obtained by defining a sequence of matrices $\mathbf{D}_n$ for $n \geq 0$. The entry $d_n(i, j)$ represents the frequency of transition from state i to state j with n arrivals. The generating function $\tilde{\mathbf{P}}(z, t)$ of the joint random variable $(A(t), J(t))$ has the same expression as given in eq. (3.68), provided that we let $\mathbf{D}(z) = \sum_{n=0}^{\infty} z^n \mathbf{D}_n$. The corresponding mean arrival rate is $\lambda_{\text{BMAP}} = \alpha \sum_{n=1}^{\infty} n\mathbf{D}_n\mathbf{e}$.

The MAP and BMAP models can be also stated in discrete time, by referring to a Markov chain as the modulating process. The corresponding models are defined by a sequence of non-negative matrices $\mathbf{D}_n$ such that $\mathbf{D} = \sum_{n=0}^{\infty} \mathbf{D}_n$ is the one-step transition probability matrix of a finite, irreducible Markov chain; hence the elements of the matrices $\mathbf{D}_n$ are probabilities, $d_n(i,j)$ being the probability of having n arrivals and a phase transition from i to j in a time slot. This kind of process is known as discrete MAP (DMAP) or discrete BMAP (DBMAP), according to whether only single arrivals or bulk arrivals are permitted.

3.4 Renewal Processes

Renewal processes have been extensively used in the reliability theory, to characterize life and replacement (*renewal*) of system components, e.g., electronic components of a circuit. This historical application has marked the terminology used in the context of renewal processes.

A renewal process is clearly a model suitable for arrival processes of queues. In the following we identify "events" as customer arrivals at a serving system and use a consequential language.

There is a huge literature and an extensive body of knowledge on point processes and specifically on renewal theory [59, 85, 184, 209]. We will confine ourselves to introducing definitions and properties of renewal point processes, that are useful for queueing systems and traffic engineering applications.

Let us consider a renewal process, i.e., a point process with inter-arrival times that are independent random variables denoted as $T_k, k \in \mathbb{Z}$. The random variable T_k denotes the time elapsing between the $(k-1)$-th and the k-th event. If $T_k \sim T$, $\forall k$, i.e., all inter-event times are identically distributed and can be described by means of a unique random variable T, the renewal process is said to be stationary. In that case, we let $G_T(x) = \mathcal{P}(T > x) = 1 - F_T(x)$ be the CCDF of the inter-arrival time random variable. When it exists, we let also $f_T(x)$ be the corresponding PDF, given by $f_T(x) = -G_T'(t)$.

In general, a point process can be described by the random variables introduced at the beginning of this chapter, namely the counting function, the inter-arrival times and the arrival times. By convention, $A(0,t)$ (sometimes abridged as $A(t)$, when there is no ambiguity) is the number of arrivals in the semi-open interval $(0, t]$. The arrival time of the n-th event is $S_n = T_1 + \cdots + T_n$, $n \geq 1$, and by extension and ease of notation, we also let $S_0 = 0$. The CDF and PDF of S_n are denoted with $F_{S_n}(t)$ and $f_{S_n}(t)$ for $t \geq 0$, respectively. For uniform notation, we also let $F_{S_0}(t) = 1$, $t \geq 0$.

It is easy to see that the event $S_k \leq t$ is equivalent to the event $A(0,t) \geq n$ for any non-negative n and t, so

$$P(A(0,t) \geq k) = P(S_k \leq t) = F_{S_k}(t), \qquad k \geq 1, \tag{3.79}$$

By taking differences, we obtain:

$$P(A(0,t) = k) = F_{S_k}(t) - F_{S_{k+1}}(t), \qquad k \geq 0. \tag{3.80}$$

The time origin can be related to the renewal process in different ways. In the following we define the *first* event time after the time origin as $T_1 = S_1$. We distinguish three cases for the probability distribution of T_1.

- *General renewal process*: The time T_1 has a probability distribution $G_{T_1}(t)$, possibly different from that of all other T_k, $k \geq 2$, namely $G_T(t)$;
- *Ordinary renewal process*: All T_k share the same probability distribution $G_T(t)$, i.e., $G_{T_1}(t) = G_T(t)$;
- *Equilibrium renewal process*: The time origin is chosen at random with respect to the renewal process events; in this case it can be shown that it is $f_{T_1}(t) = G_T(t)/E[T]$.

As for the third case, the time elapsing from a randomly chosen point until the next arrival is called the *residual inter-arrival time*, while the time elapsed from the last arrival until the randomly chosen time point is called the *inter-arrival age*.

In the general renewal arrival process, thanks to the independence of the inter-arrival times, the Laplace transform of the PDF of S_k is $\varphi_{S_k}(s) \equiv E[e^{-sS_k}] = \varphi_{T_1}(s)[\varphi_T(s)]^{k-1}$ for $k \geq 1$, where $\varphi_T(s) = E[e^{-sT}]$ and $\varphi_{T_1}(s) = E[e^{-sT_1}]$. Then

$$\int_0^\infty e^{-st} P(A(0,t) = k) \, dt = \frac{1}{s}[1 - \varphi_T(s)]\varphi_{T_1}(s)[\varphi_T(s)]^{k-1}, \qquad k \geq 1 \tag{3.81}$$

and

$$\int_0^\infty e^{-st} P(A(0,t) = 0) \, dt = \frac{1}{s}[1 - \varphi_{T_1}(s)] \tag{3.82}$$

The generating function of the probabilities appearing in the left-hand sides of the two previous equations can be found as:

$$P_A(z,s) \equiv \sum_{k=0}^\infty z^k \int_0^\infty e^{-st} P(A(0,t) = k) \, dt = \frac{1 - \varphi_{T_1}(s)}{s} + \varphi_{T_1}(s)\frac{z[1 - \varphi_T(s)]}{s[1 - z\varphi_T(s)]} \tag{3.83}$$

We can specialize this result to two special cases of renewal process enumerated above, i.e., ordinary and equilibrium renewal processes. In case of an ordinary process, $\varphi_{T_1}(s) = \varphi_T(s)$, so we find

$$P_A^{(o)}(z,s) = \frac{1 - \varphi_T(s)}{s[1 - z\varphi_T(s)]} \tag{3.84}$$

where the superscript o stands for ordinary. In case of an equilibrium renewal process it is $\varphi_{T_1}(s) = [1 - \varphi_T(s)]/(sE[T])$ (as shown in Section 3.4.1), hence we find:

$$P_A^{(e)}(z, s) = \frac{1}{s} + \frac{(z-1)[1 - \varphi_T(s)]}{s^2 E[T][1 - z\varphi_T(s)]} \tag{3.85}$$

Let us consider a general renewal process. We define the mean number of events over the interval $(0, t]$ as $M(t)$, i.e., $M(t) = E[A(0, t)]$. Then, by deriving $P_A(z, s)$ in eq. (3.83) with respect to z and setting $z = 1$, we find

$$\varphi_M(s) = \int_0^\infty M(t)e^{-st} \, dt = \frac{\varphi_{T_1}(s)}{s[1 - \varphi_T(s)]} \tag{3.86}$$

where $\varphi_M(s)$ is the Laplace transform of the average number of customers arrived in the interval $(0, t]$. Then

$$M(t) = F_{T_1}(t) + \int_0^t M(\tau)f_T(t - \tau) \, d\tau \tag{3.87}$$

The derivative of $M(t)$ is denoted with $m(t)$ and it is called the *renewal function* of the process. It provides the arrival density as a function of time and it is defined as the positive solution of:

$$m(t) = f_{T_1}(t) + \int_0^t m(\tau)f_T(t - \tau) \, d\tau \tag{3.88}$$

These are the fundamental renewal equations. Since the full knowledge of the statistics of an ordinary renewal process consists in knowing the probability distribution of the inter-arrival times, eqs. (3.87) and (3.88) show that equivalently we can reconstruct the entire renewal process statistics from either $M(t)$ or $m(t)$.

Equation (3.86) can be exploited to provide us with an estimate of $M(t)$ for large t for the ordinary renewal process. Let $\mu = E[T]$ and $\sigma^2 = E[T^2] - E[T]^2$ be short notation for the first two moments of the inter-arrival times. Then, it is

$$\varphi_T(s) = 1 - \mu s + \frac{1}{2}(\mu^2 + \sigma^2)s^2 + O(s^3), \quad s \to 0. \tag{3.89}$$

Let us consider an ordinary process with $T_1 \sim T$. Specializing the general expression of $\varphi_M(s)$ in (3.86) to this case, we get, in the limit for $s \to 0$,

$$\varphi_M^{(0)}(s) = \frac{\varphi_T(s)}{s[1 - \varphi_T(s)]} = \frac{1 - \mu s + \frac{1}{2}(\mu^2 + \sigma^2)s^2 + O(s^3)}{s[\mu s - \frac{1}{2}(\mu^2 + \sigma^2)s^2 + O(s^3)]}$$

$$= \frac{1 - \mu s + \frac{1}{2}(\mu^2 + \sigma^2)s^2 + O(s^3)}{\mu s^2 \left[1 - \frac{\mu^2 + \sigma^2}{2\mu}s + O(s^2)\right]}$$

$$= \frac{1}{\mu s^2}\left(1 - \mu s + \frac{\mu^2 + \sigma^2}{2}s^2 + O(s^3)\right)\left(1 + \frac{\mu^2 + \sigma^2}{2\mu}s + O(s^2)\right)$$

$$= \frac{1}{\mu s^2}\left(1 - \mu s + \frac{\mu^2 + \sigma^2}{2\mu}s + O(s^2)\right) = \frac{1}{\mu s^2} + \frac{\sigma^2 - \mu^2}{2\mu^2 s} + O(1) \tag{3.90}$$

hence, we find:

$$M^{(o)}(t) = \frac{t}{\mu} + \frac{\sigma^2 - \mu^2}{2\mu^2} + o(1), \quad t \to \infty \tag{3.91}$$

which gives an asymptotic expansion of the mean number of arrivals over the time window $(0, t]$ for an ordinary renewal process. The mean number of arrivals in a time window of duration t equals the mean number of arrivals (t/μ), with a constant offset given by $(C^2 - 1)/2$, where $C = \sigma/\mu$ is the coefficient of variation (COV) of the inter-arrival times.

In case of an equilibrium renewal process, $\varphi_{T_1}(s)$ has the special form $\varphi_{T_1}(s) = [1 - \varphi_T(s)]/(\mu s)$, and hence we find $M^{(e)}(s) = 1/(\mu s^2)$. This transform can be easily inverted to yield $M^{(e)}(t) = t/\mu$; correspondingly, it is $m^{(e)}(t) = 1/\mu$. This result says that the mean number of arrivals in $(0, t]$ is proportional to t with a constant equal to the reciprocal of the mean inter-arrival time. We call this quantity the mean arrival rate of the renewal process, $\lambda \equiv 1/E[T]$.

It is also possible to state a form of the central limit theorem (CLT) for renewal processes. Let us consider an ordinary renewal process. The occurrence time of the n-th event, $S_n = T_1 + \cdots + T_n$, is the sum of n i.i.d. random variables, that we assume have finite mean and variance. Then, by the CLT, it is

$$\lim_{n \to \infty} \mathcal{P}\left(\frac{S_n - n\mu}{\sigma\sqrt{n}} > u\right) = Q(u) \tag{3.92}$$

where $Q(u) = \int_u^\infty e^{-z^2/2}/\sqrt{2\pi}\, dz$ is the tail of the standard Gaussian PDF. Let $t = n\mu - u\sigma\sqrt{n}$. It is easy to check that this can be inverted to yield

$$n = \frac{t}{\mu} + u\frac{\sigma\sqrt{t}}{\mu^{3/2}}\left[\frac{c}{\sqrt{t}} + \sqrt{1 + \left(\frac{c}{\sqrt{t}}\right)^2}\right] \tag{3.93}$$

where $c = u\sigma/(2\sqrt{\mu})$. Since the event $S_n > n\mu - u\sigma\sqrt{n} = t$ is equivalent to the event $A(0, t) < n = \frac{t}{\mu} + \frac{\sigma\sqrt{t}}{\mu^{3/2}}u + O(1)$ (see (3.93)), we can rewrite eq. (3.92) as

$$\lim_{t \to \infty} \mathcal{P}\left(\frac{A(0, t) - t/\mu}{\mu^{-3/2}\sigma\sqrt{t}} < u\right) = \lim_{t \to \infty} \mathcal{P}\left(\frac{S_n - n\mu}{\sigma\sqrt{n}} > -u\right) = 1 - Q(u) \tag{3.94}$$

Therefore, we have $\mathcal{P}\left(\frac{A(0,t) - t/\mu}{\mu^{-3/2}\sigma\sqrt{t}} > u\right) \sim Q(u)$ for $t \to \infty$, which implies that $A(0, t) \sim \mathcal{N}(t/\mu, \sigma^2 t/\mu^3)$, i.e., $A(0, t)$ is (asymptotically) a Gaussian random variable with mean t/μ and standard deviation $\sigma\sqrt{t}/\mu^{3/2}$.

From this asymptotic relationship, we deduce $E[A(0, t)] \approx t/\mu$ and $\text{Var}[A(0, t)] \approx \sigma^2 t/\mu^3$ as $t \to \infty$. Then, asymptotically for large t we have

$$\frac{\text{Var}[A(0, t)]}{E[A(0, t)]} \to \frac{\sigma^2 t/\mu^3}{t/\mu} = \frac{\sigma^2}{\mu^2}, \quad t \to \infty \tag{3.95}$$

This is a generalization for a renewal process of the property of the Poisson process for which $\text{Var}[A(0, t)]/[E][A(0, t)] = 1$ for all t.

Summing up, the counting function $A(0, t)$ of a renewal point process is asymptotically distributed as a Gaussian random variable as $t \to \infty$. The asymptotic mean and variance of the counting functions are

$$E[A(0, t)] \sim \frac{t}{\mu} \qquad \text{Var}[A(0, t)] \sim \frac{\sigma^2 t}{\mu^3} \qquad (3.96)$$

respectively. This is consistent with expressions for the mean and variance of the counting function for any t given, e.g., in [60, Ch. 9].

Sometimes, the *hazard* function $h(t)$ is defined as a description of the renewal process. It is the rate of occurrence of an event just after t, given that no event has occurred until t, i.e.,

$$h(t) = \lim_{\Delta t \to 0} \frac{P(t < T \le t + \Delta t | T > t)}{\Delta t} = \frac{f_T(t)}{G_T(t)} \qquad (3.97)$$

Since $f_T(t) = -G'_T(t)$ and $G_T(0) = 1$, we can use this definition to write the differential equation $G'_T(t) = -h(t)G_T(t)$ with initial condition $G_T(0) = 1$. The solution is $G_T(t) = \exp(-\int_0^t h(u)\, du)$. This last equality shows that the entire renewal process description is given once $h(t)$ is known.

The hazard function gives an approximation of the probability that an arrival occurs within a short interval Δt, after time t, provided that no arrival has been seen until t: namely $P(t < T \le t + \Delta t | T > t) \approx h(t)\Delta t$. In general $h(t)$ varies with t, i.e., the arrival process has some form of memory. In the special case of a memoryless process we have

$$P(t < T \le t + \Delta t | T > t) = 1 - P(T > t + \Delta t | T > t) = 1 - P(T > \Delta t) \qquad (3.98)$$

where the last equality follows from the definition of the memoryless property. Hence

$$h(t) = \lim_{\Delta t \to 0} \frac{1 - G_T(\Delta t)}{\Delta t} = f_T(0) \qquad (3.99)$$

In words, the hazard function of a memoryless process is a positive constant. This result could have been found also by reminding that the only memoryless point process is the Poisson one, hence the inter-arrival times have a negative exponential probability distribution: $f_T(t) = \lambda e^{-\lambda t}$. In this case, $h(t) = \lambda$.

Some general results can be shown for a renewal process (e.g., see [196, § 9.12]).

1. It is not possible for an infinite number of renewals to occur in a finite period of time, i.e., $P(\lim_{n \to \infty} S_n \le t < \infty) = 0$ for a given finite t.
2. $\lim_{t \to \infty} A(t) = \infty$.
3. $\lim_{t \to \infty} A(t)/t = 1/E[T]$ with probability 1.
4. $\lim_{t \to \infty} M(t)/t = \lim_{t \to \infty} m(t) = 1/E[T]$.

The third property gives a sound foundation to common measurement procedures to estimate the mean arrival rate of an arrival process.

In the following we give a general result on the residual inter-arrival time of the renewal process, i.e., the amount of time elapsing from a random look at the renewal process until the next event. Since the process is time-reversible, i.e., its statistics remain exactly the same if we reverse the direction of the time axis, the residual inter-arrival time is the same as the inter-arrival age, i.e., the time since when the last event occurred from the point of view of a random observer of the renewal process.

3.4.1 Residual Inter-Event Time and Renewal Paradox

We want to characterize the residual inter-arrival time random variable, i.e., the random variable defined as the time elapsing from a randomly chosen point on the time axis to the subsequent event of the renewal process. The crucial part of the ensuing reasoning is devoted to giving an accurate notion of the statement "randomly chosen point on the time axis."

Let $\tilde{T}$ be the random variable associated to the residual inter-arrival time. We consider a sequence of n inter-arrival times, starting at the initial time 0 and lasting until time $S_n = \sum_{k=1}^{n} T_k$. Let us fix a time level x. The time interval $[0, S_n]$ can be split into two sets of points: those where the residual time until the next arrival is less than or equal to x and those such that the time until the next arrival is greater than x. To help visualize those two time point sets, we can consider the graphical construction in Figure 3.10. A sample of arrival times are highlighted by crosses. For each inter-arrival time T_k we consider a triangle with base T_k, height T_k and slope -1. We draw an horizontal line at level x (dashed line in the figure). Each interval delimited by a time point where the horizontal line intercepts an inclined line and the subsequent arrival contains points where the residual inter-arrival time is less than or equal to x. An example is shown in the figure with a dark shaded triangle.

Let us pick a time point at random, i.e., with uniform probability distribution, in the interval $[0, S_n]$. The probability that the residual inter-arrival time be less than or equal to x is the probability of the event that the randomly chosen point falls

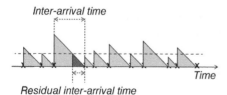

Inter-arrival time

Time

Residual inter-arrival time

Figure 3.10 Sample of an event sequence. Each triangle has slope −1, so that its height corresponds to the base, which is the inter-event time. The level of the dashed line across the triangles marks the residual times.

in one of the intervals $[\max\{t_k - x, t_k - T_k\}, t_k]$ for $k = 1, \ldots, n$. Therefore, we can write:

$$P(\tilde{T}(n) \leq x) = \frac{\sum_{k=1}^{n} \min\{x, T_k\}}{\sum_{k=1}^{n} T_k} = \frac{\frac{1}{n}\sum_{k=1}^{n} \min\{x, T_k\}}{\frac{1}{n}\sum_{k=1}^{n} T_k} \qquad (3.100)$$

where we have denoted the residual inter-arrival time random variable over n consecutive arrival times with $\tilde{T}(n)$. For a stationary renewal process, when $n \to \infty$, the right hand side has a proper limit with probability 1 and the random variable $\tilde{T}(n)$ tends to a limit r.v. $\tilde{T}$, with CDF given by:

$$F_{\tilde{T}}(x) \equiv P(\tilde{T} \leq x) = \frac{E[\min\{x, T\}]}{E[T]} \qquad (3.101)$$

It is

$$E[\min\{x, T\}] = \int_0^x t f_T(t)\, dt + \int_x^\infty x f_T(t)\, dt = \int_0^x t f_T(t) dt + x G_T(x) \qquad (3.102)$$

and, by deriving both sides with respect to x, we get

$$f_{\tilde{T}}(x) = \frac{d}{dx} F_{\tilde{T}}(x) = \frac{1}{E[T]}[x f_T(x) + G_T(x) - x f_T(x)] = \frac{G_T(x)}{E[T]} \qquad (3.103)$$

This is the PDF of the residual inter-arrival time. The mean residual inter-arrival time can be easily derived as:

$$E[\tilde{T}] = \frac{E[T^2]}{2E[T]} = E[T]\frac{1 + C_T^2}{2} \qquad (3.104)$$

where $C_T = \sigma_T/E[T]$ is the coefficient of variation of the random variable T and σ_T denotes the standard deviation of T.

For example, in case of negative exponential probability distribution, it is $C_T = 1$, so the mean residual inter-arrival time is equal to the inter-arrival time. For more variable probability distributions ($C_T > 1$), e.g., hyper-exponential ones, the mean residual inter-arrival time is *bigger* than the regular inter-arrival time. It is less in case of smoother probability distributions ($C_T < 1$), a limiting case being the deterministic distribution ($C_T = 0$), where the mean residual inter-arrival time is half the regular inter-arrival time.

As another example, let T be a gamma distributed random variable, with

$$f_T(t) = \frac{\alpha^\beta t^{\beta-1}}{\Gamma(\beta)}e^{-\alpha t}, \qquad t \geq 0,\ \alpha, \beta > 0, \qquad (3.105)$$

where $\Gamma(\cdot)$ is the Euler gamma function. It is $E[T] = \beta/\alpha$ and $\sigma_T^2 = \beta/\alpha^2$, hence $C_T = 1/\sqrt{\beta}$. It turns out that $E[\tilde{T}] = E[T](1 + 1/\sqrt{\beta})/2$. Depending on the value of β, the mean residual inter-arrival time can be either greater than the mean arrival time ($\beta < 1$) or smaller than the mean arrival time ($\beta > 1$).

The results of eqs. (3.103) and (3.104) are known as *renewal paradox*. The reason is apparent from the examples presented above. The mean residual arrival time can be *bigger* than the mean arrival time. How can something that is seemingly a "part of" be greater than the "whole"? The catch is in the subtlety of the "random look" at the arrival process. Consider a span of the time axis where a number of events occur. Some of them are widely spaced apart (those that correspond to large arrival times), while others are more tightly packed (those corresponding to small inter-arrival times). By looking at random with *uniform* probability over the considered time interval we are biased toward picking *long* inter-arrival times rather than short ones. The residual inter-arrival time is the remaining part of this *selected* interval, i.e. of longer inter-arrival times. This effect is more pronounced the bigger the variability of the inter-arrival times, i.e., the bigger the ratio of the standard deviation to the mean of the inter-arrival times.

It is easy to check that in case of negative exponential distribution of the inter-arrival time T, i.e., $G_T(t) = e^{-\lambda t}$, it is $f_{\tilde{T}}(t) = \lambda e^{-\lambda t} = f_T(t)$, that is to say, the PDF of the residual inter-arrival time is the *same* as the PDF of the entire inter-arrival time. It can be shown that the converse is true as well. Let us assume that $f_T(t) = f_{\tilde{T}}(t) = G_T(t)/E[T]$. Since $f_T(t) = -G'_T(t)$, we find $G'_T(t) = -G_T(t)/E[T]$, with initial condition $G_T(0) = 1$. From this differential equation, it is easily derived that $G_T(t) = \exp(-t/E[T])$, i.e., the PDF of T must be negative exponential. We have thus shown the following.

Theorem 3.12   The negative exponential random variable is the unique positive random variable that has the same PDF as its associated residual random variable.

In view of the fact the the negative exponential random variable is the only one exhibiting the memoryless property, Theorem 3.12 should come as no surprise.

3.4.2 Superposition of Renewal Processes

Let us consider n renewal processes with the same inter-arrival probability distribution and assume they are independent of one another. The superposition of those processes is a new point process composed of all events from each component process. Note that the superposition process is stationary, but it is not a renewal process in general.

Let $T_s(n)$ be the inter-arrival time of the superposition of n independent arrival processes and let $T^{(j)}$ denote the random variable representing the inter-arrival time of the j-th component process, $j = 1, \ldots, n$. Since the component processes are independent of one another and have same PDF of the inter-arrival times, we can focus on an arrival coming from any of them equivalently. Let us consider an arrival from process 1. The next arrival will occur after a time equal to the minimum among the full inter-arrival time $T^{(1)}$ and the *residual* inter-arrival times $\tilde{T}^{(j)}$. Then $T_s(n) = \min\{T^{(1)}, \tilde{T}^{(2)}, \dots, \tilde{T}^{(n)}\}$ and

$$
G_{T_s(n)}(t) = \mathcal{P}(T_s(n) > t) = \mathcal{P}(T^{(1)} > t) \prod_{i=2}^{n} \mathcal{P}(\tilde{T}^{(i)} > t) = G_T(t)[G_{\tilde{T}}(t)]^{n-1}
$$

(3.106)

for $n \geq 1$. The mean arrival rate of the superposition process is $n\lambda$, if λ denotes the mean arrival rate of each component process. Since we are interested in understanding what happens when n increases, and in that case the mean rate of the superposition process would grow unboundedly, we scale the component processes, by setting their respective mean arrival rates at λ/n. This corresponds to scaling the time axis by a factor $1/n$, i.e., we substitute t with t/n. In fact, the counting function becomes $A(t/n)$ and hence

$$
\lambda_{\text{scaled}} = \lim_{t\to\infty} \frac{A(t/n)}{t} = \lim_{t\to\infty} \frac{A(t/n)}{n(t/n)} = \frac{\lambda}{n}
$$

(3.107)

The CCDF of the inter-arrival times of the scaled process becomes $G_T(t/n)$. Rewriting eq. (3.106) by using the scaled component processes, we get

$$
G_{T_s(n)}(t) = G_T(t/n) \left[1 - \lambda \int_0^{t/n} G_T(x)\, dx \right]^{n-1}
$$

(3.108)

Thanks to the monotonicity of $G_T(t)$, it is easy to check that

$$
1 - \frac{\lambda t}{n} \leq 1 - \lambda \int_0^{t/n} G_T(x)\, dx \leq 1 - G_T(t/n)\frac{\lambda t}{n}
$$

(3.109)

If the function $G_T(t)$ has no jump at 0, i.e., $\lim_{t\to 0+} G_T(t) = G_T(0) = 1$, then[2]

$$
\lim_{n\to\infty} G_{T_s(n)}(t) = \lim_{n\to\infty} \left(1 - \frac{\lambda t}{n} \right)^{n-1} = e^{-\lambda t}
$$

Thus, the result of the superposition of a large number of tiny arrival processes is a point process that looks like a Poisson process *locally*, i.e., such that the inter-arrival PDF is negative exponential. Be aware, however, that the limit superposition process in *not* a Poisson process in general, since it can be correlated, i.e., inter-arrival times are not necessarily independent of one another.

3.4.3 Alternating Renewal Processes

Let us consider a point process, where the inter-arrival time PDF depends on the transitions of a finite Markov chain. Let J_k denote the state of the Markov

---

2 This means that there is probability 0 that the inter-arrival time can be 0, i.e., we consider processes with no bulk arrivals.

chain soon after the k-th transition and ℓ the number of states of the Markov chain. Let also $q_{ij} = P(J_{k+1} = j | J_k = i)$. The inter-arrival time associated with the transition from i to j denoted with T_{ij}. Its PDF is denoted with $f_{ij}(t)$ and $\varphi_{ij}(s)$ is the corresponding Laplace transform. We assume that the random variables T_{ij} are independent of one another.

Arrivals occur at transition epochs of the Markov chain. Given that the state of the Markov chain after the k-th transition is $J_k = i$ and that it moves to $J_{k+1} = j$ after the next transition, the inter-arrival time is a realization of the random variable T_{ij}. The probability distribution of T_{ij} depends on both the starting and the landing states in general; this is why we use a double subscript index.

Let us consider n successive arrivals. Let us denote with j_0 the initial state of the modulating Markov chain and let $J_k = j_k$ for the subsequent n transitions, $k = 1, \ldots, n$. The Laplace transform of the PDF of the n-th event epoch $S_n = T_1 + \cdots + T_n$ is

$$\varphi_{S_n}(s) = \mathrm{E}[e^{-s(T_1 + \cdots + T_n)}] = \sum_{j_0=1}^{\ell} \alpha_{j_0} \sum_{j_1}^{\ell} q_{j_0 j_1} \varphi_{j_0 j_1}(s) \ldots \sum_{j_n=1}^{\ell} q_{j_{n-1} j_n} \varphi_{j_{n-1} j_n}(s) \quad (3.110)$$

where $\alpha = [\alpha_1 \ldots \alpha_\ell]$ is a row vector of the probabilities of the initial state.

This result can be put in a compact form by defining the $\ell \times \ell$ matrix $\mathbf{F}(s)$, with entry (i, j) given by $q_{ij} \varphi_{ij}(s)$, and a column vector $\mathbf{e}$ of 1's of size ℓ. We can write

$$\varphi_{S_n}(s) = \alpha \mathbf{F}(s)^n \mathbf{e}, \qquad n \geq 0 \tag{3.111}$$

The probability distribution of the state (often called "phase") J_n at the end of n transitions is simply given by $\mathbf{p}(n) = \alpha \mathbf{Q}^n$, where $\mathbf{Q}$ is an $\ell \times \ell$ matrix with entry (i, j) equal to q_{ij}, i.e., it is $\mathbf{Q} = \mathbf{F}(0)$. It can be verified that the Laplace transform of the generating function of the probability distribution $P(A(0, t) = k)$, $k \geq 0$, is given by

$$P_A(z, s) = \alpha \frac{\mathbf{I} - \mathbf{F}(s)}{s} [\mathbf{I} - z\mathbf{F}(s)]^{-1} \mathbf{e} \tag{3.112}$$

with $\mathbf{I}$ denoting the identity matrix of order ℓ. This result can be derived as follows. From the very definition of the events $A(0, t) = k$ and $S_k \leq t$, we have

$$P(A(0, t) = k) = P(S_k \leq t) - P(S_{k+1} \leq t), \quad k \geq 0, t \geq 0. \tag{3.113}$$

Taking the Laplace transform of both sides, we have

$$\int_0^\infty e^{-st} P(A(0, t) = k) dt = \frac{\varphi_{S_k}(s)}{s} - \frac{\varphi_{S_{k+1}}(s)}{s} \tag{3.114}$$

Multiplying both sides by z^k and summing over $k \geq 0$, we have finally:

$$P_A(z, s) \equiv \sum_{k=0}^{\infty} z^k \int_0^\infty e^{-st} P(A(0, t) = k) dt = \frac{1}{s} \alpha \sum_{k=0}^{n} z^k [\mathbf{F}(s)^k - \mathbf{F}(s)^{k+1}] \mathbf{e} \tag{3.115}$$

which yields the desired result as we close the summation.

Such a point process is called a renewal alternating process or also semi-Markov process. It is not a renewal process, since it is correlated by virtue of the Markov underlying chain that modulates the PDFs of the independent inter-arrival times. Special cases are obtained when the PDFs of the inter-arrival times depend only on the starting or the ending state of the transition.

3.4.4 Renewal Reward Processes

Let us consider a renewal process with counting function $\{N(t), t \geq 0\}$, where X_n denotes the n-th renewal time. Assume that when an event occurs, a reward is gained (or a cost is paid). Let R_n denote the amount of reward associated with the n-th event. Consistent with the renewal character of the process, we assume that the R_n's are independent of one another and identically distributed. On the other hand, the reward R_n can depend on the renewal time X_n. We can define the cumulated reward at time t as

$$R(t) = \sum_{n=1}^{N(t)} R_n \tag{3.116}$$

In the special case that $R_n = 1, \forall n$, then $R(t) = N(t)$, i.e., the number of events in the interval $(0, t)$. As for the general renewal reward process, it can be shown that

$$\lim_{t \to \infty} \frac{R(t)}{t} = \frac{E[R]}{E[X]} \tag{3.117}$$

As a matter of fact, the reward rate can be written as

$$\frac{R(t)}{t} = \frac{\sum_{n=1}^{N(t)} R_n}{t} = \frac{\sum_{n=1}^{N(t)} R_n}{N(t)} \frac{N(t)}{t} \tag{3.118}$$

As $t \to \infty$, we know that $N(t) \to \infty$ and $N(t)/t \to 1/E[X]$ (with probability 1). With probability 1, we have also $\frac{\sum_{n=1}^{N(t)} R_n}{N(t)} \to E[R]$. These two limits prove eq. (3.117).

Example 3.3 *Application to maintenance optimization* Let us consider an application of the renewal reward theory to reliability of service systems. Any equipment (e.g., a router, a switch, a BS transceiver, a server, a machine component) has an operational lifetime, after which it is replaced, e.g., because of technological evolution or because of aging of components. Replacement can be scheduled in advance and hence made as part of ordinary management, and in that case the cost is C_1. Otherwise, the replacement is done when the equipment breaks down, i.e., it stops working at the desired level of performance/quality. In this second case, the replacement is unforeseen, and it must be carried out under exceptional procedures (e.g., to minimize the out-of-service time). We can therefore assume that the cost is bigger in this second case, i.e., it is $C_1 + C_2$.

Let Y_n be the lifetime of the piece of equipment installed after $n - 1$ replacements, $n \geq 1$. After a time Y_n the piece of equipment breaks down, if it has not been replaced before. We assume that breakdown times Y_n form a renewal process.

The variable that is under the control of the management process is the *scheduled operational lifetime*: let it be denoted with T (note that, in the assumed modeling context, there is no reason to have a scheduled replacement time depending on the index n: why?). Then, the renewal time is $X_n = \max\{Y_n, T\}$, i.e., if a breakdown intervenes before the scheduled replacement time, the equipment is substituted according to the emergency procedure; otherwise it undergoes an ordinary substitution. We can define a cost process where

$$R_n = \begin{cases} C_1 & Y_n \geq T \\ C_1 + C_2 & Y_n < T \end{cases} \tag{3.119}$$

The theory above suggests that the average long-term cost per unit time ϕ is asymptotically equal to the ratio of the mean reward to the mean renewal time. We have

$$E[X] = E[\min\{T, Y\}] = \int_0^T x f_Y(x) \, dx + T \int_T^\infty f_Y(x) \, dx$$

$$= \int_0^T x f_Y(x) \, dx + T[1 - F_Y(T)] \tag{3.120}$$

and

$$E[R] = C_1 + C_2 P(Y < T) = C_1 + C_2 F_Y(T) \tag{3.121}$$

Then

$$\phi = \lim_{t \to \infty} \frac{R(t)}{t} = \frac{C_1 + C_2 F_Y(T)}{\int_0^T x f_Y(x) \, dx + T[1 - F_Y(T)]} \tag{3.122}$$

As T gets large, ϕ approaches the limiting value $(C_1 + C_2)/E[Y]$. By contrast, for very small values of T, ϕ grows without bound. Intuitively, if we take a large scheduled time, we incur frequent if not constant equipment breakdowns and the associated supplementary replacement costs. On the other hand, if we replace the equipment frantically, we will never see breakdowns, but we will have to bear an exceedingly high replacement cost. It is intuitive that there could be an optimal value of T. From a mathematical point of view, it all depends on the shape of the CDF of the lifetime Y. If it is such that ϕ is monotonously decreasing as a function of T, then there is no optimum.

Let us assume that the lifetime be distributed uniformly over the interval $[0, U]$. This is not an especially realistic model. We adopt it because it lends itself to analytical investigation. By applying eq. (3.122) with $F_Y(x) = \min\{1, x/U\}$ for

$x \geq 0$, we get

$$\phi = \begin{cases} \dfrac{C_1 + C_2 T/U}{T - T^2/(2U)} & T \leq U \\[2mm] \dfrac{C_1 + C_2}{U/2} & T > U \end{cases} \tag{3.123}$$

With some calculations, it can be verified that ϕ has a minimum indeed and that the minimum is achieved for

$$T^* = U\left(\sqrt{\left(\dfrac{C_1}{C_2}\right)^2 + 2\dfrac{C_1}{C_2}} - \dfrac{C_1}{C_2}\right) \tag{3.124}$$

For example, if $C_1 = C_2$, we have $T^* = U(\sqrt{3} - 1) \approx 0.73 \cdot U$.

Let us now consider a different situation, where the lifetime has a negative exponential distribution. Then, it is easy to find that

$$\phi = \dfrac{1}{E[Y]}\left[\dfrac{C_1}{1 - e^{-T/E[Y]}} + C_2\right] \tag{3.125}$$

which is monotonously decreasing with T. In this case, the conclusion is that the bigger T, the smaller the average cost rate of the system. The intuition behind this result lies with the fact that the negative exponential distribution is *memoryless*. Given that an equipment has worked properly up to time t, the probability that it breaks down in the next interval of duration Δt is independent of t. In other words, aging has no effect on breakdowns. Given this behavior, there is no reason to anticipate the replacement of the equipment before it breaks. In other words, the most convenient policy with memoryless breakdown times is to set $T = \infty$ and let $X_n = Y_n$.

3.5 Birth-Death Processes

The so-called birth-death process is a commonly used stochastic model in queueing theory and network traffic engineering applications. We consider the continuous time version of this kind of process.

Let $X(t)$ be a continuous time process over the non-negative integers[3]. Let also λ_k $(k \geq 0)$ and μ_k $(k \geq 1)$ be positive constants such that:

$$P(X(t + \Delta t) = k + 1 | X(t) = k) = \lambda_k \Delta t + o(\Delta t), \quad k \geq 0 \tag{3.126}$$

$$P(X(t + \Delta t) = k - 1 | X(t) = k) = \mu_k \Delta t + o(\Delta t), \quad k \geq 1 \tag{3.127}$$

$$P(|X(t + \Delta t) - k| > 1 | X(t) = k) = o(\Delta t), \quad k \geq 0 \tag{3.128}$$

---

3 The key point is that the state space be discrete.

for $\Delta t \to 0$. Such a process is called a *birth-death* process, where λ_k represents the rate of birth in the state k and μ_k is the rate of death in the state k. Deaths are not allowed in state 0, whereas births are, i.e., it is $\lambda_0 > 0$ and $\mu_0 = 0$. The meaning of the expressions in (3.126) is that only transitions to neighboring states are likely in the short term.

The infinitesimal generator of a birth-death process is a tri-diagonal matrix:

$$\mathbf{Q} = \begin{bmatrix} -\lambda_0 & \lambda_0 & 0 & 0 & \cdots \\ \mu_1 & -(\lambda_1 + \mu_1) & \lambda_1 & 0 & \cdots \\ 0 & \mu_2 & -(\lambda_2 + \mu_2) & \lambda_2 & \cdots \\ \cdots & \cdots & \cdots & \cdots & \cdots \end{bmatrix} \tag{3.129}$$

In words, the super-diagonal contains the λ_k, $k \geq 0$, the subdiagonal contains the μ_k, $k \geq 1$, and the diagonal is set so that the sum of the elements in each row equals 0. The infinitesimal generator $\mathbf{Q}$ is irreducible if and only if $\lambda_k > 0$, $\mu_k > 0$, $\forall k$.

Let $p_n(t)$ denote the probability that the process is in state n at time t:

$$p_n(t) = \mathcal{P}(X(t) = n), \quad n \geq 0, \ t \geq 0 \tag{3.130}$$

We derive the differential equation ruling the evolution of the $p_n(t)$'s, the forward Chapman-Kolmogorov equations. For a generic state $n > 0$ we have

$$p_n(t + \Delta t) = \sum_{k=0}^{\infty} \mathcal{P}(X(t + \Delta t) = n | X(t) = k)p_k(t)$$

$$= \lambda_{n-1}\Delta t p_{n-1}(t) + \mu_{n+1}\Delta t p_{n+1}(t) + (1 - \lambda_n \Delta t - \mu_n \Delta t)p_n(t) + o(\Delta t)$$

where the only three terms of the sum in the first line that are explicitly reported in the second line are those for $k = n - 1, n, n + 1$, whereas all other terms are drowned in the $o(\Delta t)$ term, thanks to the properties (3.126). Therefore, we have

$$\frac{p_n(t + \Delta t) - p_n(t)}{\Delta t} = \lambda_{n-1}p_{n-1}(t) + \mu_{n+1}p_{n+1}(t) - (\lambda_n + \mu_n)p_n(t) + o(1) \tag{3.131}$$

as $\Delta t \to 0$. This shows that the following differential equations hold for $n \geq 1$ and $t \geq 0$:

$$\frac{dp_n(t)}{dt} = \lambda_{n-1}p_{n-1}(t) + \mu_{n+1}p_{n+1}(t) - (\lambda_n + \mu_n)p_n(t) \tag{3.132}$$

In a similar way, it is easy to derive the boundary differential equation:

$$\frac{dp_0(t)}{dt} = \mu_1 p_1(t) - \lambda_0 p_0(t) \tag{3.133}$$

The linear system (3.132) and (3.133) can be solved given an initial condition, e.g., $p_0(0) = 1$ and $p_n(0) = 0$ for $n > 0$. If the process converges toward statistical equilibrium as $t \to \infty$ and the limit $\pi_n = \lim_{t \to \infty} p_n(t) \geq 0$ is a proper probability distribution, i.e., $\sum_{n=0}^{\infty} \pi_n = 1$, then the birth-death process is said to be ergodic. If the limiting probability distribution exists, it can be found by solving (3.132) and (3.133) having set the derivatives to 0. This is nothing but the system $\pi Q = 0$ that can be written explicitly as

$$\lambda_{n-1}\pi_{n-1} + \mu_{n+1}\pi_{n+1} - (\lambda_n + \mu_n)\pi_n = 0, \quad n \geq 1 \tag{3.134}$$

$$\mu_1\pi_1 - \lambda_0\pi_0 = 0 \tag{3.135}$$

Equation (3.134) can be rearranged as

$$\mu_{n+1}\pi_{n+1} - \lambda_n\pi_n = \mu_n\pi_n - \lambda_{n-1}\pi_{n-1}, \quad n \geq 1 \tag{3.136}$$

This shows that the difference $\mu_{n+1}\pi_{n+1} - \lambda_n\pi_n$ is a constant, independent of n. By using the boundary equation (3.135), we find $\mu_{n+1}\pi_{n+1} - \lambda_n\pi_n = \mu_1\pi_1 - \lambda_0\pi_0 = 0$. Hence, we have proved that $\pi_{n+1} = \pi_n\lambda_n/\mu_{n+1}$. By repeated application of this equality, we can derive that

$$\pi_{n+1} = \frac{\lambda_n}{\mu_{n+1}}\pi_n = \frac{\lambda_n\lambda_{n-1}}{\mu_{n+1}\mu_n}\pi_{n-1} = \cdots = \pi_0 \prod_{k=0}^{n} \frac{\lambda_k}{\mu_{k+1}}, \quad n \geq 0 \tag{3.137}$$

The unknown π_0 can be found from the congruence equation, i.e., $\sum_{n=0}^{\infty} \pi_n = 1$:

$$\pi_0 + \sum_{n=0}^{\infty} \pi_0 \prod_{k=0}^{n} \frac{\lambda_k}{\mu_{k+1}} = 1 \quad \Rightarrow \quad \pi_0 = \left[1 + \sum_{n=1}^{\infty} \prod_{k=0}^{n-1} \frac{\lambda_k}{\mu_{k+1}}\right]^{-1} \tag{3.138}$$

The result holds provided that the series $\sum_{n=1}^{\infty} \prod_{k=0}^{n-1} \frac{\lambda_k}{\mu_{k+1}}$ converges. More in depth, let us define $\xi_n \equiv \prod_{k=0}^{n-1} \frac{\lambda_k}{\mu_{k+1}}$ for $n > 0$ and $\xi_0 = 1$. Further, let us define the two series

$$S_1 = \sum_{n=0}^{\infty} \xi_n \qquad S_2 = \sum_{n=0}^{\infty} \frac{1}{\lambda_n\xi_n} \tag{3.139}$$

It can be shown that the states of the birth-death process are

- *transient* if and only if $S_2 < \infty$;
- *null recurrent* if and only if $S_1 = S_2 = \infty$;
- *positive recurrent* if and only if $S_1 < \infty$.

For example, the series S_1 is convergent if the ratios λ_n/μ_{n+1} are uniformly bounded by a constant less than 1 for every n bigger than some threshold value N, i.e., if there exists N such that $\lambda_n/\mu_{n+1} \leq A < 1, \forall n \geq N$.

Example 3.4 *Application to CDMA cellular systems*   Let us consider a radio cell with n users exploiting a code division multiple access (CDMA) uplink. Let g_j denote the path gain experienced by the j-th user toward the BS and let P_j be the transmission power level of the j-th user. Let B denote the bandwidth devoted to a single user communication and W the spread spectrum bandwidth, so that $G = W/B$ is the processing gain. If N_0 denotes the thermal noise power density, the signal-to-interference-and-noise-ratio (SINR) of the j-th user uplink can be written as

$$S_j(n) = \frac{g_j P_j}{\frac{B}{W} \sum_{i \neq j} g_i P_i + N_0 B} , \quad j = 1, \dots, n. \tag{3.140}$$

Let us now assume that a perfect power control is in place, so that the *received* power level for each user is equalized, i.e., it is $g_j P_j = P$ for all $j = 1, \dots, n$. Then, eq. (3.140) can be rewritten as

$$S_j(n) = S(n) = \frac{\frac{W}{B} P}{(n-1)P + N_0 W} , \quad j = 1, \dots, n. \tag{3.141}$$

Under the usual assumption of Gaussian interference, the upper bound of the achievable information rate over the uplink CDMA channel for each user when n users are active, $C(n)$, is determined according to the Hartley-Shannon law for an additive white Gaussian noise (AWGN) channel:

$$C(n) = B \log_2 \left(1 + \frac{\frac{W}{B} P}{(n-1)P + N_0 W} \right) = B \log_2 \left(1 + \frac{\gamma}{\frac{B}{W}(n-1)\gamma + 1} \right) \tag{3.142}$$

where $\gamma \equiv P/(N_0 B)$ is the user SNR; this parameter gives the *intrinsic* quality of the communication channel of the tagged user, apart from interference. Note that, according to this ideal model, the uplink spectral efficiency, $C(n)/W$ depends only on the two parameters γ and W/B, besides the number n of users.

Let us assume that users generate globally new service requests according to a Poisson process of mean rate λ in the considered radio cell. A service request consists of the transfer of a file of length Y bit, where Y is a random variable with negative exponential probability distribution and mean value L. The mean service completion rate when n users are sharing the uplink is given by $\mu_n = nC(n)/L$, i.e.,

$$\mu_n = n \frac{B}{L} \log_2 \left(1 + \frac{\gamma}{\frac{B}{W}(n-1)\gamma + 1} \right) , \quad n > 0. \tag{3.143}$$

Under the hypotheses laid out for this example, the number of users concurrently active in the CDMA radio cell, $U(t)$, can be modeled by a birth-death process with birth rates $\lambda_n = \lambda$ and death rates μ_n.

The function μ_n is monotonously increasing with n from $\mu_1 = (B/L)\log_2(1 + \gamma)$ up to $\mu_\infty \equiv \lim_{n \to \infty} \mu_n = W/(L \log 2)$. Therefore, since $\lambda_n = \lambda, \forall n$, we can write $\lambda_n/\mu_{n+1} \leq \lambda/\mu_1 = \lambda L/(B \log_2(1 + \gamma))$, and the birth-death process governing the dynamics of the number of users active in the radio cell is ergodic if

$\lambda L < B \log_2(1 + \gamma)$. This is a sufficient condition. A more stringent condition can be found by examining the series in the denominator of π_0, i.e., $\sum_{n=1}^{\infty} \prod_{k=0}^{n-1} \frac{\lambda_k}{\mu_{k+1}}$. By applying the ratio convergence criterion, we find that the series is convergent provided that there exists N such that $\lambda/\mu_{n+1} < 1$ for all $n > N$. If it is $\lambda < \mu_\infty$, then we can always find a positive ε such that $0 < \varepsilon < 1 - \lambda/\mu_\infty$. Then, from $\mu_n \uparrow \mu_\infty$, we derive that there exists v such that $0 < \mu_\infty - \mu_{n+1} \leq \varepsilon\mu_\infty, \forall n > v$. Hence,

$$\frac{\lambda}{\mu_{n+1}} \leq \frac{\lambda}{\mu_\infty(1 - \varepsilon)} < 1, \quad \forall n > v \tag{3.144}$$

This proves that a sufficient (and in fact also necessary) condition for the series to be convergent is that $\lambda < \mu_\infty = W/(L \log 2)$. In terms of the SNR and processing gain parametrization, this condition can be rewritten as $\lambda \log(1 + \gamma) < G\mu_1$, where $G = W/B$. We rewrite the expression of the service completion rate in state n as:

$$\mu_n = n\mu_1 \frac{\log\left(1 + \frac{\gamma}{(n-1)\gamma/G+1}\right)}{\log(1 + \gamma)}, \quad n \geq 1 \tag{3.145}$$

where we have used the processing gain $G \equiv W/B$. The ratio λ/μ_n can be written as a function of three nondimensional parameters:

$$a_n \equiv \frac{\lambda}{\mu_n} = \frac{G\rho}{n \log\left(1 + \frac{\gamma}{(n-1)\gamma/G+1}\right)}, \quad n \geq 1 \tag{3.146}$$

where $\rho \equiv \lambda \log(1 + \gamma)/(G\mu_1) < 1$ is the channel utilization coefficient, referred to as *load factor* in the following. The model is parametrized by the processing gain G, the user SNR γ and the utilization factor ρ.

The steady-state probabilities can be computed numerically, with the recursion $\pi_k = \pi_{k-1}a_k$ for $k \geq 1$. The probability π_0 can be computed as $\pi_0 \approx 1/S$ where S is found with the recursion $S \leftarrow 1 + Sa_k$, per $k = N - 1, N - 2, \ldots, 1$, initialized with $S \leftarrow 1 + a_N/(1 - \rho)$. Here N is an integer big enough so that $0 < a_N - \rho < \varepsilon\rho$ for a given precision level ε (e.g., $\varepsilon = 10^{-5}$).

A minor modification of the birth-death Markov process model yields the case where only up to a finite number K of users are admitted into the system. The resulting model is obtained by truncating the infinite Markov process state space to the first $K + 1$ states, i.e., those between 0 and K. It this case, the birth-death process is a finite, irreducible Markov process, hence it is ergodic. The steady state probabilities can be calculated just as in the infinite case, starting with $\pi_0 = 1$ and using $\pi_k = \pi_{k-1}a_k$ for $k = 1, \ldots, K$. The array of numbers thus obtained is normalized so that it sums to 1 and that yields the steady state probabilities.

In the following we write expressions of performance metrics with reference to the case of a finite K. The corresponding expressions for the unrestricted system,

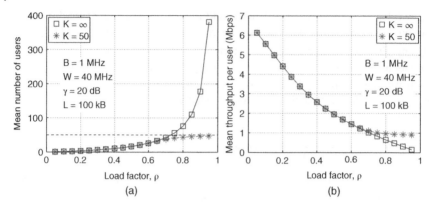

Figure 3.11 Performance metrics of the CDMA model. Mean number of backlogged users E[U] as a function of the load factor ρ (left plot). Average throughput per user Λ_1 as a function of the load factor ρ (right plot).

where there is no limit to the number of admitted users, are easily obtained by letting $K \to \infty$.

Several performance metrics can be defined. The average number of active users is simply $E[U] \equiv \sum_{n=1}^{K} n\pi_n$. The average throughput of the system is

$$\Lambda \equiv \sum_{n=1}^{K} \mu_n \pi_n = \sum_{n=1}^{K} \lambda \pi_{n-1} = \lambda(1 - \pi_K) \tag{3.147}$$

that simplifies to λ if $K = \infty$. The average throughput per user (provided there is at least one user) can be calculated as

$$\Lambda_1 = \sum_{n=1}^{K} \frac{\mu_n}{n} \frac{\pi_n}{1 - \pi_0} = \frac{\lambda}{1 - \pi_0} \sum_{n=0}^{K-1} \frac{\pi_n}{n+1} \tag{3.148}$$

As a numerical example, let us consider plausible values, e.g., $W = 40$ MHz, $B = 1$ MHz (hence $G = W/B = 40$), $\gamma = 20$ dB, and $L = 100$ kbytes. We make λ vary and use as the abscissa the normalized quantity ρ defined above. Figure 3.11 plots two performance metrics: Fig. 3.11(a) shows E[U] as a function of ρ, while Fig. 3.11(b) plots the average throughput per user Λ_1 as a function of ρ.

It is apparent that, when $K = 50$, there is a saturation phenomenon, i.e., the mean number of active users grows with ρ except that it cannot exceed K. Correspondingly, the average throughput per user decreases steadily as the number of active users grows, until the number of active users hits its cap 50. From that point on, the throughput stabilizes. The infinite size system has no upper limit. It faces a steep degradation of performance as the stability limit is approached, i.e., for $\rho \to 1$. The stability of the finite size system is paid at the price of loss. The loss probability is π_K. With the numerical values of the example, it turns out that the loss probability does not exceed 10^{-2} for $\rho \leq 0.65$, while it increases to 0.23 for $\rho = 0.95$.

3.6 Branching Processes

A branching process is a model of population growth. If defined in discrete time, successive generations are indexed by a non-negative integer. Let X_n denote the size of the population at the n-th generation, $n \geq 0$. Then X_0 is the initial population. At each step, each individual gives birth to a number of offsprings. Let ξ_k be the number of offsprings of the k-th individual. It is assumed that the ξ_k's in each generation are realizations of independent and identically distributed random variables, i.e., $\xi_k \sim \xi$, where ξ is a non-negative, discrete random variable. The variables ξ_k are also independent of X_n. In the following we use the notation:

$$a_k = \mathcal{P}(\xi = k), \quad k \geq 0. \tag{3.149}$$

The key relationship of the branching process is thus

$$X_{n+1} = \sum_{k=1}^{X_n} \xi_k, \quad n \geq 0. \tag{3.150}$$

The process X_n is a special case of a discrete-time Markov process. As a matter of fact, given the structure expressed by eq. (3.150), the probability distribution of X_{n+1} given X_n does not depend on any earlier X_t with $t < n$.

Example 3.5 *Application to message dissemination in a sensor network* Let us consider a special kind of sensor network, namely a vehicular network where each vehicle is equipped with an on-board unit. Through their sensors, vehicle collect a large amount of data that can be (at least in part) shared with other vehicles or even gathered in backhaul servers and cloud repositories.

Vehicles form collectively a Vehicular Ad-hoc Network (VANET), e.g., based on an amendment of IEEE 802.11 standard known as IEEE 802.11p. We consider dissemination of messages originating at a tagged vehicle node. The dissemination process is based on multi-hop communications, where one vehicle relays the message to other vehicles. The message is relayed hop by hop by multiple nodes, until it reaches all vehicles within the target region of interest, e.g., within a given radius relative to the position where the message originated from.

We can model with a branching process X_n the number of copies of the message that are relayed at the n dissemination step. The rule adopted by a vehicle is that it relays the message with probability p, or it cancels the message with probability $1 - p$.

If the VANET connectivity can be described with a random uniform graph (random configuration model), each vehicles sees the same distribution of the number of neighboring vehicles (two vehicles are neighbors if and only if they can communicate successfully between them over the wireless channel). Let $q_k = P(V = k)$ be the probability that a node has k neighbors, $k \geq 0$, where V is the random variable representing the number of neighbors. Under this example, the offsprings can be identified with the neighboring nodes that receive the message. When a vehicle A sends the message to a neighboring vehicle B, the number of *residual* neighbors of B (i.e., B's neighbors different from A) has the probability distribution $r_k = (k+1)q_{k+1}/E[V], k \geq 0$. The residual neighbors of B are the target of the possible relaying of the message on behalf of B. Hence, we have

$$
\begin{aligned}
a_0 &= P(\xi = 0) = 1 - p + pr_0 \\
a_k &= P(\xi = k) = pr_k \qquad k \geq 1
\end{aligned}
\tag{3.151}
$$

Given the probability distribution function of the random variable ξ, let us denote its moment generating function as $\phi(z) = E[z^\xi] = \sum_{k=0}^{\infty} a_k z^k$. Let also $\phi_n(z) = E[z^{X_n}]$ be the moment generating function of X_n. Then, from eq. (3.150) we derive

$$
\begin{aligned}
\phi_{n+1}(z) &= \sum_{h=0}^{\infty} E[z^{X_{n+1}} | X_n = h] P(X_n = h) \\
&= P(X_n = 0) + \sum_{h=1}^{\infty} E\left[\prod_{k=1}^{X_n} z^{\xi_k} | X_n = h \right] P(X_n = h) \\
&= P(X_n = 0) + \sum_{h=1}^{\infty} P(X_n = h) \prod_{k=1}^{h} E[z^{\xi_k}] \\
&= P(X_n = 0) + \sum_{h=1}^{\infty} P(X_n = h)[\phi(z)]^h = \phi_n(\phi(z))
\end{aligned}
\tag{3.152}
$$

for $n \geq 0$. By applying repeatedly eq. (3.152), we have $\phi_{n+1}(z) = \phi_n(\phi(z)) = \phi_{n-1}(\phi(\phi(z))) = \phi_{n-1}(\phi_2(z))$. The last equality holds if $X_0 = 1$. It is then easy to prove by induction the equality $\phi_n(z) = \phi_{n-k}(\phi_k(z))$, for $1 \leq k \leq n - 1$. As a special case, it is $\phi_{n+1}(z) = \phi(\phi_n(z))$.

From the relationship $\phi_n(z) = \phi_{n-1}(\phi(z))$, it is possible to derive the moments of X_n. For example, the mean value is

$$
E[X_n] = \phi_n'(1) = \phi'(1)\phi_{n-1}'(\phi(1)) = E[\xi]E[X_{n-1}], \quad n \geq 1.
\tag{3.153}
$$

where we have used the fact that $\phi(1) = 1$. By applying the iteration in eq. (3.153), we finally obtain $E[X_n] = (E[\xi])^n E[X_0]$.

In the following, we confine ourselves to the case $X_0 = 1$, since results are cleaner, yet there is little loss of generality.

One important characteristic of branching processes is the extinction probability η. As generations evolve, there is the possibility that the population size goes to 0, if $a_0 = P(\xi = 0) > 0$. It is apparent that, if $X_n = 0$, then the population size X_t remains at level 0 for all $t > n$. The time to extinction T satisfies the equality $P(T \leq n) = P(X_n = 0) \equiv \eta_n$, $n \geq 0$. We can wonder what happens as the number n of generations grows. To make this question nontrivial, we assume in the following that $0 < a_0 < 1$. With $X_0 = 1$, we have $\phi_{n+1}(z) = \phi(\phi_n(z))$, hence

$$\eta_{n+1} \equiv P(X_{n+1} = 0) = \phi_{n+1}(0) = \phi(\phi_n(0)) = \phi(\eta_n)\,, \quad n \geq 0 \qquad (3.154)$$

with $\eta_0 = 0$. Note that the generating function $\phi(z)$ is less than 1 for every $z \in [0, 1)$. Moreover, the function $\phi(z)$ is strictly increasing with z, since it is $a_0 < 1$. Then, it is $\eta_n = \phi(\eta_{n-1}) < 1$ for every $n > 0$ and $\eta_0 = 0 < 1$. Moreover,

$$\eta_1 = \phi(\eta_0) = \phi(0) = a_0 > 0 = \eta_0 \qquad (3.155)$$

In general, given the induction hypothesis $\eta_n > \eta_{n-1}$, we have

$$\eta_{n+1} = \phi(\eta_n) > \phi(\eta_{n-1}) = \eta_n \qquad (3.156)$$

Since we have already seen that the inequality holds for $n = 0$, we have proved by induction that the η_n's form a monotonously increasing sequence bounded above by 1. Therefore, the limit $\eta \equiv \lim_{n \to \infty} \eta_n$ exists and it is a probability, i.e., a real number in $[0, 1]$.

Two cases can arise: (i) $\eta = 1$, then extinction occurs eventually and T is a proper random variable, i.e., $P(T < \infty) = 1$; (ii) $\eta < 1$: then, extinction is not certain and we have $P(T = \infty) = 1 - \eta$ (i.e., T is a defective random variable).

By letting $n \to \infty$ in eq. (3.154), we get $\eta = \phi(\eta)$. We see that the extinction probability η must be a solution of the equation $z = \phi(z)$ in the interval $[0, 1]$. Actually, we can show that η is the smallest positive root of that equation.

Theorem 3.13 Let $\phi(z)$ be the probability generating function of the offsprings with $\phi(0) = a_0 > 0$. The extinction probability η is the smallest positive root of the equation $z = \phi(z)$ in the interval $[0, 1]$.

Proof: We know that $z = \phi(z)$ has at least one root in $[0, 1]$, since it is $\phi(1) = 1$. Let ζ denote the least root of $z = \phi(z)$ in $[0, 1]$. It must be $\zeta > 0$ since $\phi(0) = a_0 > 0$. Moreover, we have $\eta_1 = \phi(\eta_0) = \phi(0) < \phi(\zeta) = \zeta$, since ζ is positive and it is a fixed point. Then, by the induction hypothesis $\eta_n < \zeta$ (shown true for $n = 1$), we get $\eta_{n+1} = \phi(\eta_n) < \phi(\zeta) = \zeta$. It is therefore, $\eta_n < \zeta$, $\forall n$ and $\eta = \lim_{n \to \infty} \eta_n \leq \zeta$. Since it is also $\eta = \phi(\eta)$ and ζ is the *smallest* positive root of $z = \phi(z)$, we must have $\eta = \zeta$.

To investigate further the extinction property, let us first rule out a trivial case. Assume $a_0 + a_1 = 1$. Then, at each generation, either a new individual replaces the previous generation one or no birth occurs and the population goes to 0. In other words, this is a pure death process. When started at level 1, it goes to 0 after a geometrically distributed number of steps. Let us now move to a general case where $a_0 + a_1 < 1$, besides being $0 < a_0 < 1$. Then, the function $\phi(z)$ is strictly convex, with $0 < \phi(0) < 1$ and $\phi(1) = 1$. Only two cases are possible: either it is $\zeta = 1$ or $\zeta < 1$. Since $\phi(z)$ is strictly convex, it must lie all above any of its tangents. By taking the tangent at $z = 1$, we can write

$$\phi(z) - z > 1 + \phi'(1)(z - 1) - z = (1 - \phi'(1))(1 - z) \tag{3.157}$$

The right-hand side is non-negative for all $0 \leq z < 1$ if $\phi'(1) \leq 1$. In that case we have thus proved that $\phi(z) > z$ for $0 \leq z < 1$. Hence, it must be $\zeta = 1$. If instead it is $\phi'(1) > 1$, since $\phi'(0) = a_1 < 1$ and the derivative of $\phi(z)$ is continuous, there must exist a point $x \in (0, 1)$ such that $\phi'(x) = 1$. We can prove that it is $\phi(x) < x$. According to the average theorem, there exists a point $y \in (x, 1)$ such that $\phi(1) - \phi(x) = \phi'(y)(1 - x)$. Since it is $\phi'(y) > 1$ and $\phi(1) = 1$, we have $\phi(1) - \phi(x) = 1 - \phi(x) > 1 - x$, i.e., $\phi(x) < x$. We have thus shown that the function $\phi(z) - z$ changes sign in the interval $[0, x]$, since $\phi(0) - 0 = a_0 > 0$ and $\phi(x) - x < 0$. There must be at least one zero in $(0, x)$ with $x < 1$, and it is therefore $\zeta < 1$. Let us summarize what we have just proved.

Theorem 3.14 Let $\phi(z)$ be the generating function of a probability distribution $\{a_k\}_{k \geq 0}$ with $a_0 > 0$ and $a_0 + a_1 < 1$. Then, the smallest positive root of the equation $z = \phi(z)$ is strictly less than 1 if $\phi'(1) > 1$; it is equal to 1 if $\phi'(1) \leq 1$.

Turning from the mathematics to the branching process meaning, the result is consistent with intuition. The derivative of $\phi(z)$ at 1 is the mean number of offsprings. The result presented above says that extinction is certain, if the mean number of newborn individuals in each generation is no more than 1; otherwise, if strictly more than a single individual are born with each new generation, there is some non-null probability of the population extinction, but that outcome is not certain.

Example 3.6 (Continued from Example 3.5) Let us go back to our example of vehicular network. Let us assume that the number of neighbors of a node has a Poisson distribution, i.e., $q_k = \frac{v^k}{k!} e^{-v}$ for $k \geq 0$. Then, the probability distribution of the number of neighbors of a neighbor of a given node is $r_k = (k+1)q_{k+1}/E[V] = \frac{v^k}{k!} e^{-v}$, since the mean number of neighbors is $E[V] = v$. Then, it is $a_0 = 1 - p + pe^{-v}$ and $a_k = pr_k$, $k \geq 1$. For a working dissemination protocol it must be $p > 0$, hence it is $a_0 < 1$. It is also obviously $a_0 > 0$. It is easy to find that $\phi(z) = 1 - p + pe^{v(z-1)}$ and $\phi'(1) = pv$. If $pv \geq 1$, the message will disseminate over the whole connected component of the vehicular network with probability 1, whereas when $pv < 1$ there is a positive probability that message

Figure 3.12 Expected number of dissemination hops as a function of the relaying probability p for a branching process model of message dissemination in a vehicular network, for various values of the mean number of neighbors v.

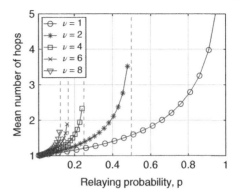

dissemination will die out after some hops. Let T be the number of hops of the message, i.e., the number of times that the message is forwarded during the dissemination process. It is $F_T(n) \equiv P(T \le n) = \eta_n$; then

$$F_T(n) = \eta_n = \phi(\eta_{n-1}) = \phi(F_T(n-1)) = 1 - p + pe^{v(F_T(n-1)-1)} \tag{3.158}$$

The CCDF $G_T(n) = 1 - F_T(n)$ is therefore the solution of the iteration

$$G_T(n) = p - pe^{-vG_T(n-1)}, \quad n \ge 1, \tag{3.159}$$

with $G_T(0) = 1$. The mean value of T can be found as $E[T] = \sum_{n=0}^{\infty} G_T(n)$. Figure 3.12 shows $E[T]$ as a function of p for various values of v. As p approaches $1/v$, $E[T]$ tends to infinity.

Branching processes can be defined in continuous time as well. Also, other variants of branching models can be analyzed, even of non-Markovian type. For a wider introduction to this topic, the interested reader can consult, e.g., [117, Ch. 11].

Summary and Takeaways

In this chapter a number of topics have been discussed, with the common denominator of providing models that are widely used in network traffic engineering and performance evaluation, to describe series or patterns of events.

The Poisson process is the most celebrated model of a point process. The key properties of the Poisson process have been presented, i.e., the negative exponential distribution of the inter-arrival times, the Poisson distribution of the counting function, the memoryless property. The extension to inhomogeneous Poisson process, i.e., one where the arrival rate is a function of time, has been introduced. A treatment of spatial point processes is given as well, with a focus on the spatial Poisson point process (PPP). A number of definitions and properties carry over from unidimensional to multi-dimensional point processes. Operations on point processes (displacement, mapping, thinning) have been defined. A generalization of the Poisson process is obtained by coupling the arrivals with a modulating Markov process. This gives rise to the seminal idea of matrix-geometric methods, introduced by M.F. Neuts [167].

Another important class of processes are renewal ones. They are the simplest possible general random process. They have no correlation structure, i.e., inter-event times form a sequence of i.i.d. random variables. Hence, the CDF of the inter-event time is sufficient to get a full knowledge of the process. The renewal paradox and the key concept of residual inter-event time, the superposition of renewal processes and renewal reward processes are discussed.

Birth-death processes are another very useful class of processes, used to capture the dynamics of systems that can be described by discrete variables that change one step at a time. The ergodicity conditions and the limiting probability distribution are derived. Birth-death processes are the basic modeling tool that allows the description of all elementary queueing models, i.e., all queues that fit into the M/M class (Poisson arrivals, negative exponential service times).

Finally, the class of branching processes is presented. This is less often used in networking application, yet it makes a useful and simple model that lends itself to a wide variety of generalization, often retaining amenability to mathematical analysis.

Problems

3.1 Consider a Poisson arrival stream with mean rate λ. In Section 3.2 you have seen that, if arriving customers are sampled with probability p, independent of one another, then the sampled sub-stream is still a Poisson process with mean rate $p\lambda$. Assume now that sampling is performed by taking exactly one arrival out of r consecutive ones. Is the resulting sampled process still a Poisson process? If not, can you characterize it? [Hint: Think in terms of inter-arrival times.]

3.2 Find the probability distribution of the minimum of M independent negative exponential random variables with mean rates λ_i, $i = 1, \dots, M$.

3.3 Repeat the task of the Problem 3.2, this time for M independent geometrically distributed non-negative random variables. Assume those M variables represent the countdown parameters of M stations competing for the access to a common channel. The winning station is the one whose countdown expires first. What is the probability that there will be a unique station winning?

3.4 Let $\{X_j\}$ be a sequence of identically distributed mutually independent random variables. Assume that X_j is a binary random variable with $P(X_j = 0) = 1 - p$ and $P(X_j = 1) = p$. Let N be a Poisson random variable with mean a. Find the probability distribution of the random variable $Y \equiv X_1 + \cdots + X_N$, i.e., the sum of a random number (with Poisson probability distribution) of Bernoulli random variables.

3.5 At a bus stop we observe that with probability 0.67 the time elapsing between two successive buses is 1 minute. If more than 1 minute goes by after the last bus arrival, we know that it will take another 19 minutes until we see the next bus coming. How long will we wait on average if we drop at the bus stop at a random time (random with respect to the bus schedule)?

3.6 You used to arrive at the bus stop near your home at a random time during the morning. After some months you find out that your average wait is 12 minutes. You learn from the time table that buses arrive one every 10 minutes *on average*, during the morning. Can you be definitely sure that the bus company is cheating on the time table? If you believe to the time table, what complaint would you submit to the bus company?

3.7 A multiplexing channel serves N packet flows. Each packet flow is generated by a user that stores new packets in a buffer as they arrive. Time on the channel is divided into frames. A frame is composed of two parts: (i) signaling slots; (ii) data slots. Signaling slots occupy a fixed interval of duration S where reservations are collected from all users. As soon as a signaling opportunity is given, the user reserves capacity for all packets waiting in its buffer at the time when the frame starts. The rest of the frame accommodates all packets that have been reserved, namely all packets waiting at the buffers of the users. If no packets are waiting, a new frame is started immediately after the signaling opportunity. The multiplexing channel capacity is C, packet lengths have a general probability distribution with mean $\overline{L}$. Packets arrive according to a Poisson process with mean rate λ_i at user i buffer ($i = 1, \ldots, N$). Find the condition for the stability of the multiplexing channel. Find the average frame duration under the stability condition.

3.8 In a VANET along a highway, a vehicle A transmits a message to those around it. The radio transmission range is R, i.e., all vehicles within a distance R of A can decode the message correctly, while more distant vehicles cannot. The highway can be modeled as a line and vehicles are distributed along the highway according to a Poisson process with mean density of λ vehicle/km. Find the PDF of the distance between vehicle A and the most

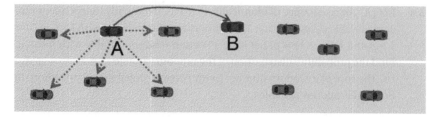

Figure 3.13 Example of vehicles along the highway, forming a VANET. Vehicle *A* sends a message that can be received up to a maximum distance *R*. *B* is the furthest vehicle reached by *A*.

distant vehicle that is reached by the message (vehicle *B* in Figure 3.13), under the condition that there is at least one vehicle within distance *R* of *A*.

3.9 A smartphone user can start a single connection at a time. While connected it cannot start another connection. Let us consider the point process defined by the start times of the user connections. Why is the Poisson process inadequate to describe such an arrival process? Now consider the population of users camping in the cell covered by a base station. The average number of users roaming in the cell is assumed to be much greater than the number of active connections through the base station at any given time. Could the Poisson process be adequate in this case? Why?

3.10 You are observing the arrivals of TCP connections at a web server. You find that such arrivals can be modeled as a Poisson process of mean rate λ. What is the probability of observing two arrivals in a time interval of duration $1/\lambda$?

3.11 An inadvertent professor gives the same meeting time to two students. The first one arrives in time, the other one is late by 10 minutes. The meeting time has a negative exponential PDF with mean 30 min. Calculate the following.
a. The probability that the second student has to wait.
b. The mean value of the time spent by the second student at the professor's office.

3.12 The packets entering a node are switched to either one of two output interfaces, according to their lengths. If the incoming packet length is less than or equal to L_0 it is sent to the interface SP, with capacity C_{SP}, otherwise the packet is routed to the interface LP, with capacity C_{LP}. Packet length have uniform probability distribution between L_{min} and L_{max}. Packet lengths are independent of the packet inter-arrival times.

If the original packet arrivals form a Poisson process with mean rate λ, prove that the two flows directed to the two output interfaces are still Poisson processes and find their respective mean rates λ_{SP} and λ_{LP}.

Then, find the output link capacities C_{SP} and C_{LP} required to obtain a link utilization coefficient equal to ρ_0 on each of the two interfaces.

In your calculations, assume $L_0 = 256$ bytes, $L_{min} = 40$ *bytes*, $L_{max} = 1500$ bytes, $\lambda = 1000$ pkts/s, $\rho_0 = 0.7$.

3.13 Customers arrive at a service facility according to a Poisson process with mean rate λ. A single server is available, with service times distributed according to a random variable X, independent of the inter-arrival times. Consider the busy period started by a given customer C_1. Find the probability P that the second arriving customer C_2 does not have to wait at all (i.e., it starts a new busy period). Calculate also the average waiting time of C_2 if X is deterministic and if it is a negative exponential random variable. Assume that the mean value of X be $1/\mu$ in both cases.

3.14 A computing cluster is equipped with M servers, completely interchangeable. Each of them can fail independently of the others. The time to failure of a running server is a negative exponential random variable with mean $1/\nu = 30$ days. There is only one technician repairing failed servers. Repair times have negative exponential PDF with mean $1/\mu = 1$ day.
   a. Define a birth-death model of the failure-restore process of the servers.
   b. Calculate the probability that no server is operating (all are down) for $M = 5$.
   c. Calculate the mean number of operational servers for $M = 5$.
   d. Find the minimum value of M that guarantees that the unavailability of the system is 10^{-9} [unavailability = probability that all servers are down].

3.15 Connection requests arrive at a fully shared channel with capacity C according to a Poisson process of mean rate λ. Each connection has a negative exponential amount of bytes to carry, with mean $\overline{Q}$. Capacity is shared equally among all ongoing connections.
   a. Identify a birth-death model of the system evolution: give expressions for the birth and death rates.
   b. Calculate the mean bit rate obtained by a connection.

3.16 Points are scattered on the plane so that for any compact (=closed and bounded) set B the probability that k point lie in B is $a^k e^{-a}/k!$, for $k \geq 0$, with $a = \lambda|B|$ ($|B|$ denotes the area of B). Let us focus on one point P. Find the probability that the point Q *closest* to P is at a distance at least $r > 0$. Try to generalize the result to find the probability density function of the distance of the n-th nearest neighbor of P for $n > 1$.

Part II

Queues

4

Single-Server Queues

If you are not too long, I will wait here for you all my life.

Oscar Wilde

4.1 Introduction and Notation

In this chapter we focus on single-server queues. The corresponding general model is noted as the $G/G/1$ queue in Kendall's notation. Unless stated otherwise, we assume the scheduling policy of the queue is first-come, first-serve (FCFS) and the server is work-conserving, i.e., it cannot stay idle if there are customers to be served, and no customer leaves the queue until it has completely received the amount of service it demands. As for the waiting line, we denote its size with K: up to $K + 1$ customers can be hosted in the queuing system, one under service, the others waiting for service. $K = \infty$ is a special case (infinite room queue). For a finite K there is the possibility that a customer arrives to find the queue full. In that case, it is lost to the queue, i.e., its service request is turned down and disappears.

Arrivals follow a stationary renewal process with inter-arrival times distributed according to a continuous positive random variable T. The mean arrival rate is denoted with λ. Service times are i.i.d. random variables with the probability distribution of the continuous positive random variable X. The mean service rate is denoted with μ. Service times are independent of inter-arrival times and of the scheduling policy. Arrivals follow a Poisson process with mean rate λ for the $M/G/1/K$ queues. The random variable X has a negative exponential probability distribution with mean $1/\mu$ in case of $G/M/1/K$ queues.

We consistently use the following notation for queue-related quantities.

$Q(t)$ number of customers in the queue at time t.
T inter-arrival time.

Network Traffic Engineering: Stochastic Models and Applications, First Edition. Andrea Baiocchi.
© 2020 John Wiley & Sons, Inc. Published 2020 by John Wiley & Sons, Inc.

X service time.

S system or response time, i.e., amount of time spent by a customer in the queue from arrival until departure.

W waiting time, i.e., the amount of time spent by a customer in the queue besides its own service time.

A number of arrivals in a service time.

B number of service completions in an inter-arrival time.

Y duration of the busy period, i.e., the time when the server is uninterruptedly busy.

M number of customers served in the busy period.

I duration of the idle time, i.e., the time when the queue is empty, between two consecutive busy periods.

The discussion of Palm's distributions in Section 2.5 points out that, for single-server, single-arrival systems, as those considered in this chapter, it is always the case that the probability distribution of the number of customers seen by an arrival that joins the system is the same as the probability distribution of the number of users left behind by a customer departing from the queue. In case of $M/G/1$ systems, the PASTA property also guarantees that the probability distribution of the number of customers seen by an arrival is the same as the random time distribution (at equilibrium). Moreover, Theorem 2.3 tells us that any of the Palm's distribution mentioned above is the same for the $M/G/1$ queue. The same theorem gives also the relationship between the probability distribution at any time or seen by an arrival and the one seen by a customer entering the queue or departing from it in case of $M/G/1/K$. In that case, departing customers do not see the same system as a random look does. For example, they can never leave a full up system, i.e., they never leave behind $K + 1$ customers, whereas it is possible for an arrival or an observer at a random time to find the system full. Only Theorem 2.2 applies to the $G/M/1$ queue, so we shall use care in deriving the probability distribution of the number of customers in the system, by specifying the point of view.

4.2 The Embedded Markov Chain Analysis of the $M/G/1$ Queue

The number $Q(t)$ of customers residing in the queue at time t does not exhaust the description of the state of the $M/G/1$ queue at time t. If we are going to be able to predict the future evolution of the system for times $t' > t$, we need to know also how long the user under service at time t (if any) has been served before of t. In other words, we need to know the amount of service it has already received, or, that is the same, the amount of service still to be delivered. The first one is the *age* of the service time, the latter is the *residual* service time. From the renewal theory, we know that the only variable that enjoys the memoryless property is the negative exponential one. Therefore, unless service times are i.i.d. negative exponential random variables, we *must* explicitly know the age of the service time to reconstruct the state of the $M/G/1$ queue. For the same reason, we do not need to know the age of the inter-arrival time: Poisson arrivals entail negative exponential inter-arrival times, hence we can predict the future of the arrival process after time t without any prior knowledge of the past before time t.

It turns out then that the couple $(Q(t), X_a(t))$ *is* indeed a full state description of the system at time t, where $X_a(t)$ is the age of the service time of the user under service at time t^1. More precisely, $(Q(t), X_a(t))$ is a Markov process. It is possible to carry out the analysis of the $M/G/1$ queue by considering the process $(Q(t), X_a(t))$, but this leads to technical difficulties and to a rather cumbersome development, given that the Markov process is bi-dimensional and it has one continuous component and one discrete component.

A different approach consists of giving up considering any time point and focusing on special time points. If suitably chosen, the sequence of random variables extracted from $Q(t)$ at those time points form an embedded process with the Markov property. The embedded Markov chain (EMC) approach is the one followed in, e.g., [130, 196]. We will follow that approach here.

Let us first introduce the notion of a *regeneration point* as done in [196]. Given a stochastic process $Y(t)$, the time point u is a regeneration point if ·

$$P(Y(t) \in I | Y(u) = y(u)) = P(Y(t) \in I | Y(\tau) = y(\tau), \tau \le u) \tag{4.1}$$

where I is a set of values of $Y(t)$ (an event), and $y(\tau)$ are given values. Equation (4.1) says that the information on $Y(u)$ summarizes anything that is worth knowing about the past of $Y(t)$ up to time u: there is no need of giving the full detail of the whole past history up to time u to be able to assess the probabilistic evolution in the future, after u. A similar notion of regeneration can be defined for discrete time processes, namely k is a regeneration time if for any positive ℓ we have

$$P(Y_{k+\ell} \in I | Y_k = y_k) = P(Y_{k+\ell} \in I | Y_h = y_h, h \le k) \tag{4.2}$$

This is but the Markov property for the discrete time random process Y_k. Now we can show the following.

Theorem 4.1 The sequence of times $\{t_{d,k}\}_{k \in \mathbb{Z}}$ of customer departures in the $M/G/1$ queue are regeneration times.

---

1 The value of $X_a(t)$ is immaterial if the queue is empty; we convene it is 0 in that case.

Proof: Let $Q_k \equiv Q(t_{d,k}^+)$ the number of customers in the $M/G/1$ queue immediately after the departure of the k-th departing customer, i.e., the number of customers left behind by the k departing customer. Assume we know that $Q_k = n$. The question is: can we write the probability distribution of future states $Q_{k+\ell}$, $\ell \geq 1$ based only on this information? Given Q_k, the value attained by $Q_{k+\ell}$ depends on the number of service completions, which are exactly ℓ by the very definition of the sequence Q_n, and on the number of arrivals. These last ones are the arrivals occurred between the k-th service completion, at time $t_{d,k}$ and $(k + \ell)$-th one, at time $t_{k+\ell}$. Since arrivals follow a Poisson process, they are independent of the state of the system. Moreover, the arrivals in $(t_k, t_{k+\ell}]$ are independent of arrival prior to time t_k. Therefore, we need not specify anything else besides Q_k. Since both k and ℓ are generic, this proves that all customer departing times $\{t_{d,k}\}_{k \in \mathbb{Z}}$, i.e., service completion times, are indeed regeneration points of the $M/G/1$ queue. ∎

Note that by $t_{d,k}$ we denote the time of departure of the k-th departing customer, irrespective of the order of service, that might well be different from FCFS. Theorem 4.1 tells us that $\{Q_n\}_{n \in \mathbb{Z}}$ is a Markov chain (an EMC indeed). We now set out to determine the limiting probability distribution of this chain, namely the probabilities

$$x_k = \lim_{n \to \infty} \mathcal{P}(Q(t_{d,n}^+) = k) = \lim_{n \to \infty} \mathcal{P}(Q_n = k) , \quad k \geq 0. \tag{4.3}$$

Let $\{t_{a,n}\}_{n \in \mathbb{Z}}$ denote the arrival times of customers at the queue and let for $k \geq 0$:

$$q_{a,k} = \lim_{n \to \infty} \mathcal{P}(Q(t_{a,n}^-) = k) \tag{4.4}$$

$$p_k = \lim_{t \to \infty} \mathcal{P}(Q(t) = k) \tag{4.5}$$

be the limiting probabilities seen by an arrival and seen at a generic time point of the statistical equilibrium. From Theorem 2.3 we know that, if those probability distributions exist, then $p_k = q_{a,k} = x_k$, $k \geq 0$.

4.2.1 Queue Length

The argument above proves that *in the $M/G/1$ queue it suffices to find the probabilities x_k to obtain any desired performance measure.* The evolution of the EMC $\{Q_n\}_{n \geq 0}$ is easily recognized to be described by

$$Q_{n+1} = \max\{0, Q_n - 1\} + A_{n+1} \tag{4.6}$$

where A_n is the number of arrivals at the queue in the time between the $(n-1)$-th and the n-th departures. Actually, if the n-th departure leaves a nonempty queue, i.e., $Q_n > 0$, then the next departing customer will leave behind the Q_n customers that were already there, except of itself, plus those arrived during its service time, namely A_{n+1}. Hence it will be $Q_{n+1} = Q_n - 1 + A_{n+1}$ for $Q_n > 0$. On the contrary, if the n-th departing customer leaves the queue empty, i.e., $Q_n = 0$, the queue will stay empty until a new customer arrives. For a work-conserving server, this newly arrived customer will enter immediately into service and will eventually leave the queue, leaving behind exactly those customers that have arrived during its service time, i.e., $Q_{n+1} = A_{n+1}$ for $Q_n = 0$.

It can be verified that the random variable A_{n+1} is independent of Q_n. In fact, A_{n+1} is the number of arrivals during the service time of the $(n + 1)$-th departing customer, while Q_n is the number of customers arrived in previous service times, up to the n-th one, and still waiting to be served. Since the arrival process is a stationary Poisson one and A_{n+1} and Q_n refer to arrivals in non overlapping intervals, they are independent of each other.

We can say even more. Since service times are i.i.d. and arrivals follow a Poisson process, the random variable A_n does not depend on the epoch n. Then, the transition mechanism described by eq. (4.6) is that of a *time-homogeneous Markov chain*, i.e., the one-step transition probabilities do not depend on the index n. Let us define

$$a_k = P(A = k), \quad k \geq 0, \tag{4.7}$$

that is to say the probability distribution of having k arrivals in a service time. From the Poisson input hypothesis and the definition of the random variable A, we have

$$P(A = k|X = t) = \frac{(\lambda t)^k}{k!} e^{-\lambda t}, \quad k \geq 0, \tag{4.8}$$

since the random variable A conditional on the service time $X = t$ is simply the number of arrivals of a Poisson process with mean arrival rate λ in a time interval of duration t. Therefore

$$a_k = P(A = k) = \int_0^\infty \frac{(\lambda t)^k}{k!} e^{-\lambda t} f_X(t) \, dt, \quad k \geq 0. \tag{4.9}$$

Note that the a_k's are all strictly positive.

The generating function of A is $\phi_A(z) = \varphi_X(\lambda - \lambda z)$, which is analytic at least for $|z| \leq 1$. In fact, we note first that

$$E[z^A|X = t] = \sum_{k=0}^\infty P(A = k|X = t)z^k = \sum_{k=0}^\infty \frac{(\lambda t)^k}{k!} e^{-\lambda t} z^k = e^{\lambda t(z-1)} \tag{4.10}$$

Then, we have

$$\phi_A(z) = E[z^A] = \int_0^\infty f_X(t)E[z^A|X = t] \, dt = \int_0^\infty f_X(t)e^{\lambda t(z-1)} \, dt = \varphi_X(\lambda - \lambda z) \tag{4.11}$$

This could have been derived directly from the definition of the generating function as a power series $\phi_A(z) = \sum_{k=0}^{\infty} a_k z^k$, by exploiting the expression of the a_k's in eq. (4.9).

The one-step transition probabilities of the EMC Q_n are given by

$$P(Q_{n+1} = j|Q_n = i) = P(A = j - i + 1) = a_{j-i+1}, \quad j = i - 1, i, \dots. \tag{4.12}$$

for $i > 0$. Starting from state i, we can end up with one less customer, if no arrival occurs within the customer departure, or with the same number of customers, if exactly one arrival compensates for the departure, etc. In the special case $i = 0$, we have

$$P(Q_{n+1} = j|Q_n = 0) = P(A = j) = a_j, \quad j = 0, 1, \dots. \tag{4.13}$$

We can display the one-step transition probabilities in matrix form as

$$\mathbf{P} = \begin{bmatrix} a_0 & a_1 & a_2 & a_3 & \cdots \\ a_0 & a_1 & a_2 & a_3 & \cdots \\ 0 & a_0 & a_1 & a_2 & \cdots \\ 0 & 0 & a_0 & a_1 & \cdots \\ 0 & 0 & 0 & a_0 & \cdots \\ \cdots & \cdots & \cdots & \cdots & \cdots \end{bmatrix} \tag{4.14}$$

The matrix $\mathbf{P}$ is a band matrix, i.e., elements along each sub- and super-diagonal on either side of the main diagonal are the same, except possibly of the elements belonging to the fist row. Those elements represent a special boundary case. Moreover, all elements below the sub-diagonal are null, which is reminiscent of the structure of the queue: since it is a single server and we sample the state of the queue at departure times, the state can at most decrease by one at each transition, whereas it can increase of whatever amount because of the Poisson arrivals. A matrix with zeros below the sub-diagonal is called a lower Hessenberg matrix. It is a generalization of a lower triangular matrix, which is obtained when also the sub-diagonal is all filled with zeros.

Since the a_k's are all positive, the Markov chain is irreducible. It is aperiodic, since $a_1 > 0$. Then, the states of the $M/G/1$ EMC are either all transient, null recurrent, or positive recurrent. In this last case, the limiting probabilities do exist and they are independent of the initial state, i.e., the EMC is ergodic. This is the condition we are most interested in. It can be shown that positive recurrence occurs if and only if

$$E[A] = \phi'_A(1) = \frac{d}{dz} \varphi_X(\lambda - \lambda z)\Big|_{z=1} = -\lambda \varphi'_X(0) = \lambda E[X] \equiv \rho < 1 \tag{4.15}$$

where ρ represents the utilization coefficient of the server. The condition $\rho < 1$ is equivalent to $\lambda < \mu$ or $E[X] < E[T]$, i.e., the mean rate that the customers arrive at the queue must be less than the maximum serving rate, the rate at which the server is able to clear the backlog. Quite a sensible result!

Theorem 4.2 Assume the probability distribution of the number of arrivals A in a service time is such that $a_0 > 0$ and $a_0 + a_1 < 1$. Assume also that the second moment of A is finite, i.e., $E[A^2] < \infty$. The EMC of the $M/G/1$ queue is positive recurrent if and only if $E[A] < 1$.

Proof: We prove only the "if" part. To that end we use the Foster-Lyapunov theorem A.6, stated in the Appendix at the end of the book.

Let the Lyapunov function be $V(x) = x^2$. The conditional drift d_k for $k \geq 1$ is

$$d_k = E[V(Q_{n+1}) - V(Q_n)|Q_n = k] = E[(k - 1 + A_{n+1})^2 - k^2] \qquad (4.16)$$

With some manipulation, it is possible to find that

$$d_k = E[(A - 1)^2] + 2k(E[A] - 1) \qquad (4.17)$$

Given $\varepsilon > 0$, the conditional drift is bounded by $-\varepsilon$ if

$$E[(A - 1)^2] + \varepsilon \leq 2k(1 - E[A]) \qquad (4.18)$$

If it is $E[A] < 1$, we see that the condition $d_k \leq -\varepsilon$ is fulfilled for all $k \geq k_{th}$ where

$$k_{th} = \left\lceil \frac{E[(A - 1)^2] + \varepsilon}{2(1 - E[A])} \right\rceil \qquad (4.19)$$

The conditions required by Theorem A.6 hold, taking the finite set $\mathcal{A}$ as the set of states $\{0, 1, \ldots, k_{th} - 1\}$.

Let us go back to the EMC Q_n and let us assume it is ergodic ($\rho < 1$): ∎

$$Q_{n+1} = \begin{cases} Q_n - 1 + A_{n+1} & Q_n > 0 \\ A_{n+1} & Q_n = 0 \end{cases} \qquad (4.20)$$

We recall that A_{n+1} is independent of Q_n, since: (i) service times are independent of the inter-arrival times; (ii) arrivals follow a Poisson process and the arrivals counted in A_{n+1} belong to an interval nonoverlapping with the intervals that the arrivals affecting Q_n belong to. It is also independent of $n + 1$, so that we can simply replace A_{n+1} with A. Then, we can write:

$$\phi_{Q_{n+1}}(z) = E[z^{Q_{n+1}}] = \sum_{h=0}^{\infty} P(Q_n = h)E[z^{Q_{n+1}}|Q_n = h]$$

$$= P(Q_n = 0)E[z^A|Q_n = 0] + \sum_{h=1}^{\infty} P(Q_n = h)E[z^{Q_n-1+A}|Q_n = h]$$

$$= P(Q_n = 0)\mathrm{E}[z^A] + z^{-1}\mathrm{E}[z^A] \sum_{h=1}^{\infty} P(Q_n = h)z^h$$

$$= P(Q_n = 0)(1 - z^{-1})\phi_A(z) + z^{-1}\phi_A(z)\phi_{Q_n}(z)$$

Taking the limit for $n \to \infty$, we get

$$\phi_Q(z) = x_0(1 - z^{-1})\phi_A(z) + z^{-1}\phi_A(z)\phi_Q(z) \tag{4.21}$$

hence

$$\phi_Q(z) = \frac{x_0(z-1)\phi_A(z)}{z - \phi_A(z)} \tag{4.22}$$

The unknown x_0 can be found by requiring that $\sum_{h=0}^{\infty} x_h = 1$, that is to say $\phi_Q(1) = 1$. By applying de L'Hôpital's rule, we find

$$1 = \phi_Q(1) = \frac{x_0}{1 - \phi_A'(1)} = \frac{x_0}{1 - \rho} \tag{4.23}$$

by eq. (4.15). Summing up, the generating function of the probability distribution of the number of customers Q in the $M/G/1$ queue is given by

$$\phi_Q(z) = \frac{(1 - \rho)(z - 1)\phi_A(z)}{z - \phi_A(z)} \tag{4.24}$$

The mean queue length at equilibrium can be found by $\mathrm{E}[Q] = \phi_Q'(1)$, that is

$$\mathrm{E}[Q] = \rho + \frac{\lambda^2 \mathrm{E}[X^2]}{2(1 - \rho)} \tag{4.25}$$

A fundamental remark is in order. The probability distribution of the number of customers in the queue given in eq. (4.24) holds for the $M/G/1$ queue at statistical equilibrium *for any queueing discipline, i.e., irrespective of the service order of the customers*. The only required conditions are that statistical equilibrium can be achieved, that is, $\rho < 1$, and that the server is work-conserving.

Before closing this section, we mention a different approach that leads to the same result (and to something more) for the analysis of the $M/G/1$. Instead of considering the EMC at departure times, we could enlarge the state space of the system, including the residual service time of the customer under service (if any), along with the number of customers.

Let us consider the sequence of arrival times $\{\tau_n\}_{n \geq 1}$. Since the arrivals follow a Poisson process, thanks to the PASTA property, we know that the probability distribution seen by arrivals is the same as the probability distribution sampled at a general time in statistical equilibrium.

We define the joint random variable (Q_n, H_n), where $Q_n = Q(\tau_{n,-})$ and H_n is the time needed to complete the current service, if any. It is $Q_n \in \mathbb{Z}^+$ and $H_n \in \mathbb{R}^+$. We let $H_n = 0$ in case it is $Q_n = 0$.

The couple $(Q_n, H_n)_{n \geq 1}$ form a Markov sequence on the state space $\mathbb{Z}^+ \times \mathbb{R}^+$. The sequence converges to statistical equilibrium if and only if $\rho < 1$. In that case, it admits a stationary joint probability distribution:

$$F_j(x) = P(Q = j, H \leq x), \quad j = 1, 2, \dots; \ x \geq 0. \tag{4.26}$$

with $F_0(x) = P(Q = 0)$, $\forall x \geq 0$. We define the double transform of the joint probability distribution:

$$\Phi(z, s) = \sum_{j=0}^{\infty} z^j \int_0^{\infty} e^{-sx} \, dF_j(x) \tag{4.27}$$

In [198] it is shown that

$$\Phi(z, s) = 1 - \rho + \frac{(1 - \rho)z\lambda(1 - z)}{z - \varphi_X(\lambda - \lambda z)} \frac{\varphi_X(s) - \varphi_X(\lambda - \lambda z)}{s - \lambda(1 - z)} \tag{4.28}$$

Equation (4.24) is obtained by letting $s = 0$ in (4.28). The joint probability distribution of the joint random variable (Q, H) and its double transform do not depend on the service order adopted by the queue. On the contrary, the waiting time *does* depend on the queueing discipline. In the rest of this chapter we consider only FCFS. Other queueing disciplines are studied along with priorities in Chapter 6.

4.2.2 Waiting Time

According to Little's law, the mean system time $E[S] = E[W] + E[X]$ can be expressed as $E[S] = E[Q]/\lambda$, so the mean waiting time is given by $E[W] = E[Q]/\lambda - E[X]$, that is:

$$E[W] = \frac{\lambda E[X^2]}{2(1 - \rho)} = E[X] \frac{\rho(1 + \sigma_X^2/E[X]^2)}{2(1 - \rho)} \tag{4.29}$$

This is the celebrated Pollaczek-Khinchine formula for the mean waiting time of the $M/G/1$ queue. As expected, $E[W]$ grows unboundedly when ρ approaches 1, i.e., the server gets saturated. Interestingly, the *mean* waiting time grows with the *variability* of service times, namely proportionally to the coefficient of variation (COV) of the service time $C_X \equiv \sigma_X/E[X]$. The best that we can do to limit the mean waiting time is to design for deterministic service times, if possible.

More deeply, waiting time in a queue is intimately related to *stochastic variability* of service and arrival times of customers. If both those time series are deterministic, i.e., inter-arrival and service times are constant with values T_0 and X_0, respectively, the condition that equilibrium be achievable implies that it must be $X_0 < T_0$, i.e., $\rho = X_0/T_0 < 1$. Since arrivals and service times are strictly deterministic, there is no queueing and waiting time is exactly 0 for every customer. This holds in spite of the fact that ρ can take any value, however close to 1. On the contrary, no matter how small ρ is, eq. (4.29) tells us that Poisson arrivals at a single-server queue will suffer some delay, the bigger the closer ρ to 1, and, for a given ρ, the bigger the more variable service times are.

The expression of the mean waiting time in eq. (4.29) is *independent* of the queue discipline. The order of service does not affect in any way the value of the mean waiting time, provided that the service times do not depend on the queue discipline and that the server is work-conserving.

For example, Figure 4.1 plots the mean waiting time, normalized with respect to the mean service time, as a function of ρ for three cases of service times: Deterministic ($C_X = 0$, dash-dot line), negative exponential ($C_X = 1$, dashed line), and a PDF estimated from measured data ($C_X = 1.247$, solid line).

The third value of C_X is taken from traffic measurement on an ethernet LAN. In that case, the service time is given by $X = L/C$, where L is the frame length. Frame lengths range from 64 up to 1518 bytes. The PDF of the frame length has few peaks and most values have negligible probabilities. Table 4.1 reports the 10 frame length values that have a probability greater than 0.01.

Contrary to the mean, the PDF of the waiting time W of the $M/G/1$ queue does depend on the service policy. In the following we refer to FCFS discipline; then the Laplace transform of the probability distribution of W can be found with the

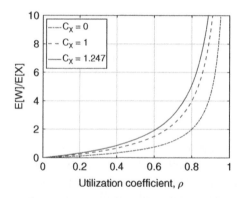

Figure 4.1 Normalized mean waiting time E[W]/E[X] as a function of ρ for three PDFs of the service times: Deterministic (dash-dot line); Negative exponential (dashed line); PDF estimated from LAN measurements (solid line).

Table 4.1 Most probable frame lengths from data captured on an ethernet LAN.

ℓ [bytes]	$P(L = \ell)$	ℓ [bytes]	$P(L = \ell)$
162	0.2031	66	0.0455
174	0.1839	142	0.0293
1090	0.1738	150	0.0152
64	0.1051	90	0.0148
1518	0.0727	570	0.0125

following argument. The number of customers left behind by a departing customer is equal to the number of customers arriving during the system time S of the departing customer. This is a consequence of the order of service: customers left behind by a departing customer are all those and only those that have arrived after it. Therefore, $\phi_Q(z) = \varphi_S(\lambda - \lambda z)$. Moreover, $S = W + X$, the waiting time W and the service time X being independent random variables. Hence

$$\varphi_S(\lambda - \lambda z) = E[e^{-S(\lambda - \lambda z)}] = E[e^{-(W+X)(\lambda - \lambda z)}] = \varphi_W(\lambda - \lambda z)\varphi_X(\lambda - \lambda z) \quad (4.30)$$

By reminding that $\phi_A(z) = \varphi_X(\lambda - \lambda z)$, putting this together with eq. (4.24), we find

$$\varphi_W(\lambda - \lambda z)\varphi_X(\lambda - \lambda z) = \phi_Q(z) = \frac{(1 - \rho)(z - 1)\varphi_X(\lambda - \lambda z)}{z - \varphi_X(\lambda - \lambda z)} \quad (4.31)$$

or

$$\varphi_W(s) = \frac{(1 - \rho)(-s/\lambda)}{1 - s/\lambda - \varphi_X(s)} = \frac{(1 - \rho)s}{s - \lambda + \lambda\varphi_X(s)} \quad (4.32)$$

This is also known as the Pollaczek-Khinchine formula for the Laplace transform of the waiting time PDF. The mean waiting time $E[W]$ can be found from $\varphi_W(s)$ by deriving with respect to s: $E[W] = -\varphi'_W(0)$. The same is true for all other moments. For example, we use the Laplace transform of the PDF of waiting time to derive the second moment of W. Instead of deriving (4.32) twice with respect to s (which may turn out to be quite cumbersome), we expand in a series of s and pick the coefficient of the s^2 term. In general, we know that the expansion of the Laplace transform $\varphi_X(s)$ of the PDF of the random variable X is $\varphi_X(s) = 1 - sE[X] + \frac{1}{2}s^2 E[X^2] - \frac{1}{6}s^3 E[X^3] + O(s^4)$ as $s \to 0$. Applying this expansion to eq. (4.32), we get

$$\varphi_W(s) = \frac{(1 - \rho)s}{s - \lambda + \lambda - s\rho + \frac{\lambda E[X^2]}{2}s^2 - \frac{\lambda E[X^3]}{6}s^3 + O(s^4)}$$

$$= \frac{1}{1 + \frac{\lambda E[X^2]}{2(1-\rho)}s - \frac{\lambda E[X^3]}{6(1-\rho)}s^2 + O(s^3)}$$

$$= \sum_{k=0}^{\infty} \left(-sE[W] + \frac{s^2}{2}\frac{\lambda E[X^3]}{3(1 - \rho)} + O(s^3) \right)^k$$

$$= 1 - sE[W] + \frac{s^2}{2}\left(\frac{\lambda E[X^3]}{3(1 - \rho)} + 2(E[W])^2 \right) + O(s^3)$$

Hence

$$E[W^2] = \frac{\lambda E[X^3]}{3(1 - \rho)} + 2(E[W])^2 = \frac{\lambda E[X^3]}{3(1 - \rho)} + \frac{(\lambda E[X^2])^2}{2(1 - \rho)^2} \quad (4.33)$$

Note that the second moment diverges faster than the mean waiting time as $\rho \to 1$. The squared coefficient of variation (SCOV) of the waiting time is:

$$C_W^2 = \frac{E[W^2]}{(E[W])^2} - 1 = 1 + \frac{4}{3}\frac{E[\hat{X}^3]}{(E[\hat{X}^2])^2}\frac{1 - \rho}{\rho} \quad (4.34)$$

where $\hat{X} \equiv X/E[X]$ is the service time normalized so that it has mean equal to 1.

As the utilization factor grows, the SCOV of the waiting time tends to 1, while for low utilization levels it can be significantly bigger than 1. The second moment of the system time S can be derived from its variance. The variance of the system time is $\sigma_S^2 = \sigma_W^2 + \sigma_X^2$.

The probability distribution of the waiting time W has a non-null mass concentrated at 0. Since an arriving customer finds the queue empty with probability $1 - \rho$, that is the probability that it waits exactly 0 before receiving service. The probability mass at 0 can be checked analytically as follows. We exploit the initial value theorem. Given an integrable function $f(t)$ with Laplace transform $\varphi(s)$, the theorem states that $\lim_{t \to 0_+} f(t) = \lim_{s \to \infty} s\varphi(s)$, provided the limits exist. The Laplace transform of the CDF of the waiting time W is $\Phi_W(s) = \varphi_W(s)/s$. Applying the initial value theorem to $\Phi_W(s)$ we find

$$F_W(0_+) = \lim_{t \to 0_+} F_W(t) = \lim_{s \to \infty} s\Phi_W(s) = \lim_{s \to \infty} \varphi_W(s) = 1 - \rho \qquad (4.35)$$

where we have used the expression of $\varphi_W(s)$ in eq. (4.32) in the last passage.

To conclude this section, we mention a result on the unfinished work of the $M/G/1$ queue, known as the Takács integro-differential equation (e.g., see [130, Ch. 5]).

First, we recall that the unfinished work $U(t)$ is the amount of time required to the server to clear the entire backlog of the queue at time t, provided no more customers arrive. Thanks to the PASTA property, this is also the system time spent by a customer arriving at the queue at time t, if FCFS discipline is used.

Let

$$F(x, t) = P(U(t) \le x | U(0) = x_0) \qquad (4.36)$$

It can be shown that $F(x, t)$ satisfies the following integro-differential equation:

$$\frac{\partial F(x, t)}{\partial t} = \frac{\partial F(x, t)}{\partial x} - \lambda F(x, t) + \lambda \int_0^\infty F_X(x - y) \frac{\partial F(y, t)}{\partial y} \, dy \qquad (4.37)$$

where $F_X(t)$ is the CDF of the service times.

The solution of this equation can be written formally in terms of transforms. Let us define the double transform of $F(x, t)$ as follows:

$$\phi(r, s) = \int_0^\infty e^{-st} \, dt \int_{0^-}^\infty e^{-rx} d_x F(x, t) \qquad (4.38)$$

The following expression of $\phi(r, s)$ can be derived

$$\phi(r, s) = \frac{(r/\eta)e^{-\eta x_0} - e^{-rx_0}}{\lambda \varphi_X(r) - \lambda - s + r} \qquad (4.39)$$

where $\eta = \eta(s)$ is the unique root of $\lambda \varphi_X(r) - \lambda - s + r = 0$, for a given s, in the region $\text{Re}[s] > 0$, $\text{Re}[r] > 0$ [27].

4.2.3 Busy Period and Idle Time

We study the busy period duration. A busy period Y starts with the arrival of a customer that finds the queue empty and stops with the first ensuing departure of a customer that leaves the queue empty. Let X_1 be the service time of the first customer of a busy period and let A_1 be the number of arrivals during that service time. Then

$$\varphi_Y(s) = E[e^{-sY}] = \sum_{n=0}^{\infty} \int_0^{\infty} f_{X_1}(t)E[e^{-sY}|A_1 = n, X_1 = t]P(A_1 = n|X_1 = t) \, dt$$

(4.40)

The busy period, conditional on $A_1 = n$, can be expressed as $Y = X_1 + Y_1 + \cdots + Y_n$, where the sub-busy periods Y_j are i.i.d. with the same probability distribution of Y. This is obtained by considering that the duration of a busy period *does not depend on the order that customers in the queue are served*, i.e., it is independent of the server policy, as long as it is work-conserving. Then, we can rearrange the customers $C_1, \ldots, C_n$ that arrive during the service time of the customer that starts the busy period: let C_1 be served first, along with all customers arriving during the service time of C_1 and so on, until the sub-busy period associated with C_1 is exhausted. Soon after that, C_2 is taken care of, along with all customers arriving during the sub-busy period started by C_2. We proceed this way until the turn of C_n comes and all customers arrive during the relevant sub-busy period. So

$$E[e^{-sY}|A_1 = n, X_1 = t] = E[e^{-s(t+Y_1+\cdots+Y_n)}] = e^{-st}[\varphi_Y(s)]^n$$

(4.41)

Putting together the last two equalities, we find

$$\begin{aligned}
\varphi_Y(s) &= \sum_{n=0}^{\infty} \int_0^{\infty} f_{X_1}(t)e^{-st}[\varphi_Y(s)]^n \frac{(\lambda t)^n}{n!} e^{-\lambda t} \, dt \\
&= \int_0^{\infty} f_{X_1}(t)e^{-t(s+\lambda-\lambda\varphi_Y(s))} \, dt \\
&= \varphi_X(s + \lambda - \lambda\varphi_Y(s))
\end{aligned}$$

(4.42)

The value of $\varphi_Y(s)$ is found as the unique solution of eq. (4.42) having modulus less than or equal to 1. In general, no closed-form expression of the PDF of the busy period can be found from eq. (4.42). In the special case of $G = M$, i.e., negative exponential service times, we have $\varphi_X(s) = \mu/(s + \mu)$ and hence

$$\varphi_Y(s) = \frac{\mu}{s + \lambda - \lambda\varphi_Y(s) + \mu}$$

(4.43)

This leads to a second-order algebraic equation in the unknown $\varphi_Y(s)$. Selecting the proper root (the one with modulus ≤ 1), we find:

$$\varphi_Y(s) = \frac{\lambda + \mu + s - \sqrt{(\lambda + \mu + s)^2 - 4\lambda\mu}}{2\lambda}$$

(4.44)

The inverse transform of (4.44) is known and gives finally the PDF of the busy period for the $M/M/1$ queue:

$$f_Y(y) = \frac{1}{y\sqrt{\rho}} e^{-(\lambda+\mu)y} I_1(2y\sqrt{\lambda\mu}), \quad y \geq 0, \tag{4.45}$$

where $I_1(\cdot)$ is the modified Bessel function of the first kind of order 1.

The mean duration of the busy period is easily obtained by derivation:

$$E[Y] = -\varphi_Y'(0) = -\varphi_X'(\lambda - \lambda\varphi_Y(0))(1 - \lambda\varphi_Y'(0)) = E[X](1 + \lambda E[Y]) \tag{4.46}$$

and it follows (remember that $\varphi_Y(0) = 1$):

$$E[Y] = \frac{E[X]}{1 - \rho} \tag{4.47}$$

The mean busy period of the $M/G/1$ queue can be found directly from first principles, by exploiting eq. (2.6). Poisson arrivals are memoryless, so the mean idle time is simply the mean residual arrival time, i.e., $E[I] = 1/\lambda$. Substituting into eq. (2.6), we find easily the result in eq. (4.47).

Note that the probability distribution of the busy period does not depend on the queue discipline. It holds in general, provided the server is work-conserving and $\rho < 1$. Similarly, it can be found that:

$$E[Y^2] = \frac{E[X^2]}{(1 - \rho)^3} \tag{4.48}$$

$$\sigma_Y^2 = \frac{\sigma_X^2 + \rho E[X]^2}{(1 - \rho)^3} \tag{4.49}$$

Following a reasoning similar to the one for the busy period duration, we can find the generating function of the probability distribution of the number of customers served during a busy period, M. A busy period starts with a customer arriving at an empty queue. During the service time of this first customer, $A_1 = n$ customers arrive. Each of them gives rise to a sub-busy period with the same probability distribution of the full one; hence the number of customers served during the i-th sub-busy period, M_i, has the same probability distribution as M, for $i = 1, \ldots, n$. Then, we have $M = 1 + M_1 + \cdots + M_n$ and

$$\phi_M(z) = E[z^{-M}] = \sum_{n=0}^{\infty} E[z^{-M} | A_1 = n] P(A_1 = n)$$

$$= \sum_{n=0}^{\infty} E[z^{-(1+M_1+\cdots+M_n)} | A_1 = n] a_n$$

$$= \sum_{n=0}^{\infty} z \prod_{i=1}^{n} E[z^{-M_i}] \int_0^{\infty} \frac{(\lambda t)^n}{n!} e^{-\lambda t} f_X(t) \, dt$$

$$= z \int_0^\infty e^{-\lambda t} f_X(t) \sum_{n=0}^\infty E[z^{-M}]^n \frac{(\lambda t)^n}{n!} dt$$

$$= z \int_0^\infty e^{-\lambda t[1-\phi_M(z)]} f_X(t) \, dt = z \varphi_X(\lambda - \lambda \phi_M(z)) \tag{4.50}$$

Again, this is a functional equation and we ought to seek the solution with modulus no bigger than 1. Also in this case the special case of the $M/M/1$ queue offers an explicit solution. In that case

$$\phi_M(z) = \frac{z\mu}{\lambda - \lambda \phi_M(z) + \mu} \tag{4.51}$$

hence

$$\phi_M(z) = \frac{1+\rho}{2\rho} \left[1 - \sqrt{1 - \frac{4\rho z}{(1+\rho)^2}} \right] \tag{4.52}$$

whose inverse transform is known and given by (see [130, Ch. 5, p. 218]):

$$P(M = n) = \frac{1}{n} \binom{2n-2}{n-1} \frac{\rho^{n-1}}{(1+\rho)^{2n-1}}, \quad n \geq 1. \tag{4.53}$$

It is possible to get an explicit expression also for another specific case, the $M/D/1$ queue, where service times are constant. In that case $\varphi_X(s) = e^{-s/\mu}$ and $\phi_M(z) = z e^{\rho(\phi_M(z)-1)}$. Amazingly, an explicit solution can be obtained for this functional equation, even if it is in series form:

$$\phi_M(z) = \sum_{n=1}^\infty \frac{(n\rho)^{n-1}}{n!} e^{-n\rho} z^n \tag{4.54}$$

This is just what we need! Remembering the very definition of moment generating function as a power series with coefficients equal to the probability distribution, we obtain readily

$$P(M = n) = \frac{(n\rho)^{n-1}}{n!} e^{-n\rho}, \quad n \geq 1. \tag{4.55}$$

Since service times are constant, we have $Y = M/\mu$. Then, we can also easily deduce the CDF of the busy period duration for the $M/D/1$ queue:

$$F_Y(y) = P(Y \leq y) = P(M \leq \mu y) = \sum_{n=1}^{\lfloor \mu y \rfloor} \frac{(n\rho)^{n-1}}{n!} e^{-n\rho} \tag{4.56}$$

where $\lfloor x \rfloor$ denotes the greatest integer not exceeding x (floor function).

Since arrivals of the M/G/1 queue follow a Poisson process, the idle time I has a negative exponential PDF with parameter λ, i.e., $G_I(t) = P(I > t) = e^{-\lambda t}$.

4.2.4 Remaining Service Time

In a work-conserving single server queue, the server may stay idle only if the queue is empty. Otherwise, a random observer (or an arrival, which for the $M/G/1$ model is the same), looking at the queue, will find the server busy. A question that turns out to be useful in many cases is the characterization of the random process defined as the remaining service time, $R(t)$, i.e., the amount of service that the server must deliver to get done with the customer under service at time t, if any. Clearly, if the queue is empty at time t, it is $R(t) = 0$. If the queue is not empty, let X denote the service time of the customer under service. Then, it must be $R(t) \leq X$. At the very moment that the customer enters service, say time t_0, it is $R(t_0) = X$. After t_0, $R(t)$ decreases at a rate of 1 s/s, until it hits zero at time $t_0 + X$. Then, the customer under service leaves the queue and a new customer is taken into service, if there are waiting customers.

An example of the time behavior of $R(t)$ is illustrated in Figure 4.2. It has a saw-tooth behavior during busy periods, while it stays at zero during idle times.

Notice that service completion epochs are regeneration points for $R(t)$, given that service times are i.i.d. random variables.

We can obtain the mean value of $R(t)$ from first principles. Let us consider a time interval $[0, t]$ during statistical equilibrium, where $Q(0) = 0$, and let $N(t)$ denote the number of customers served until time t. Let also X_i, $i = 1, \dots, N(t)$, be the service times of these $N(t)$ customers. Then, it is

$$\sum_{i=1}^{N(t)} \frac{1}{2} X_i^2 \leq \int_0^t R(u) \, du \leq \sum_{i=1}^{N(t)+1} \frac{1}{2} X_i^2 \tag{4.57}$$

Recalling that $\lim_{t \to \infty} N(t) = \infty$ and $\lim_{t \to \infty} N(t)/t = \lambda$, we have

$$\lim_{t \to \infty} \frac{1}{t} \sum_{i=1}^{N(t)} \frac{1}{2} X_i^2 = \frac{1}{2} \lim_{t \to \infty} \frac{N(t)}{t} \frac{1}{N(t)} \sum_{i=1}^{N(t)} X_i^2 = \frac{\lambda E[X^2]}{2} \tag{4.58}$$

Analogously, it can be shown that

$$\lim_{t \to \infty} \frac{1}{t} \sum_{i=1}^{N(t)+1} \frac{1}{2} X_i^2 = \frac{\lambda E[X^2]}{2} \tag{4.59}$$

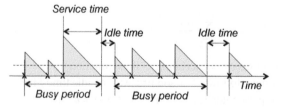

Figure 4.2 Example of realization of the remaining service time process of a single server queue.

Then, the time average of $R(t)$ is sandwiched into two sequences, having the same limit and therefore

$$\lim_{t \to \infty} \frac{1}{t} \int_0^t R(u) \, du = \frac{\lambda E[X^2]}{2} = \rho \frac{E[X^2]}{2E[X]} \tag{4.60}$$

The rightmost side of this equation reveals an intuitive interpretation of the mean remaining service time: with probability $1 - \rho$ we find the server idle and hence $R = 0$. With probability ρ we find it busy and in that case, since we behave like a random observer, the remaining service time is nothing else than the residual service time of the customer under service. In other words, the remaining service time, *conditional on a customer being under service*, is the residual service time.

We can exploit this view of the random process $R(t)$ to derive its probability distribution. Let $\tilde{X}$ be the residual service time random variable; then

$$F_R(x) = P(R(t) \le x) = (1 - \rho)P(R(t) \le x \mid idle) + \rho P(R(t) \le x \mid busy)$$

$$= 1 - \rho + \rho P(\tilde{X} \le x) = 1 - \rho + \rho \int_0^x \mu G_X(v) \, dv , \quad x \ge 0. \tag{4.61}$$

This CDF has a jump of size $1 - \rho$ at $x = 0$. The corresponding PDF can be written by using a Dirac delta function:

$$f_R(x) = (1 - \rho)\delta(x) + \lambda G_X(x) \tag{4.62}$$

Conditional on the server being busy, we obtain the random variable R_b that is just the residual service time. Hence $f_{R_b}(t) = \mu G_X(t)$. The k-th moment can be easily obtained integrating by parts the definition formula:

$$E[R_b^k] = \int_0^\infty t^k f_{R_b}(t) \, dt = \mu \int_0^\infty t^k G_X(t) \, dt$$

$$= \mu \left(\left[\frac{t^{k+1}}{k+1} G_X(t) \right]_0^\infty + \frac{1}{k+1} \int_0^\infty t^{k+1} f_X(t) \, dt \right) = \frac{E[X^{k+1}]}{(k+1)E[X]} \tag{4.63}$$

The renewal paradox of Section 3.4.1 corresponds to the case $k = 1$.

4.2.5 Output Process

The time interval between two successive departures is either a service time, if the departing customer leaves behind a nonempty queue, or it is the sum of a residual inter-arrival time and of a service time, if the departing customer leaves an empty queue. The probability of the first event is $1 - x_0 = \rho$. In the second case, the time elapsing between the departure of the customer that leaves an empty queue and the subsequent departure is the sum of two independent random variables: (i) an idle time; (ii) a service time.

Then, by denoting with D the random variable defined as the inter-departure time, its CDF is

$$F_D(t) = \rho F_X(t) + (1 - \rho) \int_0^t \lambda e^{-\lambda u} F_X(t - u)\, du \tag{4.64}$$

In terms of Laplace transform, we have

$$\varphi_D(s) = \varphi_X(s) \left(\rho + (1 - \rho)\frac{\lambda}{\lambda + s} \right) = \varphi_X(s)\frac{\lambda + \rho s}{\lambda + s} \tag{4.65}$$

Plugging into this equation the expression $\rho = \lambda/\mu$, we get

$$\varphi_D(s) = \varphi_X(s)\frac{\mu + s}{\mu}\frac{\lambda}{\lambda + s} \tag{4.66}$$

Inter-departure times of the $M/G/1$ queue do not form a sequence of i.i.d. random variables in general, i.e., the output process is not a renewal process apart from the special cases of $M/M/1$ and $M/D/1/0$ queues.

An interesting result is found in case of negative exponential PDF of the service time ($M/M/1$ queue). In that case eq. (4.66) reduces to $\varphi_D(s) = \lambda/(\lambda + s)$, that is to say, $\varphi_D(s)$ coincides with the Laplace transform of inter-arrival time PDF. The $M/M/1$ queueing system is "transparent," meaning that it does not change anything of the input random process. At the output we see a Poisson process with the *same* characteristics as the one that the queue receives as input. This means that there is no way to infer anything about the queueing system (e.g., the service rate μ) by observing *only* the departure times at the output of the queue. This property has been formally established in Burke's theorem [48].

Theorem 4.3 For an $M/M/m$ queuing system in equilibrium, if λ denotes the mean arrival rate of the input Poisson process, then:

(a) the departure process is Poisson with mean rate λ;
(b) the number of customers in the system at time t is independent of the sequence of departures prior to time t.

Specifically property (b) asserts that *we can learn nothing of the current state of an M/M queueing system, by observing its output*. Notice that the theorem has been proved for any number m of servers, including the case $m = \infty$. The key point is the memoryless property of *both* arrival and service processes.

As a matter of fact, it can be shown [94] that the maximum likelihood estimator of the service rate, given an observation interval $[0, t)$, is $\hat{\mu} = n_s(t)/T_s(t)$, where $n_s(t)$ is the number of customers served in $[0, t)$ and $T_s(t)$ is the amount of time the server is busy in $[0, t)$. The former can be observed by just counting the departing customers. The latter cannot be measured based on the departing times only. Either something is known about the "structure" of the service times, or the activity of the server should be monitored somehow.

Example 4.1 The M/G/1 model is applied to the output buffer of a packet switch. Departing customers are simply the packets sent on the output line. Service is the transmission of the packets on the link. So, $X = L/C$, where L is the packet length and C is the link capacity. If C is constant, service time variability is only due to packet length variability. In this case, a packet capture with timestamping on the link, as possible, e.g., with Wireshark, yields both departure times *and* service times, since the packet length can be read in the packet header and the link capacity is assumed to be known. Then, everything in eq. (4.66) can be reconstructed, given that λ can be estimated as the average output packet rate and the service time PDF can be estimated by a collection of packet lengths that allows an estimate of the packet length PDF.

4.2.6 Evaluation of the Probabilities $\{a_k\}_{k \in \mathbb{Z}}$

We rewrite here eq. (4.9):

$$a_k = \int_0^\infty \frac{(\lambda t)^k}{k!} e^{-\lambda t} \, dF_X(t), \quad k \geq 0. \tag{4.67}$$

In case of deterministic service times, $F_X(x) = I(x \geq 1/\mu)$, where $\mu = 1/E[X]$ and $I(E)$ is the indicator function of the event E. Then,

$$a_k = \frac{A_o^k}{k!} e^{-A_o}, \quad k \geq 0, \tag{4.68}$$

and $\phi_A(z) = e^{A_o(z-1)}$, where we define $A_o \equiv \lambda/\mu$ to be the average offered traffic. The probabilities $\{a_k\}_{k \in \mathbb{Z}}$ can be computed iteratively according to

$$\begin{cases} a_0 = e^{-A_o} \\ a_k = a_{k-1} \frac{A_o}{k}, \quad k \geq 1 \end{cases} \tag{4.69}$$

Another case where computation is simple and we need not resort to numerical integration of eq. (4.67) is when service times follow a gamma PDF, namely:

$$f_X(x) = \frac{(\alpha x)^{\beta-1}}{\Gamma(\beta)} \alpha e^{-\alpha x}, \quad x > 0, \tag{4.70}$$

where α and β are two positive real numbers and $\Gamma(\cdot)$ is the Euler gamma function defined by

$$\Gamma(z) = \int_0^\infty u^{z-1} e^{-u} \, du \tag{4.71}$$

for $z > 0$. Note that β is a nondimensional number, while α has dimension that is the reciprocal of the service time X. The Laplace transform of the gamma PDF is just

$$\varphi_X(s) = \left(\frac{\alpha}{\alpha + s} \right)^\beta \tag{4.72}$$

from which it is easily derived that $\mathrm{E}[X] = -\varphi_X'(0) = \beta/\alpha$ and $\mathrm{E}[X^2] = \varphi_X''(0) = \beta(\beta+1)/\alpha^2$. The coefficient of variation is $C_X = \sigma_X/\mathrm{E}[X] = 1/\sqrt{\beta}$. Given the desired value of the mean service time $1/\mu = \beta/\alpha$ and the value of C_X, the PDF parameters can be identified as $\beta = 1/C_X^2$ and $\alpha = \mu\beta$.

Thanks to the property $\Gamma(\beta+1) = \beta\Gamma(\beta)$, holding[2] for any real positive β, it can be checked that

$$a_k = \frac{\prod_{j=0}^{k-1}(\beta+j)}{k!}\left(\frac{\alpha}{\alpha+\lambda}\right)^{\beta}\left(\frac{\lambda}{\alpha+\lambda}\right)^k, \qquad k \geq 0, \tag{4.73}$$

where it is intended that $\prod_{j=\ell}^{u} \equiv 1$ for $\ell > u$. Also in this case, it is possible to derive a simple iteration for the numerical evaluation of the $\{a_k\}_{k\geq 0}$:

$$\begin{cases} a_0 = \left(\dfrac{\beta}{\beta+A_o}\right)^{\beta} \\[2ex] a_k = a_{k-1}\dfrac{\beta+k-1}{k}\left(\dfrac{A_o}{\beta+A_o}\right), \qquad k \geq 1 \end{cases} \tag{4.74}$$

where $A_o = \lambda/\mu$. The expression of the generating function of A is simply

$$\phi_A(z) = \left(\frac{\alpha}{\alpha+\lambda-\lambda z}\right)^{\beta} = \left(1 + \frac{A_o}{\beta} - \frac{A_o}{\beta}z\right)^{-\beta} \tag{4.75}$$

4.3 The $M/G/1/K$ Queue

For the finite size[3] $M/G/1$ queue we can resort to the EMC approach as well. Again, the departure times ar regeneration points of the process $Q(t)$. We use the same notation as in the infinite queue length case, except that now K denotes the maximum number of customers that can be standing waiting in the queue. To prevent ambiguity, we use the argument or superscript K to denote quantities relevant to the $M/G/1/K$ queue and ∞ for the $M/G/1$ queue.

Let $\{t_{a,n}\}_{n\in\mathbb{Z}}$ and $\{t_{d,n}\}_{n\in\mathbb{Z}}$ denote arrival and departure times of customers at the queue, respectively, and let

$$q_{a,k} = \lim_{n\to\infty} \mathcal{P}(Q(t_{a,n-}) = k), \quad k = 0, 1, \ldots, K+1$$

$$q_{s,k} = \lim_{n\to\infty} \mathcal{P}(Q(t_{a,n-}) = k | Q(t_{a,n-}) < K+1), \quad k = 0, 1, \ldots, K$$

---

2 It can be easily established from the definition of $\Gamma(\beta)$, by integrating by parts.
3 Recall that the convention on the notation is that K indicates the size of the *waiting line*, sometimes also referred to as the *buffer*. Therefore, up to $K + m$ customers can be accommodated in the queue, if there are m servers. It is $K \geq 0$.

$$x_k = \lim_{n \to \infty} P(Q(t_{d,n+}) = k) , \quad k = 0, 1, \dots, K$$

$$p_k = \lim_{t \to \infty} P(Q(t) = k) , \quad k = 0, 1, \dots, K+1$$

be the limiting probabilities seen by an arrival, by an arrival that joins the queue (i.e., it does not find the queue full) and at a generic time point of the statistical equilibrium. From Theorem 2.3 we know that, if those probability distributions exist, then

$$p_k = q_{a,k} , \quad k = 0, 1, \dots, K+1$$
$$x_k = q_{s,k} , \quad k = 0, 1, \dots, K$$
$$p_k = \frac{x_k}{x_0 + A_o} , \quad k = 0, 1, \dots, K; \qquad p_{K+1} = 1 - \frac{1}{x_0 + A_o} \tag{4.76}$$

where we define the mean offered traffic intensity of the $M/G/1/K$ queue as $A_o = \lambda/\mu$. We reserve the notation ρ to the utilization coefficient of the server. Since customer loss is possible in the $M/G/1/K$ queue, $A_o > \rho$, whereas in the $M/G/1$ queue they coincide.

4.3.1 Exact Solution

The evolution of the sequence Q_n of the number of customers left behind by the n-th departing one is written in this case as:

$$Q_{n+1} = \min\{K, \max\{0, Q_n - 1\} + A_{n+1}\} \tag{4.77}$$

We recall that the random variable A_{n+1}, the number of arrivals in the $(n+1)$-th service time, does not depend on n, since service times are i.i.d. random variables. Therefore, eq. (4.77) defines a homogeneous Markov chain. The one-step transition matrix of the Markov chain Q_n is almost the same as in the infinite queue case, except that the set of states is finite. It is $\{0, 1, \dots, K\}$, hence the matrix has $K+1$ rows and columns:

$$\mathbf{P} = \begin{bmatrix} a_0 & a_1 & a_2 & \cdots & a_{K-1} & \tilde{a}_K \\ a_0 & a_1 & a_2 & \cdots & a_{K-1} & \tilde{a}_K \\ 0 & a_0 & a_1 & \cdots & a_{K-2} & \tilde{a}_{K-1} \\ 0 & 0 & a_0 & \cdots & a_{K-3} & \tilde{a}_{K-2} \\ \cdots & \cdots & \cdots & \cdots & \cdots & \cdots \\ 0 & 0 & 0 & \cdots & a_0 & \tilde{a}_1 \end{bmatrix} \tag{4.78}$$

where $a_k = P(A = k)$ and $\tilde{a}_k = P(A \geq k) = \sum_{j=k}^{\infty} a_j$, for $k \geq 0$.

The EMC of the $M/G/1/K$ queue is irreducible and aperiodic, since the a_k's are positive[4]. Since the EMC is finite, irreducibility and aperiodicity guarantee that it is also ergodic and the limiting probabilities $\{x_k\}_{k=0,1,\ldots,K}$ exist. We recall that the $M/G/1$ queue EMC is ergodic under the condition $\lambda < \mu$, i.e., that the mean service time be less than the mean inter-arrival time. For a single-server queueing system this is a natural requisite to maintain stability hence to be able to reach a statistical equilibrium. For finite values of K the condition $\lambda < \mu$ is no more necessary. The finite room of the waiting line "protects" the queue from overload, since excess arrivals result in rejected customers. Therefore, statistical equilibrium can be reached anyway, even if $E[X] \geq E[T]$. Even though statistical equilibrium is not at stake, it is in any case pathological to let $E[X] \geq E[T]$, since clearly the server cannot keep up with customer demand and the probability of customer rejection achieves unbearably high values. To avoid compromising the quality of service either the system must be redesigned by improving its service capacity or the customer demand must be limited, by means of some congestion control mechanism.

The matrix $\mathbf{P}$ can be obtained directly from the evolution eq. (4.77), by the following derivation:

$$x_k(n+1) = P(Q_{n+1} = k) = \sum_{h=0}^{K} x_h(n)P(Q_{n+1} = k|Q_n = h)$$

$$= x_0(n)P(\min\{K, A_{n+1}\} = k) + \sum_{h=1}^{K} x_h(n)P(\min\{K, h-1+A_{n+1}\} = k)$$

$$= \begin{cases} x_0(n)a_k + \sum_{h=1}^{K} x_h(n)a_{k-h+1} & k < K \\ x_0(n)\tilde{a}_K + \sum_{h=1}^{K} x_h(n)\tilde{a}_{K-h+1} & k = K \end{cases}$$

where $a_k = P(A = k)$ and $\tilde{a}_k = P(A \geq k)$ for $k \geq 0$. As $n \to \infty$, the probability $x_k(n)$ tends to the limiting probability x_k that satisfies the equations:

$$x_k = \begin{cases} x_0 a_k + \sum_{h=1}^{K} x_h a_{k-h+1} & k < K \\ x_0 \tilde{a}_K + \sum_{h=1}^{K} x_h \tilde{a}_{K-h+1} & k = K \end{cases} \tag{4.79}$$

whence the matrix $\mathbf{P}$ can be recovered[5].

The probabilities x_k can be evaluated numerically by solving the linear system $\mathbf{xP} = \mathbf{x}$ and using the congruence equation $\mathbf{xe} = 1$, where $\mathbf{x} = [x_0, \ldots, x_K]$, and $\mathbf{e}$ is a column vector of 1's. More on numerical evaluation of the vector $\mathbf{x}$ will be

---

4 For a Markov chain having one-step transition probability matrix given by $\mathbf{P}$ in eq. (4.78) to be irreducible is suffices that $0 < a_0 < 1$.

5 It would be very easy to account for a different PDF of the first "service time" X' of each busy period, by just replacing the first row of this matrix with a different probability distribution $\{b_k\}_{k \in \mathbb{Z}}$, where b_k is the probability of k arrivals in the special first service time X'.

exposed in Section 4.4. In general, this problem is cast into the more general issue of finding the limiting probabilities of a finite state Markov chain. An excellent introductory treatment of this subject is given in [196, Ch. 10]

The limiting probability distribution of the number of customers found into the system at any time, p_k, can be calculated according to eq. (4.76).

As for the performance metrics, the loss probability coincides with the probability of finding the system blocked, i.e., no more room left. That is

$$P_L = p_{K+1} = 1 - \frac{1}{x_0 + \lambda/\mu} \tag{4.80}$$

The mean arrival rate of customers is $\Lambda_o = \lambda$. The mean arrival rate of customers that enter the system and then get their service is $\Lambda_s = \Lambda_o(1 - P_L) = \lambda(1 - p_{K+1}) = \lambda/(x_0 + A_o)$. The corresponding traffic intensities are $A_o = \lambda/\mu$ and $A_s = A_o(1 - P_L) = A_o/(x_0 + A_o)$. The latter is just the utilization coefficient of the server: in fact A_s is the mean number of busy servers in a queue at equilibrium. Since we have a single server in this case, then

$$\rho = A_s = \frac{A_o}{x_0 + A_o} \tag{4.81}$$

It is conversely

$$x_0 = A_o \left(\frac{1}{\rho} - 1 \right) = \frac{1 - \rho}{1 - P_L} \tag{4.82}$$

We can give expressions of the mean system time, waiting time, and of their respective PDFs (or better, the Laplace transforms of the PDFs). The mean queue length and the mean waiting line length are:

$$E[Q] = \sum_{k=0}^{K+1} k p_k = \frac{1}{x_0 + A_o} \sum_{k=1}^{K} k \, x_k + (K+1) p_{K+1} = K + 1 - \frac{K + 1 - \sum_{k=1}^{K} k \, x_k}{x_0 + A_o} \tag{4.83}$$

$$E[L] = E[Q] - \rho = E[Q] - \frac{A_o}{x_0 + A_o} \tag{4.84}$$

It is apparent that the queue length cannot exceed the upper limit $K + 1$. As the offered load A_o grows, the mean queue length gets closer and closer to this upper bound. Similarly, $\lim_{A_o \to \infty} E[L] = K$.

By applying Little's law, we obtain the mean system and waiting times as

$$E[S] = \frac{E[Q]}{\Lambda_s} = \frac{x_0 + A_o}{\lambda} E[Q] \tag{4.85}$$

$$E[W] = E[S] - E[X] \tag{4.86}$$

As A_o tends to infinity, x_0 tends to 0; hence we have the ratio $E[S]/E[Q]$ tending to $1/\mu$, i.e., as expected, the mean system time tends to $K + 1$ times the mean service time. Similarly, $\lim_{A_o \to \infty} E[W] = K/\mu$.

Although the key performance indicator expressions for the $M/G/1/K$ are less elegant than in case of the infinite queue length $M/G/1$ model and the results are less exciting, the finite queue length model can be even more useful in practice, especially if numerical evaluation of the model is required, besides qualitative understanding. First, since the state space is finite, the probability distribution can be computed by solving a linear system. Highly efficient and stable methods are known, as recalled above. Further, in many practical cases the room of the service system to accommodate customers is actually finite. The infinite queue length model is a handy approximation that can provide reasonable results as long as the loss probability is negligible in the considered context. However, in all those cases where the loss probability is a primary metric, the finite queue length model is a must.

Example 4.2   Let us consider a packet voice multiplexer, where voice packets are produced by a layered coding algorithm. Each voice packet is made up of m blocks, the j-th block containing the bits of the j-th layer. The multiplexer buffer has m thresholds $K_1 < K_2 < \cdots < K_m$. We let also $K_0 = 0$ and $K_m \equiv K$, where K is the multiplexer buffer size. It works as follows. Whenever the buffer content Q ranges in $(K_j, K_{j+1}]$, the least significant j layers are discarded before sending the packet, hence speeding up the multiplexer work. If layer j is granted ℓ_j bits in the packet format, then the packet transmission time for $Q \in (K_j, K_{j+1}]$ is $\overline{X}_j = (H + \sum_{i=j+1}^{m} \ell_i)/C$, where C is the multiplexer output link capacity and H is the packet header length, $j = 0, 1, \ldots, m - 1$. Given this coding format, the service time is a constant for each buffer operating interval.

If packets arrive at the multiplexer according to a Poisson process of rate λ, we can use an $M/G/1/K$-like model, where the EMC transition probabilities are $a_k^{(j)} = \frac{A_j^k}{k!} e^{-A_j}$, with $A_j = \lambda \overline{X}_j$, for the rows of the matrix $\mathbf{P}$ with indices r such that $K_j < r \leq K_{j+1}, j = 0, 1, \ldots, m - 1$. The 0-th row is identical to the row of index 1.

This is not a proper $M/G/1/K$ model, but the EMC being finite, there is no real difficulty obtaining the queue length PDF numerically, at least for limited values of K. The thresholds can be optimized with respect to a measure of the trade-off between quality of voice and delay through the multiplexer.

In the rest of this section we outline a result that allows the evaluation of the loss probability for large values of K, which are just those for which it might be more critical or anyway computationally expensive to find the entire probability distribution x_k.

4.3.2 Asymptotic Approximation for Large K

We start by establishing a relationship between the finite and infinite $M/G/1$ queue EMC probability distributions. Then, by exploiting this result, we develop an asymptotic property of the loss probability P_L as $K \to \infty$. To note the dependence of P_L on K we write $P_L(K)$ for the loss probability of the $M/G/1/K$ queue. In general, we use an argument K (∞) for all quantities referring specifically to the $M/G/1/K$ queue ($M/G/1$ queue).

An interesting connection can be established between the infinite queue length model $M/G/1$ probability distribution and the one of the $M/G/1/K$ queue. A comparison of the first K columns of the one-step transition probability matrices of the EMCs of the two queues shows that they are exactly the same. Then, the first K values of the vector $\mathbf{x}(\infty)$ of the EMC of the $M/G/1$ queue also form a solution of the first K equations of the system $\mathbf{x}(K)\mathbf{P}(K) = \mathbf{x}(K)$. The vector $\mathbf{x}(K)$ is the unique vector that satisfies those first K equations and the congruence equality $\mathbf{x}(K)\mathbf{e} = 1$, where $\mathbf{e}$ is a column vector of 1. Therefore, $\mathbf{x}(\infty)|_{k=0,\dots,K}$ is proportional to $\mathbf{x}(K)$. By renomalization, we conclude[6]

$$x_k(K) = \frac{x_k(\infty)}{\sum_{j=0}^{K} x_j(\infty)} = \frac{p_k(\infty)}{\sum_{j=0}^{K} p_j(\infty)}, \quad j = 0, 1, \dots, K. \tag{4.87}$$

The last equality is motivated by the fact that the queue length probability distribution at a general time coincides with that of the EMC in the $M/G/1$ queue. For the sake of notation, we let $\sigma_K = \sum_{j=K}^{\infty} p_j(\infty)$ in the following. Given eq. (4.76) and since

$$x_0(K) = \frac{p_0(\infty)}{1 - \sigma_{K+1}} = \frac{1 - A_o}{1 - \sigma_{K+1}} \tag{4.88}$$

we get

$$p_k(K) = \frac{x_k(K)}{x_0(K) + A_o} = \frac{p_k(\infty)/(1 - \sigma_{K+1})}{A_o + (1 - A_o)/(1 - \sigma_{K+1})} = \frac{p_k(\infty)}{1 - A_o\sigma_{K+1}}, \tag{4.89}$$

for $k = 0, 1, \dots, K$, and

$$p_{K+1}(K) = 1 - \frac{1}{x_0(K) + A_o} = \frac{(1 - A_o)\sigma_{K+1}}{1 - A_o\sigma_{K+1}} \tag{4.90}$$

These expressions prove that we can compute the EMC probability distribution and the queue length probability distribution at a general time for the $M/G/1$ queue as soon as we know the corresponding quantities for the $M/G/1/K$ queue and vice versa. The relationship between the EMC probabilities $x_k(K)$ and $x_k(\infty)$ shows that

$$c_k \equiv \frac{x_k(K)}{x_0(K)} = \frac{x_k(\infty)}{x_0(\infty)}, \quad x = 0, 1, \dots, K, \tag{4.91}$$

---

6 Note that it must be $A_o < 1$ for $\{p_k(\infty)\}_{k \geq 0}$ to exist.

for all $K \geq 0$. The ratios c_k do not depend on K and are defined for any value of A_o. From their definition, it follows that the vector $\mathbf{c} = [c_0, c_1, \dots]$ is the left eigenvector of $\mathbf{P}(\infty)$ having the first component equal to 1, i.e., $c_0 = 1$. The generating function of the sequence c_k is therefore the same as that of $x_k(\infty)$, apart from the factor $x_0(\infty) = 1 - A_o$. Then, from eq. (4.24) we derive

$$C(z) \equiv \sum_{k=0}^{\infty} c_k z^k = \frac{(z-1)\phi_A(z)}{z - \phi_A(z)} \tag{4.92}$$

From eq. (4.91) we can sum over k and obtain

$$d_K \equiv \sum_{k=0}^{K} c_k = \frac{1}{x_0(K)}, \quad K \geq 0. \tag{4.93}$$

Then

$$D(z) \equiv \sum_{K=0}^{\infty} d_K z^K = \frac{C(z)}{1-z} = \frac{\phi_A(z)}{\phi_A(z) - z} \tag{4.94}$$

The numerator of $D(z)$ is proportional to $\phi_A(z) = \varphi_X(\lambda - \lambda z)$, which is analytic for every z such that $|z| < 1 + \beta/\lambda$, where β is the abscissa of convergence of the Laplace transform[7] $\varphi_X(s)$. In the following we assume that $\beta > 0$, i.e., that all moments of the random variable X are finite. This is true, for example, when X has finite support, which is a reasonable model for most practical applications.

The singularities of $D(z)$ are necessarily only the zeros of the denominator. The function $\phi_A(z) - z$ is 0 for $z = 1$. Let η denote the least positive zero of $\phi_A(z) - z$ besides 1. Since $\phi_A(z)$ is monotonously increasing with z, strictly convex, $\phi_A(0) = a_0 > 0$, we can appeal to Theorem 3.14 to state that $\eta < 1$ if $\phi'_A(1) > 1$, $\eta = 1$ if $\phi'_A(1) = 1$ and $\eta > 1$ if $\phi'_A(1) < 1$. Remember that it is $\phi'_A(1) = \lambda/\mu = A_o$.

In general, the loss probability tends to 0 when $K \to \infty$ if $A_o \leq 1$; it tends instead to $(A_o - 1)/A_o$ if $A_o > 1$. This is obtained simply by observing that a single server queue can at most serve a traffic load of 1. All of the excess offered traffic cannot but be lost. Hence, we define

$$P_L(\infty) = \begin{cases} 0 & A_o \leq 1 \\ 1 - 1/A_o & A_o > 1 \end{cases} \tag{4.95}$$

From the equation above and the identity

$$P_L(K) = p_{K+1}(K) = 1 - \frac{1}{x_0(K) + A_o} \tag{4.96}$$

it is easy to derive that

$$x_0(K) - \max\{0, 1 - A_o\} = \frac{P_L(K) - P_L(\infty)}{[1 - P_L(K)][1 - P_L(\infty)]} \tag{4.97}$$

---

7 For a Laplace transform $\varphi(s) = \int_0^{\infty} e^{-st} f(t)\, dt$ the abscissa of convergence is the supremum of the set of β such that for $\text{Re}[s] > -\beta$ the integral that defines $\varphi(s)$ has a finite value.

Hence

$$\lim_{K \to \infty} \frac{P_L(K) - P_L(\infty)}{x_0(K) - \max\{0, 1 - A_o\}} = [1 - P_L(\infty)]^2 \tag{4.98}$$

for all values of A_o. Let us now consider separately the cases $A_o < 1$, $A_o > 1$ and $A_o = 1$. For $A_o < 1$, by using the identity $x_0(K) = 1/d_K$, we obtain from eqs. (4.95) and (4.98)

$$\lim_{K \to \infty} \frac{P_L(K)}{x_0(K) - (1 - A_o)} = \lim_{K \to \infty} \frac{P_L(K)}{x_0(K)(1 - A_o)\left[\frac{1}{1 - A_o} - d_K\right]} = 1 \tag{4.99}$$

whence

$$\lim_{K \to \infty} \frac{P_L(K)}{\frac{1}{1 - A_o} - d_K} = (1 - A_o)^2 \tag{4.100}$$

This equation establishes that the asymptotic behavior of $P_L(K)$ as $K \to \infty$ is the same as that of the sequence $\frac{1}{1 - A_o} - d_K$. We are left with the task of characterizing the asymptotic properties of that sequence. For that purpose we invoke a modified form of the final value theorem for generating functions. Given a sequence f_k and the transform $F(z) \equiv \sum_{k=0}^{\infty} f_k z^k$, the final value theorem asserts that

$$\lim_{k \to \infty} f_k = \lim_{z \to 1} (1 - z) F(z) \tag{4.101}$$

provided the limits exist and $F(z)$ is analytic for $|z| < 1$. If we scale the sequence by a factor γ^k, i.e., we define the new sequence $g_k = f_k \gamma^k$, we have $G(z) = \sum_{k=0}^{\infty} g_k z^k = \sum_{k=0}^{\infty} f_k \gamma^k z^k = F(\gamma z)$; then

$$\lim_{k \to \infty} f_k \gamma^k = \lim_{k \to \infty} g_k = \lim_{z \to 1} (1 - z) G(z) = \lim_{z \to 1} (1 - z) F(\gamma z) = \lim_{z \to \gamma} (1 - z/\gamma) F(z) \tag{4.102}$$

provided $F(z)$ is analytic for $|z| < \gamma$, i.e., γ is the smallest modulus singularity of the generating function $F(z)$ of the sequence f_k.

We apply the modified form of the final value theorem to the sequence $f_K = 1/(1 - A_o) - d_K$ for $K \geq 0$, so that

$$F(z) = \frac{1}{(1 - A_o)(1 - z)} - D(z) = \frac{z - (A_o + (1 - A_o)z)\phi_A(z)}{(1 - A_o)(1 - z)[z - \phi_A(z)]} \tag{4.103}$$

and we choose $\gamma = \eta$. We can do that, since the smallest modulus zeri of the denominator of $F(z)$ are 1 and $\eta > 1$. By applying twice de L'Hôpital's theorem, after tedious calculations, it can be verified that 1 is not a singularity, i.e., $F(1)$ is finite. Hence, $F(z)$ is analytic for $|z| < \eta$ and we are in a position to exploit the modified form of the final value theorem to write

$$\lim_{K\to\infty}\left(\frac{1}{1-A_0}-d_K\right)\eta^K = \lim_{z\to\eta}(1-z/\eta)F(z) = \lim_{z\to\eta}(1-z/\eta)\frac{\phi_A(z)}{z-\phi_A(z)}$$

$$= \phi_A(\eta)\lim_{z\to\eta}\frac{(1-z/\eta)}{z-\phi_A(z)} = \eta\frac{(-1/\eta)}{1-\phi'_A(\eta)} = \frac{1}{\phi'(\eta)-1} \tag{4.104}$$

where we have used $\phi_A(\eta) = \eta$ and the fact that $\phi'(\eta) > 1$. Going back to eq. (4.100) and using the last result obtained, we can write

$$\lim_{K\to\infty}\frac{P_L(K)\eta^K}{\left[\frac{1}{1-A_0}-d_K\right]\eta^K} = \lim_{K\to\infty}\frac{P_L(K)\eta^K}{1/[\phi'_A(\eta)-1]} = (1-A_0)^2 \tag{4.105}$$

and hence the final result for $A_0 < 1$:

$$\lim_{K\to\infty}P_L(K)\eta^K = \frac{(1-A_0)^2}{\phi'_A(\eta)-1} \tag{4.106}$$

Moving to the case $A_0 > 1$, we rewrite the limit of the loss probability as

$$\lim_{K\to\infty}\frac{P_L(K)-P_L(\infty)}{x_0(K)} = \lim_{K\to\infty}[P_L(K)-P_L(\infty)]d_K = \frac{1}{A_0^2} \tag{4.107}$$

We can repeat everything as above, except that in this case $\eta < 1$, $\phi'_A(\eta) < 1$, the function $D(z)$ is analytic for $|z| < \eta$ and the modified final value theorem yields

$$\lim_{K\to\infty}d_K\eta^K = \lim_{z\to\eta}(1-z/\eta)D(z) = \lim_{z\to\eta}(1-z/\eta)\frac{\phi_A(z)}{\phi_A(z)-z} = \frac{1}{1-\phi'_A(\eta)} \tag{4.108}$$

and putting together eqs. (4.107) and (4.108):

$$\lim_{K\to\infty}[P_L(K)-P_L(\infty)]\eta^{-K} = \frac{1-\phi'_A(\eta)}{A_0^2} \tag{4.109}$$

To deal with the boundary case $A_0 = 1$, we apply the final value theorem to $\int_0^z F(\zeta)\,d\zeta$, that corresponds to $f_k/(k+1)$. Then

$$\lim_{k\to\infty}\frac{f_k}{k+1} = \lim_{z\to1}(1-z)\int_0^z F(\zeta)\,d\zeta = \lim_{z\to1}(1-z)^2 F(z) \tag{4.110}$$

where the rightmost side has been obtained by applying de L'Hôpital's theorem. Therefore, for the case $A_0 = 1$, we obtain

$$\lim_{K\to\infty}\frac{d_K}{K+1} = \lim_{z\to1}(1-z)^2\frac{\phi_A(z)}{\phi_A(z)-z} = \frac{2}{\phi''_A(1)} \tag{4.111}$$

and

$$\lim_{K\to\infty}P_L(K)d_K = \lim_{K\to\infty}P_L(K)(K+1)\frac{d_K}{K+1} = 1 \tag{4.112}$$

and finally

$$\lim_{K\to\infty} P_L(K)(K+1) = \frac{\phi''_A(1)}{2} \tag{4.113}$$

We can summarize the result in the following asymptotic theorem for the loss probability of the $M/G/1/K$ queue.

Theorem 4.4 The loss probability $P_L(K)$ of the $M/G/1/K$ queue has the following asymptotic behavior as $K \to \infty$:

$$P_L(K) - P_L(\infty) = \begin{cases} \dfrac{(1-A_o)^2}{\phi'_A(\eta)-1}\eta^{-K} + o(\eta^{-K}) & A_o < 1 \\[2ex] \dfrac{1-\phi'_A(\eta)}{A_o^2}\eta^K + o(\eta^K) & A_o > 1 \\[2ex] \dfrac{\phi''_A(1)}{2}\dfrac{1}{K+1} + o\left(\dfrac{1}{K}\right) & A_o = 1 \end{cases} \tag{4.114}$$

where $A_o = \lambda/\mu$, $P_L(\infty) = \max\{0, 1 - 1/A_o\}$ and η is the smallest modulus root of the equation $z = \phi_A(z) = \varphi_X(\lambda - \lambda z)$ besides $z = 1$.

More details are provided in [15], along with extension of the result to the $G/M/1/K$ queue. Generalization to a much broader class of queues, namely the $MAP/G/1/K$ model, an instance of $M/G/1$ structured queueing systems introduced by M.F. Neuts [168], is discussed in detail in [17]. In that case the equation $z = \phi_A(z)$ turns into $\det(zI - A(z)) = 0$, where $A(z) = \int_0^\infty \exp(tD(z))\,dF_X(t)$ (see Section 3.3 for details and definitions of the MAP). The role of η is played by the least modulus root besides $z = 1$ of the equation $z = \chi(z)$, where $\chi(z)$ is the Perron-Frobenius eigenvalue of the non-negative matrix $A(z)$.

A side result of Theorem 4.4 is provided by using eq. (4.98).

Theorem 4.5 Given an $M/G/1/K$ queue, the probability $x_0(K)$, that no customer is left behind by a departure, has the following asymptotic behavior as $K \to \infty$:

$$x_0(K) = \begin{cases} 1 - A_o + \dfrac{(1-A_o)^2}{\phi'_A(\eta)-1}\eta^{-K} + o(\eta^{-K}) & A_o < 1 \\[2ex] (1-\phi'_A(\eta))\eta^K + o(\eta^K) & A_o > 1 \\[2ex] \dfrac{\phi''_A(1)}{2}\dfrac{1}{K+1} + o\left(\dfrac{1}{K}\right) & A_o = 1 \end{cases} \tag{4.115}$$

where $A_o = \lambda/\mu$ and η is the smallest modulus root of the equation $z = \phi_A(z) = \varphi_X(\lambda - \lambda z)$ besides $z = 1$.

This approximation can be useful in the numerical solution of the linear system $\mathbf{x}(K)\mathbf{P}(K) = \mathbf{x}(K)$. Let $K > 0$. By leaving out the last equation and rearranging terms, we end up writing the system as

$$
[x_1(K) \ldots x_K(K)]
\begin{bmatrix}
1 - a_0 & -a_1 & -a_2 & \ldots & -a_{K-1} \\
0 & 1 - a_0 & -a_1 & \ldots & -a_{K-2} \\
0 & 0 & 1 - a_0 & \ldots & -a_{K-3} \\
\ldots & \ldots & \ldots & \ldots & \ldots \\
0 & 0 & 0 & \ldots & 1 - a_0
\end{bmatrix}
= x_0(K)[a_0 - 1 \; a_1 \ldots a_{K-1}]
$$

If $x_0(K)$ is known, e.g., it is replaced with its asymptotic approximation as provided by Theorem 4.5, then the linear system (4.3.2) can be solved easily, since the coefficient matrix is triangular.

Another result can be derived by the asymptotic theorems. From eq. (4.88) we find for $A_o < 1$:

$$
\sigma_{K+1} = \sum_{j=K+1}^{\infty} p_j(\infty) = 1 - \frac{1 - A_o}{x_0(K)} = (1 - A_o)\left[\frac{1}{1 - A_o} - d_K\right] \tag{4.116}
$$

and therefore

$$
\lim_{K \to \infty} \sigma_{K+1} \eta^K = (1 - A_o) \lim_{K \to \infty} \left[\frac{1}{1 - A_o} - d_K\right] \eta^K = \frac{1 - A_o}{\phi_A'(\eta) - 1} \tag{4.117}
$$

This yields an asymptotic approximation of the tail of the probability distribution of the number of customers in the $M/G/1$ queue ($K = \infty$), namely

$$
\sum_{j=K+1}^{\infty} p_j(\infty) = \sum_{j=K+1}^{\infty} x_j(\infty) = \frac{1 - A_o}{\phi_A'(\eta) - 1}\eta^{-K} + o(\eta^{-K}) \tag{4.118}
$$

as $K \to \infty$, η being the least modulus root of $z = \phi_A(z)$ greater than 1.

Example 4.3 For the $M/D/1/K$ queue we have $\phi_A(z) = e^{A_o(z-1)}$, which is analytic for all finite z. Studying the plot of the function $e^{A_o(z-1)}$ it can be verified that the equation $z = e^{A_o(z-1)}$ has only two roots, 1 and η, with $\eta > 1$, if $A_o < 1$; $\eta < 1$, if $A_o > 1$ and $\eta = 1$, if $A_o = 1$ (i.e., in this last case, there is a double root at 1). Figure 4.3 plots the root η as a function of A_o.

In case of deterministic service times, the root η can be computed numerically by solving the equation $f(z) \equiv \phi_A(z) - z = e^{A_o(z-1)} - z$. The function $f(z)$ is continuous and monotonously increasing with z for $z \in [1 - \log(A_o)/A_o, \infty)$. It has opposite signs at the extremes of this interval, since $f(1 - \log(A_o)/A_o) = (1 + \log(A_o) - A_o)/A_o < 0$, while $\lim_{x \to \infty} f(x) = +\infty$. Therefore, there is a unique root η of $f(z) = 0$ besides $z = 1$. It can be approximated with the Newton-Raphson algorithm:

$$
z_{m+1} = z_m - \frac{f(z_m)}{f'(z_m)} = \frac{z_m A_o - 1}{A_o - e^{-A_o(z_m-1)}}, \qquad m \geq 0, \tag{4.119}
$$

Figure 4.3 Smallest modulus root η of $z = e^{A_o(z-1)}$, besides the root $z = 1$, as a function of A_o.

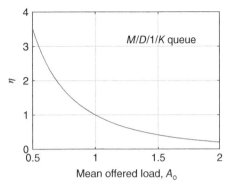

starting with a value $z_0 > \eta$. Such an initial value can be found by solving the quadratics $1 + A_o(z - 1) + A_o^2(z - 1)^2/2 - z = 0$ and taking the largest root, since it is $e^u > 1 + u + u^2/2$. The point z_0 is easily found to be $z_0 = 1 + 2(1 - A_o)/A_o^2$. This is the right initial point for $A_o < 1$. When $A_o > 1$, the initial point can be $z_0 = 0$. Convergence is very fast. Note that $\phi'_A(\eta) = A_o e^{A_o(\eta-1)} = A_o\eta$. An example of Matlab code used for the computation of η in the $M/D/1/K$ queue is given below.

```
tole = 1e-5;
erro = 1;
if Ao > 1
  ze = 0;
else
  if Ao < 1
  ze = 1+2*(1-Ao)/(Ao^2);
% obtained by approximating exp(x) to the 2nd order
  else
    ze = 1;
    erro = 0;   %if Ao=1 the cycle is skipped
  end
end
while erro > tole
  newz = (ze*Ao-1)/(Ao-exp(-Ao*(ze-1)));
  erro = abs(ze-newz)/newz;
  ze = newz;
end
eta = ze;
```

The exact value of the loss probability, computed from the solution of the linear equation system of the EMC of the $M/D/1/K$ queue, and the asymptotic

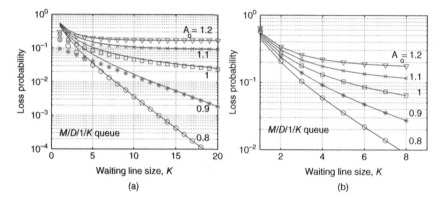

Figure 4.4 Comparison between the loss probability of the $M/D/1/K$ queue as a function of K for different values of A_o. Left plot: comparison between exact values (solid line) and the asymptotic expansion (markers). Right plot: comparison between exact values (solid line) and the approximation based on the asymptotic expansion (markers).

approximation provided by Theorem 4.4 in case of deterministic service times are compared in Figure 4.4(a) as a function of K and for five different values of A_o.

Albeit guaranteed only to be asymptotically sharp as K tends to infinity, the asymptotic approximation turns out to be accurate even for quite small values of K. Numerical experience with other service time probability distributions generally yields a similar result.

Example 4.4 As another example of computation of the root η, we discuss briefly the case of service times having a gamma distribution (see eq. (4.70) for the definition of such a distribution and the meaning of its parameters $\alpha = \mu\beta$ and β). In this case η must be found as the smallest real root of the equation $z = 1/(1 + A_o/\beta - zA_o/\beta)^\beta$. The function $f(z) = (1 + A_o/\beta - zA_o/\beta)^{-\beta}$ is monotonously increasing and convex for $z \in [0, 1 + \beta/A_o)$ and it grows from $f(0)$, with $0 < f(0) < 1$, to $+\infty$. It crosses the line $y = z$ at $z = 1$, with slope A_o. So, if $A_o < 1$, there must be one more (and only one) real root lying in the interval $(1, 1 + \beta/A_o)$. This can be efficiently approximated by, e.g., the bisection method.

Highly accurate formulas for the loss probability of the $M/G/1/K$ queue (and hence for $x_0(K)$, and ultimately for the entire probability distribution of the number of customers in the queue) can be obtained by following the approach in [16]. The basic idea is to approximate the sequence $d_K = 1/x_0(K)$ by representing its generating function $D(z) = \phi_A(z)/[\phi_A(z) - z]$ as a partial fraction expansion around its two poles 1 an η. Once an approximation of d_K is obtained, the loss probability is calculated according to

$$P_L(K) = 1 - \frac{1}{x_0(K) + A_0} = 1 - \frac{1}{1/d_K + A_0} = \frac{1 - d_K(1 - A_0)}{1 + d_K A_0} \tag{4.120}$$

As long as $A_0 \neq 1$ the two smallest modulus singularities of $D(z)$ are simple and we have

$$D(z) \approx \frac{R_1}{1 - z} - \frac{R_2}{1 - z/\eta} \tag{4.121}$$

where the residues are found from the following limits

$$R_1 = \lim_{z \to 1}(1 - z)D(z) = \lim_{z \to 1}(1 - z)\frac{\phi_A(z)}{\phi_A(z) - z} = \frac{1}{1 - A_0}$$

$$R_2 = \lim_{z \to \eta}\left(1 - \frac{z}{\eta}\right)D(z) = \lim_{z \to \eta}\frac{(1 - z/\eta)\phi_A(z)}{\phi_A(z) - z} = \frac{1}{\phi'_A(\eta) - 1}$$

The approximation of d_K is found by inverting the approximation of $D(z)$, i.e., $d_K \approx R_1 - R_2\eta^{-K}$. Plugging this into eq. (4.120), with some lengthy algebra we get for $A_0 < 1$

$$P_L(K) \approx \frac{C_1\eta^{-K}}{1 - \frac{A_0}{1-A_0}C_1\eta^{-K}} , \quad K \geq 0, \tag{4.122}$$

where $C_1 = (1 - A_0)^2/[\phi'_A(\eta) - 1]$. For $A_0 > 1$ we get

$$P_L(K) \approx 1 - \frac{1}{A_0} + \frac{C_2\eta^K}{1 - \frac{A_0}{A_0-1}C_2\eta^K} , \quad K \geq 0, \tag{4.123}$$

where $C_2 = [1 - \phi'_A(\eta)]/A_0^2$. Finally, for $A_0 = 1$ there is a double pole at 1. The approximate expansion of $D(z)$ becomes

$$D(z) \approx \frac{S_1}{1 - z} + \frac{S_2}{(1 - z)^2} = \frac{2\phi'''_A(1)}{3[\phi''_A(1)]^2}\frac{1}{1 - z} + \frac{2}{\phi''_A(1)}\frac{1}{(1 - z)^2} \tag{4.124}$$

hence

$$d_K \approx \frac{2\phi'''_A(1)}{3[\phi''_A(1)]^2} + \frac{2}{\phi''_A(1)}(K + 1) , \quad K \geq 0. \tag{4.125}$$

and finally

$$P_L(K) \approx \frac{\frac{\phi''_A(1)}{2}}{K + 1 + \frac{\phi''_A(1)}{2} + \frac{\phi'''_A(1)}{3\phi''_A(1)}} \tag{4.126}$$

A sample of numerical results for the $M/D/1/K$ queue is given in Figure 4.4(b). Note that the approximation is in excellent agreement with the exact value, except for the case $K = 0$, that is trivial to evaluate exactly: namely $P_L(0) = A_0$ $/(1 + A_0)$.

4.4 Numerical Evaluation of the Queue Length PDF

The queue length probability distribution function is the left eigenvector of **P** with respect to the maximum modulus eigenvalue 1, normalized to sum up to 1. The first K equations of the steady-state balance equations $\mathbf{xP} = \mathbf{x}$ yield:

$$x_0 a_i + \sum_{j=1}^{i+1} x_j a_{i+1-j} = x_i, \qquad i = 0, 1, \dots, K - 1 \qquad (4.127)$$

The direct numerical solution of these equations, obtained by isolating the term proportional to x_{i+1} in the left hand side and carrying everything else on the other side, suffers from numerical instability due to the subtraction of two often very close terms. Just think of the fact that the larger i becomes, the smaller the value of x_i, at least asymptotically. If one tries to evaluate such small probabilities as the difference of two close quantities, already affected by accumulated errors, numerical instability arises easily.

In [168, Ch. 1], a stable numerical procedure is given to evaluate the components of the vector **x**. The procedure is given for the infinite buffer $M/G/1$, but it readily extends to the finite buffer, as it is detailed in the following. We derive the numerical stable iteration for a slightly more general set of equations, where we allow the first row of the one-step transition probability matrix to be different from the second row:

$$x_0 b_i + \sum_{j=1}^{i+1} x_j a_{i+1-j} = x_i, \qquad i = 0, 1, \dots, K - 1 \qquad (4.128)$$

where the sequence b_i, $i \geq 0$ forms a probability distribution.

We define the CCDF of the random variable A, namely $A_k = \sum_{j=k+1}^{\infty} a_j = 1 - \sum_{j=0}^{k} a_j$, for $k \geq 0$; similarly for the b_i's, we let $B_k = \sum_{j=k+1}^{\infty} b_j = 1 - \sum_{j=0}^{k} b_j$, for $k \geq 0$. Then, $a_k = A_{k-1} - A_k$, $k \geq 1$, $a_0 = 1 - A_0$ and $b_k = B_{k-1} - B_k$, $k \geq 1$, $b_0 = 1 - B_0$.

By substituting the expression of the CCDF of A and B into eq. (4.128), it can be found that

$$x_i = x_0(B_{i-1} - B_i) + \sum_{j=1}^{i} x_j(A_{i-j} - A_{i+1-j}) + x_{i+1}(1 - A_0) \qquad (4.129)$$

whence

$$y_i \equiv x_i - x_0 B_{i-1} - \sum_{j=1}^{i-1} x_j A_{i-j} - x_i A_0 = x_{i+1}(1 - A_0) - x_0 B_i - \sum_{j=1}^{i} x_j A_{i+1-j} \qquad (4.130)$$

holding for $i = 1, \ldots, K - 1$. We recognize that this identity means that the terms of the succession y_i do not depend on i, so that they are all equal to the term y_1. Setting $i = 1$ on the left-hand side of eq. (4.130), we recover the first equation of the linear system in (4.128), hence we can see that $y_1 = 0$. Therefore, we have $y_i = 0, \forall i \geq 1$. Remembering that $1 - A_0 = a_0 > 0$, we can finally write the following iteration

$$x_i = \frac{1}{a_0} \left(x_0 B_{i-1} + \sum_{j=1}^{i-1} x_j A_{i-j} \right), \qquad i = 1, \ldots, K \tag{4.131}$$

where $a_0 = \varphi_X(\lambda)$ must be positive and less than 1, along with $\rho < 1$ in case of infinite buffer, for the $M/G/1$ system to be ergodic. The obtained iteration is highly stable, since it only involves sums and products of positive quantities. It needs to be initialized with the value of x_0. If $K = \infty$, the theory tells us that it must be $x_0 = 1 - \rho$, so it is easy to recover as many queue length probabilities as desired, by stretching the iteration in eq. (4.131) to the suitable value of K. For finite buffer $M/G/1/K$ queues, we simply set $x_0 = 1$ to carry out the iteration and find out the entire left eigenvector of **P** to within a multiplying factor. Once **x** is found, it is normalized to sum up to 1 and the resulting vector $\mathbf{x}/\sum_{k=0}^{K} x_k$ is the queue length probability distribution.

A sample Matlab code for the calculation of the probability distribution of the EMC of the $M/G/1$ queue is reported below. The first part of the code computes the probabilities $a_k = P(A = k)$, required to fill in the matrix **P**. In the example below those probabilities are computed for deterministic service times. That part of the code must be adjusted according to the specific probability distribution assumed for the service times.

```
Ao = input('Mean offered load = ');
K = input('Queue size (<=100) = ');
Kmax = 100;   % truncation index of the matrix P
% computation of P(A = k), k = 0,..,Kmax
av = zeros(1,Kmax+1);
av(1) = exp(-Ao);
for jj = 1:Kmax
  av(jj+1) = av(jj)*Ao/jj;
end
atildev = [1 1-cumsum(av)];
% computation of the probabilities x_k, k = 0,..,Kmax
xv = zeros(1,Kmax+1);   % vector of EMC probabilities
xv(1) = 1;   % x_0 is initialized at an arbitrary value
xv(2) = xv(1)*atildev(2)/av(1);
for ii = 1:Kmax-1
```

```
xv(ii+2)  =  (xv(1)*atildev(ii+2)+...
       sum(xv(2:ii+1).*atildev(ii+2:-1:3)))/av(1);
end
xinfv = xv*(1-Ao);  % M/G/1 case
xKv  = xv(1:K+1)/sum(xv(1:K+1));  % M/G/1/K case
```

4.5 A Special Case: the $M/M/1$ Queue

In the special case of negative exponential distribution of the service times, many results can be obtained in closed form. Let $G_X(t) = e^{-\mu t}$ be the CCDF of the service time, $\mu = 1/E[X]$ being the service rate of the single server. The Laplace transform of the service time PDF is $\varphi_X(s) = \mu/(s + \mu)$. By substituting it into eq. (4.32), it is found:

$$\varphi_W(s) = 1 - \rho + \rho \frac{\mu - \lambda}{s + \mu - \lambda} \tag{4.132}$$

where $\rho = \lambda/\mu$. It is easy to invert this transform, thus obtaining:

$$G_W(t) = \rho e^{-(\mu - \lambda)t} \tag{4.133}$$

i.e., a scaled negative exponential distribution with mean $\rho/(\mu - \lambda)$ and a probability mass of $1 - \rho$ concentrated in 0. The mean system time is

$$E[S] = E[X] + E[W] = \frac{1}{\mu} + \frac{\rho}{\mu - \lambda} = \frac{1}{\mu - \lambda} \tag{4.134}$$

The entire probability distribution of the system state, i.e., the number of customers in the system, can be found explicitly: it is $p_k = (1 - \rho)\rho^k$, $k \geq 0$. This can be derived in several ways. One possibility is to recognize that $Q(t)$ for the $M/M/1$ queue is a birth-death Markov process, then applying the results of Section 3.5. Alternatively, it is possible to specialize the expression of $\phi_Q(z)$ holding for the $M/G/1$ queue:

$$\phi_Q(z) = \frac{(1 - \rho)(z - 1)\phi_A(z)}{z - \phi_A(z)} = \frac{(1 - \rho)(z - 1)\mu}{z(\mu + \lambda - \lambda z) - \mu} = \frac{\mu - \lambda}{\mu - \lambda z} \tag{4.135}$$

where we have used $\phi_A(z) = \varphi_X(\lambda - \lambda z) = \mu/(\mu + \lambda - \lambda z)$.

The previous results hold under the condition $\lambda < \mu$, that guarantees that the queueing system can attain statistical equilibrium (i.e., it is stable).

In the special case at hand it is also possible to find explicit solution for the finite waiting line queueing system, namely the queue $M/M/1/K$. As usual, $K \geq 0$ denotes the maximum number of customers that can be hosted into the waiting line. Hence, the state $Q = K + 1$ is blocking. In this case, it is no more true that $\pi_{a,k} = \pi_{s,k}$, i.e., arrivals that arrive *and* enter the system do not see the same state (in a statistical sense) as customers the arrive at the system. Instead, we have

$p_k = \pi_{s,k}/(\pi_{s,0} + A)$ and conversely $\pi_{s,k} = p_k/(1 - p_{K+1})$, $k = 0, 1, \ldots, K$, where $A = \lambda/\mu$ is the mean offered load[8]. The probability distribution of the system state can be found explicitly:

$$p_k = \frac{A^k}{\sum_{n=0}^{K+1} A^n} = \pi_{a,k}, \qquad k = 0, 1, \ldots, K+1, \tag{4.136}$$

and

$$\pi_{s,k} = \frac{A^k}{\sum_{n=0}^{K} A^n} = \pi_{d,k}, \qquad k = 0, 1, \ldots, K. \tag{4.137}$$

These results hold for any value of A, since the finite queue is always stable.

As for the performance metrics, the mean arrival rate of accepted customers (i.e., the throughput of the system in terms of customers per second that are served) is

$$\Lambda_s = \lambda(1 - p_{K+1}) = \lambda \frac{\sum_{n=0}^{K} A^n}{\sum_{n=0}^{K+1} A^n} \tag{4.138}$$

Note that $\Lambda_s \to \lambda$ for $K \to \infty$, since blocking vanishes when the room inside the queuing system expands to infinity. Also, $\Lambda_s \to \mu$ when $\lambda \to \infty$ for a given K. In this last case, as the offered arrival rate increases for given queue size K and serving capacity μ, the queueing system tends to saturate its resources. In the limit the server has no more any idle time, so that it outputs served customers at its maximum allowed rate μ. This is clearly an undesirable working regime, since performance (hence QoS) offered to customers are terrible: the loss probability tends to 1 and the mean waiting time attains it maximum level K/μ.

The mean carried traffic is

$$A_s = \frac{\Lambda_s}{\mu} = \frac{\lambda}{\mu}(1 - p_{K+1}) = A(1 - p_{K+1}) \tag{4.139}$$

and the utilization coefficient is $\rho = A_s$.

The mean queue length, and waiting time are

$$E[Q] = \frac{\sum_{n=0}^{K+1} nA^n}{\sum_{n=0}^{K+1} A^n} \tag{4.140}$$

and

$$E[W] = \sum_{n=0}^{K} \frac{n}{\mu} \pi_{s,n} = \frac{\sum_{n=0}^{K} \frac{n}{\mu} A^n}{\sum_{n=0}^{K} A^n} = \frac{E[Q] - \rho}{\lambda(1 - p_{K+1})} = \frac{E[Q]}{\lambda(1 - p_{K+1})} - E[X] \tag{4.141}$$

where the last equality is nothing else than Little's law.

---

8 The subscript o has been dropped to keep notation simple.

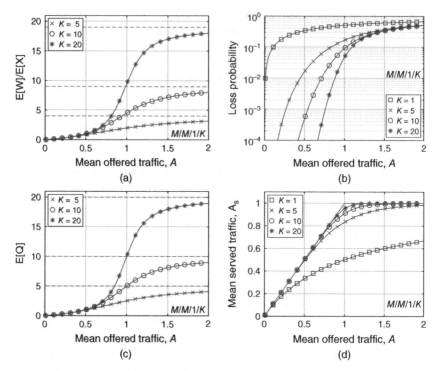

Figure 4.5 Performance metrics of the $M/M/1/K$ queue as a function of the mean offered traffic $A = \lambda/\mu$. (a) mean waiting time normalized with respect to the mean service time, $\mu E[W]$; (b) probability of loss, p_K; (c) mean queue length, $E[Q]$; (d) mean served traffic, A_s.

The performance metrics of the $M/M/1/K$ queue are plotted in Figure 4.5 as a function of A for various values of K. The asymptotic behavior of the mean waiting time and the mean queue length corresponds to the saturation of the queue space, whereas the loss probability and the mean served traffic tend to 1. Figure 4.5(d) shows the ideal behavior of a single server system, that is a straight line having slope 1 up to $A = 1$, then saturating. The actual behavior of the $M/M/1/K$ approximates the ideal one, the more the bigger K.

4.6 Optimization of a Single-Server Queue

Queueing models can be used also to state and solve optimization problems of service systems. This is often referred to as queue optimal design. In this section we give some examples, mostly aimed at showing how queueing results on single server queues derived in the previous sections can be exploited to address service system optimization for a given metric of interest.

4.6.1 Maximization of Net Profit

We consider a service system, modeled as a single server queueing facility, with possibly limited room for the waiting line. Let r denote the revenue per customer and c the cost per unit time of waiting time per customer. Then, the net balance rate, i.e., the rate of net profit, for a queue with mean arrival rate λ is

$$U = r\lambda_s - c\lambda_s E[W] \tag{4.142}$$

where λ_s denotes the mean rate of customers *admitted* into the queue.

Let us apply this setting to an $M/G/1$ queue. Then $\lambda_s = \lambda$ and $E[W]$ is given by the P-K formula:

$$U = r\lambda - c\frac{\lambda^2 E[X^2]}{2(1-\rho)} = c\frac{E[X^2]}{2E[X]^2}\rho\left(a - \frac{\rho}{1-\rho}\right) \tag{4.143}$$

with $a = 2E[X]r/(cE[X^2]) = r/(cE[\tilde{X}])$ and $\tilde{X}$ denotes the residual service time. By scaling the net profit rate by a constant factor, we can pose the problem of finding the value of ρ that maximizes the function:

$$\hat{U}(\rho) = \rho\left(a - \frac{\rho}{1-\rho}\right), \qquad \rho \in (0,1) \tag{4.144}$$

It can be checked that this function is concave and it has a unique maximum in the interval $(0,1)$ located at $\rho^* = 1 - 1/\sqrt{1+a}$. The value of the function at the maximum is $\hat{U}^* = (\sqrt{1+a} - 1)^2$.

The parameter a, appearing in the optimal level of the system utilization ρ^*, is the ratio of the revenue per customer and the cost to complete a residual service time. This is the minimum mean time that a customer has to wait when it arrives at the queue and finds the server busy. Intuitively, the term $cE[\tilde{X}]$ is a sort of "chip" cost incurred when the server is utilized. If the revenue is large compared to this minimum mean cost, then it is convenient to push the utilization factor to high levels: mathematically, when a is large, the optimal level of ρ gets close to 1. On the contrary, if the minimum mean cost $cE[\tilde{X}]$ dominates the revenue per costumer r, then it is more profitable to keep the load on the server light, i.e., for small values of a it is $\rho^* \approx a/2 \ll 1$.

Let us reconsider the optimization problem for the simpler $M/M/1$ queue: the average net utility is:

$$U = r\lambda - c\frac{\lambda}{\mu - \lambda} \tag{4.145}$$

where we have modified slightly the cost term. Now it is proportional to the *system* time of the queue rather than to the waiting time. This is consistent with assuming a cost per unit of time spent inside the service system, including both wait and service.

Given a value of the service rate, it is intuitive that there exists an optimal value of λ in the interval $(0, \mu)$. When $\lambda = 0$ the net utility is 0. When λ approaches μ the delay cost tends to infinity and exceeds definitely the maximum achievable income $r\mu$. It is easy to find that the optimal arrival rate is

$$\lambda^* = \mu - \sqrt{\frac{c\mu}{r}} \tag{4.146}$$

provided that $r \geq c\mu$, i.e., the revenue for serving one customer must at least exceed the cost incurred for serving that customer.

We consider now a fixed offered rate of service requests λ and wonder if there exists a value of μ that maximizes the net utility. We find easily that U turns out to be a monotonously increasing function of μ. This is expected, since increasing the service capacity of the system decreases the delay cost, while we are not accounting in any way for the cost of providing a given service capacity μ.

A less trivial result is obtained if we let μ grow, under the constraint of a *fixed value of the utilization coefficient* ρ. Note that increasing μ for a fixed level of ρ implies that λ grows proportionally with μ. Then, we can write

$$U = r\mu\rho - c\frac{\rho}{1-\rho} \tag{4.147}$$

It is clear that U grows without bound as μ increases. In other words, there is the advantage of *scale* of the service system: for a given level of the utilization coefficient of the serving resource, the system yields a higher gain if both μ and λ grow (bigger scale system). Even if it is extremely simple (and possibly not realistic under several aspects), this model captures the essential principles laying at the foundation of the Data Center business, as the enabling factor of cloud computing.

The utility function is no more monotonous, if we include the cost of providing the serving facility. In other words, there are two sources of cost:

1. The cost of delay incurred by customers for receiving the requested service. The source of this cost is ultimately multiplexing, i.e., the sharing of the single server facility among a population of customers.
2. The cost of providing and maintaining the single service facility with average capacity μ (and an assigned level of the quality of service provided: this is a key point of the meaning of the statement "providing service facility").

Assuming that the single server facility can be still modeled as an $M/M/1$ queue, under this new cost model the utility becomes

$$U = r\lambda - c_D\frac{\lambda}{\mu - \lambda} - c_S\mu \tag{4.148}$$

where c_D denotes the cost of delay (in monetary units per unit of time) and c_S the cost per unit of capacity of the server. It can be easily found that there exists a finite optimal value of μ in the range (λ, ∞). Setting to zero the derivative with respect to μ, it is found that $\mu^* = \lambda + \sqrt{c_D \lambda / c_S}$.

We can pose a more complex optimization problem with reference to a loss system. In this kind of systems, customer demand can be rejected, i.e., there is limited room for waiting in the system. We take the simplest model that captures this features, namely the $M/M/1/K$ queueing system.

Assume there is a fixed revenue r per served customer and that cost is associated only to waiting time, according to a fixed rate c of cost per unit time of wait. The service discipline is FCFS. Then, a customer that arrives to find n customers already into the system brings an average net profit of $r - cn/\mu$. The customer joins the queue only if the net profit is non-negative, i.e., if it finds $n \le r\mu/c$ customers into the waiting line. This is the same as saying that the queueing system under consideration is a loss system with size $K = \lfloor r\mu/c \rfloor$.

The mean net profit rate of such a system is:

$$U = r\lambda_s - c\lambda_s \mathbb{E}[W] = \lambda(1 - p_{K+1})\left(r - \frac{c}{\mu}\frac{\sum_{n=0}^{K} nA^n}{\sum_{n=0}^{K} A^n}\right)$$

$$= cA\frac{\sum_{n=0}^{K} A^n}{\sum_{n=0}^{K+1} A^n}\left(\frac{r\mu}{c} - \frac{\sum_{n=0}^{K} nA^n}{\sum_{n=0}^{K} A^n}\right)$$

with $A = \lambda/\mu$. Let us assume the serving capacity available to the system is given, i.e., μ is fixed and hence K is given as well. The only variable left is therefore λ. The intuition is as follows: as A grows, the revenue saturates, since the loss probability gets closer and closer to 1; in other words, there is a cap to the average net income rate, given by $r\mu$. On the other hand, the cost of delay increases as well, even if the mean waiting time is bounded above by K/μ. Reminding that $K = \lfloor r\mu/c \rfloor$, as $\lambda \to \infty$ we have $\lambda_s \to \mu$ and hence

$$U \to r\mu - c\mu\frac{K}{\mu} = r\mu - c\left\lfloor\frac{r\mu}{c}\right\rfloor \approx 0 \tag{4.149}$$

This suggests that it may be penalizing to let A grow too big. On the other hand, as A gets smaller, the revenue diminishes proportionally; so, even if the cost incurred because of wait is negligible, so is the realized income. Good values of A are therefore expected to be in a mid range.

Figure 4.6(a) plots U as a function of A for $\mu = 1 \text{ s}^{-1}$, $r = 10 \text{ MU/customer}$, $c = 1 \text{ MU/s}$, where MU stands for monetary unit. With these values it is found that the maximum attainable net profit rate is $U^* \approx 5.55 \text{ MU/s}$ for $\lambda^* = 0.76 \text{ s}^{-1}$.

If we instead fix λ and let A vary through the variation of μ, we know that the utility is trivially monotonously decreasing with A, since the bigger μ the less the wait and the cost incurred. To make the problem interesting, we have to account

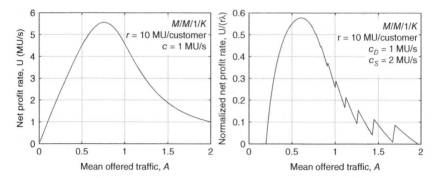

Figure 4.6 *M/M/1/K* queue optimization. Left plot: net profit rate as a function of $A = \lambda/\mu$ for varying λ and fixed $\mu = 1\ s^{-1}$. Right plot: normalized net profit rate as a function of $A = \lambda/\mu$ for varying μ and fixed $\lambda = 1\ s^{-1}$.

for the cost of the service facility[9]. Then, let us define the modified utilization as

$$\hat{U} \equiv \frac{U - c_S \mu}{r\lambda} = \frac{\sum_{n=0}^{K} A^n}{\sum_{n=0}^{K+1} A^n} \left(1 - \frac{c_D}{r\mu} \frac{\sum_{n=0}^{K} nA^n}{\sum_{n=0}^{K} A^n} \right) - \frac{c_S}{rA} \tag{4.150}$$

where $A = \lambda/\mu$, $K = \lfloor r\mu/c_D \rfloor$, and we have added subscripts to the two different costs. The utility $\hat{U}$ has been normalized by the fixed amount of the maximum possible revenue if all the potential demand were served, $r\lambda$.

Numerical results for $\lambda = 1\ s^{-1}$, $c_D = 1$ MU/s, $c_S = 2$ MU $\cdot$ s and $r = 10$ MU/customer are shown in Figure 4.6(b).

The discontinuities visible in the curve of Figure 4.6(b), for small values of μ, are due to the quantization effect of the floor function in the computation of K. Note that with varying μ we have also a variable K. We still have an optimum working point: for the given demand arrival rate $\lambda = 1\ s^{-1}$, the optimal serving rate is $\mu^* = 1.65\ s^{-1}$ and the corresponding normalized utility is $\hat{U} = 0.577$. In words, by having a service rate 65% faster then the arrival rate we get almost 58% of the overall potential revenue.

4.6.2 Minimization of Age of Information

Consider a number of agents, distributed in a given area, recording some information and reporting relevant messages to a central archival or processing facility, via a communication network. This is a scenario recurring in a number of circumstances, e.g., sensor networks, vehicular traffic monitoring, autonomous vehicle

---

9 This cost is immaterial in the optimization of U with a fixed μ, since then the serving facility cost is fixed.

supervision, software update checking, status updates of user applications, just to mention few of them.

Let us try to abstract the key common features of those disparate systems, from the point of view of a specific performance metric: the age of information (see Section 2.7.5), i.e., a measure of how old is the information collected at the central facility from each particular source. This is a most relevant metric whenever the collected data (possibly after processing) is the input to a decision-making or actuator system, whose effectiveness relies most often on the timeliness of the data.

We tackle our objective in two phases. First, we give a general expression of the mean age of information. Then, we carry out its minimization under special hypotheses that make the obtained model tractable analytically. Before that, we specify a simplified, yet general model. The material of this section is largely inspired to [119].

4.6.2.1 General Expression of the Average Age of Information

Let us consider a source that sends messages at a mean rate of λ. The messages are directed to a central facility. To reach the facility, messages go through a network, that is modeled as a service system with a single server. Service times have the same probability distribution for all messages. The central facility records only the last piece of information received from the source. As soon as a new message arrives from the source, its record is updated.

The age of information of the source record at time t is therefore $D(t) = t - T_a(t)$, where $T_a(t)$ is the last arrival time up to t of a message at the central facility. The metric of interest is the time average of the age of information, namely:

$$\langle D \rangle = \lim_{T \to \infty} \frac{1}{T} \int_0^T D(t) \, dt \qquad (4.151)$$

In the following, we assume that:

- The message generation process at the source is a stationary renewal process.
- The service system representing the network is work-conserving, has no loss and adopts the FCFS discipline.
- Service times are i.i.d. random variables, independent of inter-arrival times of the messages.

Let us follow the sample path of the age of information of the source record at the central facility. Figure 4.7 plots a sample path of $D(t)$, where it is assumed that the source record is initialized with some information at $t = 0$. Times $\{t_k\}_{k=1,...,n}$ when the source issues update messages are marked with crosses. Times $\{\theta_k\}_{k=1,...,n}$ when those messages are delivered to the central processing facility are marked with circles. The thick contour line is the sample path of the age of information. Each time a new message arrives, there is a downward jump, i.e., $D(\theta_{k+}) = \theta_k - t_k$.

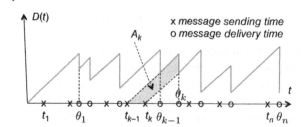

Figure 4.7 Sample path of the age of information (solid line). The source record is initialized with information collected at $t = 0$. Update messages are sent at times t_k, marked with x. Message delivery times θ_k at the central facility are marked with 'o'. The area of the shaded polygon A_k is one of the summands that composes the integral of the sample path of $D(t)$.

Then, $D(t)$ grows linearly with slope 1, until the next message delivery, i.e., $D(t) = D(\theta_{k+}) + t - \theta_k = t - t_k$ for $t \in (\theta_k, \theta_{k+1}]$.

Let T_k denote the time elapsing between the sending times of message $k - 1$ and message k, i.e., $T_k = t_k - t_{k-1}$. Let S_k denote the system time of message k through the communication network, i.e., $S_k = \theta_k - t_k$.

We need to calculate the time average of the sample path of the age of information $D(t)$. To that end, let us consider a time interval $[0, \theta_n]$, during which n update messages are delivered to the central facility. Let t_1 be the first message sending time. The area under the curve is the sum of n polygons, associated to the n delivery times $\theta_1, \ldots, \theta_n$. The first $n - 1$ polygons are trapezoids, like the one shaded in Figure 4.7, the last one is a triangle. Let A_k denote the area of the k-th polygon, $k = 1, \ldots, n$.

The area of the trapezoid k can be determined as the difference of the areas of two isosceles triangles, one with base $\theta_k - t_{k-1} = \theta_k - t_k + t_k - t_{k-1} = S_k + T_k$, the second one with base $\theta_k - t_k = S_k$. Then, it is for $k = 1, \ldots, n - 1$

$$A_k = \frac{1}{2}(S_k + T_k)^2 - \frac{1}{2}S_k^2 = S_k T_k + \frac{1}{2}T_k^2 \tag{4.152}$$

The last area is $A_n = (S_n + T_n)^2/2$.

According to this decomposition of the area under $D(t)$ over the interval $[0, \theta_n]$, we can write

$$\int_0^{\theta_n} D(t)\, dt = \sum_{k=1}^n A_k = \sum_{k=1}^n \left(S_k T_k + \frac{1}{2}T_k^2\right) + \frac{1}{2}S_n^2 \tag{4.153}$$

Let $N(t)$ denote the number of update messages generated by the source in the time interval $[0, t]$. We have $N(\theta_n) = N(t_n) + N(\theta_n) - N(t_n) = n + N(S_n)$. Since the queueing model representing the communication network is assumed to be stable, $N(S_n)$ is finite with probability 1 as $n \to \infty$. Therefore $N(\theta_n) \sim n$ as $n \to \infty$.

Using eq. (4.153), we have

$$\langle D \rangle = \lim_{n \to \infty} \frac{1}{\theta_n} \int_0^{\theta_n} D(t)\, dt = \lim_{n \to \infty} \frac{N(\theta_n)}{\theta_n} \frac{\sum_{k=1}^{n} \left(S_k T_k + \frac{1}{2} T_k^2 \right) + \frac{1}{2} S_n^2}{N(\theta_n)}$$

$$= \lim_{t \to \infty} \frac{N(t)}{t} \lim_{n \to \infty} \frac{1}{n} \sum_{k=1}^{n} \left(S_k T_k + \frac{1}{2} T_k^2 \right) = \lambda \left(E[ST] + \frac{1}{2} E[T^2] \right)$$

with probability 1. We have exploited the limiting properties of the stationary renewal arrival process $N(t)$ and the finiteness of the system time (at equilibrium), which implies that $\lim_{n \to \infty} \frac{S_n^2/2}{n} = 0$ with probability 1. Then, under the assumption of ergodicity of the system, we can write:

$$E[D] = \lambda \left(E[TS] + \frac{1}{2} E[T^2] \right) \tag{4.154}$$

This is the general result we were seeking. It holds for a broad class of service systems in which the update messages are delivered in sequence. However, the evaluation of the average age of information is not simple: the variables T and S are dependent. A large inter-arrival time T allows the queue to empty, yielding a small system time S, i.e., T and S tend to be negatively correlated. For illustrative purposes, we address next a sufficiently simple model to yield to analysis.

4.6.2.2 Minimization of the Age of Information for an $M/M/1$ Model

Let us assume that arrivals follow a Poisson process with mean rate λ; service times, i.e., the time that the communication system spends to deliver one message to the central processing facility, have a negative exponential probability distribution, with mean μ. We assume $\lambda < \mu$ and let $\rho = \lambda/\mu$. Joining these with the hypotheses of the previous section, it appears that the communication system that delivers messages can be identified as an $M/M/1$ queue.

The system time of customer i is $S_i = W_i + X_i$, where W_i is the waiting time of customer i. We have to calculate $E[S_i T_i]$. W_i is the amount of unfinished work found in the queue at the i-th arrival. With FCFS this is the sum of the work found by arrival $i - 1$ plus the work brought in by that arrival, X_{i-1}, minus the time that has elapsed between arrival $i - 1$ and i, T_i:

$$W_i = \max\{0, W_{i-1} + X_{i-1} - T_i\} = \max\{0, S_{i-1} - T_i\} \tag{4.155}$$

This is known as the Lindley's recursion for the $G/G/1$ queue. The $\max\{0, \cdot\}$ operator has been introduced to account for the fact that, if $S_{i-1} < T_i$, the i-th arrival finds the queue empty and its waiting time W_i is 0. Then

$$E[S_i T_i] = E[(W_i + X_i)T_i] = E[W_i T_i] + E[X_i T_i]$$

$$= E[\max\{0, S_{i-1} - T_i\} T_i] + E[X_i] E[T_i] \tag{4.156}$$

due to the independence of the inter-arrival and the service times of a customer. From Section 4.5, we know that S_i is a negative exponential random variable with mean $1/(\mu - \lambda)$. The inter-arrival time PDF is $\lambda e^{-\lambda t}$, since arrivals are Poisson. Then

$$E[\max\{0, S_{i-1} - T_i\} T_i] = \int_0^\infty \lambda e^{-\lambda t}\, dt \int_t^\infty (y - t)(\mu - \lambda)e^{-(\mu-\lambda)y}\, dy$$

$$= \frac{\lambda}{\mu^2(\mu - \lambda)} \tag{4.157}$$

Putting all pieces together and noting that $E[T^2] = 2/\lambda^2$, we find finally the expression of the mean age of information for an $M/M/1$ model:

$$E[D] = \lambda \left(E[\max\{0, S - T\} T] + E[X]E[T] + \frac{1}{2}E[T^2] \right)$$

$$= \lambda \left[\frac{\lambda}{\mu^2(\mu - \lambda)} + \frac{1}{\mu\lambda} + \frac{1}{\lambda^2} \right] = \frac{1}{\mu} \left[1 + \frac{1}{\rho} + \frac{\rho^2}{1 - \rho} \right] \tag{4.158}$$

for $\rho \in (0, 1)$. As expected, $E[D]$ grows both when ρ becomes small, since then the rate of message update is small and information become obsolete before being refreshed, and when ρ gets close to 1, since in that case the communication system becomes a bottleneck for the messages, again resulting in a slack refresh of information. In the special case of $M/M/1$ we are able to obtain a closed form expression for $E[D]$, thus enabling a direct computation of the optimal ρ. Yet, the trend of $E[D]$ with the load of the system has a general validity.

Since the right-hand side of eq. (4.158) is convex and goes to ∞ at both extremes of the range of ρ, there exists a unique minimum inside the interval $(0, 1)$. The optimal value ρ^* is the solution of $\rho^4 - 2\rho^3 + \rho^2 - 2\rho + 1 = 0$ belonging to $(0, 1)$. It can be verified that $\rho^* = \frac{1}{2} \left(1 + \sqrt{2} - \sqrt{2\sqrt{2} - 1} \right) \approx 0.531$.

4.7 The G/M/1 Queue

The memoryless property of the negative exponential probability distribution allows a relatively simple analysis of another single server queue, the $G/M/1$ (and the corresponding finite queue size model, the $G/M/1/K$ queue). The number of customers in the queue, $Q(t)$, is not a Markov chain, since the inter-arrival times are not memoryless in general. A viable approach to analysis goes through the EMC, in this case selecting the arrival times as the sampling epochs.

Let the inter-arrival times be i.i.d. random variables with the probability distribution $F_T(t)$ of the random variable T. The mean arrival rate is denoted with $\lambda = 1/E[T]$, as usual. Service times are exponentially distributed, hence $F_X(t) = 1 - e^{-\mu t}$ for $t \geq 0$, with $\mu = 1/E[X]$ denoting the mean service rate.

Similarly to what done for the $M/G/1$ queue, it can be shown that the arrival times of customers admitted into the queue $\{t_{s,n}\}_{n \in \mathbb{Z}}$ form a sequence of regeneration points of the process $Q(t)$. Therefore, $Q_n = Q(t_{s,n-})$ is a Markov chain. The arrival $n + 1$ finds into the system the same number of customers found by the previous arrival, plus the previous arrival, minus the overall number of service completion occurred between the two arrivals. Let B_{n+1} denote the last quantity: it is the number of service completions occurring during the inter-arrival time between arrival n and arrival $n + 1$. Then, we have

$$Q_{n+1} = \max\{0, Q_n + 1 - B_{n+1}\} \tag{4.159}$$

This holds for any value of $Q_n \geq 0$.

Since inter-arrival times are i.i.d. random variables, it follows that $B_{n+1} \sim B, \forall n$. The probability distribution of B can be found in a way entirely similar to the corresponding random variable A introduced in the analysis of the $M/G/1$ EMC:

$$P(B = k) = \int_0^\infty P(B = k \mid T = t) f_T(t) \, dt = \int_0^\infty \frac{(\mu t)^k}{k!} e^{-\mu t} f_T(t) \, dt , \quad k \geq 0. \tag{4.160}$$

The corresponding generating function is $\phi_B(z) = \mathrm{E}[z^B] = \varphi_T(\mu - \mu z)$, where $\varphi_T(s)$ is the Laplace transform of the PDF of inter-arrival times. For a short notation, let $b_k \equiv P(B = k), \ k \geq 0$. From eq. (4.159) it is easily seen that

$$p_{ij} = P(Q_{n+1} = j \mid Q_n = i) = P(B = i + 1 - j) = b_{i+1-j} \tag{4.161}$$

for $j = 1, \ldots, i + 1$ and $i \geq 0$, and

$$p_{i0} = P(Q_{n+1} = 0 \mid Q_n = i) = P(B \geq i + 1) = \sum_{h=i+1}^\infty b_h \tag{4.162}$$

for $i \geq 0$, independent of n, thanks to the stationarity of B. In matrix form, the one-step probability transition matrix of the EMC Q_n is

$$\mathbf{P} = \begin{bmatrix} \tilde{b}_0 & b_0 & 0 & 0 & 0 & \cdots \\ \tilde{b}_1 & b_1 & b_0 & 0 & 0 & \cdots \\ \tilde{b}_2 & b_2 & b_1 & b_0 & 0 & \cdots \\ \tilde{b}_3 & b_3 & b_2 & b_1 & b_0 & \cdots \\ \cdots & \cdots & \cdots & \cdots & \cdots & \cdots \end{bmatrix} \tag{4.163}$$

where $\tilde{b}_i = \sum_{h=i+1}^\infty b_h = 1 - \sum_{h=0}^i b_h$ for $i \geq 0$. The EMC is irreducible and aperiodic since $b_0 > 0$ and $b_0 + b_1 < 1$. The limiting probability distribution is the unique left eigenvector $\mathbf{x}$ of $\mathbf{P}$ corresponding to the eigenvalue 1, normalized so that its components sum to 1. Given the special structure of $\mathbf{P}$, the probability distribution $\mathbf{x}$ can be represented in a quite simple form. The analysis of sample paths of the EMC of the $G/M/1$ in [130] proves that the structure of this EMC results in constant ratio between the probabilities of adjacent states, i.e., $x_{i+1}/x_i = \xi$. Let us then consider the following structure for the limiting probability distribution:

$$x_i = C\xi^i, \quad i \geq 0, \tag{4.164}$$

where C is a normalization constant. By requiring that $\sum_{i \geq 0} x_i = 1$, we find $C = 1 - \xi$. There remains to identify the ratio $\xi \in (0, 1)$. By plugging into the i-th equation of the linear system $\mathbf{xP} = \mathbf{x}$:

$$x_i = \sum_{k=0}^{\infty} x_{i-1+k} b_k, \quad i \geq 1, \tag{4.165}$$

the expression of the probabilities $\mathbf{x}$ in eq. (4.164), we find

$$C\xi^i = C \sum_{k=0}^{\infty} \xi^{i-1+k} b_k \quad \Rightarrow \quad \xi = \sum_{k=0}^{\infty} b_k \xi^k = \phi_B(\xi) = \varphi_T(\mu - \mu\xi) \tag{4.166}$$

The parameter ξ is a solution of the fixed point equation $z = \phi_B(z)$. Since $\phi'_B(1) = \mu E[T] = 1/\rho > 1$, we can invoke Theorem 3.14, and therefore we end up proving that ξ is the unique root of $z = \phi_B(z)$ in $(0, 1)$.

It can be verified that the PDF $x_i = (1 - \xi)\xi^i$, $i \geq 0$, satisfies also the first equation of the system $\mathbf{xP} = \mathbf{x}$, namely:

$$\sum_{i=0}^{\infty} x_i \tilde{b}_i = \sum_{i=0}^{\infty} x_i \left(1 - \sum_{h=0}^{i} b_h \right) = 1 - \sum_{h=0}^{\infty} b_h \sum_{i=h}^{\infty} (1 - \xi)\xi^i$$

$$= 1 - \sum_{h=0}^{\infty} b_h \xi^h = 1 - \phi_B(\xi) = 1 - \xi = x_0$$

The amazing result for the EMC of the $G/M/1$ queue is that the probability distribution *seen by an arrival that joins the queue* is simply a *geometric* one. Since the queue is single arrival, single service, the probability distribution of the number of customers left behind by a departure is the same as x_k. In this case, however, we cannot invoke the PASTA property, hence, the probability distribution p_k of $Q(t)$ at a generic time of the equilibrium is *not* the same as x_k. The following result can be shown for the distribution p_k of the number of customers in the queue at any time at equilibrium.

Theorem 4.6 The probability distribution $p_k = P(Q(t) = k)$ of the $G/M/1$ queue is $p_k = \rho(1 - \xi)\xi^{k-1}$, $k \geq 1$ and $p_0 = 1 - \rho$, where $\rho = \lambda/\mu$.

Proof: Le us consider a generic time t_0 and let t_a be the last arrival time before t_0. $T_a = t_0 - t_a$ is the age of the inter-arrival time, whose probability distribution is the same as that of the residual inter-arrival time, i.e., $f_{T_a}(t) = \lambda G_T(t)$. For $k \geq 1$ the event $Q(t) = k$ can occur if and only if there were $k - 1 + r$ customers in the queue at t_a^-, an arrivals occurs at time t_a and r service completions are carried out during time T_a, for all possible $r = 0, 1, \ldots$. Conditioning upon $T_a = x$ and reminding that the number of service completions in a time interval x, $C(x)$, follows a Poisson law with parameter μ, we can write

$$P(Q(t) = k \mid C(T_a) = r, \ T_a = x) = P(Q(t_{a-}) = k - 1 + r) = x_{k-1+r}, \quad (4.167)$$

for $k \geq 1$. Removing the conditioning and using the PDF of T_a given by $f_{T_a}(x) = \lambda G_T(x)$ and the fact that

$$P(C(T_a) = r \mid T_a = x) = \frac{(\mu x)^r}{r!} e^{-\mu x}, \quad r \geq 0, \quad (4.168)$$

we can write

$$p_k = P(Q(t) = k)$$

$$= \int_0^\infty \sum_{r=0}^\infty P(Q(t) = k \mid C(T_a) = r, \ T_a = x) P(C(T_a) = r \mid T_a = x) f_{T_a}(t) \, dt \quad (4.169)$$

Developing this expression, we find

$$p_k = P(Q(t) = k) = \int_0^\infty \sum_{r=0}^\infty x_{k-1+r} \frac{(\mu x)^r}{r!} e^{-\mu x} \lambda G_T(x) \, dx$$

$$= \lambda \int_0^\infty \sum_{r=0}^\infty (1 - \xi) \xi^{k-1+r} \frac{(\mu x)^r}{r!} e^{-\mu x} G_T(x) \, dx$$

$$= \lambda (1 - \xi) \xi^{k-1} \int_0^\infty \sum_{r=0}^\infty \frac{(\xi \mu x)^r}{r!} e^{-\mu x} G_T(x) \, dx$$

$$= \lambda (1 - \xi) \xi^{k-1} \int_0^\infty e^{\xi \mu x - \mu x} G_T(x) \, dx$$

$$= \lambda (1 - \xi) \xi^{k-1} \left(\left[\frac{e^{\xi \mu x - \mu x}}{\xi \mu - \mu} G_T(x) \right]_0^\infty + \int_0^\infty \frac{e^{\xi \mu x - \mu x}}{\xi \mu - \mu} f_T(x) \, dx \right)$$

$$= \lambda (1 - \xi) \xi^{k-1} \left(\frac{1}{\mu - \xi \mu} + \frac{1}{\xi \mu - \mu} \varphi_T(\mu - \mu \xi) \right)$$

$$= \lambda (1 - \xi) \xi^{k-1} \left(\frac{1}{\mu - \xi \mu} - \frac{\xi}{\mu - \xi \mu} \right) = \frac{\lambda}{\mu} (1 - \xi) \xi^{k-1} \quad (4.170)$$

The probability p_0 can be found from the congruence constraint, i.e., $p_0 = 1 - \sum_{k=1}^\infty p_k = 1 - \rho$. $\blacksquare$

Again, an amazing result: it is simply $p_k = \rho x_{k-1}$ for $k \geq 1$.
Let us now derive some performance metrics.

The coefficient of utilization of the server is $P(Q(t) > 0) = \sum_{k=1}^{\infty} p_k = \rho$. This is different from the probability that an arrival finds the server busy. The latter is $P_b = P(Q_n > 0) = 1 - x_0 = \xi \neq \rho$, unless the arrivals are Poisson. In this last case, the fixed point equation for ξ becomes

$$\xi = \frac{\lambda}{\mu - \mu\xi + \lambda} = \frac{\rho}{1 - \xi + \rho} \tag{4.171}$$

It is apparent that $\xi = \rho$ is the unique solution in $(0, 1)$.

The mean number of customers in the queue is

$$E[Q(t)] = \sum_{k=0}^{\infty} k p_k = \frac{\rho}{1 - \xi} \tag{4.172}$$

As for the delay, the probability distribution we need is just the one seen by arrivals, x_k. Let us assume the FCFS discipline. A customer arriving and finding k customers, has to wait the sum of k i.i.d. negative exponential random variables with mean $1/\mu$[10]. Then:

$$\varphi_W(s) = E[e^{-sW}] = \sum_{k=0}^{\infty} x_k E[e^{-sW} \mid Q = k]$$

$$= 1 - \xi + (1 - \xi) \sum_{k=1}^{\infty} \xi^k E[e^{-s(X_1 + \cdots + X_k)}]$$

$$= 1 - \xi + (1 - \xi) \sum_{k=1}^{\infty} \xi^k [\varphi_X(s)]^k$$

$$= 1 - \xi + (1 - \xi)\frac{\xi\varphi_X(s)}{1 - \xi\varphi_X(s)} = 1 - \xi + \xi\frac{\mu(1 - \xi)}{s + \mu(1 - \xi)} \tag{4.173}$$

In the second line, it is $W = 0$ for $Q = 0$, hence $E[e^{-sW} \mid Q = 0] = 1$. In the last line $\varphi_X(s) = \frac{\mu}{s+\mu}$ has been used. The inversion of the obtained Laplace transform is elementary, yielding

$$f_W(t) = (1 - \xi)\delta(t) + \xi\mu(1 - \xi)e^{-\mu(1-\xi)t} \tag{4.174}$$

where $\delta(t)$ is a Dirac delta function. The CDF of the waiting time is:

$$F_W(t) = 1 - \xi e^{-\mu(1-\xi)t}, \quad t \geq 0. \tag{4.175}$$

The CDF has a discontinuity at $t = 0$. The size of the jump is $1 - \xi$, that is, the probability of finding the system empty upon arrival, and hence suffering no wait. The Laplace transform of the system time probability distribution is

$$\varphi_S(s) = \varphi_W(s)\varphi_X(s) = \frac{(1 - \xi)\varphi_X(s)}{1 - \xi\varphi_X(s)} = \frac{\mu(1 - \xi)}{s + \mu(1 - \xi)} \tag{4.176}$$

---

10 The residual service time of the customer under service has the same probability distribution as all other service times, thanks to the memoryless property of the negative exponential random variable.

whence the inverse transform is

$$f_S(t) = \mu(1 - \xi)e^{-\mu(1-\xi)t}, \quad t \geq 0. \tag{4.177}$$

Those are yet other amazing results of the $G/M/1$ model: the system and the waiting times have a simple negative exponential probability distribution (the latter with a jump at $t = 0$). The form of these results is exactly the same of the corresponding ones for the $M/M/1$ queue, except that ξ replaces ρ.

Finally, we consider briefly the $G/M/1/K$ queue. The same EMC at customer arrival time points can be considered as with the $G/M/1$ queue. Derivations repeat as above, except that there is a maximum number K of customers that can be accommodated in the waiting line. An arrival that finds $K + 1$ customers already into the queue is rejected. Then, the one-step probability transition matrix of the EMC is of size $(K + 2) \times (K + 2)$:

$$\mathbf{P}(K) = \begin{bmatrix} \tilde{b}_0 & b_0 & 0 & 0 & \cdots & 0 \\ \tilde{b}_1 & b_1 & b_0 & 0 & \cdots & 0 \\ \tilde{b}_2 & b_2 & b_1 & b_0 & \cdots & 0 \\ \cdots & \cdots & \cdots & \cdots & \cdots & \cdots \\ \tilde{b}_K & b_K & b_{K-1} & b_{K-2} & \cdots & b_0 \\ \tilde{b}_K & b_K & b_{K-1} & b_{K-2} & \cdots & b_0 \end{bmatrix} \tag{4.178}$$

where $\tilde{b}_k = \sum_{j=k+1}^{\infty} b_j$, $k \geq 0$.

We add an argument K to avoid confusions with the infinite $G/M/1$ queue EMC. The limiting probability distribution $\{x_k(K)\}_{k=0,1,\ldots,K+1}$ exists for whatever value of $A_o = \lambda/\mu$, provided the EMC is irreducible and aperiodic, which is the case since all b_k's are strictly positive. The limiting probabilities $\mathbf{x}(K) = [x_0(K), \ldots, x_{K+1}(K)]$ can be found by solving numerically the linear equation system $\mathbf{x}(K)\mathbf{P}(K) = \mathbf{x}(K)$, with the normalization condition $\mathbf{x}(K)\mathbf{e} = 1$, where $\mathbf{e}$ is a column vector of 1's of size $K + 2$.

Apart from delay, in the finite queue size model the loss probability is a relevant metric. It is $P_L = x_{K+1}(K)$ for the $G/M/1/K$ queue. As for the mean waiting time, we have

$$E[W] = \frac{1}{\mu} \sum_{j=0}^{K} j \frac{x_j(K)}{1 - x_{K+1}(K)} \tag{4.179}$$

Asymptotic expansions and approximations can be determined for the $G/M/1/K$ in pretty much the same way as done with the $M/G/1/K$ queue. Alternatively, a duality relationship can be noted between the two models. Given an $M/G/1/K$ queue with mean arrival rate λ and mean service rate μ, we can define a dual system by inter-changing the roles of arrivals and service processes. Namely, inter-arrival times of the dual system have the same probability distribution as service time of the original queue; similarly, service times of the dual system have the same probability distribution as the inter-arrival times of the original queue. Also, state i of the EMC at customer departure times of the original queue corresponds to state $K - i$ of the EMC at arrival times of the dual queue, $i = 0, 1, \ldots, K$. Then, it can be recognized that the dual queue is a $G/M/1/K - 1$ queue with mean arrival rate μ and mean service rate λ. With this duality definition it is possible to carry over results obtained for the $M/G/1/K$ queue to the corresponding dual $G/M/1/K - 1$ queue.

Example 4.5 Inter-arrival times form a renewal process described by the following random variable:

$$T = \begin{cases} T_0 & \text{w.p. } p \\ T_0 + \dfrac{1/\lambda - T_0}{1 - p} Y & \text{w.p. } 1 - p \end{cases} \tag{4.180}$$

where Y is a negative exponential random variable with unit mean. This is a traffic model of an intermittent source. Service requests are issued in a row, at a regular pace of 1 request every T_0 seconds. The length of the service request burst is geometrically distributed, with mean $n_b = 1/(1 - p)$. Subsequent bursts are separated by a random "silence" time of the source. The silence time is exponentially distributed. The overall mean arrival rate of request is λ. The Laplace transform of the PDF of T is

$$\varphi_T(s) = pe^{-sT_0} + (1 - p)e^{-sT_0} \frac{\alpha}{s + \alpha} \tag{4.181}$$

where $\alpha = \lambda(1 - p)/(1 - \lambda T_0)$.

The parameter ξ is found by solving numerically $\xi = \varphi_T(\mu - \mu\xi)$, that is

$$\xi = pe^{-\hat{T}_0(1-\xi)} + (1 - p)e^{-\hat{T}_0(1-\xi)} \frac{\hat{\alpha}}{1 - \xi + \hat{\alpha}} = e^{-\hat{T}_0(1-\xi)} \frac{p(1 - \xi) + \hat{\alpha}}{1 - \xi + \hat{\alpha}} \tag{4.182}$$

where $\hat{\alpha} = A_0(1 - p)/(1 - \lambda T_0)$ and $\hat{T}_0 = \lambda T_0/A_0$. The coefficients b_k forming the one-step transition probability matrix of the EMC for this example can be computed by using the following formula:

$$b_k = p\frac{\hat{T}_0^k}{k!}e^{-\hat{T}_0} + (1 - p)e^{-\hat{T}_0} \frac{\hat{\alpha}}{\hat{\alpha} + 1}\left(\frac{1}{\hat{\alpha} + 1}\right)^k \sum_{h=0}^{k} \frac{[(\hat{\alpha} + 1)\hat{T}_0]^h}{h!} \tag{4.183}$$

for $k \geq 0$. It is apparent that everything depends only on three nondimensional parameters: λT_0, A_0, and p. In this example we assume $A_0 = \lambda/\mu = 0.7$, $\lambda T_0 = 1/25$. We consider three different values of the mean number of requests in a burst, $n_b = 5, 10, 20$; the probability p is calculated as $p = 1 - 1/n_b$.

The mean waiting time for the infinite queue size case is simply $E[W(\infty)] = \frac{1/\mu}{1-\xi}$. For the finite case with queue size K, first the state probability distributions are

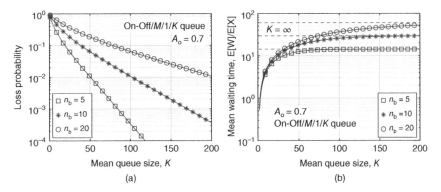

Figure 4.8 Performance metrics of the On-Off/M/1/K queue as a function of the queue size K. Left plot: loss probability $P_L(K)$. Right plot: mean waiting time normalized with respect to mean service time.

found, then eq. (4.179) is applied. The loss probability is found directly from the state probabilities: $P_L(K) = x_{K+1}(K)$.

Figure 4.8 shows the plots of $P_L(K)$ and $E[W(K)]/E[X]$ versus K. In the right plot, the dashed lines show the asymptotic values of the mean waiting time, when the queue size grows to infinity.

To gain insight into the numerical values displayed, note that the range of queue sizes in the plot of Figure 4.8 goes up to 20 times the maximum considered burst size. In spite of that, it is apparent that the loss probability is quite high (in the order of 0.01 for $n_b = 20$ even for $K = 200$.) The mean waiting time grows quickly close its upper limit. The critical performance shown by these results are directly related to the burstiness of the traffic demand. Arrivals of service requests tend to cluster as n_b gets bigger. This leads to a degradation of the performance, both in terms of loss probability and mean waiting time. Above all, it entails a quite unfavorable trade-off on K. If we keep K small to limit the mean delay, we end up with an excessively high loss probability. If, on the contrary, we boost K to pull down the loss probability, we cannot avoid having a mean waiting time as large as we would experience with an infinite queue size system.

4.8 Matrix-Geometric Queues

This section is devoted to hint at a quite general model stemming from the $M/G/1$ and the $G/M/1$ queues. The main contributor in this area has been M.F. Neuts [167, 168]. The reader wishing to learn more and take a systematic introduction to the matrix-geometric methods in queueing theory can consult his books. Here we aim at grasping the potential offered by this large class of models.

The key idea is to describe the state of the queueing system with a joint discrete variable $(Q(t), J(t))$, where $Q(t)$ is the number of customers in the queue and $J(t)$ is the *phase* of the process, at time t. The variable $J(t)$ can take a finite number r of values, say the integers in the set $\{1, \ldots, r\}$. The models considered in this section are those for which the sampled process $(Q_n, J_n) \equiv (Q(t_n), J(t_n))$ in discrete time is a Markov chain[11]. The one-step transition probability matrix $\mathbf{P}$ of the Markov chain can be organized so as to highlight its structure. While the limiting probability state vector can be found by using general numerical methods, there are cases of practical interest where the special structure of the matrix $\mathbf{P}$ allows a deep analytical development, with very efficient numerical methods even for large size systems.

In the rest of this section we introduce of the most common structures and the relevant basic results.

4.8.1 Quasi Birth-Death (QBD) Processes

This is the case where Q_n can make transitions only between adjacent levels. In other words, if the number of customers at time t_n is equal to k, at the next time t_{n+1} it can only move to k, $k+1$ or $k-1$. During the same transition, the phase J_n can change from whatever initial state i to any other final state j. For concreteness, we assume that the time points t_n refer to customer departures. As a result, the matrix $\mathbf{P}$ has the following structure:

$$
\mathbf{P} = \begin{bmatrix}
\mathbf{B}_{00} & \mathbf{B}_{01} & \mathbf{0} & \mathbf{0} & \mathbf{0} & \mathbf{0} \\
\mathbf{B}_{10} & \mathbf{A}_1 & \mathbf{A}_2 & \mathbf{0} & \mathbf{0} & \mathbf{0} \\
\mathbf{0} & \mathbf{A}_0 & \mathbf{A}_1 & \mathbf{A}_2 & \mathbf{0} & \mathbf{0} \\
\mathbf{0} & \mathbf{0} & \mathbf{A}_0 & \mathbf{A}_1 & \mathbf{A}_2 & \mathbf{0} \\
\cdots & \cdots & \cdots & \cdots & \cdots & \cdots
\end{bmatrix}
\tag{4.184}
$$

where $\mathbf{A}_s$ are square matrices of size $r \times r$ for $s = 0, 1, 2$. $\mathbf{B}_{00}$, $\mathbf{B}_{01}$, and $\mathbf{B}_{10}$ are boundary matrices of size $\ell \times \ell$, $\ell \times r$, and $r \times \ell$, respectively, where ℓ can be different from r in general. For $\mathbf{P}$ to be stochastic, each component block of $\mathbf{P}$ must be non-negative, $\mathbf{A} = \mathbf{A}_0 + \mathbf{A}_1 + \mathbf{A}_2$ must be a stochastic matrix and similar conditions for the first $\ell + r$ boundary rows of the matrix must hold.

The boundary matrices are useful to allow the possibility of a special queue behavior in the boundary level $Q = 0$. The meaning of the entries of the various matrices is intuitive. For example, the (i, j) entry of $\mathbf{A}_s$ is

$$
a_s(i, j) = P(A = s, J_{n+1} = j \mid J_n = i), \quad s = 0, 1, 2
\tag{4.185}
$$

---

[11] In the following we refer to discrete time, e.g., the time point series defined by an EMC. Everything can be quite easily mapped to Markov processes, if continuous time is considered. In that case probabilities are replaced by transition rates in the matrix entries.

where A denotes the number of arrivals during a service time. Thus, $a_0(i,j)$ is the probability that no arrival takes place in a service time and the phase moves from i to j. Since there is a departure, but no arrival, the state Q goes from its current level n to $n - 1$. This is why the matrix $\mathbf{A}_0$ appears in the subdiagonal of $\mathbf{P}$.

A process with a one-step transition probability matrix like $\mathbf{P}$ in eq. (4.184) is named a quasi birth-death (QBD) process. The reason of the name is evident from the fact that $\mathbf{P}$ shares the same tri-diagonal structure of the transition rate matrix of a birth-death process. We have encountered a similar matrix structure when dealing with the $M/M/1$ queue. In that case the phase is a constant ($r = 1$) and the limiting state probability distribution is simply geometric. Therefore, we are motivated to search for an analogous distribution, though with a matrix structure. We start by considering a block vector structure for the limiting probabilities:

$$\mathbf{x} = [\mathbf{x}_0 \quad \mathbf{x}_1 \quad \mathbf{x}_2 \dots] \tag{4.186}$$

where each block $\mathbf{x}_k$ is a row vector of size r, except of $\mathbf{x}_0$, that has length ℓ. Then, we seek a solution having the following property:

$$\mathbf{x}_k = \mathbf{x}_{k-1}\mathbf{R} , \quad k \geq 2, \tag{4.187}$$

where $\mathbf{R}$ is a suitable matrix yet to be determined. Equation (4.187) implies that

$$\mathbf{x}_k = \mathbf{x}_1\mathbf{R}^{k-1} , \quad k \geq 1. \tag{4.188}$$

The linear system $\mathbf{x}\mathbf{P} = \mathbf{x}$ can be written explicitly as follows:

$$\mathbf{x}_0\mathbf{B}_{00} + \mathbf{x}_1\mathbf{B}_{10} = \mathbf{x}_0 \tag{4.189}$$

$$\mathbf{x}_0\mathbf{B}_{01} + \mathbf{x}_1\mathbf{A}_1 + \mathbf{x}_2\mathbf{A}_0 = \mathbf{x}_1 \tag{4.190}$$

$$\mathbf{x}_{k-1}\mathbf{A}_0 + \mathbf{x}_k\mathbf{A}_1 + \mathbf{x}_{k+1}\mathbf{A}_2 = \mathbf{x}_k , \quad k \geq 2 \tag{4.191}$$

Plugging into these equations the hypothesized solution form, we find that all equations for $k \geq 2$ are simultaneously satisfied if $\mathbf{R}$ solves the following quadratic matrix equation:

$$\mathbf{A}_0 + \mathbf{R}\mathbf{A}_1 + \mathbf{R}^2\mathbf{A}_2 = \mathbf{R} \tag{4.192}$$

It can be shown that under stability conditions that will be discussed soon, the sought-for matrix $\mathbf{R}$ is the minimal non-negative solution of eq. (4.192). *Minimal* means that for any other matrix $\mathbf{S}$ solving eq. (4.192), we have $\mathbf{0} \leq \mathbf{R} \leq \mathbf{S}$. Moreover, $\mathbf{R}$ is sub-stochastic. Hence there exists the inverse of the matrix $\mathbf{I} - \mathbf{R}$, where $\mathbf{I}$ is the $r \times r$ identity matrix.

As for the boundary equations (4.189) and (4.190), by using eq. (4.187), they can be written as

$$[\mathbf{x}_0 \quad \mathbf{x}_1] \begin{bmatrix} \mathbf{B}_{00} - \mathbf{I} & \mathbf{B}_{01} \\ \mathbf{B}_{10} & \mathbf{A}_1 + \mathbf{R}\mathbf{A}_0 - \mathbf{I} \end{bmatrix} = [\mathbf{0} \quad \mathbf{0}] \tag{4.193}$$

We have to find a nontrivial solution of this homogeneous linear system, under the normalization condition obtained by requiring that the sum of all probabilities be 1, i.e.,

$$\sum_{k=0}^{\infty} \mathbf{x}_k \mathbf{e} = \mathbf{x}_0 \mathbf{e} + \mathbf{x}_1 \sum_{k=1}^{\infty} \mathbf{R}^{k-1} \mathbf{e} = \mathbf{x}_0 \mathbf{e} + \mathbf{x}_1 (\mathbf{I} - \mathbf{R})^{-1} \mathbf{e} = 1 \tag{4.194}$$

where $\mathbf{e}$ is a column vector of 1's and we have used the identity

$$\sum_{i=0}^{\infty} \mathbf{R}_i = (\mathbf{I} - \mathbf{R})^{-1} \tag{4.195}$$

that holds because the spectral radius[12] of $\mathbf{R}$ is less than 1, $\mathbf{R}$ being a sub-stochastic matrix. The linear system (4.193), together with the normalization condition (4.194) and the knowledge of the matrix $\mathbf{R}$, yields the entire state probability distribution.

Finally, we need to determine when the limiting probability distribution does exist, i.e., the stability conditions of the system. First, we assume that the matrix $\mathbf{P}$ is irreducible and aperiodic. This is true if the matrix $\mathbf{A} = \mathbf{A}_0 + \mathbf{A}_1 + \mathbf{A}_2$ is the one-step transition probability matrix of an irreducible and aperiodic Markov chain. Since this last Markov chain has a finite number of states, then its limiting state probability distribution exists. We denote it with θ. Then, it can be proved that the condition for stability is (see [167, Th. 3.1.1, p. 82])

$$\theta \mathbf{A}_2 \mathbf{e} < \theta \mathbf{A}_0 \mathbf{e} \tag{4.196}$$

The stability condition has an intuitive interpretation. By observing the structure of $\mathbf{P}$, the inequality (4.196) can be read as the requirement that the drift of the Markov chain be negative, i.e., the probability of making a transition up (one more customer into the queue) be strictly less than the probability of making a transition down (one less customer into the queue).

To obtain the probability distribution vector $\mathbf{x}$ numerically, we need to compute $\mathbf{R}$. This can be done simply by using the following iteration

$$\mathbf{R}_{m+1} = -\mathbf{A}_0 (\mathbf{I} - \mathbf{A}_1)^{-1} - \mathbf{R}_m^2 \mathbf{A}_2 (\mathbf{I} - \mathbf{A}_1)^{-1}, \qquad m \geq 0, \tag{4.197}$$

initialized with $\mathbf{R}_0 = \mathbf{0}$. This iteration is not particularly efficient. A much more efficient algorithm has been defined by Latouche and Ramaswami [140].

4.8.2 M/G/1 and G/M/1 Structured Processes

An immediate generalization of a QBD process is to have a block Hessenberg structure for $\mathbf{P}$. We define a *G/M/1* structured Markov chain as one having the

---

12 The spectral radius of a matrix $\mathbf{C}$, denoted with $sp(\mathbf{C})$, is the maximum of the moduli of its eigenvalues.

following form of the one-step transition probability matrix:

$$P = \begin{bmatrix} \mathbf{B}_0 & \mathbf{A}_0 & 0 & 0 & 0 & 0 & \cdots \\ \mathbf{B}_1 & \mathbf{A}_1 & \mathbf{A}_0 & 0 & 0 & 0 & \cdots \\ \mathbf{B}_2 & \mathbf{A}_2 & \mathbf{A}_1 & \mathbf{A}_0 & 0 & 0 & \cdots \\ \mathbf{B}_3 & \mathbf{A}_3 & \mathbf{A}_2 & \mathbf{A}_1 & \mathbf{A}_0 & 0 & \cdots \\ \cdots & \cdots & \cdots & \cdots & \cdots & \cdots & \cdots \end{bmatrix} \tag{4.198}$$

where $\mathbf{A}_s$ are square matrices of size $r \times r$ for $s \geq 0$, and the $\mathbf{B}_i$ are boundary matrices of appropriate sizes. For $\mathbf{P}$ to be stochastic, we have that all component block matrices are non-negative and

$$\mathbf{B}_k \mathbf{e} + \sum_{i=0}^{k} \mathbf{A}_i \mathbf{e} = \mathbf{e} , \qquad k \geq 0, \tag{4.199}$$

where $\mathbf{e}$ is a column vector of 1's. As a consequence of eq. (4.199), the matrix $\mathbf{A} = \sum_{k=0}^{\infty} \mathbf{A}_k$ is sub-stochastic. In most applications, it turns out to be irreducible.

The block lower Hessenberg structure is clearly reminiscent of that of the EMC of the $G/M/1$ queue. Therefore, we expect that a geometric structure still applies to the probability distribution, when we consider its block structure as in eq. (4.186). We assume a solution structure given by

$$\mathbf{x}_k = \mathbf{x}_0 \mathbf{R}^k , \qquad k \geq 0, \tag{4.200}$$

We can find out the fundamental equation that the matrix $\mathbf{R}$ must satisfy by substituting the expression in eq. (4.200) for $\mathbf{x}$ in the linear system $\mathbf{x}\mathbf{P} = \mathbf{x}$. It is

$$\sum_{i=0}^{\infty} \mathbf{R}^i \mathbf{A}_i = \mathbf{R} \tag{4.201}$$

In [167, Ch. 1] there is a detailed discussion of the solution of the equation (4.201). The entries of the matrix $\mathbf{R}$ have a probabilistic interpretation: the element R_{ij} is the probability that the state moves from (ℓ, i) to $(\ell + 1, j)$, without ever returning to level ℓ, for $i, j = 1, \ldots, r$. Such probabilities do not depend on the starting level ℓ, given the special structure of $\mathbf{P}$. It is proved that, if the Markov chain $\mathbf{P}$ is positive recurrent, then $\mathbf{R}$ is the minimal non-negative solution of the equation $\sum_{i=0}^{\infty} \mathbf{X}^i \mathbf{A}_i = \mathbf{X}$, and it has a spectral radius $sp(\mathbf{R}) \leq sp(\mathbf{A}) \equiv \xi < 1$. As for the stability of the Markov chain, following is proved:

Theorem 4.7 If $\mathbf{A}$ is irreducible, the Markov chain $\mathbf{P}$ is positive recurrent if and only if: (i) $\theta \sum_{k=0}^{\infty} k\mathbf{A}_k \mathbf{e} > 1$, where θ is the left eigenvector of $\mathbf{A}$ corresponding to its maximum modulus eigenvalue ξ, normalized so that $\theta \mathbf{e} = 1$; (ii) the stochastic matrix $\sum_{k=0}^{\infty} \mathbf{R}^k \mathbf{B}_k$ has a strictly positive left invariant vector for the eigenvalue 1.

Under the hypotheses of Theorem 4.7, the boundary block probabilities $\mathbf{x}_0$ can be found as the suitably normalized left invariant vector of a matrix depending on the boundary matrices $\mathbf{B}_k$'s. In formulas

$$\mathbf{x}_0 = \mathbf{x}_0 \sum_{k=0}^{\infty} \mathbf{R}^k \mathbf{B}_k \qquad \mathbf{x}_0 (\mathbf{I} - \mathbf{R})^{-1} \mathbf{e} = 1 \tag{4.202}$$

Then, the entire probability distribution is known, as given in eq. (4.200), where $\mathbf{R}$ is the minimal non-negative solution of the eq. (4.201). Even in this apparently much more general case, a simple matrix-geometric solution is found. This is somewhat less surprising than it might be, in view of the result obtained for the ordinary $G/M/1$ queue. There it is shown that the probability distribution of the number of customers seen by an arrival has a simple geometric form. The result given above for the $G/M/1$ structured queues is a natural generalization of the one holding for the simple $G/M/1$ queue.

Note that the ergodicity condition $\theta \sum_{k=0}^{\infty} k \mathbf{A}_k \mathbf{e} > 1$ reduces to the negativity of the drift for QBD. This can be checked by remembering that $\mathbf{A}_k = \mathbf{0}$ for $k > 2$ and $\mathbf{A}_1 \mathbf{e} = (\mathbf{A} - \mathbf{A}_0 - \mathbf{A}_2)\mathbf{e} = \mathbf{e} - \mathbf{A}_0 \mathbf{e} - \mathbf{A}_2 \mathbf{e}$ for a QBD process[13] .

The other class of structured processes corresponds to an upper Hessenberg structure of the one-step transition probability matrix, that is an $M/G/1$ structured process:

$$\mathbf{P} = \begin{bmatrix} \mathbf{B}_0 & \mathbf{B}_1 & \mathbf{B}_2 & \mathbf{B}_3 & \cdots \\ \mathbf{C}_0 & \mathbf{A}_1 & \mathbf{A}_2 & \mathbf{A}_3 & \cdots \\ \mathbf{0} & \mathbf{A}_0 & \mathbf{A}_1 & \mathbf{A}_2 & \cdots \\ \mathbf{0} & \mathbf{0} & \mathbf{A}_0 & \mathbf{A}_1 & \cdots \\ \mathbf{0} & \mathbf{0} & \mathbf{0} & \mathbf{A}_0 & \cdots \\ \cdots & \cdots & \cdots & \cdots & \cdots \end{bmatrix} \tag{4.203}$$

where $\mathbf{A}_s$ are square matrices of size $r \times r$ for $s \geq 0$, and the $\mathbf{B}_i$ and $\mathbf{C}_0$ are boundary matrices of appropriate sizes. For $\mathbf{P}$ to be stochastic, we have that all component block matrices are non-negative and that both $\mathbf{B} = \sum_{k=0}^{\infty} \mathbf{B}_k$ and $\mathbf{A} = \sum_{k=0}^{\infty} \mathbf{A}_k$ are stochastic matrices.

This time, if $\mathbf{A}$ is irreducible, the Markov chain is positive and recurrent if $\theta \sum_{k=0}^{\infty} k \mathbf{A}_k \mathbf{e} < 1$, where θ is the limiting state probability vector associated to the irreducible and finite (hence ergodic) Markov chain $\mathbf{A}$.

The key role is played by the matrix $\mathbf{G}$ that is shown to be the minimal non-negative solution of the matrix equation

$$\mathbf{G} = \sum_{k=0}^{\infty} \mathbf{A}_k \mathbf{G}^k \tag{4.204}$$

---

13 Note the different numbering convention adopted here with respect to the numbering used in the definition of the QBD one-step probability transition matrix.

The entry (i, j) of $\mathbf{G}$ has a probabilistic interpretation. G_{ij} is the probability that the process enters level $n - 1$ for the first time by touching phase j, given that the process started from level n in phase i, for $i, j = 1, \ldots, r$ and for any level $n \geq 2$.

As usual, we decompose the probability vector of the Markov chain $\mathbf{P}$ in blocks, as shown in eq. (4.186). Analogously to the $M/G/1$ queue, for which no simple closed form is known for the limiting state probabilities, also in the case of $M/G/1$ structured processes we cannot give an elegant matrix-geometric solution. Nevertheless, it is possible to lay out a numerically feasible procedure to evaluate the probabilities $\mathbf{x}$ even in the case of an $M/G/1$ structured process. The derivation is quite lengthy and we skip the details. The interested reader can find an extensive discussion in [168] and a synthetic account in [196, Ch. 10]. We can summarize the steps required to compute the $\mathbf{x}_k$'s as follows:

1. Compute the matrix $\mathbf{G}$, by solving eq. (4.204).
2. Compute the boundary probability vector $\mathbf{x}_0$ as the solution of the linear system

$$\mathbf{x}_0(\mathbf{I} - \mathbf{B}_0 - \mathbf{B}_1^*[\mathbf{A}_1^*]^{-1}\mathbf{C}_0) = \mathbf{0} \tag{4.205}$$

   subject to the normalization condition

$$\mathbf{x}_0\mathbf{e} + \mathbf{x}_0 \left(\sum_{i=0}^{\infty} \mathbf{B}_i^* \right) \left(\sum_{i=0}^{\infty} \mathbf{A}_i^* \right) \mathbf{e} = 1 \tag{4.206}$$

3. Compute $\mathbf{x}_1$ from $\mathbf{x}_1 = \mathbf{x}_0\mathbf{B}_1^*[\mathbf{A}_1^*]^{-1}$.
4. Compute $\mathbf{x}_i$, $i \geq 2$, from

$$\mathbf{x}_i = \left(\mathbf{x}_0\mathbf{B}_i^* - \sum_{j=1}^{i-1} \mathbf{x}_j\mathbf{A}_{i-j+1}^* \right) [\mathbf{A}_1^*]^{-1} \tag{4.207}$$

In the algorithm above we used the following definitions:

$$\mathbf{B}_i^* = \sum_{k=i}^{\infty} \mathbf{B}_k\mathbf{G}^{k-i}, \qquad i \geq 1;$$

$$\mathbf{A}_i^* = -\sum_{k=i}^{\infty} \mathbf{A}_k\mathbf{G}^{k-i}, \qquad i \geq 2;$$

$$\mathbf{A}_1^* = \mathbf{I} - \sum_{k=1}^{\infty} \mathbf{A}_k\mathbf{G}^{k-1}$$

The major numerical issues with the algorithms in this section stem from the computation of the key matrices $\mathbf{R}$ or $\mathbf{G}$, given that the summations appearing in the respective equations to be solved are infinite in general. This requires truncation and an accurate control of error propagation in the iterative solution procedure. Another numerical issue lies with the size of the matrices, namely r, which might give rise to both a storage and computational complexity problem. In many applications those matrices are relatively sparse, so that efficient storage and computation algorithm for specific cases can be devised.

4.9   A General Result on Single-Server Queues

Let us consider a single-server queue with a work-conserving server character-ized by an average serving rate C. The queue has infinite storage, so that new service requests are always admitted into the queue. Arrivals are described in a general way, by means of the arrival process counting function. Each arrival brings an amount of workload, say L, such that its service time is L/C. Let $L(t)$ denote the amount of workload arrived at the queue in $[0, t]$, where the origin of the time axis is set so that $L(0_-) = 0$[14]. $L(t)$ is a nondecreasing, monotonic function of time. With discrete arrivals, $L(t)$ is a step-wise function, with jumps at arrival times. If L_i is the workload of the i-th arriving customer and $A(t)$ is the count-ing function of the arrival process, we have $L(t) = \sum_{i=1}^{A(t)} L_i$, with the understand-ing that $A(0_-) = 0$ and that $\sum_{i=a}^{b} \equiv 0$ if $a > b$. We assume $L(t)$ as well as $A(t)$ to be right-continuous. Jump sizes are equal to the amount of workload brought in by the arriving customer. Under a fluid approximation, where the individual customers are neglected, the workload function $L(t)$ is a continuous function of time.

The so-called Reich's formula (sometimes referred to as Reich's theorem) gives a general expression for the amount of workload into the queue at a generic time $t > 0$, denoted with $Q(t)$. It is

$$Q(t) = \sup_{0_- \le s \le t} [L(t) - L(s) - C(t - s)] \tag{4.208}$$

Note that $L(t) - L(s)$ is the amount of work entering the queue in the time inter-val $(s, t]$, while $C(t - s)$ is the maximum workload that can be served during the same time interval.

Proof: The equality (4.208) can be proved as follows. Let $D(t)$ be the output work-load, i.e., the amount of work actually done by the server up to time t. For any s it is $D(t) - D(s) \le C(t - s)$. By the very definitions of the involved quantities, we have

$$L(t) - L(s) = Q(t) + D(t) - D(s) \le Q(t) + C(t - s) \tag{4.209}$$

---

14 Since $L(t)$ has jumps, we are accounting for the fact that the first arrival could occur just at time $t = 0$, so that $L(0_+) > 0$.

for any $s \in [0_-, t]$. Hence $Q(t) \geq L(t) - L(s) - C(t-s)$ for any $s \in [0_-, t]$, which leads to

$$Q(t) \geq \sup_{0_- \leq s \leq t} [L(t) - L(s) - C(t-s)] \tag{4.210}$$

On the other hand, let us consider the set $\{\tau : 0_- \leq \tau < t, Q(\tau) = 0\}$. This is non empty, since at least it is $Q(0_-) = 0$. Therefore we can define the last time that the queue was empty, namely $v = \sup\{\tau : 0_- \leq \tau < t, Q(\tau) = 0\}$. The queue is continuously not empty, hence the server is continuously busy, in the interval (v, t). Then, $D(t) - D(v) = C(t-v)$ and all of the workload arrived in $[v, t]$ is either in the queue at time t or has been already served, i.e., $L(t) - L(v) = D(t) - D(v) + Q(t) = Q(t) + C(t-v)$. We can write

$$Q(t) = L(t) - L(v) - C(t-v) \leq \sup_{0_- \leq s \leq t} [L(t) - L(s) - C(t-s)] \tag{4.211}$$

which completes the proof. ∎

Reich's formula is a very general result. It has mainly a theoretical impact, yet it is sometimes useful, since it can be applied broadly, with minimal assumptions on the arrival and service processes and even on the stability of the queue.

An example sketch of the quantities involved in Reich's formula is illustrated in Figure 4.9. The step-wise function represents an instance of workload arrival function $L(t)$. The solid line represents the work done by the server. It has slope C (the server capacity) as long as there is work to serve in the queue. It is flat when the queue is empty (idle time). At a generic time t, $L(t)$ is simply the sum of $Q(t)$ and $D(t)$. This is simply a balance of the input workload up to time t with the amount of work done up to time t plus the backlogged work at time t. In fact, $Q(t)$ is just the gap between the step-wise function representing the incoming workload and the solid line representing the amount of served workload. Note that it must be $D(t) \leq L(t)$, for all $t > 0$, given that the system starts empty at time $t = 0$.

Reich's formula can be generalized even further, to account for servers with time varying service capacity. Let $C(t)$ be the maximum amount of work that can be

Figure 4.9 Illustration of Reich's formula. The queue length at a generic time t is shown to be the difference between $L(t)$ and $D(t)$. The thick black segment represents the difference $L(t) - [L(s) + (t-s)C]$ for a given value of s.

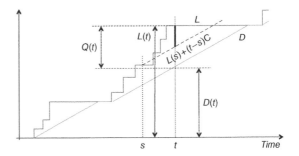

done by the server up to time t. Then, it is

$$Q(t) = \sup_{0_- \leq s \leq t} [L(t) - L(s) - (C(t) - C(s))] \tag{4.212}$$

i.e., the amount of workload to be found in the queue at time t depends on the history of the amount of workload received by the queue and the potential work that the server could have performed over the same time interval.

Summary and Takeaways

Single-server queueing systems are the focus of this chapter. In general, this is noted as a $G/G/1$ queue. We have relaxed this extreme generality by first considering *renewal* arrival and service processes. Moreover, we have simplified further the model by assuming that the input process be a particular point process, the Poisson process, that is amenable to extensive analysis. This has brought us to the $M/G/1$ model. The analysis of this model has been instrumental to introducing one of the most seminal approaches for the performance modeling and analysis of service systems: the EMC. It is important to grasp the essential idea of sampling variables describing the state of a system at suitably chosen time points, so as to obtain a Markov chain, i.e., a process whose future in stochastically independent of the past, given its present state. It is also important to carefully use the results of the EMC analysis, remembering that it gives the probability distribution of the state *at the chosen sampling time points*.

The $M/G/1/K$ queue is then considered. Asymptotically sharp, simple approximations of the probability of empty system, and of the loss probability are discussed in detail. Along with an algorithm for the stable computation of the probability distribution of the number of customers in the system, these tools provide very efficient and stable means to use the $M/G/1/K$ models.

A section is devoted to the optimization of the parameters of the queue. This lays a bridge between the mathematical queueing model and its applications in practical problems.

To complete the picture, a review of the $G/M/1$ queue is provided and a detailed analysis of the special case of $M/M/1$ queue is given as well. The latter is a useful model in that it provides explicit closed formulas for any desired performance metric, including those of the finite queue size $M/M/1/K$ model.

Finally, the Markov structured queueing system is outlined. While keeping tractability, at least from a numerical point of view, those models allow a vast generalization of single server queueing models, with respect to all those considered in the rest of the chapter. In fact, while for the whole chapter we constantly addressed renewal arrival and service processes, with structured $M/G/1$ and $G/M/1$ queues we allow correlated sequences of inter-arrival times or service times. The correlation structure is the one offered by the associated Markov phase process.

Problems

4.1 Consider the following variation of the standard $M/G/1$ queueing system. As soon as the queue becomes empty, the server goes on vacation for a time V, where V is a positive random variable with PDF $f_V(x)$. At the end of the vacation, the server samples the queue to see if any customer has arrived. If it finds the queue still empty, it immediately goes for another vacation. On the contrary, if anyone shows up, the server starts a regular service cycle and does not stop until the queue becomes empty again.

Give the EMC model of this $M/G/1$ queue. Calculate the mean waiting time. Calculate also the mean waiting time of the first customer that arrives at an empty queue.

4.2 A firewall has a processing capability of μ pkts/s. The time required to analyze a packet can be assumed to be a constant. Packets arrive according to a Poisson process from a link with bit rate R. The mean packet length is $\bar{L}$. The utilization of the link is ρ. The firewall has a buffer that can store up to K packets.

a) Define a model of the firewall processing unit.

b) Compute the loss probability of the firewall buffer.

c) Give a graph of the packet loss probability of the firewall as a function of R for R ranging between 2 Mbit/s up to 1 Gbit/s.

(Assume the following numerical values: $\bar{L} = 1250$ bytes, $\rho = 0.7$, $\mu = 10000$ pkts/sec, $K = 100$.)

4.3 Consider the variant of the $M/G/1$ queueing system in which the server can fail, but only when busy. The time to failure, when the server is busy, has an exponential distribution with mean $1/v$. When a failure occurs, the customer in service is dropped, and the server enters a repair period R with probability distribution $F_R(x)$. After repair, the server is immediately operational and takes a new customer from the waiting line, if one is present. Show that this variant is equivalent to an ordinary $M/G/1$ queue with a modified service time probability distribution. Determine the equilibrium condition and the mean waiting time of a customer.

4.4 Consider the variant of $M/G/1$ described in Problem 4.3, where a service, interrupted by a failure, is resumed whenever the service comes back into operation. Successive repair times are independent and identically distributed and completion of a repair restores the server in the original condition. Discuss the service model under this failure-resume variant.

4.5 **The $M/G/1$ queue with N-policy**

In this variant of the $M/G/1$ queue an idle server does not start serving customers unless there are at least $N \geq 1$ waiting for service. Once the server starts a busy period, it behaves as usual, until the queue becomes empty again and the server enters a new idle time. The case for $N = 1$ is the ordinary $M/G/1$ queue. This policy trades off longer busy periods against possibly longer waiting times.

Write the one step evolution equation of the EMC of the queue at departure times for this variant. Derive the mean number of customers in the queue and the mean system time.

4.6 **The $M/G/1$ queue with batch arrivals**

Consider an $M/G/1$ queue where arrivals are in batch of fixed size g. Batches arrive according to a Poisson process of mean rate λ. Find the equilibrium condition of the queue. Write and solve the one-step evolution equations of the EMC at departure times. Derive the mean number of customers in the system and the mean waiting time.

4.7 The aggregated traffic offered to a packet multiplexer can be modeled as a Poisson process, with mean rate 1000 pkts/sec. The mean packet length is 500 bytes and the standard deviation is 300 bytes. Calculate the minimum multiplexer output capacity such that the mean time through the multiplexer be no more than 1 ms.

4.8 A statistical multiplexer receives the traffic of 40 users, each sending a Poisson flow of packets at a mean rate of 300 pkts/s. Packet lengths have negative exponential PDF with mean 500 bytes. Calculate the minimum multiplexer output capacity such that the probability that the time through the multiplexer be more than 10 ms be less than 1%.

4.9 In a finite queue length system you know that the mean queue length is 12 customers, the probability that a customer, that joins the queue, finds an empty queue is 0.361, the mean service time is 90 s. If the average rate of arriving customer is 0.5 min^{-1}, calculate the average rate of customers out of the queuing system and the mean waiting time.

4.10 A Stop&Wait ARQ protocol is operated on an unreliable channel of capacity C. The PDUs have fixed payload length L bytes. PDU header and ACK

PDUs are H byte long. The unreliable channel propagation delay is fixed and equal to τ. There is a probability p that a data PDU is found with errors at the receiver and hence discarded. The probability that an ACK is errored can be neglected. Assume also that the timeout is set equal to the fixed RTT value and that the number of attempts for a same packet is unlimited. The ARQ protocol entity serves a network layer entity that offers packets at a mean rate λ.

1. Define a queueing model to represent the delivery of packets from the sending network entity to the receiving network entity. Identify arrival and service processes.
2. What is the upper bound of λ that guarantees equilibrium is achieved?
3. Calculate the mean time required for a network layer packet to be delivered through the channel at equilibrium (time elapsing since the packet arrives until when it is successfully delivered to the destination).

4.11 A tagged packet flow of mean rate λ can cross an ISP network through two alternative paths. The bottleneck capacities on the two paths are C_1 and C_2, respectively. The mean rate of packets offered to the two paths (background traffic) are λ_1 and λ_2 respectively. Assume all packet flows can be modeled as Poisson processes. Packet lengths are independent of packet inter-arrival times and have a general cumulative probability function $F_L(x)$ for $x \geq 0$.

1. Define a queueing model of the ISP network from the point of view of the tagged flow.
2. Find the optimal routing, i.e., the mean fraction α of packets of the tagged flow that are routed to path 1, so as to minimize the average delay of the tagged flow packets through the ISP network.

4.12 Discuss the scaling of $E[W]$ and $E[Q]$ of an $M/M/1$ queue as λ and μ are both multiplied by a common factor b.

4.13 In a wireless network time is divided into intervals of duration T. The capacity of the wireless channel allows the transmission of N packets per interval. Packets destined to a tagged user terminal arrive at the access point (AP). The AP stores any packet arrived during the interval k in a buffer. The buffer can store up to N packets. At the beginning of interval $k + 1$ the user terminal sends a message to the AP to inquire whether there are packets pending for it. If no packet has arrived during interval k, the user terminal can go to sleep for the interval $k + 1$. If instead some packets have arrived, the user terminal must stay awaken for the whole duration of the interval $k + 1$ and receive any packet addressed to it. Packets destined to the user terminal arrive at the AP according to a Poisson process of mean rate λ. Assume $N = 4$, $T = 1$ s.

a) Calculate the probability of packet loss for $\lambda = 2$ pkts/s and for $\lambda = 8$ pkts/s. For which values of λ is the system stable?

b) The user terminal drains a power level P whenever it is active. Calculate the mean consumed power under the sleep mode operation described above ($P = 100$ mW, $\lambda = 2$ pkts/s).

4.14 *Balking* in a queuing system refers to an upcoming customer deciding not to join the queue. Reasons can be external or internal to the queueing system. In the latter case, they result in a state dependent process of arrivals that actually join the queue.

Consider an $M/M/1$ queue with service rate μ and Poisson arrivals with mean rate λ. Newly arriving customers join the queue with a probability that is inversely proportional to the number of customers already in the queue plus the customer itself. So, for example, a customer finding 3 other customers in the queue joins the queue with probability $1/4$.

Find the mean waiting time of such a queuing system and compare it with that of the regular $M/M/1$ queue with the same mean arrival and service rates. Find also the probability that a customer joins the queue.

4.15 *Reneging* in a queuing system refers to customer inside the system that decide to leave it before receiving service.

Consider an $M/M/1$ queue with service rate μ and Poisson arrivals with mean rate λ. Customers that have joined the queue and are waiting for service can decide to leave. A waiting customer leaves the queue after a negative exponential time, with mean $1/\beta$.

Give a birth-death process model of the queue with reneging. Find the mean waiting time of a customer that is actually served in such a queuing system and compare it with that of the regular $M/M/1$ queue with the same mean arrival and service rates.

4.16 Consider a service system that fits into an $M/M/1$ model, i.e., you can safely assume that arrivals be Poisson, service time have negative exponential PDF, and arrival and service times be independent of one another. You can only observe customers leaving the queue. Hence, record their inter-departure times. Is it possible to define an estimator of the mean service rate of the queue server using that information? Explain your answer.

5

Multi-Server Queues

The queue besides you is always faster.

Arthur Bloch

5.1 Introduction

There is no simple general solution for a $G/G/m/K$ queueing system. Many special cases can be analyzed, though. Among them, the $M/M/m/K$ (both with finite and infinite K), the $M/G/m/0$, the $G/M/m$, and many cases of infinite server queues.

In the following we address the elementary $M/M/m/K$ queueing system, the pure loss system with Poisson arrivals, and the infinite server case. A complete treatment of the $G/M/m$ model can be found in [130]. Rather than being exhaustive we aim at understanding the impact of having multiple servers on traffic models.

First, some system remarks are in order. Single-server queues model those systems where the serving capacity is granted as a whole to a customer, as long as it needs it to complete its service demand. We have already encountered first-come, first-served (FCFS) discipline; we will see further queueing disciplines in Chapter 6, i.e., policies according to which the capacity is assigned to the customers waiting for service. In many cases, the serving capacity cannot be split. At any time the server is either idle (only if there is no customer waiting for work-conserving servers) or it is assigned *fully* to some customer. For example, the capacity of the output link of an interface card of an IP router is given as a whole to the head-of-the-line packet waiting in the interface buffer. While the packet is engaging the output line, being sent at link speed, no other packet can use the same link (in the same direction). This is definitely different from the

Network Traffic Engineering: Stochastic Models and Applications, First Edition. Andrea Baiocchi.

situation of a number of user terminals sharing the capacity of a cellular base station. In this last case, *simultaneous* data transfer can take place, e.g., in the downlink direction, between the base stations and the user terminals. Then, *more than one* packet at the same time can be sent over the air in the radio channel. The base station radio channel capacity cannot be modeled as a single server. Actually, it can serve up to a given number of packets *in parallel*.

The examples above point out that those systems where it is structurally possible to serve multiple customers at the same time are to be modeled as multi-server queues. A number of structural features should now be specified.

- *Interchangeability.* Servers might be equivalent, i.e., each of them provides the same kind of service, or they can have different capabilities.
- *Accessibility.* Customers may access only some of the servers, depending on their class or their specific service demand.
- *Distinguishability.* A customer may choose one of the available servers with equal probability or join a specific server according to some criterion. This implies that servers are labeled in some unique way.

Examples of the above categories are as follows.

Consider a connection that is routed through an ISP network, from an edge node A to another edge node B. In general, there exist multiple alternative paths between A and B. Assume the ISP offers a protection service for paths through its own network. Two protection levels are possible: (i) unprotected, for interruptible traffic; (ii) $1 + 1$ protection, for high dependability traffic. A model of the connectivity service offered by the ISP can be defined as a pool of servers, each server representing the network resource that support an end-to-end connection between A and B. In this model, multiple "servers" are available, but they are not fully equivalent. A given connection request will be served by an unprotected or a protected path, according to the requirements of the carried traffic.

As another example, a base station (BS) has in general a number m of channels to accomodate requests of connection of customers roaming in the area covered by the base station. Requests from customers can be divided into two types: (i) new connection requests; and (ii) incoming handoff requests, i.e., connections already ongoing for customers that are moving from a nearby BS toward the tagged BS. For a customer it is in general much more annoying to have its ongoing connection interrupted, because of a failed handoff, rather than being rejected when submitting a request for opening a new connection. This induces a need to deal with customers requesting a channel in different ways, according to their kind of request (new or handoff): n_{HO} channels can be reserved for incoming handoffs. Therefore, an incoming handoff sees the tagged BS as a serving system with m servers (channels), while a newborn connection experiences a system with $m - n_{HO}$ servers. This means that, as long as there are at least $m - n_{HO}$ channels already busy serving ongoing connections, no new connection requests in the service area of the tagged BS will be accepted. However, incoming handoffs *will* be accepted as long as there is at least one channel available.

Frequency channels that suffer different attenuation levels are yet another example of noninterchangeable servers.

Finally, suppose tasks are offered to a data center where a certain number m of virtual machines (VMs) are available to execute the tasks. We can order the VMs, by labeling them with numbers from 1 up to m. Then, we can stipulate that a new arriving task is assigned to the *first* available VM, according to the label order. As a consequence, the VMs with lower rank labels will be most heavily loaded (i.e., they will be busy for a fraction of time bigger than others). This unbalance of the workload splitting among the alternative servers (the VMs) could be desirable, e.g., since it facilitates energy saving policies obtained by switching off physical machines hosting idle VMs.

In the following, unless otherwise stated, we assume that the m servers are indistinguishable, fully accessible and interchangeable. This leads to the simplest possible multi-server queueing system description, since the only quantity that matters is the *number* of servers.

5.2 The Erlang Loss System

The multi-server queueing system with no wait goes under the name of Erlang loss system, in honor of Agner Krarup Erlang, the founder of the theory of telecommunications traffic. It corresponds to the $M/G/m/0$ queue, with m servers and no waiting line. Thus, an arriving customer either finds an idle server and hence starts being served right away or is rejected by the system.

Arrivals follow a Poisson process with mean rate λ. The mean service time is denoted with $1/\mu$.

We begin with the simpler system $M/M/m/0$. The number of customers $Q(t)$ in the system at time t comprises the set $\{0, 1, \dots, m\}$. With negative exponential inter-arrival and service times, $Q(t)$ is the state of a Markov process. Given $Q(t)$, no other information on the system state needs be specified to be able to predict the future evolution of the system. This is due to the memoryless property of the negative exponential PDF.

Since arrival and service completions take place one at a time (no bulk arrival or service), only a transition to nearby states is possible within vanishingly small time intervals. Formally, it is easy to see that for $0 < h < m$

$$P(Q(t + \Delta t) = k \mid Q(t) = h) = \begin{cases} \lambda \Delta t + o(\Delta t) & k = h + 1 \\ h\mu\Delta t + o(\Delta t) & k = h - 1 \\ o(\Delta t) & |k - h| > 1 \end{cases} \tag{5.1}$$

with obvious adaptations for $h = 0$ and $h = m$.

The transition rate matrix of the Markov process $Q(t)$ is then given by

$$
\mathbf{Q} = \begin{bmatrix}
-\lambda & \lambda & 0 & 0 & \cdots & 0 & 0 & 0 \\
\mu & -\mu - \lambda & \lambda & 0 & \cdots & 0 & 0 & 0 \\
0 & 2\mu & -2\mu - \lambda & \lambda & \cdots & 0 & 0 & 0 \\
\cdots & \cdots & \cdots & \cdots & \cdots & \cdots & \cdots & \cdots \\
0 & 0 & 0 & 0 & \cdots & -(m-2)\mu - \lambda & \lambda & 0 \\
0 & 0 & 0 & 0 & \cdots & (m-1)\mu & -(m-1)\mu - \lambda & \lambda \\
0 & 0 & 0 & 0 & \cdots & 0 & m\mu & -m\mu
\end{bmatrix}
\tag{5.2}
$$

This is a tri-diagonal matrix. It is then recognized that $Q(t)$ is a finite, irreducible Markov process of birth-death type. Since it is irreducible and finite, $Q(t)$ is ergodic for every positive value of λ and μ. The limiting state probabilities of the number of customers seen by an arrival or at a general time during equilibrium (which is the same due to the Poisson arrival and PASTA property) can be found by applying the general formulas given in Section 3.5 for birth-death processes:

$$
p_k = p_0 \frac{A^k}{k!} = \frac{A^k/k!}{\sum_{j=0}^{m} A^j/j!}, \qquad k = 0, 1, \ldots, m,
\tag{5.3}
$$

where $A \equiv \lambda/\mu$ is the mean offered traffic intensity[1]. p_0 is the probability to find an empty system. It can be found from the normalization condition $\sum_{k=0}^{m} p_k = 1$. The probability distribution in eq. (5.3) is known as the Erlang distribution.

As for the performance metrics, there is no wait in this queue. So only server utilization and the probability of being rejected are interesting quantities. The mean number of busy servers is simply

$$
A_s = \sum_{k=1}^{m} k p_k = A \frac{\sum_{k=1}^{m} A^{k-1}/(k-1)!}{\sum_{j=0}^{m} A^j/j!} = A \left[1 - \frac{A^m/m!}{\sum_{j=0}^{m} A^j/j!} \right]
\tag{5.4}
$$

As for the rejection, we distinguish between the *blocking* probability P_B and the loss probability P_L. The first one is the probability that the service system is in a blocking state, i.e., a state where all its resources are saturated. With the Erlang loss system, the only blocking state is $Q = m$. When the system is blocked, there is rejection of a service request, if one is submitted to the system. Otherwise, no rejection takes place. The loss probability is the probability that the system is blocked, *given that a service request arrives at the system.* Formally

$$
P_L = \lim_{\Delta t \to 0} \mathcal{P}(Q(t) = m \mid A(t, t + \Delta t) > 0) = \frac{\lambda p_m}{\lambda} = P_B
\tag{5.5}
$$

---

1 In other parts of the book the mean offered traffic intensity is denoted with A_o. For the sake of a simple notation, here we drop the subscript o.

It turns out that $P_B = P_L$. This identity holds for Poisson arrivals, since the limiting probability of finding the system in a given state at a general time (from which P_B is calculated) is the same as the probability that an arriving customer finds the system in that state (from which the loss probability is obtained). In general, they are not the same. For example, let us consider a $D/D/1/0$ queue, where the inter-arrival time is fixed and equal to T_0 and the service time is fixed as well and equal to $X_0 < T_0$. The blocking probability is $P_B = X_0/T_0$, while the loss probability is $P_L = 0$. A more complex example is one where the population of customers that offer service demands is finite, say N customers. Then, if m of them are inside the service system (and hence make the system blocked), there remain only $N - m$ customers that can still pose service demands to the system and eventually get rejected. If $N \gg m$ we fall back toward the Poisson case, whereas if N is comparable with m there is definitely a difference between P_B and P_L.

In the Erlang loss system, the blocking and loss probabilities coincide and they are equal to p_m, namely the celebrated Erlang B-formula.

$$P_L = P_B = p_m = \frac{A^m/m!}{\sum_{k=0}^m A^k/k!} \equiv B(m, A) \tag{5.6}$$

The Erlang B-formula, or briefly, $B(m, A)$ depends only on the positive integer parameter m and on the non-negative real A. It can be extended to real non-negative values of m by using the Fortet's formula.

$$[B(m, A)]^{-1} = \frac{m!}{A^m} \sum_{k=0}^m \frac{A^{m-k}}{(m-k)!} = \sum_{k=0}^m \binom{m}{k} \frac{k!}{A^k} = \tag{5.7}$$

$$= \sum_{k=0}^m \binom{m}{k} \frac{1}{A^k} \int_0^\infty u^k e^{-u} \, du = \int_0^\infty \left(1 + \frac{u}{A}\right)^m e^{-u} \, du \tag{5.8}$$

where the last but one equality is based on the definition the Euler gamma function $\Gamma(x) = \int_0^\infty u^{x-1} e^{-u} \, du$ and on the property $\Gamma(k+1) = k!$ for integer k. The expression (5.7) is meaningful also for noninteger m, so we define:

$$[B(x, A)]^{-1} = \int_0^\infty \left(1 + \frac{u}{A}\right)^x e^{-u} \, du \,, \qquad x, A \geq 0. \tag{5.9}$$

For numerical evaluation, it is best to write

$$B(x, A) = \frac{A^x e^{-A}}{\Gamma(A, x+1)} \tag{5.10}$$

where $\Gamma(y, z) = \int_y^\infty u^{z-1} e^{-u} \, du$ is the upper incomplete Euler gamma function.

The integral expression of $B(x, A)$ lends itself also to derive asymptotics. For example, assume $x = \beta A$, where β is a coefficient. Since $\lim_{A \to \infty} (1 + u/A)^{\beta A} = e^{\beta u}$, it is easy to derive that

$$\lim_{A \to \infty} B(\beta A, A) = \begin{cases} 0 & \beta \geq 1 \\ 1 - \beta & \beta < 1 \end{cases} \tag{5.11}$$

This result is one face of the *trunking efficiency* phenomenon, typical of the Erlang loss system, due to the nonlinear character of the Erlang B-formula.

Let $\phi(u) = e^{-u^2/2}/\sqrt{2\pi}$ be the PDF of the standard Gaussian random variable and $\Phi(u) = \int_{-\infty}^{u} \phi(y)\,dy$ the corresponding CDF. A more sophisticated result can be stated as follows (see [209]):

Theorem 5.1   In an $M/G/m/0$ system

$$\lim_{A \to \infty} \sqrt{A}\, B([A + \beta\sqrt{A}], A) = \phi(\beta)/\Phi(\beta) \tag{5.12}$$

where $[x]$ is the greatest integer less than or equal to x and β is a positive parameter.

This result again shows that the loss probability tends asymptotically to 0 for very large systems, i.e., scaling improves loss performance. The utilization coefficient with the scaling of Theorem 5.1 tends to 1 as $A \to \infty$:

$$\rho = \frac{A}{A + \beta\sqrt{A}}[1 - B(A + \beta\sqrt{A}, A)] \sim \frac{A}{A + \beta\sqrt{A}}\left[1 - \frac{\phi(\beta)}{\Phi(\beta)\sqrt{A}}\right] \to 1 \tag{5.13}$$

This shows further that scaling to large systems it is possible to achieve high utilization and low loss probability simultaneously.

A different scaling can be found if we fix the ratio $r = A/m$. Let $m = N$ and $A = rN$, with $0 < r < 1$. The denominator of the Erlang-B function can be written as

$$\sum_{k=0}^{N} \frac{(rN)^k}{k!} = e^{rN} \int_{rN}^{\infty} \frac{u^{N-1}}{(N-1)!} e^{-u}\, du \tag{5.14}$$

The function $f_N(u) = \frac{u^{N-1}}{(N-1)!} e^{-u}$ is the PDF of a gamma random variable Y, which can be expressed as the sum of N i.i.d. negative exponential random variables X_i with mean 1, i.e., $Y = X_1 + \cdots + X_N$, with $X_i \sim \mathrm{Exp}(1)$, $i = 1, \ldots, N$. Then, $E[Y] = N$ and $\mathrm{Var}(Y) = N$. By virtue of the CLT, we have $P\left(\frac{Y-N}{\sqrt{N}} > x\right) \to 1 - \Phi(x)$ as $N \to \infty$, where $\Phi(x)$ is the CDF of the standard Gaussian random variable. Then

$$\sum_{k=0}^{N} \frac{(rN)^k}{k!} = e^{rN} P(Y > rN) = e^{rN} P\left(\frac{Y-N}{\sqrt{N}} > -\sqrt{N}(1 - r)\right) \sim e^{rN}\, \Phi(\sqrt{N}(1 - r))$$

Using Stirling's formula, the numerator of the Erlang-B function can be manipulated as follows:

$$\frac{(rN)^N}{N!} \sim \frac{r^N N^N}{(N/e)^N \sqrt{2\pi N}} = \frac{r^N e^N}{\sqrt{2\pi N}} \tag{5.15}$$

Using the asymptotic expressions of the numerator and denominator, we find

$$B(N, rN) \sim \frac{r^N e^N e^{-rN}}{\sqrt{2\pi N} \ \Phi(\sqrt{N}(1-r))} = \frac{(re^{1-r})^N}{\sqrt{2\pi N} \ \Phi(\sqrt{N}(1-r))} \qquad (5.16)$$

If $0 < r < 1$, this expansion tends to 0 as $N \to \infty$, exponentially fast. The decay rate is re^{1-r}. Note that for $r = 1$ we get $B(N, N) \sim \sqrt{2/(\pi N)}$, which agrees with the statement of Theorem 5.1 if $\beta = 0$ and $A = N$.

The Erlang B-formula can be calculated iteratively by using the recursion

$$y_m = 1 + \frac{m}{A} y_{m-1}, \qquad m \geq 1 \qquad (5.17)$$

initialized with $y_0 = 1$. Then $B(m, A) = 1/y_m$. A different recursion is

$$B(m, A) = \frac{AB(m-1, A)}{m + AB(m-1, A)}, \qquad m \geq 1, \qquad (5.18)$$

initialized with $B(0, A) = 1$. The advantage of the latter recursion is that it deals only with quantities between 0 and 1.

Inverse problems can be posed, e.g., find the maximum A such that $B(m, A) \leq B_{th}$, for a given m. B_{th} represents a performance requirement, i.e., the maximum tolerable loss probability. Since $B(m, A)$ is continuous and monotonously increasing from 0 to 1 when A goes from 0 to ∞, the required maximum level of A, A_{max}, exists, it is unique and is such that $B(m, A_{max}) = B_{th}$. Another inverse problem consists in finding the minimum integer m such that $B(m, A) \leq B_{th}$ for a given A. Since $B(m, A)$ is monotonously decreasing with m and it tends to 0 as $m \to \infty$, for any given A, the searched m_{min} exists and it is unique. In general, it is $B(m_{min}, A) \leq B_{th}$.

A simple code to calculate the Erlang B-formula and solve the two inverse problems is shown below.

```
% Computation of B(m,A)
m = input('Number of servers m = ');
A = input('Mean intensity of offered traffic A = ');
p = 1;
for n = 1:m
  p = A*p/(n+A*p);
end
ErlangB = p;


% inverse ErlangB w.r.t. number of servers for given A
Bth = input('Loss probability constraint = ');
A = input('Mean intensity of offered traffic A = ');
n = 1;
p = A/(1+A);
while p > Bth
```

```
   n = n+1;
   p = A*p/(n+A*p);
end
m_min = n;

% inverse ErlangB w.r.t. traffic for given m
Bth = input('Loss probability constraint p = ');
m = input('Number of servers m = ');
tole = 1e-5;
Ainf = 0;
Asup = m/(1-Bth);
Amed = 0.5*(Ainf+Asup);
erro = (Asup-Ainf)/Amed;
while erro > tole
   if B(m,Amed) > Bth
      Asup = Amed;
   else
      Ainf = Amed;
   end
   Amed = 0.5*(Ainf+Asup);
   erro = (Asup-Ainf)/Amed;
end
Amax = Amed;
```

The behavior of $B(m, A)$ as a function of A for various values of m is plotted in Figure 5.1(a). For a given m, $B(m, A)$ grows quickly, in a concave way, tending to

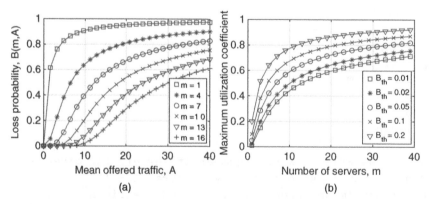

(a)   (b)

Figure 5.1 Loss probability of the Erlang queue. Left plot: $B(m, A)$ as a function of A for various values of m. Right plot: maximum value of the utilization coefficient $\rho = A[1 - B(m, A)]/m$ under the constraint that $B(m, A) \leq B_{th}$ as a function of m.

1 as A tends to ∞. The bigger m, the more the curve is shifted to the right, i.e., a smaller loss is experienced for low up to moderate traffic intensity levels.

Figure 5.1(b) shows the maximum utilization coefficient that can be sustained under a given loss probability requirement, as a function of m. The utilization coefficient of any server is

$$\rho = \frac{A}{m}[1 - B(m, A)] \tag{5.19}$$

It turns out that ρ is increasing with m for any given loss probability requirement. In other words, the system resources are used *more efficiently* as the system *scale* grows. This is another landmark of the trunking efficiency phenomenon or of the *economies of scale*. Moving to a larger system is convenient, i.e., less resources are required to provide a given level of quality of service (loss probability in the system at hand). In other words, a bigger utilization per server can be realized for a fixed level of quality of service.

Equation (5.16) shows that the dream of a traffic engineer is not impossible (at least asymptotically). Letting $r = A/m < 1$, we have $\rho = r[1 - B(m, mr)] \sim r$ as $m \to \infty$. This hold for any $r < 1$, however close to 1. We can have *both* server utilization as close to 1 as we want and a negligible loss probability.

Example 5.1 Let us consider an LTE cell served by a BS, the evolved Node B (eNB) in the LTE jargon. The radio resource is organized in resource blocks (RBs). An RB corresponds to a number of OFDM sub-carriers for a time slot. Therefore, it provides a given bit rate to a given user, depending on the characteristics of the radio channel between the user and the eNB. Let $\bar{r}$ denote the average bit rate obtained by a user for one RB. Let g be the number of RBs allocated to a single user connection. Connection requests arrive according to a Poisson process of mean rate λ. A connection aims at carrying an average amount Q of bits. The eNB has K RBs overall, out of which it takes the RBs to allocate for each served connection. If a connection request arrives at the eNB when there are less than g RBs available, the connection is rejected.

We model the eNB as an Erlang loss system with $\lfloor K/g \rfloor$ servers. The mean offered traffic is the product of the mean arrival rate of the connection requests λ and the mean duration of a connection, $\overline{X}$. Since Q bits must be transferred in a connection and the available bit rate is $g\bar{r}$ on average, then $\overline{X} = Q/(g\bar{r})$. Hence, $A = \lambda Q/(g\bar{r})$ and the loss probability is

$$P_L = B\left(\left\lfloor \frac{K}{g} \right\rfloor, \frac{\lambda Q}{g\bar{r}}\right) \tag{5.20}$$

We can find the maximum value of λQ (mean offered bit rate) that meets the requirement $P_L \le B_{th}$. Another problem consists in determining the maximum value of g that is compatible with the requirement $P_L \le B_{th}$, for a given level of

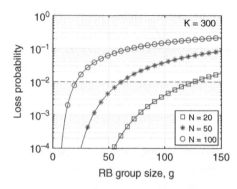

Figure 5.2 Probability of loss of connection requests in an LTE cell with RB grouping. The mean offered load per user is 8 Mbit/s, the mean rate supported by an RB is $\bar{r} = 500$ kbit/s. The overall number of RBs in the cell is $K = 300$.

the offered bit rate. Note that, as g increases, the mean duration of a connection decreases, i.e., service is faster and thus the load is relieved, but, on the other side, the number of servers decreases. This worsens the loss performance eventually, given the trunking efficiency effect (increasing g means moving toward a smaller-scale system).

Numerical evaluation of the loss probability is plotted in Figure 5.2. We assume N users are sharing the cell capacity, each offering a mean rate of connection requests of λ_1. The numerical values of the parameters are $K = 300$ RBs, $\lambda_1 = 0.1$ s^{-1}, $Q = 1$ Mbyte, $\bar{r} = 500$ kbit/s. The curves of P_L as a function of g for three values of N have been obtained by using the generalization of the Erlang formula for a non integer number of servers (Fortet's formula). This gives an upper bound of the true loss performance. The approximation is only relevant for very large values of g.

The loss probability grows with g, as well as the average allocated bit rate (which is $g\bar{r}$). The most convenient dimensioning of the system consists of finding the maximum level of g that is compatible with a requirement on the loss probability. The line at level 10^{-2} is shown in the Figure 5.2. The corresponding maximum value of g compatible with a requirement that the loss probability be no more than 10^{-2} is $g_{max} = 125,\ 61,\ 20$ for $N = 20,\ 50,\ 100$, respectively.

The behavior of g_{max} as a function of N for the same parameter values as in Figure 5.2 is shown in Figure 5.3(a), under the requirement that the loss probability be no more than $B_{th} = 0.01$. g_{max} is monotonously decreasing with N, which is consistent with intuition: the more users are contending for the system resources, the bigger the parallelism of the service system should be to maintain the same level of rejection of arriving connection requests.

Figure 5.3(b) plots the throughput delay trade-off of the LTE-cell. The ordinate represents the mean overall throughput of the cell, i.e., $TH = N_{max}\lambda_1 Q$, where N_{max} is the maximum number of users compatible with the constraint $B_{th} \leq 0.01$.

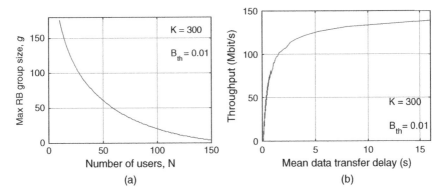

Figure 5.3 Left plot: maximum value of the RB group size g as a function of N, under the constraint that the loss probability be no more than 10^{-2}. Right plot: throughput-delay trade-off for data transfer in the LTE cell, under the requirement that the loss probability be no more than 10^{-2}. The mean offered load per user is 8 Mbit/s, the mean rate supported by an RB is $\bar{r} = 500$ kbit/s. The overall number of RBs in the cell is $K = 300$.

The abscissa reports the mean time required for a user to complete its data transfer, i.e., $D = Q/(g\bar{r})$. The trade-off is generated by letting the RB group size g vary from 1 to K. The trade-off is monotonously increasing. The larger the tolerable delay, the bigger the throughput that can be sustained.

Example 5.2 *Random server hunting* Assume servers are searched at random as a new request arrives at an Erlang loss system. That is to say, a permutation of the integers in the range $[1, m]$ is generated uniformly at random among all possible permutations, say it is $(s_1, \ldots, s_m)$. Starting with $j = 1$, the server s_j is checked: if it is idle, it is assigned to the pending service request, otherwise the server s_{j+1} is explored, and so on, until either an available server is found or all servers are found busy. In this last case the pending service request is rejected and lost to the system. We ask what is the probability $P(i)$ that the first i explored servers are found busy.

The problem can be solved by writing:

$$P(i) = \sum_{k=i}^{m} p_k \mathcal{P}(\mathcal{E}_i \mid Q = k) \tag{5.21}$$

where $\mathcal{E}_i$ is the event that the first i servers explored are found busy. Since server hunting is done completely at random, all possible configurations of server engagement are equiprobable. The number of configurations of k busy servers out of m is $\binom{m}{k}$. After i servers have been explored and found busy, the number of

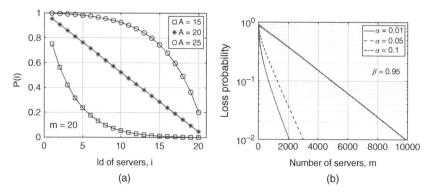

Figure 5.4 Left plot: probability of finding the first i explored servers busy, $P(i)$, in an Erlang loss system with $m = 20$ servers and random server exploration, for three values of the mean offered load. Right plot: probability of finding the first d explored servers busy as a function of the size of the system (number of servers).

possible configurations of the remaining $k - i$ busy servers out of the remaining $m - i$ servers is $\binom{m-i}{k-i}$. Therefore

$$P(i) = \sum_{k=i}^{m} p_0 \frac{A^k}{k!} \frac{\binom{m-i}{k-i}}{\binom{m}{k}} = \frac{B(m, A)}{B(m - i, A)}, \qquad i = 1, \ldots, m. \tag{5.22}$$

Figure 5.4(a) plots $P(i)$ for an Erlang loss system with $m = 20$ servers and for three values of the mean offered load, $A = 15, 20$, and 25 Erl. The probability $P(i)$ is monotonously decreasing with the index i of the server. When the system bears a heavy load ($A = 25$ Erl), it exhibits a concave behavior; on the contrary for a lighter load ($A = 15$ Erl) it decreases in a convex way, quite sharply, e.g., the probability that more than 5 servers need be explored to find one idle is 0.2. The special case $A = 20$ Erl $= m$ lies in between.

We can use the result (5.22) to evaluate how the loss probability scales with the system size. For this purpose, we let $m = N$ and $A = \beta N$ with $\beta = 0.95$. In a large system (e.g., N in the order of thousands) it could be too costly to explore all servers. It makes sense to limit the server exploration to a small fraction of the overall set of available servers. Let $d = \alpha m = \alpha N$, with $\alpha \ll 1$. Since we stop the exploration when up to d servers have been probed, the probability of blocking is

$$P(N) = \frac{B(N, \beta N)}{B(N - \alpha N, \beta N)} \tag{5.23}$$

Figure 5.4(b) plots $P(N)$ as a function of N for three values of α and $\beta = 0.95$. It is evident that $P(N)$ decreases exponentially with N (the plot is on semi-logarithmic scale). The slope is greater for bigger values of α, as expected. It is interesting to note that performance improve rapidly as α grows, so that a large part of the performance gain is reaped with relatively small levels of α. For example, exploring only 5% of servers makes the blocking probability fall below to 0.01 if a system is composed of 3000 servers or more. This number is not so unrealistic. It can be found in real large-scale systems such as data centers of the public cloud.

Example 5.3 *Sequential server hunting* We consider now an Erlang loss system where the availability of servers is checked sequentially. Let the servers be labeled with integers from 1 to m. Assume servers are explored sequentially as a new request arrives and the first idle server, if any, is assigned to the request. If no idle server is found, the request is rejected and lost.

Under this setting, the first i servers form an $M/G/i/0$ queueing system with mean offered traffic equal to A, the mean offered traffic offered to the entire original system. The probability $\rho(i)$ that server i is busy coincides with the mean carried traffic $A_c(i)$ of that server. The mean carried traffic can be found as the difference between the mean traffic offered to server i and the mean traffic rejected by server i, $A_r(i)$. The former is the mean traffic rejected by server $i - 1$. Hence $\rho(i) = A_c(i) = A_r(i - 1) - A_r(i)$. The mean intensity of the rejected traffic of an Erlang system with i servers and mean offered traffic intensity equal to A is $A_r(i) = AB(i, A)$, for $i = 1, \ldots, m$. We let $A_r(0) = A$ for ease of notation.

The final result is $\rho(i) = A[B(i - 1, A) - B(i, A)]$, for $i = 1, \ldots, m$. This is also the probability that server i is found busy. It can be seen that it is a monotonously decreasing function of i.

Sequential hunting allocates work in an unequal way among servers, the lower index ones being much more loaded than the last ones. Each new server added to the system can be conceived as a "second-choice" alternative, to which only traffic overflown from first-choice servers is offered.

To appreciate the differentiated load of the servers, if $m = 100$ and $A = 80$ it can be computed that $\rho(1) = 0.9877$ and $\rho(100) = 0.0814$.

5.2.1 Insensitivity Property of the Erlang Loss System

The Erlang loss system as described above corresponds to the $M/M/m/0$ queue. The same limiting probability distribution applies to a more general system, namely the $M/G/m/0$ queue. It is possible to prove the following.

Theorem 5.2 The limiting probability distribution of the $M/G/m/0$ exists (provided the mean of the service times is finite) and is given by the Erlang probability distribution (5.3).

Proof: We outline a sketch of the proof[2] . Let $f_k(x_1, \ldots, x_k)$ denote the joint probability density of the number of customers in the queue and of the age of service of the k customers in the queue at a generic time of the equilibrium. Formally, if $Y_i(t)$ denotes the age of the service of customer i in the queue at time t, we let

$$F_k(x_1, \ldots, x_k) = P(Q(t) = k, Y_1(t) \le x_1, \ldots, Y_k(t) \le x_k), \qquad k = 1, \ldots, m,$$

(5.24)

Then

$$f_k(x_1, \ldots, x_k) = \frac{\partial^k F_k(x_1, \ldots, x_k)}{\partial x_1 \ldots \partial x_k}$$

(5.25)

and $f_0 = P(Q(t) = 0) = p_0$. Since the service times are i.i.d. random variables, independent of the number of customers in the queue, and the PDF of the age of the service time is $f_Y(x) = \mu G_X(x)$, where $G_X(x)$ is the CCDF of the service time, we have

$$F_k(x_1, \ldots, x_k) = P(Y_1(t) \le x_1, \ldots, Y_k(t) \le x_k \mid Q(t) = k)P(Q(t) = k) = p_k \prod_{i=1}^{k} F_{Y_i}(x_i)$$

(5.26)

hence

$$f_k(x_1, \ldots, x_k) = p_k \prod_{i=1}^{k} f_{Y_i}(x_i) = p_k \mu^k \prod_{i=1}^{k} G_X(x_i)$$

(5.27)

Moreover, upon an arrival that finds k customers already in the queue ($k < m$), the service ages change from $x_1, \ldots, x_k$ to $x_1, \ldots, x_k, 0$, i.e., the newly arrived customer has an age of 0. For symmetry reasons, there are $k + 1$ configurations of the arguments of the function $f_{k+1}(x_1, \ldots, x_k, 0)$ that are fully equivalent. In other words, the arrival causes the transition from the state $(k, x_1, \ldots, x_k)$ to anyone of the $k + 1$ micro-states $(k + 1, x_1, \ldots, x_{j-1}, 0, x_{j+1}, \ldots, x_{k+1})$. We collapse all of those micro-states into the unique state $(k + 1, x_1, \ldots, x_k, 0)$. Then, we have

$$\lambda f_k(x_1, \ldots, x_k) = (k + 1)f_{k+1}(x_1, \ldots, x_k, 0), \qquad k = 1, \ldots, m - 1.$$

(5.28)

Considering the arrival of a new customer at an empty queue, we can add the boundary equation $\lambda p_0 = f_1(x_1)$. Putting all these equations together, we see that

$$f_k(0, \ldots, 0) = \frac{\lambda}{k} f_{k-1}(0, \ldots, 0) = \cdots = \frac{\lambda^k}{k!} p_0$$

(5.29)

---

2 See Takács [199]. He also mentions that the first correct proof of the insensitivity property was given by B. A. Sevastyanov in 1957.

Applying eq. (5.27) for $x_1 = x_2 = \cdots = x_k = 0$ and using eq. (5.29), we obtain

$$p_k \mu^k \prod_{i=1}^{k} G_X(0) = \frac{\lambda^k}{k!} p_0 \quad \Rightarrow \quad p_k = p_0 \frac{A^k}{k!}, \quad k = 1, \ldots, m, \tag{5.30}$$

where $A = \lambda/\mu$. Then, p_0 is found from the congruence relationship $\sum_{k=0}^{m} p_k = 1$. ∎

The result stated in Theorem 5.2 is a manifestation of the so called *insensitivity* property of the Erlang loss system. Amazingly, the only thing that matters of the service time is its mean value!

5.2.2 A Finite Population Model

In the Erlang model the mean rate of arrival of service requests is constant, no matter what the state of the system is. That is, the "pressure" of the demand posed by customers outside the system is the same independently of the number of customers residing into the system. Such a model entails that the population of customers from which the service demand comes is infinite (or, from a practical point of view, of much larger size than the number of servers).

When that is not the case, different models must be used. A simple variant of the Erlang model consists of a population of N customers. When outside the service system, a customer takes an exponentially distributed time with mean $1/\gamma$ to make a service request to the system. If accepted, the service time has a general PDF with mean $1/\mu$. If rejected, the request is canceled and a new timer starts with the same negative exponential PDF until the next service request. The service system has m identical, fully accessible servers and no room for a waiting line (pure loss system). This model is called the Engset system, after T. O. Engset, a Norwegian mathematician and engineer who did pioneering work in the field of telephone traffic. We assume obviously $N > m$, otherwise the system has no blocking (it would become equivalent to an infinite server system).

If the service system has negative exponential PDF, it is easy to see that the number $Q(t)$ of customers in the system at time t is a birth-death Markov process. The birth rate in state k is $(N - k)\gamma$ for $k = 0, \ldots, m - 1$. The death rate is $k\mu$ for $k = 1, \ldots, m$. The Markov process is finite and irreducible for positive γ and μ, then the limiting state probabilities exist. They are

$$p_k = p_0 \binom{N}{k} \left(\frac{\gamma}{\mu}\right)^k = \frac{\binom{N}{k} \left(\frac{\gamma}{\mu}\right)^k}{\sum_{j=0}^{m} \binom{N}{j} \left(\frac{\gamma}{\mu}\right)^j}, \quad k = 0, \ldots, m. \tag{5.31}$$

Let $a = \gamma/\mu$, a measure of mean offered load of a single customer. The mean rate of service request offered to the system is by its very definition

$$\Lambda_o = \sum_{k=0}^{m} \lambda_k p_k = \sum_{k=0}^{m} (N - k)\gamma p_k = \gamma(N - E[Q]) \tag{5.32}$$

The mean offered traffic is $A_o = \Lambda_o/\mu = a(N - E[Q])$. Simple algebra yields:

$$E[Q] = Na \frac{\sum_{j=0}^{m-1} \binom{N-1}{j} a^j}{\sum_{j=0}^{m} \binom{N}{j} a^j} \tag{5.33}$$

The mean rate of *accepted* service requests is obtained by observing that the only blocking state is m:

$$\Lambda_s = \sum_{k=0}^{m-1} (N - k)\gamma p_k \tag{5.34}$$

The difference $\Lambda_o - \Lambda_s$ is the mean rate of lost requests of service. The probability of loss of a service request is:

$$P_L = \frac{\Lambda_o - \Lambda_s}{\Lambda_o} = \frac{(N - m)\gamma p_m}{\Lambda_o} = \frac{(N - m)p_m}{N - E[Q]} = \frac{\binom{N-1}{m} a^m}{\sum_{j=0}^{m} \binom{N-1}{j} a^j} \tag{5.35}$$

Note that this is different from the blocking probability, i.e., the probability the system is the blocking state, namely $P_B = p_m$. This is not so surprising: after all arrivals at this system are *not* Poisson[3]. The blocking probability is named Engset function and here denoted with $E(m, N, a)$:

$$P_B = p_m \equiv E(m, N, a) = \frac{\binom{N}{m} a^m}{\sum_{j=0}^{m} \binom{N}{j} a^j} \tag{5.36}$$

5.2.3 Non-Poisson Input Traffic

We have defined the offered traffic process as the number $N(t)$ of servers engaged by the service request flow in a test system made of infinite servers.

For a stationary Poisson traffic it is $P(N(t) = k) = \frac{A^k}{k!} e^{-A}$, $k \geq 0$, where $A = E[N(t)]$ is the mean traffic intensity.

The variance of the traffic process is $V = E[N(t)^2] - (E[N(t)])^2$. In case of Poisson traffic it is $V = A$.

---

3 The concise statement that "traffic is Poisson" means that arrivals of service requests follow a Poisson process. The time required to serve each service request must be independent of the arrival process, but it can have a general PDF.

We define also the *peakedness factor* $z = V/A$. It equals 1 for a Poisson traffic. A traffic process for which $z < 1$ is said to be a *smoothed* traffic. If it is $z > 1$, we call it a *peaked* traffic. Smoothed traffic appears to be more regular, since fluctuations around the mean value are limited, whereas peaked traffic is expected to present more burstiness, given the large variance with respect to the mean.

We have seen that a pure loss queueing system is amenable to analysis as soon as we assume that the input traffic is of Poisson type. This is a reasonable hypothesis for an isolated system that receives its traffic directly from a population of sources made of a large number of entities, each contributing a small fraction of the overall traffic.

Things are different if we look at a network. Pure loss systems are apt to describe *circuit-switched* networks. Let us consider a network modeled as a graph of n nodes and ℓ links. Assume that communications through the network is based on end-to-end connections, each requiring a same amount of capacity. Link i can multiplex up to m_i connections. We say that link i has m_i *circuits*, using a terminology that is reminiscent of telephone networks (hence the name of circuit-switched networks). The concepts developed here apply to a broad set of technological networks, e.g., transport networks based on SDH or on DWDM optical technologies, cellular access network based on radio channel reservation and individual channel assignment to traffic sources. We will see in Chapter 11 that the techniques developed for circuit-switched network analysis carry over also to packet network analysis at flow level, if we manage network resources based on the effective bandwidth concept or taking a deterministic traffic model approach.

In a circuit-switched network, before communication can take place over a connection, an end-to-end path must be found between the originating node and the destination node. This task is accomplished by the routing function, exploring the network graph and reserving one circuit on each link that composes the end-to-end path. In general, more than a single path exists in the network topology for a given origin-destination pair. If none of those paths provides the required circuits end-to-end, connection set-up fails and the set-up request is rejected, i.e., there is usually no waiting option. This makes pure loss models suitable for describing such a traffic handling approach.

Set-up requests move forward node after node as the routing progresses inside the network, eventually completing the end-to-end path successfully, if one circuit per crossed link is available. Otherwise, if the connection setup is blocked due to a congested link, alternate routes are explored, e.g., according to a predefined sequential search plan, established for each origin-destination pair. Once alternatives are exhausted, the connection set-up request is rejected and lost. As a consequence of this operation mode, traffic is "filtered" by network nodes. Even if it is modeled as a Poisson process natively at network edge, once it goes through nodes that forward connection requests or block them, according to the congestion state of links, the traffic process changes.

To understand quantitatively what happens, let us consider an isolated Erlang system, i.e., an $M/M/m/0$ queue. The carried traffic is by definition the number of busy servers $Q(t)$. In the Erlang queueing system $Q(t)$ has the Erlang probability distribution. Mean and variance of $Q(t)$, denoted with A_c and V_c, respectively, are easily found to be:

$$A_c = A\,(1 - P_L)$$
$$V_c = A_c - A\,P_L\,(m - A_c) \tag{5.37}$$

where $P_L = B(m, A)$ is the loss probability.

The lost traffic, i.e., the service demands that are rejected by the Erlang system, form a modulated Poisson process, whose mean rate is 0 as long as $Q(t) < m$, and it is equal to the offered rate λ, when $Q(t) = m$. It can be verified that the first two moments of the lost traffic, denoted with A_l and V_l, respectively, are:

$$A_l = A\,P_L$$
$$V_l = A_l\left(1 - A_l + \frac{A}{m + 1 + A_l - A}\right) = A_l\,\frac{m + 1 - A_l(m - A_c)}{m + 1 - A_c} \tag{5.38}$$

It is easy to verify that $z_c = V_c/A_c < 1$ and $z_l = V_l/A_l > 1$.

The carried traffic is therefore *smoother* than the offered Poisson traffic, for which it is $z = 1$. This matches the intuition that the more extreme peaks of the offered traffic are clipped by the loss-biased service system. The lost traffic instead turns out to be *peaked*. Again, this is consistent with intuition. The traffic rejected by the Erlang system is made up of those peaks that do not fit into the service capacity of the system, hence we expect that the lost traffic is more stochastically variable than the offered traffic.

We have established that the traffic at the output of the Erlang system (the carried traffic) and the traffic overflown from the Erlang system (the lost traffic) are both non-Poisson traffic processes. Analysis of loss systems that receive as input one of those processes cannot be carried out using the classic Erlang model. This is exactly what happens when considering a network of loss systems, rather than an isolated loss system. A communication network operated according to the circuit-switching paradigm is modeled as a network of loss system each loss system corresponding to one link of the circuit-switched network. When analyzing circuit switched networks, we are faced with the issue of determining the probability that a connection set-up fails because of congestion of some link. The traffic offered to a link can be the composition of traffic flows that are carried by or overflow from upstream links. These traffic processes are not Poisson, as we have seen. We cannot apply the Erlang model exactly any more.

As an example, Figure 5.5 illustrates a four node network. Poisson traffic flows are offered at nodes A and B, both directed to node C. The routing of the origin-destination (OD) pair AC is so defined. As a first choice the direct link AC is

Figure 5.5 Example of circuit-switched network. Connection requests are routed from origin to destination according to alternative paths, e.g., traffic from A to C is carried on link 1 as a first choice; if that is blocked, on the path made by links 3 and 5 as a second choice.

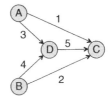

preferred. If no available capacity can be found in that link, the alternative routing AD-DC s explored. If either of the two last links is saturated, the connection request is rejected and lost. Similarly, for the BC traffic flow, direct path routing is preferred over the two-link alternative routing through node D.

Links AD and BD receive non-Poisson traffic flows, made up by those connection requests that were rejected by link AC and link BC, respectively. The link DC input is composed of the superposition of two independent non Poisson traffic flows, that are the traffic carried by the links AD and BD. We can associate multi-server pure loss models to each of the network links. However, it is clear that we need to generalize the Erlang model so as to be able to deal with traffic flows of any kind.

Approximate methods have been defined to tackle analysis and dimensioning of circuit-switched networks. Most of them limit the description of the considered processes to the first two moments and reduce in some way the evaluation of the loss probability to the calculation of the Erlang-B formula, applied to a suitable equivalent system. We will see two of these methods in the following.

The approximations presented in this section have been motivated initially by the need of analyzing the telephone network, but they hold in general for any setting where resources are pre-allocated to connections during set-up phase and multiplexing is done according to resource partitioning on network links.

5.2.3.1 Wilkinson's Method

The approximation proposed by Wilkinson [208] applies only to peaked traffic, i.e., we assume $z > 1$. Let A and V be the mean and variance of the tagged peaked traffic, which is offered to a loss system with m servers. We assume that the tagged traffic results from the overflow of an equivalent Erlang loss system with mean offered traffic A_E having m_E servers (see Figure 5.6).

Figure 5.6 Scheme of Wilkinson's approximation. The non-Poisson traffic is deemed to be the overflow traffic of an equivalent Erlang loss system of m_E servers to which a Poisson traffic of intensity A_E is offered. The traffic lost by the original loss system of m servers can be obtained as the traffic lost by the extended system made of $m_E + m$ servers with Poisson offered traffic A_E.

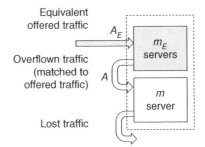

The parameters A_E and m_E are determined so as to match the known mean and variance of the considered traffic, considered as the lost traffic of the equivalent Erlang system. Using eq. (5.38) we have

$$A = A_E B(m_E, A_E)$$
$$V = A \frac{m_E + 1 - A(m_E + A - A_E)}{m_E + 1 + A - A_E} \tag{5.39}$$

This nonlinear system in the unknowns A_E and m_E can always be solved, provided we relax m_E to be a real number. This is possible thanks to Fortet's formula. Once the parameters of the auxiliary Erlang system have been found, the blocking probability of the original system, made up by m servers to which the peaked traffic of mean intensity A and peakedness z is offered, can be calculated as the loss probability of an equivalent system made up of $m + m_E$ servers, to which a Poisson traffic of mean intensity A_E is offered. Then

$$P_L = B(m + m_E, A_E) \tag{5.40}$$

Analogously, the first two moments of the lost traffic of the original system can be calculated using formulas (5.38), applied to the equivalent system with $m + m_E$ servers and Poisson input traffic of mean intensity A_E.

The first two moments of the carried traffic can be calculated as follows. Let $A_{c,E}$ and $V_{c,E}$ be the mean and the variance of the carried traffic of the equivalent loss system comprising $m + m_E$ servers. Let further $A_{c1,E}$ and $V_{c1,E}$ be the first two moments of the traffic carried by the loss system consisting of the additional m_E servers, to which a Poisson traffic of intensity A_E is offered. The first two moments of the traffic carried by the original system of m servers, to which the original non-Poisson traffic is offered, can be estimated as $A_c = A_{c,E} - A_{c1,E}$ and $V_c = V_{c,E} - V_{c1,E}$.

5.2.3.2 Fredericks' Method

An alternative method has been developed by Fredericks [83] and applies to any kind of traffic, either smoothed or peaked. The idea is to replace the original non-Poisson traffic with a Poisson batch arrival process of service requests having the same mean and variance of the original traffic process. Let the batch have fixed size equal to J. The moment-generating function (MGF) of the offered traffic is then $\phi_N(z) = e^{A(z^J - 1)}$. The first two moments are $E[N] = A J$ and $\sigma_N^2 = A J^2$. Their ratio is $Var(N)/E[N] = J$. We can therefore identify J with the peakedness factor z (we assume for the time being that z is an integer).

The key idea of the approximation, suggested by the reasoning above, is to replace the original loss system composed of m servers, to which a non-Poisson traffic of moments A and $V = zA$ is offered, with an equivalent system made of m/z servers, to which a Poisson traffic of mean intensity A/z is offered. It is as if the customers of this new system were identified with the batches of the batch Poisson arrival process.

With this approximation, we can apply the analysis technique holding for the Erlang loss system to the equivalent system with m/z servers and mean offered traffic A/z. We calculate the loss probability using Erlang-B formula, and the moments of the carried and lost traffic processes using formulas (5.37) and (5.38). To recover the first two moments of the carried and lost traffic processes of the original loss system, we account for the fact that we have divided the offered traffic by z. Hence, we multiply the mean values of the carried and lost traffic by z and the variances by z^2. The resulting expressions are usually referred to as Fredericks-Lindberger equations.

The blocking probability is approximated by using the Erlang B-formula applied to the equivalent system:

$$P_L = B\left(\frac{m}{z}, \frac{A}{z}\right) \tag{5.41}$$

where the extension of Erlang B-formula to noninteger values of the number of servers must be used in general.

The first two moments of the carried traffic are given by the following expressions:

$$\begin{aligned} A_c &= A(1 - P_L) \\ V_c &= zA_c - AP_L(m - A_c) \end{aligned} \tag{5.42}$$

while the following equations hold for the lost traffic:

$$\begin{aligned} A_l &= AP_L \\ V_l &= zA_l\left(1 - \frac{A_l}{z} + \frac{A}{m + z + A_l - A}\right) \end{aligned} \tag{5.43}$$

where the loss probability is computed according to eq. (5.41).

This approximation can be used both for peaked ($z > 1$) and for smooth traffic ($z < 1$). Equations (5.42) and (5.43) are exact in case of Poisson input traffic ($z = 1$).

As a last tile to complete the puzzle of two moments traffic analysis, when merging traffic flows, the mean values and the variances sum up, i.e., traffic flow components are assumed to be independent of one another. The loss probability experienced by the service request of the i-th component flow at a link where the overall loss probability is P_L is given by

$$P_{L,i} = \frac{z_i}{z} P_L \tag{5.44}$$

where z_i is the peakedness coefficient of the i-th traffic flow.

Example 5.4 We apply the method of Fredericks-Lindberger to the network in Figure 5.5. We consider the Poisson traffic flows AC and BC with mean intensities A_{AC} and A_{BC} ranging from 80 Erl to 100 Erl. The alternative routing is defined by the direct path as a first choice, the path through D as a second choice. In this example, all quantities referred to a link are labeled with a subscript number corresponding to the link, as shown in Figure 5.5. The number of circuits available for the network links are $m_1 = 64$, $m_2 = 96$, $m_3 = m_4 = m_5 = 32$.

Since the offered traffic is assumed to be Poisson for the two considered traffic flows, we have

$$P_{L,1} = B(m_1, A_{AC})$$
$$A_3 = A_{l,1} = A_{AC}P_{L,1}$$
$$V_3 = V_{l,1} = z_1 A_{l,1} \left(1 - \frac{A_{l,1}}{z_1} + \frac{A_{AC}}{m_1 + z_1 + A_{l,1} - A_{AC}} \right)$$
$$z_3 = V_3/A_3$$
$$P_{L,3} = B\left(\frac{m_3}{z_3}, \frac{A_3}{z_3} \right)$$
$$A_{c,3} = A_3(1 - P_{L,3})$$
$$V_{c,3} = z_3 A_{c,3} - A_3 P_{L,3}(m_3 - A_{c,3})$$

for the traffic flow AC, and

$$P_{L,2} = B(m_2, A_{BC})$$
$$A_4 = A_{l,2} = A_{BC}P_{L,2}$$
$$V_4 = V_{l,2} = z_2 A_{l,2} \left(1 - \frac{A_{l,2}}{z_2} + \frac{A_{BC}}{m_2 + z_2 + A_{l,2} - A_{BC}} \right)$$
$$z_4 = V_4/A_4$$
$$P_{L,4} = B\left(\frac{m_4}{z_4}, \frac{A_4}{z_4} \right)$$
$$A_{c,4} = A_4(1 - P_{L,4})$$
$$V_{c,4} = z_4 A_{c,4} - A_4 P_{L,4}(m_4 - A_{c,4})$$

for the traffic flow BC. The traffic offered to link 5 is

$$A_5 = A_{c,3} + A_{c,4}$$
$$V_5 = V_{c,3} + V_{c,4}$$
$$P_{L,5} = B\left(\frac{m_5}{z_5}, \frac{A_5}{z_5} \right)$$

The end-to-end loss probabilities for the two considered traffic flows are:

$$z_{5,AC} = V_{c,3}/A_{c,3}$$
$$z_{5,BC} = V_{c,4}/A_{c,4}$$
$$P_L(A,C) = P_{L,1} \left[P_{L,3} + (1 - P_{L,3}) \frac{z_{5,AC}}{z_5} P_{L,5} \right]$$
$$P_L(B,C) = P_{L,2} \left[P_{L,4} + (1 - P_{L,4}) \frac{z_{5,BC}}{z_5} P_{L,5} \right]$$

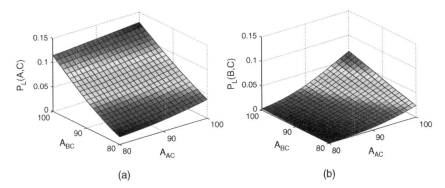

Figure 5.7 End-to-end loss probability of the traffic flows AC (left) and BC (right) as a function of their respective mean traffic intensities. Alternative routing on two paths is assumed (direct path and alternate path through D).

The loss probabilities $P_L(A, C)$ and $P_L(B, C)$ are shown in Figure 5.7 as a function of A_{AC} and A_{BC}. As expected, the probability $P_L(A, C)$ is bigger than $P_L(B, C)$ since the link 1 has 64 circuits against the 96 circuits of link 2.

5.2.4 Multi-Class Erlang Loss System

A last generalization addressed here is a multi-class scenario where the loss system has m servers and the offered traffic is made of the superposition of demands from c classes of customers. A class i customer needs $b_i \in [1, m]$ servers to accommodate its request. Hence, it gets blocked if $m - b_i + 1$ or more servers are already busy. In general, the state of the system is described by a c-tuple of nonnegative integers $(Q_1, \ldots, Q_c)$, where Q_i is the number of class i customers in the queue at a general time of the equilibrium. Feasible vectors $(n_1, \ldots, n_c)$ are all and only those such that $\sum_{i=1}^{c} b_i n_i \leq m$. If arrivals follow a Poisson law, with mean rate λ_i for class i, a very efficient algorithm has been devised to find the blocking probability of each class, without the need to calculate the limiting probabilities of the joint random variable $(Q_1, \ldots, Q_c)$ [118, 183].

In those papers it is shown that the probability distribution of the number of servers that are busy in equilibrium can be obtained by means of the following simple recursion:

$$p_k = \frac{1}{k} \sum_{i=1}^{c} A_i b_i p_{k-b_i}, \qquad k = 1, \ldots, m, \tag{5.45}$$

where $A_i = \lambda_i/\mu_i$ and $1/\mu_i$ is the mean service time of class i. The recursion is used along with the congruence condition $\sum_{k=0}^{m} p_k = 1$ and the boundary conditions $p_x = 0$ for $x < 0$.

This model holds for general service times (insensitivity property) and the complete sharing policy, i.e., any server can be allocated to any service demand, only provided there are enough to completely match the demand. The loss probability of class i, $P_L(i)$, is found as

$$P_L(i) = \sum_{j=0}^{b_i-1} p_{m-j}, \qquad i = 1, \ldots, c. \tag{5.46}$$

The state probability $p(\mathbf{n}) \equiv P(Q_i = n_i, \ i = 1, \ldots, c)$ for $\mathbf{n} \equiv (n_1, \ldots, n_c) \in \Omega$, $\Omega = \{\mathbf{n} : \sum_{i=1}^{c} n_i b_i \leq m\}$, can be calculated as follows:

$$p(\mathbf{n}) = G(\Omega)^{-1} \prod_{i=1}^{c} \frac{A_i^{n_i}}{n_i!}, \qquad \mathbf{n} \in \Omega, \tag{5.47}$$

where the normalization constant is

$$G(\Omega) = \sum_{\mathbf{n} \in \Omega} \prod_{i=1}^{c} \frac{A_i^{n_i}}{n_i!} \tag{5.48}$$

It is easy to recognize that $G(\Omega) = 1/p_0$, where p_0 is obtained from the recursion (5.45).

Example 5.5 Let us consider an LTE base station giving access to two classes of customers. The average amount of data that is transferred over a class 1 connection is $B_1 = 1$ Mbyte, it is $B_2 = 10$ Mbyte for a class 2 connection. The BS has $m = 100$ Resource Blocks (RB) to assign to customers for carrying traffic. A radio channel consisting of one RB per frame can sustain a bit rate of $r = 500$ kbit/s. Class 1 customers are assigned a single RB per frame ($b_1 = 1$), while class 2 customers are allocated 10 RBs per frame ($b_2 = 10$).

We assume that class 1 customers account for a fraction $\xi = 0.8$ of the overall connection requests. If λ denotes the mean rate of offered connection requests, the mean offered traffic intensity is:

$$A_o = \lambda(\xi b_1 E[X_1] + (1 - \xi) b_2 E[X_2]) \tag{5.49}$$

where $E[X_i] = 8 B_i/(b_i r)$.

The utilization coefficient of the multi-class system is

$$\rho = \frac{\lambda}{m}[\xi b_1 E[X_1](1 - P_{L,1}) + (1 - \xi) b_2 E[X_2](1 - P_{L,2})] \tag{5.50}$$

where $P_{L,i}$ is the loss probability of class i customers.

This system can be modeled with the multi-class Erlang model, provided that the connection request arrivals can be modeled as a Poisson process. Figure 5.8(a) plots the loss probability of the two classes as a function of the utilization coefficient. As expected, class 2 customers suffer a larger loss probability. Whenever less than 10 RBs are idle, a class 2 customer is rejected, whereas a class 1 customer

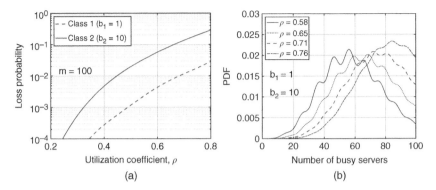

Figure 5.8 Multi-class Erlang system modeling an LTE radio access. Left plot: loss probability of the two classes as a function of the utilization coefficient. Right plot: probability distribution of the number of busy RBs.

can still find room to accommodate its connection, as long as there is at least one idle RB.

Four probability distributions of the number of busy RBs are plotted in Figure 5.8(b) for four values of the utilization coefficient, respectively. The peculiar feature of this distribution is the oscillating behavior, induced by the interplay of the different amounts of resource required by customers belonging to the two classes. This kind of behavior is characteristic of the probability distribution of busy servers in any multi-class Erlang loss system.

Example 5.6 As another example, we consider a large system, comprising $m = 10000$ servers. A set of 10000 servers that are assigned to jobs having different requirements in terms of how many servers they should be allocated is easily found in the cloud.

Incidentally, this example proves also the extreme numerical efficiency of the iteration used to compute the probability distribution of the number of busy servers and the loss probabilities of the different traffic classes.

We consider four classes of customers, requesting 1, 20, 50, and 100 servers respectively. Figure 5.9(a) shows the loss probability of the four classes versus the utilization coefficient ρ. The probability distribution of the number of busy servers for $\rho \approx 0.9$ is shown in Figure 5.9(b).

The oscillating behavior of the probability distribution has disappeared, due to the very large size of the considered system. The crucial point is the high ratio between the number of available servers ($m = 10000$) and the amount of servers allocated to service requests (at most $b_4 = 100$).

As for the loss probability, customers belonging to high demand classes (e.g., class 4, having $b_4 = 100$) suffer a strong degradation of their loss performance at

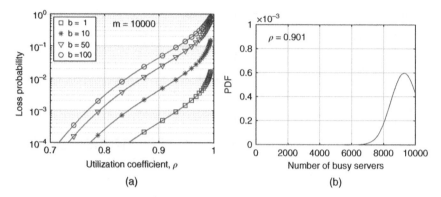

Figure 5.9 Multi-class Erlang system modeling a large server cluster. Left plot: loss probability of the four classes as a function of the utilization coefficient. Right plot: probability distribution of the number of busy servers.

high utilization, while less demanding customers still enjoy a relatively low loss probability, even for utilization levels as high as 0.9. In general, we can push utilization to quite high levels before impairing loss performance significantly. This is yet another evidence that large scale system offer a good trade-off between utilization and performance.

5.3 Application of the Erlang Loss Model to Cellular Radio Access Network

Let us consider the radio access network (RAN) of a cellular network. A cell is an area served by a radio station equipment referred to as a base station. The specific technical name of the BS depends on the technology. It is BTS in GSM, NodeB in UMTS, evolved NodeB (eNB) in LTE, next generation NodeB (gNB) in 5G. In the following sub-sections, we will address performance evaluation and dimensioning problems related to a cellular RAN. For this purpose, we make a simplified model of the RAN. We assume that the coverage area of base stations (BSs) has a hexagonal shape with radius R. The coverage area of the cellular network is therefore tiled with a hexagonal grid. We use the following notation in the rest of this section.

N number of BSs in the coverage area.

K cluster size, i.e, the number of BSs (cells) making up a reuse cluster.

M overall number of channels available to the cellular operator.

m number of channels assigned to a BS: $m = \lfloor M/K \rfloor$.

D reuse distance, i.e., the minimum distance between two BSs using the same channels.

R cell radius.

P_{tx} BS transmission power level in a single channel.

U mean number of users in a cell.

δ user density; it is $\delta = U/S_c$, where S_c is the area of the cell.

a mean offered traffic intensity of a user.

Y connection holding time.

μ connection completion rate; it is $E[Y] = 1/\mu$.

V sojourn time of a user in a cell.

New connections are generated by a user according to a Poisson process of mean rate $\lambda = \mu a$. The connection holding time Y is distributed according to a negative exponential random variable with mean $1/\mu$, unless stated otherwise.

5.3.1 Cell Dimensioning under Quality of Service Constraints

A BS is assigned a number of channels to serve users roaming in the relevant cell. Channels are *reused spatially*, i.e., the same channel can be used by different BSs provided that their distance makes the reciprocal interference tolerable. The largest compact set of contiguous cells that are assigned different channels (no reuse) is referred to as *reuse cluster*.

We refer to a simple geometrical model of the BS layout, namely, BS are assumed to be spread uniformly over the plane, according to a hexagonal grid. Figure 5.10 illustrates two examples of hexagonal cell grids and cell clusters of sizes $K = 7$ and $K = 4$. The reuse distance D is defined as the distance between the two closest BSs that reuse the same radio channel. With the hexagonal geometry of cells, it can be proved that $D = R\sqrt{3K}$.

Densifying the BSs' layout, the capacity of the RAN grows. The cost of the RAN infrastructure grows as well. As a matter of fact, for each BS the operator must provide power supply and interconnection to the backhaul network, equipment installation, configuration, management, and maintenance (updates, repairing),

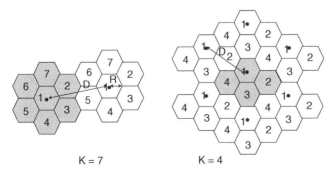

Figure 5.10 Examples of cluster and reuse distance in a cellular network coverage.

cell site planning, permissions, right-of-way, rent. The tipping point where there is no more incentive to push densification further depends on the density of the offered traffic.

For example, consider a scenario where a region of area $S = 20$ km^2 is to be covered with a RAN. Density of users is $\delta = 3000$ users/km^2 and each user offers an amount of traffic $a = 0.1$ Erl. The overall offered traffic intensity is $A = a\delta S = 6000$ Erl. Assume N BSs are used to cover the tagged region, with a cluster size of $K = 9$. Let $r = 1$ be the revenue per Erlang of carried traffic and $c = 5$ be the cost per installed BS. The net profit of the RAN can be expressed as:

$$P = r\,N\frac{A}{N}\left[1 - B\left(\frac{M}{\min\{N,K\}}, \frac{A}{N}\right)\right] - c\,N = r\,A\left[1 - B\left(\frac{M}{\min\{N,K\}}, \frac{A}{N}\right)\right] - c\,N$$

$$(5.51)$$

for $N \geq 1$. It is easy to find that the net profit is maximized at $N^* = 569$. This corresponds to a coverage radius per BS in the order of 100 m.

Going back to our radio access model, we aim at connecting the dimensioning of the RAN to requirement of radio channel quality at physical layer and requirements of user perceived quality at application level. The former is measured by means of the signal to noise and interference ratio (SNIR) of the radio channel. The latter is quantified by means of blocking of new connection attempts.

Let us start with modeling the impact of a requirement on SNIR, e.g., $SNIR \geq \gamma$.

We assume a very simple propagation model, with a power law attenuation with decay exponent α. Then, the received power level at distance d from the emitting source is

$$P_{rx} = P_{tx}G(d) = \frac{\kappa P_{tx}}{d^\alpha} \tag{5.52}$$

where $G(d)$ is the path gain at distance d.

Let us focus on the downlink (transmission of data from the BS toward the user terminals). A user is served by the BS from which it receives the strongest signal level. According to the assumed path loss model, the serving BS is the one closest to the user terminal.

We consider a worst-case scenario, where the tagged user is at the biggest possible distance from the serving BS, i.e., at distance R. Therefore, the power level of the signal received from the serving BS is $P_{tx}G(R)$. On top of this useful signal component, the user terminal receives interference from other BSs and background noise. Interfering BSs are only those that reuse the same radio channel as the tagged terminal. Interfering BSs form concentric tiers around the serving BS, at distance of multiples of the reuse distance D. There are $6j$ interfering BSs on tier j at distance jD, for $j \geq 1$. Taking all of these into account, the SNIR can be written as

$$SNIR = \frac{\kappa P_{tx}/R^\alpha}{P_N + \sum_{j=1}^{\infty} 6j\,\kappa P_{tx}/(j\,D)^\alpha} \approx \frac{D^\alpha}{6R^\alpha} \geq \gamma$$

where P_N is the background noise power. In the rightmost side we have simplified the expression of *SNIR* neglecting the background noise power level P_N with respect to interference and the interfering BSs further than the first tier ones.

From the SINR requirement we find $\frac{D}{R} \geq (6\gamma)^{1/\alpha}$. On the other hand, it can be shown that $\frac{D}{R} = \sqrt{3K}$. Hence, it must be

$$K \geq \frac{1}{3}(6\gamma)^{2/\alpha} \tag{5.53}$$

This constraint ties together the cell layout parameter K, the physical layer requirement γ and the only parameter of the physical channel, α.

We can model the offered traffic process as follows:

- Each user generates channel requests at a mean rate λ.
- Once a users gets a channel, it holds the channel for the whole duration of its connection; let $1/\mu$ be the average holding time.
- When all m channels of a cell are used and a new user request arrives at that cell, the new request is rejected with no wait.

If we assume that user requests arrive as a Poisson process then we can model the cell from the point of view of channel usage as an $M/G/m/0$ queue, i.e., Poisson input, general service times, m inter-changeable and equivalent servers, m customers at most can be admitted into the queue (no wait). This is but the Erlang queueing model.

Note that we are not accounting for handoffs, hence we are neglecting user mobility (i.e., we are assuming $V = \infty$ with probability 1). We will relax this restriction in next subsections.

Given the Erlang model, the blocking probability experienced by a user is $P_B = B(m, A)$, where A is the mean intensity of the traffic offered to a BS and m is the number of radio channels assigned to the BS. We set a requirement p on the blocking probability, i.e., we require that $P_B \leq p$.

Given the overall number of radio channels available to the operator for cell planning, M, and the reuse cluster size, we have $m = \lfloor M/K \rfloor$. The offered traffic can be expressed as a function of the mean number of users in the cell, U, and the mean intensity of traffic offered by a user, a: $A = aU$. In turn, $U = \delta \pi R^2$, where δ is the user spatial density and the cell shape has been approximated as a circle.

Summing up, the dimensioning evolves through the following steps:

1. Evaluate the minimum cluster size to maintain the requirement on the quality of physical radio channels, i.e., $SNIR \geq \gamma$;
2. Evaluate the maximum traffic intensity that can be sustained by the BS under the requirement of quality of service provided to users, i.e., $P_B \leq p$.
3. Given user density and traffic model, derive the maximum feasible cell radius R.

Table 5.1 Example of cell coverage dimensioning in a cellular RAN: reference case.

Symbol	Value	Dimensioning
γ	9 dB (7.94)	$K \geq \frac{1}{3}(6\gamma)^{2/\alpha} = 2.29 \rightarrow K = 3$
M	90	$A \leq B^{-1}(M/K, p) = B^{-1}(90/3, 0.02)$
p	0.02	$A_{\max} = 21.93$ Erl
λ	0.6 req/h	$U\lambda/\mu = A \leq A_{\max} \rightarrow U_{\max} = \dfrac{A_{\max}}{\lambda/\mu}$
$1/\mu$	1.76 min	$U_{\max} = 1246$ users/cell
δ	1600 users/km^2	$\pi R^2 \delta = U \leq U_{\max} \rightarrow R \leq \sqrt{\dfrac{U_{\max}}{\pi \delta}} \approx 498$ m

Table 5.2 Example of cell coverage dimensioning in a cellular RAN: low user density.

Symbol	Value	Dimensioning
γ	9 dB (7.94)	$K \geq \frac{1}{3}(6\gamma)^{2/\alpha} = 2.29 \rightarrow K = 3$
M	90	$A \leq B^{-1}(M/K, p) = B^{-1}(90/3, 0.02)$
p	0.02	$A_{\max} = 21.9$ Erl
λ	0.6 req/h	$U\lambda/\mu = A \leq A_{\max} \rightarrow U_{\max} = \dfrac{A_{\max}}{\lambda/\mu}$
$1/\mu$	1.76 min	$U_{\max} = 1246$ users/cell
δ	160 users/km^2	$\pi R^2 \delta = U \leq U_{\max} \rightarrow R \leq \sqrt{\dfrac{U_{\max}}{\pi \delta}} \approx 1574$ m

The maximum cell radius is related to the density of BS deployment required to sustain the offered traffic under the given quality constraints, hence it is an indirect measure of the cost of the RAN infrastructure.

Let us give some numerical examples in Tables 5.1, 5.2, and 5.3. The path gain exponent is $\alpha = 4$ in all cases.

The first table sets a reference case, with a relatively high user density. As a result, the cell radius compatible with all constraints turns out to be around 500 m (still not a really small cell). If we consider lower values of the user density, the required cell radius increases. With the same density, a more stringent requirement on the quality of the radio link in terms of SNIR brings the radius to an intermediate value (about 1300 m). The reason is that less users can be accommodated because of the increase of the cluster size required to meet the radio link quality constraint.

Table 5.3 Example of cell coverage dimensioning in a cellular RAN: high channel quality requirement.

Symbol	Value	Dimensioning
γ	12 dB (15.8)	$K \geq \frac{1}{3}(6\gamma)^{2/\alpha} = 3.25 \rightarrow K = 4$
M	90	$A \leq B^{-1}(M/K, p) = B^{-1}(90/4, 0.02)$
p	0.02	$A_{\max} = 14.9$ Erl
λ	0.6 req/h	$U\lambda/\mu = A \leq A_{\max} \rightarrow U_{\max} = \dfrac{A_{\max}}{\lambda/\mu}$
$1/\mu$	1.76 min	$U_{\max} = 847$ users/cell
δ	1600 users/km^2	$\pi R^2 \delta = U \leq U_{\max} \rightarrow R \leq \sqrt{\dfrac{U_{\max}}{\pi\delta}} \approx 1298$ m

The antennas of the BS can be designed to split the cell area into *sectors*. As an example, with 120° antennas three sectors are defined. Sectorization can bring advantages, since interfering BSs reduce to only those whose signals are received in the tagged cell sector. However, channels must be partitioned over sectors, hence smaller-scale systems are designed (less users in a sector with respect to a cell, less channels assigned to a sector with respect to a cell). The final balance that the system strikes is to be checked.

Let n denote the number of antenna sectors per BS. We compare three designs, with $n = 1$ (no sectorization), $n = 3$ and $n = 6$. Numerical results are provided in Table 5.4 for the following numerical values of the parameters: $\gamma = 12$ dB, $\alpha = 4$, $M = 120$, $p = 0.02$, $\lambda = 0,6$ req/h, $1/\mu = 1.76$ min, $\delta = 1600$ users/km^2. In

Table 5.4 Example of cell coverage dimensioning in a cellular RAN with antenna sectorization. Parameter values set as in Table 5.1.

	$n = 1$ (no sect.)	$n = 3$ (120°)	$n = 6$ (60°)
$K \geq \frac{1}{3}\left(\frac{6}{n}\gamma\right)^{2/\alpha}$	$K \geq 3.2 \rightarrow K = 4$	$K \geq 1.8 \rightarrow K = 3$	$K \geq 1.3 \rightarrow K = 3$
$m = \left\lfloor \dfrac{M}{nK} \right\rfloor$	$m = 30$	$m = 13$	$m = 6$
$A_{\max} = B^{-1}\left(\dfrac{M}{nK}, p\right)$	$A_{\max} = 21.9$ Erl	$A_{\max} = 7.4$ Erl	$A_{\max} = 2.27$ Erl
$U_{\max} = \dfrac{A_{\max}}{\lambda/\mu}$	$U_{\max} = 1246$	$U_{\max} = 420$	$U_{\max} = 129$
$R = \sqrt{\dfrac{nU_{\max}}{\pi\delta}}$	$R = 498$ m	$R = 500$ m	$R = 393$ m

Table 5.5 Example of cell coverage dimensioning in a cellular RAN with antenna sectorization. Parameter values set as in Table 5.2.

	$n = 1$ (no sect.)	$n = 3$ (120°)	$n = 6$ (60°)
$K \geq \frac{1}{3}\left(\frac{6}{n}\gamma\right)^{2/\alpha}$	$K \geq 3.2 \to K = 4$	$K \geq 1.8 \to K = 3$	$K \geq 1.3 \to K = 3$
$m = \left\lfloor \frac{M}{nK} \right\rfloor$	$m = 30$	$m = 13$	$m = 6$
$A_{max} = B^{-1}\left(\frac{M}{nK}, p\right)$	$A_{max} = 21.9$ Erl	$A_{max} = 7.4$ Erl	$A_{max} = 2.27$ Erl
$U_{max} = \frac{A_{max}}{\lambda/\mu}$	$U_{max} = 1246$	$U_{max} = 420$	$U_{max} = 129$
$R = \sqrt{\frac{nU_{max}}{\pi\delta}}$	$R = 2226$ m	$R = 2240$ m	$R = 1757$ m

Table 5.6 Example of cell coverage dimensioning in a cellular RAN with antenna sectorization. Parameter values set as in Table 5.3.

	$n = 1$ (no sect.)	$n = 3$ (120°)	$n = 6$ (60°)
$K \geq \frac{1}{3}\left(\frac{6}{n}\gamma\right)^{2/\alpha}$	$K \geq 6.48 \to K = 7$	$K \geq 3.7 \to K = 4$	$K \geq 2.6 \to K = 3$
$m = \left\lfloor \frac{M}{nK} \right\rfloor$	$m = 21$	$m = 12$	$m = 8$
$A_{max} = B^{-1}\left(\frac{M}{nK}, p\right)$	$A_{max} = 14.0$ Erl	$A_{max} = 6.6$ Erl	$A_{max} = 3.6$ Erl
$U_{max} = \frac{A_{max}}{\lambda/\mu}$	$U_{max} = 797$	$U_{max} = 375$	$U_{max} = 206$
$R = \sqrt{\frac{nU_{max}}{\pi\delta}}$	$R = 1781$ m	$R = 2118$ m	$R = 2218$ m

Table 5.5 the user density has been varied to $\delta = 80$ users/km^2, while all other numerical parameters are the same as in Table 5.4. Finally, in Table 5.6 the SNIR requirement is changed to $\gamma = 18$ dB, all other parameter values being the same as in Table 5.4.

It can be observed that the best design is obtained with $n = 3$ sectors in the first two cases, while $n = 6$ is better in the third one.

5.3.2 Number of Handoffs in a Connection Lifetime

In the following we model the user sojourn time within a cell as a random variable V with continuous CDF $F_V(t)$. The Laplace transform of the corresponding PDF is denoted with $\varphi_V(s)$. We assume users move according to a constant-speed linear trajectory, with speed v. The direction of motion is taken at random. To keep the analysis simple, we approximate the cell shape with a circle of radius R. Then, the mean sojourn time in a cell with the assumed mobility model is $E[V] = 4R/\pi v$.

Alternatively, we could model the user movement by using a planar random walk. Then the sojourn time can be identified with the time to absorption of the random walk process, given the initial position where the user enters the cell, the absorbing barrier being the cell border. For example, we could model user mobility by means of a two-dimensional Brownian motion. If the drift is 0 and the variance coefficient is σ^2, assuming a square cell shape of side $2R$, the time to absorption V starting from the center of the cell has the following CCDF (see [60, Ch. 5, p. 222]):

$$G_V(t) = \frac{4}{\pi} \sum_{k=1}^{\infty} \frac{(-1)^{k-1}}{2k-1} \exp\left(-\frac{(2k-1)^2 \pi^2 \sigma^2}{8R^2} t \right) \tag{5.54}$$

The m channels available to a BS are assigned on demand to support user connections, as long as a user is visiting the cell or the connection is up. The channel holding time is therefore $X = \min\{Y, V\}$, where Y is the connection holding time and V is the user sojourn time in a cell, under the service of the BS covering that cell. We assume that cell sojourn times are independent of the connection holding times. Therefore, the CCDF of X is $G_X(t) = P(X > t) = P(Y > t, V > t) = G_Y(t)G_V(t)$. If Y has a negative exponential PDF with parameter μ, we have

$$E[X] = \int_0^\infty G_X(t)\, dt = \int_0^\infty e^{-\mu t} G_V(t)\, dt = \frac{1 - \varphi_V(\mu)}{\mu} \tag{5.55}$$

Connection set-up requests arriving at a congested BS (i.e., having no idle channel) are rejected immediately, with no wait. Then, the congestion at connection level can be modeled by an Erlang loss system with m servers, service time X and arrival process defined by the connection set-up request arrivals. We assume this process is Poisson with rate λ. The mean traffic intensity offered to the BS is $A = \lambda E[X]$.

We evaluate the probability distribution of making k handoffs during a connection lifetime. Let H be the random variable denoting the number of handoffs during a connection lifetime, $H \geq 0$. We have

$$P(H \geq k) = P(V_1 + \cdots + V_k \leq Y) = \int_0^\infty f_Y(t) P(V_1 + \cdots + V_k \leq t)\, dt \tag{5.56}$$

If Y has negative exponential PDF with parameter μ, the rightmost side of eq. (5.57) is recognized as the Laplace transform of the CDF of the random variable $V_1 + \cdots + V_k$. If sojourn times in different cells are independent of one another, we can write:

$$P(H \geq k) = \mu \int_0^\infty e^{-\mu t} P(V_1 + \cdots + V_k \leq t)\, dt = \mu \left. \frac{\varphi_V^k(s)}{s} \right|_{s=\mu} = \varphi_V^k(\mu) \tag{5.57}$$

Hence, H has a geometric probability distribution with ratio $\varphi_V(\mu) < 1$:

$$P(H = k) = [1 - \varphi_V(\mu)]\varphi_V^k(\mu), \qquad k \geq 0 \tag{5.58}$$

It is $E[H] = \varphi_V(\mu)/[1 - \varphi_V(\mu)]$ with $\mu = 1/E[Y]$. If V has negative exponential PDF as well, it is easy to find $E[H] = E[Y]/E[V]$, which is a rather intuitive result. The probability that a radio channel is released because the connection has been torn down is $P(Y < V) = E_Y[G_V(Y)]$. This is just $1 - \varphi_V(\mu)$.

5.3.3 Blocking in a Cell with User Mobility

Let us now turn to the evaluation of the traffic load of a cell where both new connection requests and handoff requests from neighboring cells are offered. Connection holding time is distributed according to a negative exponential random variable Y with mean $E[Y] = 1/\mu$. We assume the same mobility model as in previous section. Hence, the mean channel holding time $E[X]$ is given in eq. (5.55) and the probability that a connection is handed off, before it terminates, is $\varphi_V(\mu)$.

Let λ_{lo} be the average rate of the Poisson process describing new connection requests and λ_{ho} the average incoming handoff rate. If the cellular network coverage area is modeled with a uniform hexagonal grid, the average rate of radio channels being released in a given cell is $\lambda_s = \lambda(1 - p)$, where λ is the overall channel requests rate and p is the loss probability. Of these requests, a fraction $\varphi_V(\mu)$ leaves the current cell while the connection is still up, hence gives rise to a handoff toward one of the six neighboring cells. It motion is at random, 1/6 of the outgoing handoffs will be directed to a specific neighboring cell. Hence, the average rate of incoming handoffs in a cell is $\lambda_{ho} = 6 \cdot 1/6 \cdot \varphi_V(\mu)\lambda_s = \varphi_V(\mu)\lambda(1 - p)$. Further, we have $p = B(m, A) = B(m, \lambda E[X]) = B(m, \lambda[1 - \varphi_V(\mu)]/\mu)$, where $B(\cdot, \cdot)$ is the Erlang B function. Reminding that $\lambda = \lambda_{lo} + \lambda_{ho}$, we end up with the following equation system with unknowns λ_{ho} and p:

$$\begin{cases} \lambda_{ho} = \varphi_V(\mu)(\lambda_{lo} + \lambda_{ho})(1 - p) \\ p = B\left(m, (\lambda_{lo} + \lambda_{ho})\dfrac{1 - \varphi_V(\mu)}{\mu}\right) \end{cases} \tag{5.59}$$

If we let $q \equiv \varphi_V(\mu)$ and normalize average arrival rates with μ, so as to turn to traffic intensities, the system can be rewritten as:

$$\begin{cases} A_{ho} = \dfrac{q(1 - p)}{1 - q(1 - p)}A_{lo} \\ p = B(m, (A_{lo} + A_{ho})(1 - q)) = B\left(m, A_{lo}\dfrac{1 - q}{1 - q + qp}\right) \end{cases} \tag{5.60}$$

The latter equation can be solved numerically for p, once the fresh traffic intensity A_{lo}, the number of radio channels m and the probability q are given. Then the

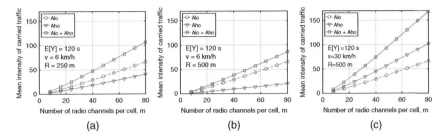

Figure 5.11 Mean intensity of carried traffic as a function of the number of radio channels per cell, with three different parameter settings for the cell radius R and the user speed v (the mean connection duration is $E[Y] = 120$ s).

incoming handoff rate is found as $\lambda_{ho} = A_{ho}\mu$. The average rate of handoffs per unit covered area is therefore $\Lambda_{houa} = \lambda_{ho}/(\pi R^2)$. Note that it is also $A_{lo} = a\delta\pi R^2$, where a is the user mean offered traffic intensity (typically 25 mErl) and δ is the user average spatial density.

Note that the above calculations assume that the incoming handoff flow can be dealt with as a Poisson process. This is an approximation, since the handoff flow is the carried traffic of an Erlang loss system; it is therefore a smoothed traffic. Poisson handoff traffic is a good approximation as far as sojourn times have negative exponential PDFs and loss is small (e.g., less than 0.01), as it should be in a well-designed cellular network.

Numerical results are plotted in Figure 5.11. The mean connection holding time is kept fixed at $E[Y] = 1/\mu = 120$ s, while cell radius R and user speed v are varied, so as to experiment different sojourn times. We assume that sojourn times have a negative exponential PDF with mean $E[V] = 4R/(\pi v)$. The probability q is given by $q = \varphi_V(\mu) = 1/(1 + \mu E[V])$.

The relative weight of the handoff traffic intensity and of the local traffic intensity changes according to the mobility of users. In case of "speedy" users, the handoff induced traffic is overwhelming, whereas local traffic prevails for slower users. The critical parameter is the ratio of the mean connection holding time with the mean cell sojourn time.

The cell model yields another performance measure of interest: the handoff rate. Handoffs imply an intense exchange of signaling messages over the interfaces of the RAN and a significant amount of processing in all systems ranging from user terminals up to core network access nodes. Therefore, it is key to assess how frequently the handoff procedure is invoked in the face of user mobility and connection holding times.

Figure 5.12 plots the handoff rate as a function of the number m of channels assigned to each cell (left plot) and as a function of the cell radius R (right plot). The requirement on connection blocking probability is $p = 0.01$.

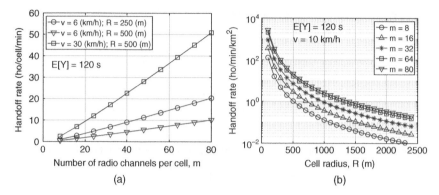

Figure 5.12 Incoming handoff rate as a function of the number of radio channels per cell (left plot) and as a function of the cell radius (right plot). The requirement on connection blocking probability is $p = 0.01$.

As for Figure 5.12(a), the slower the user mobility, the larger the sojourn times, and the more probable that few handoffs, if any, are performed during the connection lifetime. As users move faster, cell border crossing is more frequent and the handoff rate grows. Whatever the mobility setting, the handoff rate is obviously an increasing function of the number of radio channel assigned to a cell. This behavior is a consequence of the requirement on connection blocking probability. The amount of traffic that can be handled by a cell within the given loss requirement grows with the number of channels that the cell is assigned, hence the handoff rate grows as well.

Handoff rate per unit coverage area is plotted in Figure 5.12(b). It is apparent that reducing cell sizes boosts the number of handoffs per unit time and unit area toward large values, with a steep growth (note that the ordinate scale is logarithmic). Along with the reduced effect of multiplexing, this is one critical issue of BS densification.

5.3.4 Trade-off between Location Updating and Paging

The model developed under user mobility can be used to evaluate another typical trade-off of cellular networks. While a user equipment (or better, a subscriber identity module, SIM) roams in the cellular network coverage area, if it is not engaged in a connection, it updates the central network database whenever it crosses the border of a Localization Area (LA). An LA is a cluster of contiguous cells that defines the granularity of the spatial localization of a SIM. When there is an incoming call to a SIM and the LA where the SIM is currently roaming is known, a notification message (paging) is sent downlink in each cell making up the LA where the SIM is deemed to be.

The cost of updating the localization of the SIM grows as LAs are made smaller. On the other hand, the cost of paging grows as the LA size grows.

In our simple geometrical model of the cellular network coverage, where cells are all equal and shaped like a hexagon of radius R, the size of an LA can be measured by the number of cells making up the LA. To tile the entire plane, LAs must be configured as cell clusters. Hence, the LA size must be chosen among the cluster sizes K given by the formula $K = i^2 + j^2 + ij$, where i and j take non-negative integer values. These are the so called rhomboidal numbers.

Let us consider a coverage area comprising N cells. Let K denote the number of cells of an LA. We approximate the perimeter of the LA as $P(K) = 2\pi D$, where D is the re-use distance, given by $D = R\sqrt{3K}$ in the hexagonal grid. The mean rate at which a user crosses the cell border is $1/E[V]$, where V is the user sojourn time in a cell. The mean rate of cell border crossing is $\lambda_{\text{cross}} = U/E[V]$, where U is the mean number of users per cell. The mean number of crossings per unit of length of the cell border is $\lambda_{\text{cross}}/(2\pi R)$. Then, the mean rate of border crossing for an LA is $2\pi D\lambda_{\text{cross}}/(2\pi R) = \lambda_{\text{cross}}\sqrt{3K} = U\sqrt{3K}/E[V]$.

Denoting the mean offered traffic intensity of connection requests originating in a cell with A_{lo}, we have $U = A_{lo}/a = \lambda_{lo}/(\mu a)$. The overall mean rate of location updating in the entire coverage area is therefore

$$\lambda_{\text{LU}} = \frac{N}{K} \cdot \frac{U\sqrt{3K}}{E[V]} = \frac{N\lambda_{lo}\sqrt{3}}{\mu a E[V]} \frac{1}{\sqrt{K}} \tag{5.61}$$

As for paging, new connection requests arrive with a mean rate of λ_{lo} per cell. Each incoming call must be paged over all cells making up the LA where the addressed user is roaming. Hence, the mean rate of paging procedures triggered in a cell is $\lambda_{lo}K$. The overall number of paging procedures per unit time in the entire coverage area is $\lambda_{\text{PG}} = N\lambda_{lo}K$.

Let c_{LU} and c_{PG} denote the cost in terms of complexity and resource consumption of the location updating and the paging procedures. The overall cost per unit time and per cell for the given coverage area composed of N cells is

$$c = \frac{1}{N}(c_{\text{LU}}\lambda_{\text{LU}} + c_{\text{PG}}\lambda_{\text{PG}}) = \frac{1}{N}\left(c_{\text{LU}}\frac{N\lambda_{lo}\sqrt{3}}{\mu a E[V]}\frac{1}{\sqrt{K}} + c_{\text{PG}}N\lambda_{lo}K\right) = \frac{w_{\text{LU}}}{\sqrt{K}} + w_{\text{PG}}K \tag{5.62}$$

where w_{LU} and w_{PG} are suitable constants, depending on the parameters of the cellular network. There is clearly an optimal size of the LA. It is

$$K^* = \left(\frac{w_{\text{LU}}}{2w_{\text{PG}}}\right)^{2/3} = \left(\frac{c_{\text{LU}}\sqrt{3}}{2c_{\text{PG}}a\mu E[V]}\right)^{2/3} \tag{5.63}$$

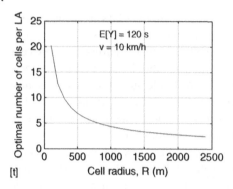

Figure 5.13 Optimal number of cells making up a location area (LA) versus the cell radius. The cost of the location updating procedure is set to twice the cost of the paging procedure.

Figure 5.13 plots the optimal number of cells for an LA, to minimize the cost of the mobility management procedures, as a function of the cell radius. The cost of location updating has been assumed as double of the cost of paging, i.e., $c_{LU}/c_{PG} = 2$, while the mean offered traffic per user is $a = 0.05$ Erl and the mean connection holding time is $E[Y] = 1/\mu = 120$ s. We assume that $E[V] = 4R/(\pi v)$, where v is the user speed. We let $v = 10$ km/h. Using the numerical values set for the parameters, we have, $K^* \approx 0.7734 \cdot \left(\frac{c_{LU}v}{c_{PG}a\mu R}\right)^{2/3} \approx \left(\frac{9069}{R}\right)^{2/3}$, where R is measured in meters.

As R grows, the mean sojourn time in a cell grows as well, and the optimal number of cells that make up an LA cluster gets smaller. Then, the cost of paging prevails, given that larger cells imply a lower border crossing rate, hence a lower rate of location updating.

5.3.5 Dimensioning of a Cell with Two Service Classes

Let us consider a BS provided with two kinds of radio channel: full rate (FR) and half rate (HR) channels. FR channels offer adequate communication QoS if the SNR at reception is no less than γ_F. HR channels require a higher SNR γ_H to provide adequate QoS, since they use stronger source coding to shrink the source bit rate to a half of that used in the FR channel, hence the bit error ratio tolerable for the HR channels is lower than that for FR channels.

We assume a power law with exponent α for the geometric attenuation, so that the SNR γ is inversely proportional to d^α. To emphasize the dependence of the SNR on the distance from the BS, we write $\gamma(d)$. Let R be the coverage radius of the BS, i.e., the maximum value of d such that $\gamma(d) \geq \gamma_F$. Then, the maximum distance from BS where HR channels are allowed is the maximum value of d such that $\gamma(d) \geq \gamma_H$. Let this value be r. It must be $\gamma_H = \gamma(r) = \gamma(R)(R/r)^\alpha = \gamma_F(R/r)^\alpha$, from which r can be found as a function of the SNR requirements, i.e., $r = R(\gamma_F/\gamma_H)^{1/\alpha}$.

Assume users are uniformly spread over the coverage area, so that the average number of connection attempts per unit time is proportional to the considered area. We can consider the average offered traffic from the inner circle around the

BS, with radius r, let it be A_i. Let also A_e denote the average offered traffic in the outermost ring ranging from distance r up to distance R from the BS. The overall offered traffic is $A = A_i + A_e$, where $A_i = A\pi r^2/(\pi R^2) = A(\gamma_F/\gamma_H)^{2/\alpha} = A\xi$ and $A_e = A(1 - \xi)$, with $\xi = (\gamma_F/\gamma_H)^{2/\alpha}$.

The spectral resource assigned to the BS is equivalent to m FR channels. It can be configured by using part of it as FR channels and the remaining as HR channels. Let m_F and m_H denote the number of FR and HR channels, respectively. The resource constraint is therefore $m_F + 2m_H = m$ for a given value of m.

We can state the following optimization problem: maximize the mean offered traffic intensity A subject to the following constraints.

1. *Traffic QoS requirement.* The connection blocking probability is no more than p.
2. *Radio channel QoS requirement.* The SNR of FR (HR) radio channel is no less than γ_F (γ_H).
3. *Resource allowance.* The number of FR and HR channel must satisfy $m_F + 2m_H \leq m$.

We assume that user mobility is negligible during a connection lifetime. This leads to reliable results if there is little chance that users move from the outer to the inner cell area or vice versa during a connection lifetime.

The optimization strategy is based on the remark that connection requests offered in the inner circle can be served by either kind of channel, whereas outermost connection requests must use a FR channel. Let q be the probability that a connection request originating from the inner circle takes an HR channel. The part of the overall average offered traffic loading HR channels is therefore $A_H = q\xi A$, while $A_F = (1 - q)\xi A + (1 - \xi)A = (1 - q\xi)A$ is the average offered traffic loading the FR channels, that is to say the sum of traffic coming from the outermost part of the cell plus a fraction $1 - q$ of the traffic of the inner circle. By assuming Poisson connection request arrivals and a pure loss system, the blocking probability can be expressed as

$$P = \frac{A_F B(m_F, A_F) + A_H B(m_H, A_H)}{A} = (1 - q\xi)B(m_F, (1 - q\xi)A) + q\xi B(m_H, q\xi A) \tag{5.64}$$

Thanks to the monotonicity of the Erlang loss function, the constraint on the loss probability must be saturated, i.e., we set $P = p$. For each fixed m_F from 1 up to m, we search for the value of q that maximizes A, under the constraint:

$$(1 - q\xi)B(m_F, (1 - q\xi)A) + q\xi B(2(m - m_F), q\xi A) = p \tag{5.65}$$

where m and p come from constraints and ξ is a known quantity. Let A^* denote the maximum value of the mean offered traffic and q^* the optimal value of the channel selection probability that achieves that maximum.

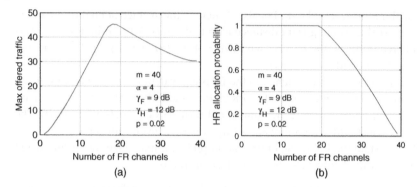

Figure 5.14 Maximum value of the cell offered traffic (left plot) and optimum value of the probability of choosing a half-rate channel for inner circle connections (right plot) as a function of the number of full-rate channels for given constraints on the connection blocking probability, on the quality of radio channels, and on the overall number m of FR channels.

Figure 5.14(a) shows the plot of A^* as a function of m_F for $m = 40$, $p = 0.02$ and $\xi = 1/\sqrt{2}$. The optimum value q^* is plotted against m_F in Figure 5.14(b). As long as the number of FR channels is below a threshold, it is not convenient to select FR channels in the inner circle around the BS, that is, $q^* = 1$. When there are plenty of FR channels, q^* becomes less than 1, i.e., it is best to select a FR channel, rather than an HR one, in the inner circle with probability $1 - q^*$. The optimal channel configuration corresponds to the breakpoint of the curve of q^* in Figure 5.14(b).

5.4 The $M/M/m$ Queue

In the $M/M/m$ queue customers arrive according to a Poisson process with mean rate λ. Service times are i.i.d. random variables having a negative exponential PDF with mean $1/\mu$, *independently of the server that takes care of the service*. Service and inter-arrival times are independent of one another.

The number $Q(t)$ of customers in the queue at time t is a Markov process under the assumptions holding for the $M/M/m$ queue. No information is required on the age of the service under way or the inter-arrival time age at time t to be able to predict the future evolution of the system after time t, once the number $Q(t)$ of customers residing in the system is assigned.

The Markov process $Q(t)$ is a birth-death one. Since no bulk arrival or bulk service are allowed, when a transition out of the state $Q = k$ occurs, it can only lead to either $k + 1$ or $k - 1$. We could derive results for the $M/M/m$ queue directly as special instance of birth-death process. We prefer instead to give a full account, by deriving the transition rates from first principles, for illustrative purposes.

Let $A(t_1, t_2)$ and $D(t_1, t_2)$ denote the number of arrivals and departures in $(t_1, t_2]$, respectively. The birth rate can be found from

$$P(Q(t + \Delta t) = k + 1 \mid Q(t) = k) =$$

$$= \sum_{j=1}^{\infty} P(A(t, t + \Delta t) = j, D(t, t + \Delta t) = j - 1 \mid Q(t) = k)$$

$$= \sum_{j=1}^{\infty} P(A(t, t + \Delta t) = j) P(D(t, t + \Delta t) = j - 1 \mid Q(t) = k)$$

$$= \sum_{j=1}^{\infty} \frac{(\lambda \Delta t)^j}{j!} e^{-\lambda \Delta t} P(D(t, t + \Delta t) = j - 1 \mid Q(t) = k)$$

$$= \lambda \Delta t e^{-\lambda \Delta t} e^{-\min\{k,m\}\mu \Delta t} + o(\Delta t) = \lambda \Delta t + o(\Delta t)$$

for $\Delta t \to 0$ and all $k \geq 0$. The second equality is a consequence of the independence of service and inter-arrival times.

The probability of seeing a departure from the queue is a function of the state of the queue: if k customers are in the queue, there are $\min\{k, m\}$ servers working simultaneously. Let $\tilde{X}_j(t)$ be the residual service time of the j-th server at time t, for $j = 1, \ldots, \min\{k, m\}$. Then

$$P(D(t, t + \Delta t) = 0 \mid Q(t) = k) = P(\tilde{X}_j(t) > \Delta t, j = 1, \ldots, \min\{k, m\})$$

$$= \prod_{j=1}^{\min\{k,m\}} P(\tilde{X}_j(t) > \Delta t)$$

$$= \prod_{j=1}^{\min\{k,m\}} e^{-\mu \Delta t} = e^{-\min\{k,m\}\mu \Delta t}$$

Hence, we conclude for all $k \geq 0$:

$$\lim_{\Delta t \to 0} \frac{P(Q(t + \Delta t) = k + 1 \mid Q(t) = k)}{\Delta t} = \lambda \tag{5.66}$$

Similarly, it is easy to derive that for all $k \geq 0$ and $h \geq 2$:

$$\lim_{\Delta t \to 0} \frac{P(Q(t + \Delta t) = k + h \mid Q(t) = k)}{\Delta t} = 0 \tag{5.67}$$

As for the death rates, we have

$$P(Q(t + \Delta t) = k - 1 \mid Q(t) = k) =$$

$$= \sum_{j=1}^{\infty} P(A(t, t + \Delta t) = j - 1, D(t, t + \Delta t) = j \mid Q(t) = k)$$

$$= \sum_{j=1}^{\infty} P(A(t, t + \Delta t) = j - 1) P(D(t, t + \Delta t) = j \mid Q(t) = k)$$

$$= e^{-\lambda \Delta t} P_1 + \lambda \Delta t e^{-\lambda \Delta t} P_2 + o(\Delta t)$$

where

$$P_1 = P(D(t, t + \Delta t) = 1 \mid Q(t) = k)$$
$$= P(\exists j_1 : \tilde{X}_{j_1}(t) \le \Delta t, \tilde{X}_j(t) > \Delta t, j \ne j_1, j \in [1, \min\{k, m\}])$$
$$= \binom{\min\{k, m\}}{1}(1 - e^{-\mu\Delta t})e^{-(\min\{k,m\}-1)\mu\Delta t} = \min\{k, m\}\mu\Delta t + o(\Delta t)$$

and

$$P_2 = P(D(t, t + \Delta t) = 2 \mid Q(t) = k)$$
$$\le 1 - P(D(t, t + \Delta t) = 1 \mid Q(t) = k) - P(D(t, t + \Delta t) = 0 \mid Q(t) = k)$$
$$= 1 - e^{-\min\{k,m\}\mu\Delta t} - \min\{k, m\}\mu\Delta t + o(\Delta t) = o(\Delta t)$$

Summing it up, we get for all $k \ge 1$

$$P(Q(t + \Delta t) = k - 1 \mid Q(t) = k) = \min\{k, m\}\mu\Delta t + o(\Delta t) \tag{5.68}$$

Hence

$$\lim_{\Delta t \to 0} \frac{P(Q(t + \Delta t) = k - 1 \mid Q(t) = k)}{\Delta t} = \min\{k, m\}\mu, \quad k \ge 1. \tag{5.69}$$

It can easily be shown also that

$$\lim_{\Delta t \to 0} \frac{P(Q(t + \Delta t) = k - h \mid Q(t) = k)}{\Delta t} = 0, \tag{5.70}$$

for all $k \ge h$ and all $h \ge 2$. We have now all the information required to fill up the transition rate matrix $\mathbf{Q}$ of the continuous time Markov process $Q(t)$:

$$Q_{kj} = \begin{cases} \lambda & j = k + 1, \ k \ge 0 \\ \min\{m, k\}\mu & j = k - 1, \ k \ge 1 \\ -\lambda - \min\{m, k\}\mu & j = k, \ k \ge 0 \\ 0 & |j - k| > 1 \end{cases} \tag{5.71}$$

Since $\lambda, \mu > 0$, the Markov chain is irreducible. The limiting state probabilities $x_i \equiv P(Q(t) = i)$, when they exist, are the unique nontrivial solution of the linear system $\mathbf{x}\mathbf{Q} = \mathbf{0}$, with the normalization condition $\mathbf{x}\mathbf{e} = 1$. Here $\mathbf{x} = [x_0 \ x_1 \ x_2 \dots]$ and $\mathbf{e}$ is a column vector of 1's. The linear system can be written in a compact form by introducing the notation $\mu(k) \equiv \min\{k, m\}\mu$. We get:

$$\lambda x_0 = \mu(1)x_1, \quad k = 0 \tag{5.72}$$
$$\lambda x_{k-1} + \mu(k + 1)x_{k+1} = [\lambda + \mu(k)]x_k, \quad k > 0 \tag{5.73}$$

Let $A_o \equiv \lambda/\mu$ be the mean intensity of the offered traffic. Since there is no loss, this is also the mean carried traffic. Therefore the utilization factor is $\rho = A_c/m = A_o/m = \lambda/(m\mu)$. In the sequel, to simplify notation, we drop the subscript o of the mean offered traffic intensity.

We could apply the general formulas holding for the birth-death processes to eqs. (5.72) and (5.73). It is however more instructive to solve them directly. By rearranging terms, we get:

$$\mu(1)x_1 - \lambda x_0 = 0, \qquad k = 0 \tag{5.74}$$

$$\mu(k)x_k - \lambda x_{k-1} = \mu(k+1)x_{k+1} - \lambda x_k, \qquad k > 0 \tag{5.75}$$

It is apparent that the differences $\mu(k)x_k - \lambda x_{k-1}$, $k \geq 1$, are independent of k and all equal to 0. Therefore, we have $x_k = \frac{\lambda}{\mu(k)}x_{k-1}$, for $k \geq 1$. In the end

$$x_k = x_0 \prod_{j=1}^{k} \frac{\lambda}{\mu(j)} = \begin{cases} x_0 \dfrac{A^k}{k!} & 1 \leq k \leq m-1 \\[2mm] x_0 \dfrac{A^k}{m!m^{k-m}} & m \leq k \end{cases} \tag{5.76}$$

where $A = \lambda/\mu$. The probability x_0 is found by the normalization condition:

$$1 = \sum_{k=0}^{\infty} x_k = x_0 \left[\sum_{k=0}^{m-1} \frac{A^k}{k!} + \frac{A^m}{m!} \sum_{k=m}^{\infty} \frac{A^{k-m}}{m^{k-m}} \right] = x_0 \left[\sum_{k=0}^{m-1} \frac{A^k}{k!} + \frac{A^m}{m!} \frac{1}{1-A/m} \right] \tag{5.77}$$

hence

$$x_0 = \frac{1}{\sum_{k=0}^{m-1} \frac{A^k}{k!} + \frac{A^m}{m!} \frac{1}{1-A/m}} \tag{5.78}$$

and finally

$$x_k = \begin{cases} \dfrac{\frac{A^k}{k!}}{\sum_{k=0}^{m-1} \frac{A^k}{k!} + \frac{A^m}{m!} \frac{1}{1-A/m}} & 0 \leq k \leq m-1 \\[6mm] \dfrac{\frac{A^k}{m!m^{k-m}}}{\sum_{k=0}^{m-1} \frac{A^k}{k!} + \frac{A^m}{m!} \frac{1}{1-A/m}} & k \geq m \end{cases} \tag{5.79}$$

This result holds provided that the series in eq. (5.77) converges. This is true if and only if $A/m < 1$, i.e., $\rho < 1$. This is not a surprising condition for an infinite size queueing system. It says that the mean arrival rate shall be less than the maximum service rate of the system for stability.

We assume this condition holds and proceed with the derivation of performance metrics.

The mean number of customers in the queue is $E[Q] = \sum_{k=1}^{\infty} kx_k$. From this we derive the mean system time by applying Little's law.

The main performance metrics of the $M/M/m$ queue concern delays. A customer arriving at a queue where there are less than m customers will experience no wait, since there is at least one available server. Therefore, the probability of delay, i.e., $P_d \equiv P(W > 0)$ is

$$P_d = \sum_{k=m}^{\infty} x_k = \frac{\frac{A^m}{m!} \frac{1}{1-A/m}}{\sum_{k=0}^{m-1} \frac{A^k}{k!} + \frac{A^m}{m!} \frac{1}{1-A/m}} \equiv C(m, A) \tag{5.80}$$

This is a function of $A \geq 0$ and $m \geq 1$. It is called the Erlang-C formula and often denoted with $C(m, A)$. We have

$$C(m, A) = \frac{B(m, A)}{1 - \frac{A}{m} + \frac{A}{m} B(m, A)} \tag{5.81}$$

which yields to numerical iterative calculation based on the recursions for the Erlang B-formula. Then, the probability of delay is $P_d = C(m, A)$.

The waiting time for FCFS service is 0 if the arriving customer finds the system in any of the states between 0 and $m - 1$. Otherwise, the customer has to wait the residual service time of the customer that ends its service first, then the time required for assigning servers to the $k - m$ customers waiting in front of it. Thanks to the memoryless property of the negative exponential PDF, all of these times share the same negative exponential PDF with mean $1/(m\mu)$. Therefore,

$$\varphi_W(s) = \sum_{k=0}^{m-1} x_k + \sum_{k=m}^{\infty} x_k \left(\frac{m\mu}{s+m\mu} \right)^{k-m+1} \tag{5.82}$$

$$= 1 - P_d + P_d \sum_{k=m}^{\infty} \left(\frac{m\mu}{s+m\mu} \right)^{k-m+1} \left(1 - \frac{A}{m} \right) \frac{A^{k-m}}{m^{k-m}} \tag{5.83}$$

$$= 1 - P_d + P_d \left(\frac{m\mu}{s+m\mu} \right) \left(1 - \frac{A}{m} \right) \sum_{h=0}^{\infty} \left(\frac{A\mu}{s+m\mu} \right)^h \tag{5.84}$$

$$= 1 - P_d + P_d \left(1 - \frac{A}{m} \right) \frac{\left(\frac{m\mu}{s+m\mu} \right)}{1 - \frac{\lambda}{s+m\mu}} \tag{5.85}$$

$$= 1 - P_d + P_d \frac{m\mu - \lambda}{s + m\mu - \lambda} \tag{5.86}$$

This can be easily inverted, and the following CDF is found:

$$F_W(t) = 1 - P_d e^{-(m\mu-\lambda)t}, \qquad t \geq 0. \tag{5.87}$$

The obtained expression is extremely simple: it resembles the expression of the CDF of the $M/M/1$ (to which it reduces for $m = 1$), the only difference being that the serving rate is replaced by the collective serving rate of all m servers and the probability of delay has a more complex expression ($C(m, A)$ instead of simply ρ).

The mean waiting time is

$$E[W] = \frac{P_d}{m\mu - \lambda} \tag{5.88}$$

The system time can be obtained by noting that $S = W + X$ and W and X are independent random variables. Then, $\varphi_S(s) = \varphi_W(s)\frac{\mu}{s+\mu}$. The mean system time is

$$E[S] = E[W] + E[X] = \frac{P_d}{m\mu - \lambda} + \frac{1}{\mu} \tag{5.89}$$

It is interesting to study the delay performance of the multi-server queue for large systems. Let the number of servers be $m = N + \beta\sqrt{N}$ and the offered traffic $A = N$. With this scaling, the utilization coefficient is $\rho = A/m = \frac{1}{1+\beta/\sqrt{N}}$. Using the result of Theorem 5.1, it can be verified that:

$$C(N + \beta\sqrt{N}, N) = \frac{B(N + \beta\sqrt{N}, N)}{1 - \rho + \rho B(N + \beta\sqrt{N}, N)} \sim \left(1 + \frac{\beta}{\sqrt{N}}\right)\frac{\phi(\beta)}{\beta\Phi(\beta) + \phi(\beta)} \tag{5.90}$$

as $N \to \infty$. Rewriting eq. (5.88) with the scaling expression of m and A, we have

$$E[W] = E[X]\frac{C(N + \beta\sqrt{N})}{N + \beta\sqrt{N} - N} \sim E[X]\frac{\phi(\beta)}{\beta[\beta\Phi(\beta) + \phi(\beta)]}\frac{1}{\sqrt{N}} \tag{5.91}$$

We see that large systems bring about a significant delay performance advantage: the utilization coefficient of the servers tends to 1, while the mean waiting time tends to 0. The performance benefit of large-scale systems motivates the success of the cloud paradigm, based on data centers that concentrate from several tens of thousands to hundreds of thousands of servers, to which a large flow of jobs is submitted.

5.4.1 Finite Queue Size Model

The $M/M/m$ model can be easily extended to account for finite room in the waiting line. The $M/M/m/K$ model can accommodate at most $m + K$ customers, m of which under service, the others waiting for their turn. Customers arriving when the queue is blocked, i.e., $Q(t) = m + K$, are rejected and lost to the system.

This queue is described by a birth-death process on the state space $\{0, \ldots, m + K\}$. The transition rates are exactly the same as for the $M/M/m$ queue, except that in the present case the birth-death process is truncated to $m + K + 1$ states.

Going through all steps we have done with the $M/M/m$ model, it is easy to recognize that the Markov process limiting state probabilities are given by:

$$x_k = \begin{cases} x_0\dfrac{A^k}{k!} & 0 \le k \le m \\[2mm] x_0\dfrac{A^k}{m!m^{k-m}} & m + 1 \le k \le m + K \end{cases} \tag{5.92}$$

The probability x_0 is found using the congruence equation $\sum_{k=0}^{m+K} x_k = 1$.

The probability distribution $\{x_k\}_{0 \le k \le m+K}$ refers to a general time during statistical equilibrium. Since this is a loss system, the probability distribution seen by arrivals and departures may be different. Specifically, thanks to the Poisson arrivals, the probability distribution seen by arrivals is the same as $\{x_k\}_{0 \le k \le m+K}$.

The probability distribution seen by an arrival that joins the queue (or, which is the same, by a departing customer) is given by $\pi_k = \lambda x_k / \Lambda_s$, where $\Lambda_s = \sum_{k=0}^{m+K-1} \lambda x_k = \lambda(1 - x_{m+K})$. Hence, it is $\pi_k = x_k/(1 - x_{m+K})$ for $k = 0, \dots, m+K-1$.

As for metrics, we have:

$$P_d = \sum_{k=m}^{m+K-1} \pi_k$$

$$P_L = x_{m+K}$$

$$E[Q] = \sum_{k=0}^{m+K} k x_k \tag{5.93}$$

$$E[W] = \frac{1}{m\mu} \sum_{k=m}^{m+K-1} (k - m + 1) \pi_k$$

$$E[S] = E[W] + \frac{1}{\mu}$$

Probability distribution can also be found in an elementary way, though leading to cumbersome expressions. For example, the Laplace transform of the waiting time PDF in the case of FCFS service is given by:

$$\varphi_W(s) = 1 - P_d + \sum_{k=m}^{m+K-1} \pi_k \left(\frac{m\mu}{s + m\mu} \right)^{k-m+1} \tag{5.94}$$

Finally, note that the spacial case $K = 0$ corresponds to the Erlang loss system with negative exponential service times.

5.4.2 Resource Sharing versus Isolation

We compare different ways of allocating serving capacity to a flow of service requests. We consider m flows of service requests arriving according to a Poisson process, each one at a mean rate λ. An overall serving capacity with mean serving rate $m\mu$ is available.

We compare three different arrangements of the service system (a sketch of the three system configurations is shown in Figure 5.15):

1. m different queues are provided, one for each service request flow; each queue is equipped with an equal share of the serving capacity, i.e., serving rate μ (separate queueing).
2. A single queue with m different servers, each having serving rate μ (serving capacity slicing and parallel service).
3. A single queue with the whole serving capacity, with mean rate $m\mu$ (fully shared system).

Figure 5.15 Diagram of three alternative service system configurations: (1) m separate queues; (2) a multi-server queue (aggregation of arrivals); (3) a single queue (aggregation of arrivals and of service capacity).

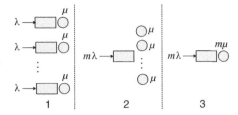

The last approach corresponds to assigning the whole serving capacity in the system to one service demand at a time. In the first approach both the arrival flow of service demands and the service capacity are split in advance, so as to manage m parallel queueing systems. The second solution is an intermediate one: the flow of service demand is dealt with as a whole, but the serving capacity is sliced into m portions that are assigned to pending service requests, so that possibly more than one request can be served in parallel.

In the first case, the system model consists of m $M/M/1$ queues, each of which has a Poisson arrival process of mean rate λ and a single server with mean serving rate μ. Then, the mean system time is

$$E[S_1] = \frac{1}{\mu - \lambda} \tag{5.95}$$

The second system configuration is modeled as an $M/M/m$ queue with a single Poisson arrival process with mean rate $m\lambda$ and m servers, each one having a mean serving rate of μ. The mean system time is then

$$E[S_2] = \frac{P_d}{m\mu - m\lambda} + \frac{1}{\mu} = \frac{C(m, m\lambda/\mu)}{m(\mu - \lambda)} + \frac{1}{\mu} \tag{5.96}$$

where $C(\cdot, \cdot)$ is the Erlang-C formula.

Finally, in the third system configuration, we have a unique $M/M/1$ with Poisson input with a mean rate $m\lambda$ and a single server with serving capacity $m\mu$. Then, the mean system time is

$$E[S_3] = \frac{1}{m\mu - m\lambda} \tag{5.97}$$

The normalized mean system times of the three arrangements are compared in Figure 5.16(a) as a function of the load coefficient a, defined as $a = \lambda/\mu$. The normalization is done with respect with $1/\mu$. Then, all three expressions depend only on the two parameters a and m. We assume $m = 10$ and let a vary between 0 and 1 (the upper limit of the load factor to guarantee the stability of the considered queueing systems).

It is apparent that the aggregated system of the third approach offers the best performance, the array of m single-server queues performs worst, and the multi-server system lies in between. In other words, complete sharing of

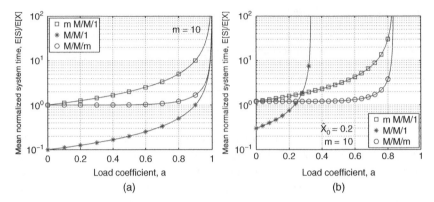

Figure 5.16 Comparison of the mean system time with different arrangements of the serving capacity; the load factor is $a = \lambda/\mu$. Left plot: fully scaling service times. Right plot: service time including a nonscaling overhead component.

resources and aggregation of demand appear to improve delay performance of service systems. As the scale of systems grows up (that is to say, the service demand rate and the serving rate grow proportionally), we can expect that the performance improve, for the same level of utilization of the resources.

The obtained result depends on the crucial hypothesis that the time to serve a request in inversely proportional to m, i.e., moving from a system made up of m separate servers of mean rate μ to a single server with serving rate $m\mu$, the mean service time gets divided by m. Let us now examine a variation of the setting above, where the serving time has a component that *does not scale with m*. Let X be a negative exponential random variable that represents the service time. Then the mean value of X depends on which system configuration is considered:

1. With the array of m $M/M/1$ queues, we let $E[X] = X_0 + 1/\mu$ for each queue.
2. With the multi-server queue $M/M/m$, we let $E[X] = X_0 + 1/\mu$ for each server.
3. With the single $M/M/1$ queue, we let $E[X] = X_0 + 1/(m\mu)$.

Note that the mean service time has a constant contribution, that does not scale with m, along with a second component, that behaves as before.

It is not difficult to find the mean system times in this new scenario. It is

$$E[S_1] = \frac{1 + \mu X_0}{\mu - \lambda(1 + \mu X_0)} \tag{5.98}$$

$$E[S_2] = \frac{(1 + \mu X_0)C(m, m\lambda(1 + \mu X_0)/\mu)}{m[\mu - \lambda(1 + \mu X_0)]} + \frac{1}{\mu} + X_0 \tag{5.99}$$

$$E[S_3] = \frac{1 + m\mu X_0}{m[\mu - \lambda(1 + m\mu X_0)]} \tag{5.100}$$

where $C(\cdot, \cdot)$ is the Erlang-C formula. It is apparent that the mean system times normalized with respect to $1/\mu$ depend on the three nondimensional parameters m, $a = \lambda/\mu$, $\hat{X}_0 = \mu X_0$. The upper limit of the load coefficient for the first two approaches is $1/(1 + \hat{X}_0)$, whereas it is smaller and equal to $1/(1 + m\hat{X}_0)$ for the single queue.

For the numerical evaluation we set $\hat{X}_0 = 0.2$. Figure 5.16(b) shows that the aggregated serving system outperforms the others as long as the load coefficient is small enough. For bigger loads, the multi-server system configuration becomes definitely the best solution, as far as the mean delay is concerned.

The lesson learned from this example is that aggregation of the demand and of the service capacity is beneficial (i.e., it offers better delay performance) as long as the service time of customers scales in inverse proportion with the aggregation parameter (m in our case). If instead there is some nonscaling part of the service time, it is not necessarily convenient to collapse the serving system in a unique super-powerful serving entity. In that case it might be more effective to aggregate the demand (sharing of the waiting line), but to keep serving capacity split into a number of separate servers that can run in parallel. Slower servers are less penalized by the "overhead" imposed by the nonscaling service component.

WiFi is a real system where the technology evolution is demonstrating tangibly the finding of this example. The air bit rate offered by the PHY layer has been multiplied by a factor of about 3500 over 15 years (from 2 Mbps of the original IEEE 802.11 in 1999 up to the almost 7 Gbps of the IEEE 802.11ac version in 2014). However, the effective realized throughput has not kept that pace, due to inter-frame spaces, back-off slots, MAC level ACK time and other MAC and physical related overhead, that do not scale as the air bit rate grows.

5.5 Infinite Server Queues

Infinite server queues aim to model service systems where customers do not "interfere" with one another, i.e., there is no competition for the allocation of the serving capacity, given that there is an unlimited amount of it. The time that a customer spends in the system amounts to just its own service time, independently of how many other customers are visiting the system. This kind of models is especially useful to represent systems that impose a delay on the customers in transit, yet we can neglect any form of serving resource limitation. As an example, let us consider a network composed on N links. Every now and then links go out-of-service (e.g., for maintenance or upgrade purposes, for a failure). Assume up-times are i.i.d. random variables. Then, each running link is "waiting" for an out-of-service to terminate its own up-time. This kind of system lends itself quite naturally to an infinite server queueing model where customers are links and the service time of the queue is the up-time. As another example, the number of users camping the area covered by a base station of a cellular network can be identified as the number of customers in an infinite server queue. Here the arrival process represents user arrivals within the coverage area and the service time models the amount of time that a user spends roaming within the base station coverage area, which depends on the user mobility. To a good extent, we can safely assume that users do not interact among themselves (e.g., in a macro-cell). The unlimited "serving capability" of the first example is the appearance of out-of-service events, while in the second example it is the space where users move. This last case points out that the resource need not be strictly "unlimited" (the amount of space covered by a given base station is not). What matters is that there is "enough" resource for making contentions among customers negligible (in our example, it is unrealistic that a new arriving user does not get into the area covered by the base station because it sees too many people already there, unless we are referring to very small cells, e.g., femto-cells).

We focus on infinite server queues with Poisson input and general i.i.d. service times, denoted with $M/G/\infty$. This simplifying hypothesis makes them nicely tractable. Let $P_k(t) = P(Q(t) = k)$ denote the probability that there are $Q(t) = k$ customers in the $M/G/\infty$ queue at time t, given that at time $t = 0$ the queue was empty. If there are n arrivals in $(0, t]$, then the event of finding k customers in the queue at t is equivalent to the event that exactly k out of the n customers are still in service at t. Since customers are independent of one another, given the probability $s(t)$ that a customer arriving during the interval $(0, t]$ is still in service at t, we have

$$P(Q(t) = k | Q(0) = 0, A(0, t) = n) = \binom{n}{k} [s(t)]^k [1 - s(t)]^{n-k}, \quad n \geq k \geq 0$$

(5.101)

Thanks to a well-known property of Poisson arrivals (see Theorem 3.6), given that there are n arrivals in $(0, t]$, they are uniformly distributed in the considered interval. Let $U = u \in (0, t)$ be the arrival time of a customer with service time X. The probability that it is still in service at t, given the arrival time u, is

$$s(t|u) = P(X > t - u) = 1 - F_X(t - u)$$

(5.102)

By removing the conditioning on the random variable U, that is uniformly distributed over $(0, t]$, we get

$$s(t) = \frac{1}{t} \int_0^t s(t|u) \, du = \frac{1}{t} \int_0^t [1 - F_X(u)] \, du$$

(5.103)

By introducing the expression of $s(t)$ into eq. (5.101) and removing the conditioning on the number of arrivals we get finally for $k \geq 0$

$$P_k(t) = \sum_{n=k}^{\infty} \frac{(\lambda t)^n}{n!} e^{-\lambda t} \binom{n}{k} \left[\frac{1}{t} \int_0^t F_X(u) \, du \right]^{n-k} \left[\frac{1}{t} \int_0^t (1 - F_X(u)) \, du \right]^k$$

$$= \frac{\left(\lambda \int_0^t (1 - F_X(u)) \, du \right)^k}{k!} e^{-\lambda t} \sum_{n=k}^{\infty} \frac{1}{(n-k)!} \left[\lambda \int_0^t F_X(u) \, du \right]^{n-k}$$

$$= \frac{\left(\lambda \int_0^t (1 - F_X(u)) \, du \right)^k}{k!} \exp\left(-\lambda \int_0^t (1 - F_X(u)) \, du \right) \qquad (5.104)$$

This is nothing but a Poisson probability distribution with mean $\lambda \int_0^t (1 - F_X(u)) \, du$. For the $M/G/\infty$ queue we can therefore obtain a simple, explicit form even of the *transient* probability distribution of the number of customers in the queue.

As $t \to \infty$, the integral $\int_0^t (1 - F_X(u)) \, du$ tends to E[X]. Hence we get

$$p_k \equiv \lim_{t \to \infty} P_k(t) = \frac{(\lambda E[X])^k}{k!} \exp(-\lambda E[X]) \qquad (5.105)$$

The main result is that the limiting probability distribution of the number of customers in an $M/G/\infty$ queue is simply a Poisson one, with mean $\lambda E[X]$. It is insensitive to the form of the PDF of service times. It depends only on the mean of the service time. The number of busy servers is by definition the offered traffic. Hence, its mean value is the mean intensity of the offered traffic, $A_o = \lambda E[X]$.

It is possible to find an expression also for the Laplace transform of the PDF of the busy period even for a more general system than $M/G/\infty$. Let $C = Y + I$ be the busy cycle, i.e., the sum of the busy period Y and the ensuing idle time I. For a $G/G/\infty$ queue. The sequence of the busy cycles forms a renewal process. Then, we can show the following.

Theorem 5.3   In the $G/G/\infty$ queue, assuming the system is empty initially, the Laplace transform of the probability of empty system $P_0(t) = \mathcal{P}(Q(t) = 0 \mid Q(0) = 0)$ is

$$P_0^*(s) = \int_0^\infty P_0(t) e^{-st} \, dt = \frac{1}{s} - \frac{\varphi_T(s)}{s} \frac{1 - \varphi_Y(s)}{1 - \varphi_C(s)} \qquad (5.106)$$

Proof: Given that the system starts empty at time 0, the probability that it is empty at time t can be expressed as the sum of the probabilities of two disjoint events: (i) the first arrival occurs after time t; (ii) the first arrival occurs at $x \in (0, t)$, but its busy period and any other ensuing busy periods in the interval $(0, t)$ are over by time t. In this second case, we use the probability $Q_0(t)$ that the a system is found empty at time t, given that a busy period starts at 0. Then:

$$P_0(t) = P(T > t) + \int_0^t Q_0(t - x) f_T(x)\, dx \tag{5.107}$$

The probability $Q_0(t)$ can be found with a regenerative argument as well. Two cases are possible, given that at time 0 a busy period starts: (i) t falls immediately after the end of the busy period and before the next busy period starts, i.e., $Y \le t < C$; and (ii) the first busy cycle lasts a time x; then, we can invoke the probability $Q_0(t - x)$, since $C = x$ means that at time x a new busy period starts. Summing up, we have

$$Q_0(t) = P(Y \le t < C) + \int_0^t Q_0(t - x) f_C(x)\, dx$$

$$= G_C(t) - G_Y(t) + \int_0^t Q_0(t - x) f_C(x)\, dx$$

where $G_V(t) \equiv P(V > t)$ denotes the CCDF of the random variable V. Taking the Laplace transform, we get

$$Q_0^*(s) = \int_0^\infty Q_0(t) e^{-st}\, dt = \frac{1 - \varphi_C(s)}{s} - \frac{1 - \varphi_Y(s)}{s} + Q_0^*(s)\varphi_C(s) \tag{5.108}$$

hence

$$Q_0^*(s) = \frac{\varphi_Y(s) - \varphi_C(s)}{s[1 - \varphi_C(s)]} \tag{5.109}$$

Going back to eq. (5.107), we have

$$P_0^*(s) = \frac{1 - \varphi_T(s)}{s} + Q_0^*(s)\varphi_T(s) = \frac{1}{s} - \frac{\varphi_T(s)}{s}\frac{1 - \varphi_Y(s)}{1 - \varphi_C(s)} \tag{5.110}$$

∎

In the special case of $M/G/\infty$, we have $\varphi_C(s) = \varphi_I(s)\varphi_Y(s) = \varphi_Y(s)\lambda/(s + \lambda)$ and $\varphi_T(s) = \lambda/(s + \lambda)$. Then, the result of Theorem 5.3 can be used to yield

$$\varphi_Y(s) = 1 + \frac{s}{\lambda} - \frac{1}{\lambda P_0^*(s)} \tag{5.111}$$

where $P_0^*(s) = \int_0^\infty e^{-st} P_0(t)\, dt$ and $P_0(t) = \exp\left(-\lambda \int_0^t [1 - F_X(u)]\, du\right)$.

We can calculate the mean busy period $E[Y]$ from these expressions. Applying the final value theorem and using the expression of $P_0(t)$, we find

$$\lim_{s \to 0} s P_0^*(s) = \lim_{t \to \infty} P_0(t) = \lim_{t \to \infty} e^{-\lambda \int_0^t [1 - F_X(u)]\, du} = e^{-\lambda E[X]} \tag{5.112}$$

From eq. (5.111) we derive $P_0^*(s) = [s + \lambda - \lambda\varphi_Y(s)]^{-1}$ and hence, applying de l'Hôpital rule,

$$\lim_{s \to 0} s P_0^*(s) = \lim_{s \to 0} \frac{s}{s + \lambda - \lambda\varphi_Y(s)} = \lim_{s \to 0} \frac{1}{1 - \lambda\varphi_Y'(s)} = \frac{1}{1 + \lambda E[Y]} \tag{5.113}$$

Equating the two expression we find finally:

$$E[Y] = \frac{e^{\lambda E[X]} - 1}{\lambda} \tag{5.114}$$

In the special case of deterministic service times, a closed form expression can be given for $P_0^*(s)$ and for the Laplace transform of the PDF of the busy period:

$$P_0^*(s)|_{M/D/\infty} = \frac{s + \lambda e^{-(s+\lambda)D}}{s(s + \lambda)} \tag{5.115}$$

$$\varphi_Y(s)|_{M/D/\infty} = \frac{(s + \lambda)e^{-(s+\lambda)D}}{s + \lambda e^{-(s+\lambda)D}} \tag{5.116}$$

Another special case leads to a manageable result, namely the $G/D/\infty$ queue. It is instructive to develop an ad hoc analysis, rather than specializing the result above. The busy period can be described as follows:

$$Y = \begin{cases} D & T > D \\ T + Y' & T \le D \end{cases} \tag{5.117}$$

where $Y' \sim Y$, thanks to the renewal property of the arrivals and the infinite server structure of the queue. Then

$$\varphi_Y(s) = E[e^{-sY}] = e^{-sD}G_T(D) + \int_0^D E[e^{-s(t+Y)}]f_T(t) \, dt$$

$$= e^{-sD}G_T(D) + \varphi_Y(s)\int_0^D e^{-st}f_T(t) \, dt$$

Then

$$\varphi_Y(s) = \frac{e^{-sD}G_T(D)}{1 - \int_0^D e^{-st}f_T(t) \, dt} \tag{5.118}$$

Also the probability distribution of the number M of customers served in a busy period can be determined in a closed form. Adapting the approach used above for the busy period duration, we can write

$$M = \begin{cases} 1 & T > D \\ 1 + M' & T \le D \end{cases} \tag{5.119}$$

where $M' \sim M$. Hence $\phi_M(z) = E[z^M] = z(1 - p) + zE[z^{M'}]p = z(1 - p) + zp\phi_M(z)$, where $p = P(T \le D) = F_T(D)$. Then $\phi_M(z) = z\frac{1-p}{1-pz}$, that can be easily inverted to yield $P(M = k) = (1 - p)p^{k-1}, k \ge 1$.

As for the output process, in the special case of $M/G/\infty$ (Poisson input process), the queueing system is essentially a displacer, i.e., the effect on the arrival point process of going through the $M/G/\infty$ is to displace every point by an amount equal to a service time, independently of any other point. Thank to the property of a stationary Poisson process with respect to i.i.d. displacements (see Sec. 3.2.3.1), we conclude that the output process of the $M/G/\infty$ queue is a Poisson process with the same statistical characteristics of the input process.

Extensions and further results can be found, e.g., in [149].

5.5.1 Analysis of Message Propagation in a Linear Network

The material in this example is inspired and largely based on the work by Miorandi and Altman [161].

An interesting analogy can be established between the $G/G/\infty$ queue and the problem of connectivity in a one dimensional vehicular network where message passing is based on multi-hop networking. Let us consider a road and vehicles scattered along the road. Assume that vehicles are equipped with on-board units (OBUs) capable of sending and receiving messages in dedicated radio channels. An example is provided by the dedicated short range communication (DSRC) technology [165]. Messages are propagated among vehicles by means of dissemination protocols, i.e., they are broadcasted and forwarded from vehicle to vehicle according to some policy that aims at maximizing the distance covered by the message while keeping the number of forwarding operations (hence the number of copies of the same message received by each OBU) at a minimum.

Multi-hop message dissemination is necessary to extend the message coverage as far as possible, given the limited range of the OBU radio signals. Let R denote the distance within which reliable communication is possible, i.e., such that $SNR(R) = G(R)P_{tx}/P_N > \gamma_{th}$, where P_{tx} is the transmission power level, P_N is the background noise power at the receiver, $G(x)$ is the path gain at distance x and γ_{th} is the threshold of the SNR required to make the reception of a message successful with some (high) probability for the chosen modulation and coding set. In the following we assume that reception is reliable within distance R. In general, R is a random variable, since the path gain is made up of a deterministic component plus shadowing and fading. Those last two components are modeled by means of probability distributions of the relevant path gain component, e.g., typically a log normal PDF for the shadowing and a Rayleigh PDF for the fading component. Since typically the average covered distance with DSRC equipment is in the order of several hundred meters, much bigger than typical road widths, we neglect the road width and assimilate the road to a straight line. Hence, vehicle positions are identified by means of a single coordinate. Let x_k denote the position of the k-th vehicle moving along the positive x semi-axis (hence $x_k < x_{k+1}$). We assume that vehicle spatial distribution is stationary, i.e., $x_{k+1} - x_k$ are i.i.d. random variables. Let $f_Y(x)$ and $F_Y(x)$ denote their common PDF and CDF, respectively; that is to say $x_{k+1} - x_k \sim Y, \forall k \geq 0$. In spite of vehicle mobility, we assume their positions as frozen, since the time taken for message forwarding through radio communications is about three orders of magnitude less than the time required for vehicle to

Figure 5.17 Example of the evolution of the $G/G/\infty$ queue: Y is a busy period, C is a busy cycle (busy period plus the ensuing idle time) and X is a service time.

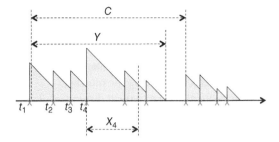

move significantly (e.g, tens of ms for message delay vs. typical speed values in the order of 10 m/s).

Assume that at least one vehicle among those that receive the message from the originating node forwards it further. As an example, the 'Furthest Forwarder' rule can be used, i.e., if nodes $N_1, N_2, \ldots, N_k$ receive the message from a node N_0, the one that will forward the message is the node furthest away from N_0 among $N_1, N_2, \ldots, N_k$. It is easy to see that the message keeps propagating along the positive direction of the x-axis, as long as the coverage regions of the forwarding nodes keep overlapping (see Figure 5.17).

We can interpret the vehicular network model in terms of an infinite server queue. Arrivals correspond to vehicles' positions (and hence the "time" axis is in fact the x-axis in this case: one-dimensional space takes the place of time). Service time corresponds to the distance covered by the OBU radio equipment, i.e., R. Since there is no limit to the number of vehicles whose communication ranges may overlap[4], the queueing model in this analogy has infinite servers.

Let Y denote the spacing between subsequent vehicles along the road line (it corresponds to the inter-arrival time of the queueing model) and let $\lambda = 1/E[Y]$ be the mean spatial density of vehicles. Vehicles in the same lane interact with one another. If the road has more than one lane, the dependency among spacings of nearby vehicles weakens. We neglect it, so that we assume that "inter-arrival times" are i.i.d. random variables.

Let R denote the range of the radio communication of a vehicle: it plays the role of the service time of the queueing model. In general R can be characterized by a random variable, depending on the statistical characteristics of multi-path propagation and obstacles. We neglect path loss due to vehicles acting as obstacles and the correlation of shadowing of nearby vehicles, so that we assume that

---

4 We are clearly neglecting the finite size of vehicles: infinite servers may be seen as a limiting approximation to a model with a large, but finite number of servers, that accounts for the length of vehicles. Given that a vehicle occupation of the road is in the order of 10 m and that there can be more than one lane in the road, the number of vehicles within E[R] $\approx$ several hundreds meters can scale up to hundreds.

"service times" are i.i.d. random variables and they are independent of the "inter-arrival times."

We identify the following metrics of interest:

- Coverage distance, D_c: the distance between the originator of the message and the last vehicle receiving the message[5].
- Probability of connectivity at distance x, $P_c(x)$: the probability that a given vehicle can send a message up to a distance x from itself, i.e., it can cover up to a distance x (obviously by means of multi-hop message passing); it is $P_c(x) = P(D_c > x)$.
- The number of vehicles that receive the message, N_c: it is the number of vehicles covered by the propagation of the message.

It is easy to recognize that D_c corresponds to the busy period of the queueing model. N_c is the number of customers served in a busy period. The coverage probability at distance x is the probability that the busy period exceeds x.

Let us begin with a simple scenario where only deterministic path loss is accounted for. We can then assume that the path gain $G(x)$ is a monotonously decreasing function of the distance x between the transmitter and the receiver. The requirement for successful message decoding is that the *SNR* at the receiver be greater than a threshold γ_{th}, hence reception is successful at distance d if $G(d)P_{tx}/P_N \geq \gamma_{th}$. The maximum allowed range is therefore $R = G^{-1}(\gamma_{th}P_N/P_{tx})$. Since the transmission range is a constant, the queueing model of the vehicular network casts into a $G/D/\infty$ queue.

The Laplace transform of the busy period PDF, hence of the coverage distance PDF, is

$$\varphi_{D_c}(s) = \frac{e^{-sR}(1 - F_Y(R))}{1 - \int_0^R e^{-st} f_Y(t)\, dt} \tag{5.120}$$

where the random variable Y represents the distance between subsequent vehicles. The mean covered distance is found by deriving $\varphi_{D_c}(s)$:

$$E[D_c] = -\varphi'_{D_c}(0) = R + \frac{\int_0^R t f_Y(t)\, dt}{1 - F_Y(R)} \tag{5.121}$$

To be concrete, we assume that vehicle spacing follows a gamma law, hence $f_Y(t) = a \frac{(at)^{b-1}}{\Gamma(b)} e^{-at}$, with $1/\lambda = E[Y] = b/a$ and $\sigma_Y = \sqrt{b}/a$. Plugging this function into eq. (5.120) we find

$$\varphi_{D_c}(s) = \frac{e^{-sR}[1 - \Gamma(aR, b)]}{1 - \left(\frac{a}{s+a}\right)^b \Gamma((a+s)R, b)} \tag{5.122}$$

---

[5] In the calculations below the coverage distance is extended by the transmission range, to simplify formulas in the queueing analogy.

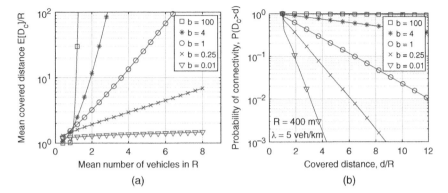

Figure 5.18 Queue model for a linear vehicular network. Left plot: mean covered distance (normalized to the transmission range R) as a function of the mean number of vehicles in a transmission range. Right plot: probability of connectivity at distance d as a function of the normalized covered distance d/R.

where $\Gamma(x,b) = \frac{1}{\Gamma(b)} \int_0^x u^{b-1} e^{-u} \, du$ is the incomplete gamma function. As for the mean value, noting that $a = b\lambda$, we have

$$E[D_c] = R + \frac{b}{a} \frac{\Gamma(aR, b+1)}{1 - \Gamma(aR, b)} = R \left[1 + \frac{1}{\lambda R} \frac{\Gamma(\lambda Rb, b+1)}{1 - \Gamma(\lambda Rb, b)} \right] \tag{5.123}$$

The probability of connectivity at a distance d is simply $P_c(d) = P(D_c > d)$. The number N_c of vehicles covered by the message dissemination has a geometric probability distribution, i.e., $P(N_c = k) = (1-p)p^{k-1}$ for $k \geq 1$, where $p = P(Y \leq R) = F_Y(R)$. In our specific case, we have $p = \Gamma(\lambda Rb, b)$.

All of the metrics depend on two nondimensional parameters, λR and b, in the specific considered case. The first one is the mean number of vehicles found in a transmission range. The second one is the reciprocal of the squared coefficient of variation[6] of the PDF of the inter-vehicle spacing.

The mean covered distance (normalized with respect to R) and the probability of connectivity are plotted as a function of λR and of the distance d (normalized with respect to R) in Figure 5.18, for five values of b. Those values correspond to a coefficient of variation of Y equal to $1/\sqrt{b}$.

It is apparent that as the product λR grows, the mean covered distance improves, the steeper the bigger b (i.e., the smaller the variance of the inter-vehicle spacing). The meaning of the result can be easily unveiled: laying out vehicles along the linear road with a standard deviation of the inter-vehicle spacing high with respect to the mean vehicle spacing increases the probability that the chain of multi-hop

---

6 The coefficient of variation of a random variable X is defined as the ratio of the standard deviation to the mean of X.

message passing gets interrupted because of a gap bigger than the transmission range. This probability becomes smaller as the mean vehicle density in increased.

The same insight can be gained from the plot of the probability of connectivity at a given distance. That plot is done for $\lambda = 5$ veh/km and $R = 400$ m, hence $\lambda R = 2$. With increasing variability of OBUs' radio range, the probability of connectivity over a distance d plunges to very small values as soon as the distance d moves away from R. On the contrary, for a low variability radio range of OBUs, the probability of connectivity is high even for distance levels much bigger than R.

By exploiting the analogy with the infinite server queue, we can analyze the linear vehicular network for a variable transmission range, e.g., as due to a random path loss. A common stochastic model of path loss with obstructions consists of multiplying the deterministic power law component by a log-normal random variable with standard deviation σ. In this case, the CDF of the transmission range is

$$
F_R(a) = P(R \leq a) = P\left(\frac{S e^{-\sigma^2/2} G_d(a) P_{tx}}{P_N} \leq \gamma_{th} \right) =
$$

$$
= P\left(S \leq \frac{\gamma_{th} e^{\sigma^2/2} P_N}{G_d(a) P_{tx}} \right)
$$

$$
= \Phi\left(\frac{\log(\gamma_{th} e^{\sigma^2/2} / SNR_d(a))}{\sigma} \right) \tag{5.124}
$$

where S is the log-normal gain component, $\Phi(x)$ is the CDF of the standard Gaussian random variable, and $SNR_d(x) \equiv G_d(x) P_{tx}/P_N$ is the average SNR at the receiver when it is at a distance x from the transmitter. The factor $e^{-\sigma^2/2}$ is introduced to make the mean path gain equal to the deterministic component of the path gain, $G_d(x)$. In the following we assume a simple power law for this gains, namely, $G_d(x) = \kappa/x^\alpha$. The exponent α takes values typically bigger than 2 and up to 4. With this path gain model, with some calculations it can be found that

$$
E[R] = \left(\frac{\kappa P_{tx}}{\gamma_{th} P_N} \right)^{1/\alpha} e^{-\frac{\sigma^2}{2\alpha}\left(1 - \frac{1}{\alpha}\right)} \tag{5.125}
$$

We can use the queue analogy and identify the linear vehicular network with an $M/G/\infty$ model, if we can assume that vehicles are scattered along the road according to a 1-D Poisson process with mean density λ. Then, we can exploit the results above and write

$$
\varphi_{D_c}(s) = 1 + \frac{s}{\lambda} - \frac{1}{\lambda P_0^*(s)} \tag{5.126}
$$

where $P_0^*(s)$ is the Laplace transform of $P_0(t)$ defined as $P_0(t) = e^{-\lambda \int_0^t [1 - F_R(a)]\, da}$. Then

$$
E[D_c] = \frac{e^{\lambda E[R]} - 1}{\lambda} \tag{5.127}
$$

It is apparent that the normalized mean covered distance $E[D_c]/E[R]$ depends only on the parameter $\lambda E[R]$. Under the modeling assumption for the path loss, this parameter is equal to the power $1/\alpha$ of the SNR at a distance $1/\lambda$ (mean inter-vehicle spacing) discounted by the exponential factor that accounts for the log-normal shadowing:

$$\lambda E[R] = \left[\frac{\kappa \lambda^{\alpha} P_{tx}}{\gamma_{th} P_N} e^{-\frac{\sigma^2}{2}\left(1-\frac{1}{\alpha}\right)}\right]^{1/\alpha} = \left[SNR_d(1/\lambda) e^{-\frac{\sigma^2}{2}\left(1-\frac{1}{\alpha}\right)}\right]^{1/\alpha} \tag{5.128}$$

Eventually, a shorter distance is covered by the multi-hop message passing, as a consequence of the randomness of the path gain, expressed through the parameter σ. We recover the deterministic case (with Poisson vehicle spacing) in the limit for $\sigma \to 0$.

Summary and Takeaways

Queueing models with multiple servers have been investigated in this chapter. Out of the general $G/G/m/K$ class of queues, we have focused our attention on pure loss systems with Poisson input, also known as Erlang system; pure wait systems with Poisson input and negative exponential service times; and so called "ample" systems, i.e., those with an unlimited number of serving seats.

In many networking and computing systems the whole available communication and/or processing resource is partitioned in shares and each of them can be allocated to a service request. In all those cases, multi-server systems arise naturally.

The first kind of model is very useful in circuit-switching networks and in channel assignment based cellular networks. In those cases contention for resources among competing users are generally solved by rejecting the excess traffic. The key metric here is the loss probability. A major result is the insensitivity of the Erlang model to the probability distribution of the service times, so that results hold for the whole class of $M/G/m/0$ queues. Generalizations of the Erlang loss model are addressed as well, namely the multi-class Erlang loss model, for which a smart numerical solution exists, and approximations to deal with non-Poisson input traffic to loss systems.

The second kind of models arises when there is room for keeping service requests while they wait to be served. In this case, key metrics concern the delay performance of the system. An important result is about scaling of performance with the number of servers. We have compared service systems with the same overall input rate and the same overall serving rate, but with different structures: (i) separate queues; (ii) shared waiting line, but separate servers (here is where the multi-server queueing model appears); (iii) shared waiting line and serving capacity. In general, delay performance improve with the scale of the system, i.e., shared systems perform best. This might turn out not to hold if there are service time components that do not scale with the serving capacity.

Finally, the third kind of model (infinite server queues) can have disparate applications. We have focused on a specific application case, i.e., modeling multi-hop message passing in a one-dimensional network.

Problems

5.1 Consider a private telephone exchange serving 100 users, each offering an average traffic equal to 0.05 Erl during the peak hour. Blocked calls are lost and the target loss probability must satisfy the constraint $P \leq 0.01$. Assume calls are offered according to a Poisson process.

(a) Find the required number of lines connecting the private branch exchange to the outside network (one line can be used to carry a single call at a time).

(b) What is the average utilization factor of the lines?

(c) Find the blocking probability and the average utilization factor if 10% of the lines are out of service.

(d) Identify the blocking probability in case the mean offered traffic is increased by 10%.

5.2 You have to plan the upgrade of the connections of a network access node of the telephone mobile network over a time period of 3 years. You can assume that: (1) the offered traffic is of Poisson type, with an average offered call rate that grows by a constant factor each year. In the first year it is $\lambda = 2000$ calls/h and in the last year it is $\lambda = 10000$ calls/h; (2) the average duration of a call is 1.5 min; (3) the target loss probability is 0.01; (4) the upgrade cards support up to 32 connections each.

Calculate the minimum number of cards that need be procured each year, to maintain the target QoS requirements of the loss probability.

5.3 Consider a multi-service facility with $m = 5$ servers, all having a negative exponential service time probability distribution with mean value $1/\mu = 10$ min. Assume they are all busy at time $t_0 = 0$ and no more customers enter the system. Calculate the following:

(a) The mean time required for the first customer completing its service and leaving the system.

(b) The mean time required for the last customer completing its service and leaving the system.

   (c) The probability distribution function of the times considered in point (a) and (b) above; check that the corresponding mean values are consistent with those found in points (a) and (b).

   (d) The probability that no customer has left the system after 10 min.

   (e) The probability that all customers have left the system within 20 min.

5.4 The University of Utopia plans to establish a call center to provide students with information relating their careers and administrative issues. The University comprises 10 schools, half of which are big ones, with 15,000 students each; the others are medium-size schools with 8000 students each. During the enrollment time (one month) each student has a chance of 50% to make one phone call to the information help desk. Calls are uniformly spread over the entire month (20 working days) and the office time of the help desk (7 hours per working day). The mean duration of a call is 4 min.

   (a) Calculate the number of call-center operators required to guarantee that the blocking probability be no more than 5%, if the service is organized independently for each school.

   (b) Redo the calculation of point (a) if a centralized service is set up for the whole University.

5.5 The front-end desk of a company is served by a single operator, collecting all incoming calls to deliver them to the appropriate internal extension. Calls arrive according to a Poisson process with average rate 1 call/min and the operator takes 40 s to serve a call.

   (a) Calculate the blocking probability of the front desk in case of lost blocked calls.

   (b) If a second operater can take the call when the first one is busy, how much is the blocking probability reduced? What is the average busy time of the second operator?

5.6 You must design the interconnection system of three private telephone exchanges of a company among themselves and to the outside network. Let A, B, and C denote the three sites, to be inter-connected. The mean offered traffic in Erlang between the three sites and toward the outside network is given in Table 5.7.

The target loss probability (probability that a connection setup is rejected) must be no more than 0.01. The costs of the connection line lease and of the traffic is as follows:

C_s cost per month of a subscriber line to the external network

C_l cost per month of a leased line to interconnect the company sites

Table 5.7 Mean offered traffic among the company sites and with the outside network (ext).

in/out	A	B	C	ext
A	–	5	3	30
B	10	–	2	12
C	10	2	–	12
ext	30	12	12	–

Compare the following solutions by assuming $C_s = 50$ Euro/month and $C_l = 750$ Euro/month.

(a) All inter-site traffic is carried on leased lines.

(b) All inter-site traffic is carried through leased lines with the additional constraint that sites B and C are not directly connected; calls between B and C are switched through A.

(c) All traffic is carried on the public network with no leased lines.

(d) Find the break-even point ratio of C_s/C_l that makes the costs of solution (a) and (c) equal.

5.7 Let us consider the dimensioning of the voice interconnection network of a company having four different sites. The average offered traffic matrix among the four sites is given by (the entry A_{ij} of the matrix represents the average offered traffic intensity from site i to site j, $i,j = 1, 2, 3, 4$).

$$
\mathbf{A} = \begin{bmatrix}
- & 4 & 4 & 3 \\
5 & - & 2 & 1 \\
5 & 2 & - & 2 \\
4 & 1 & 2 & -
\end{bmatrix}
\tag{5.129}
$$

You can assume that offered traffic follows a Poisson statistics and that blocked calls are lost.

(a) Calculate the number of voice connections required to meet a blocking probability requirement of 0.01, if the inter-connection topology is a full mesh.

(b) Repeat the calculation of point (a) for a star topology centered on site 1.

(c) Suppose that voice connections are provided by means of VoIP, with an average bit rate of 25 kbit/s, and that the link dedicated to VoIP must be loaded at most up to 70%. Link capacity can be leased in multiples of 64 kbit/s. Find the minimum capacity that must be rented to sustain a full mesh topology.

5.8 Customer arrivals to a store follow a Poisson law with an average arrival rate of 40 customer/hour. On the average, each customer spends 30 min in the store. 90% of the customer come to the store by car. A parking lot is reserved at the store, where there is space for up to 25 cars. People arriving by car at the store and unable to find room for their car in the parking lot, leave immediately.

    (a) Find the probability that a customer has to give up going to the store because it cannot park.

    (b) How many places would you suggest the parking lot be sized for to guarantee that customers arriving by car find parking place with probability at least 99%?

5.9 An m server facility receives two streams of service requests that can be modeled as independent Poisson processes of mean rate λ_1 and λ_2, respectively. A customer of stream 1 requires a single server and has mean service time equal to $1/\mu_1$. Customer of type 1 that cannot find an available server upon arrival are rejected and lost to the system. A customer of type 2 requires b servers, $1 < b \leq m$, and has mean service time $1/\mu_2$. Customers of type 2 that cannot find b available servers upon their arrivals are rejected and lost. Find the average fraction of customer of either type that are rejected with the following numerical values: $m = 100$, $b = 8$, $\lambda_1/\mu_1 = 40$ and $\lambda_2/\mu_2 = 5$.

5.10 The owner of a big store hires clerks for the registers at a cost (wages plus overhead) of 25 MU/hour (MU stands for monetary unit). Customers come to the store according to a Poisson process with mean rate of 100 customer/hour. The service time at the register is exponentially distributed with mean 4 min. Customers lining up at the registers form a unique waiting line. They pick the earliest available register. The amount of net profit per customer is a random variable with uniform probability distribution between 0 and 10 MU.

    (a) Compute the mean waiting time and the probability of a positive waiting time in case the minimum number of clerks to guarantee finite delays is hired.

    (b) The owner, in an effort to improve his customer care and win the preference of the customers, offers a discount of 1 MU/min of wait time for a customer. What is then the most profitable number of clerks to hire?

5.11 The access quality offered by a network is constantly monitored by agents that collect data each day. The agents upload their data to a central database server, by establishing a connection. Each connection has a fixed throughput of R bit/s. The amount of data to be uploaded has a mean size of $M = 8$ Mbytes. Connection requests to the server are so scheduled that they can be deemed to arrive according to a Poisson process of rate $\lambda = 24$ connections/hour. The server is linked to the monitored network with an access capacity $C = 1$ Mbit/s, entirely devoted to uploading connections. When an agent tries to connect to the server and the connection cannot be established because of lack of capacity (i.e., the capacity is already taken up completely by ongoing connections), the agent gives up uploading its daily report and throws away the data. Find the probability that data is lost as a function of the ratio R/C. Why it is not advisable to make R too small?

5.12 A video distribution service company sets up a server with an access bandwidth capable to maintain up to $m = 100$ simultaneous video streaming flows. Customers of this service pay $E = 0.01$ MU/min of connection to the video server. Their request arrive at the server according to a Poisson process with mean rate $\lambda = 80$ req/h. If the maximum number of streaming flows is on, further requests are turned down and lost. The average duration of a video session is $1/\mu = 120$ min. The requests are spread uniformly over 4 hour per day, 30 days per month and are negligible for the rest of the time. The cost of renting more access capacity for the video server is $C = 100$ MU per each additional streaming flow per month.

(a) Find the average amount of MUs lost per month with the initial setting of the video distribution service.

(b) Is there any convenience for the service provider to add more access capacity? If there is, how many additional capacity slices pay off on average (1 capacity slice = one additional streaming flow admitted)?

(c) Does the answer to point (b) change if the price structure is no more time dependent, i.e., a video session cost is fixed to 1.2 MU?

(d) Suppose now that $\lambda = N\lambda_0$, N being the number of the video service subscribers. Assume that a customer pays P MU per month for $\lambda_0 = 5$ req/month, and customers are guaranteed a blocking probability not exceeding 0.01. Determine the relationship among P, N and the access capacity renting cost mC. What is the minimum value of N for given C and P that makes the service profitable? Given $m_{max} = 500$ for the server (bigger values of m imply a substantial upgrade of the server capabilities), find the minimum subscription fee that makes the service profitable as a function of N.

5.13 The ticketing office of an airline company is connected to the telephone network with 8 lines, served each by a clerk. During working days, there is a peak time of four hours when the mean arrival rate of service requests is 100 calls/h, calls lasting 5 min on average. Customers finding all lines busy are lost. 75% of them will purchase a ticket from another company, spending 150 MU on average. Lost calls are negligible during off-peak hour. A clerk costs the company 4000 MU/month (including overhead costs), the cost of additional lines is comparatively negligible. Would you advice the company to hire more clerks? In that case, how many?

5.14 Consider a queueing system with m servers and a infinite waiting line. Customer arrive according to a Poisson process with mean rate λ and service times have negative exponential probability distribution with mean $1/\mu_1$ so long as there are no more than m customers in the system. The mean service time becomes $1/\mu_2$ when the number of customers in the system exceeds m. Find the probability that an arriving customer has to wait for service as a function of the ratio μ_1/μ_2.

5.15 A server is connected to the network with a link of capacity C. Users download files with mean size Q and share fairly the overall capacity of the server, i.e., if n users are downloading simultaneously, each one gets a capacity C/n. An arriving customer joins the server and starts downloading a file with probability $1/(n + 1)$ if there are already n users downloading (this is an approximate representation of the fact that an arriving user starts downloading, then it immediately gives up if its connection is too slow). Find the steady-state probability p_n of finding n users in the system. Evaluate the mean number of users that are downloading simultaneously for $\lambda = 3$ req/s, $Q = 4$ Mbyte, $C = 100$ Mbit/s.

5.16 A base station has $m = 20$ transmitter/receiver. When powered up, the base stations demand a fixed power P_0 plus an amount that is proportional to the number of active transmitter/receivers $N(t)$. So, the power required at time t is $P(t) = P_0 + P_1 N(t)$. During the peak hour, new connections arrive according to a Poisson process with mean rate λ, the connection holding time is a negative exponential random variable with mean value $1/\mu$. Calculate the average amount of energy consumed during the peak hour [$P_0 = 50$ W, $P_1 = 1$ W, $\lambda/\mu = 18$ Erl].

5.17 A content delivery provider has set up an essentially nonblocking service by using a conveniently large number of servers with adequate access capacity. Download requests come from the target user population according to a

Poisson process of mean rate λ. Given the bandwidth allocation policy, the average amount of time required to download a content is $1/\mu$.

(a) Define a model of the interaction between the content delivery network and the user population.

(b) What is the probability that there are more then 100 connections active at any given time if $\lambda = 2400/h$ and $1/\mu = 2$ min?

5.18 A call center has 20 responders. Call attempts arrive according to Poisson process with mean rate 10 calls/min. Call holding time is distributed according to a negative exponential random variable with mean 3 min. At most K customers can be put on wait, if all responders are busy. Further call attempts will receive a courtesy message, inviting the customer to try later.

Find the minimum value of K that makes the probability of rejection of the call attempt less than 0.01.

6

Priorities and Scheduling

Tout vient à point à celui qui sait attendre.
Said by Kutuzov in "War and peace," by Lev Nikolajevič Tolstoj

6.1   Introduction

Service systems often discriminate among customers, offering different levels of quality of service to different customers. In general, getting served entails some kind of benefit for the customer that asks for service. The benefit has a cost or might be associated with a penalty, e.g., in terms of waiting time suffered before being served, service time (depending on the server being slow or fast), possibility of being rejected and having to try later on, and the like.

Let us consider a single server system where a number of customers are waiting to be served. As soon as the server is done with the customer currently being served, the question becomes: who's next? The policy according to which the server decides which customer is taken care of next defines the *queueing discipline*. A queueing discipline can be based on arrival times, service demand, feature associated with customers (e.g., priorities), or quality of service requirements.

We will use also the terms *priority* and *scheduling*. The former arises specifically when a label is assigned to each customer, specifying a level of preference for service. In general, priority queueing is often extended to several queueing disciplines where preference for service is based on aspects other than static labels assigned to customers as they join the queue. Scheduling is usually referred to policies to serve concurrent *flows* of customers. We could as well encompass all priority and scheduling algorithms under the umbrella of queueing disciplines. To keep consistent with terminology used extensively in most theoretical and

Network Traffic Engineering: Stochastic Models and Applications, First Edition. Andrea Baiocchi.
© 2020 John Wiley & Sons, Inc. Published 2020 by John Wiley & Sons, Inc.

application-oriented contexts, we will use either priority or scheduling, as appropriate, when referring to specific algorithms.

The following aspects qualify different queueing disciplines:

- Whether customers are served one at a time, without interruptions between the beginning and the end of service (head-of-line service), or interruptions are possible.
- Whether parallel service is possible, or only a single customer at a time can be under service (possibly interrupted and resumed later).

We can distinguish three categories of queueing disciplines, according to the following modes of operation:

1. Customers are served one at a time, without interruptions.
2. Customers are served one at a time, but interruptions are possible.
3. More than a single customer can be simultaneously under service.

At most a single customer can be partially served with the first category, whereas more than a single customer could be partially served with the other two categories. The third category is usually referred to as the family of *processor-sharing* policies. It is important to observe that most practical implementation of processor-sharing policies actually work according to serial service. In other words, work is done by the server in small chunks moving from one customer to the other one. An example of this mechanisms is round robin, e.g., as implemented by polling system. Time can be slotted. In each time slot one customer receives a chunk of service. Different customers are polled from time slot to time slot. The true parallel processor-sharing policy is obtained as the limit of the round robin as the time slot duration shrinks to zero.

The simplest case of priority corresponds to customers being labeled[1]. Let labels be $1, 2, \ldots, P$, the smallest index corresponding to the highest priority level. An example priority rule could state that the server selects the customer with the smallest label among the waiting customers. Ties are broken according to time of arrival (first come, first served, FCFS). This simple rule identifies the next customer to be served with no ambiguity, as long as there are no bulk arrivals. As another example, each customer could carry a counter that grows linearly at a rate specific of the class the customer belongs to. Then, after having waited for a time T, the customer has accumulated a count $c_0 + c_1 T$, where c_0 is the initial count at customer arrival at the service system and c_1 is the count growth rate. Both parameters can be specific of the class the customer belongs to. When ready to serve, the server picks the customer with the highest count.

---

1 Either customers carry a label already assigned when they arrive or the label is assigned as soon as they are admitted to the system, according to some predefined characteristic of the customer.

The example assumes a *centralized* control of the queue, with a queue manager that knows the priority level and the time of arrival of each customer. The knowledge can be stored in several ways, e.g., tags appended to customers, or some physical arrangement of the service system. As an example, the output of a packet multiplexer can be equipped with P buffers, each one associated to one level of priority. Packets with priority level j are stored sequentially in the j-th buffer, according to their arrival time. The server scans the buffers sequentially, starting from 1, and stops at the first nonempty buffer, where it picks the head-of-line packet. The priority rule is implemented thanks to the structure of the system.

More complex disciplines can be defined. A key generalization is that service need not be supplied fully in a single interaction. The server could do some work for one customer, without necessarily completing the demand of the customer. Then it takes care of other customers, eventually getting back to the first customer to carry on its service. After being admitted to the system, a customer experiences waiting times alternating with service times, until its service demand has been matched completely and the customer can leave the system. As it completes a slice of service, the server must answer the usual question: who's next? Except that now the next customer to be served will not be served completely, but it will receive a slice of service covering part of its demand. The policy according to which the server capacity is shared among the customers goes under the name of *scheduling*. A scheduling policy requires defining the priority among customers, and how much service is supplied in each instance. The priority may also depend on the amount of service already attained (or still missing to complete the demand of the customer).

A second key issue is whether the scheduler is centralized (a queue manager that knows everything that is needed to implement the scheduling policy) or distributed (customers interact in some way, following an algorithm that takes as input events detectable by the customers and rules/parameters programmed in advance).

Before delving into the mathematical modeling and performance evaluation of some priority and scheduling policies, let us ask one more question: why priorities? In networked service systems, different flows of customers interact and contend for the service resource. Different customers may have different quality of service requirements. As an example, packets sharing a link capacity can belong to a Voice over Internet Protocol (VoIP) application, to a video streaming, to an HTTP session, to a file download, to an email. The first two applications have stringent requirements on delay and delay variation, less on information integrity. The last three applications have strict requirements on the integrity of the transferred data, but looser requirements as for delays. We could even differentiate among the last three applications: to be effective, HTTP needs smaller delays than email.

As long as the service system has plenty of resources for any request, no issue arises. If resources are not so abundant and contention occurs, even if temporarily, decisions must be taken as to who shall be penalized and how much. More critical or demanding or valuable traffic flows get better treatment with respect to others, i.e., they must be *protected* from overload caused by other flows.

Priority and scheduling serve two purposes: (i) creating *differentiated service*; and (ii) protecting customers against excessive demands of other customers that could hog the service capacity thus degrading unacceptably the quality of service experienced by less overwhelming customers.

In the following, we give an overview of priority queueing systems with Poisson input traffic. Then, we turn to server capacity sharing, by analyzing the processor sharing policy, its generalization and possible practical algorithms that implement that policy. Finally, we touch the least attained service (LAS) discipline and then review known results for the optimality of serving policies in single server queues. There are certainly several other topics that deserve attention. The interested reader can find extensive and excellent treatment in [131, Ch. 3], or, in the more recent in [86, Ch. 5] and reference therein. A very insightful introduction to scheduling, both in single server and multi-server queuing systems, is available in [98]. Polling systems are especially tackled in the evergreen text by Hideaki Takagi [200]. An extensive amount of research work has been carried out to apply scheduling policies to the sharing of wireless link capacity, e.g., see [13] or for computer systems, e.g., see [207].

6.2 Conservation Law

The mean waiting times of the priority classes are the result of a trade-off. We expect that any improvement of the mean waiting time of a class should come at the detriment of the performance achieved by some other classes. Nothing comes for free. The conservation law for $M/G/1$ priority systems states formally this intuition.

We assume the following hypotheses:

1. The service system is work-conserving.
2. The scheduling is non preemptive[2].
3. The service times do not depend on the scheduling discipline, i.e., each customer arrives at the system with a given service demand that is not going to change according to the way that the service is provided to the customer.

We further assume that the system is in equilibrium and evaluate the mean unfinished work at a general time t in the equilibrium regime. The unfinished

---

2 Preemptive resume scheduling is allowed, if the service times are i.i.d. negative exponential variables for all customers.

work at time t is the sum of the service times of all customers waiting for service plus the residual service time of the customer under service, denoted with X_0. From section 4.2.4, we know that the mean remaining service time seen by a Poisson arrival is $\lambda E[X^2]/2$. This expression must be generalized to account for the different service times of the P classes. At equilibrium, a Poisson arrival sees a customer of class j in service with probability[3] ρ_j. The mean residual service time of a customer of class j is $E[X_j^2]/(2E[X_j])$. Putting pieces together:

$$W_0 \equiv E[X_0] = \sum_{j=1}^{P} \rho_j \frac{E[X_j^2]}{2E[X_j]} = \sum_{j=1}^{P} \frac{\lambda_j E[X_j^2]}{2} = \frac{\lambda}{2} \sum_{j=1}^{P} \frac{\lambda_j}{\lambda} E[X_j^2] = \frac{\lambda E[X^2]}{2}$$

(6.1)

where we have used the definition $E[X^2] = \sum_{j=1}^{P} \lambda_j E[X_j^2]/\lambda$, with $\lambda = \lambda_1 + \cdots + \lambda_P$.

The unfinished work can be written as

$$U(t) = X_0 + \sum_{j=1}^{P} \sum_{k=1}^{N_j(t)} X_j(k)$$

(6.2)

where $X_j(k)$ is the service time of the k-th customer of class j. The mean, at equilibrium, is independent of t:

$$E[U] = W_0 + \sum_{j=1}^{P} E[N_j]E[X_j] = W_0 + \sum_{j=1}^{P} \lambda_j E[W_j]E[X_j] = W_0 + \sum_{j=1}^{P} \rho_j E[W_j]$$

(6.3)

where Little's law was applied, thanks to the equilibrium hypothesis. The mean unfinished work is *independent* of the service discipline, so long as the assumptions at the beginning of this section hold. Under these assumptions, we can evaluate $E[U]$ by using any serving policy within our hypotheses, including the special case of FCFS. The resulting mean unfinished work coincides with the mean waiting time, expressed by the P-K formula:

$$E[U] = \frac{W_0}{1 - \rho}$$

(6.4)

Finally, merging eqs. (6.3) and (6.4), we get for $\rho < 1$:

$$\sum_{j=1}^{P} \rho_j E[W_j] = \frac{\rho W_0}{1 - \rho}$$

(6.5)

Equation (6.5) is the quantitative statement of the intuitive fact that any gain on the mean waiting time of a given class of customers comes at the expense of a wait penalty of some other classes.

---

3 Note that $1 - \sum_{j=1}^{P} \rho_j = 1 - \rho$ is the probability that an arriving customer finds an empty system and therefore does not suffer any wait.

Example 6.1 In a two-class system the mean waiting time of the high-priority class is required to be a fraction β of the low-priority class mean waiting time, i.e., $E[W_1] = \beta E[W_2]$. The conservation law tells us that

$$\rho_1 E[W_1] + \rho_2 E[W_2] = \frac{(\rho_1 + \rho_2)W_0}{1 - \rho_1 - \rho_2} \tag{6.6}$$

Then, substituting $E[W_1] = \beta E[W_2]$, we obtain

$$E[W_1] = \beta E[W_2] = \frac{\beta(\rho_1 + \rho_2)(\rho_1 E[\tilde{X}_1] + \rho_2 E[\tilde{X}_2])}{(\rho_1 \beta + \rho_2)(1 - \rho_1 - \rho_2)} \tag{6.7}$$

where $E[\tilde{X}_j] = E[X_j^2]/(2E[X_j])$, $j = 1, 2$.

The plots of Figure 6.1 refer to the requirement that the mean waiting time of the high-priority class be a given fraction β of the mean wait of the low-priority class. In those figures we have assumed $\beta = 0.1$. In these two graphs $E[W_1]/E[X_1]$ is plotted as a function of ρ_2/ρ (left plot) and of $C_{X_2} \equiv \sigma_2/E[X_2]$ (right plot). The first case illustrates the impact of the load of the low-priority class. Since the system is designed according to a *differentiated service* principle, both mean waiting times diverge as the overall load tends to saturate the system resource, i.e., $\rho_1 + \rho_2 \to 1$. $E[W_1]$ grows as the low-priority traffic load is increased, still maintaining that it be only 10% of $E[W_2]$. Ultimately, this undesired effect comes from sharing a same server between traffic classes with very different service times and using a differentiated service approach (i.e., there is no absolute requirement on the mean wait of the high-priority class, only that high-priority customers should be better off with respect to low-priority ones in a relative way).

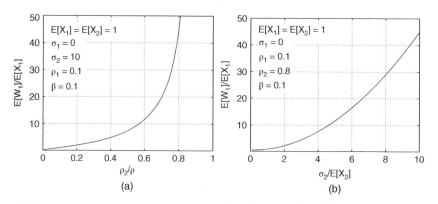

Figure 6.1 Two-class priority system with a fixed ratio β of the mean waiting times of high- and low-priority classes: normalized mean waiting time of the higher-priority class as a function of the load of the lower-priority class (left plot) and as a function of the coefficient of variation (COV) of the service time of the lower-priority class (right plot).

The right graph points out that the mean wait of the high-priority class grows also with the variability of the service times of the low-priority class, for a *fixed* load level ρ_2 of that class. This is a specific trait of nonpreemptive priority. A high-priority customer arriving at the queue must wait for the current service to be completed before it can access the server (if there is no other high-priority customer waiting). Thus, the mean remaining service time includes also a term that accounts for the second moment of low-priority class service times.

Example 6.2 In the same setting as the previous example, namely a two-priority class system of $M/G/1$ type, if the mean waiting time of the high-priority class is assigned a required value $\overline{W}_1$, the mean wait of the low-priority class turns out to be

$$E[W_2] = \frac{1}{\rho_2} \left[\frac{(\rho_1 + \rho_2)(\rho_1 E[\tilde{X}_1] + \rho_2 E[\tilde{X}_2])}{1 - \rho_1 - \rho_2} - \rho_1 \overline{W}_1 \right] \tag{6.8}$$

Figure 6.2 shows $E[W_2]/E[W_1]$ as a function of $E[W_1]/E[X_1]$, for the same mean service time of the two classes, null standard deviation of the high-priority class, standard deviation 10 times the mean service time in case of low-priority class, and $\rho_1 = 0.1$, $\rho_2 = 0.8$. As for class 1, we set the constraint $E[W_1] = \overline{W}_1$. The parameter choice corresponds to a light load high-priority traffic flow, with low variability service times, mixed together with a heavily loaded low-priority flow with high variability service times. Since a fixed level of mean wait is desired for the high-priority class, quite a big penalty is imposed on the low-priority class, even when the requirement $\overline{W}_1$ is relaxed. For example, if we maintain that $\overline{W}_1 = 4E[X_1]$, then it must be $E[W_2] = 100E[W_1]$.

Example 6.3 Let us consider an $M/G/1$ system with mean arrival rate λ, where customers are assigned a random label L, distributed on $[0, 1]$ with CDF

Figure 6.2 Two-class priority system with a fixed requirement for the higher-priority class mean waiting time: ratio of the mean waiting times of the lower- to higher-priority class versus the normalized mean waiting time of the higher-priority class.

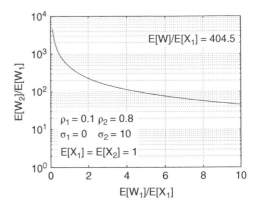

$F(x) \equiv P(L \leq x)$. The smaller the label, the higher the service priority. Service times are i.i.d. random variables with the same PDF across all "priority" classes. Let us start with a discrete probability distribution $q_i = P(L = i/n)$, with $i = 1, \ldots, n$. The CDF is $F_i = \sum_{j=1}^{i} q_j$. We can map this system into an n priority classes system. The i-th class mean arrival rate is $\lambda_i = q_i \lambda$. The conservation law application yields:

$$\sum_{j=1}^{n} \rho_j E[W_j] = \rho \sum_{j=1}^{n} q_j E[W_j] = \frac{\rho W_0}{1 - \rho} \tag{6.9}$$

In the limit for $n \to \infty$, we move toward a continuous probability distribution, with PDF $f(x)$. Letting $\overline{W}(x)$ denote the mean waiting time of those customers that are assigned label x, we can restate the conservation law in this case as follows:

$$\int_0^1 f(x)\overline{W}(x) \, dx = \frac{W_0}{1 - \rho} \tag{6.10}$$

We will resume this example when we analyze the head-of-line (HOL) priority case.

6.3 M/G/1 Priority Queueing

The $M/G/1$ priority queueing model consists of a single server system fed by customers bearing a priority. The queueing system has infinite room for waiting customers, so no loss occurs. We assume that there exist P priority classes. For the time being, we do not specify a particular priority rule. All we subscribe is that: (i) any conceivable priority rule shall be a function of the labels $1, \ldots, P$; (ii) HOL service is adopted, i.e., a single customer at a time is served; and (iii) service time is independent of the priority rule. More subtly, the key point lies with *how* we decide to assign a given label to a given traffic flow (e.g., based on its offered load or service requirement characteristics).

Class j customers arrive at the system according to a Poisson process with mean rate λ_j. The CDF of their service demand X_j is denoted with $F_{X_j}(x)$, the mean service time with $E[X_j] = 1/\mu_j$. Let also $N_j(t)$ be the number of class j customers residing in the system at time t. The overall number $Q(t)$ is the sum of the $N_j(t)$'s: $Q(t) = N_1(t) + \cdots + N_P(t)$. Finally, let W_j denote the waiting time of the class j customers. The priority scheme based on labels needs two further specifications:

- What to do with ties, i.e., customer having the same label;
- Whether service can be interrupted.

As for the first issue, we assume an FCFS rule, unless otherwise stated. As for the second point, two kinds of priorities are defined.

- *Nonpreemptive* priority: Once a customer is allocated the server, the server stays with that customer until its demand has been fully met, no matter of the priority level of customers arriving *after* the beginning of the current service; in other words, service is not interruptible.
- *Preemptive* priority: If a customer of class j is under service and a customer with priority level $i < j$ enters into the system, the service is interrupted immediately and the new, higher-priority customer is taken into service, releasing the customer that was being served; two kinds of resume mechanisms can be defined.
  - *Preemptive-resume*: When the server is available again and the interrupted customer has the right to get it, according to the priority rules, service starts off just from where it had been stopped on the last interruption.
  - *Preemptive-repeat*: Whenever resuming service after an interruption, a customer starts all over from scratch; thus the work done in previous partial service times is lost

A key observation is that the probability distribution of the number $Q(t)$ of customers in the system at equilibrium does not depend on the service priority as long as the server is work-conserving, i.e., all the workload admitted into the queue is carried out, sooner or later, *and* no time is wasted by the server, i.e., the server is always serving one customer as long as there is at least one in the system (no server vacations allowed).

This is evident by reviewing the derivation of the z-transform $\phi_Q(z)$ in Section 4.2.1 or even by reading the one-step evolution equation of the $M/G/1$ EMC Q_n. The subscript n refers to the n-th *departing* customer, whatever the order of service be. Note that also the duration of the busy period and its probability distribution are independent of the order of service for a work-conserving server. On the other hand, the *probability distribution* of the waiting time *does depend* on the order of service.

As a result of the independence of the queue length PDF of the order of service, it follows that statistical equilibrium of the system is achieved under the same conditions as in the plain $M/G/1$ queue, namely it must be $\rho < 1$, where $\rho = \rho_1 + \cdots + \rho_P = \sum_{i=1}^{P} \lambda_i / \mu_i$ is the overall server utilization coefficient.

6.3.1 Non-FCFS Queueing Disciplines

Before moving to specific priority schemes and scheduling algorithms, we devote this section to analyzing two classic queueing disciplines alternative to FCFS, namely last-come, first-served (LCFS) and random order (RO). Both of them are addressed in [198].

With LCFS, the server picks the last-arrived customer to serve whenever it has to start a new service. No interruption is allowed. This kind of discipline is typically implemented in a stack data structure, as the one used to manage subroutine calls in a computer program.

Let us consider an $M/G/1$ queue, with mean arrival rate λ. We assume that $\rho < 1$ so steady-state exists and we refer to a general time during statistical equilibrium.

A customer that joins the queue can find it empty, with probability $1 - \rho$. In this case it will be served immediately, i.e., the waiting time is 0.

If the arriving customers finds the server busy, they must wait until the current service ends, i.e., for a residual service time $\tilde{X}$. Let N be the number of customers arriving *after* the tagged customer, during the same residual service time. They will be served before the tagged customer, according to the LCFS discipline, along with any other customer arriving during the sub-busy periods starting with their service. Let us denote with Y_k the sub-busy period generated by the k-th arriving customer, $k = 1, \dots, N$.

Translating in equations the above reasoning, we can write the following expression of the Laplace transform of the PDF of the waiting time, conditional on $\tilde{X} = x$ and $N = n$:

$$E[e^{-sW}|\tilde{X} = x, N = n] = 1 - \rho + \rho \, E[e^{-s(x + \sum_{k=1}^{n} Y_k)}] = 1 - \rho + \rho \, e^{-sx}[\varphi_Y(s)]^n$$

$$(6.11)$$

where $\varphi_Y(s)$ is the Laplace transform of the busy period of the $M/G/1$ queue as given in eq. (4.42). Removing the conditioning, we have

$$\varphi_W(s) = \int_0^\infty \sum_{n=0}^\infty \frac{(\lambda x)^n}{n!} e^{-\lambda x} E[e^{-sW}|\tilde{X} = x, N = n] f_{\tilde{X}}(x) \, dx$$

$$= 1 - \rho + \rho \int_0^\infty \sum_{n=0}^\infty \frac{(\lambda x)^n}{n!} e^{-\lambda x} e^{-sx}[\varphi_Y(s)]^n f_{\tilde{X}}(x) \, dx$$

$$= 1 - \rho + \rho \int_0^\infty e^{-x[s + \lambda - \lambda \varphi_Y(s)]} f_{\tilde{X}}(x) \, dx$$

$$= 1 - \rho + \rho \, \varphi_{\tilde{X}}(s + \lambda - \lambda \varphi_Y(s))$$

$$= 1 - \rho + \lambda \frac{1 - \varphi_X(s + \lambda - \lambda \varphi_Y(s))}{s + \lambda - \lambda \varphi_Y(s)}$$

$$= 1 - \rho + \lambda \frac{1 - \varphi_Y(s)}{s + \lambda - \lambda \varphi_Y(s)}$$

where we have used the expression of the Laplace transform of the random variable $\tilde{X}$, i.e., $\varphi_{\tilde{X}}(s) = \frac{1 - \varphi_X(s)}{s E[X]}$, in the fourth line, and eq. (4.42) in the sixth line.

It can be verified that the mean waiting time of the LCFS $M/G/1$ queue is the same as the FCFS. In fact, the mean waiting time in *independent* of the queueing discipline, as long as it serves one customer at a time and the server is work-conserving. The probability distribution and higher-order moments are instead different from the FCFS case.

As for random order (RO) service, whenever a new service is started, the server picks one of the customers waiting for service, if any, uniformly at random. A customer is selected with probability $1/n$ if $Q = n$.

In [198] it is shown that it is

$$\varphi_W(s) = (1 - \rho)\left[1 + \frac{\lambda}{s}\int_{\varphi_Y(s)}^1 A(u)e^{-B(u)}\,du\right] \tag{6.12}$$

where

$$A(u) = 1 + \frac{u-1}{u - \varphi_X(\lambda - \lambda u)} \qquad B(u) = \int_u^1 \frac{1}{w - \varphi_X(s + \lambda - \lambda w)}\,dw \tag{6.13}$$

By deriving the Laplace transforms of the waiting time it is possible to calculate the moments of the waiting time. As we know, the mean waiting time is independent of the queueing discipline and given by the Pollaczek-Khinchine formula:

$$E[W] = \frac{\lambda E[X^2]}{2(1-\rho)} \tag{6.14}$$

As for the second moment, we have

$$E[W_{FCFS}^2] = \frac{\lambda E[X^3]}{3(1-\rho)} + \frac{\lambda^2(E[X^2])^2}{2(1-\rho)^2} \tag{6.15}$$

for FCFS service,

$$E[W_{LCFS}^2] = \frac{\lambda E[X^3]}{3(1-\rho)^2} + \frac{\lambda^2(E[X^2])^2}{2(1-\rho)^3} = \frac{1}{1-\rho}E[W_{FCFS}^2] \tag{6.16}$$

for LCFS service, and

$$E[W_{RO}^2] = \frac{2\lambda E[X^3]}{3(1-\rho)(2-\rho)} + \frac{\lambda^2(E[X^2])^2}{(1-\rho)^2(2-\rho)} = \frac{2}{2-\rho}E[W_{FCFS}^2] \tag{6.17}$$

for RO service.

Figure 6.3 compares the COV of the waiting time C_W for the three queueing disciplines as a function of ρ for Pareto distributed service time (left plot) and deterministic service time (right plot). We assume $E[X] = 1$, i.e., the mean service time is taken as the unit of time. We recall that $C_W = \sigma_W/E[W]$. The Pareto CCDF is given by $G_X(t) = (\theta/t)^\alpha$, where the parameter θ is set so that $E[X] = \alpha\theta/(\alpha - 1)$ is equal to a prescribed value. We set $\alpha = 3.2$. Note that it must be $\alpha > 3$ for the first three moments of the service time to exist.

The plots are similar, except that the values of the COV are much higher in case of Pareto service times for the same value of ρ. The COV appears to diverge both for low and for high load levels. This means that a large jitter, with respect to the average of the waiting time, is to be expected in those two cases. Since the mean waiting time tends to 0 as $\rho \to 0$, the diverging COV means simply that the

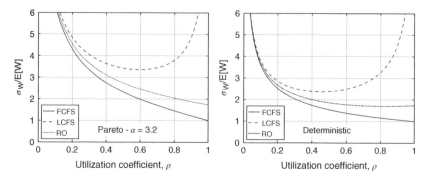

Figure 6.3 Coefficient of variation of waiting time as a function of the utilization coefficient for an M/G/1 queue: comparison among FCFS, LCFS, and RO queueing disciplines. Left plot: Pareto distributed service time. Right plot: deterministic service time.

standard deviation of the waiting time tends to 0 slower than the mean. Yet, for low load values, delay is not an issue. As ρ approaches 1, the plots highlight that the jitter of the waiting time is growing even faster than the mean waiting time. Worse delay jitter performance is to be expected with LCFS, while RO results in values intermediate between LCFS and FCFS.

6.3.2 Head-of-Line (HOL) Priorities

We specialize the general label priority model by requiring that class j customers have priority over class i customers for any $i > j$. HOL priority consists of this simple rule: the server, whenever ready to serve a customer waiting in the system, chooses the one having the least label. Among those with the least label, it chooses the one that has waited longest. Preemption is not allowed.

The waiting time of a class j customer can be written as

$$W_j = X_0 + \sum_{i=1}^{j} \sum_{k=1}^{N_i} X_i(k) + \sum_{i=1}^{j-1} \sum_{k=1}^{M_i} Y_i(k) \tag{6.18}$$

where N_i is the number of customers of class i found in the system by the tagged arrival, $X_i(k)$ is the service time of the k-th of such customers, M_i is the number of customers arriving at the system *after* the arrival of the tagged customer, $Y_i(k)$ is the service time of the k-th of those last customers.

Equation (6.18) is self-explanatory: the wait of the class j tagged customer is made up of three components:

1. The time required for the server to clear its current residual service time (no preemption is allowed).
2. The service times of customers having higher or *the same* priority as the tagged customer, already in the system upon the arrival of the tagged customer.

3. The service times of customers of higher priority than the tagged customer, arriving after its arrival, since those higher-priority arrivals jump ahead of the tagged customer according to HOL scheme.

At equilibrium we can apply Little's law and thus obtain the following expression of the mean wait of class j customers:

$$
\begin{aligned}
E[W_j] &= W_0 + \sum_{i=1}^{j} E[N_i]E[X_i] + \sum_{i=1}^{j-1} E[M_i]E[X_i] \\
&= W_0 + \sum_{i=1}^{j} \lambda_i E[W_i]E[X_i] + \sum_{i=1}^{j-1} \lambda_i E[W_j]E[X_i] \\
&= W_0 + \sum_{i=1}^{j} \rho_i E[W_i] + E[W_j]\sum_{i=1}^{j-1} \rho_i
\end{aligned}
\tag{6.19}
$$

for $j = 2, \ldots, P$. For $j = 1$ it is easy to find that

$$
E[W_1] = W_0 + E[N_1]E[X_1] = W_0 + \rho_1 E[W_1] \tag{6.20}
$$

In this derivation we have used the fact that, at equilibrium, the mean number of class i customers residing in the waiting line (i.e., in the system, but not under service) is $E[N_i] = \lambda_i E[W_i]$ by Little's law. The mean number of customers of class i arriving at the queue during the tagged class j customer wait are $E[M_i] = \lambda_i E[W_j]$.

Equations (6.19) and (6.20) can be rearranged as follows:

$$
E[W_1] = \frac{W_0}{1 - \rho_1} \tag{6.21}
$$

$$
E[W_j]\left(1 - \sum_{i=1}^{j} \rho_i\right) = W_0 + \sum_{i=1}^{j-1} \rho_i E[W_i] \tag{6.22}
$$

Let $S_j = \sum_{i=1}^{j} \rho_i$, $j = 1, \ldots, P$, and $S_0 = 0$, to simplify notation. Subtracting the $(j-1)$-th equation from the j-th one, we get $E[W_j]\left(1 - S_j\right) = E[W_{j-1}]\left(1 - S_{j-2}\right)$. Multiplying by $1 - S_{j-1}$ we obtain finally

$$
E[W_j](1 - S_j)(1 - S_{j-1}) = E[W_{j-1}](1 - S_{j-1})(1 - S_{j-2}) \tag{6.23}
$$

for $j = 2, \ldots, P$. Therefore, $Z_j = E[W_j](1 - S_j)(1 - S_{j-1})$ is a constant, independent of j, whose value is found by letting $j = 1$ and noting that

$$
Z_1 = E[W_1](1 - S_1) = E[W_1](1 - \rho_1) = W_0 \tag{6.24}
$$

The final result is

$$
E[W_j] = \frac{W_0}{(1 - S_j)(1 - S_{j-1})}, \qquad j = 1, \ldots, P, \tag{6.25}
$$

with $S_0 \equiv 0$ and $S_j = \sum_{i=1}^{j} \rho_i$ for $j = 1, \ldots, P$. The overall mean waiting time is $E[W] = \sum_{i=1}^{P} \frac{\lambda_i}{\lambda} E[W_i]$, where $\lambda = \lambda_1 + \cdots + \lambda_P$.

The form of the result is consistent with intuition of the HOL scheme working. The mean waiting time of class j depends on the residual time (hence on the variability of the service times of *all* classes: no preemption!) and on the mean load of all classes from 1 (highest priority) down to j itself. It does not depend on the mean load of classes with priority lower than j. This is the kind of protection offered by the HOL scheme to higher-priority classes against possible overloads caused by lower priority classes. In turn, this gives an indication of when HOL could possibly be used.

Example 6.4   Consider the output link of a router or a cellular network base station, where VoIP, interactive transaction (e.g., chatting), video streaming, web browsing, email, cloud storage, and file-sharing applications traffic flows are multiplexed. The first three kinds of traffic have stringent delay requirements. The first two are expected to offer a little fraction of the overall traffic (tens of kbps for each VoIP connection and few kbps for chatting), while video streaming is more demanding as for throughput, ranging from few hundreds kbps up to few Mbps. Web browsing is somewhat more elastic than real-time traffic. The last three categories have definitely loose requirements on delay. The last two can produce massive data transfer, typically much more intense than email or web browsing (up to Gbytes as opposed to typically few Mbytes or less data to move).

If a HOL scheme is applied at the output link, it is natural to assign top priority to small, inelastic flows such as VoIP and transactional traffic. Immediately below this category there comes video streaming, since it needs low mean delay and delay jitter and it has a capped throughput requirement. Low priority should be given to email, cloud storage and file sharing, especially the last one. File sharing is usually executed by greedy applications that try to take as much bandwidth as possible (even if parameters are provided to configure how aggressive they should be). It is therefore likely that such applications can cause overloads. Assigning them the lowest priority guarantees that they will not disturb more critical and delay demanding applications. On the other hand, high-priority applications are expected to leave enough room for the low priority ones to survive well off, since they do not demand a big throughput.

Example 6.5   To give a numerical example, let us assume $P = 4$. The mean service times, coefficients of variation (ratio of the standard deviation to the mean) of the service times and the fraction of the overall load ρ of each class are listed in Table 6.1.

Figure 6.4 illustrates $E[W_j]$ as a function of ρ for a four classes priority queueing system. It compares HOL priority (left plot) with preemptive-resume priority (right plot; see section 6.3.3 for the analysis of the latter queueing discipline). For

Table 6.1 Numerical values for Example 6.5.

class	1	2	3	4
$E[X_j]$	1	1	2	10
C_{X_j}	0	0.5	1	2
ρ_j/ρ	0.1	0.1	0.3	0.5

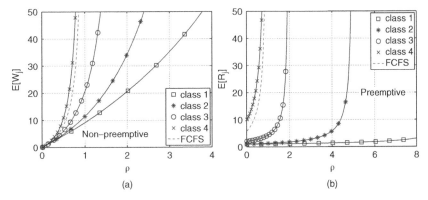

Figure 6.4 Example of HOL priority: mean waiting time as a function of load for nonpreemptive priority (left) and mean response time as a function of load for preemptive resume priority (right). The values of the mean service times, coefficient of variation of the service time and fraction of the overall load of each class are listed in Table 6.1.

comparison purposes, the mean waiting time of a plain FCFS queueing system with the same load is shown as a dashed line.

The different limits of stability of the priority classes are apparent, the lowest priority class having the same limit as the equivalent FCFS queueing system. The mean waiting times highlight that the system offers quite a good gain for a relatively small penalty of the lowest priority class. This is consistent with the assignment of priorities, i.e., the highest priority goes to the light traffic flows, with smoothest service times, while the least priority is given to the bulk of traffic, with highly variable service times.

Comparing HOL and preemptive-resume priorities, it can be seen that the latter improves the mean waiting time experienced by higher-priority classes at low to moderate loads, with respect to the former policy. The price to be paid is worsening of low priority mean waiting time even at very low loads. As suggested by intuition, at very low loads, HOL priority implies little difference with respect to a plain FCFS system, i.e., all classes undergo the same mean waiting time, dominated by the mean residual unfinished work W_0. In the preemptive-resume policy, differentiation is already evident even at very low loads. The added value of this policy is that high-priority classes do not suffer in case of high variance of service times of low-priority class.

Example 6.6 (Continued from Example 6.3)   The random labeling priority service system can be assimilated to a HOL scheme, where class j customers are selected with probability q_j from the whole stream of arriving customers. Then

$$E[W_j] = \frac{W_0}{(1 - \rho F_j)(1 - \rho F_{j-1})} \tag{6.26}$$

where $W_0 = \lambda E[X^2]/2$. In the limit for a continuous probability distribution, we have

$$\overline{W}(x) = \frac{W_0}{[1 - \rho F(x)]^2} \ , \quad x \in [0, 1], \tag{6.27}$$

where $F(x)$ is the CDF of the random label. Using eq. (6.27) it can be checked that the conservation law holds. The minimum possible waiting time is $\overline{W}(0) = W_0$, while the upper bound is $\overline{W}(1) = W_0/(1 - \rho)^2$.

We can compare $\overline{W}(x)$ with the mean waiting time $E[W] = W_0/(1 - \rho)$ and ask what is the probability that an arriving customer will experience a wait lower than the mean. The inequality $\overline{W}(x) \le E[W]$ yields $F(x) \le (1 - \sqrt{1 - \rho})/\rho$, i.e., the probability that a lucky customer waits less than the mean waiting time is just $p = (1 - \sqrt{1 - \rho})/\rho \in (1/2, 1)$ as ρ ranges in $(0, 1)$. For $\rho = 0.8$, we obtain $p \approx 0.69$. Note that, as ρ gets closer to 1, the mean waiting time raises fast. Experiencing a mean waiting time less than the overall mean wait does not necessarily imply that the customer waits for a *short* time, as compared to its mean service time. In formulas, $\overline{W}(x) < E[W]$ does not imply necessarily that $\overline{W}(x) \sim E[X]$.

We calculate the probability p that $\overline{W}(x) < D$ for a given fixed value D. It is apparent that it must be $D > W_0$. The mean wait $\overline{W}(x) = W_0/[1 - \rho F(x)]^2$ is less than the prescribed threshold D for $x \le x_D = F^{-1}((1 - \sqrt{W_0/D})/\rho)$. Then $p = P(\overline{W}(x) \le D) = P(L \le x_D) = F(x_D) = (1 - \sqrt{W_0/D})/\rho$. For $\rho = 0.9$ and $D = 2W_0$, we have $p \approx 0.325$. About one third of customers experiences a mean wait of no more than twice the minimum mean wait W_0, while the overall mean wait, averaged over all customers, is $10W_0$ for $\rho = 0.9$.

Note that the probability p does not depend on the form of $F(x)$, even if the value of $\overline{W}(x)$ does depend on the specific behavior of $F(x)$.

As a last remark, the random labeling model could account also for customer rejection. It can be stipulated that with some probability q arriving customers are dropped. Then, random labeling applies only to surviving customers. The flow of admitted customers is still a Poisson process, if the original arrival process is, since dropping is made at random, independently of arrival times (it may well depend though on service times: in that case the PDF of the service times of the admitted customers should be adjusted accordingly).

The result on HOL mean waiting times has been derived by assuming $\rho < 1$ to make statistical equilibrium of the *whole* system achievable. It can be seen, however, that the mean waiting time of customers up to class j does exist finite as long as $S_j < 1$, even if $S_i \geq 1$ for $i > j$. In this last case, customers of classes $i > j$ will experience a disaster, the service system appearing overloaded and ineffective at accommodating their service demand.

This is a first sample of a deep and key issue in resource sharing, dubbed under the name of *fairness*. It has to do with the criteria that rule resource sharing among contending customers, when the amount of available resource falls short of the overall demand[4]. Fairness is usually associated to some form of guarantee that every demand presented at the system will be taken care of in some way, maybe with delay or allowing only part of the required service capacity. There is no general rule as to what is the "ideal" fairness, since this is actually a "political" issue, in the sense that it depends on the management policies of the service system. For an interesting discussion of the fairness concept see, e.g., [45]. We will tackle this fundamental topic in Section 10.7, in conjunction with congestion control.

Example 6.7 Priorities and human activity. An unusual instance of priority queueing emerges from the analysis and modeling of human activities [23]. Several measurements give evidence for non-Poisson activity patterns in individual human behavior. In Barabási's paper the email sending times of an individual user are taken as an example. It appears that the best fit for the PDF of the inter-departure times of sent emails is heavy-tailed rather than exponential-tailed. It is also reported that similar patterns have been found in other measurements: examples of heavy-tail distribution are revealed in the distribution of the time differences between consecutive instant messages sent by individuals during online discussions. The timing of job submissions on a supercomputer, directory listing and file transfers (FTP request) initiated by individual users, or the timing of printing jobs submitted by users are mentioned in [23] to have been reported to display non-Poisson features.

The explanation model proposed by Barabási is extremely simple and inspired by the everyday management of tasks involving human individual activity. A list of size L of prioritized tasks is given. Each task in the list is assigned a priority level,

---

4 As long as the overall demand can be fully served, i.e., there is service capacity for all, no fairness issue arises; any work-conserving system just goes ahead to serve everyone. This does not imply that everyone gets the *same* share of the service capacity, since service requests may well differ among themselves. This is not an issue in a plentiful world (or under a modest demand).

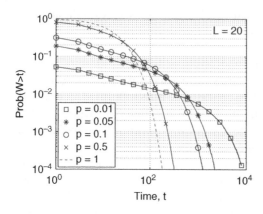

Figure 6.5 CCDF of the task waiting time in the human activity model of Barabási [23]. The task list length is $L = 20$.

x_i for task i. The time axis is discretized. In each time slot a new task is selected and worked out. In the same slot a new task replaces the selected one. The priority level of the new task is chosen from a given PDF, independently of all other tasks.

The selection of a task from the list is biased according to priorities. With probability p (another parameter of the model) the task is chosen at random, otherwise the highest-priority task among the L in the list is selected. This simple model tries to capture a schedule based on priorities, yet allowing occasional violations. For $p \to 1$ the model converges to a pure random order serving system (though with limited room for arriving tasks and saturated load). In that case, it is easy to see that the number of slots W that a task inserted in the list has to wait before being taken care of has a simple geometric PDF:

$$P(W = k) = \left(1 - \frac{1}{L}\right)^{k-1} \frac{1}{L}, \qquad k \geq 1. \tag{6.28}$$

On the contrary, as $p \to 0$ the model moves toward a strict priority scheduling. The PDF for assigning priority levels to new tasks is assumed to be uniform in the interval $[0, 1]$. Simulations have been run to estimate the CCDF of the task waiting time W for various levels of p. The estimated CCDF are plotted in Figure 6.5 along with the CCDF of the pure random selection of eq. (6.28) (dashed line).

In the log-log scale of the graph, the straight initial part of the CCDF denotes the emergence of a heavy-tail behavior. Eventually the extreme tail is exponential, as it is the case for $p > 0$. A heavy tail behavior is seen for about three orders of magnitude of the abscissa if $p = 0.01$. The high variability of the waiting times of this apparently simple model lies in the priority schedule: high-priority tasks joining the list are cleared quickly. Low–priority ones are continuously overrun by the higher-priority tasks and fall heavily behind schedule, accumulating a large delay. A question arises at this point: Why do HOL priorities and other priority queueing systems exhibit *exponential* tails of the waiting times, rather than heavy tails? The key to answer this question is noting two specific feature of the human activity model:

- The priority list is finite; usually queueing priority models envisage an infinite buffer.
- The list is always full, i.e., the system is *saturated* ; priority queueing models with infinite room for customers are assumed to be stable, i.e., the mean arrival rate of tasks is strictly less than the mean rate of task clearing. This guarantees that the system converges to a statistical equilibrium, under stationary task arrivals. Then, all tasks are eventually served, busy period are finite with probability one. As a consequence, the tail of the waiting time CCDF is exponential as long as service times have finite moments.

6.3.3 Preempt-Resume Priorities

The preempt-resume policy models systems where service is interruptible without harm. Only when no higher-priority customer is in the queue can the interrupted customer resume service from where it left off. An example that could fit this scheme is sharing of processing capabilities (either physical servers or virtual machines) in a data center among competing tasks.

We will evaluate the mean of the *system time* (also known as *response time*), i.e., $R_j = X_j + W_j$ for class j customers. This is the relevant quantity with preempt-resume, since service can be interrupted.

Mimicking the approach of Section 6.3.2, we consider a tagged customer of class j arriving at the system, that is supposed to be at equilibrium. The response time of the tagged customer is the sum of the time required to clear the backlog it finds in the system upon its arrival plus the time it takes to complete the tagged customer's service, including possible interruptions:

$$R_j = U_j + \sum_{i=1}^{j-1} \sum_{k=1}^{M_i} X_i(k) + X_j \tag{6.29}$$

where U_j is the unfinished work of customers with priority level up to j found by the tagged class j customer in the system, M_i is the number of priority i customers arriving *during* the tagged customer response time R_j. By using Little's law, we have $E[M_i] = \lambda_i E[R_j]$. Then

$$E[R_j] = E[U_j] + \sum_{i=1}^{j-1} \lambda_i E[R_j] E[X_i] + E[X_j] = E[U_j] + E[R_j] \sum_{i=1}^{j-1} \rho_i + E[X_j] \tag{6.30}$$

The mean unfinished work $E[U_j]$ can be determined thanks to the following remark: a class j customer *does not see* lower-priority customers in any way.

It jumps ahead of those found waiting, and it can interrupt the service of any lower-class customer under service. From the point of view of a class j customer, lower-priority customers do not exist. What does a class j customer see then? The *mean* amount of work it finds in the queue upon arrival is nothing else than an instance of a plain $M/G/1$ mean unfinished work, thanks to the work-conserving property of the server. The specific $M/G/1$ queue backlog seen by the class j customer corresponds to a system where only customers of classes from 1 to j exist. Therefore, invoking the classic P-K formula for the mean waiting time of the $M/G/1$ queue fed by customers of classes $i = 1, \ldots, j$, we have

$$E[U_j] = \frac{W_{0j}}{1 - \sum\limits_{i=1}^{j} \rho_i} \tag{6.31}$$

where W_{0j} is the mean residual service time of customers of class up to j:

$$W_{0j} = \sum_{i=1}^{j} \frac{\lambda_i E[X_i^2]}{2} \tag{6.32}$$

Summing it all up, we get finally

$$E[R_j] = \frac{E[U_j] + E[X_j]}{1 - S_{j-1}} = \frac{1}{1 - S_{j-1}} \left[E[X_j] + \frac{W_{0j}}{1 - S_j} \right] \tag{6.33}$$

where $S_j = \sum_{i=1}^{j} \rho_i, j = 1, \ldots, P$.

Figure 6.4(b) illustrates $E[R_j]$ as a function of ρ. The different limits of stability of the priority classes are apparent, the lowest-priority class having the same limit as the equivalent FCFS queueing system. Numerical values of the parameters are as listed in Table 6.1.

It is apparent that preemptive-resume HOL is even more effective than plain HOL in favoring higher-priority classes (note the different x-axis scale of the two graphs in Figs. 6.4(b) and 6.4(a)). As expected, the penalty paid by the lowest-priority class is heavier than in the HOL case. To gauge correctly the results displayed in these graphs it is noted that the mean response time shown in Figure 6.4(b) includes the service time. Service times in this numerical example vary significantly among classes. This explains the big gap for small levels of ρ.

6.3.4 Shortest Job First

The HOL model can be exploited to derive the mean waiting time of the shortest job first (SJF) policy. With this policy a customer has a higher priority the smaller its service demand. Let us start by a discrete model, obtained by binning the range of the service time. Class j is defined as all service times lying between x_{j-1} and

x_j, where $x_0 = 0 < x_1 < \cdots < x_P$. The mean arrival rate of class j customers is $\lambda_j = \lambda \int_{x_{j-1}}^{x_j} f_X(t)\, dt$, $j = 1, \ldots, P$, where $f_X(x)$ is the PDF of the service times. Therefore

$$S_j = \sum_{i=1}^{j} \rho_i = \sum_{i=1}^{j} \lambda \int_{x_{i-1}}^{x_i} f_X(t)\, dt \int_{x_{i-1}}^{x_i} t\, \frac{f_X(t)}{\int_{x_{i-1}}^{x_i} f_X(t)\, dt}\, dt = \lambda \int_{0}^{x_j} t f_X(t)\, dt$$

(6.34)

Letting the set of discrete points splitting the range of the service time become dense, we may pass to a continuous spectrum of classes. The mean waiting time of class $x \in (0, \infty)$ is then

$$\overline{W}(x) = \frac{W_0}{\left(1 - \lambda \int_0^x t f_X(t)\, dt\right)^2}$$

(6.35)

where $W_0 = \lambda E[X^2]/2$. The mean waiting time ranges from W_0 when $x \to 0$ to $W_0/(1-\rho)^2$ for $x \to \infty$. The overall mean waiting time is $E[W] = \int_0^\infty f_X(x) \overline{W}(x)\, dx$. It can be shown that the conservation law holds for the SJF discipline [196, Ch. 14, p. 541].

Example 6.8 Let the service time PDF be Pareto, with PDF $f_X(x) = \alpha \theta^\alpha / x^{\alpha+1}$, $x \in [\theta, \infty)$. If $\alpha > 2$ at least the first two moments exist and the $M/G/1$ SJF model can be applied. We have

$$\rho = \frac{\alpha}{\alpha - 1} \lambda \theta$$

(6.36)

$$W_0 = \frac{1}{2} \lambda \frac{\alpha}{\alpha - 2} \theta^2$$

(6.37)

and

$$\overline{W}(x) = \frac{W_0}{\left[1 - \rho + \rho\left(\frac{\theta}{x}\right)^{\alpha-1}\right]^2}, \qquad x \geq \theta.$$

(6.38)

With a fair amount of calculations, it can be found that $\overline{W}(x) \leq W_0/(1-\rho)$ for $x \leq x_{sup} = \theta\left(1 + \frac{1}{\sqrt{1-\rho}}\right)^{1/(\alpha-1)}$, which occurs with probability

$$p = \int_0^{x_{sup}} f_X(x)\, dx = 1 - \left(\frac{\theta}{x_{sup}}\right)^\alpha = 1 - \left(\frac{\sqrt{1-\rho}}{1 + \sqrt{1-\rho}}\right)^{\frac{\alpha}{\alpha-1}}$$

(6.39)

For $\rho = 0.9$ and $\alpha = 3$, we have $p \approx 0.88$. It is a typical effect of power laws that samples stay below the mean with high probability. Here we see another manifestation of the power law effect. There is a remarkably high probability that a customer arriving at a SJF service system will undergo a mean waiting time less than the FCFS mean waiting time. The conservation law reminds us that the counterpart of this nice behavior is that the remaining customers (with probability ≈ 0.12 in our numerical example) will suffer rather large delays. However, note that the overall mean waiting time will improve over FCFS.

6.3.5 Shortest Remaining Processing Time

The SJF discipline discriminates in favor of short jobs, yet maintaining that at most a single job can be partially served at any time. In other words, the server capacity cannot be shared concurrently among different jobs. If we relax this constraint and let more jobs be partially served, along the same lines of SJF, another queueing discipline arises quite naturally, namely the so called shortest remaining processing time (SRPT). We still assume that the server is work-conserving. Hence, it can go idle only if no customer is waiting for service in the queueing system. According to SRPT, when the server becomes idle, if any customer is waiting to be served, the server selects the one whose *residual* service time is minimum. In other words, the server gives preference to that job that requires the minimum amount of time to be completely served among those that are waiting for service. When a new customer joins the queue, if its service demand is less than the residual service time of the customer under service, the newly arrived customer will preempt the server. A resume kind of preemption is here assumed. In other words, no work is lost when the server leaves a customer not fully served, to start serving a higher-priority customer, i.e., one whose remaining service time is the least among all customers residing in the queue. As with other queueing disciplines, SRPT can be implemented by defining a quantum of service. After having completed each service quantum, the server checks among all customers in the queue at that time and selects the one with the least remaining service time for the next quantum of service.

In the following we analyze the performance of SRPT as the quantum size tends to 0. We consider a single, work-conserving server. Customers arrive according to a Poisson process of mean rate λ and service times are i.i.d. random variables with CDF $F_X(x)$. We offer an informal argument. For a more formal proof, the interested reader can consult, e.g., [86].

The mean response time conditional on the tagged customer requesting a service time of x is denoted with $R(x)$. It can be decomposed into the sum of two terms:

1. The mean time $V(x)$ since arrival until the tagged customer is served for the first time.
2. The mean time $S(x)$ elapsing since when service of the tagged customer starts until when it leaves the queueing system having completed its service, including all interruptions due to arriving customers with shorter remaining service time.

As for $V(x)$, we recognize that SRPT effectively corresponds to a priority queueing system, where customers with service demand $< x$ have higher priority than the tagged customer. Upon arrival, the tagged customer with service demand x has to wait for the server to complete the current service, provided that service is given to a customer whose residual service demand is less than x. Then, the mean time for the server to complete the current service is

$$W_0(x) = \frac{1}{2}\lambda \left[\int_0^x t^2 \, dF_X(t) + x^2(1 - F_X(x)) \right] \tag{6.40}$$

The mean time the tagged customer has to wait before getting to the server for the first time can be calculated as the mean waiting time of an $M/G/1$ priority system, whose effective arrival rate is $\lambda(x) = \lambda F_X(x)$, i.e., it accounts only for customers having service times smaller than x, and service time PDF equal to $f_X(t)/F_X(x)$ for $t \in [0, x)$. Since the mean time for the server to become empty upon arrival of the tagged customer is $W_0(x)$, we obtain

$$V(x) = \frac{W_0(x)}{\left(1 - \lambda(x)\int_0^x t \, dF_X(t)/F_X(x)\right)^2} = \frac{\lambda \int_0^x t^2 \, dF_X(t) + \lambda x^2[1 - F_X(x)]}{2\left(1 - \lambda \int_0^x t \, dF_X(t)\right)^2} \tag{6.41}$$

As for $S(x)$, let us consider the interval $[t, t + dt]$ during the service (possibly interrupted) of the tagged customer, with $t \leq x$, i.e., let us consider the evolution of the *residual* service time of the tagged customer when decreasing from $t + dt$ to t. The tagged customer can be preempted if a customer with service demand less than t arrives in the considered time interval. In that case the interruption lasts for the whole mini-busy period generated by the new arriving customer and all those possibly arriving *and* having service time less than t. Those form a Poisson process with mean rate $\lambda(t) = \lambda F_X(t)$. Their service time is a random variable X_t, whose PDF is $f_X(u)/F_X(t)$, for $u \in [0, t)$. The mean duration of such mini-busy periods is unaffected by the queueing discipline (since the server is work-conserving), hence we can calculate it by using the known expression holding for the FCFS queue. Let B denote the random variable representing the duration of one such mini-busy period. Then

$$E[B] = \frac{E[X_t]}{1 - \lambda(t)E[X_t]} = \frac{\int_0^t u \, dF_X(u)/F_X(t)}{1 - \lambda \int_0^t u \, dF_X(u)} \tag{6.42}$$

The mean number of such mini-busy periods in the reference time interval $[t, t + dt]$ is $\lambda(t)dt$. Therefore, the tagged customer spends time dt plus the preemption time it suffers, i.e., $\lambda(t)dt \cdot E[B] = \lambda F_X(t)dt\frac{\int_0^t u \, dF_X(u)/F_X(t)}{1-\lambda\int_0^t u \, dF_X(u)}$. Summing it up, we have

$$dt + dt\lambda \frac{\int_0^t u \, dF_X(u)}{1 - \lambda \int_0^t u \, dF_X(u)} = \frac{dt}{1 - \lambda \int_0^t u \, dF_X(u)}. \text{ Integrating over } t, \text{ we get finally}$$

$$S(x) = \int_0^x \frac{dt}{1 - \lambda \int_0^t u \, dF_X(u)} \tag{6.43}$$

The mean conditional response time of the $M/G/1$ queue run according to the SRPT queueing discipline is $R(x) = V(x) + S(x)$, i.e.,

$$R(x) = \frac{\lambda \int_0^x u^2 \, dF_X(u) + \lambda x^2 [1 - F_X(x)]}{2\left(1 - \lambda \int_0^x u \, dF_X(u)\right)^2} + \int_0^x \frac{dt}{1 - \lambda \int_0^t u \, dF_X(u)} \tag{6.44}$$

The unconditional mean response time is obtained simply as $\overline{R} = \int_0^\infty R(x) \, dF_X(x)$.

6.3.6 The μC Rule

Let us pose the following optimization problem, following closely [131, Ch. 3, p. 125]. Consider a service system accessed by P customer classes, each one characterized by its own service time probability distribution and mean arrival rate. Assume that the system can be modeled as an $M/G/1$ priority queue. Let C_j be the cost per unit time that a customer of class j spends in the system. The overall mean cost rate for running the system is then

$$\overline{C} = \sum_{j=1}^P C_j E[N_j] = \sum_{j=1}^P C_j \lambda_j (E[X_j] + E[W_j]) = \sum_{j=1}^P C_j \rho_j + \sum_{j=1}^P C_j \mu_j \rho_j E[W_j] \tag{6.45}$$

The first term on the right-hand side sum is a constant, once the arrival and service processes parameters are given. We should then minimize the sum $\sum_{j=1}^P C_j \mu_j \rho_j E[W_j]$ by choosing among all possible work-conserving priority disciplines. If we restrict ourselves to those priority schemes for which the conservation law holds, the optimization problem becomes:

$$\max_{\pi \in D} \sum_{j=1}^P \mu_j C_j \rho_j E[W_j] \tag{6.46}$$

subject to:

$$\sum_{j=1}^P \rho_j E[W_j] = \frac{\rho W_0}{1 - \rho} \tag{6.47}$$

where π is a policy and D is the set of all work-conserving disciplines for which the conservation law holds.

Since the sum of the variables $\rho_j E[W_j]$ must be a constant by the constraint (6.47), the optimum is achieved by ordering the "weights" $\mu_j C_j$, $j = 1, \ldots, P$, in descending order. We choose the policy that makes $E[W_1]$ attain the least possible value, since it is weighted by the biggest value $\mu_1 C_1$, $E[W_2]$ attain the second smallest value and so on so forth. This is but the HOL priority scheme. We have therefore a rule for assigning labels to the classes, so as to set up the HOL priority. The rule is: once the indices are given so that $\mu_1 C_1 \geq \mu_2 C_2 \geq \cdots \geq \mu_P C_P$, the highest priority is given to class 1, the lowest to class P. The result is therefore that HOL priority is the solution of our optimization problem.

A little thought reveals that the so called "μC" rule is sensible. If mean service times were the same for all customers, priority would be simply proportional to the cost per unit time of staying in the system. The more a customer sojourn in the system costs, the fastest it should be served and sent away. Different values of the mean service times come to change somewhat this picture. To be concrete, suppose one class of customers has a cost of $C_1 = 10$ and a mean service time $1/\mu_1 = 10$, while another class has $C_2 = 2$ and $1/\mu_2 = 1$. Type 1 customers have a high cost; however, they tend to hog the server, since their mean service time is 10 times the other class mean service time. Suppose there are only two customers to serve, one for each class. If class 1 customer is preferred, the average cost for waiting time is $1/\mu_1 \times C_2 = 20$. If on the contrary type 2 customer is selected for service first, the mean cost for the waiting time is $1/\mu_2 \times C_1 = 10$. It is more convenient to serve first the customer whose cost per unit time in the system is *lower* in this case. It can be verified that the μC-rule gives the correct answer: $\mu_1 C_1 = 1 < \mu_2 C_2 = 2$.

Even this very simple example points out that a compromise must be found between cost and mean service time. The outcome is strikingly simple: just assign priorities in descending order of the products $\mu_j C_j$, $j = 1, \ldots, P$.

6.4 Processor Sharing

We touch here the key topic of modeling systems where the service facility provided by a single server is shared *dynamically* among customers visiting the system. The ability to serve simultaneously all pending customers by assigning a portion of the service capability to each one motivates the term *sharing*. The word *processor* refers to the service capability itself. This terminology has become commonplace since the first application of this model, which addressed the analysis of computing facilities.

After introducing the classic $M/G/1$ processor sharing model, we outline the generalized processor sharing model. Then we turn to ways to realize in practice those theoretical models in a packet multiplexer, e.g., at the output line of a router in the Internet.

6.4.1 The $M/G/1$ Processor Sharing Model

The processor sharing model is an ordinary $M/G/1$ queue except that the service discipline is special: the server offers its capacity to all awaiting customers, in equal parts. Let μ denote the service rate of the queue, i.e., the reciprocal of the mean service time. When the number of customers in the queue is $Q(t) = n$, the service rate allocated to each customer is μ/n, i.e., the service potential of the server is equally shared by the n customers. Processor sharing can be considered as the limit for $\delta \to 0$ of a round robin scheduler that cycles through customers and gives no more than an amount δ of service per round (service quantum) to each customer in the queue.

Note that with the processor sharing model there is no clear-cut separation between the waiting time and the service time. A customer starts being served as soon as it enters the queue. The customer leaves the queue when its service demand has been completely served. Therefore, an $M/G/1$ processor-sharing server is work-conserving, but there is in general more than a single customer that has been partially served.

Let λ denote the mean customer arrival rate and let $\rho = \lambda/\mu$ be the utilization coefficient of the server. The system is stable, hence it admits a steady state, if and only if $\rho < 1$. A rigorous derivation of the steady-state probability distribution of the $M/G/1$ processor sharing model can be found in many textbooks, e.g., in [86, Ch. 4, p. 221]. Here we offer an informal derivation.

Let $Q(t)$ denote the number of customers in the system at time t, and let $R_i(t)$ be the remaining service time of the i-th customer at time t, $i = 1, \ldots, n$. It can be seen that $(Q(t), R_1(t), \ldots, R_{Q(t)}(t))$ is a Markov process that gives a full description of the $M/G/1$ processor sharing model. For any positive n, let $F_n(y_1, \ldots, y_n)$ be a joint CDF defined as

$$F_n(y_1, \ldots, y_n) = P(Q = n, R_1 \le y_1, \ldots, R_n \le y_n) \tag{6.48}$$

where the random variables $Q, R_1, \ldots, R_n$ are referred to a generic time in the steady state. The corresponding joint probability density function is

$$f_n(y_1, \ldots, y_n) = \frac{\partial^n F_n(y_1, \ldots, y_n)}{\partial y_1 \, \partial y_2 \ldots \partial y_n} \tag{6.49}$$

It can be shown that

$$f_n(y_1, \ldots, y_n) = (1 - \rho)\lambda^n \prod_{i=1}^{n} [1 - F_X(y_i)] \tag{6.50}$$

where $F_X(\cdot)$ is the CDF of the service times. The steady state probability distribution of Q can be found by integrating $f_n(y_1, \ldots, y_n)$ with respect to $y_i, i = 1, \ldots, n$ over $(0, \infty)$. Since $\int_0^\infty [1 - F_X(y_i)]dy_i = 1/\mu$, it follows

$$P(Q = n) = (1 - \rho)\lambda^n \frac{1}{\mu^n} = (1 - \rho)\rho^n , \qquad n \ge 0. \tag{6.51}$$

The striking result is that the probability distribution of the number of customers residing in the $M/G/1$ processor sharing system in steady state is the same as the one of the $M/M/1$ queue, *irrespective of the probability distribution of the service times!*

In the sequel, we derive this result and other special ones by resorting to a non-rigorous yet insightful reasoning. We start by finding the average response time $R(x) \equiv E[S|X = x]$, conditional on the service requirement being x. This is the average of the time that an arriving customer requesting a service time x remains in the system before completing its service requirement.

Let t_a denote the arrival time of a tagged customer and S be its sojourn time into the system. Then, according to the processor sharing model of operation, we have:

$$\int_{t_a}^{t_a+S} \frac{1}{1 + Q(\tau)} \, d\tau = x \tag{6.52}$$

Note that $Q(t_a)$ is the number of customers found in the system by the tagged customer and $Q(t_a + S)$ the number of customers left behind by the tagged customer upon departure. Taking derivatives with respect to x, we get

$$\frac{dS}{dx} \frac{1}{1 + Q(t_a + S)} = 1 \tag{6.53}$$

The random variable $Q(t_a + S)$ is nothing but the random variable $Q_d =$ number of customers left behind by a departing one, that in turn has the same PDF of $Q_a =$ number of customers found by the tagged customer in the queue, and the latter is obviously independent of the service time x. Then, we have $Q(t_a + S) \sim Q(t_a) \sim Q$, where Q denotes a random variable that has the probability distribution of the number of customers found in the system at a general time of the steady state. The last statement is a consequence of Poisson arrivals and of the PASTA property. Then, $1 + Q(t_a + S) \sim 1 + Q$, and it is independent of x and of S. Therefore, integrating eq. (6.53), we have

$$S = x(1 + Q) \qquad \Rightarrow \qquad R(x) = E[S|X = x] = x(1 + E[Q]) \tag{6.54}$$

By removing the conditioning on the service time, we get

$$E[S] = E[X](1 + E[Q]) = \frac{E[Q]}{\lambda} \tag{6.55}$$

where the last equality is just Little's law. Then

$$E[Q] = \frac{\rho}{1 - \rho} \tag{6.56}$$

where $\rho = \lambda/\mu < 1$ to guarantee the existence of statistical equilibrium. Getting back to the conditional sojourn time, we find finally

$$R(x) = \frac{x}{1 - \rho} \tag{6.57}$$

This result illustrates the nice property that the response time of the $M/G/1$ processor sharing system is insensitive to the specific form of the service time PDF, it only depends on the service demand. Moreover, the response time is simply proportional to the service demand of the tagged customer. This is an easy-to-understand form of fairness among customers having different service requirements.

A different derivation of the main result of the processor sharing $M/G/1$ queue is as follows. Let us assume that service is scheduled in quanta of time Δt, that is each of the $Q(t)$ customers in the queue at time t are given an amount of service (up to) Δt per round. Let us consider a tagged customer arriving at time t_0 at the processor sharing queue with a service demand of x seconds, that finds $Q(t_0)$ customers already in the queue. We assume x is an integer multiple of Δt, which is not restrictive in view of the fact that the processor sharing discipline is obtained by letting Δt shrink to 0. Then, the random variable $S(x)$, defined as the system response time of the tagged customer, conditional on the service request being x, is

$$S(x) = \sum_{k=0}^{x/\Delta t - 1} \left[\Delta t + \Delta t\, Q(t_k) \right] = x + \Delta t \sum_{k=0}^{x/\Delta t - 1} Q(t_k) \tag{6.58}$$

By taking expectations of both sides and assuming statistical equilibrium exists, so that $Q(t)$ is a stationary process whose mean is independent of time, we obtain

$$E[S|X = x] = x + x\, E[Q|X = x] \tag{6.59}$$

The random variable Q is independent of the service time of the tagged customer, hence $E[Q|X = x] = E[Q]$. By Little's law, we get $E[Q] = \lambda E[S]$. So, taking expectation with respect to service time on both sides of eq. (6.59), we have

$$E[S] = \frac{E[Q]}{\lambda} = E[X](1 + E[Q]) \quad \Rightarrow \quad E[Q] = \frac{\rho}{1 - \rho} \tag{6.60}$$

Getting back to eq. (6.59) we have $E[Q|X = x] = E[Q]$, hence

$$R(x) = E[S|X = x] = \frac{x}{1 - \rho} \tag{6.61}$$

Processor sharing is a limiting case of round-robin service, where customers in the queueing system get equal chunks of service in turn, until their service demand is met. It is an example of server policy that offers customers *isolation*. Assume customers represent jobs or packet flows. The time to complete service of a customer depends only on the service demand of that customer, i.e., the amount of work that the server should spend on the customer to fulfill its service demand, and on the *average* server load, ρ. It does not depend in any way on the specific values of demands of other competing customers, nor is it affected by the distribution characteristics of other customers' demand. As an example, if a big job is sharing the server facility with a tiny, small job, they get equal server shares so long as they carry on, irrespective of their "mass." Thus, the small job does not get choked by the overwhelming one, suffering a long delay.

6.4.2 Generalized Processor Sharing

The ability of the server to give attention to all pending jobs, without being swamped by long ones, can be biased, by defining the generalized processor sharing (GPS) policy for a single-server queue.

Let the arriving customers be divided into N classes. No a priori priority level is tied to the class concept here. Classes are introduced only to define weights and to be able to associate them to arriving customers. Let ϕ_j be a positive, real number, representing the weight of class $j, j = 1, \dots, N$. Informally, the service rate provided to class j customers is proportional to their weight. If all weights are integer valued, we could roughly say that the server spends up to ϕ_j units of service for customer of class j, moving in turn from class j to class $(j \bmod N) + 1$. In other words, a *weighted round robin* policy is realized. GPS represents the limit of that policy when the duration of the unit of service shrinks to zero.

Let μ denote the service rate of the server. We assume the server is work-conserving.

Let $\overline{\mu}_j[\tau, t]$ be the average service rate obtained by class j customers in the time interval $(\tau, t]$. Traffic class j is said to be *backlogged* at time t, if there is at least one customer belonging to class j that has unfinished work in the queue at time t. We say that the server operates according to the GPS policy, if

$$\frac{\overline{\mu}_j[\tau, t]}{\phi_j} \geq \frac{\overline{\mu}_i[\tau, t]}{\phi_i}, \quad \forall i \tag{6.62}$$

for any class j that is *continuously* backlogged in $(\tau, t]$. Note that the definition implies that equality sign holds for classes that are continuously backlogged in $(\tau, t]$. Then, we see that those classes receive an amount of work that is proportional to the respective weights.

We define the following discrete valued stochastic processes:

$Z_j(t)$ It is equal to 1 if there is at least one customer of class j in the queue; otherwise it is 0.

$Z(t)$ It is equal to 1 if there at least one customer, of whatever class, into the queue; otherwise it is 0.

At least conceptually, we can think of the queueing system as being organized into N (virtual) queues, the j-th of which hosts customers of class j, while the server spends its service effort for the head-of-line customers of all virtual queues simultaneously, according to the weights ϕ_j.

Formally, the serving rate of a customer belonging to class j at time u can be expressed as follows:

$$\mu_j(u) = \frac{\phi_j Z_j(u)}{\sum\limits_{i=1}^{N} \phi_i Z_i(u)} \, \mu \tag{6.63}$$

Here and in the following we interpret as 0 any expression resulting in the ratio 0/0. For example, if only two classes are populated at time u, e.g., 1 and 2, then we have $\mu_1(u) = \frac{\phi_1}{\phi_1 + \phi_2} \mu$, $\mu_2(u) = \frac{\phi_2}{\phi_1 + \phi_2} \mu$ and $\mu_j(u) = 0$, for $j = 3, \dots, N$.

The cumulative service provided to class j over the interval $[0, t]$ is given by

$$\int_0^t \mu_j(u) \, du = \int_0^t \frac{\phi_j Z_j(u)}{\sum\limits_{i=1}^{N} \phi_i Z_i(u)} \, \mu \, du \tag{6.64}$$

The average service rate experienced by customers of class j in the time interval $[0, t]$, $\overline{\mu}_j[0, t]$, is the ratio of the amount of service received by those customers over $[0, t]$ and the overall time that class j has been backlogged in $[0, t]$. Therefore

$$\overline{\mu}_j[0, t] = \frac{\int_0^t \frac{\phi_j Z_j(u)}{\sum\limits_{i=1}^{N} \phi_i Z_i(u)} \, \mu \, du}{\int_0^t Z_j(u) \, du} \tag{6.65}$$

Let us consider a time interval $[0, t]$, where class j is backlogged, i.e., such that $Z_j(u) = 1$ for $u \in [0, t]$. Since $Z_i(u) \leq 1$ for all u, we can write

$$\overline{\mu}_j[0, t] \geq \frac{\int_0^t \frac{\phi_j Z_j(u)}{\sum\limits_{i=1}^{N} \phi_i} \, \mu \, du}{\int_0^t Z_j(u) \, du} = \frac{\phi_j}{\sum\limits_{i=1}^{N} \phi_i} \, \mu \, , \qquad \forall t > 0 \tag{6.66}$$

This inequality states the most important property of GPS. So long as class j is backlogged, it is guaranteed to receive a minimum rate of service, depending on the ratio of its own weight to the sum of all weights. GPS realizes sort of partition of the server capacity among the contending customer classes. It provides also isolation of each class, in that each class is guaranteed to benefit of at least a given share of the server capacity. Obviously, when some class is not backlogged, more capacity can be assigned to the others. In other words, we are assuming that the server is work-conserving. Its entire service capacity is spent to serve customers in the queue, no matter what class they belong to. The policy that dictates how the capacity is shared however yields the minimum guaranteed average rate per class, as given by eq. (6.66).

Example 6.9   Let us consider a traffic flow described by means of the counting function $A(t)$, defined as the amount of workload offered by the arrival stream over the interval $[0, t]$. We assume that the workload offered by the traffic flow is controlled at the source, so that for any $0 \le v < u$ it holds that $A(u) - A(v) \le b + c \cdot (u - v)$. The amount of workload emitted over the time interval $[v, u]$ is bounded above by a linear function. Here c represents the long-term average emission rate of the traffic flow, while b accounts for the allowed maximum burstiness. This linear constraint on the workload arrival function of a traffic flow is a classic way of defining a traffic description for a variable bit rate traffic source (see Section 11.2).

Assume the tagged traffic flow goes through a buffered link of capacity C that can be modeled as a GPS system. The packets of the tagged traffic source form a customer class of the system, say class 1. The weight of class 1 is chosen so that $\frac{\phi_1}{\sum_{i=1}^{N} \phi_i} C = c$, where c is the long-term average rate parameter of the traffic flow constraint. By using Reich's formula (see Section 4.9), since a class 1 customer (a packet of the tagged traffic flow) is guaranteed to receive a serving rate of at least c as long as it is backlogged, we can write the following bound for the amount of workload belonging to the tagged traffic flow and residing into the GPS system at time t:

$$Q(t) = \sup_{0 \le s \le t} \{A(t) - A(s) - c(t - s)\} \tag{6.67}$$

Thanks to the constraint on the offered traffic flow $A(t)$, we can set an upper bound on the queue length. It is $Q(t) \le \sup_{0 \le s \le t} \{b + c(t - s) - c(t - s)\} = b$.

Let $D(t)$ be the amount of workload that leaves the GPS system over the time interval $[0, t]$. The amount of workload of the tagged flow leaving the system in the interval $[v, u]$, $D(u) - D(v)$, can be upper bounded by the sum of the amount of workload arrived over that interval, $A(u) - A(v)$, and the amount of workload $Q(v)$ residing in the system at the start time v of the interval. Then, by using eq. (6.67), we obtain

$$
\begin{aligned}
D(u) - D(v) &\le Q(v) + A(u) - A(v) \\
&\le \sup_{0 \le s \le v} \{A(v) - A(s) - c(v - s)\} + A(u) - A(v) \\
&= \sup_{0 \le s \le v} \{A(u) - A(s) - c(v - s)\} \\
&\le \sup_{0 \le s \le v} \{b + c(u - s) - c(v - s)\} = b + c(u - v)
\end{aligned}
$$

where the last inequality holds by virtue of the constraint on the arrival function.

We thus get a nice result, which is a cornerstone of deterministic traffic analysis (see Chapter 11): a traffic flow, shaped so as to match a linear constraint, maintains this property when going through a buffered multiplexer run according to a GPS policy. This property, extensively discussed by Parekh and Gallager [173], has underpinned the Integrated Services (IntServ) and Differentiated Services (Diff-Serv) traffic engineering approach since the late 1990s.

A useful function tied to the GPS model is *virtual time*. With the same agreement that $0/0$ must be interpreted as 0, we let the virtual time $V(t)$ be defined as follows:

$$V(t) = \int_0^t \frac{\mu Z(u)}{\sum_{i=1}^N \phi_i Z_i(u)} \, du \tag{6.68}$$

This is a non-negative, monotonously nondecreasing function of the "wall-clock" time t. It is $V(0) = 0$ by definition. It is piecewise linear function of time, with a slope $\mu / \sum_i \phi_i$, where the sum is taken over backlogged classes. The slope of $V(t)$ is 0 when the system is empty.

An interesting property of virtual time $V(t)$ is that it identifies the order of service completion of customers of a given class. Let us first consider an ordinary FCFS single server queue and let us introduce some notation.

a_k Time of the k-th arrival.

d_k Departure time of the k-th arriving customer.

L_k Amount of work associated with the k-th arriving customer (e.g., number of bits of a packet, if customers correspond to packets in a communication network and the FCFS server models a link capacity equipped with a FIFO buffer).

μ Serving rate of the FCFS server; it is measured in units of work per unit time. So, the service time of the k-th arriving customer is L_k/μ.

In this simple case, it is easy to establish a connection among arrival and departure times, assuming as usual that the server is work-conserving. Specifically, we have

$$d_{k+1} = \max\{d_k, a_{k+1}\} + \frac{L_{k+1}}{\mu}, \qquad k \geq 0 \tag{6.69}$$

The sequence is initialized with $d_0 = 0$.

The recurrence in eq. (6.69) is easily explained. During a busy period, i.e., when the k-th departing customer leaves behind a nonempty system, the departing time of the next customer occurs just after its service time, which is L_{k+1}/μ. If instead the k-th departing customer leaves behind an empty system, after an idle time there will be a new arrival at time a_{k+1}. Then, the departing time will occur after the service time of such newly arrived customer.

Given the sequence of arrival times, $\{a_k\}_{k\geq 1}$, and that of workload, $\{L_k\}_{k\geq 1}$, the sequence of departing times can be easily calculated. Actually, the calculation of the departing time of a customer can be done upon its arrival, just keeping in memory the last calculated departing time.

The recurrence in eq. (6.69) does not hold any more for a GPS system. Let us introduce some new notation, generalizing the previous one to account for different classes.

$a_k^{(j)}$ Time of the k-th arrival belonging to the j-th class.
$s_k^{(j)}$ Time when the service of the k-th arriving customer of class j starts.
$d_k^{(j)}$ Departure time of the k-th arriving customer of class j.
$L_k^{(j)}$ Amount of work associated with the k-th arriving customer of class j.
μ Serving rate of the GPS server; it is measured in units of work per unit time. So, the service time of the k-th arriving customer of class j is $L_k^{(j)}/\mu$.

We assume that service order within a class is according to FCFS. Then

$$V\left(d_{k+1}^{(j)}\right) = \int_0^{s_{k+1}^{(j)}} \frac{\mu Z(u)}{\displaystyle\sum_{i=1}^N \phi_i Z_i(u)}\, du + \int_{s_{k+1}^{(j)}}^{d_{k+1}^{(j)}} \frac{\mu Z(u)}{\displaystyle\sum_{i=1}^N \phi_i Z_i(u)}\, du$$

$$= \int_0^{s_{k+1}^{(j)}} \frac{\mu Z(u)}{\displaystyle\sum_{i=1}^N \phi_i Z_i(u)}\, du + \frac{1}{\phi_j}\int_{s_{k+1}^{(j)}}^{d_{k+1}^{(j)}} \frac{\phi_j \mu}{\displaystyle\sum_{i=1}^N \phi_i Z_i(u)}\, du$$

$$= V(s_{k+1}^{(j)}) + \frac{L_{k+1}^{(j)}}{\phi_j}$$

The last equality derives from the very definition of the workload $L_{k+1}^{(j)}$ and observing that $Z_j(u) = 1$ (hence also $Z(u) = 1$) for $u \in [s_{k+1}^{(j)}, d_{k+1}^{(j)}]$. Since $\frac{\phi_j}{\sum_{i=1}^N \phi_i Z_i(u)}\mu$ is just the serving rate experienced by the class j customer $k+1$, the integral of such serving rate over the entire time interval when that customer gets served amounts to its workload.

Moreover, it is $s_{k+1}^{(j)} = \max\{d_k^{(j)}, a_{k+1}^{(j)}\}$, since the GPS server is work-conserving, the serving policy within a given class is FCFS, and under GPS an arriving customers starts being served immediately as it arrives, provided no other customer of the same class is in the queue. Remembering that $V(t)$ is monotonously nondecreasing, we have

$$V\left(s_{k+1}^{(j)}\right) = V\left(\max\{d_k^{(j)}, a_{k+1}^{(j)}\}\right) = \max\left\{V\left(d_k^{(j)}\right), V\left(a_{k+1}^{(j)}\right)\right\} \tag{6.70}$$

Putting results together, we find finally

$$V\left(d_{k+1}^{(j)}\right) = \max\left\{V\left(d_k^{(j)}\right), V\left(a_{k+1}^{(j)}\right)\right\} + \frac{L_{k+1}^{(j)}}{\phi_j}; \qquad k \geq 0 \tag{6.71}$$

Here $d_k^{(j)}$ denotes the departing times of k-th arriving customer of class j, so it is different from the departing time that could be calculated according to the FCFS policy, even for the same arrival pattern.

Equation (6.71) exhibits exactly the same pattern as eq. (6.69), except that the virtual time replaces plain time. Therefore, we can use virtual time as a "clock" to sort customers according to their order of departure.

The virtual time function provides us with a tool to associate a sequence of increasing values to customers, according to their order of service completion. According to the recurrence in eq. (6.71), the quantity $V\left(d_{k+1}^{(j)}\right)$ can be calculated upon the arrival of the $(k+1)$-th customer of class j, just maintaining in memory the values of the virtual times associated to the most recent arrivals for each class, i.e., $\left\{V\left(d_{k_i}^{(i)}\right)\right\}_{i=1,\dots,N}$. The importance of eq. (6.71) lies in the fact that it could be used as a timestamp to append on each arriving customer. Then, we can state that the ideal GPS would serve customers so that they would leave the queue according the increasing order of their respective timestamps.

6.4.3 Weighted Fair Queueing

Weighted fair queueing (WFQ) is a practical implementation of GPS, first proposed by Demers, Keshav, and Shenker [66] and further discussed and analyzed, under the name of Packet-by-packet GPS (PGPS), by Parekh and Gallager [173]. PGPS can be considered as the closest approximation to GPS. Its implementation proves somewhat complex. A simpler algorithm, which is less precise, but much easier to realize, was proposed independently by Golestani, under the name of self-clocked fair queueing (SCFQ) [91], and by Roberts, under the name virtual spacing [182]. A preliminary version of this simplified algorithm was first proposed by Davin and Heybey in an early work on fair queueing performed at MIT [64].

Note that WFQ has been conceived in the context of packet multiplexing. However, the scheduling concept can be adapted to any other setting.

We have seen at the end of the previous section that the virtual time function $V(t)$, computed according to the iteration (6.71), provides us with a sequence of numbers, associated to the customers served by the GPS server, which is monotonously increasing and corresponds to the order of service offered by the GPS discipline. For ease of notation, let d_k be the time at which the k-th departing customer will depart (finish service) under GPS. A good approximation of GPS would be a work-conserving server that serves customers in increasing order of d_k. Now, suppose that the server completes the ongoing service at time t, thus being available to serve the next customer, if any. The point is that the next customer to depart under GPS may not have arrived at time t. The server has no knowledge of when this customer could arrive, so there is no way for the server to be both work-conserving and serve the customers in increasing order of d_k. The server is therefore left with the option of picking the first customer that would complete service in the GPS simulation among those currently in the queue.

Summing up, WFQ works as follows:

1. The server keeps track of the virtual time $V(t)$ for $t \geq 0$ (see details of the computation of virtual time later in this section).
2. Upon arrival at time t of customer k belonging to flow i, a *finish tag* $F_{i,k}$ is appended to the customer. The finish tag is computed as follows (compare with eq. (6.71)):

$$F_{i,k} = \max\{F_{i,k-1}, V(t)\} + \frac{L_{i,k}}{\phi_i}, \qquad k \geq 1, \tag{6.72}$$

where $L_{i,k}$ is the amount of work brought in by the k-th customer and $F_{i,0} = 0$.
3. When the server is ready to select a new customer (i.e., it has completed serving the previous customer), it chooses the one with the smallest finish tag among all customers currently in the queue.

Let $L_{\max}$ denote the maximum amount of work brought into the system by a single customer (maximum packet length in case of packet multiplexing). That WFQ is a good approximation of the ideal GPS is proved by the two following theorems. Theorem 6.1 proves that there exists a constant offset between the delay suffered by a customer under the ideal, fluid GPS and its practical implementation WFQ. Theorem 6.2 shows that the amount of service received by class j customers in the time t with WFQ is offset only by a constant quantity with respect to the ideal value achieved under GPS.

Theorem 6.1   Let $\hat{d}_k$ be the departure time of the k-th departing customer out of the WFQ system and let d_k be the departing time of the same customer in the ideal GPS system. It is

$$\hat{d}_k \leq d_k + \frac{L_{\max}}{\mu} \tag{6.73}$$

Proof: We number customers according to the order of departure from the WFQ queue, which is the same as the order of service start for this system (HOL property). Note that this property does not necessarily hold for GPS

Since the server is work-conserving, busy periods coincide in the GPS and WFQ systems. Let us focus on one busy period and let $t = 0$ be the initial time of the busy period.

Let us consider customer k. If $d_j < d_k$ for $j = 1, \ldots, k-1$, all customers numbered from 1 to k have been completely served in the GPS system by time d_k. Then

$$d_k \geq \sum_{j=1}^{k} \frac{L_j}{\mu} = \hat{d}_k \tag{6.74}$$

where the last equality is due to the definition of k-th customer and to the work-conserving property of the server.

If for some j it is $d_j > d_k$, let m denote the largest index in the set $\{1, \ldots, k-1\}$ such that $d_m > d_k$. Then, it is $d_j < d_k$ for $j = m+1, \ldots, k-1$, by definition of m. Let also $\hat{s}_m$ be the time by which customer m starts service in WFQ. We claim that all customers indexed from $m+1$ to k are served in the interval $[\hat{s}_m, d_k]$. This conclusion stems from the two following facts:

1. Customers $m+1, \ldots, k$ must have arrived no earlier than $\hat{s}_m$, otherwise the WFQ system would not have started serving customer m at time $\hat{s}_m$, since the finish tags of customer m is bigger than any of the finish tags of customers $m+1, \ldots, k$.
2. Customers $m+1, \ldots, k$ depart no later than d_k by the definition of m.

Then, we have

$$d_k \geq \hat{s}_m + \sum_{j=m+1}^{k} \frac{L_j}{\mu} \geq \sum_{j=1}^{m-1} \frac{L_j}{\mu} + \sum_{j=m+1}^{k} \frac{L_j}{\mu} = \hat{d}_k - \frac{L_m}{\mu} \tag{6.75}$$

Equations (6.74) and (6.75), along with the inequality $L_m \leq L_{\max}$, complete the proof of the theorem. ∎

Theorem 6.2 Let $S_j(\tau, t)$ and $\hat{S}_j(\tau, t)$ denote the amount of work received by customers of class j in the time interval $[\tau, t]$ in the GPS and WFQ systems, respectively. It is

$$S_j[0, t] - \hat{S}_j[0, t] \leq L_{\max} \tag{6.76}$$

where $t = 0$ marks the beginning of a busy period.

Proof: Let us prove this statement by contradiction. Assume that for some t it is

$$\hat{S}_j[0, t] < S_j[0, t] - L_{\max} \tag{6.77}$$

This implies that at least one customer has been completely served in GPS, that is not in the WFQ system. Let k be the index of that customer and let $\tilde{L}_k$ be the remaining work to be done at time t for that customer in the WFQ system. That work has been carried out in the GPS system, along with an additional amount $\Delta L = L_{\max} - \tilde{L}_k$ of work, due to the assumption (6.77). The additional work ΔL must have been done *after* customer k departure in the GPS system, since service order within a flow is FCFS. Then it must be

$$d_k < t - \frac{L_{\max} - \tilde{L}_k}{\mu} \tag{6.78}$$

On the other hand, departure of customer k in the WFQ system requires at least completing the residual work. Then

$$\hat{d}_k \geq t + \frac{\tilde{L}_k}{\mu} \tag{6.79}$$

The two inequalities imply

$$\hat{d}_k - d_k > t + \frac{\tilde{L}_k}{\mu} - \left(t - \frac{L_{max} - \tilde{L}_k}{\mu}\right) = \frac{L_{max}}{\mu} \tag{6.80}$$

which contradicts Theorem 6.1. ∎

The implementation of WFQ is based on the virtual time function, $V(t)$. By observing arrival times of customers and knowing the amount of work they bring into the system, it is possible to calculate $V(d_k^{(j)})$, where $d_k^{(j)}$ is the departing time of the k-th arriving customer of class j. To this end, we exploit eq. (6.71) and we observe that it is enough to evaluate $V(t)$ for each busy period. Let the starting time of a generic busy period be $t_1 = 0$. Let also t_j, for $j \geq 2$, denote the time epoch of the j-th event occurring at the queue, either a customer arrival or a customer departure. The t_j's are ordered in increasing order[5]. Then, we set $V(0) = 0$. The expression of $V(t)$ for $t > 0$ is

$$V(t_{j-1} + \tau) = V(t_{j-1}) + \frac{\tau}{\sum_{i \in B_j} \phi_i}, \qquad t_{j-1} \leq \tau < t_j, \quad j \geq 2, \tag{6.81}$$

where B_j denotes the set of backlogged traffic classes in the time interval $[t_{j-1}, t_j)$. Given $V(t)$, WFQ works according to steps (1)-(3) listed above.

WFQ requires the (nontrivial) task of computing the virtual time function. It also requires the amount of work of a newly arrived customer to be known. Once those are given, deciding who's next has the complexity of choosing the minimum in a finite set.

The main source of complexity of the WFQ scheduling comes from the necessity to track the virtual time function. This is a piecewise linear function, whose break-points correspond to arrivals and departures. As the frequency of these events grows, the computational complexity of $V(t)$ can become incompatible with the requirement of performing the computation in real time.

Golestani [91] and Roberts [182] proposed an approximation to the virtual time function, to obtain a strong simplification, without giving up to much performance. The practical implementation proposed by Golestani is the SCFQ, while Roberts dubbed his algorithm virtual spacing. Apart from notation, the algorithm defined in Section 3.3 of [182] coincides with SCFQ.

---

5 It is immaterial how ties, if any, are broken.

With reference to a generic busy period of the server, SCFQ is defined by the following algorithm:

1. The k-th arriving customer of class j is tagged with the service tag $F_k^{(j)}$, before it is placed in the queue. The customers in the queue are picked up for service in increasing order of the associated service tags.
2. Let $F(t)$ be the service tag of the customer receiving service at time t. It is initialized at 0 at the start of any new busy period.
3. The service tag of customer k of class j, arriving at time t, is

$$
F_k^{(j)} = \max\{F_{k-1}^{(j)}, F(t)\} + \frac{L_k^{(j)}}{\phi_j} , \qquad k \geq 1, \tag{6.82}
$$

initialized with $F_0^{(j)} = 0$ at the beginning of each busy period. Here $L_k^{(j)}$ denotes the amount of work brought into the system by the k-th arriving customer of class j.

The key point is the second step. Instead of simulating the virtual time function of the theoretical GPS system, the virtual time function is immediately derived by the real system. As a new customer arrives at time t, the service tag of the customer being served is sampled and the sampled value is assigned to $F(t)$. In his work Golestani discusses the performance of SCFQ, showing that the disparity of the amounts of service received by a class under SCFQ and the theoretical GPS is bounded.

6.4.4 Credit-Based Scheduling

Credit-based fair queueing (CBFQ) is a practical implementation of GPS proposed for packet switched network multiplexers [29].

We consider a single server of capacity μ and N classes of customers. Each class is characterized by a weight, ϕ_j for class j, that is related to the share of the server capacity that the class is entitled to. Within each class, the order of service is according to FCFS discipline. Only a single customer partially served can exist at any given time in the system, i.e., preemption is not allowed. Let also L_j denote the amount of work requested by the head-of-line customer of class j (if any). The corresponding service time is $X_j = L_j/\mu$. Finally, we define a counter K_j for class $j, j = 1, \ldots, N$. This counter is used to store the number of serving credits earned by the corresponding class.

Based on the values of the counters, and the sizes of the head-of-line workloads, the algorithm decides which customer is to be served next. In other words, instead of waiting until a class earns enough credits to have its head-of-line customer served, which would result in a waste of server capacity, the class that needs the shortest time to earn enough credits to have a customer served will be served first. Each time a customer is served, the corresponding counter is reset to zero and other classes counters are updated.

Formally, the CBFQ algorithm is stated as follows:

1. Let $t = 0, h = 1$, and $K_j = 0$ for $j = 1, \dots, N$.
2. Let $J(t)$ be the set of classes backlogged at time t, i.e., those that have at least one customer waiting.
3. Let $y_j \equiv \frac{L_j - K_j}{\phi_j}$, $j \in J(t)$ and let j^* denote the index of the minimum among the y_j, $j \in J(t)$.
4. Serve the HOL customer of class j^*, set $X_h = L_{j^*}/\mu$ and update counters as follows:

$$K_j \leftarrow K_j + \phi_j \frac{L_{j^*} - K_{j^*}}{\phi_{j^*}}, \qquad j \in J(t)\backslash\{j^*\} \tag{6.83}$$

and

$$K_{j^*} \leftarrow 0 \tag{6.84}$$

5. Set $t \leftarrow t + X_h$, $h \leftarrow h + 1$, and go back to step 2

We can define a fairness bound as follows. Let $S_j(u, v)$ be the amount of work of class j served in the time interval $[u, v]$. Given a non-negative constant B, we say that the service system is fair if, for any two classes i and j that are constantly backlogged in the time interval $[t_1, t_2]$, we have

$$\left| \frac{S_i(t_1, t_2)}{\phi_i} - \frac{S_j(t_1, t_2)}{\phi_j} \right| \leq B \tag{6.85}$$

The meaning is that classes i and j could possibly receive different (weighted) amounts of service, yet their offset can never exceed a finite, constant bound.

Assume that a single customer service demand is bounded above by a constant $L_{\max}$, which represents the maximum amount of demand that can be serviced on a single scheduling decision. Then, it can be shown that CBFQ can guarantee a fairness bound [29]. Let us go through the proof, which gives insight into the meaning of the credits K_j. To that purpose we establish a Lemma first.

Lemma 6.1 Flow credit counters are limited within the range $0 \leq K_j \leq L_j \leq L_{\max}$ for $j = 1, \dots, N$.

Proof: We prove the result by induction.

Let $K_j(t)$ be the credit count of class j at step t and $L_j(n_j(t))$ the amount of demand of the customer with sequence number $n_j(t)$, which denotes the number of sequence of the HOL customer of class j at time t.

At $t = 0$, we have $K_j(0) = 0$ and $n_j(0) = 1$ for all j. Let

$$Z_j(t) = \frac{L_j(n_j(t)) - K_j(t)}{\phi_j} \tag{6.86}$$

and

$$Z^*(t) = \min\{Z_1(t), \dots, Z_N(t)\} \tag{6.87}$$

We prove by induction that $0 \le Z_j(t) \le L_j(n_j(t))/\phi_j$, which is readily seen to yield the property stated in the Lemma.

The inequalities are true at initialization, since than $Z_j(0) = L_j(1)/\phi_j$, $j = 1, \dots, N$.

The updating equations can be written as follows:

$$Z_j(t+1) = \begin{cases} L_j(n_j(t) + 1)/\phi_j & \text{if class } j \text{ is scheduled,} \\ Z_j(t) - Z^*(t) & \text{otherwise.} \end{cases} \tag{6.88}$$

It is evident that $0 \le Z_j(t+1) \le Z_j(t)$ for all nonscheduled classes. By the induction hypothesis we have then $0 \le Z_j(t+1) \le L_j(n_j(t))/\phi_j$ for nonscheduled flows. The same inequalities hold evidently also for the scheduled flow, which completes the proof. ∎

Lemma 6.2 Let $Z_j(t) = (L_j(n_j(t)) - K_j(t))/\phi_j$. After a scheduling step is completed, for any two backlogged classes i and j it is $Z_i(t+1) - Z_j(t+1) = Z_i(t) - Z_j(t)$ if neither i nor j are scheduled, it is $Z_i(t+1) - Z_j(t+1) = Z_i(t) - Z_j(t) + L_i(n_i(t+1))/\phi_i$ if class i is scheduled.

Proof: If both considered classes are non-scheduled, both $Z_i(t)$ and $Z_j(t)$ are decreased by the same amount. Then, their difference remains constant.

If instead class i is scheduled it is $Z^*(t) = Z_i(t)$ and hence we have $Z_i(t+1) = L_i(n_i(t+1))/\phi_i$, while $Z_j(t+1) = Z_j(t) - Z^*(t) = Z_j(t) - Z_i(t)$. These relationships yield immediately the result stated in the Lemma. ∎

We are now ready to prove the main theorem on the fairness of CBFQ.

Theorem 6.3 Let $J(u, v)$ be the set of classes that are continuously backlogged from scheduling step u to step v. Then, $\forall t_1, t_2$, $\forall i, j \in J(t_1, t_2)$, we have

$$\left| \frac{S_i(t_1, t_2)}{\phi_i} - \frac{S_j(t_1, t_2)}{\phi_j} \right| \le \frac{L_{\max}}{\phi_i} + \frac{L_{\max}}{\phi_j} \tag{6.89}$$

where $S_k(t_1, t_2)$ is the overall amount of demand of class k that has been served in the interval $[t_1, t_2]$.

Proof: Let

$$B_j(t) = \begin{cases} L_j(n_j(t)) & \text{if class } j \text{ is scheduled at step } t, \\ 0 & \text{otherwise.} \end{cases} \tag{6.90}$$

The total amount of service received by class j in the interval $[t_1, t_2]$ is therefore

$$S_j(t_1, t_2) = \sum_{t=t_1}^{t_2} B_j(t) \tag{6.91}$$

According to the CBFQ algorithm, we have for a flow backlogged at step t:

$$K_j(t+1) = \begin{cases} K_j(t) + \phi_j Z^*(t) & \text{if a flow other than } j \text{ is scheduled,} \\ 0 & \text{if flow } j \text{ is scheduled,} \end{cases}$$

$$= \begin{cases} K_j(t) + \phi_j Z^*(t) & \text{if a flow other than } j \text{ is scheduled,} \\ K_j(t) + \phi_j Z^*(t) - L_j(n_j(t)) & \text{if flow } j \text{ is scheduled,} \end{cases}$$

$$= K_j(t) + \phi_j Z^*(t) - B_j(t)$$

Summing over t, we get

$$K_j(t_2) = K_j(t_1) + \phi_j \sum_{t=t_1}^{t_2} Z^*(t) - S(t_1, t_2) \tag{6.92}$$

Note that the equality sign holds only if class j is continuously backlogged in $[t_1, t_2]$. Therefore

$$\left| \frac{S_i(t_1, t_2)}{\phi_i} - \frac{S_j(t_1, t_2)}{\phi_j} \right| = \left| \frac{K_i(t_1) - K_i(t_2)}{\phi_i} - \frac{K_j(t_1) - K_j(t_2)}{\phi_j} \right|$$

$$\leq \left| \frac{K_i(t_1) - K_i(t_2)}{\phi_i} \right| + \left| \frac{K_j(t_1) - K_j(t_2)}{\phi_j} \right|$$

Thanks to Lemma 6.1, we have $|K_i(t_1) - K_i(t_2)| \leq L_{\max}$, which completes the proof. ∎

We can give also a bound for Jain's fairness index [113]. Let $x_1, \ldots, x_N$ be the amount of resource allocated to users in a shared resource system with N users. Jain's fairness index is defined as follows:

$$F = \frac{\left(\sum_{i=1}^{N} x_i \right)^2}{N \sum_{i=1}^{N} x_i^2} \tag{6.93}$$

It is easy to verify that $0 < F \leq 1$, with $F = 1$ if and only if $x_1 = \cdots = x_N$. We apply the index to the ratios $x_i = S_i(t_1, t_2)/\phi_i$, $i = 1, \ldots, N$. We have

$$0 \leq \sum_{i=1}^{N} \sum_{j=1}^{N} (x_i - x_j)^2 = 2N \sum_{i=1}^{N} x_i^2 - 2 \sum_{i=1}^{N} \sum_{j=1}^{N} x_i x_j = 2N \sum_{i=1}^{N} x_i^2 - 2 \left(\sum_{i=1}^{N} x_i \right)^2 \tag{6.94}$$

From Theorem 6.3 we have:

$$(x_i - x_j)^2 = \left(\frac{S_i(t_1, t_2)}{\phi_i} - \frac{S_j(t_1, t_2)}{\phi_j} \right)^2 \leq L_{max}^2 \left(\frac{1}{\phi_i} + \frac{1}{\phi_j} \right)^2 \leq 4\frac{L_{max}^2}{\phi_{min}^2} \quad (6.95)$$

where ϕ_{min} is the minimum among the weights ϕ_i, $i = 1, \dots, N$. From the last two relationships we obtain:

$$0 \leq 2N \sum_{i=1}^{N} x_i^2 - 2\left(\sum_{i=1}^{N} x_i \right)^2 \leq 4N^2 \frac{L_{max}^2}{\phi_{min}^2} \quad (6.96)$$

whence, letting $Y = \sum_{i=1}^{N} x_i$,

$$F \geq \frac{Y^2}{Y^2 + \frac{2N^2 L_{max}^2}{\phi_{min}^2}} \quad (6.97)$$

The bound is an increasing function of Y, so that we can strengthen it by finding a lower bound of Y. We have

$$Y = \sum_{i=1}^{N} x_i = \sum_{i=1}^{N} \frac{S_i(t_1, t_2)}{\phi_i} \geq \frac{1}{\phi_{max}} \sum_{i=1}^{N} S_i(t_1, t_2) = \frac{(t_2 - t_1)\mu}{\phi_{max}} \quad (6.98)$$

The last equality is a consequence of the fact that customer classes are continuously backlogged during the interval $[t_1, t_2]$ and the server is work-conserving. Finally, we have

$$F \geq \frac{(t_2 - t_1)^2 \mu^2}{(t_2 - t_1)^2 \mu^2 + 2N^2 L_{max}^2 \left(\frac{\phi_{max}}{\phi_{min}} \right)^2} \quad (6.99)$$

We see that Jain's fairness index converges to 1 as time goes by (i.e., as $t_2 - t_1 \to \infty$). This proves that CBFQ is long-term fair among continuously backlogged traffic flows, in the sense of Jain's index. Note that fairness refers to the ratios $S_i(t_1, t_2)/\phi_i$, that is to say, it is weighted.

6.4.5 Deficit Round Robin Scheduling

Deficit round robin (DRR) [190] is yet another form of weighted round robin scheduling algorithm. It has been introduced to share capacity among backlogged traffic flows, accounting for the actual amount of bytes each flow sends out of its queue at each link usage opportunity. A pseudo-code of the DRR algorithm is given in the following algorithm.

The algorithm assumes N flows share a link for sending packets. A service quantum Q_i is defined for flow i, representing the maximum amount of bytes the i-th flow can send at each opportunity. A list of backlogged flows is maintained, *ActiveList*. It contains the indexes of backlogged flows only. A nonbacklogged flow

---

Algorithm Pseudo-code of DRR: dequeue algorithm.

1: *ActiveList* ← List of backlogged flows
2: $DC_i = 0$, $i = 1, \ldots, N$
3: **while** *ActiveList* $\neq \emptyset$ **do**
4:     $i \leftarrow$ extract_head(*ActiveList*)
5:     $DC_i \leftarrow DC_i + Q_i$
6:     *done* ← FALSE
7:     **while** $(DC_i > 0)$**and** $(done == \text{FALSE})$ **do**
8:       $pkt =$ dequeue(*queue$_i$*)
9:       $L \leftarrow pkt.size$
10:       **if** $L \leq DC_i$ **then**
11:         $DC_i \leftarrow DC_i - L$
12:         send(*pkt*)
13:         **if** is_empty(*queue$_i$*) == TRUE **then**
14:           *done* ← TRUE
15:         **end if**
16:       **else**
17:         *done* ← TRUE
18:       **end if**
19:     **end while**
20:     **if** is_empty(*queue$_i$*) == TRUE **then**
21:       $DC_i \leftarrow 0$
22:     **else**
23:       *ActiveList* ← *ActiveList* $\bigcup \{i\}$
24:     **end if**
25: **end while**

---

is appended at the end of the list when new packets arrive at its buffer and hence it becomes backlogged again. If, after having being served, a flow buffer becomes empty, the flow is taken out of the *ActiveList*. If instead, the flow buffer has still packets waiting after having exhausted the service quantum, the corresponding flow is moved to the end of the *ActiveList*.

The deficit count variable DC is initialized to 0. It is incremented by Q_i at each opportunity of the i-the flow and decremented by the length of packets of the i-th flow that get transmitted at that opportunity. If any leftover arises, it is summed to the byte quantum at the next opportunity. The deficit count is reset if the queue has been emptied.

The algorithm can be adapted to realize a fair sharing of usage time of the link, rather than byte counts. It suffices to substitute the byte quantum with a time quantum and the packet length with the packet time.

6.4.6 Least Attained Service Scheduling

Round-robin and generalized processor sharing policies aim also at protecting short jobs from long ones, that could hog the server capacity for a long time and thus penalize the delay performance of possibly several other jobs. We might ask ourselves what is the most discriminatory service policy to protect and give advantage to short jobs. The answer is simple: a policy that shares the full capacity of the server among those customers that have so far received the least amount of service. Such a policy is called *least attained service* (LAS), also known as *foreground-background* (FB) .

The policy can be described by defining a quantum of service q. There are N queues, each one run according to FCFS discipline. The server gives its attention to customers in queue k only if queues from 1 up to $k-1$ are empty. The service given to the HOL customer of the served queue consists of one quantum. A new arriving customer joins queue 1, where it eventually receives one quantum of service. If the customer is done, it leaves the system. Otherwise, the customer joins queue 2. In general, after having received one quantum of service through the queue k, the customer leaves the system if its work demand has been met, i.e., if the work demand is no more than kq. Otherwise, the customer joins the subsequent queue $k+1$. When at the last queue, the N-th one, the customer rejoins the same queue, if its service has not been completed yet.

The simpler version of this scheduling is obtained with $N = 2$. In that case, jobs in queue 1 are considered to be in foreground. They are served with priority over the jobs enqueued in queue 2, which is deemed to be the background queue. Hence the FB name used for this scheduling policy.

In the following we consider a processor sharing model of LAS. This can be thought of as the limit for $N \to \infty$ and $q \to 0$ of the N queue system with quantized service described above. To gain insight into the way (processor sharing) LAS works, let us give a small example. Consider a customer arriving at time t_1 at a server with capacity μ, requiring a service of X_1. After an interval of time Δ_1 customer 2 arrives, requiring a service amount X_2. In the meantime, customer 1 has already received Δ_1 seconds of service time . The customer having so far received the least attention by the server is then customer 2, who has zero achieved service time. Then, customer 2 is assigned the full capacity of the service and customer 1 is set in stand-by for a time Δ_1. If no other customer arrives by then, customers 1 and 2 have equalized the amount of service they have received. Then, they can carry on by sharing equally the server capacity, until either they complete service or new customers arrive. The basic behavior of this policy envisages that a customer in "foreground", i.e., being served (customer 1 in the time interval $[t_1, t_1 + \Delta_1)$ in our example) can be temporarily moved in "background," i.e., not being served (customer 1 in the time interval $[t_1 + \Delta_1, t_1 + 2\Delta_1)$ in our example).

Let us now study the processor sharing LAS model. We will derive the mean system response time conditional on the service time demand of a tagged arriving customer. The derivation holds for an $M/G/1$ type of queue, i.e., arrivals follow a Poisson process with mean rate λ and service times are distributed according to a renewal process with CDF $F_X(x)$. When a tagged customer requiring x seconds of work joins the queue, it has received 0 service time, hence the server will devote its attention to that customer immediately. To all purposes, any customer that has already received up to x seconds of service, while the tagged customer is sojourning in the queue, is nonexistent from the point of view of the tagged customer. In other words, service demand exceeding x "disappears" effectively for the tagged customer. We can therefore define a modified service time CDF, denoted with $F_X(u; x)$, as follows:

$$F_X(u; x) = \begin{cases} F_X(u) & u < x \\ 1 & u \geq x \end{cases} \tag{6.100}$$

Service times less than x have the same probability distribution as the original one. Those exceeding x are replaced instead by simply x and given a mass probability equal to $1 - F_X(x) = P(X > x)$. For ease of notation, we let $m_1(x)$, $m_2(x)$ and $\sigma(x)$ denote the mean, second moment and standard deviation of the random variable $X_x \equiv \min\{x, X\}$, representing the modified service time with CDF $F_X(\cdot; x)$. We have

$$m_h(x) = \int_0^x t^h \, dF_X(t) + x^h[1 - F_X(x)], \qquad h = 1, 2. \tag{6.101}$$

We assume the queueing system is in equilibrium, which is the case if $\rho = \lambda E[X] < 1$, where λ is the mean arrival rate of customers at the queue. The tagged customer finds an average amount of "effective" workload in the queue equal to

$$E[W(x)] = \frac{\lambda m_2(x)}{2[1 - \rho(x)]} \tag{6.102}$$

where $\rho(x) = \lambda m_1(x)$. This is the workload of a regular $M/G/1$ queueing system with the modified service time probability distribution. Note that the mean amount of workload in a single server queueing system in equilibrium, with a work-conserving server, is independent of the service discipline. It depends only on arrivals and service demand, not on the service order. All of the workload expressed by eq. (6.102) must be completed before the tagged customer can leave the queueing system. Then, the mean conditional response time $R(x)$ of the tagged customer, i.e., the mean value of the time that the tagged customer with service demand x spends into the system before leaving, is the sum of $E[W(x)]$ and the service time x itself. We need to add still one more contribution, namely the amount of service carried out for customers joining the queue *after* the tagged customer. The mean number of such customers is $\lambda R(x)$. Each of them brings an "effective" work drawn from the modified CDF $F_X(u; x)$. Therefore, the mean amount of work due to customers arrived after the tagged one is $\lambda R(x) m_1(x) = R(x)\rho(x)$. Summing up, we get

$$R(x) = E[W(x)] + x + R(x)\rho(x) \quad \Rightarrow \quad R(x) = \frac{E[W(x)] + x}{1 - \rho(x)} \tag{6.103}$$

More explicitly, we can write

$$R(x) = \frac{\lambda m_2(x)}{2[1 - \lambda m_1(x)]^2} + \frac{x}{1 - \lambda m_1(x)} \tag{6.104}$$

The unconditional mean response time is obtained simply as $\overline{R} = \int_0^\infty R(x)\,dF_X(x)$.

It is possible to derive the Laplace transform of the PDF of the conditional response time as well. Let $W(x) = w$. The amount of workload to be worked out before the tagged customer leaves the queue is then $w + x$ plus the amount of work due to customers arriving after the tagged one. Those customers are $N(w + x)$, where $N(t)$ denotes the number of arrivals according to the Poisson process of mean rate λ. Each arriving customer gives rise to a busy period of customers characterized by the 'effective' service time PDF. Let $\varphi_Y(s; x)$ denote the Laplace transform of the PDF of the busy period $Y(x)$ conditional on the service time x of the tagged customer. Let also $\varphi_X(s; x)$ be the analogous Laplace transform for the PDF of the effective service times, i.e.,

$$\varphi_X(s; x) = \int_0^x e^{-su}dF_X(u) + e^{-sx}[1 - F_X(x)] \tag{6.105}$$

These two transforms are related as established for the ordinary $M/G/1$ queue:

$$\varphi_Y(s; x) = \varphi_X(s + \lambda - \lambda\varphi_Y(s; x) \, ; \, x) \tag{6.106}$$

Then

$$E\left[e^{-sR(x)} | W(x) = w\right] = e^{-s(w+x)} E\left[e^{-s \sum\limits_{j=1}^{N(w+x)} Y_j(x)}\right]$$

$$= e^{-s(w+x)} \sum_{n=0}^\infty \frac{[\lambda(w+x)]^n}{n!} e^{-\lambda(w+x)} \prod_{j=1}^n E[e^{-sY_j(x)}]$$

$$= e^{-s(w+x)} e^{-\lambda(w+x)} \sum_{n=0}^\infty \frac{[\lambda(w+x)\varphi_Y(s; x)]^n}{n!}$$

$$= e^{-(w+x)[s+\lambda-\lambda\varphi_Y(s;x)]}$$

By removing the condition on $W(x)$, we get finally

$$\varphi_R(s;x) = \varphi_W(s + \lambda - \lambda\varphi_Y(s;x);x) \, e^{-x[s+\lambda-\lambda\varphi_Y(s;x)]} \tag{6.107}$$

The Laplace transform of the PDF of $W(x)$ is the same as for the ordinary $M/G/1$ queue, given that the workload is not affected by the queueing discipline, for a work-conserving server. Then

$$\varphi_W(s;x) = \frac{s[1 - \rho(x)]}{s - \lambda + \lambda\varphi_X(s;x)} \tag{6.108}$$

Wrapping up, we get:

$$\varphi_R(s;x) = \frac{[1 - \rho(x)][s + \lambda - \lambda\varphi_Y(s;x)] \, e^{-x(s+\lambda-\lambda\varphi_Y(s;x))}}{s + \lambda - \lambda\varphi_Y(s;x) - \lambda + \lambda\varphi_X(s + \lambda - \lambda\varphi_Y(s;x);x)}$$

$$= \frac{1 - \rho(x)}{s} \, [s + \lambda - \lambda\varphi_Y(s;x)] \, e^{-x(s+\lambda-\lambda\varphi_Y(s;x))}$$

where we have used the identity $\varphi_Y(s;x) = \varphi_X(s + \lambda - \lambda\varphi_Y(s;x);x)$.

Example 6.10 Let us assume a Pareto PDF for the service time: $F_X(x) = 1 - (\theta/x)^\alpha$, for $x \geq \theta > 0$. The first two moments are finite provided $\alpha > 2$. In this example we set $\alpha = 2.5$ and $E[X] = 1$, so that $\theta = E[X](1 - 1/\alpha) = 0.6$.

Simple algebra yields

$$m_h(x) = \frac{\theta^h}{\alpha - h} \left[\alpha - h \left(\frac{\theta}{x} \right)^{\alpha-h} \right], \qquad h = 1, 2. \tag{6.109}$$

By using eq. (6.104) and the expressions above, we can evaluate numerically the conditional response time as a function of x. Figure 6.6 plots $R(x)$ for three values of $\lambda = \rho/E[X]$. The conditional response time turns out to be increasing in a concave way. The rate of increase gets bigger as the utilization coefficient grows.

Figure 6.6 Mean conditional response time of the $M/G/1$ queue with LAS scheduling as a function of the service demand of the tagged customer for three values of the utilization factor of the queue.

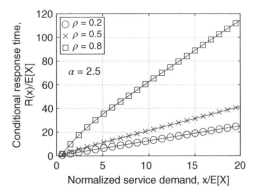

As the server becomes more and more busy, the discriminatory power of the LAS discipline is stronger.

6.5 Miscellaneous Scheduling

In this section we review two popular applications of scheduling algorithms, that have received a large attention in the technical literature. First we discuss scheduling of radio resources in the downlink of a cellular access network. Then we turn to job dispatching in a cluster of computing servers.

6.5.1 Scheduling on a Radio Link

We give an example of scheduling algorithm applied to wireless cellular networks. There is a vast literature and several different approaches. We consider the downlink of a single cell where radio resources are assigned to a number of contending flows according to an orthogonal multiplexing principle, with the aim to maximize throughput (sum-rate), under fairness and power budget constraints.

6.5.1.1 Proportional Fairness

Given a set of N contending flows that share a link (the wireless downlink of a cellular base station (BS) in our case), a rate allocation R_i, $i = 1, \dots, N$ is said to be proportionally fair if for any other allocation $\tilde{R}_i$, $i = 1, \dots, N$ we have

$$\sum_{i=1}^{N} \frac{\tilde{R}_i - R_i}{R_i} \leq 0 \tag{6.110}$$

i.e., the sum of the relative increments of flow rates is non positive. It can be shown that proportional fair (PF) allocation maximizes $\sum_i \log(R_i)$ under the constraint imposed by the overall capacity of the shared link. With reference to the downlink of a BS, finding proportionally fair rates for the N contending flows, under a power budget P, can be stated as the following optimization problem:

$$\max \sum_{k=1}^{N} \log R_k$$

$$[R_1, \dots, R_N] \in C(P)$$

$$R_k \geq 0, \quad k = 1, \dots, N,$$

where $R_k = \mathrm{E}[r_k(t)], k = 1, \dots, N$, are the long-term average rates, $(r_1(t), \dots, r_N(t)) \in C(t, P)$ are the instantaneous rates at time t, $C(t, P)$ is the capacity region at time t (determined by the power constraints and the current channel state), $C(P)$ is the ergodic capacity region.

It can be shown [141] that a scheduler is proportionally fair if the instantaneous rates $\{r_1(t), \ldots, r_N(t)\}$ maximize the following weighted sum:

$$\sum_{k=1}^{N} \frac{r_k(t)}{R_k} \tag{6.112}$$

Hence, our goal is to maximize the weighted sum rate $\sum_k w_k r_k$, where the weight w_k assigned to flow k is an estimate of $1/E[r_k(t)]$. This is an example of a max-weight scheduler.

In the following we estimate R_k in frame $t + 1$ by means of a simple exponentially weighted smoothing to deal with a time varying channel, i.e.,

$$\hat{R}_k(t + 1) = \beta \hat{R}_k(t) + (1 - \beta) r_k(t) \tag{6.113}$$

where $r_k(t)$ is the rate assigned in frame t to the k-th flow and $\hat{R}_k(\cdot)$ denotes the estimate of the long term average rate of the k-th flow.

So, the optimization problem solved in each allocation interval is of the general form:

$$\max_{\mathbf{p}} \sum_{k=1}^{N} \frac{r_k}{\hat{R}_k} \tag{6.114a}$$

$$[r_1, \ldots, r_N] \in C(t, \mathbf{p}) \tag{6.114b}$$

$$\sum_{k=1}^{N} p_k \leq P \tag{6.114c}$$

$$r_k \geq 0, \quad k = 1, \ldots, N, \tag{6.114d}$$

where $\mathbf{p} = [p_1, \ldots, p_N]$ are the transmission powers. Equation (6.114b) is the communication channel capacity constraint, while (6.114c) states the constraint on the power budget of the BS.

6.5.1.2 Multi-rate Orthogonal Multiplexing

We focus on orthogonal multiplexing. We assume that time is slotted and the frequency band is divided into sub-bands (e.g., this is the multiple access scheme of LTE). The resource unit made up of a single sub-band for a single time slot is called a resource block (RB). An RB can be assigned to at most a single flow at a time (orthogonal multiplexing). We assume an ideal link adaptation mechanism where the capacity of an RB is a function of the link signal-to-noise ratio (SNR) according to the Hartley-Shannon law. Let K be the number of RBs available per time slot. RB allocation decisions are taken frame by frame, where a frame is made up of a fixed number of time slots.

The rate assignment to contending flows in time slot t is the solution of the following optimization problem:

$$\max_{\mathbf{x}, \mathbf{p}} \sum_{k=1}^{N} w_k r_k$$

$$r_k = \sum_{j=1}^{K} x_k(j) \log_2 \left(1 + p_k(j)\alpha_k(j)\right) \qquad k = 1, \ldots, N$$

$$\sum_{k=1}^{N} x_k(j) \leq 1 \qquad j = 1, \ldots, K$$

$$\sum_{k=1}^{N} \sum_{j=1}^{K} x_k(j)p_k(j) \leq P$$

$$x_k(j) \in \{0, 1\} \quad k = 1, \ldots, N \quad j = 1, \ldots, K \qquad (6.115)$$

The binary variable $x_k(j)$ determines the allocation of the j-th RB to flow k. The coefficient $\alpha_k(j)$ is proportional to the path gain of flow k in RB j. Given a flow allocation vector $\mathbf{x}$, the Lagrangian associated to problem (6.115) is:

$$L_{\mathbf{x}}(\mathbf{p}, \mu) = \sum_{j=1}^{K} \sum_{k=1}^{N} x_k(j)w_k \log_2(1 + \alpha_k(j)p_k(j)) - \mu \sum_{j=1}^{K} \sum_{k=1}^{N} x_k(j)p_k(j) \qquad (6.116)$$

For a given RB allocation $\mathbf{x}$, the Lagrangian is a function of $\mathbf{p}$, with a feasible domain for $\mathbf{p}$ that is closed, bounded and convex. Moreover, the Hessian matrix of the Lagrangian in strictly negative definite. Then, there is a unique maximizer $\mathbf{p}^*$ of the Lagrangian, that can be determined by setting the gradient of the Lagrangian to 0.

The resulting optimal vector $\mathbf{p}^*$ is calculated according to the water-filling formula:

$$p_k^*(j) = \max \left\{ 0, \frac{w_k}{\mu} - \frac{1}{\alpha_k(j)} \right\}, \qquad j = 1, \ldots, K. \qquad (6.117)$$

Then, we have

$$L_{\mathbf{x}}(\mathbf{p}^*, \mu) = \sum_{j=1}^{K} \sum_{k=1}^{N} x_k(j) \left[w_k \log_2(1 + \alpha_k(j)p_k^*(j)) - \mu p_k^*(j) \right]$$

$$= \sum_{j=1}^{K} \sum_{k=1}^{N} x_j(k)w_k f\left(\frac{w_k \alpha_k(j)}{\mu} \right) \leq \sum_{j=1}^{K} \max_k w_k f\left(\frac{w_k \alpha_k(j)}{\mu} \right) \qquad (6.118)$$

where $f(z)$ is defined as follows:

$$f(z) = \begin{cases} 0 & 0 \leq z \leq 1, \\ \log_2 z - 1 + 1/z & z \geq 1. \end{cases} \qquad (6.119)$$

Let k_j^* denote the flow index that maximizes the j-th term appearing inside the sum of eq. (6.118). Formally, for $j = 1, \ldots, K$, we have

$$
k_j^* =
\begin{cases}
\arg\max\limits_{k} w_k f\left(\dfrac{w_k \alpha_k(j)}{\mu}\right) & \exists k \in (1, \ldots, N) : \dfrac{w_k \alpha_k(j)}{\mu} > 1, \\
0 & \text{otherwise.}
\end{cases}
\tag{6.120}
$$

The multiplier μ is computed from

$$
\sum_{j:\,k_j^* > 0} \max\left\{0, \frac{w_{k_j^*}}{\mu} - \frac{1}{\alpha_{k_j^*}(j)}\right\} = P
\tag{6.121}
$$

Equations (6.120) and (6.121) yield the maximum of the Lagrangian under the power constraint, hence the optimal solution to problem (6.115). The j-th RB is assigned to flow k_j^*, if $k_j^* > 0$, or it is left unused otherwise. That is to say, in case $k_j^* > 0$, the optimal assignment is such that $x_j(k_j^*) = 1$ and $x_j(k) = 0$, $k \neq k_j^*$, for $j = 1, \cdots, K$. The value of μ solving both eqs. (6.120) and (6.121) lies in the range $(0, M)$, where $M = \max_j \max_k w_k \alpha_k(j)$ and it can be found, e.g., by means of a simple bisection algorithm, based on the fact that the left hand side of (6.121) is a monotonously decreasing function of μ for $\mu > 0$.

As a special case, the flat fading channel, where a single radio resource is to be assigned as a whole ($K = 1$), yields a very simple result. By dropping the sub-band index j, the problem statement can be given as

$$
k^*(t) = \arg\max_{k} w_k(t) \log_2\left(1 + \alpha_k(t)P\right)
\tag{6.122}
$$

$$
r_k(t) =
\begin{cases}
\log_2\left(1 + \alpha_k(t)P\right) & k = k^*(t), \\
0 & k \neq k^*(t).
\end{cases}
\tag{6.123}
$$

$$
w_k(t+1) = \frac{w_k(t)}{\beta + (1 - \beta)w_k(t)r_k(t)}, \quad k = 1, \ldots, N
\tag{6.124}
$$

where we explicitly introduce the time index t referring to the t-th allocation interval. In the flat fading case, orthogonal access implies flows are given capacity in turn. The time dependent weights are updated according to the obtained rate.

The resource allocation driven by the PF allocation criterion can be compared with a Maximum Throughput (MT) algorithm, based on the same snapshot optimized resource allocation frame by frame except weights are all equal (set to 1). In the general case of frequency selective channel ($K > 1$), the sub-band allocation and power level setting as a solution of the sum rate optimization with sum power constraint can be easily obtained. The sub-band j goes to the user k_j with the biggest channel coefficient in that sub-band, i.e., $\alpha_{k_j}(j) \geq \alpha_k(j), \forall k$. Once all sub-bands are assigned, power levels are determined according to the water-filling principle with power constraint P and channel coefficients $\alpha_{k_j}(j), j = 1, \ldots, K$. Formally, the MT resource assignment for a generic allocation frame is

$$k_j^* = \arg\max_k \alpha_k(j)$$

$$p(j; \mu) = \max\left\{ 0, \frac{1}{\mu} - \frac{1}{\alpha_{k_j^*}(j)} \right\}, \qquad j = 1, \dots, K$$

$$\sum_{j=1}^{K} p(j; \mu) = P$$

Example 6.11 Let us consider a BS serving N user terminals. In this numerical example, we set the frame duration T_f equal to 20 time slots and the slot time to 0.5 ms, hence it is $T_f = 10$ ms. Resource allocation decisions are taken once every frame.

We assume a flat fading channel ($K = 1$), with time varying gain frame by frame. The radio channel to a single user is modulated by a two state Markov chain $Z(t)$. State 1 corresponds to a bad channel (high attenuation), while state 2 represents a good channel (low attenuation). The SNR of the i-th flow in frame t is given by

$$SNR_i(t) = G_f(t)G_s(Z(t))\frac{\kappa}{d^\alpha}\frac{P_{tx}}{P_N} = G_f(t)G_s(Z(t))SNR_0 \tag{6.125}$$

where $SNR_0 = 24.78$ dB, $G_f(t) \sim \text{Exp}(1)$ is the fast fading gain component, independently drawn frame by frame as a negative exponential random variable of mean 1, and $G_s(x)$ is the shadowing gain in state x of the Markov chain $Z(t)$. We set $G_s(1) = -15$ dB and $G_s(2) = 0$ dB. The mean time spent in state 1 is 20 frames, the mean time in state 2 is 30 frames. The base SNR_0 corresponds to a user terminal at 100 m from the BS, background thermal noise over 1 MHz bandwidth with noise figure 10 dB, and 10 W transmitted power of the BS.

The capacity realized by a user terminal in a frame is expressed by the Hartley-Shannon law:

$$C_i(t) = C_0 \log_2\left(1 + \frac{SNR_i(t)}{\Gamma}\right) \tag{6.126}$$

where $\Gamma = 9.78$ is the gap factor and $C_0 = 1.08$ $Mbit/s$. The role of the term $\alpha_k(t)P$ for flow k in eq. (6.123) is played here by the ratio $SNR_k(t)/\Gamma$.

The weights are initialized to 1 and the algorithm described by eqs. (6.122), (6.123), (6.124) is used. For comparison purposes, we consider two other allocation mechanisms: (i) max throughput, which is obtained from proportional fair allocation by simply setting all weights equal to 1; (ii) round robin, which is obtained by assigning the radio resource to each user terminal cyclically, independently of the radio channel quality.

Figure 6.7(a) shows the achieved average throughput as a function of the number N of contending flows. The throughput is obtained as the sum of the long-term

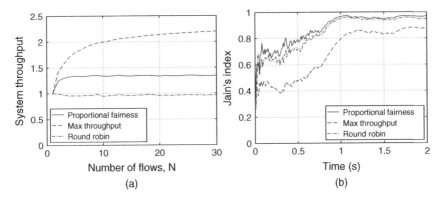

Figure 6.7   Proportional fair radio resource allocation in a cellular downlink. Left plot: average throughput as a function of the number of contending flows. Right plot: evolution over time of Jain's index of fairness.

average throughputs of all flows, normalized to the overall throughput realized in case of a single user terminal. Note that the average throughput achieved by each flow is $1/N$-th of the overall throughput.

Round robin performance do not vary with N, hence the throughput per flow is inversely proportional to N. This is not surprising, since round robin does not exploit any diversity gain. On the contrary, max throughput takes the maximum advantage of the diversity gain. Proportional fair allocation obtains an intermediate performance. Since it imposes a fair allocation in the long term, it loses some of the diversity gain. Yet its overall throughput grows, at least for lower values of N.

The other end of the story is shown in Figure 6.7(b), where Jain's fairness index of the transmitted data is plotted versus time. Let $x_i(t)$ be the amount of bits transmitted by flow i up to frame t. We define the Jain index as

$$J(t) = \frac{\left(\sum\limits_{i=1}^{N} x_i(t) \right)^2}{N \sum\limits_{i=1}^{N} x_i^2(t)} \tag{6.127}$$

This index ranges between 0 and 1, the closer to 1, the better from fairness point of view. It is apparent from Figure 6.7(b) that max throughput achieves a high overall throughput at the expense of fairness, especially in the short-term. On the contrary, round robin and proportional fair allocation schemes are both fair, except a short initial transient. Proportional fair is even fairer than round robin in this plot, even though it achieves a much better overall throughput.

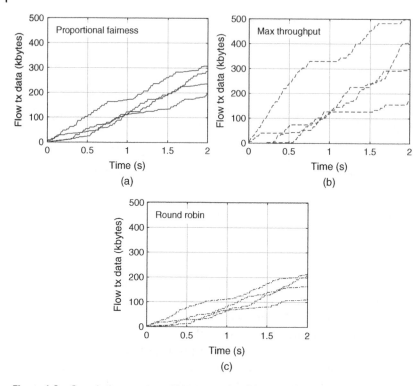

Figure 6.8 Cumulative number of bits transmitted by each flow in a cellular downlink with $N = 4$ flows. (a): Proportional fairness. (b): Max throughput. (c): Round robin.

The behavior of the cumulative amount of data transmitted by each flow over time is plotted in Figure 6.8 for $N = 4$ flows, comparing the three considered scheduling algorithms. The spreading of the curves in case of max throughput illustrates how this algorithm achieves better throughput, i.e., essentially favoring the user terminal that experiences the best radio channel quality, but penalizing those experiencing a bad channel.

6.5.2 Job Dispatching

Computing systems are made up of a large number of servers. Jobs offered to a computing cluster must be dispatched to servers for execution. The typical architecture envisages a number of load balancers connected to the server cluster. Load balancers are the front end versus user requests. They collect job request and distribute them over servers.

Figure 6.9   Scheme of a load balancer
feeding a cluster of servers.

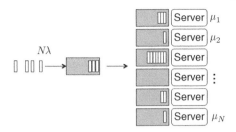

A scheme of a prototypical system of this kind is shown in Figure 6.9, where a single load balancer acts as a job dispatcher toward N servers.

The aim of the job dispatcher is to minimize the overall response time to jobs. The response time is the sum of the time required for the job dispatcher to take its decision, the waiting time of the job at the selected server queue, and the time required for the selected server to execute the job. Servers can have different serving capacities.

We denote the mean job arrival rate at the dispatcher with $N\lambda$. The mean serving rate of the j-th server is denoted with μ_j for $j = 1, \dots, N$. For the stability of the system, it must be $N\lambda < \sum_{j=1}^{N} \mu_j$, i.e., the mean arrival rate must be less than the overall serving capacity available to the server cluster. We introduce the utilization coefficient defined as $\rho = \lambda/\overline{\mu}$, where $\overline{\mu} = \frac{1}{N} \sum_{j=1}^{N} \mu_j$.

We know from Section 5.4.2 that the most efficient approach is to collapse the whole serving capacity in a unique super-server. However, this ideal serving system could be technologically unfeasible. Cloud computing data centers comprise tens or even hundreds thousands servers. Multiple servers are therefore a must or at least the most convenient solution in most settings.

We identify two categories of load balancing algorithms:

- *Pull schemes*, where the servers themselves signal to the dispatcher their status (e.g., that they are ready for a new job; servers "pull the job").
- *Push*: the dispatcher polls servers to inquire about their status and take the scheduling decision (the dispatcher "pushes the job").

The aim of the load balancer is to get jobs to be served in the shortest possible time, i.e., delay minimization is the target. Delay is the sum of the dispatching delay and the response time of the selected server. A second relevant aspect is overhead, i.e., the amount of data that is exchanged between the server cluster and the dispatcher, as well as the data structures that are stored in the dispatcher itself.

We say that a job dispatcher is delay-optimal if it achieves the same performance as the ideal single queue having a service capacity equal to the sum of the capacities of the N servers (complete resource-pooling system). More in depth, let us assume that the system admits a statistical equilibrium regime. Let Q_j denote the queue length at server j and let Q_e denote the queue length of an equivalent single-server system having serving rate $N\mu_e = \sum_{j=1}^{N} \mu_j$ and the same job arrival process as the dispatcher. Let also $\epsilon = \sum_{j=1}^{N} \mu_j - N\lambda = N(\mu_e - \lambda) > 0$, the last inequality holding to guarantee stability and hence the existence of statistical equilibrium. We say that a scheduling algorithm is heavy-traffic delay optimal if

$$\lim_{\epsilon \to 0} \epsilon \sum_{j=1}^{N} E[Q_j] = \lim_{\epsilon \to 0} \epsilon E[Q_e] \tag{6.128}$$

Foschini and Salz [80] proved that the join shortest queue (JSQ) policy is asymptotically delay-optimal in heavy-traffic. JSQ consists in polling the N queues upon each job arrival at the dispatcher and then selecting the queue with the shortest line. Ties are broken at random. The crux of JSQ is that it requires exchanging $2N$ messages between the dispatcher and the server cluster for each arrival, that is, the overhead message rate is $2N^2\lambda$, if we let $N\lambda$ to be the arrival rate of jobs at the system.

A simplified version of JSQ is obtained by limiting the number of servers that are polled to $d \geq 2$. This kind of policy goes under the name of power-of-d (Pod). With $d = N$ we recover JSQ. The interesting point is that good performance can be obtained with $d \ll N$. Even $d = 2$ reaps a big part of the performance offered by JSQ. It has been shown that Pod is heavy-traffic delay-optimal [55,157]. Moreover, the probability distribution of the queue length of servers decays very fast (in a double-exponential form) for large N [203, 162].

JSQ and Pod are examples of push policies. The drawback of this class of policies lies in the fact that they imply nonzero dispatching delay and potentially high message overhead.

Kelly and Laws [120] have proposed a simplified scheduling policy, known as join below threshold (JBT). An integer r is added to the policy name, to specify the threshold value. The idea of the policy is that the dispatcher allocates an arriving job to a server having less than r jobs already assigned.

More precisely, the dispatcher maintains an N bit data structure. Initially, all servers are idle (no job enqueued) and all bits of the data structure are set to 1. When a new job arrives, the dispatcher does the following:

1. If not all bits are 0, it picks one server at random among those marked with a bit equal to 1 and sends the job request to that server; the corresponding bit is reset to 0.
2. If all bits of the data structure are 0, the dispatcher selects a server at random and sends the job to that server.

On their part, servers signal their status to the dispatcher, according to the following rules. A server reports its identity to the dispatcher, whenever it completes serving a job, *if*

1. The queue length drops below the threshold r after the served job departure.
2. At least one new job has been assigned to the server since the last time it reported its identity to the dispatcher.

The second rule avoids the server reporting its identity multiple times, if its corresponding bit in the dispatcher data structure is already set to 1.

This algorithm reduces drastically the number of messages exchanged between the dispatcher and the server cluster, with respect to JSQ. It is an example of a pull policy. The main advantage of this class of policies is to have zero dispatching delay and to entail a relatively low message overhead. In [214] it has been shown that JBT(r) is heavy-traffic delay-optimal. More in depth, it is shown that JBT(r) is heavy-traffic delay-optimal in steady state provided that $r \geq K \log(1/\epsilon)$, for some constant K, and $r = o(1/\epsilon)$ as $\epsilon \to 0$. If instead $r = 1/\epsilon^{1+\alpha}$ for any positive α, JBT(r) is asymptotically the same as a pure random scheduler in heavy-traffic. Intuitively, this means that r should grow as the system moves toward heavy-traffic (i.e., as $\epsilon \to 0$). It should grow fast enough, at least as $\log(1/\epsilon)$, but not too fast, i.e., it must be slower than $1/\epsilon$. Moreover, extensive simulations point out that JBT(r) offers delay performance comparable with JSQ in various system settings.

A special case of JBT(r) is obtained for $r = 1$: it is called join idle queue (JIQ). Though it has zero dispatching delay, JIQ suffers from a strong degradation of delay performance in heavy-traffic.

To conclude this section, we compare a memoryless random dispatcher with an ideal dispatcher that knows the exact workload at each queue. The former represents a baseline algorithm that does not require any information on the status of the queues, nor does it imply any knowledge of the job service times. The latter entails a full knowledge of the status of the queue. Moreover, it is anticipative, since it requires knowledge of the service time of each arriving job.

Let $U_n(t)$ be the workload of queue n at time t. Upon a job arrival at time $t = a$, with service time X, the dispatcher evaluates the minimum workload, i.e., $U^* = \min_{0 \leq n \leq N} U_n(a)$. Then, it selects uniformly at random a server whose workload equals U^* and sends the job request to that server. Say n^* is the index of the selected queue. Then, we have $U_n(a^+) = U_n(a^-)$ for $n \neq n^*$, and $U_{n^*}(a^+) = U_{n^*}(a^-) + X$.

As for the memoryless dispatching, let β_n be the probability that a job is sent to queue n. Assume job arrivals follow a Poisson process with mean rate $N\lambda$. Sampling this arrival flow randomly and independently of service times preserves its Poisson character. Hence, each server can be modeled as an $M/G/1$ queue. Let L denote the workload brought by an arriving job. Let C_n be the capacity of the n-th server. Then, the service time of a job executed at server n is $X_n = L/C_n$. The mean response time can be written as

$$E[R] = \sum_{n=1}^{N} \beta_n \left(\frac{\beta_n N \lambda E[L^2]/C_n^2}{2(1 - \beta_n N \lambda E[L]/C_n)} + \frac{E[L]}{C_n} \right) \tag{6.129}$$

We will find the probability distribution β_n, $n = 1, \ldots, N$, that minimizes the response time. The minimization problem can be restated in the following way:

$$\min_{\mathbf{x}} \sum_{n=1}^{N} \left(\frac{c^2 x_n^2}{1 - x_n} + x_n \right) \tag{6.130}$$

subject to:

$$\sum_{n=1}^{N} a_n x_n = 1 \tag{6.131a}$$

$$x_n \equiv \beta_n \frac{N \lambda E[L]}{C_n} \geq 0, \quad n = 1, \ldots, N, \tag{6.131b}$$

where

$$c^2 = \frac{E[L^2]}{2(E[L])^2} = \frac{C_L^2 + 1}{2}$$

$$a_n = \frac{C_n}{N \lambda E[L]}, \quad n = 1, \ldots, N.$$

where C_L^2 is the SCOV of the workload associated with a job request. The target function is separable, i.e., it is the sum of N functions $f(x_n)$, each depending only on a single variable. The function $f(\cdot)$ is non-negative, strictly increasing and convex. Hence the Hessian of the target function if strictly positive definite. The minimizer $\mathbf{x}^*$ can be found by setting to 0 the gradient of the Lagrangian:

$$L(\mathbf{x}, \psi) = \sum_{n=1}^{N} \left(\frac{c^2 x_n^2}{1 - x_n} + x_n \right) - \psi \sum_{n=1}^{N} a_n x_n \tag{6.132}$$

The resulting optimal routing probabilities β_n^* are

$$\beta_n^* = \max \left\{ 0, \min \left\{ 1, a_n \left(1 - \frac{1}{\sqrt{1 - \frac{1}{c^2} + \psi \frac{a_n}{c^2}}} \right) \right\} \right\} \tag{6.133}$$

where we have accounted for the limited range allowed to the probabilities β_n^*. The multiplier ψ is found as the unique solution of the normalization equation $\sum_{n=1}^{N} \beta_n^* = 1$.

As a numerical example, let us consider the case where the workload has a Pareto probability distribution with mean 1, i.e., it is $P(L > y) = \min\{1, (\theta/y)^\alpha\}$, with $E[L] = \alpha\theta/(\alpha - 1) = 1$. The parameter θ is determined from this condition. We let $\alpha = 2.1$.

As for the server capacities, we let

$$C_n = C_{\min} + (C_{\max} - C_{\min})\frac{n-1}{N-1}, \quad n = 1, \dots, N. \tag{6.134}$$

with $C_{\min} = 1$ and $C_{\max} = 10$. The utilization coefficient of the servers is defined as

$$\rho = \frac{N\lambda E[L]}{\sum_{n=1}^{N} C_n} \tag{6.135}$$

Figure 6.10 plots the mean response time for $N = 4$ servers (left plot) and for $N = 100$ servers (right plot) as a function of ρ. Ideal minimum backlog dispatching and random memoryless dispatching, optimized as described above, are compared. It is evident that, in spite of the optimization of the routing probabilities β_n, random job dispatching is definitely inferior to the ideal policy. This is the more remarkable as the number of server grows. Moving from $N = 4$ to $N = 100$ response time with random dispatching exhibits little change, while ideal dispatching results in negligible response time for loads as high as $\rho = 0.95$.

The takeaway of this simple comparison is that there is a huge performance gain with exploiting information on the status of queues at servers. This motivates the research of policies that offer a good trade-off between status report overhead, dispatching delay and response time.

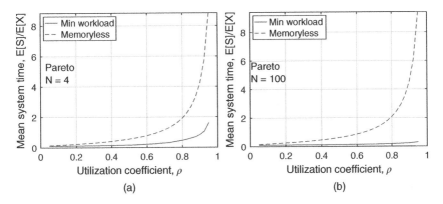

Figure 6.10 Mean response time of job dispatching to a cluster of N unequal servers ($N = 4$ in the left plot, $N = 100$ in the right plot) as a function of the utilization coefficient of the servers. Comparison between ideal dispatching to the minimum workload queue and optimized random memoryless job dispatching.

6.6 Optimal Scheduling

Before closing this introduction of scheduling policies in queues, we hint at *optimal* scheduling. The general setting consists of a service system, with possibly multiple flows of customers, with different statistical characteristics both as for arrival patterns and service demand. Some target function of the queueing system performance is defined, e.g., the mean number of customers residing in the system, or the weighted sum of the mean delays of different classes of customers, or a quantile of the customer sojourn time. Then, a serving policy is searched for within a set of admissible policies, such that it optimizes the target function.

In the following, we confine ourselves to single-class systems, i.e., a serving system visited by customers belonging to a same class. Many more details and pointers to specialized references can be found, e.g., in [131, Ch. 4][86]. Hence, the service demand of all customers is characterized by a single statistical description. We assume service times of arriving customers form a renewal process with CDF $F_X(x)$. We also assume consistently a single, work-conserving server.

We can identify two main classification criteria that affect the admissible set of scheduling policies:

1. Whether at most a single customer with partially completed service is allowed or not;
2. Whether the service demand of an arriving customer is known when it joins the queueing system or not.

We will refer to a queueing system allowing more than a single partially served customer as *server-sharing* system. On the contrary, a queueing system where at any time at most a single customer can be partially served is defined to be non-server-sharing. As for the second classification feature, we call *anticipative* a queueing system where the service demand of an arriving customer is declared explicitly and hence known to the server upon arrival. Nonanticipative means that only the probability distribution is possibly known, not the specific sample of the service time of an arriving customer.

For example, FCFS is a non-server-sharing, nonanticipative queueing discipline. Processor sharing and the generalized version thereof are examples of server sharing, non-anticipative disciplines. Finally a representative of the server-sharing, anticipative category is SRPT.

Let us discuss what the optimal scheduling is for single-class, single work-conserving server queueing systems, by considering separately three cases, according to the two criteria stated above. We assume consistently that the target function to be optimized is the overall mean number of customers residing in the queueing system. In equilibrium, Little's law guarantees that this target is equivalent to the mean sojourn time of a customer inside the system, that is to say, the response time.

6.6.1 Anticipative Systems

In this case, SRPT can be shown to be the optimal scheduling policy. We can sketch the proof briefly as follows. Let $Q_S(t)$ and $Q_A(t)$ be the number of customers in the system at time t under the SRPT policy and under any other policy different from SRPT. Let $q(t) = \min\{Q_S(t), Q_A(t)\}$ and consider the sum of the $q(t)$ largest remaining processing times in the queue, say $W_S(t)$ and $W_A(t)$ for the SRPT and the "any-other" queueing policies. Since SRPT favors customers with the least remaining processing times, it must be $W_S(t) \geq W_A(t)$. On the other hand, for a work-conserving server in equilibrium the *overall* amount of unfinished work at any time is invariant with respect to the serving policy, hence it must be the same for SRPT and the "any-other" policies. This is only possible if $Q_S(t) \leq Q_A(t)$. Moving from this sample-path argument to time-averages, it follows that the mean number of customers residing in the system attains its minimum level under SRPT policy.

The argument outlined above applies to general arrival processes, not necessarily the Poisson process. Note that if preemption is not allowed, it can be shown that SJF is optimal instead of SRPT.

6.6.2 Server-Sharing, Nonanticipative Systems

In this case only the CDF of the service times, $F_X(x)$, is assumed as known. The serving policy minimizing the mean number of customers in the system or equivalently the response time of the system is the *Gittins index*. To introduce that policy, we need some definitions.

Let a denote the attained service of a customer. We let

$$\kappa(a,x) = \frac{\int_a^{a+x} dF_X(u)}{\int_a^{a+x} [1 - F_X(u)]du}, \qquad x > 0, a \geq 0. \tag{6.136}$$

For $x \to 0$ the Gittins index $\kappa(a,x)$ tends to the hazard function of the probability distribution of the service times, calculated at a, i.e., $\lim_{x\to 0}\kappa(a,x) = h(a) \equiv f_X(a)/[1 - F_X(a)]$. As $x \to \infty$ we have $\kappa(a,x) \to G_X(a)/\int_a^\infty G_X(x)dx < \infty$. Hence $\kappa(a,x)$ is continuous and limited in any interval of the positive semi-axis as a function of x. Then, we can find a finite supremum, i.e., we can define $\kappa(a) = \sup_{x\geq 0}\kappa(a,x)$ for any positive a. Let also $x^* = \arg\max_{x\geq 0}\kappa(a,x)$, including the possibility that it is $x^* = \infty$.

The Gittins index policy can be stated as follows. At any time, serve that customer that has the largest level of $\kappa(a)$ among all those within the queueing system. Continue serving that customer until one of the following conditions is met, whichever comes first:

1. The service demand of the customer is fully served.
2. The customer has received up to x^* time units of service.
3. A new customer arrives with a higher Gittins index.

The idea underlying this policy is to select a customer with a high probability of completing service within time x and at the same time having the least expected remaining service time. The Gittins index aims at capturing this selection criterion, thus it does its best to behave like SRPT does.

Example 6.12 In case of negative exponential service times, we have $F_X(x) = 1 - e^{-\mu x}$. It is easy to find that $\kappa(a, x) = \mu$ for any x and a. Then, it is also $\kappa(a) = \mu$. In this special case, all customers are at the same level. It is an optimal policy to select whichever customer out of those that are in the queue when the server is available to start service. Nor is there any reason to interrupt service. We can therefore deem FCFS to be optimal in this special case. This is not completely surprising, given the memoryless property of the negative exponential probability distribution.

Example 6.13 Let the service times have a Pareto probability distribution given by $F_X(x) = 1 - (\theta/x)^\alpha$ for $x \geq \theta$. It can be found that

$$\kappa(a, x) = \frac{\alpha - 1}{a} \frac{1 - \left(\frac{a}{a+x}\right)^\alpha}{1 - \left(\frac{a}{a+x}\right)^{\alpha-1}} \tag{6.137}$$

for $a + x \geq \theta$ and $\alpha > 1$. This is a monotonously increasing function of x, whose supremum is $(\alpha - 1)/a$, attained as $x \to \infty$. Hence it is $x^* = \infty$. According to the Gittins index policy, we should select the customer with the minimum level of a, given that $\kappa(a)$ is inversely proportional to a. Since a is the attained service, it turns out that Gittins index coincides with LAS in case of Pareto service times.

6.6.3 Non-Server-Sharing, Nonanticipative Systems

In the special framework where service times of arriving customers are not known to the server and at most one customer with partial service accomplished is allowed, the conservation law for work-conserving queueing discipline applies with the same PDF of service times for all classes. Then, we know that all such serving policies achieve the *same* mean response time, hence the same mean number of customers residing in the system, by Little's law. Any such discipline is therefore optimal. It is customary to adopt FCFS in most cases, since this is well understood and usually accepted by competing customers.

Summary and Takeaways

Differentiated treatment of customers and sharing of the server capacity are the topics of this chapter. A vast amount of literature, techniques, algorithms has been developed and applied to a wide variety of systems, ranging from telecommunication networks, to transportation systems, computing systems, data centers, workflow management.

We have seen a basic model derived from the $M/G/1$ queue. It is useful to understand the fundamental trade-offs, particularly the conservation law, a mathematical statement of the fact that nothing comes for free. Discrete levels and continuous classes of customers are studied, thus deriving the well known shortest job first and shortest remaining processing time disciplines. Preemption of the server is discussed as well. This introduces the idea of sharing the server capacity among many customers simultaneously, i.e., at any time there can exist more than one customer that has been served only partially. The most famous processor sharing model, the $M/G/1$ processor sharing is studied. A generalization of the processor sharing system is presented, paving the way to practical algorithms that allow weighted sharing of the server capacity. Two specific examples from the communication networks realm are presented: Weighted fair queueing and credit-based scheduling. As hinted at by the name, the key issue these algorithm address is to share the server capacity efficiently while protecting customers demanding little amount of work from heavy load customers, that could otherwise hog the server. In fact, the scheduling problem is intimately connected with the fairness issue.

Finally, we review queueing disciplines for single-server queues under the perspective of optimality, specifically, disciplines that minimize the mean number of customers in the queue. To that purpose, we categorize queueing disciplines according to server-sharing and service time anticipation properties.

Problems

6.1   In a priority queueing system of $M/G/1$ type, there are two priority classes. We are given the fact that class two mean waiting time is

$$E[W_2] = \frac{W_0}{1 - \alpha\rho_1 - \beta\rho_2} \tag{6.138}$$

where $\rho_j = \lambda_j E[X_j]$, for $j = 1, 2$; α, β lie in the interval $(0, 1)$, and W_0 is the residual service time found by an arriving customer.

Find the mean waiting time of class one, $E[W_1]$.

6.2 A statistical multiplexer with an output capacity $C = 10$ Mbit/s receives a Poisson flow of packets of mean rate $\lambda = 1000$ pkts/s. The packet length distribution is a discrete one, with three values, as detailed below:

$$
\begin{aligned}
p_1 &= P(L = 100 \text{ bytes}) = 0.4 \\
p_2 &= P(L = 1000 \text{ bytes}) = 0.2 \\
p_3 &= P(L = 1500 \text{ bytes}) = 0.4
\end{aligned} \qquad (6.139)
$$

Assume the multiplexer adopts a queueing discipline that assigns priority according to packet lengths. Specifically, shortest packets have highest priority, longest packets lowest priority. Calculate the mean waiting time of the three packet types.

6.3 The packet flow out of a domestic network is sent to a link of capacity $C = 1$ *Mbit/s*. The packet flow is composed of packets belonging to two classes: VoIP packets with fixed length of 80 bytes, and data packets, with negative exponential lengths with mean value 1500 bytes. It is known that 20% of the overall link load is due to VoIP packets. The overall average link load (=utilization factor of the link) is equal to 0.9.

(a) Is it possible to devise a non preemptive, nonanticipative priority rule so that the mean waiting time of both classes falls below 80 ms?

(b) Calculate the mean waiting time of the data packets, if we impose that it must be twice the mean waiting time of the VoIP packets.

6.4 Consider a work-conserving priority-based statistical packet multiplexer. There are traffic flows arriving at the multiplexer: (i) voice packets, with constant length $L_1 = 100$ bytes; (ii) background traffic packets, having a negative exponential probability distribution of packet lengths with mean value $E[L_2] = 800$ bytes. The capacity of the multiplexer is $C = 1$ Mbit/s. The utilization coefficient of the multiplexer output link is 0.8; 10% of that load is due to voice packets, the remaining 90% of the load is made by background traffic packets.

(a) Calculate the mean waiting times of packets of either flow, under a HOL priority scheme, where the top priority class is given to voice packets.

(b) Does there exists any work-conserving priority scheme that can guarantee a reduction by 5% for *both* classes of the mean waiting times obtained in the previous point?

(c) Suppose that voice packets are guaranteed a mean delay of no more than 1 ms by using some work-conserving priority scheme. What is the minimum value of the mean waiting time for the background traffic packets?

6.5 A file server receives requests of file download according to a Poisson process of mean rate $\lambda = 0.2$ requests/s. Requested files are of two types: the short ones have constant length equal to $L_1 = 100$ kbyte, the long ones have variable length with mean value $E[L_2] = 1$ Mbyte and standard deviation $\sigma_{L_2} = 2$ Mbyte. 50% of the requests are for short files. The server has a link with capacity $C = 1$ Mbit/s.

(a) Calculate the mean system time if there is no priority.

(b) Calculate the mean system time if HOL priorities are used, giving higher priority to the short file download.

(c) Calculate the mean system time of a job of the class "short file" and of a job of the class "long file" under the processor sharing policy. Compare the results with those of points (a) and (b) above and discuss the reason why of the different values.

7

Queueing Networks

Overheard while waiting in a long ticket line, from a disgruntled customer who was heading toward the back of the queue: "The only thing worse than waiting in line is waiting in the wrong line!"

7.1 Structure of a Queueing Network and Notation

A queueing network is a system of J queues interconnected so that customers leaving one queue can possibly visit other queues. Queueing networks provide a wide class of models that encompass isolated queues as a special case. The key feature of the queueing network is its topological structure, i.e., the relationship among the queues established by the flows of customers moving from one queue to another one. The topology is defined by a weighted directed graph[1] , where nodes are queues. Let the queues be numbered with positive integers, from 1 to J. An edge is directed from queue i to queue j and it is labeled with $r_{ij} \in [0, 1]$ if a customer leaving queue i joins queue j with probability r_{ij}. These probabilities form a matrix $\mathbf{R}$, whose row sums are less than or equal to 1.

For ease of notation we introduce one more "queue" of the network, representing the outside world. Let it be labeled with 0. Define $r_{i0} = 1 - \sum_{j=1}^{J} r_{ij}$. A queueing network is said to be *open* if $r_{i0} > 0$ for at least one index i. In the opposite case, it is referred to as *closed*, and the corresponding matrix $\mathbf{R}$ is stochastic.

If $r_{i0} > 0$ for at least one index i the matrix $\mathbf{R}$ is sub-stochastic and exit out of the queueing network is allowed. In that case there must be an external arrival process as well. Let γ_i be the average rate of customers arriving at queue i. We can think

---

1 Basic definitions of graph theory and how to determine whether a graph is connected are reviewed in the Appendix at the end of this chapter.

Network Traffic Engineering: Stochastic Models and Applications, First Edition. Andrea Baiocchi.
© 2020 John Wiley & Sons, Inc. Published 2020 by John Wiley & Sons, Inc.

of these arrivals as belonging to a single external flow with rate $\gamma = \sum_{j=1}^{J} \gamma_j$. New arrivals join the i-th queue with probability r_{0i}, hence $\gamma_i = \gamma r_{0i}$ for $i = 1, \ldots, J$.

7.2 Open Queueing Networks

We focus on open queueing networks. Queues are assumed to have infinite size, so that no customer is rejected. External arrivals are assumed to follow a Poisson process. The serving facility at a queue has a state-dependent average service rate. The service rate of the j-th queue is $\mu_j(x_j)$, if the queue length is x_j. It must be $\mu_j(x) > 0$ for $x > 0$ and $\mu_j(0) = 0$. In the ensuing analysis, we assume that service times have negative exponential PDF with mean $1/\mu_j(x_j)$ at queue j, when its state is x_j. For example, if m_j equivalent servers are available at queue j, each with mean service time $1/\mu_j$, we have $\mu_j(x_j) = \min\{x_j, m_j\}\mu_j$ for $x_j \geq 0$.

The average flow through the queue i, λ_i, can be found by summing up all contributions at the input of queue i and taking into account that the average flow λ_i into queue i coincides with the average flow out of queue i. This is true at statistical equilibrium, if the network can achieve this regime. Then, we have

$$\lambda_i = \gamma_i + \sum_{j=1}^{J} \lambda_j r_{ji}, \quad i = 1, \ldots, J. \tag{7.1}$$

This is a nonhomogeneous, linear equation system, with a unique positive solution, provided the matrix $\mathbf{R}$ is irreducible. Since $\mathbf{R}$ is also sub-stochastic for open networks, it follows that $\mathbf{I} - \mathbf{R}$ is invertible[2], $\mathbf{I}$ being the identity matrix. In matrix form we can write $\lambda = \gamma(\mathbf{I} - \mathbf{R})^{-1}$, with $\lambda = [\lambda_1, \ldots, \lambda_J]$ and $\gamma = [\gamma_1, \ldots, \gamma_J]$.

We can give a meaning to the matrix $(\mathbf{I} - \mathbf{R})^{-1}$, by considering the discrete-time, discrete state Markov process $Y(\tau)$ defined by customer movements through the network. Let $Y(\tau) = i$ if at step τ queue i is visited. Then, with probability r_{ij}, queue j will be visited at step $\tau + 1$. If the process is started at queue i with probability $q_0(i)$, we have $\mathbf{q}_\tau = \mathbf{q}_0 \mathbf{R}^\tau$, where $\mathbf{q}_\tau$ is a row vector. The entry $q_\tau(i)$ is the probability that the state at time τ is i, i.e., $q_\tau(i) = P(Y(\tau) = i)$. Let $Z_i(\tau)$ be a binary random variable equal to 1, if $Y(\tau) = i$, 0 otherwise. The number of visits to queue i up to time T is $V_i(T) = \sum_{\tau=0}^{T} Z_i(\tau)$. The mean number of visits is therefore $E[V_i(T)] = \sum_{\tau=0}^{T} P(Y(\tau) = i)$. In vector form, $E[\mathbf{V}(T)] = \mathbf{q}_0 \sum_{\tau=0}^{T} \mathbf{R}^\tau$. As $T \to \infty$ we get $E[\mathbf{V}] = \mathbf{q}_0(\mathbf{I} - \mathbf{R})^{-1}$. If we choose $q_0(k) = 1$ and $q_0(i) = 0, \forall i \neq k$ for a given k, we find that the entry (k, j) of the matrix $(\mathbf{I} - \mathbf{R})^{-1}$ represents the mean number of visits at queue j, given that we start from queue k.

The same result could be found more directly by considering that the mean number of visits $E[V_j]$ to queue j can be written as the probability of starting from queue

---

2 It is actually an M-matrix, so that its inverse has positive entries.

j itself plus the mean number of visits to any queue i that is a neighbor of queue j multiplied by the probability of making a transition from i to j. Formally $E[V_j] = q_0(j) + \sum_{i=1}^{J} E[V_i] r_{ij}$, for $j = 1, \dots, J$. This linear equation system yields immediately $E[\mathbf{V}] = \mathbf{q}_0(I - R)^{-1}$.

Another interesting interpretation of the matrix $(I - R)^{-1}$ is related to the number of hops that a customer makes through the network. Let us consider the vector $\mathbf{r} = (I - R)\mathbf{e}$. The i-th entry of the vector $\mathbf{r}$ is r_{i0}, the probability of leaving the queueing network from queue i. Let $\mathbf{D}$ be a diagonal matrix having the elements of $\mathbf{r}$ on the diagonal. We can consider a transient Markov chain with one-step transition probability matrix given by

$$\mathbf{P}_e = \begin{bmatrix} \mathbf{R} & \mathbf{D} \\ \mathbf{0} & \mathbf{I} \end{bmatrix}. \tag{7.2}$$

The first J states correspond to the queues of the queueing network. The last J states are absorbing. The i-th absorbing state corresponds to exit out of the queueing network through queue i. Let T be the number of hops to absorption from one queue to another queue and let $Y(T)$ denote the queue from which we leave the queueing network. It can be verified that

$$P(T = n, Y(T) = j | Y(0) = i) = [\mathbf{R}^{n-1}\mathbf{D}]_{ij}, \qquad n \geq 1, \tag{7.3}$$

where $[\mathbf{A}]_{ij}$ denotes the entry (i, j) of the matrix $\mathbf{A}$. Let h_{ij} be the mean number of hops given that we enter the queueing network at queue i and we leave it from queue j. Let $\mathbf{H}$ be the matrix with entries h_{ij}. It is $h_{ij} = \sum_{n=1}^{\infty} n [\mathbf{R}^{n-1}\mathbf{D}]_{ij}$ and $\mathbf{H} = (I - R)^{-2}\mathbf{D}$. A customer enters the network at queue i with probability γ_i/Γ, where $\Gamma = \gamma_1 + \cdots + \gamma_J$. Removing the conditioning on the initial state and saturating the event $Y(T) = j$, we obtain the mean number of hops through the network $\bar{h} = \gamma(I - R)^{-1}\mathbf{e}/\Gamma = \Lambda/\Gamma$, where $\Lambda = \lambda_1 + \cdots + \lambda_J$. The interesting result is that $\Lambda = \bar{h}\Gamma$. The sum of the link loads equals the sum of the exogenous input rates multiplied by the mean number of hops. A customer making h hops "consumes" capacity on h links. This remark makes the formula intuitively appealing.

The assumptions leading to Jackson's model of queueing network are summarized below:

(a) External arrivals follow a Poisson process.
(b) The service time at every queue is a negative exponential random variable with a possibly state-dependent mean value, and it is independent of the external arrival process.
(c) All customers arriving at a queue can join the queue, i.e., queue sizes are infinite.

(d) Routing is memoryless: each customer chooses the next queue, when leaving the currently visited queue, independently of all other customers.

(e) The $J \times J$ routing matrix $\mathbf{R}$, with entries r_{ij}, $i,j = 1,\ldots,J$, is irreducible[3].

Let $Q_j(t)$ be the number of customers in queue j at time t. The state of the network at time t is described by the vector $\mathbf{Q}(t) = [Q_1(t)\ldots Q_J(t)]$. Under the hypotheses (a)-(e) the network state $\mathbf{Q}(t)$ defines a Markov process over the state space Ω. Since $Q_j \in \mathbb{Z}^+$, i.e., it takes integer non-negative values, the state space Ω is made up of all non-negative integer valued vectors of dimension J, i.e., $\Omega = \{\mathbf{x} = [x_1,\ldots,x_J]$: $x_j \in \mathbb{Z}^+\}$. If the Markov process is ergodic, we denote the steady-state random variable $\mathbf{Q}(\infty)$ simply with $\mathbf{Q}$. Similarly, $Q_j = Q_j(\infty)$ denotes the steady-state random variables describing the number of customers in queue j.

Let $\pi(\mathbf{x})$ be the limiting state probabilities of the Markov process, assumed to be ergodic. We will see later the conditions under which ergodicity holds. The vector $\pi(\mathbf{x})$ can be computed from the homogeneous linear equation system $\pi\mathbf{M} = \mathbf{0}$ and the equation $\pi\mathbf{e} = 1$, where $\mathbf{e}$ denotes a column vector of 1's and $\mathbf{M}$ is the infinitesimal generator of the Markov process $\mathbf{Q}(t)$. The equations of this system can be written down by balancing the probability flow in and out of each state $\mathbf{x}$, yielding

$$\pi(\mathbf{x})\sum_{j=1}^{J}[\mu_j(x_j)(1 - r_{jj}) + \gamma_j] = \sum_{j=1}^{J}\pi(\mathbf{x} - \mathbf{e}_j)\gamma_j + \sum_{j=1}^{J}\pi(\mathbf{x} + \mathbf{e}_j)\mu_j(x_j + 1)r_{j0}$$

$$+ \sum_{j=1}^{J}\sum_{i\neq j,i=1}^{J}\pi(\mathbf{x} - \mathbf{e}_j + \mathbf{e}_i)\mu_i(x_i + 1)r_{ij} \qquad (7.4)$$

where $\mathbf{e}_i$ indicates a vector with all elements equal to 0 except the i-th one, which is 1. This balance equation holds for $\mathbf{x} > 0$ and it is modified in an obvious way for states with some null entries (e.g., no departure can take place out of a queue, if its state is 0)[4].

Let us define the non-negative, discrete random variable Y_i with PDF given by

$$P(Y_i = k) = P(Y_i = 0)\frac{\lambda_i^k}{M_i(k)}, \qquad k \geq 0 \qquad (7.5)$$

with

$$M_i(k) = \begin{cases} \prod_{h=1}^{k}\mu_i(h), & k \geq 1, \\ 1 & k = 0. \end{cases} \qquad (7.6)$$

---

3 This is equivalent to stating that the graph associated to the queueing network is strongly connected (see the Appendix of this chapter).
4 It suffices to let $\pi(\mathbf{x}) = 0$ if any component of $\mathbf{x}$ is negative.

where $\mu_i(h)$ is the serving rate of queue i when there are $Q_i = h$ customers in that queue, $i = 1, \ldots, J$. Note that $\mu_i(0) = 0$, while it must be $\mu_i(h) > 0$ for any $h > 0$.

The Y_i's are properly defined random variables, provided that the infinite sums $\sum_{k=0}^{\infty} \frac{\lambda_i^k}{M_i(k)}$, for $i = 1, \ldots, J$, converge to positive, finite values. In that case the probabilities in eq. (7.5) can be normalized to sum up to 1 by letting:

$$
P(Y_i = 0) = \left(1 + \sum_{k=1}^{\infty} \frac{\lambda_i^k}{M_i(k)}\right)^{-1}, \qquad i = 1, \ldots, J, \tag{7.7}
$$

The conditions for the stability of the queueing network are therefore:

$$
\sum_{k=0}^{\infty} \frac{\lambda_i^k}{M_i(k)} < \infty, \qquad i = 1, \ldots, J, \tag{7.8}
$$

The conditions in eq. (7.8) are indeed necessary and sufficient for the ergodicity of the Markov process $\mathbf{Q}(t)$.

The result due to Jackson is stated as follows.

Theorem 7.1 (*Jackson, [111]*) If an open queueing network satisfies the hypotheses (a)-(e) above and the J conditions in eq. (7.8) are satisfied, the limiting PDF of the Markov process $\mathbf{Q}(t)$ exists and it is given by a product form as

$$
\pi(\mathbf{x}) = \prod_{j=1}^{J} P(Y_j = x_j) \tag{7.9}
$$

where the marginal PDFs are given in eq. (7.5).

Proof: We will check that the probabilities in eq. (7.9) satisfy the rate-balance equations (7.4). According to the expression of the limiting state probabilities given in the theorem statement, we have:

$$
\pi(\mathbf{x} - \mathbf{e}_j) = \pi(\mathbf{x})\frac{\mu_j(x_j)}{\lambda_j}, \qquad x_j > 0
$$

$$
\pi(\mathbf{x} + \mathbf{e}_j) = \pi(\mathbf{x})\frac{\lambda_j}{\mu_j(x_j + 1)},
$$

$$
\pi(\mathbf{x} - \mathbf{e}_j + \mathbf{e}_i) = \pi(\mathbf{x})\frac{\mu_j(x_j)\lambda_i}{\lambda_j\mu_i(x_i + 1)}, \qquad x_j > 0
$$

Then, by substituting into eq. (7.4) and canceling out $\pi(\mathbf{x})$, we get

$$
\sum_{j=1}^{J}[\mu_j(x_j)(1 - r_{jj}) + \gamma_j] =
$$

$$
= \sum_{j=1}^{J} \frac{\mu_j(x_j)}{\lambda_j}\gamma_j + + \sum_{j=1}^{J} \frac{\lambda_j}{\mu_j(x_j + 1)}\mu_j(x_j + 1)r_{j0} + \sum_{j=1}^{J}\sum_{i \neq j, i=1}^{J} \frac{\mu_j(x_j)\lambda_i}{\lambda_j\mu_i(x_i + 1)}\mu_i(x_i + 1)r_{ij}
$$

$$= \sum_{j=1}^{J} \frac{\mu_j(x_j)}{\lambda_j} \left(\gamma_j + \sum_{i \neq j, i=1}^{J} \lambda_i r_{ij} \right) + \sum_{j=1}^{J} \lambda_j r_{j0} = \sum_{j=1}^{J} \mu_j(x_j)(1 - r_{jj}) + \sum_{j=1}^{J} \lambda_j r_{j0} \quad (7.10)$$

The last equality is a consequence of the flow-balance equations (7.1). Equation (7.10) reduces to

$$\sum_{j=1}^{J} \gamma_j = \sum_{j=1}^{J} \lambda_j r_{j0} \quad (7.11)$$

It is easily checked that eq. (7.11) is true by summing up both sides of eq. (7.1) over i and taking into account that $\sum_{i=1}^{J} r_{ji} = 1 - r_{j0}$. Note that (7.11) amounts to the statement that the overall mean input rate to the network equals the overall mean output rate of the network, which is obviously true at equilibrium. ∎

Jackson's result points out that the queue states are independent random variables. Their respective marginal probability distributions are just those that could be computed for each queue as if it were in isolation, according to an M/M/1 model with state-dependent serving rates. To compute the marginal PDF at each queue we have to use the value of λ as obtained by the balance equations (7.1). This holds even though the superposition at the input of each queue of the external input and of the customer flows coming from other queues is not necessarily a Poisson process.

In the special case of state independent, negative exponential service times (i.e., single server), namely $\mu_j(k) = \mu_j \ \forall k > 0$, we find easily that

$$P(Y_j = k) = (1 - \rho_j)\rho_j^k, \qquad k \geq 0 \quad (7.12)$$

where $\rho_j = \lambda_j/\mu_j$. The ergodicity condition is that $\rho_j < 1$ for $j = 1, \dots, J$.

Moments of the queue length can be easily found. The average delay through the whole network can be evaluated easily as well. By Little's law, we have $E[D] = E[Q]/\gamma$, where $Q = \sum_{j=1}^{J} Q_j$ is the overall number of customers inside the network (waiting or receiving service at any queue), $\gamma = \sum_{j=1}^{J} \gamma_j$ and D is the network delay, i.e., the delay suffered by a generic customer going through the network. We have $E[Q] = \sum_{j=1}^{J} E[Q_j]$. In the special case of single server queues, a simple explicit form of $E[Q_j]$ is obtained, namely $E[Q_j] = \frac{\rho_j}{1-\rho_j}$. Therefore

$$E[D] = \frac{1}{\gamma} \sum_{j=1}^{J} \frac{\lambda_j}{\mu_j - \lambda_j} \quad (7.13)$$

This expression depends on the overall offered traffic rate, on the mean serving rate of the single server at each queue and on routing through the flow rates $\lambda_j, j = 1, \dots, J$.

Example 7.1 *Kleinrock's model of a packet network* A classic application of the open queueing network model is Kleinrock's model of a packet network [131, Ch.5]. Let us consider a packet network. The traffic offered to the network can be described by the *traffic matrix*. The entry (i, j) of this matrix gives the mean rate of the end-to-end (e2e) traffic flow entering the network at node i and leaving the network from node j. The nodes of the packet network are *routers* , since their main function is to route packets through the network path from entrance point i to exit point j. Routers are interconnected through subnets, that we can think of as direct links (e.g., fiber-optic links, as typical of core networks). We can abstract a view of the network at packet level as a graph, where nodes correspond to routers and an arc is placed between two nodes if a packet can be sent directly between the corresponding routers, without the help of any other router. Let **V** denote the adjacency matrix of the graph. Then, $v_{ij} = 1$ if and only if there is a link between router i and router j. In a typical packet network links are bidirectional, hence the graph model is undirected. A discussion of the properties of the adjacency matrix **V** that guarantee the network graph is connected is given in the Appendix to this Chapter. We assume the graph is connected (actually, it is multi-connected for reliability reasons).

We assume further that a single route is defined for any e2e flow through the network (e.g., shortest path routing according to some link metric, without load sharing or multi-path). To apply the open queueing network model, we focus on network links. A link connecting router R_1 to router R_2 is equipped with a buffer on the sender side. Packets dispatched by router R_1 to router R_2 via the link are sent immediately, if possible. Otherwise, they are stored in the buffer at R_1 output port, waiting for the first opportunity for being sent out to the link. The typical serving discipline at router output buffers is FCFS.

Let $\mathcal{F}$ denote the set of e2e flows and $\mathcal{L}$ the set of network links. Let x_s be the offered packet rate of e2e flow $s \in \mathcal{F}$. Let a_{sk} be a binary coefficient, equal to 1 if flow s is routed through link k, 0 otherwise[5] . Let b_{sk} be 1, if flow s is injected into the network through link k. Only a single b_{sk} is equal to 1 for any given s. We denote the matrices collecting those coefficients as **A** and **B** respectively. Let C_k be the bit rate of link $k \in \mathcal{L}$. If the mean packet length is $\overline{L}$, then the serving rate of link k is $\mu_k = C_k/\overline{L}$.

The link load rates λ_k are most often found as follows in a packet network:

$$\lambda_k = \sum_{s \in \mathcal{F}} x_s a_{sk} , \qquad k \in \mathcal{L}. \tag{7.14}$$

---

5 If the coefficients a_{sk} are generalized to real values in $[0, 1]$, multi-path routing can be modeled.

Similarly, the mean external offered packet rate at link k can be calculated as

$$\gamma_k = \sum_{s \in \mathcal{F}} x_s b_{sk} , \qquad k \in \mathcal{L}. \tag{7.15}$$

Note that the sum of all input flows equals the sum of all flows x_s, i.e., $\gamma = \sum_{k \in \mathcal{L}} \gamma_k = \sum_{s \in \mathcal{F}} x_s$, since $\sum_{k \in \mathcal{L}} b_{sk} = 1, \forall s$.

If we can assume memoryless routing[6] , the probability r_{hk} of moving from link h to link k can be expressed as follows, provided there is a non-null flow on link h:

$$r_{hk} = \frac{\sum_{s \in \mathcal{F}} x_s a_{sh} v_{hk} a_{sk}}{\sum_{s \in \mathcal{F}} x_s a_{sh}} , \qquad h, k \in \mathcal{L}. \tag{7.16}$$

The expression of r_{hk} consists of the ratio between the mean rate of those flows whose routing contains *both* links h and k, divided by the mean rate of flows coming from link h. In other words, r_{hk} is the mean fraction of flow shifted from link h to link k. The factor v_{hk} is there to make the ratio equal to 0 for nonadjacent links (h and k could belong simultaneously to the route of a same flow, even if they are not adjacent). It can be verified that the flow rates λ_k, the external flow arrival rates γ_k, and the routing probabilities r_{hk} defined above satisfy the flow balance equations of the queueing network, i.e., $\lambda_k = \gamma_k + \sum_{h \in \mathcal{L}} \lambda_h r_{hk}$.

The mean delay through the network is therefore

$$E[D] = \sum_{k \in \mathcal{L}} \frac{\lambda_k}{\gamma} \left(\frac{1}{C_k / \overline{L} - \lambda_k} + \tau_k \right) = \frac{1}{\gamma} \sum_{k \in \mathcal{L}} \left(\frac{\lambda_k \overline{L}}{C_k - \lambda_k \overline{L}} + \lambda_k \tau_k \right) \tag{7.17}$$

where we have introduced the propagation delay τ_k associated with link k. More generally, τ_k could be deemed as the mean delay through the subnet corresponding to link k. This result holds under the stability conditions $\lambda_k \overline{L} < C_k, \forall k$, that is

$$\sum_{k \in \mathcal{F}} x_s a_{sk} \leq C_k / \overline{L} \tag{7.18}$$

The end-to-end delay experienced by a flow, say flow s, can be calculated as follows:

$$E[D_s] = \sum_{k \in \mathcal{L}} a_{sk} \left(\frac{1}{C_k / \overline{L} - \lambda_k} + \tau_k \right) \tag{7.19}$$

---

6 This is definitely an approximation for a packet network: routers do not switch packets randomly, rather they use the destination address field carried inside each packet, consult their forwarding tables and dispatch the packet to the appropriate output accordingly. Moreover, loops are not possible, whereas there may be a non-null probability of loop routing with a general open network and memoryless routing. Yet, the approximation can be expected to provide reasonable results if the degree of multiplexing in the router is high enough, i.e., if packets of many different flows get mixed and interleaved as they are processed and eventually forwarded by the router.

The validity of those results requires also that the hypotheses underlying Jackson's networks hold.

If we are modeling a portion of the backbone Internet, external packet flows come from other parts of the Internet, hence they are the result of multiplexing of the individual packet flows of many traffic sources. Therefore, assuming those flows behave as a Poisson process (hypothesis (a)) is reasonable. The assumption is more questionable if we are modeling the network of an access ISP.

The hypothesis (b) on service times is hard to justify. Packet length probability distribution in real networks is very different from a negative exponential PDF. Packets have a minimum nonzero length and a maximum length. Besides, only few special sizes have significant probability, most of the admissible lengths being relatively improbable. Moreover, since packets arrive at routers through serial communication links and the transmission time is proportional to the packet length, there is a clear dependency between the packet length and packet inter-arrival times. If only a small fraction of packets coming from a same input link is routed to a given output link, the dependency is weak and could be neglected without impairing the model effectiveness.

As for hypothesis (c), routers have finite buffer sizes. In normal operational conditions, buffer overflow is a relatively rare event (less than a few percent of the packets get dropped). Therefore, infinite queue size is an acceptable approximation, especially if we aim at estimating the network delays.

Memoryless routing (hypothesis (d)) is not true in a packet network. Packets are routed according to their destination address (and possibly other parameters read from packet header), so routing is not random at all. Packets belonging to the same end-to-end flow are routed according to the same rules and hence follow the same network path (so long as the forwarding tables are not updated). If we consider a backbone router, where input traffic has a high degree of multiplexing, packets belonging to a large number of end-to-end flows are mixed up. Therefore, it is a reasonable approximation to assume that the link where each packet is routed can be described as the realization of a random variable. In fact, the probability that consecutive packets belong to the same end-to-end flow is very low.

Finally, hypothesis (e) corresponds to assuming that the topology of the considered packet network is connected, which is certainly true for an operational communication network.

Therefore, the result (7.17) for the mean delay through a packet network is only an approximation. The approximation is expected to be reliable, when there is a high degree of multiplexing of many flows in each link and many flows get mixed and interleaved within routers. This is most probable in core network routers, rather than in (small) routers at the edge of the network, especially those that connect small office and home networks.

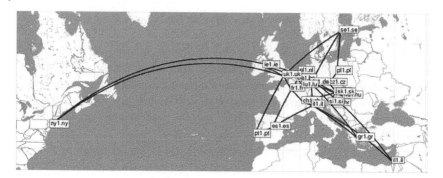

Figure 7.1 Picture of the Geant topology. It comprises 22 nodes and 72 links (image from `http://sndlib.zib.de`).

We close the example of application of the open network model with a numerical application to a specific network, namely Geant, the backbone of European national research networks. A large amount of experimental data on Geant traffic has been collected in the framework of the project Survivable fixed telecommunication Network Design[7]. The topology of Geant is illustrated in Figure 7.1. It comprises 22 nodes, and 72 (unidirectional[8]) links. Traffic demand matrices have been collected over a period of 4 months during 2005. The $22 \cdot 21 = 462$ off-diagonal elements of those matrices, arranged in a stacked vector, form the e2e flow vector denoted with $\mathbf{x} \equiv [x_s, s \in \mathcal{F}]$.

By using shortest path routing, it is possible to derive the 462×72 matrix $\mathbf{A}$ for Geant. Propagation delays have been computed by assuming a unit delay of 5 μs/km, typical of fiber-optic links. Link capacities have been set to the minimum among $\{2.5, 10, 20\}$ Gbit/s that guarantees link stability in the face of offered link traffic. A sequence of traffic matrices has been considered, corresponding to about 700 hours (one matrix every 15 min). The resulting time series of the overall mean network delay is shown in Figure 7.2(a).

An in-depth analysis of the numerical results points out that most of the delay is accounted for by the propagation delay, except for few links that get occasionally highly loaded and give rise to delay spikes. This can be seen in Figure 7.3, where we have plotted a sample of the the queueing delays, obtained by stripping off propagation delays, of the paths originating from 9 of the 22 nodes of the Geant network. For each originating node the plot shows the 21 time plots of the end-to-end queueing delay from the tagged origin node and every other node of the Geant network.

---

7 For more information and download resources refer to `http://sndlib.zib.de`.
8 Links between routers provide transfer capacity in both directions, independently, i.e., they are full-duplex. To account for that, each link between two routers is replaced with two directional arcs between the two corresponding nodes in the network graph.

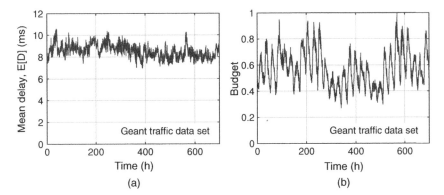

Figure 7.2 Left plot: mean delay through the Geant network for a time series of 2900 traffic demand matrices, one every 15 min. Right plot: budget as a function of time for the link capacity-optimized Geant network, under the constraint that the mean e2e delay be no greater than 12 *ms*.

Note that the measure unit on the y-axis is microseconds here. The overall mean network delay in Figure 7.2(a) is surprisingly stable, in spite of the wide time window observed. This is also a consequence of the averaging over the entire network traffic. It is also consistent with the fact the end-to-end delays are dominated by the propagation delays.

Example 7.2 *Vehicular traffic flows on a highway* We consider a unidirectional highway span. There are r ramps providing entrance/exit points to/from the highway. The considered highway span is divided up into r highway *links* , labeled with integers ranging from 1 to r. We use an integer index to refer to the k-th entrance point and the k-th exit point as well. The former is always assumed to be located upstream with respect to the latter. Link k starts at the k-th entrance point and ends at the k-th exit point and has length L_k ($k = 1, \ldots, r$). The Figure 7.4 illustrates an example of a linear highway with four entrance and exit points. In fact, the model applied in this example captures any connected topology of highway links.

The example could be easily generalized to a highway system comprising a set $\mathcal{L}$ of links, where each highway link has an entrance/exit ramp. Vehicular traffic offered to the highway system is made up by end-to-end vehicle flows (akin to packet flows in a communication network). We denote the intensity of the vehicular flow accessing ramp k by γ_k.

As for vehicle mobility, we denote the mean velocity of vehicle on link k as v_k. The maximum allowed velocity is denoted with $v_{\max}$. According to the fundamental transportation equation applied to a highway link (assumed to be in equilibrium), the mean vehicular traffic flow ϕ, density δ and velocity v are related as

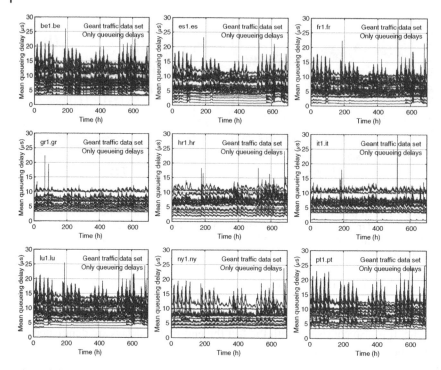

Figure 7.3 Mean end-to-end queueing delay (no propagation delay) for the paths originated from nine nodes of the Geant network; the label of the originating node is reported in the upper-left corner of each plot.

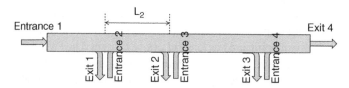

Figure 7.4 Example of linear highway topology with four entrance points and exits.

follows: $\phi = \delta \cdot v$. The mean density is actually a function of the velocity. According to Greenberg flow model, we can write $\delta = \delta_{max}e^{-v/v_c}$, where v_c is the critical velocity, at which the mean vehicular flow attains its maximum. Therefore, the mean sustainable flow rate on a highway link under equilibrium conditions can be written as $\phi(v) = \delta_{max}ve^{-v/v_c}$.

Finally, we let p_{hk} be the probability that a vehicle leaving the highway link h will enter the highway link k. The mean flow rate on highway link k is denoted with λ_k. The link flow rates can be found by solving the balance equations:

$$\lambda_k = \gamma_k + \sum_{h \in \mathcal{L}} \lambda_h p_{hk} , \qquad k \in \mathcal{L}. \tag{7.20}$$

We model the highway as an open network of queues. Single-server queues are used to model ramps. Arrivals of vehicle at those ramps (external flows) are assumed to follow a Poisson process. Serving at the access ramp consists of merging into the flow of vehicles traveling on the highway. We assume the time it takes for the head-of-line vehicle to merge into the highway flow can be represented by a negative exponential random variable.

Let μ_k denote the serving rate of the highway link k, i.e., the mean number of vehicles per unit time that can travel on that link at the assigned mean velocity in equilibrium. In the highway framework, the serving rate of link k is just the mean number of vehicles that can be sustained at equilibrium at the assigned mean velocity level, that is $\mu_k = \phi(v_k)$. Vehicles arriving at a mean rate γ_k at the ramp k see a *reduced* serving rate $\mu_k^{(r)}$ on the relevant highway link. Given that on average $\lambda_k - \gamma_k$ vehicles per unit time carry on to link k, arriving from upstream links, the mean available serving rate for vehicle at ramp k to merge is $\mu_k^{(r)} = \mu_k - (\lambda_k - \gamma_k)$. Then, the mean delay through the ramp k is

$$E[D_{\text{ramp},k}] = \frac{1}{\mu_k^{(r)} - \gamma_k} = \frac{1}{\mu_k - \lambda_k} \tag{7.21}$$

The mean number of vehicles on the k-th ramp can be found by using Little's law as

$$E[Q_{\text{ramp},k}] = \gamma_k E[D_{\text{ramp},k}] \tag{7.22}$$

As for the link k, we have:

$$E[D_{\text{link},k}] = \frac{L_k}{v_k} \qquad E[Q_{\text{link},k}] = \lambda_k E[D_{\text{link},k}] \tag{7.23}$$

Each highway link can be modeled as an infinite server queue with general service time. Here service time is just the time it takes the vehicle to travel the entire link, i.e., L_k/v_k, where L_k denotes the length of link k.

The mean delay through the highway system is obtained by applying Little's law to the open queueing network and calculating the mean number of vehicle residing in the system at equilibrium (ramps plus highway links).

The stability conditions are written as $\lambda_k < \phi(v_k), \forall k$.

Summing up, the steps for the delay analysis are as follows.

1. Given the external flow rates offered to the ramps $\gamma_k, k \in \mathcal{L}$, and the matrix $\mathbf{P}$ whose entries are the probabilities $p_{hk}, h, k \in \mathcal{L}$, calculate the link flow rates by solving the linear equation system:

$$\lambda_k = \gamma_k + \sum_{h \in \mathcal{L}} \lambda_h p_{hk} , \qquad k \in \mathcal{L}. \tag{7.24}$$

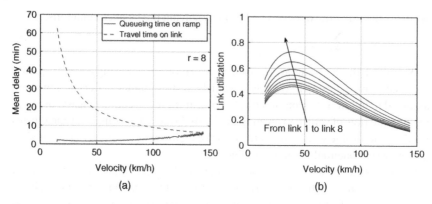

Figure 7.5 Performance of a linear highway span with random vehicle add-drop model. Left plot: mean delay through ramps (solid line) and on highway links (dashed line). Right plot: mean vehicle flow rate normalized with respect to the maximum capacity on each link.

Note that in the linear highway case, we have $p_{hk} = 0$ except for $h = k - 1$. Then, it is $\lambda_k = \gamma_k + \lambda_{k-1}p_{k-1,k}$, for $k \geq 2$ and $\lambda_1 = \gamma_1$.

2. Given the highway link velocity v_k, calculate the mean serving rate of highway link k:

$$\mu_k = \phi(v_k) = \delta_{\max}v_k e^{-v_k/v_c}, \qquad k \in \mathcal{L}. \tag{7.25}$$

3. Given the highway link lengths L_k, $k \in \mathcal{L}$, calculate the mean delay through the highway system by using eqs. (7.22) and (7.23) and Little's law. The result is:

$$E[D] = \frac{\sum\limits_{k\in\mathcal{L}}(E[Q_{\mathrm{ramp},k}] + E[Q_{\mathrm{link},k}])}{\sum\limits_{k\in\mathcal{L}}\gamma_k} = \frac{1}{\sum\limits_{k\in\mathcal{L}}\gamma_k}\sum\limits_{k\in\mathcal{L}}\left(\frac{\gamma_k}{\phi(v_k) - \lambda_k} + \lambda_k\frac{L_k}{v_k}\right) \tag{7.26}$$

To give numerical examples, we consider a probabilistic vehicle routing, where a fraction p of vehicles leaves at each exit ramp. Then, $p_{k-1,k} = 1 - p$ for all k, and all other entries of the matrix $\mathbf{P}$ are 0.

In Figure 7.5 numerical results for the highway example are given. We set a same velocity for all links. As for the numerical values of the parameters, we have assumed $v_c = 38.7$ km/h, $v_{\max} = 144$ km/h, $1/\delta_{\max} = 8$ m, $r = 8$, $L_k = 5000$ m, $\forall k$, $p = 0.25$.

Figure 7.5(a) shows the mean delay through ramps (solid line curve) and the mean delay through highway links (dashed line curve) as a function of the velocity. It is apparent that the transit delay through highway links is dominant, unless at very high velocity levels.

Figure 7.5(b) illustrates the link rate λ_k, normalized with respect to the link serving rate $\mu_k(v)$, as a function of the velocity v, i.e., the link utilization $\rho_k = \lambda_k/\mu_k(v)$. There is an optimal velocity level, i.e., one that maximizes the achieved link utilization. This is consistent with the fundamental transportation equation that gives a maximum flow rate at the critical velocity, while the flow rate decays to small values when the velocity is very low or very high.

7.2.1 Optimization of Network Capacities

As an example of optimization of a queueing network, we present here the classic link capacity optimization problem. Let us consider an open Jackson network of single server queues. We have discussed in Example 7.1 how this model can be used to study a packet network. For a given traffic demand offered to the network, we are interested in assigning the link capacities to minimize the mean delay through the network. By link capacity of the j-th link, C_j, here we refer to the serving rate of the single server of queue j, μ_j. The change of notation is related to the motivating context of the optimization problem. For example, minimizing the mean delay is a sensible performance target in the dimensioning of a communication or a transportation network. In these cases, serving rates are identified as the capacity of network links (communication links, roads).

Since the mean delay through the network $E[D]$ is monotonously decreasing with the link capacities $C_j, j = 1, \ldots, J$, the optimization is nontrivial only under suitable constraints. A natural constraint is to set a cap to the infrastructure cost, i.e., to the cost of installing link capacities. Let us assume that the cost of having a capacity C_j on link j is linear with C_j. Let the cost of link j capacity be $b_j C_j$ and let B denote the overall allowed cost budget.

The optimization problem can be stated as follows:

$$\textbf{Minimize:} \quad E[D] = \frac{1}{\gamma} \sum_{j=1}^{J} \frac{\lambda_j}{C_j - \lambda_j} \tag{7.27}$$

$$\textbf{Subject to:} \quad C_j > \lambda_j, \qquad j = 1, \ldots, J, \tag{7.28}$$

$$\sum_{j=1}^{J} b_j C_j \leq B \tag{7.29}$$

We assume that the external demand is given in terms of end-to-end flow rates $x_s, s \in \mathcal{F}$, and that also routing is assigned. Then, the link rates are determined as $\lambda_j = \sum_{s \in \mathcal{F}} x_s a_{sj}$, where a_{sj} is equal to 1 if and only if link j belongs to the route of flow s. Moreover, the overall offered rate γ is given by $\gamma = \sum_{s \in \mathcal{F}} x_s$.

The objective function is convex in the variables C_j, while the constraints define a bounded convex feasible set. Provided the feasible set is nonempty, the solution to the optimization problem is unique and it can be found by using Lagrange multipliers. The Lagrangian is

$$L(C_1, \ldots, C_J, \psi) = \sum_{j=1}^{J} \frac{\lambda_j}{C_j - \lambda_j} + \psi \sum_{j=1}^{J} b_j C_j \tag{7.30}$$

where we have dropped the positive factor $1/\gamma$. Setting the derivatives to 0, we find

$$\frac{\partial L}{\partial C_j} = -\frac{\lambda_j}{(C_j - \lambda_j)^2} + \psi b_j = 0 \quad \Rightarrow \quad C_j = \lambda_j + \sqrt{\frac{\lambda_j}{\psi b_j}} \tag{7.31}$$

Using these values for the C_j's into the constraint[9] we have

$$\sum_{j=1}^{J} b_j \lambda_j + \frac{1}{\sqrt{\psi}} \sum_{j=1}^{J} b_j \sqrt{\frac{\lambda_j}{b_j}} = B \tag{7.32}$$

from which we derive an expression of ψ. The final expression of the link capacity at the optimum, denoted with a superscript asterisk, is

$$C_j^* = \lambda_j + \frac{\sqrt{b_j \lambda_j}}{\sum\limits_{k=1}^{J} \sqrt{b_k \lambda_k}} \frac{1}{b_j} \left(B - \sum_{k=1}^{J} b_k \lambda_k \right), \qquad j = 1, \ldots, J. \tag{7.33}$$

Plugging the optimal capacities back into the expression of the mean delay, we obtain

$$E[D^*] = \frac{1}{\gamma} \frac{\left(\sum\limits_{j=1}^{J} \sqrt{b_j \lambda_j} \right)^2}{B - \sum\limits_{j=1}^{J} b_j \lambda_j} \tag{7.34}$$

The final result has a nice interpretation. The optimal capacity of link j is the sum of two terms. The first term, λ_j, is the bare minimum capacity required to keep up with the traffic load of link j (completely determined by the assigned external demand and the routing). The second one is proportional to the *excess* budget $B - \sum_{k=1}^{J} b_k \lambda_k$, left over once the minimum requirement to guarantee stability has been met. That excess budget is distributed over the links according to weights proportional to the square root of the link load and of the unit link cost.

As clear from (7.33), the optimization problem is feasible (i.e., the search region for the capacities C_j is not empty) only if $B > \sum_{k=1}^{J} b_k \lambda_k$.

---

9 Since the objective function is monotonously decreasing with the C_j's, the constraint is attained with equality sign at the optimum. From a mathematical point of view, the complementary slackness conditions imply that the constraint is satisfied with equality sign, since the Lagrange multiplier ψ cannot be 0.

Example 7.3 *Optimization of a backbone communication network* We apply the capacity optimization to the Geant network, introduced in Example 7.1. We assume the cost of a link is proportional to the link capacity C_j and to the distance between the two end-points of the link. The weight b_j is therefore identified with the geographical distance between the two cities where the link is terminated. To fix a budget, we require that the mean end-to-end delay through the network be no more than a required level D_{req}. Using the expression (7.34) of the mean end-to-end queueing delay and accounting for the propagation delays, we find

$$B = B_{req} = \sum_{j=1}^{J} b_j \lambda_j + \frac{\left(\sum_{j=1}^{J} \sqrt{b_j \lambda_j} \right)^2}{\gamma D_{req} - \sum_{j=1}^{J} \lambda_j \tau_j} \tag{7.35}$$

Since we consider a sequence of 2900 traffic matrices (one every 15 minutes), we can evaluate a time sequence of required budget values. The series of budgets obtained for the Geant network is plotted in Figure 7.2(b). Budget values are normalized with respect to the maximum required budget among all considered traffic matrices. The average budget value over the inspected time horizon is about 0.52. This makes a strong case for being able to adapt capacities to the offered traffic flow. Going for the worst-case dimensioning appears to be highly wasteful of communication resources.

Note that, having link capacities optimized at each time for the given budget implies from a system point of view that link capacities can be reconfigured on a time scale of 15 minutes (the time period of traffic demand matrix sampling). This is possible with a number of technologies, i.e., if links are obtained by using multi-protocol label switching (MPLS) over a transport network, so that a link is an MPLS path, whose capacity can be reconfigured by means of the traffic engineering tools of MPLS. Another example of rearrangeable link capacities is offered by optical networks based on dense wavelength division multiplexing (DWDM), where links correspond to light paths, that can be configured via a centralized software control of the network, e.g., according to the software defined network (SDN) paradigm.

7.2.2 Optimal Routing

We discuss here a fundamental network optimization problem: given a set of nodes and links, forming a connected network topology, given a traffic matrix expressing the amount of traffic demand from node i to node j, for all couples (i, j), find traffic flow routes that minimize the overall mean end-to-end delay of packets through the network. In this problem, link capacities are assumed as given.

We introduce notation to give a formal statement. Let S denote the set of traffic origin-destination pairs of the network, R denote the set of routes of the network, and L the set of links of the network. We denote the flow on route j with x_j. Let $\mathbf{x}$ be the column vector collecting all x_j's.

We define also the matrices $\mathbf{A}$, with entry $a_{kj} = 1$ if route j uses link k, 0 otherwise, and $\mathbf{B}$, with entry $b_{sj} = 1$ if route j is used by the origin-destination pair s, 0 otherwise.

The vector $\mathbf{f} = \mathbf{Bx}$ gives the end-to-end traffic flows of the origin-destination pairs. The traffic rates loading network links are given by $\mathbf{y} = \mathbf{Ax}$.

We assume that Kleinrock's model applies to the packet network. Let C_k and τ_k be the capacity and the propagation delay of link k. Let $\gamma = \sum_{s \in S} f_s$ be the overall traffic flow rate offered to the network. We can express the mean end-to-end delay through the network as

$$\overline{D} = \frac{1}{\gamma} \sum_{k \in L} \frac{y_k}{C_k - y_k} + \tau_k y_k \tag{7.36}$$

provided that $y_k < C_k$, $\forall k \in L$. We define the link "cost" D_k as a function proportional to the contribution of link k to the overall mean delay:

$$D_k(y) = \frac{y}{C_k - y} + \tau_k y \tag{7.37}$$

Routing the traffic demand f_s, $s \in S$, means determining how much of that traffic demand is to be sent on route $r \in R$. Optimal routing is obtained as the solution of the following optimization problem:

$$\textbf{Minimize}: \quad \sum_{k \in L} D_k \left(\sum_{j \in R} a_{kj} x_j \right) \tag{7.38}$$

$$\textbf{Subject to}: \quad \sum_{j \in R} b_{sj} x_j = f_s, \quad s \in S \tag{7.39}$$

$$\sum_{j \in R} a_{kj} x_j \leq C_k, \quad k \in L \tag{7.40}$$

$$x_j \geq 0, \quad j \in R. \tag{7.41}$$

The target function is separable. Each component under the summation sign is a strictly increasing, convex function. The domain of the variables is bounded and convex. Therefore, if the feasible domain is not empty, i.e., if there exists at least one flow vector $\mathbf{x}$ satisfying all constraints, there is a unique minimizer $\mathbf{x}^*$. Any algorithm that leads to finding one minimizer, will find *the* global minimum.

Let us characterize the optimal flows. First we observe that

$$\frac{\partial \overline{D}}{\partial x_j} = \sum_{k \in L} D_k'(y_k) \frac{\partial y_k}{\partial x_j} = \sum_{k \in L} a_{kj} D_k'(y_k) \tag{7.42}$$

We see that the derivative of the target function $\overline{D}$ with respect to the route flow x_j is the *sum of the derivatives of the link cost functions over all links making up the route j*. We will refer to this sum as the *route length* or route cost. Thus, the shortest path for the origin-destination pair s is any route j, belonging to the set of routes usable by pair s, i.e., such that $b_{sj} = 1$, having the least value of the derivative (7.42).

Let us consider the optimal flow x_j^* on route j, used by origin-destination pair s. If we shift a small amount δx of flow form route j to another route i, also used by the same origin-destination pair s, we cannot improve the target function, otherwise the flow x_j^* would not be optimal.

The variation of the target function, to first order, is given by

$$\delta \overline{D} = \delta x \frac{\partial \overline{D}}{\partial x_i} - \delta x \frac{\partial \overline{D}}{\partial x_j} \geq 0 \tag{7.43}$$

where the derivatives are calculated at $\mathbf{x}^*$. We conclude that

$$x_j^* > 0 \quad \Rightarrow \quad \frac{\partial \overline{D}}{\partial x_j} \leq \frac{\partial \overline{D}}{\partial x_i}, \quad \forall i \in \mathcal{R} \text{ such that } b_{si} = b_{sj} = 1. \tag{7.44}$$

In words, the end-to-end flow s is routed on the least cost routes, i.e., those that have the smallest possible length among all routes available to flow s. Optimal routing consists of sending the traffic flow offered to the origin-destination pair s via the shortest routes available to that pair. Note however that route lengths depend on the flow distribution over routes. Therefore, it is not known a priori which routes will be the shortest ones.

The optimization can be pursued by a gradient descent method. Let $\mathbf{x}$ be a feasible flow allocation to routes, i.e., a flow allocation that satisfies all constraints. A feasible variation of the flow assignment is a vector $\Delta \mathbf{x}$ such that $\mathbf{x} + \Delta \mathbf{x}$ still satisfies all constraint. Note that this implies that $\Delta x_j \geq 0$ if it is $x_j = 0$. Moreover, since it must be $\sum_{j \in \mathcal{R}} b_{sj} x_j = f_s$, we have $\sum_{j \in \mathcal{R}} b_{sj} \Delta x_j = 0$ for a feasible route flow update.

The direction of the route flow update should be such that the inner product of the increment and of the gradient of the target function is negative, i.e.,

$$\nabla \overline{D} \cdot \Delta \mathbf{x} = \sum_{j \in \mathcal{R}} \sum_{k \in \mathcal{L}} a_{kj} \frac{\partial D_k}{\partial x_j} \Delta x_j \leq 0 \tag{7.45}$$

The inner product can be obtained as the derivative of the function $G(\alpha) = \overline{D}(\mathbf{x} + \alpha \Delta \mathbf{x})$ with respect to the scalar parameter α.

The remarks above have lead to the definition of the famous *flow deviation* algorithm [81]. This is a special case of a multi-commodity flow problem [82], specifically of the Frank-Wolfe approach to the minimization of convex function over a convex feasible domain.

Let us assume we are given a feasible route flow assignment $\mathbf{x}$. We can then compute the route lengths (7.42). For each origin-destination pair s we can thus determine the shortest path, say it is j_s. Let $\overline{x}_j$ denote the route flows obtained by assigning the whole flow of the origin-destination pair s to its shortest path j_s as determined by the route length calculated for the current flow assignment $\mathbf{x}$, for $s \in S$. Formally, $\overline{x}_{j_s} = f_s$ and $\overline{x}_j = 0$ for all $j \neq j_s$ and such that $b_{sj} = 1$.

The route flows are updated as follows:

$$\mathbf{x}' = \mathbf{x} + \alpha^*(\overline{\mathbf{x}} - \mathbf{x}) \tag{7.46}$$

where α^* is the value of $\alpha \in [0, 1]$ that minimizes $G(\alpha) = \overline{D}(\mathbf{x} + \alpha(\overline{\mathbf{x}} - \mathbf{x}))$. The rationale of (7.46) is clarified by observing that, for those j such that $\overline{x}_j = 0$ (non-shortest paths) it is $x'_j = x_j(1 - \alpha^*)$, i.e., some flow is shifted to those paths. For the shortest path j_s, the new assigned flow is *less* than $\overline{x}_{j_s} = f_s$.

It appears that the updated point $\mathbf{x}'$ is obtained from the shortest path flow allocation by deviating some flow amount to nonshortest paths, in search of an improvement of the overall network cost (descent direction along the line $\mathbf{x} + \alpha(\overline{\mathbf{x}} - \mathbf{x})$).

The flow deviation algorithm is carried out until the decrease of the target function drops below a prescribed threshold.

To run the flow deviation algorithm we need:

- An efficient way to find the shortest paths on a weighted network graphs (the weight of link k is given by $D'_k(y_k)$). This is accomplished, e.g, by using Dijkstra's algorithm.
- An efficient way to find α^*. This requires the solution of a unidimensional optimization problem. The optimal α can be located efficiently by the bisection method or the Fibonacci search algorithm.
- An efficient way to determine an initial feasible route flow allocation, if one exists. This is not straightforward. An efficient algorithm is described in [81,82].

7.2.3 Braess Paradox

Queues are nonlinear systems and much of the interest and nontrivial results are rooted in this character of the queueing models. Networks of queues add a topological structure, represented by a graph, to the nonlinearity of the queues. It should be expected that queueing networks may sometimes defeat intuition and give rise to (apparently) paradoxical results. One of the most famous is the Braess paradox, originally introduced in [42] with reference to transportation engineering.

Let us consider a four-node, four-link topology, described by a directed graph (see Figure 7.6). Nodes are labeled with capital letters. A single flow of customers arrives at the network at node A and it is addressed to node D. We assume that arrivals follow a Poisson process with mean rate 2λ. As for the network, we assume that links AB and CD are equipped with a single-server queue, with serving rate μ

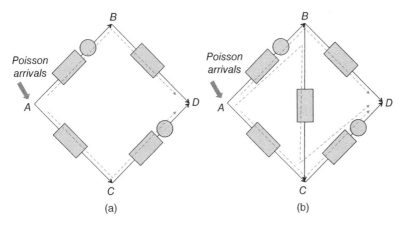

Figure 7.6 Example network of Braess paradox. Left plot: original network. Right plot: network after the addition of a new link.

and negative exponential service times. The links BD and AC impose a fixed delay. We can think of them as $M/D/\infty$ queueing systems.

A queueing network model of this system is therefore made of four queues, as sketched in Figure 7.6(a). It is a special case of queueing network, since customers can never loop back into an already-visited queue. Such networks are called feed-forward. They lend themselves to a nice analysis, especially in the present case. In fact, the input process of queues AB and AC is a Poisson process, if splitting occurs randomly at A, independently of the inter-arrival times and the queue states. We assume this is the case. Moreover, the output process of an $M/G/\infty$ queue is still a Poisson process. This is a consequence of the general result proved in § 3.2.3.1, stating that independent, random displacement of a Poisson point process yields yet another Poisson point process with the same mean rate. Then, also the input processes of queues BD and CD are Poisson. Summing up, the two queues AB and CD are $M/M/1$, while AC and BD are $M/D/\infty$ queues.

Let the constant service time of the infinite server queues be 2 (in some given time unit). Let also $\mu > \lambda + 1$.

The topology of the Braess network offers two alternative paths for customers to move from A to D, namely ABD and ACD. The key point is the criterion to select one of the two paths. We assume that customers act randomly, independently on one another. They prefer the path with the least expected mean travel time T. At equilibrium the mean travel times of the two paths must be the same, i.e., $T_{ABD} = T_{ACD}$. In fact, if there were any mismatch between the travel times of the two paths, more customers would choose the "shorter" path, i.e., the one with the smaller travel time, until the congestion level is such that the travel time of one path equals the travel time of the other path.

Let λ_{ABD} and λ_{ACD} be the mean rate of customers selecting either path. At equilibrium we have

$$T_{ABD} = T_{ACD} \qquad \Rightarrow \qquad 2 + \frac{1}{\mu - \lambda_{ABD}} = 2 + \frac{1}{\mu - \lambda_{ACD}}$$

which implies $\lambda_{ABD} = \lambda_{ACD}$. Since $\lambda_{ABD} + \lambda_{ACD} = 2\lambda$, we find $\lambda_{ABD} = \lambda_{ACD} = \lambda$, as expected because of the symmetry of the network.

Let us now add one more link, from B to C, to the topology, that is, more capacity for customers to move from A to D (see Figure 7.6(b)). The new link BC is equipped with an infinite server queue with constant service times equal to 1. If customer splitting at B among the alternative paths ABD and $ABCD$ takes place randomly, the queue corresponding to link BC is modeled as an $M/D/\infty$ queue.

To understand how the customer preferences adjust with the new link, consider the situation at the time the new link has just been opened. A new customer arriving at A can choose one of the old paths, being guaranteed an average travel time of $2 + 1/(\mu - \lambda)$. If the customer ventures into the new path, it will experience a travel time of $T_{ABCD} = 1 + 2/(\mu - \lambda)$. It can be verified that $T_{ABCD} < T_{ABD} = T_{ACD}$ since $\mu > \lambda + 1$. Therefore, the tagged customer will certainly choose the newly opened path. This "migration bias" of customers from old paths to the new path will go on until a new equilibrium is achieved. Given the selfish nature of independent customer choices, the statistical equilibrium is achieved only when the travel times of all three paths are the same. Let $\lambda_1 = \lambda_{ABD} = \lambda_{ACD}$, by symmetry, and $\lambda_2 = \lambda_{ABCD}$. The rates λ_1 and λ_2 satisfy the equations:

$$\begin{aligned} 2 + \frac{1}{\mu - \lambda_1 - \lambda_2} &= 1 + \frac{2}{\mu - \lambda_1 - \lambda_2} \\ 2\lambda_1 + \lambda_2 &= 2\lambda \end{aligned} \qquad (7.47)$$

which yields $\lambda_1 = 2\lambda + 1 - \mu$ and $\lambda_2 = 2\mu - 2 - 2\lambda$, assuming $\mu \leq 1 + 2\lambda$. Since $1/(\mu - \lambda_1 - \lambda_2) = 1$, it turns out that $T_{ABD} = T_{ACD} = T_{ABCD} = 3$. This is *larger* than in the case without link BC, since $3 > 2 + 1/(\mu - \lambda)$, due to the assumption $\mu > \lambda + 1$.

The appalling conclusion is that, having added more capacity to the network, the average end-to-end travel time of the customer flow has *increased* with respect to the original situation.

The basic motivation of this paradoxical conclusion is the selfish path selection criterion of customers. After all, Braess paradox is yet another manifestation of the nonoptimality of the equilibrium outcomes of games among selfish players. In other words, it may well be the case that the social utility is not maximized by letting selfish independent entities make their individual choices. The explanation of the paradox and an approach to add capacity to a network without incurring into the effects of Braess paradox are discussed in [133].

An optimization framework insight of Braess paradox is given in [122]. Given a network of directed links, forming a set $\mathcal{L}$, we define a routing optimization problem as follows. We consider the source-destination pair set S and let f_s denote the mean rate of the flow associated with the source-destination pair s, $s \in S$. Let also $\mathcal{R}$ denote the set of routes or paths connecting source-destination pairs and x_r the mean rate of the flow routed on $r \in \mathcal{R}$. Finally, let y_j be the link rate of link $j \in \mathcal{L}$. We define two incidence matrices:

- The matrix $\mathbf{H}$ is such that $h_{sr} = 1$ if and only if route $r \in \mathcal{R}$ connects the source-destination pair s.
- The matrix $\mathbf{A}$ is such that $a_{jr} = 1$ if and only if link $j \in \mathcal{L}$ belongs to route r.

The vectors $\mathbf{f}$ and $\mathbf{x}$ of flow rates and the vector $\mathbf{y}$ of link rates are related by $\mathbf{f} = \mathbf{Hx}$ and $\mathbf{y} = \mathbf{Ax}$.

We assume the topology and routing information condensed into $\mathbf{H}$ and $\mathbf{A}$ is assigned, as well as the source-destination flows $\mathbf{f}$. Our degree of freedom is the vector $\mathbf{x}$, i.e., the way the end-to-end flow f_s is split among the possible alternative routes serving it. To guide the splitting, we define the delay of link j as a function of the link load, i.e., $D_j(y_j), j \in \mathcal{L}$. This is a monotonously increasing function of y_j, typically convex and strictly increasing. It might have also a vertical asymptote at some critical link load level.

We can state the following optimization problem:

$$\textbf{Minimize}: \quad \sum_{j \in \mathcal{L}} \int_0^{y_j} D_j(u) \, du \tag{7.48}$$

$$\textbf{Subject to}: \quad \mathbf{f} = \mathbf{Hx}, \mathbf{y} = \mathbf{Ax} \tag{7.49}$$

$$\mathbf{x} \geq \mathbf{0}, \quad \mathbf{y} \tag{7.50}$$

The Lagrangian is

$$L(\mathbf{x}, \mathbf{y}, \xi, \psi) = \sum_{j \in \mathcal{L}} \int_0^{y_j} D_j(u) \, du + \xi(\mathbf{f} - \mathbf{Hx}) - \psi(\mathbf{y} - \mathbf{Ax}) \tag{7.51}$$

We write down the Karush-Kuhn-Tucker (KKT) conditions, namely

$$\frac{\partial L}{\partial x_i} = -\xi_{s(i)} + \sum_{j \in \mathcal{L}} \psi_j a_{ji} \quad i \in \mathcal{R}, \qquad \frac{\partial L}{\partial y_j} = D_j(y_j) - \psi_j, \quad j \in \mathcal{L}, \tag{7.52}$$

where $s(i)$ is the index of the origin-destination pair served by route i. Setting the derivatives to 0^{10}, it follows that it must be $\psi_j = D_j(y_j)$ and

$$\begin{cases} \xi_{s(i)} = \sum_{j \in \mathcal{L}} \psi_j a_{ji} = \sum_{j \in \mathcal{L}} D_j(y_j) a_{ji} & x_i > 0 \\ \xi_{s(i)} \leq \sum_{j \in \mathcal{L}} \psi_j a_{ji} = \sum_{j \in \mathcal{L}} D_j(y_j) a_{ji} & x_i = 0 \end{cases} \tag{7.53}$$

---

10 More precisely, the derivative with respect to x_i must be 0 if $x_i > 0$; if $x_i = 0$, the condition is that the derivative must be non-negative.

The sum $\sum_{j \in \mathcal{L}} D_j(y_j) a_{ji}$ is therefore minimal where $x_i > 0$ for the source-destination pair $s(i)$. It turns out that the optimal routes are those yielding the minimum delay, i.e., those that minimize $\sum_{j \in \mathcal{L}} D_j(y_j) a_{jr}$ for r varying over $\mathcal{R}$. Thus, the optimization problem stated above is the one solved by a customer selecting its path so that everyone gets an end-to-end delay that it cannot improve with unilateral selection changes, given the selections of other customers. This selfish behavior translates into the optimization problem stated above, which makes end-to-end flows go through routes offering a delay smaller than any alternative route. Then, there is no incentive to change for anyone. Why is this a bad strategy? It is apparent that what is getting optimized has little to do with a measure of the mean "social delay" suffered by customers traveling through the network.

The optimization problem that makes sense to minimize the "social" mean delay through the network could be stated instead as follows:

$$\textbf{Minimize}: \quad \sum_{j \in \mathcal{L}} y_j D_j(y_j) \tag{7.54}$$

$$\textbf{Subject to}: \quad \mathbf{f} = \mathbf{Hx}, \mathbf{y} = \mathbf{Ax} \tag{7.55}$$

$$\mathbf{x} \geq \mathbf{0}, \quad \mathbf{y} \tag{7.56}$$

Here the target function is the mean number of customers found in the network at equilibrium. By Little's law, this is proportional to the mean delay through the network.

When writing the KKT conditions for this last optimization problem, one finds out that the selected routes are those having minimum path cost, where now the link cost is not the sheer delay $D_j(y_j)$; rather, it is $D_j(y_j) + y_j D'_j(y_j)$. The second term can be considered a "toll" cost introduced to obtain an incentive to achieve the overall network target of minimum average transit delay.

If we write the mean delay for Braess network (actually we write the mean number of customers found in the network at equilibrium) we have

$$\lambda_1 T_{ABD} + \lambda_1 T_{ACD} + \lambda_2 T_{ABCD} = 2\lambda_1 \left(2 + \frac{1}{\mu - \lambda_1 - \lambda_2} \right) + \lambda_2 \left(1 + \frac{2}{\mu - \lambda_1 - \lambda_2} \right)$$

$$= 4\lambda_1 + \lambda_2 + \frac{2(\lambda_1 + \lambda_2)}{\mu - \lambda_1 - \lambda_2} \tag{7.57}$$

under the constraint $2\lambda_1 + \lambda_2 = 2\lambda$. Substituting λ_1 from the constraint, we get

$$4\lambda - \lambda_2 + \frac{2\lambda + \lambda_2}{\mu - \lambda - \lambda_2/2}, \quad 0 \leq \lambda_2 \leq 2\lambda. \tag{7.58}$$

It can be verified that this function is monotonously increasing with λ_2 under the condition $\mu > \lambda + 1$. The minimum is attained for $\lambda_2 = 0$, which means that the new added link of Braess network shall not be used (hence we would have better not to add it at all).

7.3 Closed Queueing Networks

A closed queueing network is similar to an open one, except that no external arrival is allowed and, correspondingly, no exit out of the network can take place. The number of customers in the network is fixed, denoted by N in the following. The N customers move around the J queues (service stations) all the time. The assumptions closely match those made for open queueing networks:

(a) Negative exponential service time, possibly state-dependent.
(b) Independence of inter-arrival and service times.
(c) Infinite queue lengths, i.e., all customers arriving at a queue can join the queue.
(d) Memoryless routing: a customer leaving one queue chooses the next one independently of all other customers.
(e) The $J \times J$ routing matrix $\mathbf{R}$ whose entries are the r_{ij}'s is irreducible.

We still consider the solution of the linear equation system:

$$v_j = \sum_{i=1}^{J} v_i r_{ij}, \qquad j = 1, \dots, J \tag{7.59}$$

This is a homogeneous system, still it admits a nontrivial solution. In fact, the routing matrix $\mathbf{R}$ is a $J \times J$ irreducible, stochastic matrix, hence it corresponds to an ergodic finite Markov chain with a positive stationary vector, that is the left eigenvector of $\mathbf{R}$ corresponding to the eigenvalue 1. The vector $\mathbf{v} = [v_1, \dots, v_J]$ is proportional to such a stationary vector. For our purposes, in the closed queueing network, the vector $\mathbf{v}$ can be defined up to a multiplicative constant. The constant can be fixed, e.g., by setting the sum of the components of $\mathbf{v}$ equal to 1 or letting one of its components be equal to 1.

Let us define the state of the queueing network as the vector $\mathbf{Q}(t) = [Q_1(t), \dots, Q_J(t)]$, where $Q_j(t)$ is the number of customers residing into queue j at time t.

Under the hypotheses (a)-(e) listed above, the state $\mathbf{Q}(t)$ is a Markov chain on the non-negative integer vectors $\mathbf{x} = [x_1, \dots, x_J]$ satisfying the constraint $|\mathbf{x}| \equiv x_1 + \cdots + x_J = N$. This is a finite state space and the continuous time Markov chain $\mathbf{Q}(t)$ is irreducible since the routing matrix $\mathbf{R}$ is irreducible itself. So, a limiting stationary probability vector exists, let it be $\pi(\mathbf{x}) = P(\mathbf{Q} = \mathbf{x})$. Let us define

$$P(Y_i = k) = P(Y_i = 0) \frac{v_i^k}{M_i(k)}, \qquad 0 \le k \le N \tag{7.60}$$

with

$$M_i(k) = \begin{cases} 1 & k = 0 \\ \prod_{h=1}^{k} \mu_i(h) & k \ge 1 \end{cases} \tag{7.61}$$

where $\mu_j(h)$ is the serving rate of queue j when there are $Q_j = h$ customers in that queue, $j = 1, \ldots, J$. Note that $\mu_j(0) = 0$, while it must be $\mu_j(h) > 0$ for any $h > 0$.

The key result for Jackson-type closed queueing networks is stated as follows:

Theorem 7.2 (Gordon-Newell [92]) If a closed queueing network satisfies the hypotheses (a)-(e) listed above, the limiting PDF of the Markov process $\mathbf{Q}(t)$ exists and it is given by a product form as

$$\pi(\mathbf{x}) = \frac{\prod_{j=1}^{J} P(Y_j = x_j)}{P(|\mathbf{Y}| = N)} = \frac{1}{G} \prod_{j=1}^{J} \frac{v_j^{x_j}}{M_j(x_j)} \tag{7.62}$$

for all $\mathbf{x}$ such that $\mathbf{x} \geq 0$ and $|\mathbf{x}| = N$, where G is a normalization constant.

Remark: The queue lengths of the closed networks are not independent random variables as in the open network, and this is apparent from the expression of $\pi(\mathbf{x})$ in eq. (7.62).

Proof: The expression of the balance equations for the closed queueing network is given by:

$$\pi(\mathbf{x}) \sum_{j=1}^{J} \mu_j(x_j)(1 - r_{jj}) = \sum_{j=1}^{J} \sum_{i \neq j, i=1}^{J} \pi(\mathbf{x} - \mathbf{e}_j + \mathbf{e}_i)\mu_i(x_i + 1)r_{ij} \tag{7.63}$$

According to the expression in eq. (7.60), it is

$$\pi(\mathbf{x} - \mathbf{e}_j + \mathbf{e}_i) = \pi(\mathbf{x}) \frac{\mu_j(x_j)v_i}{v_j \mu_i(x_i + 1)}, \qquad x_j > 0 \tag{7.64}$$

Then, by substituting into eq. (7.4) and canceling out $\pi(\mathbf{x})$, we get

$$\sum_{j=1}^{J} \mu_j(x_j)(1 - r_{jj}) = \sum_{j=1}^{J} \sum_{i \neq j, i=1}^{J} \frac{\mu_j(x_j)v_i}{v_j \mu_i(x_i + 1)} \mu_i(x_i + 1)r_{ij} =$$

$$= \sum_{j=1}^{J} \frac{\mu_j(x_j)}{v_j} \sum_{i \neq j, i=1}^{J} v_i r_{ij} = \sum_{j=1}^{J} \mu_j(x_j)(1 - r_{jj}) \tag{7.65}$$

The last equality is a consequence of eq. (7.59). This proves the theorem. ∎

The normalizing constant $G \equiv G(J, N)$ can be expressed as:

$$G = \sum_{\mathbf{x} \geq 0 \,:\, |\mathbf{x}|=N} \prod_{j=1}^{J} \frac{v_j^{x_j}}{M_j(x_j)} \tag{7.66}$$

Note that any scaling factor multiplying the vector **v** cancels out in eq. (7.62) since it appears with the same power N both in the v_j's in the numerator and in the expression of the constant G.

The computation of the constant G requires summing $\binom{N+J-1}{N}$ terms[11], so that efficient algorithms are required unless the closed network is very small, i.e., N and J reduce to a few units. We will address this issue later on. Now we illustrate how the constant G appears in some useful performance metrics.

The marginal probability distribution of Q_j is found to be

$$P(Q_j = k) = P(Y_j = k) \sum_{\mathbf{x}_{-j} \geq 0 \,:\, |\mathbf{x}_{-j}| = N-k} \frac{\prod\limits_{i \neq j, i=1}^{J} P(Y_i = x_i)}{P(|\mathbf{Y}| = N)}$$

$$= P(Y_j = k) \frac{P(|\mathbf{Y}_{-j}| = N - k)}{P(|\mathbf{Y}| = N)}$$

where $\mathbf{Y}_{-j}$ is a row vector of the random variables Y_i's, except the j-th one, Y_j, i.e., $\mathbf{Y}_{-j} = [Y_1, \ldots, Y_{j-1}, Y_{j+1}, \ldots, Y_J]$. An analogous definition holds for $\mathbf{x}_{-j}$.

The throughput of queue j is by definition

$$E[\mu_j(Q_j)] = \sum_{k=1}^{N} \mu_j(k) P(Q_j = k) \tag{7.67}$$

so we have

$$E[\mu_j(Q_j)] = \sum_{k=1}^{N} \mu_j(k) P(Y_j = k) \frac{P(|\mathbf{Y}_{-j}| = N - k)}{P(|\mathbf{Y}| = N)}$$

$$= \sum_{k=0}^{N-1} \mu_j(k+1) P(Y_j = k+1) \frac{P(|\mathbf{Y}_{-j}| = N - 1 - k)}{P(|\mathbf{Y}| = N)}$$

$$= \sum_{k=0}^{N-1} v_j P(Y_j = k) \frac{P(|\mathbf{Y}_{-j}| = N - 1 - k)}{P(|\mathbf{Y}| = N)}$$

$$= v_j \frac{P(|\mathbf{Y}| = N - 1)}{P(|\mathbf{Y}| = N)} \sum_{k=0}^{N-1} P(Y_j = k) \frac{P(|\mathbf{Y}_{-j}| = N - 1 - k)}{P(|\mathbf{Y}| = N - 1)}$$

$$= v_j \frac{P(|\mathbf{Y}| = N - 1)}{P(|\mathbf{Y}| = N)} \tag{7.68}$$

---

11 Counting the states can be done as follows. To distribute N customers over J queues, imagine aligning all customers over a straight line and set $J - 1$ barriers, separating them into J groups. A group is made by all customers lying within two barriers, including the boundary groups at the two ends of the line. To realize this layout, we can provide a line with $N + J - 1$ places and mark each place with a C for a customer, with a B for a barrier. The number of choices of the $J - 1$ positions of the barriers over the available $N + J - 1$ positions, i.e., $\binom{N+J-1}{J-1}$, is just the number of ways of forming J groups of customers.

since the summation on the right-hand side in the last but one passage is just the sum of the marginal PDF of the random variable Q_j in a closed queueing network identical to the given one except that there are $N-1$ customers. The expression of the throughput reveals that the only part that depends on the specific queue is the parameter v_j. This leads to the definition of a global network parameter that goes under the name of *network throughput*:

$$TH(N) = \frac{P(|\mathbf{Y}| = N-1)}{P(|\mathbf{Y}| = N)} \tag{7.69}$$

Since the vector $\mathbf{v}$ is defined up to a multiplicative constant, we can set $|\mathbf{v}| = 1$ and thus we see that the network throughput is the sum of the throughputs of queues making up the network.

7.3.1 Arrivals See Time Averages (ASTA)

We have already discussed the PASTA property in queues with Poisson arrivals. An arriving customer belonging to a Poisson flow finds the queue in state x with its steady-state probability distribution $\pi(x)$ (provided the visited queue is at equilibrium). Arrivals in a closed queueing network do not consist of Poisson processes. Still, an analogous property holds for the class of queueing networks we are considering, often called Jackson networks (e.g., see [86, Ch. 6]).

Let $\pi^{(N)}(\mathbf{x})$ be the probability that a closed Jackson network with J queues and N customers is in state $\mathbf{x}$. We highlight the number of customers circulating in the network in the superscript of the steady-state probabilities. Note that the elements of $\mathbf{x}$ sum up to N. Let also $\mathbf{z}$ be a vector of non-negative integers summing to $N-1$ and $p_j(\mathbf{z})$ be the probability that a customer arriving at queue j sees z_k customers in queue k for $k = 1, \dots, J$. This probability can be obtained by considering the transition of the customer from its originating queue, say i, to queue j, in a time interval h shrinking to 0:

$$p_j(\mathbf{z}) = \lim_{h \to 0} \frac{\displaystyle\sum_{i=1}^{J} \pi^{(N)}(\mathbf{z} + \mathbf{e}_i) r_{ij} [\mu_i(z_i + 1)h + o(h)]}{\displaystyle\sum_{\mathbf{y}} \sum_{i=1}^{J} \pi^{(N)}(\mathbf{y} + \mathbf{e}_i) r_{ij} [\mu_i(y_i + 1)h + o(h)]} \tag{7.70}$$

where the sum over $\mathbf{y}$ is extended to all J-dimensional vectors of non-negative integers that sum up to $N-1$ (all the customers circulating in the considered network, except of the tagged one, arriving at queue j).

For ease of notation, let $\beta_k(x) = v_k^x / M_k(x)$ for $k = 1, \ldots, J$ and $0 \le x \le N$. Taking the limit for $h \to 0$ in eq. (7.70), we find

$$p_j(\mathbf{z}) = \frac{\displaystyle\sum_{i=1}^{J} \pi^{(N)}(\mathbf{z} + \mathbf{e}_i) r_{ij} \mu_i(z_i + 1)}{\displaystyle\sum_{\mathbf{y}} \sum_{i=1}^{J} \pi^{(N)}(\mathbf{y} + \mathbf{e}_i) r_{ij} \mu_i(y_i + 1)}$$

$$= \frac{\displaystyle\sum_{i=1}^{J} [G(N)]^{-1} \left[\prod_{k \ne i} \beta_k(z_k) \right] \beta_i(z_i + 1) \mu_i(z_i + 1) r_{ij}}{\displaystyle\sum_{\mathbf{y}} \sum_{i=1}^{J} [G(N)]^{-1} \left[\prod_{k \ne i} \beta_k(y_k) \right] \beta_i(y_i + 1) \mu_i(y_i + 1) r_{ij}}$$

$$= \frac{\displaystyle\sum_{i=1}^{J} \left[\prod_{k \ne i} \beta_k(z_k) \right] \beta_i(z_i) v_i r_{ij}}{\displaystyle\sum_{\mathbf{y}} \sum_{i=1}^{J} \left[\prod_{k \ne i} \beta_k(y_k) \right] \beta_i(y_i) v_i r_{ij}}$$

$$= \frac{\left[\prod_{k=1}^{J} \beta_k(z_k) \right] \displaystyle\sum_{i=1}^{J} v_i r_{ij}}{\displaystyle\sum_{\mathbf{y}} \left[\prod_{k=1}^{J} \beta_k(y_k) \right] \displaystyle\sum_{i=1}^{J} v_i r_{ij}}$$

$$= \frac{\left[\prod_{k=1}^{J} \beta_k(z_k) \right] v_j}{\displaystyle\sum_{\mathbf{y}} \left[\prod_{k=1}^{J} \beta_k(y_k) \right] v_j} = \pi^{(N-1)}(\mathbf{z})$$

where we have used the balance equations of the closed queueing network and eq. (7.59). The meaning of the result is as follows: a tagged customer, moving to a given queue j, sees a probability distribution of the remaining $N - 1$ customers over the J queues, which is just the same that one could observe on the *same* queueing network at equilibrium, if $N - 1$ customers were circulating in it. This is the "meaning" of the ASTA property.

7.3.2 Buzen's Algorithm for the Computation of the Normalization Constant

In general, the computation of the normalization constant G, hence of the entire PDF of the queueing network state, can be reduced to a number of convolution sums. Let $G(J, N)$ denote the constant for a queueing network with J queues and N customers. Let also

$$g_j(k) = \frac{v_j^k}{M_j(k)}, \qquad k \geq 0, j = 1, \dots, J. \tag{7.71}$$

Then

$$G(j+1, n) = \sum_{|x|+x_{j+1}=n} \frac{v_{j+1}^{x_{j+1}}}{M_{j+1}(x_{j+1})} \prod_{i=1}^{j} \frac{v_i^{x_i}}{M_i(x_i)}$$

$$= \sum_{k=0}^{n} \frac{v_{j+1}^k}{M_{j+1}(k)} \sum_{|x|=n-k} \prod_{i=1}^{j} \frac{v_i^{x_i}}{M_i(x_i)} = \sum_{k=0}^{n} g_{j+1}(k) G(j, n-k)$$

for $j = 1, \dots, J-1$ and $n \geq 0$, with the boundary conditions $G(j, 0) = 1, \ j = 1, \dots, J$ and $G(1, n) = g_1(n), \ 0 \leq n \leq N$. The sums in eq. (7.72) are discrete convolutions and can be computed efficiently, e.g., by resorting to the discrete Fourier transforms. This is known as *Buzen's algorithm*.

A nicer version of this algorithm can be defined if service rates do not depend on queue states, that is with single server queues. In that case $g_j(k) = a_j^k$ with $a_j = v_j/\mu_j$. Then, from eq. (7.72) we get

$$G(j+1, n) = \sum_{k=0}^{n} a_{j+1}^k G(j, n-k) = G(j, n) + \sum_{k=1}^{n} a_{j+1}^k G(j, n-k)$$

$$= G(j, n) + a_{j+1} \sum_{k=0}^{n-1} a_{j+1}^k G(j, n-1-k)$$

$$= G(j, n) + a_{j+1} G(j+1, n-1) \tag{7.72}$$

for $n = 1, \dots, N$ and $j = 1, \dots, J-1$. The boundary conditions are $G(1, n) = a_1^n, \ n \geq 0$ and $G(j, 0) = 1, \ j = 1, \dots, J$. Once the sequence $G(j, n)$ is computed for $j = 1, \dots, J$ and $n = 0, \dots, N$, the network throughput can be computed as $TH(N) = G(J, N-1)/G(J, N)$.

7.3.3 Mean Value Analysis

If the service rates are all independent of the queue state, we have $\mu_j(k) = \mu_j$, so

$$\pi(\mathbf{x}) = \frac{1}{G} \prod_{j=1}^{J} a_j^{x_j} \tag{7.73}$$

with $a_j = v_j/\mu_j$ and G computed with the iteration (7.72).

In this case, we can find a simple expression of the average queue lengths. Let us denote with $Q_j^{(N)}$ the random variable representing the j-th queue length for the queueing network with N customers. Then

$$L_j(N) \equiv \mathrm{E}[Q_j^{(N)}] = \sum_{k=0}^{N} kP(Q_j^{(N)} = k)$$

$$= \sum_{k=1}^{N} kP(Y_j = k) \frac{P(|\mathbf{Y}_{-j}| = N - k)}{P(|\mathbf{Y}| = N)}$$

$$= \sum_{k=0}^{N-1} (k+1)P(Y_j = k+1) \frac{P(|\mathbf{Y}_{-j}| = N - k - 1)}{P(|\mathbf{Y}| = N)}$$

$$= \frac{v_j}{\mu_j} \sum_{k=0}^{N-1} (k+1)P(Y_j = k) \frac{P(|\mathbf{Y}_{-j}| = N - k - 1)}{P(|\mathbf{Y}| = N)}$$

$$= \frac{v_j}{\mu_j} \left[\sum_{k=0}^{N-1} kP(Q_j^{(N-1)} = k) + \sum_{k=0}^{N-1} P(Q_j^{(N-1)} = k) \right] \frac{P(|\mathbf{Y}| = N - 1)}{P(|\mathbf{Y}| = N)}$$

$$= \frac{v_j}{\mu_j} [L_j(N-1) + 1]TH(N)$$

for $N \geq 1$, with $L_j(0) = 0$. The throughput of the j-th queue is given in eq. (7.68). Applying Little's law to the j-th queue, we obtain

$$L_j(N) = v_j TH(N)W_j(N) \tag{7.74}$$

where $W_j(N)$ is the mean waiting time at queue j. Since we are dealing with a closed queueing network, we have $\sum_{j=1}^{J} L_j(N) = N$, so

$$N = TH(N) \sum_{j=1}^{J} v_j W_j(N) \tag{7.75}$$

whence

$$L_j(N) = N \frac{v_j W_j(N)}{\sum_{i=1}^{J} v_i W_i(N)} \tag{7.76}$$

An explicit expression the mean delay of queue j can be derived from the iteration on $L_j(N)$ and eq. (7.74), as:

$$W_j(N) = \frac{1}{\mu_j} [1 + L_j(N-1)] \tag{7.77}$$

Summing up, we can write a set of iterative equations for $j = 1, \ldots, J$ and $N \geq 1$ as follows:

$$
\begin{cases}
W_j(N) = \dfrac{1}{\mu_j}[1 + L_j(N-1)] \\[2ex]
L_j(N) = N\dfrac{v_j W_j(N)}{\displaystyle\sum_{i=1}^{J} v_i W_i(N)}
\end{cases}
\tag{7.78}
$$

initialized with $L_j(0) = 0$ for $j = 1, \ldots, J$. The network throughput can be computed as $TH(N) = N/\sum_{j=1}^{J} v_j W_j(N)$. Iteration (7.78) yields performance metrics of the closed queueing network directly. This is known as *mean value analysis* (MVA).

An example Matlab script code that implements MVA is provided below. The code exploits the vector operations available in Matlab.

```
% The script assumes that the following quantities are given
% number of queues J
% number of customers N
% the routing matrix Rmat
% the vector of serving rates muv
ev=ones(J,1);
Remat=eye(J,J)-Rmat;
Remat=[Remat(1:J,1:J-1) ev];
bv=[zeros(1,J-1) 1];
vv=bv/Remat;
Lv=zeros(1,J);
for k=1:N
  Wv=(1+Lv)./muv;
  TH=k/sum(vv.*Wv);
  Lv=TH*vv.*Wv;
end
```

Example 7.4   Let us consider a window flow-controlled connection, crossing M links, with constant window size equal to K. The capacity of links is managed according to dynamic multiplexing, as in a packet network. Output contention is solved by means of delay, storing packets in link buffers, assumed to be sufficiently large so that overflow probability is negligible. The j-th link has capacity C_j available to the considered connection. Packet length is exponentially distributed, with mean $\bar{L}$. Then, service times have a negative exponential PDF with service rate $\mu_j = C_j/\bar{L}$ at link j. We assume service times are independent of one another. As soon as a packet is delivered to the destination, an ACK message is issued and it goes back to the source. We assume congestion is negligible in the reverse path.

The flow-controlled connection can be cast into a closed queueing network model [164]. The model applies also to a network where the connection path is loaded with exogenous traffic at each link. A general model for such a network consists of a mixed multi-class queueing network where a class of customer models the flow-controlled connection packets, while several open customer classes model cross traffic. The statistics of the closed class modeling the flow-controlled packets can be obtained directly from the analysis of a network construct which is identical to the general multi-class network except that: (i) all open classes are deleted; and (ii) the nodal service rates are redefined to be the reduced service rates [179]. Assuming a balanced model (i.e., the reduced capacity is the same on all connection path links), we let $\mu_1 = \cdots = \mu_M = \mu$. This construct is called the *reduced closed network*.

A second important fact of product-form networks is that even though the propagation delays are distributed over the links of the connection path, the calculation of many quantities of interest may be done as if there were only one propagation delay with mean equal to the cumulative propagation delay of the connection path (the so called base round trip time). Let T_0 be the overall propagation delay over the whole connection path.

The network state is given by the number of packets of the tagged connection stored in each link buffer (Q_k for the k-th link) plus the Q_0 packets in flight into the pipe with propagation delay T_0. The model is composed of $M + 1$ queues, M single-server queues, modeling the link buffers, and an infinite server queue with deterministic service times, modeling the propagation delay of the pipe. The routing matrix entries are $r_{j,j+1} = 1$ for $j = 0, \ldots, M - 1$, and $r_{M,0} = 1$, all other $r_{i,j}$'s being equal to 0. It is easy then to find that the quantities v_j are all equal to 1. Then

$$\pi(\mathbf{n}) \equiv P(\mathbf{Q} = \mathbf{n}) = \frac{1}{G'} \frac{1}{n_0!(1/T_0)^{n_0}} \frac{1}{\mu^{n_1+\cdots+n_M}} = \frac{1}{G} \frac{1}{n_0!} \frac{1}{a^{n_1+\cdots+n_M}} \qquad (7.79)$$

where $a = T_0\mu$ is the bandwidth-delay product of the connection. The normalizing constant is found as

$$G \equiv G(K, M) = \sum_{|\mathbf{n}|=K} \frac{1}{n_0!} \frac{1}{a^{n_1+\cdots+n_M}}$$

$$= \sum_{k=0}^{K} \binom{M-1+k}{M-1} \frac{1}{a^k} \frac{1}{(K-k)!}$$

$$= \frac{1}{K!(M-1)!} \int_0^\infty e^{-u} u^{M-1} \left(1 + \frac{u}{a}\right)^K du$$

where we have exploited the identity

$$n! = \int_0^\infty x^n e^{-x} dx, \qquad n \geq 0. \qquad (7.80)$$

A deep asymptotic analysis for $a \to \infty$ (large bandwidth-delay product networks, also referred to sometimes as "fat pipes") is presented in [164]. That paper gives a deep insight into the effect of scaling the window size K with the bandwidth-delay product a. It provides very useful results to dimension window-based flow control algorithms. First, following Kleinrock, we define the power as the ratio of the achieved throughput TH to the mean delay D. At equilibrium, Little's law suggests that in our case it is $D = K/TH$, since there are constantly K packets in flight for a greedy window flow-controlled connection. Then, in the case at hand, the power is $P(K) = TH(K)/D(K) = TH(K)^2/K$, where we have emphasized the dependence on the number of in-flight packets K. Mitra [164] defines the scaling $K \sim \Gamma a - \alpha \sqrt{a}$, as $a \to \infty$, and three asymptotic regimes as a function of the values of the constant Γ: (i) $\Gamma > 1$: high usage; (ii) $\Gamma < 1$: light usage; (iii) $\Gamma = 1$: moderate usage. It turns out that the connection power is maximized for $K^* \sim a - \alpha^* \sqrt{a}$.

Due to symmetry of the network, the statistics of the number of customers (packets) in each queue is the same as all other queues. Under the optimal asymptotic regime, Mitra finds that the mean and the standard deviation of the number of customers in a queue are $E[Q_1] \sim \beta^* \sqrt{a}$ and $\sigma_{Q_1} \sim \gamma^* \sqrt{a}$.

The constants α^*, β^*, and γ^* can be expressed (approximately) as a function of the number of queues M:

$$\alpha^* \approx -\frac{1}{2\sqrt{M}} \qquad \beta^* \approx \frac{1}{\sqrt{M}} \qquad \gamma^* \approx \frac{1}{\sqrt{M+1}} \tag{7.81}$$

This is a beautiful result in itself and a very useful one in networking applications. It tells us that for large bandwidth-delay product networks, we can expect to reap high throughput levels with marginal queueing (the number of packets in flight grows as a, whereas the queue length grows as $\sqrt{a}$).

This promise is what is currently hardly sought for in new congestion control designs promoted within the IETF and at major companies, to overcome what is known as bufferbloat (see Section 10.9.3), i.e., the huge increase of queueing delays showing up especially in access routers, due to excessively large FIFO buffers coupled with loss-based congestion control mechanisms that fill up queues greedily until packet overflow occurs (see Chapter 10).

To conclude the analysis of this example, let us consider a simplified case, where $M = 1$, i.e., there is a single bottleneck queue. Then, the normalization constant is

$$G = \sum_{k=0}^{K} \frac{1}{(K-k)!a^k} = \frac{1}{K!B(K,a)} \tag{7.82}$$

where $B(\cdot, \cdot)$ is the Erlang-B formula (see Section 5.2). The state reduces to the two variables (Q_0, Q_1), where Q_0 denotes the packets in flight in the network pipe, while Q_1 refers to the packets stored in the bottleneck buffer. The state probabilities are

$$\pi(K-k,k) = P(Q_0 = K - k, Q_1 = k) = \frac{K!B(K,a)}{(K-k)!a^k}, \qquad k = 0, \dots, K.$$

Figure 7.7 Analysis of the bottleneck buffer of a window-based flow control algorithm. Buffer overflow probability as a function of the normalized threshold H/K for three values of the bandwidth-delay product a.

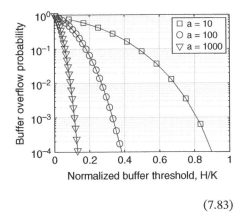

(7.83)

The probability that more than $H < K$ packets be stored in the buffer is

$$P = \sum_{k=H+1}^{K} \pi(K-k,k) = \frac{K(K-1)\cdots(K-H)}{a^{H+1}} \frac{B(K,a)}{B(K-H-1,a)} \qquad (7.84)$$

The buffer overflow probability P is plotted against the normalized threshold, H/K, for three values of the bandwidth-delay product a in Figure 7.7.

For a given a, the window size K is chosen as $K = \lceil a + \sqrt{a/2} \rceil$, which is the optimal value suggested by the asymptotic study for $M = 1$. It is apparent that most packets within the window K are actually in flight in the pipe rather than stored in the buffer as the bandwidth-delay product a grows. Under that regime, with overwhelming probability most of the packets travel through the network, and only a relatively small fraction is in the bottleneck buffer with non-negligible probability. For example, choosing $H = \sqrt{K}(1 + 3/\sqrt{2})$, the probability that this threshold H is exceeded is less than 10^{-3}. For $a = 1000$ that threshold is about 10% of the window size K. The threshold has been fixed as the expected mean queue length plus three times the standard deviation, as provided by the optimal asymptotic design for $M = 1$.

The lesson learned from this simple example is that one can expect to maintain a high throughput (full bottleneck utilization) with a quite small buffer as compared to the bandwidth-delay product, as the product grows, i.e., for "fat" networks. In other words, there is no reason to insist on the rule of thumb for providing buffer sizes in the order of the bandwidth-delay product to reap the maximum throughput.

Example 7.5 In this example, we consider a factory running M machines, that need be all operational for the factory to carry on its production. Machines go off service independently of one another, according to a memoryless process with failure rate λ for each machine. Once a machine is down, it is taken care of by local

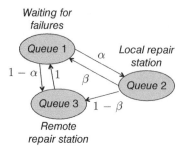

Waiting for
failures

Queue 1

α

Local repair
station

$1 - \alpha$ 1

β

Queue 2

Queue 3 $1 - \beta$

Remote
repair station

Figure 7.8   Sketch of the closed queueing network of Example 7.5.

repairmen groups. There are m_A groups. The time required to fix a machine is assumed to be a negative exponential random variable, independent of the failure process and of other repair times, with mean value $1/\mu_A$. With probability β the machine gets properly repaired and goes back into operation. With probability $1 - \beta$ it needs further assistance and it is sent to a remote maintenance center, where m_B groups of repairmen work and take on average a time $1/\mu_B$ to fix the machine. The remote center repairs the machine with probability 1. We also assume that the machine can be sent directly to the remote center in the first place with probability $1 - \alpha$ and it is handed to local repairmen only with probability α. Overall there are $M + S$ machines, where M are those required to be in service and S are spare ones that replace failed machines until they are restored back into service.

With the hypotheses laid down above, the machine maintenance process can be modeled as a closed queueing network, with $J = 3$ nodes and $N = M + S$ "customers" that represent the machines. A first node represents the working status of machines, where each one of them can fail with individual rate λ. So this queue is an infinite server one, with service time having negative exponential PDF with mean value $1/\lambda$. When leaving this queue, a machine can move to the second queue with probability α or to the third one with probability $1 - \alpha$. The second queue models the repairmen groups of the local maintenance center. So this queue has m_A servers and negative exponential service times with mean $1/\mu_A$. The third queue models the repairmen of the remote maintenance center. Hence it consists of m_B servers with negative exponential service times having a mean value of $1/\mu_B$. The layout of the resulting closed queueing network model is depicted in Figure 7.8.

The linear equation system to calculate the v's is given by $\mathbf{v} = \mathbf{vR}$, where the routing matrix $\mathbf{R}$ is

$$\mathbf{R} = \begin{bmatrix} 0 & \alpha & 1-\alpha \\ \beta & 0 & 1-\beta \\ 1 & 0 & 0 \end{bmatrix} \tag{7.85}$$

The linear system for the v's is therefore:

$$v_1 = \beta v_2 + v_3, \qquad v_2 = \alpha v_1, \qquad v_3 = (1 - \alpha)v_1 + (1 - \beta)v_2 \qquad (7.86)$$

Up to a factor, the solution is $v_3 = v_1(1 - \alpha\beta)$ and $v_2 = \alpha v_1$. Letting $v_1 = 1$, we obtain $v_2 = \alpha, v_3 = 1 - \alpha\beta$.

Queue 1 has infinite servers, while queues 2 and 3 have m_A and m_B servers, respectively. To simplify the resulting expression, assume $m_A = m_B = 1$. Then

$$\pi(M + S - k - h, k, h) = \frac{1}{G} \cdot \frac{(1/\lambda)^{M+S-k-h)}}{(M + S - k - h)!} \left(\frac{\alpha}{\mu_A}\right)^k \left(\frac{1 - \alpha\beta}{\mu_B}\right)^h \qquad (7.87)$$

for $k = 0, \ldots, M + S$ and $h = 0, \ldots, M + S - k$. Scaling the constant by the factor λ^{M+S}, we can rewrite the probabilities conveniently as

$$\pi(x_1, x_2, x_3) = \frac{1}{G} \cdot \frac{1}{x_1!} \left(\frac{\lambda\alpha}{\mu_A}\right)^{x_2} \left(\frac{\lambda(1 - \alpha\beta)}{\mu_B}\right)^{x_3} \qquad (7.88)$$

The normalization constant is

$$G = \sum_{x_1+x_2+x_3=M+S} \frac{1}{x_1!} a_A^{x_2} a_B^{x_3} = \sum_{j=0}^{M+S} \frac{1}{j!} \sum_{i=0}^{M+S-j} a_A^i a_B^{M+S-j-i}$$

$$= \sum_{j=0}^{M+S} \frac{1}{j!} \frac{a_B^{M+S+1-j} - a_A^{M+S+1-j}}{a_B - a_A} \qquad (7.89)$$

where $a_A \equiv \lambda\alpha/\mu_A, a_B \equiv \lambda(1 - \alpha\beta)/\mu_B$.

The probability P that at least M machines are operating is the probability that $Q_1 \geq M$, i.e.,

$$P = \sum_{j=M}^{M+S} \mathcal{P}(Q_1 = j) = \sum_{j=M}^{M+S} \sum_{k=0}^{M+S-j} \pi(j, k, M + S - j - k) \qquad (7.90)$$

This can be found most simply by first writing down the marginal probability distribution of Q_1:

$$\mathcal{P}(Q_1 = j) = \frac{1}{G} \frac{1}{j!} \sum_{k=0}^{M+S-j} a_A^k a_B^{M+S-j-k} = \frac{1}{G} \frac{1}{j!} \frac{a_B^{M+S+1-j} - a_A^{M+S+1-j}}{a_B - a_A} \qquad (7.91)$$

for $j = 0, \ldots, M + S$.

Let $b = 1/\max\{a_A, a_B\}$ and $c = \min\{a_A, a_B\}/\max\{a_A, a_B\} < 1$. For numerical evaluation purposes, the marginal probability distribution of Q_1 can be conveniently rewritten as

$$\mathcal{P}(Q_1 = j) = \frac{\frac{b^j}{j!} \frac{1-c^{M+S+1-j}}{1-c}}{\sum_{i=0}^{M+S} \frac{b^i}{i!} \frac{1-c^{M+S+1-i}}{1-c}}, \qquad j = 0, \ldots, M + S. \qquad (7.92)$$

We are interested in finding the minimum S that can guarantee that at least M machines are operational at any given time with some prescribed level of probability $1 - \varepsilon$, that is the minimum S such that the unavailability U is not larger than ε:

$$U(S) = 1 - P = \sum_{j=0}^{M-1} P(Q_1 = j) \le \varepsilon \qquad (7.93)$$

We add the argument S to stress the dependence of the unavailability U on the number of spare machines S.

As a numerical example, let us assume $1/\lambda = 90$ days, $1/\mu_A = 1$ day, $1/\mu_B = 7$ days, $\alpha = 0.8$, $\beta = 0.95$, $\varepsilon = 0.1$. We find $a_A = 0.0089$ and $a_B = 0.0187$. For M up to 13, it suffices to provide $S = 1$ spare machine to guarantee that the probability of at least M operational machines exceeds 0.9. For M ranging between 14 and 21, we need $S = 2$, that must be raised to $S = 3$ for $22 \le M \le 27$, to $S = 4$ for $28 \le M \le 31$. Going on with bigger and bigger M, we end up with $S = 17$ for $M = 44$, but after that, for $M \ge 45$, no value for S can be found such that $U(S) \le 0.1$.

To understand why this target seems to be unattainable, let $S \to \infty$ and find the least achievable level of $U(S)$ for any finite S:

$$U(S) > U(\infty) = \sum_{j=0}^{M-1} \frac{b^j}{j!} e^{-b} = \int_b^{\infty} \frac{u^{M-1}}{(M-1)!} e^{-u} \, du \qquad (7.94)$$

The last equality reveals that $U(\infty)$ is the tail of the PDF of a random variable X that is identified as the sum of M i.i.d. negative exponential random variables with mean 1. Hence, $E[X] = M$ and $\sigma_X = \sqrt{M}$. For large M we can invoke the central limit theorem and we approximate the tail of the PDF of X with the tail of a Gaussian random variable, i.e.,

$$U(\infty) \sim 1 - \Phi\left(\frac{b - M}{\sqrt{M}}\right) \qquad (7.95)$$

where $\Phi(x) = \int_{-\infty}^{x} \frac{1}{\sqrt{2\pi}} e^{-u^2/2} du$ is the CDF of the standard Gaussian variable. To meet the requirement $U(S) \le \varepsilon$, it must be at least $U(\infty) = \varepsilon$, i.e., $\Phi((b - M)/\sqrt{M}) = 1 - \varepsilon$. Inverting the Gaussian CDF, we find $b \approx M + \gamma\sqrt{M}$, where $\gamma = \Phi^{-1}(1 - \varepsilon)$. For $\varepsilon = 0.1$ it is $\gamma \approx 1.282$. With the numerical values of the parameters of this example, we have $b = 53.57$. For $M = 45$ is it $M + \gamma\sqrt{M} \approx 53.597$, thus exceeding the value of b. In words, the target $U(S) \le 0.1$ is not achievable for $M \ge 45$ with the numerical values of the parameters chosen for this example, since it is $U(S) > U(\infty) \approx 0.105$ for any finite S.

The asymptotic analysis presented above hints at a way to size b so that the requirement on the number of operational machines can be met. For example, we can select machines with an average lifetime $1/\lambda \approx \max\{\frac{\alpha}{\mu_A}, \frac{1-\alpha\beta}{\mu_B}\}(M + \gamma\sqrt{M})$, with $\gamma = 1.282$. Alternatively, the repair-chain organization can be rearranged so as to increase μ_A and μ_B.

7.4 Loss Networks

In previous models of network of queues we have consistently assumed infinite room so that no arriving customer is ever rejected at any queue[12] .

Here we introduce a structurally different model, where a queue has *no* waiting line, hence an arriving customer is served immediately, if there is at least one available server, otherwise it is rejected right away.

We will give a concise account of this subject. An elegant and in-depth treatment can be found in [122, Ch. 3].

The network can be thought of as a graph, made up of a set $\mathcal{N}$ of nodes, interconnected by directed links, forming a set $\mathcal{L}$. Let $\ell = |\mathcal{L}|$ denote the number of links in the network. We assume the network graph is strongly connected.

Since this kind of model arises naturally in circuit-switched networks (more generally, in connection-oriented communication systems), where we can identify customers as connection requests, or *calls*, we will use this last term instead of customer.

An arriving call is characterized by the originating node, the destination node, and one or more alternative *routes* or *paths* through the network that connect the origin with the destination. Since the network graph is strongly connected, there is at least one route for each origin-destination pair. In general, there could exist a multiplicity of routes connecting an origin-destination pair. Let $\mathcal{R}$ be the set of routes used in the network and let $n = |\mathcal{R}|$ be the number of routes. A route is made of a sequence of links and nodes, starting from the origin node up to the destination node, such that any two consecutive nodes are connected by a link.

To be established successfully a call needs one resource (i.e., one circuit, to be consistent with the circuit-switched terminology) on each link of its path. Each link of the network is labeled with a positive integer, representing the number of circuits of that link, hence the maximum number of simultaneous calls that can be established through that link. This is called the *capacity* of the link. Let C_k denote the capacity of link k. Let also $\mathbf{c} = [C_1 \dots C_\ell]$ be the row vector of the link capacities.

If an arriving call cannot be accommodated in any of its associated routes because of lack of circuits along those routes, the call is rejected (no wait is allowed). Unless explicitly stated, in the following we assume that a single route

---

12 For closed queueing networks, a finite room, at least equal to the number of customers in the network, is enough to avoid customer rejection.

per origin-destination pair is selected. Calls for a given origin-destination pair i-j are associated with the predetermined route from i to j. Therefore, in the sequel we refer to calls as arriving to route $s \in \mathcal{R}$.

We assume calls arrive at route s according to a Poisson process with mean rate λ_s. We assume that call holding times are exponentially distributed and independent among one another and of the inter-arrival times. Let $1/\mu_s$ be the mean holding time of calls arriving at route s. We denote the mean offered traffic intensity to route s as $a_s \equiv \lambda_s/\mu_s$.

The state of the network is defined by the number of calls set up for each route in $\mathcal{R}$. Let $\mathbf{x} = [x_1 \ldots x_n]$ be a row vector of non-negative integers, with x_s being the number of calls active on route $s \in \mathcal{R}$. Let also r_{sk} be binary 0–1 coefficients such that $r_{sk} = 1$ if and only if link k belongs to route s. We denote the $n \times \ell$ matrix collecting all coefficients $r_{sk}, s \in \mathcal{R}, k \in \mathcal{L}$ as $\mathbf{R}$.

The capacity constraint that defines the loss network can be summarized by the matrix inequality $\mathbf{xR} \le \mathbf{c}$. The state space of the loss network is defined by $S = \{\mathbf{x} \in \mathbb{Z}^{+n} | \mathbf{xR} \le \mathbf{c}\}$.

The random process $\mathbf{X}(t)$, defined as the state $\mathbf{x}$ of the loss network at time t, is a Markov process on the state space S. In fact, arrivals from outside the network occur according to Poisson processes. Holding times have a negative exponential probability distribution. Inter-arrival times and holding times are independent of each other and form renewal sequences. Therefore, once the state description $\mathbf{X}(t)$ is given for some t, it is possible to predict the behavior of the process for any $\tau > t$ entirely, no matter what its past states have been.

The Markov process $\mathbf{X}(t)$ is stable for whatever value of the mean arrival rates and mean call holding times. This is due to the lossy character of the network, which "protects" the network from overload.

The exact limiting probability distribution $\pi(\mathbf{x}) = \mathcal{P}(\mathbf{X}(t) = \mathbf{x}), \mathbf{x} \in S$ can be found explicitly by using a property of time-reversible Markov processes. A Markov process X with state space $\mathcal{A}$ and transition rates $q_{ij}, i, j \in \mathcal{A}$ is said to be time-reversible if the detailed balance equations hold, i.e.,

$$\pi_j q_{ji} = \pi_i q_{ij}, \qquad \forall i, j \in \mathcal{A}. \tag{7.96}$$

The Markov process Y is said to be obtained by truncating X to the subspace $\mathcal{B} \subset \mathcal{A}$ if

- The transition rates of Y are equal to $q_{ij}, i, j \in \mathcal{B}$, while it is $q_{ik} = 0$ for all pairs (i, k) such that $i \in \mathcal{B}$ and $k \in \mathcal{A} \setminus \mathcal{B}$.
- The resulting process Y is irreducible.

It is recognized easily that the truncated process Y is still time-reversible, if the original process X is. Hence, the limiting probability distribution of Y satisfies the detailed balance equations as prescribed in eq. (7.96). Then, it is also possible to verify that the limiting probability distribution of Y, $\{\pi_i^Y\}_{i \in B}$, can be expressed as a function of that of X, $\{\pi_i^X\}_{i \in A}$, as follows:

$$\pi_i^Y = \frac{\pi_i^X}{\sum_{j \in B} \pi_j^X} \tag{7.97}$$

In other words, the probability distribution of Y is obtained from that of X simply by re-normalization.

Let us apply this result to loss networks. Given a loss network, let us replace it with the same network (same route matrix, same network topology, same arrival and service processes), except that now capacities are unbounded. There is no capacity constraint and the state space is now the entire set of non-negative integer n-dimensional vectors. The model turns out to be a Jackson open queueing network, made of n queues, representing the service offered by routes. Each route is modeled as an infinite-server queue, since the unbounded capacity assumption poses no limit on the number of admissible calls per route. Arrivals at queue s follow a Poisson process with mean rate λ_s and have mean service time $1/\mu_s$. By applying the open queueing network formulas to this special case, we end up with the following simple expression:

$$\pi^{(\infty)}(x_1, \dots, x_n) = \prod_{s=1}^{n} e^{-a_s} \frac{a_s^{x_s}}{x_s!}, \qquad x_s \geq 0, s = 1, \dots, n, \tag{7.98}$$

where $a_s = \lambda_s/\mu_s$ and the superscript ∞ reminds us that the probability distribution corresponds to the modified model with unbounded link capacities.

The original loss network can be recovered by reintroducing finite capacity values for the links and hence truncating the state space $\mathbb{Z}^{+n}$ to $S(\mathbf{c}) = \{\mathbf{x} \in \mathbb{Z}^{+n} \mid \mathbf{xR} \leq \mathbf{c}\}$, where we have highlighted the dependency of the loss network state space on the capacity vector $\mathbf{c}$.

The resulting probability distribution for the loss network is

$$\pi(x_1, \dots, x_n) = G(\mathbf{c}) \prod_{s=1}^{n} \frac{a_s^{x_s}}{x_s!}, \qquad \mathbf{x} \in S(\mathbf{c}), \tag{7.99}$$

where $G(\mathbf{c})$ is the normalization constant given by

$$G(\mathbf{c}) = \left(\sum_{\mathbf{x} \in S(\mathbf{c})} \prod_{s=1}^{n} \frac{a_s^{x_s}}{x_s!} \right)^{-1} \tag{7.100}$$

Once the limiting probability distribution of $\mathbf{X}(t)$ is found, we can evaluate performance metrics. The key metric of the loss network is of course the loss probability, i.e., the probability that an arriving call is rejected because there is no available room in its route. Since arrivals follow a Poisson process, invoking the PASTA property, we know that this is the same as the blocking probability. Let L_s denote the loss probability of route s, briefly referred to as *route loss*. This is the probability that route s is blocked, i.e., the capacity of at least one link belonging to route s is saturated. Let $\mathbf{e}_s$ denote an n-dimensional row vector with entries defined as $e_s(s) = 1$ and $e_s(r) = 0, \forall r \neq s$. It corresponds to a state where there is just one call through route s and nothing else. So long as the state of the loss network is within the set $\overline{B}_s \equiv S(\mathbf{c} - \mathbf{e}_s \mathbf{R})$, there is room for at least one call on route s, i.e., route s is not blocked. Then, we can write

$$L_s = 1 - \sum_{\mathbf{x} \in \overline{B}_s} \pi(\mathbf{x}) = 1 - \frac{G(\mathbf{c})}{G(\mathbf{c} - \mathbf{e}_s \mathbf{R})}, \qquad s = 1, \dots, n. \qquad (7.101)$$

It should come as no surprise, then, that the numerical evaluation of the probability distribution of the loss network and its key performance metrics depend crucially on computing the normalization constant efficiently. This is hard to do for large state spaces (large capacities and/or number of routes). This makes the obtained results of little help, except for relatively small networks. However, powerful approximations can be obtained, as explained in depth in [122, Ch. 3]. In the next section we introduce the famous Erlang fixed-point approximation.

Example 7.6 Consider a cellular network with a spatial uniform layout. We model the cell coverage by using a hexagonal grid, where each hexagon corresponds to a cell served by a base station (BS)[13] . Multiple access of user terminals is obtained by dividing the available spectral radio resource into orthogonal channels. Say C channels are available to the system. Channels are spatially reused, i.e., a same channel can be in use in multiple cells at the same time. Co-channel interference constraints limit the number of channels that can be used in neighboring cells. To this end, the reuse cluster is defined as the least compact set of neighboring cells where all channels are assigned once, i.e., no channel reuse is possible. Figure 7.9 shows two coverage examples with reuse cluster sizes 4 and 7 cells, respectively. Two sample clusters, one for each size, are colored to highlight them. The reuse distance D is the distance between the two nearest BSs that can use the same radio channel. It can be shown that $D = R\sqrt{3K}$, where R is the cell radius and K is the cluster size. This result holds for the hexagonal coverage model.

A connection request from a user terminal needs one channel to be set up. If no channel is available, user requests are typically rejected (no wait). We can use a loss network model for this system, if connection requests arrive at each BS according to a Poisson process. The capacity constraints can be written as $\sum_{j \in \mathcal{K}_k} x_j \leq C$, where x_j is the number of connections in cell j and $\mathcal{K}_k$ represents the cell cluster k. The

---

13 The technical name given to this equipment varies with the technology standard, e.g., it is eNB for LTE.

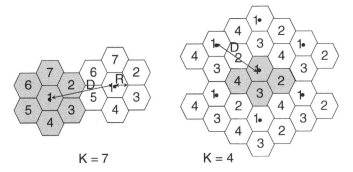

Figure 7.9 Example of spatial reuse clusters in a cellular network model. The shape of the cluster is highlighted. K denotes the cluster size, D is the reuse distance.

constraints must be applied to all possible clusters. For each given cell, we must consider all clusters of the given size K that it belongs to.

For example, for clusters of size $K = 7$, we can define one cluster per cell as the cluster centered on that cell. Each cell belongs to 7 clusters, namely, the one centered on itself and the clusters centered on each of its 6 neighboring cells.

In this example there is no routing of connections in the classical sense. We can associate a sort of "virtual route" to each BS by considering the BS itself and the BSs that belong to all clusters that the tagged BS belongs to. It is as if a new arriving connection requires a "link usage" out of its own BS and of all its neighboring BSs.

7.4.1 Erlang Fixed-Point Approximation

The basic idea of this approximation is simple. Let B_j denote the blocking probability at link j, i.e., the probability that all C_j resources of link j are being used. Then, a new arriving request for any route s that uses link j will be lost.

If the link j could be studied *in isolation* with respect to the rest of the network, and if arrivals at that link formed a Poisson process with mean intensity A_j, we could apply the Erlang loss model to link j and conclude that $B_j = B(C_j, A_j)$, where $B(m, A)$ is the Erlang B-formula for m servers and a mean offered traffic A (see Section 5.2).

The approximation lies in *assuming* that the two conditions mentioned above hold, in spite of the fact that they do not in general.

What remains is to find the mean intensity of the traffic offered to link j. This is the traffic originated by call requests that survive the rejection of other links belonging to the same route as j. For each given route $s \in \mathcal{R}$, the mean rate of offered calls is λ_s. Those calls are offered to link j

- If $r_{sj} = 1$, i.e., if link j belongs to route s;
- With probability $1 - B_i$ for all other links i on the same route s.

The mean *reduced* rate of calls belonging to route s that are offered to link j in this approximation is given by $\lambda_s r_{sj} \prod_{i \,:\, r_{si}=1; i \neq j}(1 - B_i)$. Summing up over all routes, we find finally

$$A_j = \sum_{s \in R} \frac{\lambda_s}{\mu_s} r_{sj} \prod_{i \neq j}(1 - B_i)^{r_{si}} \tag{7.102}$$

The derivation of (7.102) entails that blocking at each link is assumed to be *independent* of all other links, which is not true in general.

Writing the link-blocking probability by modeling each link as an Elrang loss system in isolation, we end up with the following nonlinear system of ℓ equations in the ℓ unknowns E_j:

$$E_j = B\left(C_j, \frac{1}{1-E_j} \sum_{s \in R} a_s r_{sj} \prod_{i=1}^{\ell}(1 - E_i)^{r_{si}}\right), \qquad j = 1, \ldots, \ell. \tag{7.103}$$

The arguments of the Erlang B formula $B(\cdot, \cdot)$ in eq. (7.103) are the number of circuits of link j and the mean intensity of the traffic offered to link j. The latter is found as the ratio of the mean intensity of the traffic carried by link j and (1 − loss probability of link j). In turn, the mean intensity of the traffic carried by link j is found as the sum of all traffic offered to routes that use link j, thinned by the loss incurred on other links.

The nonlinear equation system (7.103) defines a continuous mapping of $[0, 1]^\ell$ onto itself. Since the unit hypercube is compact and convex, we can apply Brouwer's theorem and conclude that there must exist a fixed point. In [122] and [137, Ch. 6] it is shown that a *unique* solution to the nonlinear equation system (7.103) exists in $[0, 1]^\ell$.

The Erlang fixed-point approximation is intimately related with the link blocking probabilities in the asymptotic regime for large networks [122]. In other words, it can be expected that the Erlang fixed-point method yields accurate predictions for those networks that can be considered as "large", i.e., such that there is a small probability that two connections share more than a single link.

Let us consider a sequence of networks indexed by an integer N. We will take the limit as N grows according to the following asymptotic regime: the call arrival rates $\lambda_s(N)$, $s \in R$, and the link capacities $C_j(N)$, $j \in L$, diverge as $N \to \infty$, in such a way that

$$\lim_{N \to \infty} \frac{1}{N} \lambda_s(N) = \bar{\lambda}_s > 0, \qquad \lim_{N \to \infty} \frac{1}{N} C_j(N) = \bar{C}_j > 0. \tag{7.104}$$

We define also $a_s(N) = \lambda_s(N)/\mu_s$. According to the scaling assumed in this analysis, we have $\lim_{N \to \infty} \frac{1}{N} a_s(N) = \bar{a}_s$. Let $\xi_s = x_s/N$, where x_s is the number of circuits engaged on route s. We rewrite the expression of the steady-state probability distribution of the loss network:

$$\log \pi(x_1, \dots, x_n) = \log G(\mathbf{c}) + \sum_{s \in \mathcal{R}} (x_s \log a_s - \log x_s!) \tag{7.105}$$

Let us focus on the summand inside the summation sign and use Stirling formula $\log n! \sim n \log n - n + O(\log n)$. We get

$$
\begin{aligned}
x_s \log a_s - \log x_s! &= N\xi_s \log(N\bar{a}_s) - N\xi_s \log(N\xi_s) + N\xi_s + O(\log N) \\
&= N\xi_s \log(\bar{a}_s) - N\xi_s \log(\xi_s) + N\xi_s + O(\log N) \\
&= Ng(\xi_s, \bar{a}_s) + O(\log N)
\end{aligned}
$$

where $g(y, a) = y \log a - y \log y + y$. Let us drop the $O(\cdot)$ contribution in the limit for $N \to \infty$, and let us consider two states $\mathbf{x}^{(1)} = N\xi^{(1)}$ and $\mathbf{x}^{(2)} = N\xi^{(2)}$. Assume it is $\sum_s g(\xi_s^{(1)}, \bar{a}_s) > \sum_s g(\xi_s^{(2)}, \bar{a}_s)$. Then

$$\frac{\pi(\mathbf{x}^{(2)})}{\pi(\mathbf{x}^{(1)})} \sim \exp\left(N \sum_{s \in \mathcal{R}} g(\xi_s^{(2)}, \bar{a}_s) - N \sum_{s \in \mathcal{R}} g(\xi_s^{(1)}, \bar{a}_s) \right) = e^{-N\Delta} \tag{7.106}$$

where Δ is a positive constant. It turns out the steady-state probabilities become negligible when compared to the probability of a special state, the one that maximizes $\sum_s g(\xi_s, \bar{a}_s)$. We can identify the "most" probable state of the limiting regime of the sequence of loss networks parametrized by N as the solution of the following optimization problem:

$$\textbf{Maximize}: \ \sum_{s \in \mathcal{R}} (y_s \log \bar{a}_s - y_s \log y_s + y_s) \tag{7.107}$$

$$\textbf{Subject to}: \ \mathbf{y}R \le \bar{c} \tag{7.108}$$

$$\mathbf{y} \ge 0 \tag{7.109}$$

where $\bar{\mathbf{c}} = [\bar{C}_1, \dots, \bar{C}_\ell]$.

The target function is separable, each summand being a strictly concave function. The feasible set is closed, bounded, and convex. Hence, there exists a unique maximizer $\mathbf{y}^*$ of this problem.

The steady-state probability distribution $\pi(N\mathbf{y})$ tends to concentrate on a single value as $N \to \infty$, namely, we have $\lim_{N \to \infty} \pi(N\mathbf{y})/\pi(N\mathbf{y}^*) = 0$ for any $\mathbf{y} \ne \mathbf{y}^*$. This is a typical example of *state space collapse* under an asymptotic regime. When scaling a system to large sizes of its parameters, a typical outcome is that the variability due to statistical fluctuations tends to disappear, that is, it has a relatively negligible value with respect to the mean. Under such regime, the stochastic process that described the dynamics of the system tends to a deterministic system. This is closely related to the *fluid approximation* (see Section 8.6).

Let us make the limit of the asymptotic regime for large network precise. Let $B_j(N)$ be the blocking probability of link j for the N-th network. It can be shown (e.g., see [122]) that $B_j(N) \to B_j$ as $N \to \infty$, where the quantities $B_j, j \in \mathcal{L}$ are any solution to:

$$
\begin{cases}
\sum_{s \in \mathcal{R}} \bar{a}_s r_{sj} \prod_{i \in \mathcal{L}} (1 - B_i)^{r_{si}} = \bar{C}_j & \text{if } B_j > 0 \\
\sum_{s \in \mathcal{R}} \bar{a}_s r_{sj} \prod_{i \in \mathcal{L}} (1 - B_i)^{r_{si}} < \bar{C}_j & \text{if } B_j = 0
\end{cases}
\tag{7.110}
$$

It can be shown that there always exists a solution to (7.110). The solution is unique, if the matrix $\mathbf{R}$ has rank ℓ, i.e., if it is full rank. It is also possible to prove that

$$
\lim_{N \to \infty} \frac{1}{N} E[X_s(N)] = \bar{x}_s , \qquad s \in \mathcal{R},
\tag{7.111}
$$

where $X_s(N)$ is the random variable defined as the state of route s of the N-th loss network of the considered sequence, and

$$
\bar{x}_s = \lim_{N \to \infty} \frac{1}{N} a_s(N) \prod_{j \in \mathcal{L}} [1 - B_j(N)]^{r_{sj}} = \bar{a}_s \prod_{j \in \mathcal{L}} [1 - B_j]^{r_{sj}}
\tag{7.112}
$$

where the B_j's are a solution to (7.110).

The limiting probability distributions of $U_s(N) \equiv [X_s(N) - \bar{x}_s]/\sqrt{N}$ can be characterized as well. It turns out that the probability distribution of $\mathbf{U}(N)$ converges to that of a random vector $\mathbf{U}$, such that $U_s \sim \mathcal{N}(0, \bar{x}_s)$ and $\mathbf{UR}_B = \mathbf{0}$, where $\mathbf{R}_B$ is the matrix obtained from $\mathbf{R}$ by deleting all columns j for which $B_j = 0$.

The loss probability in the limiting regime is obtained as the solution of (7.110). That system has an intuitive interpretation: for those links where the sum of average traffic intensities loading the link is less than the link capacity, the link blocking probability is 0 (underloaded links). The loss probability has a positive value for overloaded links, where the mean offered traffic is bigger than the capacity. This is reminiscent of a fluid approximation and it is a consequence of the large traffic, large link capacity asymptotic regime.

The link of the Erlang fixed-point method with the asymptotic regime of large loss networks is established by the fact, proved in [122, § 3.6], that $E_j(N) \to B_j$, $j \in \mathcal{L}$, where $E_j(N)$ is the solution to the Erlang fixed-point equation system (7.103) for the N-th network and B_j is a solution to (7.110), i.e., the limit of the blocking probability $B_j(N)$ of the N-th network.

In essence, the limit theorem shows that for large networks the probability distribution of the state space of the loss networks is concentrated around the mean. In turn, the mean can be determined by solving for the blocking probabilities B_j from (7.110). But for large systems, those B_j values are close to what the Erlang fixed-point method yields.

The limit theorem provides thus a theoretical underpinning to the Erlang fixed-point algorithm. Actually, it turns out that the Erlang fixed-point algorithm is quite accurate for most cases.

Once the link blocking probabilities have been found, the route loss can be calculated again resorting to the independence assumption:

$$L_s = 1 - \prod_{j=1}^{\ell} (1 - B_j)^{r_{sj}} \tag{7.113}$$

The Erlang fixed-point algorithm yields generally good results when the independence assumption it is based on leads to a reasonable approximation. In turn, this is expected when there is a negligible probability that two routes share more than one link. This is essentially what happens for large networks.

Example 7.7 *Fixed vs. dynamic channel assignment in a cellular network*
Let us consider a cellular network model with hexagonal cells. Let K denote the size of the reuse cluster, i.e., the number of closest cells that must be assigned different radio channels to keep co-channel interference within acceptable limits. Assume the cellular network operator has M radio channels overall. Connection requests to a base station (BS) arrive according to a Poisson process with mean rate λ. Let $1/\mu$ be the mean channel holding time. The mean offered traffic offered to a cell is $A = \lambda/\mu$. Channels are allocated to BSs according to two different strategies:

- *Fixed channel allocation* (FCA): Each BS is assigned its own radio channels, for exclusive use of users connected to that BS. We assume that each BS gets the same number of channels, M/K.
- *Dynamic channel allocation* (DCA): The M channels are shared dynamically by BSs. A BS takes a channel when it is required to set up a new connection. To keep interference within acceptable limits, a BS must check that the number of channels in use in each reuse cluster it belongs to is no more than M.

In case of FCA, to account for the fact that M/K may not be integer, we assign $\lceil M/K \rceil - 1$ channels to N BSs and $\lceil M/K \rceil$ channels to the remaining $K - N$ BSs. N is chosen so that

$$N \left(\left\lceil \frac{M}{K} \right\rceil - 1 \right) + (K - N) \left\lceil \frac{M}{K} \right\rceil = M \tag{7.114}$$

whence $N = K \lceil M/K \rceil - M$. The blocking probability with FCA is

$$P_{FCA} = \left(\left\lceil \frac{M}{K} \right\rceil - \frac{M}{K} \right) B \left(\left\lceil \frac{M}{K} \right\rceil - 1, A \right) + \left(1 + \frac{M}{K} - \left\lceil \frac{M}{K} \right\rceil \right) B \left(\left\lceil \frac{M}{K} \right\rceil, A \right) \tag{7.115}$$

where $B(\cdot, \cdot)$ is the Erlang B formula.

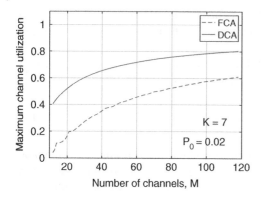

Figure 7.10 Comparison between the maximum channel utilization obtained with fixed and dynamic channel allocations (FCA and DCA, respectively) under the requirement that the blocking probability of a new connection request P_0 be no more than 0.02.

As for DCA, a BS belongs to K clusters, i.e., it can take the role of cell $1, 2, \ldots, K$ of the cluster. By applying the Erlang fixed-point method, thanks to the symmetry of the system (all BSs are equivalent), we obtain the blocking probability of DCA as $P_{DCA} = 1 - (1 - \beta)^K$, where β is the unique solution to

$$\beta = B(M, KA(1 - \beta)^{K-1}) \tag{7.116}$$

Figure 7.10 compares the maximum channel utilization achievable under a requirement on the blocking probability P, namely that it be $P \leq P_0 = 0.02$. The analysis is carried out for $K = 7$.

The gain offered by DCA over FCA is apparent. Even greater advantage of DCA over FCA can be observed by increasing the cluster size K. The price to pay for the increased efficiency of DCA is the implementation of coordination function among the BSs to share the channels dynamically.

7.4.2 Alternate Routing

Up until here we have assumed single path routing. It is often the case that multiple routes can be selected and used alternatively. This has been current practice in circuit switched networks (like the telephone network) for decades and is used as well in cellular networks, where diversity routing is possible whenever more than a single BS covers a user terminal, or in optical networks, where calls correspond to lightpath setup requests.

In general, graphs describing communication networks are multi-connected. Not only are they connected, to guarantee reachability of any node, but they usually provide diverse routing with disjoint end-to-end paths, to be robust in the face of link or node failures. Given that multiple paths exist to connect two given nodes i and j, it is natural to conceive *alternate routing* strategies, where a first choice path is explored. If that is found to be unavailable, a second choice path is explored,

Figure 7.11 Example of alternate routing: the first choice is the direct path; the second and last choice is a two-link route through a relay node.

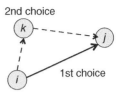

and so on, until all alternatives have been checked. If all alternatives for the traffic relation i-j fail, the call is rejected.

When alternate routing is used there might arise difficulties leading to multiple regimes of the loss network. To illustrate this point, let us consider a very simple example (see [122]). Given N nodes, we consider a network providing direct links between any pair of nodes. Thus the number of links is $\ell = N(N-1)/2$. Each link is equipped with a capacity C, i.e., it can sustain up to C simultaneous calls.

Routing exploits the direct path as a first choice. If the direct link is saturated, an alternate path made up of two links through a randomly selected relay node is explored. If that fails as well, the offered call is lost.

The "route" of the loss network model for calls between nodes i and j is made up of two elements: the direct path i-j and a randomly selected path i-k-j, where $k \neq i,j$ is an intermediate relay node. Figure 7.11 gives an example scheme of the two alternate paths: the direct path (first choice) and the two-link path through a randomly selected relay node (second and last choice).

By symmetry, the blocking probability of all links is the same. Let the link blocking probability be denoted with β, while L denotes the route blocking probability. Applying the independence assumption of the Erlang fixed-point method, the probability that a route is not blocked is given by $1 - L = (1 - \beta) + \beta(1 - \beta)^2$. The first term is the probability that the direct path is available $(1 - \beta)$. The second term is the probability that the first choice route is blocked (β) multiplied by the probability that the randomly selected two-link route is not blocked, i.e., $(1 - \beta)^2$, since this is the joint event that both i-k and k-j are not blocked.

To write down the Erlang fixed-point equation $\beta = B(C, A_o)$, we have to find the mean offered traffic A_o of a generic link. This is related to the mean carried traffic A_c by $A_o = A_c/(1 - \beta)$. In turn, the mean carried traffic equals the mean number of busy circuits on the link. If λ is the mean arrival rate of calls at any "route," $1/\mu$ is the mean call holding time, and $a \equiv \lambda/\mu$, we have $A_c = a(1 - \beta) + 2a\beta(1 - \beta)^2$. This is obtained by considering that one circuit is used with probability $1 - \beta$, if the first choice path is available, while two circuits are required if the first choice path is found busy and the alternate two-link path is available. Therefore $A_o = a + 2a\beta(1 - \beta)$ and the fixed-point equation can be written as

$$\beta = B(C, a(1 + 2\beta - 2\beta^2)) \tag{7.117}$$

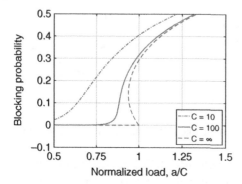

Figure 7.12 Blocking probability of the symmetric alternate routing network as a function of the normalized traffic load a/C, for various values of the link capacity C.

The loss probability of a route (made up by the direct link and the alternate two-link routing) is:

$$L = 1 - [1 - \beta + \beta(1 - \beta)^2] = \beta^2(2 - \beta) \tag{7.118}$$

We expect that the expressions in eqs. (7.117) and (7.118) provide accurate results for $N \to \infty$, since it is in that regime that the independence assumption underlying the fixed-point Erlang approximation holds.

The curves in Figure 7.12 show the behavior of the blocking probability β as a function of the normalized traffic load a/C for three values of the link capacity C.

The limit for large capacity values is obtained by scaling both C and a by a same factor s and letting $s \to \infty$. We know from eq. (5.11) that

$$\lim_{s \to \infty} B(s\hat{C}, s\hat{a}) = \max\left\{0, 1 - \frac{\hat{C}}{\hat{a}}\right\} \tag{7.119}$$

In the limiting case, β is the solution of

$$\beta = \max\left\{0, 1 - \frac{\hat{C}}{\hat{a}(1 + 2\beta - 2\beta^2)}\right\} \tag{7.120}$$

As shown by the dashed line curve in Figure 7.12, this equation has multiple solutions for[14] $\frac{\hat{a}}{\hat{C}} \geq \frac{5\sqrt{10}-13}{3} \approx 0.9371$.

The behavior of the curves shown in Figure 7.12 can be interpreted as follows. For moderate levels of $\hat{C}$, as the normalized load grows, the blocking probability gets larger and larger, in a slow way at first, then with a steep growth slope in a transition region that goes across $\hat{a}/\hat{C} = 1$. The transition region is narrower as $\hat{C}$ increases. In the limit for very large values of $\hat{C}$ a bi-stable behavior arises. As the

___

14 This value can be found by considering the function $y = [(1 - \beta)(1 + 2\beta - 2\beta^2)]^{-1}$, for $\beta \in (0, 1)$. Searching for the minimum of $y(\beta)$ in $[0, 1)$, it is easy to check that there is a unique minimum at $\beta = \frac{4-\sqrt{10}}{6} \approx 0.1396$. The lower limit of the ratio $\hat{a}/\hat{C}$ is found as the value of $y(\beta)$ at its minimum.

normalized load increases the blocking probability can shift abruptly from a relatively small level to a high level (high blocking regime). This corresponds to an increasing number of calls finding their direct path blocked, hence being routed on a two-link alternative. The rerouting doubles the amount of circuits used by a single call. If a series of rerouting is performed, the network is likely driven into a high congestion level, where blocking (both of direct path and alternate path) is more probable. As the load is relieved, circuits come back to idle state and the network moves back from the high congestion level to a low congestion level where blocking is relatively rare.

What is affected is the dynamic behavior of the network, while the Markov process describing the state of the network might still admit a steady state probability distribution, that exhibits two separate peaks (bimodal PDF). The state tends to oscillate stochastically between the two congestion regimes corresponding to high and low blocking levels, even if it admits a unique steady-state probability distribution. This phenomenon is reminiscent of what happens with the nonstabilized version of ALOHA in random access protocols (see Section 9.2).

A way to protect the network from such oscillations, and yet allow alternate routing, is to introduce dumping. For example, *sticky routing* [89] is defined by the following two provisions:

1. An incoming call is assigned a circuit on a target link, if a circuit is available on the direct link.
2. For any couple of nodes, a random relay node is selected ahead of call arrival time, and stored in node tables. Let $k(i,j)$ denote the relay selected for the couple i-j. $k(i,j)$ is maintained as long as call requests arising at node i are successfully routed, either directly to j or through the intermediate relay $k(i,j)$. Whenever call routing fails for lack of circuits also on the alternate route, the call is lost and the relay node is randomly reselected.

It is shown that sticky routing is successful in avoiding bi-stability and hence it improves the loss performance.

7.5 Stability of Queueing Networks

Up to this point, we have not given too much emphasis to the issue of stability of a queueing network. We hinted at the fact that the solutions we explored refer to steady-state analysis, which can exist only if the queueing network is stable. We devote this section to giving a concise introduction to the vast topic of defining and assessing stability of a network of queues. This aspect has received a special attention, since it turns out that surprising results can arise. More on stability can be found in [86, Ch. 8] or in specialized textbooks, e.g., [63, 44, 160, 54]. We confine ourselves to some basic definitions and an illustrative example. It is, however, enough to grasp where the catch can be when dealing with stability of a network of queues and not just an isolated queue.

To illustrate the stability issue, we refer to a generalized queueing network model, formed by nodes (also called sometimes stations), connected by links, forming a connected graph. Each node is provided with a single server facility and it can host one or more queues (also called buffers). The service time of a customer depends in general on the buffer it joins. Buffers are sometimes associated with specific classes of customers, so that a customer of class x entering a node will join the buffer reserved for class x, where service time has a probability distribution specific of that class. This can model successive phases of a workflow, where a "customer" (piece of product or a job) undergoes different working phases, each one requiring service times that can be modeled as drawn from a random variabile specific of that working phase. Nodes are equipped with a single server, whose serving capacity is shared among the queues hosted in the node. Thus, the description of this model entails assigning

- The statistical characteristics of the external arrivals at the nodes (external stands for outside of the network);
- The statistical description of service times of customers at each buffer;
- The routing rules, i.e., which buffer is joined by a customer leaving its current buffer; if none is chosen, the customer leaves the network;
- The single server sharing policy at each node.

The stochastic model defined by such a structure is by its nature a discrete one, i.e., its state includes the description of the number of customers sojourning in each buffer.

We give first an example to illustrate the nontriviality of the stability issue in this kind of networks. Then, we will introduce the formal definitions of stability for a stochastic discrete network and its fluid approximation. The reason to involve the fluid approximation is that it can be used to show whether a stochastic discrete queueing network is stable, i.e., stability can be investigated by reducing the stochastic discrete network to a fluid queueing network, that is generally simpler to study. To complete the topic, we detail how to identify the fluid approximation of a stochastic discrete queueing network of the type dealt with in this section.

Example 7.8 We provide a classic example for the discussion of stability of stochastic queueing networks. The example has been introduced concurrently in two papers and brings the name of the researchers that have designed the example, namely Rybko and Stolyar [185], Kumar and Seidman [136].

The network comprises two nodes, denoted with A and B (see Figure 7.13).

Figure 7.13 Stochastic queueing network example for the discussion of stability (also known as Rybko-Stolyar-Kumar-Seidman network).

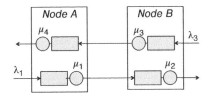

Two external flows of customers are offered to the network. Class 1 customers arrive at node A with a mean rate λ_1. They require a mean service time $1/\mu_1$ and are stored in buffer 1. When they are done they leave node A to join node B as class 2 customers. At node B class 2 customers are served at a mean rate μ_2. After service completion, class 2 customers leave the network.

A symmetric customer flow is defined, in the reverse direction. Namely, class 3 customers arrive at node B at a mean rate λ_3, are served at a mean rate μ_3, then join node A as class 4 customers, receiving service at a mean rate μ_4. After service completion, class 4 customers leave the network.

Node A and B are equipped with a single server each. The service capacity of the single server is shared among the buffers hosted in each node, according to the pre-emptive-resume priority discipline. Specifically, class 2 customers have priority over class 3 ones in node B, while class 4 customers have priority over class 1 customers at node A.

Obvious necessary conditions for the stability of the queueing network is that the mean utilization of the single servers at node A and B must be less than 1, i.e., it must be

$$\frac{\lambda_1}{\mu_1} + \frac{\lambda_3}{\mu_4} < 1 \qquad \frac{\lambda_1}{\mu_2} + \frac{\lambda_3}{\mu_3} < 1 \tag{7.121}$$

Remarkably, it turns out that these necessary conditions are *not sufficient* .

We can experimentally check this by simulating the network, starting from an empty state, with the following numerical values of the network parameters (normalized to some arbitrary unit): $\lambda_1 = \lambda_3 = 1$, $\mu_1 = 5$, $\mu_2 = 10/7$, $\mu_3 = 4$, $\mu_4 = 4/3$. The necessary conditions for stability are verified with these numerical values, since

$$\frac{\lambda_1}{\mu_1} + \frac{\lambda_3}{\mu_4} = 0.95 < 1 \qquad \frac{\lambda_1}{\mu_2} + \frac{\lambda}{\mu_3} = 0.95 < 1 \tag{7.122}$$

The results of a simulation experiment are plotted in Figure 7.14. On the left the content of each individual buffer is plotted as a function of time. On the right, the sum of the number of customers in all buffers of the network is plotted against time.

It is apparent that there is a growth trend of the number of customers in the buffers, which is a clear sign of instability of the network.

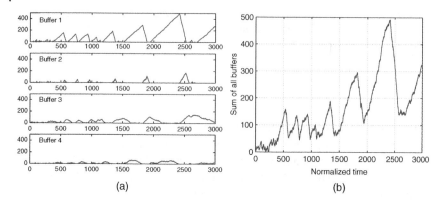

(a)   (b)

Figure 7.14 Number of customers in each buffer (left plots) and sum of all buffers (right plot) from a simulation experiment with the stochastic discrete queueing network of Figure 7.13.

The explanation of the instability resides in the operational structure of the system. Let $X_i(t)$ denote the number of customers in buffer i at time t, $i = 1, 2, 3, 4$. We can prove that $X_2(t)X_4(t) = 0$ at any time $t \geq 0$ if the system starts empty at time $t = 0$.

Assume that immediately before some time t_1 it is $X_2(t_1) > 0$ and $X_4(t_1) = 0$. Then, no arrival at buffer 4 can occur at t_1, since the only way that an arrival can show up at buffer 4 is after a service completion at buffer 3. However, no service can be completed at buffer 3 so long as buffer 2 is nonempty due to the preemptive priority policy of the single server of node B. Therefore, we have shown that $X_4(t)$ is forced to stay at 0 as long as $X_2(t)$ is positive.

A completely symmetric argument shows that also $X_2(t)$ must remain at 0, if initially at 0, until buffer 4 is nonempty. This is again due to the priority policy at node A and the fact that arrivals at buffer 2 correspond to customers having completed their service at buffer 1.

Since buffers 2 and 4 start out empty, by assumption, and we have shown that if one of them is nonempty, the other one is forced to stay empty, it turns out that the event that *both* buffers are nonempty at some time is not possible[15] .

As a consequence, the system as a whole must have capacity enough to devote a fraction λ_1/μ_2 to the service of customers of class 2 plus another fraction λ_3/μ_4 for customers of class 4. There cannot be any overlap between these two service time sequences, hence it must be

$$\frac{\lambda_1}{\mu_2} + \frac{\lambda_3}{\mu_4} < 1 \tag{7.123}$$

---

15 Note that simultaneous events occur with probability 0, as usual in single arrival, single head-of-line service, continuous-time service systems.

This is yet another condition for stability to be added to the necessary conditions in eq. (7.121). It is as if there exists a virtual node made up of buffers 2 and 4, equipped with a virtual single server.

In this numerical example we have $\lambda_1/\mu_2 + \lambda_3/\mu_4 = 1.45$, hence the additional condition for stability is not met.

Other examples can be provided, e.g., with systems where the serving policy at nodes does not use any priority, just FCFS. In that case *reentrant flows* are present, i.e., customers flows that enter a buffer of a node after having completed their service in another buffer of the *same* node.

7.5.1 Definition of Stability

Let us now give a precise notion of what is meant by stability for a stochastic queueing network. The pathway that can be followed to assess the stability of a given stochastic network is to transform it into a deterministic equivalent fluid network and then to show that the latter is stable. This motivates us to give stability definitions for both stochastic discrete and deterministic fluid queueing networks.

Let b be the number of buffers in the network. Let $X_i(t)$ denote the number of customers in buffer i at time t. We can state two definitions of stability.

Definition 7.1 (Stability)　A stochastic discrete queueing network is said to be stable if $\sum_i X_i(t) < \infty$ for all t with probability 1, including in the limit for $t \to \infty$. This is typically shown by proving that the stochastic process associated with the queueing network state is positive recurrent.

For the next definition, let $\boldsymbol{\gamma}$ denote a row vector whose i-th element γ_i denotes the mean external arrival rate at buffer i. Let r_{ij} be the probability of moving from buffer i to buffer j, when leaving buffer i. Let $\mathbf{R}$ denote the square matrix collecting the routing probabilities. Note that the row sums of $\mathbf{R}$ are less than or equal to 1, with at least one of them being strictly less than 1, since customers must eventually leave the network. The effective arrival rate at buffer i is given by a_i, where the vector $\mathbf{a} = [a_1, a_2, \ldots, a_b]$ is given by $\mathbf{a} = \boldsymbol{\gamma}(\mathbf{I} - \mathbf{R})^{-1}$, $\mathbf{I}$ being the identity matrix.

Definition 7.2 (Rate stability)　A stochastic discrete queueing network is said to be rate stable if the steady-state departure rate equals the steady-state effective arrival rate (the arrival rate of the customer flow coming from the outside of the network and those arriving from other buffers of the network) for all buffers in the network. If $D_i(t)$ denotes the cumulative number of departures from buffer i in $[0, t]$, rate stability means that $D_i(t)/t \to a_i, \forall i$, as $t \to \infty$, almost surely.

The first definition is stronger, while the second one requires only a balance between the input and output customer flows at each buffer, but it does not imply that buffers should ever be empty. On the contrary, with the first definition, buffers empty infinitely often.

The stability can be assessed via a fluid model of the stochastic discrete queueing network. The fluid approximation will be presented in Section 8.6. Here it suffices to say that, given a counting process $A(t)$, having discrete unit jumps at time when events occur, the fluid approximation of that process is obtained as the limit for $n \to \infty$ of $A(nt)/n$. The meaning of the scaling and of the limit is considering the original stochastic discrete process on a wider and wider time horizon, so that discrete jumps become negligible with respect to the overall trend of the process. Since we know that for a stationary process with mean rate λ it is $A(t)/t \to \lambda$ as $t \to \infty$, it is easy to check that the scaled fluid approximation tends to behave like λt if $A(t)$ is a stationary process.

The reason why we are interested here in the fluid approximation of the stochastic discrete queueing network is that the stability of the latter can be studied by observing suitable stability definition applied to the fluid model, which usually turns out to be much simpler than the original problem.

Specifically, let $x_i(t) = \lim_{n \to \infty} X_i(nt)/n$ be the fluid limit of the content of buffer i for any $t \geq 0$. We give two definitions for the stability of the fluid equivalent of a stochastic discrete queueing network, leaving the detailed description of how to identify the fluid equivalent network to the next section.

Definition 7.3 (Stability of a fluid network) A deterministic fluid queueing network is said to be stable if there exists a finite $T > 0$ such that $\sum_{i=1}^{b} x_i(t) = 0$ for all $t \geq T$, for any initial finite fluid level $x_i(0)$, $\forall i$. That means that all fluid queues will eventually drain out whatever the initial fluid level be.

Definition 7.4 (Weak stability of a fluid network) A deterministic fluid queueing network is called weakly stable if $\sum_{i=1}^{b} x_i(t) = 0$ for all $t > 0$, given that initially the network is empty, i.e., $\sum_i x_i(0) = 0$.

The second definition is weaker. It implies that the fluid queueing network remains empty if it starts out empty. Saying that a fluid buffer is empty does not mean that no fluid is going through the buffer. It means that the rate of the flow at the input of the buffer is less than or equal to the drain rate out of the buffer. Then, all the fluid arriving at the buffer immediately leaves the buffer and no accumulation occurs, i.e., the buffer remains empty.

The second definition of stability states that full drainage is maintained over time. So, if all buffers in the network start out empty, they will remain empty throughout. The first definition says something more. Provided the queueing network buffers start with any finite backlog, it can be drained out completely (along with the new fluid arriving at the nework) over a finite time horizon. That is, the serving rate at the buffers is strictly greater than the arrival rate, so that there is a positive excess serving capacity that is used over a finite transient to get rid of the initial backlog.

The key result that can be shown and that motivates the whole introduction of the deterministic fluid model corresponding to a stochastic discrete network is the following: a stochastic discrete queueing network is stable (rate stable) if the equivalent deterministic fluid network is stable (resp., weakly stable).

Note the strength of the result. All that is required to construct the equivalent fluid model, as detailed in the next section, is the mean arrival and service rate at each buffer, besides the structure of the network, the routing and the serving policy at each node. Stability results do not depend on the statistical distributions of the arrival and service processes, apart from their respective mean rates.

Next, we complete the argument by giving details on the construction of the equivalent fluid model of a stochastic discrete network. For further details and proofs, the interested reader can refer to, e.g., [54,63].

7.5.2 Turning a Stochastic Discrete Queueing Network into a Deterministic Fluid Network

Let us summarize the full definition of the kind of stochastic discrete queueing network we are considering here, introducing some notation as well. The elements constituting a stochastic discrete queueing network are the following:

1. N nodes, that provide service to customers.
2. Each node has a single server shared by customers of possibly different classes served by that node.
3. The network contains b buffers, at least one in each node. Customers residing in buffer i are identified as composing class i, $i = 1, \dots, b$. We define an incidence matrix for the association buffer-node: let $b_{ij} = 1$ if buffer i belongs to node j; let $\mathbf{B}$ be the matrix collecting the 1-0 coefficients b_{ij}.
4. $1/\mu_i$ denotes the mean service time of a customer at buffer i, if the customer is served in isolation. In other words, the service rate for a customer served at buffer i is μ_i, if that is the only customer the server of the node containing buffer i is processing.
5. Each node has a policy for sharing its single server capability among the buffers belonging to the node. Policy requirements are only: (a) it must be work-conserving; (b) head-of-the-line service at every buffer, i.e., there can be at most one partially completed service in a buffer at any time. Note the

condition (b) does not preclude having processor sharing policies across buffers.

6. There is infinite waiting room at each buffer.
7. External arrivals at buffer i occur at an average rate of γ_i. Arrivals are independent of each other and of the network state and service times.
8. When a customer completes service at buffer i, it either goes to buffer $j \neq i$ (then we set $r_{ij} = 1$), or leaves the network (in that case $\sum_{j=1}^{b} r_{ij} = 0$). Routing is deterministic, defined by the 1–0 matrix $\mathbf{R}$, made up with the entries $r_{ij}, i,j = 1, \dots, b$. We assume that $\mathbf{I} - \mathbf{R}$ is invertible.

The elements above give a full description of the stochastic discrete queueing network, as far as stability matters are of interest. If we intended to simulate the network, we should also specify the statistical characteristics of the external arrival processes at buffers and of the service times of each customer class, besides assigning numerical values to all parameters.

Confining ourselves to the study of stability, the deterministic fluid equivalent network is obtained by replacing the discrete flows of customers with fluids and describing buffers as liquid containers, where fluid is poured into and drained out, according to the given mean arrival and serving rates. To be more precise, the only quantities that matter are the average arrival rate and the average serving rate at each buffer. Those are converted into the input fluid rate and the drain rate out of the buffer in the fluid network.

Example 7.9 Let us identify the deterministic fluid network associated with the Rybko-Stolyar-Kumar-Seidman network of Example 7.8. There are $N = 2$ nodes and $b = 4$ buffers. Following the same labels used in Example 7.8, we obtain the 4×2 incidence matrix $\mathbf{B}$ as

$$\mathbf{B} = \begin{bmatrix} 1 & 0 \\ 0 & 1 \\ 0 & 1 \\ 1 & 0 \end{bmatrix} \tag{7.124}$$

The routing matrix is easily derived from the description of the network:

$$\mathbf{R} = \begin{bmatrix} 0 & 1 & 0 & 0 \\ 0 & 0 & 0 & 0 \\ 0 & 0 & 0 & 1 \\ 0 & 0 & 0 & 0 \end{bmatrix} \tag{7.125}$$

The input arrival rates at the four buffers are $[\lambda_1 \ 0 \ \lambda_3 \ 0]$. and the drain rates of fluid out of the buffers are $[\mu_1 \ \mu_2 \ \mu_3 \ \mu_4]$.

Finally, as for the serving policy at node 1, the single server is assigned to drain buffer 4 whenever there is fluid in that buffer. If buffer 4 is empty, that does not mean that no fluid of class 4 is being processed by node 1, just that the rate of that fluid is less than the full serving capacity of the node. Then, whatever is left unused by class 4 fluid of the single server capacity of node 1, is exploited by fluid of class 1. As a consequence, the single server capacity at node 1 can be underutilized (as a special case, idle) only if the demand of fluid summing up class 1 and class 4 is less than the server capacity. An entirely similar argument applies to node 3, hosting buffers 2 and 3, where 2 has preemptive priority over 3.

Given a stochastic discrete network with all its parameters, we convert it into a deterministic fluid network by preserving its structure (nodes, buffers), the deterministic routing rules and specifying only the input and drain rates of each buffer.

When the only fluid at a tagged node is in one of its buffers, say buffer i, the drain rate of that buffer coincides with the mean service rate of the node. To account for sharing of the node single server among buffers within the node, let $z_i(t)$ denote the amount of time allocated by the node hosting buffer i to that buffer in the time interval $[0, t]$. We define the right-continuous derivative of $z_i(t)$, as $\zeta_i(t) = \frac{d}{dt} z_i(t)$. The function $\zeta_i(t)$ gives the fraction of the server that is devoted to buffer i at time t. Therefore, the sum of all such fractions over the buffers of a node cannot exceed 1, i.e.,

$$\sum_{j=1}^{b} b_{ij} \zeta_j(t) \leq 1 , \qquad i = 1 \ldots , b \tag{7.126}$$

The sum is equal to 1, if at least one buffer of the node is nonempty, given that the node service policy is work-conserving. The fraction $\zeta_j(t)$ depends on: (i) the input rate at buffer j; (ii) the state of buffer j (empty or nonempty); (iii) the policy of the server. For example, if the state of buffer j is $X_j(t) = 0$, then it must be $\mu_j \zeta_j(t) \leq a_j(t)$, i.e., the service rate granted to buffer j cannot be larger than its input rate.

Taking the instantaneous balance of input and output flow rates at buffer j, the input fluid rate a_j at buffer j is obtained as

$$a_j(t) = \gamma_j + \sum_{i=1}^{b} \mu_i \zeta_i(t) r_{ij} \qquad j = 1, \ldots, b. \tag{7.127}$$

The drain rate at buffer j in general can be identified as $\mu_j \zeta_j(t), j = 1 \ldots , b$.

This completes the identification of the deterministic fluid network equivalent to a given stochastic discrete queueing network.

A last remark concerns how to assess whether a given deterministic fluid network is stable or weakly stable. Essentially, it is a case-by-case study. There is no general approach. Several examples are provided, e.g., in [86, Ch. 8]. A practical approach consists of simulating the fluid version of the queueing network and checking that queues get drained out of whatever initial backlog is assigned.

7.6 Further Readings

The kind of open network considered in this chapter is often referred to as Jackson network; sometimes it is referred to as *migration process*, e.g., see [122, Ch. 2]. The relatively simple product form yielding the steady-state probability distribution of the number of customers in each queue of the network holds if the queueing network nodes offer: (i) state-dependent service (e.g., multi-server queues with a finite number of servers) with negative exponential service times; and (ii) infinite server queues and general renewal service times [86, Ch. 6]. Several generalizations are possible, while still retaining the strong property of a product-form solution for the joint state probability distribution. Some of the main generalizations still retaining the product form solution are:

- *Deterministic routing.* Unlike the memoryless routing model, we define a set $\mathcal{R}$ of routes through the network. Arriving customers are associated with one route of $\mathcal{R}$. The route specifies the sequence of queues that the associated customer visits orderly before leaving the queueing network.
- *Multi-class networks.* K different classes of customers are defined, each one characterized by its own arrival rate and routing probabilities at each queue. When leaving queue i a customer of class k can either leave the network with probability $r_{i0}(k)$, or it can join queue j as a customer of class h with probability $r_{ij}(k, h)$. Admissible serving policies at the queues are FCFS, processor sharing or LCFS with preemptive resume policy.
- *BCMP networks.* This concept was originally introduced in the paper by Baskett, Chandy, Muntz, and Palacios [25]. A summary can also be found in [196, Ch. 15].

Combinations of features from various generalization paradigms are possible as well, e.g., Kelly networks [121].

Moving from exact solutions to approximation, a wealth of models has been analyzed, including G/G queueing networks, i.e., networks where component queues exhibit general i.i.d. service times and external arrivals form general renewal processes. Fluid approximations are also available for even more general distributional assumptions. Another generalization that can be dealt with by means of approximation is priority queueing networks, both global (priorities are assigned to customers when they arrive at the network and they do not change as customers move inside the network) or local priorities (a customer can change its priority level at any visited queue). For more details, the interested reader can refer to [86] for an excellent introductory, yet extensive treatment, with many numerical examples. Specialized references for approximate solution of queueing network are [37, 206, 54].

The product-form solution is intimately related to a property of the underlying Markov process described by the queueing network state $\mathbf{Q}(t)$. Specifically, the key property is the reversibility, which amounts to saying that, if $\mathbf{Q}(t)$ is a Markov process and it is reversible, then $\mathbf{Q}(-t)$ is a Markov process as well.

A deep theoretical foundation of queueing networks along with many general results is available in Chen and Yao [54]. Besides [54], recent results are reviewed also in Serfozo [188] and Chao, Miyazawa, and Pinedo [53]. Gautam [86] gives a quite extensive account of the queueing network topic, at a more accessible level. The deep connection between reversibility of Markov chains and queueing network is explored in depth in Kelly [121].

Appendix

Let us consider a graph $\mathcal{G}$. Let V denote the set of vertices or nodes of the graph and E the set of edges (also referred to as links or arcs) of the graph. E is a subset of the Cartesian product $V \times V$. A graph is called undirected if $(i,j) \in E \Rightarrow (j,i) \in E$. It is called directed in the opposite case, i.e., if there exists even a single pair $(i,j) \in E$ for which $(j,i) \notin E$. It is customary to draw the graph by representing nodes with circles and edges with lines connecting the circles. In the case of directed graphs the lines are actually arrows, to represent the direction of the edge (from i to j). With undirected graphs there is no need to draw arrows. Let $n = |V|$ be the number of nodes and $\ell = |E|$ be the number of edges.

A graph is a very useful model for representing interactions. In the framework of communication networks, it is useful to describe connectivity as seen at a given architectural level. The interpretation of nodes and edges depends on the considered architectural level, e.g., nodes can be routers and edges subnets connecting them at IP level or switches connected with physical links at layer 2 (e.g., in an ethernet network).

If a function $w : E \mapsto \mathbb{R}$ is defined, we say the graph is weighted. The weight of arc (i,j) is denoted with w_{ij}.

A path from i to j is an ordered sequence of nodes $k_1, k_2, \ldots, k_m$ such that the edges $(i,k_1), (k_1,k_2), \ldots, (k_m,j)$ all belong to E. A path from node i to node j is denoted with $P(i,j)$. A path with $i = j$ is called a loop. In networking applications usually one is interested in paths between nodes i and j, with $i \neq j$ that do not include loops, i.e., such that there are not two indices a and b so that $k_a = k_b$. For ease of notation, we let $a, b = 0, 1, \ldots, m+1$, extending the definition of k_j with $k_0 = i$ and $k_{m+1} = j$.

A graph is strongly connected if there exists a path $P(i,j)$ for all $i,j \in V$ with $i \neq j$. To assess the connectivity of an undirected graph, it suffices to check that there exists a node i such that the paths $P(i,j), \forall j \in V \setminus \{i\}$ can be found. Given any two other nodes, h and k, a path connecting them can be obtained as the union of the two paths $P(k, i)$ and $P(i, h)$. These last two paths exist necessarily, since $P(k, i) = P(i, k)$, the graph being undirected. In words, if a node x communicates with any other node of the graph, any two nodes communicate through x. The node x plays the role of a "hub" node. Note that it is key that edges are not directed for this property to hold.

In an undirected graph the degree of node i is the number of edges $(i, j) \in E$ for all $j \in V$, i.e., the number of edges outgoing from node i. This coincides with the number of edges pointing at i from any other node of the graph. We denote the degree of i with d_i. It is a non-negative integer number. If $d_i = 0$, node i is isolated. In a directed graph we distinguish between in-degree of node i, i.e., the number of edges pointing at node i, and out-degree, i.e., the number of edges coming out of node i.

Let $\mathbf{A}$ denote a 1-0 matrix, called the adjacency matrix, defined as follows: $a_{ij} = 1$ if and only if $(i, j) \in E$. For an undirected graph $a_{ij} = 1 \Rightarrow a_{ji} = 1$, i.e., the matrix $\mathbf{A}$ is symmetric.

The Laplacian of a graph is an $n \times n$ matrix $\mathbf{L}$ defined as follows:

- $L_{ii} = d_i$, where d_i is the degree of node i;
- $L_{ij} = -1$ if and only if $(i, j) \in E$.

We can write:

$$\mathbf{L} = \mathbf{D} - \mathbf{A} \tag{7.128}$$

where $\mathbf{D}$ is a diagonal matrix whose i-th diagonal element is d_i and $\mathbf{L}$ is the Laplacian matrix. $\mathbf{L}$ is a symmetric matrix, if the graph is undirected. The sum of each row and each column of $\mathbf{L}$ is 0. Therefore, 0 is always an eigenvalue of $\mathbf{L}$.

The incidence matrix of a graph is an $\ell \times n$ matrix $\mathbf{M}$ defined as follows:

- $M_{ki} = 1$, if link k is outgoing from link i;
- $M_{ki} = -1$, if link k is incoming into node i;
- $M_{ki} = 0$, otherwise.

It can be verified that:

$$\mathbf{L} = \frac{1}{2}\mathbf{M}^T\mathbf{M}$$

Hence $\mathbf{L}$ is a positive semi-definite matrix, i.e., $\mathbf{x}^T\mathbf{L}\mathbf{x} \geq 0$ for any vector $\mathbf{x}$. Being a positive semi-definite and symmetric matrix (for an undirected graph), the eigenvalues of $\mathbf{L}$ are real and non-negative. We already know that $\mathbf{L}$ has an eigenvalue equal to 0. Let the algebraic multiplicity of the eigenvalue 0 be v. It can be shown that the graph $\mathcal{G}$ breaks up into v connected components. Therefore, the graph is *connected* if $v = 1$. This, in turn, implies that the second smallest eigenvalue of the

---

Algorithm Pseudo-code of the breadth-first-search algorithm.

1: **for** each node $n \in V$ **do**
2:   $n.distance = \infty$;
3:   $n.parent = NIL$;
4: **end for**
5: create empty queue Q;
6: $root.distance = 0$;
7: $Q.enqueue(root)$;
8: **while** $Q \neq \emptyset$ **do**
9:   $c = Q.dequeue()$;
10:   **for** each node $n : a_{cn} = 1$ **do**
11:     **if** $n.distance = \infty$ **then**
12:       $n.distance = c.distance + 1$;
13:       $n.parent = c$;
14:       $Q.enqueue(n)$;
15:     **end if**
16:   **end for**
17: **end while**

Laplacian must be positive. This property can be exploited to set up a test for graph connectivity.

Alternatively, it can be shown that the graph is connected if the adjacency matrix is irreducible. This means that there does not exist a relabeling of nodes such that the adjacency matrix can be put into the following structure:

$$A = \begin{bmatrix} A_{11} & 0 \\ A_{21} & A_{22} \end{bmatrix}$$

It can be verified that the $n \times n$ matrix A is irreducible if and only if

$$I + A + A^2 + \cdots + A^{n-1} > 0$$

where the inequality is understood as entry-wise. Finally, checking the connectivity of an undirected graph can be carried out also by resorting to the breadth-first-search algorithm. A pseudo-code of the breadth-first-search algorithm is shown above. It assumes that a graph (V, E) and a node "root" are assigned. Starting from the root node, the algorithm visits neighboring nodes, then neighbors of neighbors and so on, until no new node can be visited. To know if the graph is connected it suffices to check whether the number of visited nodes, other than the root node, is equal to $n - 1$.

Numerical experience with growing n shows that the breadth-first-search algorithm is by far the most computationally efficient algorithm among those presented in this Appendix.

Summary and Takeaways

This chapter provides an introduction to a powerful class of models, namely networks of queues. The topic has received wide attention over several decades, and many results and generalizations are available. Here we focus on Jackson-type networks, both open and closed, and on loss networks. Even though those are rather simple models, they provide useful tools for modeling service systems networks and protocols. We have addressed optimization over a queueing network (link capacity dimensioning to minimize the mean end-to-end delay, optimal routing) and numerical issues, related to the evaluation of the normalization constant of a closed queueing network. Queueing networks are a rich enough model to easily give rise to counterintuitive results. We have presented the Braess paradox, which warns about improving performance of a network by adding capacity. Another interesting issue is stability of a queueing network. In Jackson-type networks we have seen that stability (hence the existence of a statistical equilibrium, steady-state regime) depends only on the utilization coefficient of each queue of the network being less than 1. This is not necessarily the case in general queueing networks, where priorities and/or reentrant customer flows can be defined. Interested readers can further investigate the references discussed in the Further Readings section of this chapter.

Problems

7.1 In an archival system, files undergo two kinds of processing. Task 1 is performed by server 1, to which files are offered according to a Poisson process of mean rate $\lambda = 0.3 \ s^{-1}$. After completing task 1, files are submitted to server 2 that is in charge of running task 2. Service times at the two servers have negative exponential PDFs with mean $E[X_1] = 1 \ s$ and $E[X_2] = 2 \ s$, respectively.

(a) Evaluate the mean response time of the whole system.

(b) The system is rearranged by having a unique server do both tasks. The mean service time of the new server is the sum of the mean service times of the two servers. Compare the mean system response time of this new solution with the previous one. Give an intuitive explanation of the result.

7.2   An ISP Point of Presence (PoP) is made up of N routers. Time to off-line maintenance of a router has a negative exponential probability distribution with mean value 300 days. There is only one technician in service at the PoP, at any time (multiple workers alternate on daily shifts). The maintenance time of a router has a negative exponential probability distribution with mean 4 hours. Find the minimum value of N that guarantees that there are at least six routers operational at any given time with probability no less than 0.999.

7.3   An urban area is connected by a ring network comprising $N = 7$ nodes. Node 1 is a gateway to the Internet, while the other six nodes are the access nodes of six different local networks. All the traffic from local networks (upstream traffic) is addressed to the Internet: the corresponding mean bit rate for each local network is $R_{out} = 100$ Mbit/s. The traffic coming from the Internet and directed to the local networks (downstream traffic) is uniformly distributed among them. The overall mean bit rate of that traffic is $R_{in} = 1$ Gbit/s. Both traffic flows can be modeled as Poisson processes at the packet level. Packet lengths are random variables, with negative exponential distribution, and the mean packet length is $L = 500$ bytes. The links connecting neighboring nodes are bidirectional, with a capacity C in each direction.
(a) Identify a queueing model of the ring network at packet level.
(b) Find the minimum value of C that guarantees that the utilization coefficient in each link is not greater than 0.8.
(c) Find the mean delay from node 1 to each other node with the dimensioning of point (b).

7.4   Consider an ISP network with J links connecting routers. The offered traffic matrix and the routing in the network are such that it is possible to maintain link utilization equal to a constant value $\rho_0 = 0.7$ for all links. The volume of the overall input traffic is increased by 10% (i.e., the mean rate of each flow of packets offered to the network is increased by 10%). Calculate the percentage increase of the mean delay of a packet through the network.

7.5   A packet network is made up of ℓ links. The mean offered packet rate to the whole network Γ and the internal network routing are so engineered that the mean load on each link has a same value ρ_0. The propagation delay of the i-th link is τ_i, for $i = 1, \dots, \ell$.
(a) Define an open queueing network model of the network and give the expression of the mean delay through the i-th link, *by accounting for the*

propagation delay. [Hint: the mean delay through a link is the sum of the queueing and the propagation delays.]

(b) Write the expression of the mean delay through the network E[D], by using the result obtained in point (a).

(c) Find the admissible value of ρ_0 under the requirement that $E[D] \leq 4\ell/\Gamma$, if the average bandwidth delay product of the network links is three times the mean packet length.

7.6 A factory is made up of M machines. Times to failure and times to repair can be modeled as random variables with negative exponential probability distribution, independent of one another for all machines. Their respective mean values are $\lambda = 1$ failure/year and $1/\mu = 1$ week. A single repairman is available.

(a) Define a closed queueing network model of the system for a generic M and calculate the state probabilities at equilibrium.

(b) Calculate the probability $P(\beta)$ that at least a fraction $\beta \in (0, 1)$ of the M machines are working. Find the numerical value for $\beta = 0.9$ and $M = 100$.

7.7 Let us consider an asymmetric bidirectional communication system between two endpoints A and B (e.g., A is a server and B is a host). A TCP connection is set up between A and B. A sends data segments of length L_1 to B. B sends ACK segments of length L_2 back to A, one ACK per data segment. The link from A to B has a capacity C_1, the other link has a capacity C_2. At any given time there are W data segments in flight, i.e., A has sent them and is still waiting for their ACKs.

(a) Define a closed queueing network model of the communication system and find the state probabilities at equilibrium.

(b) Calculate the utilization coefficients of link 1 (from A to B) and link 2 (from B to A) as a function of the variables L_1, L_2, C_1, C_2, and W.

(c) Find the value of C_2 such that the two utilization coefficients are equal by assuming $L_1 = 1500$ bytes, $L_2 = 50$ bytes and $C_1 = 30$ Mbit/s. Then, calculate the utilization coefficient as a function of W and for $W = 7$.

7.8 Two communication paths connect the same two end points. They have the same capacity C and they are used in load balancing mode. Each of them can fail independently of the other: the time to failure of a working path has a negative exponential PDF with parameter ν. A failed path goes into a repair process that restores the path in a time R, where R is a negative exponential random variable with mean $1/\mu$. A *single* repair team operates at any time, when required. Assume $1/v = 4$ days, $1/\mu = 4$ hours and $C = 1$ Gbit/s.

(a) At time t_0 one link fails, while the other one is still operational. What is the probability that the failed link is restored before the other one fails as well?

(b) Define a queueing model to describe the failure-restore evolution of the two paths system [Hint: you can use a closed queueing network; alternatively, you could use a simple birth-death Markov process].

(c) Given the model defined in point (b), calculate the unavailability P_U of the two paths system [unavailability=probability that the two end points are disconnected].

(d) Calculate the average available capacity to connect the two end points.

8

Bounds and Approximations

It is the mark of an educated mind to rest satisfied with the degree of precision which the nature of the subject admits and not to seek exactness where only an approximation is possible.

Aristotle

8.1 Introduction

In spite of an impressive amount of exact and elegant results for stochastic models and, specifically, for queues, it is infrequently the case that a service system can be analyzed with a model yielding to exact analysis. Even if the aim of an analytic model is to grasp the key properties of a system and to gain insight in the fundamental trade-offs, achieving this target may require accounting for more details or stochastic characteristics than those that could feasibly be captured by a model solvable in closed form. Hence, approximations are commonplace in the application of network traffic engineering.

The first part of the chapter is concerned with approximations for general queueing models, namely $G/G/1$ and $G/G/m$. We restrict our discussion to the case of renewal arrival and service processes. Both bounds and approximations are discussed, mainly for the mean waiting time (or system time). We also provide an upper bound of the probability distribution of the system time of a $G/G/1$ queue. We discuss a continuous-state approximation of the $G/G/1$ queue, based on the reflected Brownian motion process. This approach stems as an example of a general approach where a strong simplification of the analysis can be obtained by forgetting about the discrete nature of the queueing system.

Network Traffic Engineering: Stochastic Models and Applications, First Edition. Andrea Baiocchi.
© 2020 John Wiley & Sons, Inc. Published 2020 by John Wiley & Sons, Inc.

The second part of the chapter is devoted to fluid models. The common feature of this kind of modeling is to use continuous-state processes to describe service systems, neglecting the discrete character of the customers. More generally, there are systems that are naturally modeled as having a continuous state varying over continuous time (e.g., dams or water basins). In other contexts (e.g., computer, communications, and transportation systems) discrete models are usually the most natural approach to describe the operation of the service system. Continuous-state processes arise as approximations. Their main merit is to defeat the state-space explosion that affects many discrete-state models for practical parameter ranges.

For example, let us consider a Markov modulated arrival process, where the modulating phase process has J states. The arrival process can be a Poisson process, whose mean rate is a function of an irreducible, positive recurrent continuous-time Markov chain with J states. If the arrival stream (e.g., packets) is offered to a buffer of size K, the entire system could be described by a two-dimensional discrete process (j, k). The first component $j \in \{0, 1, \ldots, J\}$ represents the state of the modulating phase process. The second component $k \in \{0, 1, \ldots, K\}$ represents the number of packets in the buffer (including the one possibly under service). The state space has size $(J + 1) \times (K + 1)$. If both J and K are in the order of 100, the state space size is in the order of 10^4 states, which makes it hard to deal with it as a Markov chain. If modeling is done via Markov chain of dimensionality higher than two, the explosion of the state space can easily forbid the practical implementation of the model analysis.

A fluid model can circumvent this issue by modeling the content of the buffer as a continuous level x ranging from 0 up to K. The resulting description falls into the category of a stochastic fluid flow model. The state space size reduces to $J + 1$, the size of the modulating process. The complexity of the two-dimensional big state space is shifted to the mathematical relationship that involves the probability distribution of the joint variables (j, x). As we will see, such joint probability distribution is the solution of a system of first-order differential equations, with suitable boundary conditions. The key points here are:

- In various interesting cases an analytical solution can be derived for the system of differential equations.
- If a closed-form solution cannot be found, the implicit description of the joint probability distribution by means of the system of differential equations can provide useful qualitative information on the behavior of the solution or for developing approximations.
- There exist many well-established and reliable numerical solvers for systems of differential equations of the first-order, so that the probability distribution can be obtained at least numerically with relative ease and often at a very reasonable computational cost.

We present the fluid approximation both as a limit of a suitably scaled version of the original queueing process and as a direct approximation of the original discrete-state process describing the service system.

With reference to the latter use of the fluid model, we introduce a stochastic fluid flow model, where the rate of input fluid is modulated by a finite-state Markov process. This is a simple example of how fluid modeling leads to tractable models under quite sophisticated queueing settings. Even more general frameworks yield to analysis via fluid modeling, e.g., queues where the input or service processes are time-varying, or driven by semi-Markov processes, or (possibly nonlinear) functions of system state. Fluid modeling is also useful for transient analysis.

There is a vast literature on bounds and approximations, encompassing introductory material and specialized monographs and papers. Good introductory material can be found in [131, Ch. 2] and [94, Ch. 7]. The former gives an excellent account of the diffusion approximation. An extensive treatment of the fluid approximation and of stochastic fluid models can be found in [86, Chs. 8, 9, 10]. The same book offers also a good introduction to approximate analysis of queueing networks. Pointers to specific in-depth sources are given in Section 8.5.

One aspect of solving models to obtain performance measures is numerical evaluation of the model. This is a crucial and essential part of network traffic engineering. It is often the case that issues can be dealt with by resorting to standard numerical analysis tools (e.g., numerical integration, differential equation solvers, linear algebra numerical tools). Two specialized topics emerging in network traffic engineering model evaluation are: (i) numerical inversion of Laplace transforms of PDFs; and (ii) numerical solution of a Markov chain, i.e., computation of the limiting state probabilities. As for the first topic, an excellent survey of best methods is provided by Abate and Whitt [1–3]. Numerical evaluation of Markov chain is covered in a neat and extensive way in [196, Ch. 10].

8.2 Bounds for the *G/G/1* Queue

We consider a single server queue with independent, general renewal processes for arrivals and service times. The $G/G/1$ queue can be considered the simplest model which is beyond easy mathematical analysis. Anyway, it is rather straightforward to generate a sequence of waiting or system times for this queue, once sequences of inter-arrival times and service times are given, either estimated from measurements or drawn from assigned probability distributions.

The main notation used in this chapter, specifically for the analysis of $G/G/1$ queues, are listed in Table 8.1.

Let $\{T_n\}_{n\geq 1}$ and $\{X_n\}_{n\geq 1}$ denote the sequence of inter-arrival times and service times, respectively. Let also A_n and D_n denote the arrival and departure time of the

Table 8.1 Main notation for the $G/G/1$ queue.

Symbol	Definition
A_n	arrival time of the n-th customer
D_n	departing time of the n-th customer
T_n	inter-arrival time between customer $n-1$ and n, i.e., $T_n = A_n - A_{n-1}$.
X_n	service time of the n-th customer.
U_n	inter-departure time of the n-customer, i.e., $U_n = D_n - D_{n-1}$.
W_n	waiting time of the n-th customer.
S_n	system time of the n-th customer, i.e., $S_n = W_n + X_n$.
λ	mean arrival rate, assuming the arrival process is (wide-sense) stationary: $\lambda = 1/E[T_n]$.
σ_T	standard deviation of arrival times, assuming the arrival process is (wide-sense) stationary.
C_T^2	squared coefficient of variation (SCOV) of arrival times: $C_T^2 = \lambda^2 \sigma_T^2$.
μ	mean service rate, assuming the service process is (wide-sense) stationary: $\mu = 1/E[X_n]$.
σ_X	standard deviation of service times, assuming the service process is (wide-sense) stationary.
C_X^2	squared coefficient of variation (SCOV) of service times: $C_X^2 = \mu^2 \sigma_X^2$.

n-th customer. We assume that no customer is rejected. We assume also first-come, first-served (FCFS) service order. Setting initially $A_0 = D_0 = 0$, we derive easily the following iterations for $n \geq 0$:

$$A_{n+1} = A_n + T_{n+1} \tag{8.1}$$

$$D_{n+1} = \max\{D_n, A_{n+1}\} + X_{n+1} \tag{8.2}$$

$$W_{n+1} = \max\{0, D_n - A_{n+1}\} \tag{8.3}$$

$$S_{n+1} = D_{n+1} - A_{n+1} = W_{n+1} + X_{n+1} \tag{8.4}$$

where W_n and S_n are the waiting and system times of the n-th customer, respectively. For those iterations to hold, we require that the single server is *work-conserving* and offers *head-of-line* (HOL) service. Work-conserving means that the server cannot stay idle if there is at least one customer in the queueing system. HOL service means that there cannot be more than one customer under service at any given time.

From eqs. (8.1)–(8.4) the well known Lindley's recursion for a single-server queue can be deduced:

$$W_{n+1} = \max\{0, W_n + X_n - T_{n+1}\}, \quad n \geq 1, \tag{8.5}$$

initialized with $W_1 = 0$. W_n is the amount of workload found in the queue by the n-th customer upon arrival.

Note that eqs. (8.2)–(8.4) hold for any work-conserving, HOT serving discipline, provided that the subscript n denotes the n-th *departing* customer. For eq. (8.1) to hold as well, service must be offered according to FCFS queueing policy. In that case, the n-th departing customer is also the n-th arriving customer. In the rest of the chapter, we assume FCFS service, unless stated otherwise. We also assume that service times are positive with probability 1 and that the arrival process is nondegenerate and both stochastic sequences are wide-sense stationary, with finite mean and variance.

Equation (8.5) generates the sequence of waiting times, even for nonstationary or unstable $G/G/1$ queues, for any finite time horizon. Obviously, the sequence $\{W_n\}_{n\geq 1}$ provides an estimate of E[W] (or of any other steady-state statistics of the random variable W) only if the steady state exists, i.e., the arrival and service process are stationary and the queue is stable. The condition for the $G/G/1$ queue to be stable is E[X] < E[T]. If we introduce the arrival and service rates, namely $\lambda = 1/E[T]$ and $\mu = 1/E[X]$, this is stated as $\lambda < \mu$ or $\rho < 1$, where $\rho \equiv \lambda/\mu$ is the utilization coefficient of the server.

Example 8.1 As a numerical example, we assume inter-arrival times uniformly distributed in the interval $[T_{\min}, T_{\max}]$ and service times with a Pareto probability distribution with complementary cumulative distribution function (CCDF) $G_X(x) = \min\{1, (\theta/x)^\beta\}$.

The first two moments of the inter-arrival times are

$$E[T] = \frac{1}{\lambda} = \frac{T_{\min} + T_{\max}}{2} \qquad \sigma_T^2 = \frac{(T_{\max} - T_{\min})^2}{6} \tag{8.6}$$

As for the service times, we have ($\beta > 2$)

$$E[X] = \frac{\beta\theta}{\beta - 1} \qquad E[X^2] = \frac{\beta\theta^2}{\beta - 2} \tag{8.7}$$

We set $T_{\min} = 0$, $T_{\max} = 2$, so that E[T] = 1 is assumed to be the time unit. Moreover, we let $\beta = 2.25$, while θ is found by assigning a value of $\rho = \frac{\beta}{\beta-1}\lambda\theta$. For this numerical example, we let $\rho = 0.8$, hence $\theta = 0.444$.

The mean and squared coefficient of variation (SCOV) of the inter-arrival and service times are E[T] = 1, $C_T^2 = 0.667$, E[X] = 0.8, $C_X^2 = 1.778$.

The first 1000 samples of the sequence of waiting times W_n are plotted against time in Figure 8.1.

The high variability of waiting times, due to the high SCOV of the service times, is apparent from the sequences of relatively low levels of waiting times interleaved by sudden peaks.

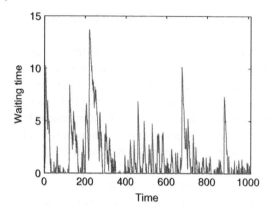

Figure 8.1 Sample path of the waiting time of a $G/G/1$ queue. The utilization coefficient is $\rho = 0.8$. Inter-arrival times are uniformly distributed over $[0, 2]$; service times follow a Pareto distribution with CCDF $\min\{1, (0.444/x)^{2.25}\}$.

8.2.1 Mean Value Analysis

Let us consider a stochastic sequence X_n converging to a proper random variable X as $n \to \infty$. A mean value analysis (MVA) consists of a set of formulas that can be used to derive the mean value of X, i.e., $\lim_{n\to\infty} E[X_n]$.

We apply the MVA to waiting time of a $G/G/1$ queue, thus deriving an exact result. The exact result is however of little use in practice, except that it can be the starting point for deriving approximations and bounds of the mean waiting time.

We assume that the limits of the first two moments of W_n as $n \to \infty$ exist and are positive and finite. Notation is as defined in Table 8.1.

The time spent by customer $n + 1$ in the queue is the sum of the waiting time W_n and of the service time S_n:

$$S_{n+1} = X_{n+1} + W_{n+1} = X_{n+1} + \max\{0, D_n - A_{n+1}\} \tag{8.8}$$

The last equality follows from the fact that customer $n + 1$ has to wait only if its arrival time A_{n+1} occurs before the departure time of the previous customer n. Note that, by definition, it is $S_n = D_n - A_n$; then

$$D_n - A_{n+1} = D_n - A_n + A_n - A_{n+1} = S_n + A_n - A_{n+1} = S_n - T_{n+1} \tag{8.9}$$

Moreover

$$\max\{0, A_{n+1} - D_n\} \equiv I_{n+1} \tag{8.10}$$

represents the idle time occurring between the departure of customer n and the arrival of the following customer $n + 1$. The idle time I_{n+1} is 0, if customer $n + 1$ is already in the queue when customer n departs.

Exploiting those definitions, we can modify (8.8), thus obtaining

$$S_{n+1} = X_{n+1} + D_n - A_{n+1} + \max\{0, A_{n+1} - D_n\} = X_{n+1} + S_n - T_{n+1} + I_{n+1} \tag{8.11}$$

Taking expectations of (8.11) and then the limit for $n \to \infty$, we get

$$E[S_{n+1}] = E[X_{n+1}] + E[S_n] - E[T_{n+1}] + E[I_{n+1}], \quad \forall n \geq 1, \tag{8.12}$$

and

$$\lim_{n \to \infty} E[I_{n+1}] = E[I] = E[T] - E[X] = \frac{1}{\lambda} - \frac{1}{\mu} \tag{8.13}$$

Under the hypotheses required for this limit to exist (queue stability), it is $\lambda < \mu$, hence $E[I] > 0$. This is but one more manifestation of a general trait of stable, ergodic system, namely they must visit all possible states with positive probability, including the idle state, where the queue is empty. Then, the mean fraction of time spent in the idle state in the statistical equilibrium regime must be positive.

To obtain the mean of S_n we first square eq. (8.11), take the expectation and then the limit for $n \to \infty$. Before proceeding, we rearrange (8.11) as follows:

$$S_{n+1} - X_{n+1} - I_{n+1} = S_n - T_{n+1} \tag{8.14}$$

In doing this we account for four facts:

1. $S_{n+1} - X_{n+1} = W_{n+1}$ is independent of X_{n+1}.
2. $(S_{n+1} - X_{n+1})I_{n+1} = W_{n+1}I_{n+1} = 0$, since, if customer $n+1$ suffers a nonzero wait, the idle time preceding its service must be 0; on the contrary, if the idle time preceding the service time of customer $n+1$ is positive, it means that the server went idle after completing service of customer n, hence customer $n+1$ finds the server idle upon its arrival and suffers no delay.
3. S_n and T_{n+1} are independent of each other, since in the $G/G/1$ queue arrivals and service times form independent renewal processes.
4. S_n depends only on inter-arrival and service times up to those of the n-th customer.

Then, we have

$$E[(S_{n+1} - X_{n+1})^2] + E[I_{n+1}^2] = E[S_n^2] + E[T_{n+1}^2] - 2E[S_n]E[T_{n+1}] \tag{8.15}$$

whence

$$E[W_{n+1}^2] + E[I_{n+1}^2] = E[W_n^2] + E[X_n^2] + 2E[W_n]E[X_n] + E[T_{n+1}^2] - 2E[S_n]E[T_{n+1}] \tag{8.16}$$

where we have used $S_n = W_n + X_n$. By taking the limit for $n \to \infty$, we find

$$E[I^2] = E[X^2] + E[T^2] + 2E[W](E[X] - E[T]) - 2E[X]E[T] \tag{8.17}$$

Rearranging terms, we derive an expression for the mean waiting time

$$
\begin{aligned}
E[W] &= \frac{E[X^2] + E[T^2] - 2E[X]E[T] - E[I^2]}{2\left(\frac{1}{\lambda} - \frac{1}{\mu}\right)} \\
&= \lambda \frac{\frac{1+C_X^2}{\mu^2} + \frac{1+C_T^2}{\lambda^2} - \frac{2}{\mu\lambda} - E[I^2]}{2(1-\rho)} \\
&= \frac{\rho^2 C_X^2 + C_T^2 + (1-\rho)^2 - \lambda^2 E[I^2]}{2\lambda(1-\rho)}
\end{aligned}
\tag{8.18}
$$

Since we have shown that $E[I] = \frac{1}{\lambda} - \frac{1}{\mu} = (1-\rho)/\lambda$ and $\sigma_I^2 = E[I^2] - E[I]^2 \geq 0$, we have $E[I^2] \geq E[I]^2 = (1-\rho)^2/\lambda^2$. Therefore, we have $(1-\rho)^2 - \lambda^2 E[I^2] \leq 0$ and an upper bound is easily derived from (8.18):

$$
E[W] = \frac{\rho^2 C_X^2 + C_T^2 + (1-\rho)^2 - \lambda^2 E[I^2]}{2\lambda(1-\rho)} \leq \frac{\rho^2 C_X^2 + C_T^2}{2\lambda(1-\rho)} = \frac{\lambda(\sigma_X^2 + \sigma_T^2)}{2(1-\rho)}
\tag{8.19}
$$

The upper bound on the rightmost side of eq. (8.19) is also known as Kingman's bound of the mean waiting time of the $G/G/1$ queue. We will derive it in a different way in Section 8.2.3. It is remarkable that we can express a meaningful upper bound of the $G/G/1$ queue with renewal arrival and service times by means only of the first two moments of the arrival and service processes.

8.2.2 Output Process

In a subsequent section we will present an approximate analysis of a network of $G/G/1$ queues. A piece of that analysis relies on the expression of the first two moments of the output process of a $G/G/1$ queue, given the first two moments of the input process and of the service times.

Let $U_{n+1} = D_{n+1} - D_n$ denote the inter-departure time of the $(n+1)$-customer. We know that $D_{n+1} = \max\{A_{n+1}, D_n\} + X_{n+1}$. Then

$$
U_{n+1} = \max\{A_{n+1}, D_n\} + X_{n+1} - D_n = \max\{A_{n+1} - D_n, 0\} + X_{n+1} = I_{n+1} + X_{n+1}
$$

It is easy to derive the mean of the inter-departure time

$$
E[U] = \lim_{n\to\infty} E[U_n] = E[I] + E[X] = \frac{1}{\lambda}
\tag{8.20}
$$

This is an expected result for a lossless, stable queueing system.

As for the second moment of U, we exploit the independence of X_{n+1} and I_{n+1} to find:

$$
E[U^2] = \lim_{n\to\infty} E[U_n^2] = E[I^2] + E[X^2] + 2E[I]E[X] =
$$

$$
= E[I^2] + \frac{1+C_X^2}{\mu^2} + 2\frac{1-\rho}{\lambda}\frac{1}{\mu} = E[I^2] - \left(\frac{1}{\lambda} - \frac{1}{\mu}\right)^2 + \frac{1}{\lambda^2} + \frac{C_X^2}{\mu^2}
$$

Reminding eq. (8.18) we obtain

$$C_U^2 = \frac{\sigma_U^2}{\mathrm{E}[U]^2} = \lambda^2\mathrm{E}[U^2] - 1 = \rho^2 C_X^2 + \lambda^2\mathrm{E}[I^2] - (1 - \rho)^2$$
$$= C_T^2 + 2\rho^2 C_X^2 - 2\lambda(1 - \rho)\mathrm{E}[W] \tag{8.21}$$

The exact knowledge of the second moment of the output process is hence equivalent to knowing the exact expression of the mean waiting time.

Summing up, the first two moments of the output process of the $G/G/1$ queue are given by:

$$\begin{cases} \mathrm{E}[U] = \frac{1}{\lambda} \\ C_U^2 = C_T^2 + 2\rho^2 C_X^2 - 2\lambda(1 - \rho)\mathrm{E}[W] \end{cases} \tag{8.22}$$

8.2.3 Upper and Lower Bounds of the Mean Waiting Time

We can derive an upper bound for the mean waiting time of the $G/G/1$ queue in a different way, with respect to the bound found at the end of Section 8.2.1. Beside its own interest, the new approach leads us to derive also a lower bound for the mean waiting time of the $G/G/1$ queue. We maintain the same notation and assumption as in the previous subsections. The subscript n denotes the n-th departing customer. We also introduce the sequence of random variables Z_n defined as $Z_n = X_n - T_{n+1}$. Since inter-arrival and service times form renewal processes, we have $Z_n \sim Z$, $\forall n$, where Z is a random variable with first two moments $\mathrm{E}[Z] = \mathrm{E}[X] - \mathrm{E}[T] = 1/\mu - 1/\lambda < 0$ and $\sigma_Z^2 = \sigma_X^2 + \sigma_T^2$.

We know that the sequence of waiting times W_n follows Lindley's recursion in eq. (8.5), that we rewrite here as

$$W_{n+1} = \max\{0, W_n + Z_n\}, \qquad n \geq 1 \tag{8.23}$$

where $Z_n = X_n - T_{n+1}$, $n \geq 1$. We define also

$$W_{n+1}^- = -\min\{0, W_n + Z_n\} \tag{8.24}$$

It is easy to check that $W_{n+1}W_{n+1}^- = 0$ and

$$W_{n+1} - W_{n+1}^- = W_n + Z_n, \qquad n \geq 1 \tag{8.25}$$

Note that W_n is independent of Z_n. Taking squares of both sides and expectations we get

$$\mathrm{E}[W_{n+1}^2] + \mathrm{E}[(W_{n+1}^-)^2] = \mathrm{E}[W_n^2] + \mathrm{E}[Z_n^2] + 2\mathrm{E}[W_n]\mathrm{E}[Z_n] \tag{8.26}$$

Taking expectations of eq. (8.25) yields

$$\mathrm{E}[W_{n+1}] - \mathrm{E}[W_{n+1}^-] = \mathrm{E}[W_n] + \mathrm{E}[Z_n] \tag{8.27}$$

Assume the steady state exists, for which it is necessary and sufficient that $\rho = \lambda/\mu < 1$. From eqs. (8.27) and (8.26), we derive as $n \to \infty$:

$$E[W^-] = -E[Z] \tag{8.28}$$

and

$$E[(W^-)^2] = E[Z^2] + 2E[W]E[Z] \tag{8.29}$$

Putting these two equations together, we find

$$E[W] = -\frac{E[Z^2]}{2E[Z]} + \frac{E[(W^-)^2]}{2E[Z]} = \frac{\sigma_Z^2 - \sigma_{W^-}^2}{-2E[Z]} = \frac{\sigma_X^2 + \sigma_T^2 - \sigma_{W^-}^2}{2(1/\lambda - 1/\mu)} \tag{8.30}$$

Since variances are non-negative, an upper bound of $E[W]$ is easily obtained as:

$$E[W] \le \frac{\lambda(\sigma_X^2 + \sigma_T^2)}{2(1 - \rho)} \tag{8.31}$$

Equation (8.31) yields again Kingman's bound of eq. (8.19).

The approach used to derive the upper bound can be used to obtain a lower bound as well. First, note that $W^- = \max\{0, -W - Z\} = \max\{0, T - (W + X)\} \le T$. Hence, it follows that $E[(W^-)^2] \le E[T^2]$. Plugging this inequality into eq. (8.30), we get

$$E[W] = -\frac{E[Z^2]}{2E[Z]} + \frac{E[(W^-)^2]}{2E[Z]} = \frac{E[Z^2] - E[(W^-)^2]}{2\left(\frac{1}{\lambda} - \frac{1}{\mu}\right)} \ge$$

$$\ge \lambda\frac{\sigma_Z^2 + E[Z]^2 - E[T^2]}{2(1 - \rho)} = \lambda\frac{\sigma_X^2 + \sigma_T^2 + \frac{1}{\lambda^2} + \frac{1}{\mu^2} - \frac{2}{\lambda\mu} - \sigma_T^2 - \frac{1}{\lambda^2}}{2(1 - \rho)} =$$

$$= \frac{\lambda^2\sigma_X^2 + \rho(\rho - 2)}{2\lambda(1 - \rho)}$$

The lower bound can become negative and in that case it is of no use.

Summing up, we have the following bounds for the mean waiting time of the $G/G/1$ queue:

$$\max\left\{0, \frac{\lambda^2\sigma_X^2 + \rho(\rho - 2)}{2\lambda(1 - \rho)}\right\} \le E[W] \le \frac{\lambda(\sigma_X^2 + \sigma_T^2)}{2(1 - \rho)} \tag{8.32}$$

Example 8.2   With the same probability distributions and numerical values as in Example 8.1 we estimate the mean waiting time from a simulation of the $G/G/1$ queue (10^6 samples have been generated). The mean waiting time estimated from the simulation is compared with upper and lower bounds as provided by eq. (8.32). This has been done for several values of ρ and the results are plotted in Figure 8.2 as a function of ρ.

Figure 8.2 Comparison among the mean waiting time and its upper and lower bounds for a $G/G/1$ queue as a function of the utilization coefficient ρ. The inter-arrival times are uniformly distributed over $[0, 2]$; service times follow a Pareto distribution with CCDF $\min\{1, [(\rho/1.8)/x]^{2.25}\}$.

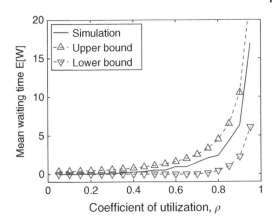

The bounds are not especially tight, still they provide useful information. While the lower bound is very loose, the upper bound gives a reasonable overestimate of the mean waiting time and becomes asymptotically tight as $\rho \to 1$.

It is to be noted that estimating the mean waiting time of this particular queue is tricky. Since the variance of the waiting time depends on the third moment of the service time, but in this case that moment is not finite (β should be greater than 3 to yield a finite third moment), it follows that the waiting time has infinite variance. It is therefore an extremely "unstable" random variable. Running different simulations can provide estimates that depart significantly, possibly even breaking the belt traced by the upper and lower bounds. In that case, the problem lies with estimating the exact mean waiting time, not with the bounds. Conversely, having firm bounds removes the need to run such computationally expensive simulations. On the other hand, the difficulties experienced with simulations, ultimately due to the infinite variance of the waiting time, point out that the steady state mean could be a poor performance metric from the point of view of a customer, concealing the high variability of waiting times.

8.2.4 Upper Bound of the Waiting Time Probability Distribution

A simple exponential bound can be derived for the $G/G/1$ queue by adding the assumption that the Laplace-Stiltjes transforms of the inter-arrival and service time CDFs be analytic for $\text{Re}[s] > -\beta$, where β is a positive number. This implies that all moments of the inter-arrival and service times are finite.

Let $F_T(\cdot), F_X(\cdot)$ and $F_Z(\cdot)$ denote the CDFs of the random variables T, X and $Z = X - T$. While T and X are non-negative random variables, Z takes all real values, hence its transform is bilateral, i.e., $\varphi_Z(s) = \text{E}[e^{-sZ}] = \int_{-\infty}^{\infty} e^{-su} dF_Z(u)$. Since we assume that inter-arrival and service times are independent random variables, we have $\varphi_Z(s) = \text{E}[e^{-s(X-T)}] = \text{E}[e^{-sX}]\text{E}[e^{sT}] = \varphi_X(s)\varphi_T(-s)$. If there exists a positive

β_T (β_X) such that $\varphi_T(s)$ (resp., $\varphi_X(s)$) is analytic for $\text{Re}[s] > -\beta_T$ (resp., $\text{Re}[s] > -\beta_X$), then the function $\varphi_Z(s)$ is analytic for $-\beta_X < \text{Re}[s] < \beta_T$.

Let us restrict our attention to real values of s. The function $\varphi_Z(s)$ is defined for $s \in (-\beta_X, \beta_T)$, it is continuous, strictly positive, and convex in that interval and unbounded at the extremes. It is also $\varphi_Z(0) = 1$ and $\varphi_Z'(0) = -E[Z] = E[T] - E[X] > 0$. The last inequality is a consequence of the queue stability assumption. These properties imply that $\varphi_Z(s)$ is strictly less than 1 for $s < 0$ and close to the origin. The following definition is therefore well grounded:

$$s_0 = \sup\{s > 0 \ : \ \varphi_Z(-s) \le 1\} \tag{8.33}$$

and it must be $s_0 < \beta_X$. A sketch graph is displayed in Figure 8.3 to illustrate the behavior of $\varphi_Z(s)$ and the definition of s_0.

We recall Lindley's recursion (8.5), i.e., $W_{n+1} = \max\{0, W_n + Z_n\}$, $n \ge 1$. Let us assume that the initial value of the waiting time, W_1, is chosen so that $P(W_1 \ge y) \le e^{-s_0 y}$. In fact, it is $W_1 = 0$, then $P(W_1 \ge y) = 0$ for any $y > 0$ and $P(W_1 \ge 0) = 1$. Then, it is actually $P(W_1 \ge y) \le e^{-s_0 y}$ for any $s_0 \ge 0$ and $y \ge 0$.

Given that $P(W_1 \ge y) \le e^{-s_0 y}$, we prove by induction that $P(W_n \ge y) \le e^{-s_0 y}$ holds for any positive n.

Suppose the inequality holds up to the index n and let us prove that the inequality carries over to $n + 1$. We have for any positive y:

$$P(W_{n+1} \ge y) = P(W_n + Z_n \ge y) = \int_{-\infty}^{\infty} P(W_n \ge y - u)\, dF_Z(u) \tag{8.34}$$

since Z_n is independent of W_n. Then,

$$P(W_{n+1} \ge y) = \int_{-\infty}^{y} P(W_n \ge y - u)\, dF_Z(u) + \int_{y}^{\infty} P(W_n \ge y - u)\, dF_Z(u)$$

$$\le \int_{-\infty}^{y} e^{-s_0(y-u)}\, dF_Z(u) + 1 - F_Z(y)$$

$$\le \int_{-\infty}^{y} e^{-s_0(y-u)}\, dF_Z(u) + \int_{y}^{\infty} e^{-s_0(y-u)}\, dF_Z(u)$$

$$= \int_{-\infty}^{\infty} e^{-s_0(y-u)}\, dF_Z(u) = e^{-s_0 y}\varphi_Z(-s_0) \le e^{-s_0 y}$$

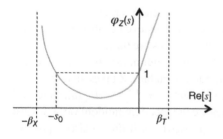

Figure 8.3 Example of $\varphi_Z(s)$ plot and definition of s_0.

since $P(W_n \geq t) \leq e^{-s_0 t}$ for non-negative t, by induction hypothesis, $P(W_n \geq t) = 1$ for any non positive t, $e^{-s_0(y-u)} \geq 1$ for $u \geq y$, since $s_0 > 0$, and $\varphi_Z(-s_0) \leq 1$ by construction of s_0.

In the limit for $n \to \infty$, this proves the exponential bound of the tail of waiting time probability distribution:

$$P(W \geq y) \leq e^{-s_0 y}, \qquad y \geq 0, \tag{8.35}$$

The exponential cutoff exhibited by the waiting time CDF is characteristic of those $G/G/1$ queues where inter-arrival and service times probability distributions have a light tail, i.e., they have finite moments of all orders. The exponential cutoff property breaks down if heavy-tailed probability distributions are used. In that case it is no more true that $\varphi_Z(s)$ is analytic for negative values of $\text{Re}[s]$.

Example 8.3 Let us consider a $G/G/1$ queue with inter-arrival times uniformly distributed in $[T_{\min}, T_{\max}]$ and service times with a Rayleigh probability distribution, i.e., the CCDF of the service times is $G_X(t) = e^{-(\alpha x)^2}$.

Given the mean inter-arrival time $E[T] = T_0$, we have $T_{\max} = 2T_0 - T_{\min}$ and $C_T^2 = \frac{2}{3}\left(1 - \frac{T_{\min}}{T_0}\right)^2$. The Laplace transform of the PDF of T is:

$$\varphi_T(s) = e^{-sT_{\min}} \frac{1 - e^{-2sT_0}}{2sT_0} \tag{8.36}$$

As for the service times, it is $\alpha = \frac{1}{E[X]}\sqrt{\frac{\pi}{4}}$ and $C_X^2 = 4/\pi - 1$, with $E[X] = \rho T_0$. The Laplace transform of the service times PDF can be written as follows:

$$\varphi_X(s) = 1 - sE[X]f(sE[X]/\sqrt{\pi}) \tag{8.37}$$

where $f(x) \equiv e^{x^2}\frac{2}{\sqrt{\pi}}\int_x^\infty e^{-t^2}\,dt$ is the scaled complementary error function.

To carry over the numerical example, let us assume $T_{\min} = 0$, $E[T] = T_0 = 1$ (i.e., the mean interarrival time is taken as the unit of time). It is then $C_T \approx 0.8165$, $E[X] = \rho$, $C_X \approx 0.5227$. Moreover, it is

$$\varphi_Z(s) = \frac{e^{2s} - 1}{2s} \cdot [1 - s\rho f(s\rho/\sqrt{\pi})] \tag{8.38}$$

with $\beta_T = \beta_X = \infty$ in this case.

Figure 8.4 plots the actual (simulated) CCDF of the waiting times (solid line) and the exponential upper bound (dashed line) for two values of the utilization coefficient, $\rho = 0.6$ and $\rho = 0.8$. The corresponding values of the exponent coefficient of the bound are $s_0 \approx 1.9475$ for $\rho = 0.6$, and $s_0 \approx 0.7785$ for $\rho = 0.8$.

It can be appreciated that the upper bound offers a quite tight approximation, only offset by a constant factor with respect to the actual values. The good estimate provided by the bound can be ascribed to the "good" behavior of the inter-arrival and service time PDFs, both of which have coefficient of variation less than 1. They

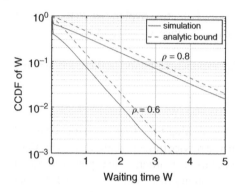

Figure 8.4 Comparison between the simulated CCDF of the waiting times of the $G/G/1$ queue of Example 8.3 (solid line) and the exponential upper bound (dashed line).

are even more concentrated around the mean than the negative exponential PDF, hence large deviations with respect to the mean value occur with small probability. The exponential decay of the CCDF of the waiting time is the mark of such a regular behavior of the inter-arrival and service times.

8.3 Bounds for the $G/G/m$ Queue

We consider a $G/G/m$ queue, where inter-arrival and service times are independent, stationary renewal processes. Multiple identical and fully accessible servers are provided, as well as infinite room for arriving customer (pure delay system, no loss).

Exact results are known for few special cases. Bounds and approximations are therefore especially useful for this general model.

Following [94], we obtain bounds for the $G/G/m$ queue by considering two other queueing systems, obtained by transforming the original $G/G/m$ queue.

As for the upper bound, assume that arriving customers are orderly assigned to a server upon their arrival, by scanning servers cyclically in a round robin fashion. This corresponds to having m separate $G/G/1$ queues. The mean waiting time experienced by a customer joining this modified system, W_1, is stochastically greater than the waiting time of the original queueing system. In fact, a customer could be waiting in front of the server that it has been assigned to in spite of other servers being idle, which cannot occur in the original multi-server queueing system. Therefore, we can state that $E[W] \leq E[W_1]$. The latter can be found exactly, if the equivalent $G/G/1$ can be solved. Otherwise, we can use an upper bound holding for the $G/G/1$ queue. The service times of the modified single-server system have the same probability distribution as the original one, whereas the arrival times are obtained as sums of m consecutive arrival times of the original process. Hence, their PDF is the m-fold convolution of the PDF of the original inter-arrival times. To apply the bounds we are actually interested only in the first two moments. The argument above implies that $E[X]$ and σ_X^2 remain the same, while the mean arrival rate and the variance of the inter-arrival times become, respectively, λ/m and $m\sigma_T^2$ for the $G/G/1$ system.

By applying the upper bound from eq. (8.32), we have

$$E[W] \leq E[W_1] \leq \frac{(\lambda/m)(m\sigma_T^2 + \sigma_X^2)}{2(1 - (\lambda/m)/\mu)} = \frac{\lambda(\sigma_T^2 + \sigma_X^2/m)}{2(1 - \rho)} \tag{8.39}$$

where $\rho = \lambda/(m\mu)$ in this case.

A tighter upper bound can be found as follows. Let us rewrite the upper bound of the mean waiting time as found for the $G/G/1$ queue in terms of the SCOV of arrivals and service times:

$$E[W] \leq \frac{C_T^2 + \rho^2 C_X^2}{2\lambda(1 - \rho)} \tag{8.40}$$

We could guess that this be also a bound for the mean waiting time of the $G/G/m$, where now ρ should be read as $\lambda/(m\mu)$. If that were true, we would have the bound

$$E[W] \leq \frac{\lambda(\sigma_T^2 + \sigma_X^2/m^2)}{2(1 - \rho)} \tag{8.41}$$

which is tighter than (8.39). It can be proved that the bound in eq. (8.41) holds [69].

As for the lower bound, we consider a single server queueing system where the server has a mean serving rate m times bigger than each server of the original multi-server queueing system, i.e., for the same arrival process, we replace the m servers with serving rate μ with a single server with serving rate $m\mu$.

This is the same as saying that each arriving customer brings into the system the same amount of work as before ($1/\mu$ on the average), but the single server available to the system is m times faster than each server of the original system. Therefore, the service time of this single server, X_2 is distributed as X/m. The first two moments of the modified service time random variable are $E[X]/m$ and σ_X^2/m^2.

In [47] it is shown that the mean unfinished work in steady state of the $G/G/m$ queue is lower bounded by the mean unfinished work of the modified single server queue. As a consequence, it is possible to derive a lower bound for the mean waiting time $E[W]$ of the $G/G/m$ queue as a function of the mean waiting time of the $G/G/1$ queue whose server has a speed-up factor of m.

To that end, we exploit a general relationship between the mean unfinished work $E[U]$ and the mean waiting time $E[W]$ of a queue. Let $E[Q]$ denote the mean queue length. The unfinished work of the $G/G/m$ queue is:

$$E[U] = E[Q]E[X] + m\rho\frac{E[X^2]}{2E[X]} = \lambda E[W]E[X] + \frac{\lambda E[X^2]}{2} \tag{8.42}$$

where the first term is the mean amount of work due to customers in the waiting line and the second term is the product of the mean number of busy servers by the mean residual service time. In the first term, we use Little's law to express the mean number of customers in the waiting line as a function of the mean waiting time.

Applying this expression to the $G/G/1$ queue whose server has a speed-up factor of m, we find

$$E[U_2] = \lambda E[W_2]E[X_2] + \frac{\lambda E[X_2^2]}{2} = \lambda E[W_2]\frac{E[X]}{m} + \frac{\lambda E[X^2]}{2m^2} \tag{8.43}$$

To compare the mean unfinished work of the original $G/G/m$ queue and that of the fast server $G/G/1$ queue, we have to account for the speed-up factor m. In other words, we can compare the mean unfinished works normalized with respect to the respective mean service times of the two queues. We can write

$$\frac{E[U]}{E[X]} \geq \frac{E[U_2]}{E[X_2]} \quad \Rightarrow \quad \lambda E[W] + \frac{\lambda E[X^2]}{2E[X]} \geq \lambda E[W_2] + \frac{\lambda E[X^2]}{2mE[X]} \tag{8.44}$$

Rearranging terms, we end up with the following expression:

$$E[W] \geq E[W_2] - \frac{m-1}{m}\frac{C_X^2+1}{2\mu} \tag{8.45}$$

If the exact result for $E[W_2]$ were known, it could be substituted. Otherwise, we can resort to the general lower bound we have found for a $G/G/1$ queue, thus writing

$$E[W_2] \geq \frac{\lambda^2\sigma_X^2/m^2 + \rho(\rho-2)}{2\lambda(1-\rho)} = \frac{\rho^2C_X^2 + \rho(\rho-2)}{2\lambda(1-\rho)} \tag{8.46}$$

thus obtaining finally the following Brumelle-Marchal's lower bound

$$E[W] \geq \frac{\rho^2C_X^2 + \rho(\rho-2)}{2\lambda(1-\rho)} - \frac{m-1}{m}\frac{C_X^2+1}{2\mu} \tag{8.47}$$

where $\rho = \lambda/(m\mu)$.

Summing up we have for the $G/G/m$ queue:

$$\max\left\{0, \frac{\rho^2C_X^2 + \rho(\rho-2)}{2\lambda(1-\rho)} - \frac{m-1}{m}\frac{C_X^2+1}{2\mu}\right\} \leq E[W] \leq \frac{\lambda(\sigma_T^2 + \sigma_X^2/m^2)}{2(1-\rho)} \tag{8.48}$$

Example 8.4 Let us consider a multi-server queue with m servers, arrivals and services following renewal processes with inter-event probability distributions of Gamma and Weibull type, respectively. The PDF of the inter-arrival times is $f_T(t) = \frac{a_T^{\beta_T} t^{\beta_T-1}}{\Gamma(\beta_T)}e^{-\alpha_T t}$, $t \geq 0$, where $\Gamma(z) = \int_0^\infty u^{z-1}e^{-u}du$ is the Euler Gamma function. The positive parameters α_T and β_T can be determined once the first two moments of the PDF are given, namely $\beta_T = 1/C_T^2$ and $\alpha_T = \beta_T/E[T]$. As for the service times, the CCDF of the Weibull distribution can be written as $G_X(t) = e^{-(\alpha_X t)^{\beta_X}}$, $t \geq 0$. The parameters α_X and β_X can be determined from the first two moments. It is

$$C_X^2 = \frac{\Gamma(2/\beta_X + 1)}{\Gamma(1/\beta_X + 1)^2}$$

which can be inverted since the right-hand side is a monotonously strictly decreasing function of β_X. Then, we use $\alpha_X = \Gamma(1/\beta_X + 1)/E[X]$. Moreover, the mean service time is obtained once the mean inter-arrival time and the utilization coefficient ρ are given, i.e., $E[X] = E[T]m\rho$.

In this numerical example we let $E[T] = 1$, i.e., time is measured in units of mean inter-arrival time, and $C_T = 1.5$, i.e., the standard deviation of the inter-arrival time is 1.5 times the mean inter-arrival time. The default queue configuration corresponds to $\rho = 0.8$, $m = 3$, and $C_X = 3$. We let each of those parameters vary over a range, keeping the other two fixed, to explore the behavior of the mean waiting time $E[W]$ as a function of ρ, m, and C_X^2.

Figure 8.5(a) plots the mean waiting time as a function of the utilization coefficient. Besides the upper and lower bounds given in eq. (8.48), Figure 8.5 plots also the approximation that will be introduced in the next section (see eq. (8.51)). It is apparent that the bounds are rather crude, yet they capture the qualitative behavior of the mean waiting time and give the correct order of magnitude. As expected, both bounds are more interesting at high loads, while for low levels of ρ they give highly incorrect or not significant results.

Similar comment apply to Figs 8.5(b) and 8.5(c), where $E[W]$ is plotted against the number of servers and the SCOV of service times, respectively.

It can be observed also that $E[W]$ is decreasing with the number of servers, for a fixed utilization coefficient, i.e., there is a scaling phenomenon. For the same mean load, larger systems perform better. The mean waiting time increases as the

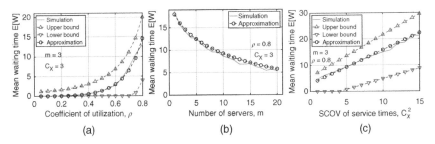

Figure 8.5 Mean waiting time of the $G/G/m$ queue as a function of: (a) utilization coefficient ρ; (b) number of servers m; (c) SCOV of service times C_X^2. Inter-arrival times have a Gamma PDF, service times are distributed according to a Weibull distribution. Time is normalized with respect to the mean inter-arrival time.

service time variability grows, as shown in Figure 8.5(c). $E[W]$ scales linearly with C_X^2. This behavior is captured correctly by the upper bound.

Finally, the plots compare the estimates of the mean waiting time obtained through simulation with the approximation of eq. (8.51). It turns out that the approximation is extremely accurate for this class of multi-server queues (i.e., Gamma inter-arrival times and Weibull service times).

8.4 Approximate Analysis of Isolated G/G Queues

A number of different approaches can be used to provide approximate expressions of performance measures of $G/G/m$ queues. Some approaches are best suited for special ranges of parameters, specifically of the utilization coefficient of the server, e.g., the heavy-traffic approximation, which gives accurate predictions in the limit $\rho \to 1$, or the diffusion approximation, which holds on time scales much bigger than the average inter-arrival and service times.

The presentation is organized in sub-sections according to the approach used to derive the approximation.

8.4.1 Approximations from Bounds

The bounds can be used to yield approximations. We discuss specifically the scaling proposed by Marchal of Kingman's upper bound for the $G/G/1$ queue, namely $E[W] \le \frac{\lambda(\sigma_T^2 + \sigma_X^2)}{2(1-\rho)}$. By scaling that bound by the factor $\frac{\rho^2 \sigma_T^2 + \sigma_X^2}{\sigma_T^2 + \sigma_X^2}$, the following results is obtained

$$E[W^{G/G/1}] \approx \overline{W}_1 = \frac{\lambda(\rho^2 \sigma_T^2 + \sigma_X^2)}{2(1-\rho)} = E[X] \frac{\rho}{1-\rho} \frac{C_X^2 + C_T^2}{2} \qquad (8.49)$$

This is quite an elegant result. It decomposes the mean waiting time into the product of three factors: (i) the mean service time; (ii) a factor depending only on the server utilization coefficient and shaping the asymptotic behavior of the mean waiting time in heavy traffic (as $\rho \to 1$); and (iii) a factor depending on the variability of the arrival and service processes.

The approximation is exact for the $M/G/1$ queue. It is also sharp in heavy traffic.

It is noted that $\overline{W}_1$ can be thought as the mean waiting time of the $M/M/1$ queue multiplied by the factor $(C_T^2 + C_X^2)/2$ that accounts for the non-Poisson character of the arrival and service processes. Inspired by this remark, we can define an approximation for the multi-server queue as

$$E[W^{G/G/m}] \approx \overline{W}_m = \frac{C_X^2 + C_T^2}{2} E[W^{M/M/m}] \qquad (8.50)$$

It is $E[W^{M/M/m}] = E[X]C(m, m\rho)/[m(1 - \rho)]$, where $C(m, m\rho)$ is the Erlang-C function and gives the probability that all servers are busy[1]. Then, we obtain the following approximation for the mean waiting time of the $G/G/m$ queue

$$E[W^{G/G/m}] \approx \overline{W}_m = E[X] \frac{C(m, m\rho)}{1 - \rho} \frac{C_X^2 + C_T^2}{2m} \tag{8.51}$$

This expression reduces to the approximation for the $G/G/1$ queue in the special case $m = 1$. The approximation can be further simplified by using the approximation

$$C(m, m\rho) \approx \begin{cases} \frac{1}{2}(\rho + \rho^m) & \rho > 0.7 \\ \rho^{\frac{m+1}{2}} & \rho < 0.7 \end{cases} \tag{8.52}$$

8.4.2 Approximation of the Arrival or Service Process

Another approach to derive approximations of the $G/G/1$ queue performance measures is to substitute the general renewal process with a tractable process whose probability distribution function approximates the original one.

A class of processes that yields to analysis via Laplace transforms is the class of renewal processes whose renewal time PDF has a rational Laplace transform, often referred to as the class of Coxian PDFs, after the name of Donald C. Cox, a most prominent contributor to the theory of point processes.

Another class that yields to effective numerical implementation and has a vast potential for approximating any renewal process is the class of phase-type probability distribution (PH distributions). An extensive account of the PH-type probability distributions can be found, e.g., in [167, Ch. 2]. Here we give just the definition of a PH-type distribution.

Given a continuous-time Markov chain on $n + 1$ states, n of which are transient and one is absorbing, the absorption time (first-passage time to the absorbing state $n + 1$), given that the chain is initialized according to the probability vector $\mathbf{q}$, defines a random variable whose probability distribution is said to be phase-type. The number of phases is n.

Formally, let us define the random variable T as $T = \min\{t \geq 0 : X(t) = n + 1\}$, where $X(t)$ denotes the state of the Markov chain over the state space $\{1, \ldots, n, n + 1\}$, initialized according to the probability vector $\mathbf{q}$ at $t = 0$. The CCDF of T is given by $G_T(t) = \mathbf{q}\exp(\mathbf{Q}t)\mathbf{e}$, where $\mathbf{e}$ is a column vector of ones and

---

1 We recall that $C(m, A)$ can be computed efficiently using the identity $C(m, A) = \frac{B(m,A)}{1-A/m+B(m,A)A/m}$, where $B(m, A)$ is the Erlang-B formula. The latter can be efficiently computed by using the recursion $B(m, A) = \frac{AB(m-1,A)}{m+AB(m-1,A)}$, for $m \geq 1$, initialized with $B(0, A) = 1$.

the infinitesimal generator of the Markov chain is

$$\mathbf{P} = \begin{bmatrix} \mathbf{Q} & -\mathbf{Qe} \\ \mathbf{0} & 0 \end{bmatrix} \tag{8.53}$$

with $\mathbf{Q}$ an $n \times n$ matrix. Note that $\mathbf{Q}$ is such that $\mathbf{Qe} \leq \mathbf{0}$, with strict inequality for at least one row. This guarantees that all eigenvalues of $\mathbf{Q}$ have negative real part, so that $G_T(t)$ decays exponentially fast to 0 as $t \to \infty$ (light-tailed probability distribution). More in depth, $G_T(t) \sim \kappa e^{-\eta t}$ as $t \to \infty$, where η is the maximal real part eigenvalue of $\mathbf{Q}$, $\mathbf{v}$, and $\mathbf{u}$ are its associated left and right eigenvectors, and $\kappa = \mathbf{quve}$.

According to this last approach, the original $G/G/m$ queue is approximated with a $PH_n/PH_k/m$ queue, where n (resp., k) is the number of "phases" of the PH-type distribution approximating the arrival (resp., service) process. For the use of a $\sum_i PH(i)/PH/m$ queue to fit a general $G/G/m$ model the interested reader can consult also [86, § 4.4].

A versatile class of probability distributions of positive random variables was introduced by Luchak [153]. These are mixtures of Erlang K probability distributions of the form

$$\sum_{j=1}^{\infty} p_j f_{K_j}(x; \lambda) \tag{8.54}$$

where $p_j \geq 0$, $\forall j$ and $\sum_{j=1}^{\infty} p_j = 1$.

The PDF of the Erlang K random variable is

$$f_K(x; \lambda) = \frac{(K\lambda)^K x^{K-1}}{(K-1)!} e^{-K\lambda x}, \quad x \geq 0, \ K \geq 1. \tag{8.55}$$

This is the convolution of K negative exponential PDFs with mean $1/(K\lambda)$. The Erlang K random variable Y is therefore generated as $Y = (X_1 + \cdots + X_K)/(K\lambda)$, with $X_j \sim \text{Exp}(1)$ for $j = 1, \dots, K$.

As a special case, we consider the mixture obtained with $K_j = j$. It can be proved that this mixture is dense in the set of the PDFs on $[0, \infty)$, i.e., it is possible to fit any given PDF of a non-negative random variable by means of a suitable mixture of Erlang K PDFs. If the probability distribution p_j has a finite support, the probability distribution obtained with the mixture of eq. (8.54) is of phase type. It can be proved that also the class of PH probability distributions is dense in the set of PDFs on $[0, \infty)$.

8.4.3 Reflected Brownian Motion Approximation

In this section we introduce the reflected Brownian motion (RBM) representation of the for $G/G/1$ queues. Under heavy traffic conditions, i.e., as the utilization coefficient of the server tends to 1, and at steady state (large time values after the queueing system has started its operation) it provides a useful approximation of the $G/G/1$ queue, also known as heavy-traffic limit or diffusion approximation.

Let us consider a $G/G/1$ queue and let us introduce the following notation:

$A(t)$ counting function of the arrival process, i.e., the number of customers arrived in $[0, t]$.

$S(t)$ counting function of the service process, i.e., number of service completions in $[0, t]$ if the server is continuously busy in $[0, t]$.

$Q(t)$ number of customers in the queue at time t.

$I(t)$ cumulative amount of idle time in $[0, t]$.

$B(t)$ cumulative amount of busy time in $[0, t]$,

By definition, it is $B(t) + I(t) = t, \forall t \geq 0$. The number of customers in the system at time t can be written as

$$Q(t) = Q(0) + A(t) - S(B(t)) \tag{8.56}$$

which is nothing more than a balance between arrivals and departures, taking into account that the server is actually busy only for an overall amount of time $B(t)$ in $[0, t]$.

Let us assume that there exists $u \in [0, t]$ such that $Q(u) = 0$. Let then t_0 be defined as the largest time that the queue was empty in $[0, t]$, i.e., $t_0 = \max\{0 \leq u \leq t : Q(u) = 0\}$. By definition it is $I(t) = I(t_0) \geq I(u), \forall u \in [0, t]$. Let us consider the function $Y(u) = \mu I(u) - Q(u)$. Since it is $Q(t) \geq 0$, we have $Y(u) \leq \mu I(t_0) = \mu I(t), \forall u \in [0, t]$. In words, $Y(t)$ is the sum of two components: (a) $\mu I(t)$, which is monotonically nondecreasing with t; and (b) $-Q(t)$, which is always nonpositive; it is 0 (i.e., it attains its maximum) only when $I(t)$ is strictly increasing. Then, we have $\sup_{0 \leq u \leq t} Y(u) = \mu I(t_0) = \mu I(t)$.

If the queue is never empty in $[0, t]$, it is $I(t) = 0$ and hence $Y(u) = -Q(u) < 0, \forall u \in [0, t]$.

Putting together the two cases, we can say that $\sup_{0 \leq u \leq t} \max\{0, Y(u)\} = \mu I(t)$. Using this result, we can represent $Q(t)$ as follows:

$$Q(t) = U(t) + V(t) \tag{8.57}$$

where

$$U(t) \equiv Q(t) - \mu I(t) = Q(0) + (\lambda - \mu)t + A(t) - \lambda t - [S(B(t)) - \mu B(t)] \tag{8.58}$$

$$V(t) \equiv \mu I(t) = \sup_{0 \leq \tau \leq t} \max\{0, \mu I(\tau) - Q(\tau)\} = \sup_{0 \leq \tau \leq t} \max\{0, -U(\tau)\} \tag{8.59}$$

We arrive thus at our final result, the decomposition of the number of customers in the $G/G/1$ queue as follows:

$$Q(t) = U(t) + \sup_{0 \leq \tau \leq t} \max\{0, -U(\tau)\} \tag{8.60}$$

with $U(t)$ given in eq. (8.58).

In [86], it is shown that, given $U(t)$, there exists a unique pair of functions $X(t)$ and $V(t)$ such that $X(t) = U(t) + V(t)$ and the following three conditions are satisfied:

1. $X(t) \geq 0$.
2. $V(t)$ is nondecreasing for $t \geq 0$ with $V(0) = 0$.
3. $X(t) \frac{dV}{dt} = 0$, i.e., $V(t)$ is strictly increasing whenever $X(t)$ is equal to 0, while it is flat whenever $X(t) > 0$.

What is this decomposition good for? It is the basic relationship that paves the way to heavy-traffic approximation of the state of a $G/G/1$ queue through the reflected Brownian motion (RBM) process.

Intuitively, for ρ less than 1, yet close to 1, i.e., in the limit $\rho \to 1$, $I(t)$ is negligible since the queue is (almost) never empty. Then, we can write $Q(t) \approx U(t)$.

For large t, we know that the counting process associated with a renewal process is approximatively normally distributed with mean $t/\overline{T}$ and variance $\sigma^2 t/\overline{T}^3$, where $\overline{T}$ and σ^2 are the mean and variance of the inter-event times.

Applying this result to the arrival counting function $A(t)$ we get $A(t) \sim \mathcal{N}(\lambda t, \lambda C_T^2 t)$ for large t (formally, as $t \to \infty$)[2] .

As for the service counting process, since we deal with $S(B(t))$ and $B(t)$ is itself a random process, the argument is a bit trickier. The mean can be evaluated exactly as $E[E[S(B(t))|B(t)]] = E[\mu B(t)] = \mu \rho t = \lambda t$ (not surprisingly, since the mean amount of service in $[0, t]$ equals the mean amount of work arrived at the queue in the same time interval). As for the variance, we replace $B(t)$ by its mean (which is asymptotically correct for t tending to infinity) and then evaluate $Var(S(\rho t)) = \mu C_X^2 \rho t = \lambda C_X^2 t$. Also for the service counting process, we can say that $S(B(t))$ is asymptotically normally distributed, i.e., $S(B(t)) \sim \mathcal{N}(\lambda t, \lambda C_X^2 t)$ for large t.

The argument developed so far shows that, for large t, i.e., in the limit for $t \to \infty$, we can state that $U(t)$ in eq. (8.58) is itself normally distributed with mean and variance given by

$$E[U(t)] = Q(0) + (\lambda - \mu)t \qquad \sigma_{U(t)}^2 = \lambda(C_T^2 + C_X^2)t \qquad (8.61)$$

Let us assume (with no limitation) that $Q(0) = 0$. We see that, if $A(t)$ and $S(t)$ are stationary renewal processes, it turns out that $U(t)$ tends asymptotically to a Brownian motion process with drift $\alpha = \lambda - \mu$ (which is < 0 if $\rho < 1$, i.e., the queue is stable) and variance coefficient $\beta^2 = \lambda(C_T^2 + C_X^2)$. The definition and some useful properties of the Brownian motion process (also known as Wiener process) are given in the Appendix.

---

2 We recall that the notation $Y \sim \mathcal{N}(\mu, \sigma^2)$ means that Y is a Gaussian random variable with mean μ and variance σ^2.

We can conclude saying that, in heavy-traffic, i.e., when $\rho \to 1$, we can approximate the state of the $G/G/1$ queue $Q(t)$ by means of a real-valued, continuous time random process $U(t)$, i.e., the queue content level of a stationary $G/G/1$ queue in heavy traffic behaves approximately as a Brownian motion process with drift $\lambda - \mu$ and variance parameter $\lambda(C_T^2 + C_X^2)$.

If we consider also the term $V(t)$ in the decomposition of $Q(t)$, we obtain the reflected Brownian motion approximation for $Q(t)$, namely $Q(t) \approx Q_{RBM}(t) = U(t) + V(t) = U(t) + \sup_{0 \leq \tau \leq t} \max\{0, -U(\tau)\}$, where $U(t)$ is a Brownian motion process with mean $(\lambda - \mu)t$ and variance $\lambda(C_T^2 + C_X^2)t$. The preceding argument points out that the RBM approximation is asymptotically sharp as $\rho \to 1$.

The role of $V(t)$ is to avoid that $Q(t)$ attains negative values (which is instead possible for $U(t)$). We can think of the RBM process as a Brownian motion process constrained by a reflecting barrier at 0. If the drift is negative, the RBM process impacts the time axis infinitely often. In terms of queueing, this means that the queue becomes empty infinitely often, which is the mark of a stable system (the state 0 must have positive probability in steady-state).

Given $Q(t)$ we can identify the workload of the $G/G/1$ queue at t as $W(t) = Q(t)/\mu$. This is also a reflected Brownian motion process with drift $(\lambda - \mu)/\mu = \rho - 1$ and variance parameter $\lambda(C_T^2 + C_X^2)/\mu^2$.

The probability distribution of a RBM process can be found by solving a differential equation under proper boundary conditions. Let $F(t, x; w_0)$ be the CDF of $W(t)$ conditional on its initial state, i.e.,

$$F(x, t; w_0) = \mathcal{P}(W(t) \leq x | W(0) = w_0) \tag{8.62}$$

It is possible to derive a partial differential equation that involves $F(x, t; w_0)$ (e.g., see [131, Ch. 2]). The equation is[3]

$$\frac{\partial F}{\partial t} = -\alpha \frac{\partial F}{\partial x} + \frac{1}{2} \beta^2 \frac{\partial^2 F}{\partial x^2} \tag{8.63}$$

where α is the drift and β^2 is the variance parameter of the RBM process. In our case, it is $\alpha = \rho - 1$ and $\beta^2 = \lambda(C_T^2 + C_X^2)/\mu^2$.

The boundary conditions for the RBM process are $F(x, t; w_0) = 0$, $x < 0$, $F(\infty, t; w_0) = 1$, for $t > 0$, $F(x, 0; w_0) = 0$ for $x < w_0$, and $F(x, 0; w_0) = 1$, for $x \geq w_0$.

For $\rho < 1$ there exists a steady state, i.e., the limit of $F(x, t; w_0)$ for $t \to \infty$ is a proper, nondegenerate CDF $F(x)$, independent of the initial state. The corresponding PDF $f(x) = F'(x)$ satisfies the following differential equation (obtained

---

3 This is but the Fokker-Planck equation for the diffusion process with homogeneous and uniform drift coefficient α and diffusion parameter β^2. The general equation for time and size dependent drift coefficient $\alpha(x, t)$ and diffusion parameter $D(x, t)$ is written as $\frac{\partial}{\partial t} F(x, t) = -\frac{\partial}{\partial x} [\alpha(x, t) F(x, t)] + \frac{1}{2} \frac{\partial^2}{\partial x^2} [D(x, t) F(x, t)]$.

from (8.63) by setting to 0 the derivative of $F(x)$ with respect to time and taking into account that $f(x)$ is the derivative of $F(x)$ with respect to x):

$$\frac{df}{dx} = \frac{2\alpha}{\beta^2} f(x) \tag{8.64}$$

The solution is a negative exponential PDF $f(x) = \frac{1}{\overline{W}} e^{-x/\overline{W}}$, where

$$\overline{W} = -\frac{\beta^2}{2\alpha} = \frac{\lambda(C_T^2 + C_X^2)/\mu^2}{2(1-\rho)} = E[X]\frac{\rho(C_T^2 + C_X^2)}{2(1-\rho)} \tag{8.65}$$

It turns out that the approximation $\overline{W}$ in (8.65) leads to the exact result of $M/G/1$ queue, when $C_T^2 = 1$. In the heavy-traffic limit $\rho \to 1$, the right-hand side of eq. (8.65) coincides with the upper bound of the mean waiting time of the $G/G/1$ queue found in eq. (8.39). It coincides with Marchal's approximation (8.49) as well.

To sum up, $Q(t) \sim Q_{RBM}(t) = U(t) + \sup_{0 \le \tau \le t} \max\{0, -U(\tau)\}$ for large t, i.e., on time scales much bigger than the mean inter-arrival and service times. An even simpler approximation holds under heavy-traffic, namely $Q(t) \sim Q_{BM}(t) = U(t)$. This does not lead to an especially accurate approximation, yet it can be useful to gain insight into the qualitative behavior of the queueing process.

Example 8.5   Figure 8.6 plots a sample path of RBM process for $\rho = 0.8$, $C_T = 1.5$, and $C_X = 3$. The time axis is normalized with respect to the mean service time. The BM process is generated through i.i.d. Gaussian increments, i.e.,

$$U(k\Delta t) = \sum_{j=1}^{k} \left[\Delta t(\rho - 1) + Z_j \sqrt{\Delta t \rho(C_T^2 + C_X^2)} \right], \qquad k \ge 1, \tag{8.66}$$

with $Z_j \sim \mathcal{N}(0, 1), \forall j$.

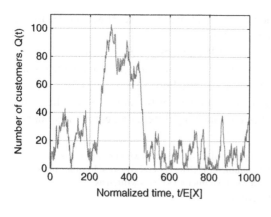

Figure 8.6   Sample path of the number of customers in the queue according to the RBM approximation. Time is normalized with respect to the mean service time. $\rho = 0.8$, $C_T = 1.5$, $C_X = 3$.

Table 8.2 Comparison between simulations (95% confidence intervals) and RBM approximation for a $G/G/1$ queue with Gamma distributed inter-arrival times and Weibull distributed service times. Time is normalized with respect to the mean service time. The utilization coefficient is $\rho = 0.9$.

C_T	C_X	Simulations (95% conf. int.)	RBM approximation
0.5	0.5	[2.060,2.064]	2.250
1.0	0.5	[5.591,5.601]	5.625
2.0	0.5	[19.940,19.975]	19.125
0.5	1.0	[5.346,5.357]	5.625
1.0	1.0	[9.009,9.026]	9.000
2.0	1.0	[23.383,23.426]	22.500
0.5	2.0	[18.766,18.806]	19.125
1.0	2.0	[22.271,22.316]	22.500
2.0	2.0	[37.833,37.909]	36.000

Example 8.6 We consider a $G/G/1$ queue with Gamma distributed inter-arrival times and Weibull distributed service times. The mean service time is set to 1 (i.e., it is chosen as the unit of time), the mean inter-arrival time is $E[T] = E[X]/\rho$. In this example, we set $\rho = 0.9$. The coefficients of variation of inter-arrival and service times range between 0.5 and 2. Table 8.2 compares the mean waiting time obtained with simulations versus the RBM approximation of eq. (8.65). Simulations are displayed by giving the 95% confidence level interval. It appears that the RBM approximation is quite accurate. Note also that it does not behave consistently, i.e., sometimes it provides an overestimate sometimes an underestimate.

8.4.4 Heavy-traffic Approximation

To shed more light into the heavy-traffic regime of the $G/G/1$ queue, we start from Lindley's recursion, which we rewrite here

$$W_{n+1} = \max\{0, W_n + X_n - T_{n+1}\} = \max\{0, W_n + Z_n\}, \quad n \geq 1, \tag{8.67}$$

where $Z_n \equiv X_n - T_{n+1}$ and W_n is the workload found by the n-th arriving customer in the queue, $n \geq 1$. In the renewal $G/G/1$ queue the random sequence Z_n is stationary. In a stable queue, it is $E[Z] = -(1 - \rho)/\lambda < 0$, while $\sigma_Z^2 = \sigma_X^2 + \sigma_T^2$.

Applying the recursion (8.67) with the initial condition $W_1 = 0$, we obtain

$$W_{n+1} = \max\{0, W_n + Z_n\} = \max\{0, Z_n, W_{n-1} + Z_{n-1} + Z_n\} =$$
$$= \max\{0, Z_n, Z_{n-1} + Z_n, \ldots, Z_1 + Z_2 + \cdots + Z_n\} = \max_{0 \le k \le n} U_k \quad (8.68)$$

where $U_0 = 0$ and $U_k = \sum_{j=1}^{k} Z_{n+1-j}$, $k = 1, \ldots, n$. The sequence U_k is a discrete-time random walk with i.i.d. steps $Z_j \sim Z = X - T$, starting from 0. For a stable queue, the sequence of random variables $\{W_n\}_{n \ge 1}$ converges in distribution to W, the waiting time of customers joining the queue at steady state. From eq. (8.68), we obtain $W = \sup_{k \ge 0} U_k$, i.e., the waiting time is the supremum attained by the random walk. This is finite, given that $E[Z] = -(1 - \rho)/\lambda < 0$, and $\sigma_Z^2 < \infty$, i.e., the random walk $\{U_k\}_{k \ge 0}$ has a negative drift and finite variance steps.

Normalizing eq. (8.68) by the mean waiting time $\overline{W}$ given in eq. (8.49), we have $\hat{W} = W/\overline{W} = \sup_{k \ge 0} \hat{U}_k$, with $\hat{U}_k = U_k/\overline{W}$. The step of the normalized random walk is distributed according to the random variable $\hat{Z} = (X - T)/\overline{W}$. The first two moments of $\hat{Z}$ are $E[\hat{Z}] = -B(1 - \rho)^2$ and $\sigma_{\hat{Z}}^2 = 2B(1 - \rho)^2$, where $B = 2/[\lambda^2(\sigma_X^2 + \sigma_T^2)]$. Let us consider the limit as $\rho \to 1$ (heavy-traffic regime).

The normalized random walk increments have negative drift, vanishing as $(1 - \rho)^2$, and standard deviation vanishing as $(1 - \rho)$. Thus, all conditions recur for the random walk U_k to converge to a standard Brownian motion (BM) process with drift $\alpha = -B(1 - \rho)^2$ and variance coefficient $\beta^2 = 2B(1 - \rho)^2$. The probability that the supremum of that random walk exceeds a threshold x equals the probability of absorption of the BM process starting from 0 with an absorbing barrier at x. The absorption probability for the BM process is known to be $\exp(\beta^2 x/(2\alpha))$ if $\alpha < 0$. Applying the result to our case, we find $\frac{\beta^2}{2\alpha} = -\frac{2B(1-\rho)^2}{2 \cdot B(1-\rho)^2} = -1$ and hence the probability that $\hat{W} = \sup_{k \ge 0} U_k$ exceeds x tends to e^{-x} in the limit for $\rho \to 1$. Denormalizing, we have that the heavy-traffic limit of the waiting time of the $G/G/1$ queue is a negative exponential random variable with mean $\overline{W}$.

An alternative reasoning is based on the application of Wald's identity (see [60, Chapter 2]). Let X_i denote i.i.d. random variables with $X_i \sim X$, where X is a continuous random variable and $f(x)$ its PDF. Given an absorbing barrier at $a > 0$ and a random walk $\{Y_n, n \ge 1\}$, with $Y_0 = 0$ and $Y_n = X_1 + X_2 + \cdots + X_n, n \ge 1$, let $N = \min\{n : n \ge 1, Y_n \ge a\}$. This is the absorption time. Wald's identity states that $E[e^{-\theta Y_N}(\phi(\theta))^{-N}] = 1$, where $\phi(\theta)$ is the generating function of the random walk step random variable, i.e., $\phi(\theta) = \int_{-\infty}^{\infty} e^{-\theta x} f(x) \, dx$. Making the absorption probability p_a appear explicitly, we can restate Wald's identity as follows:

$$p_a E[e^{-\theta Y_N}(\phi(\theta))^{-N} | Y_N \ge a] = 1 \quad (8.69)$$

Now we exploit a property of generating functions. Since the step random variable X is continuous, there exists a unique nonzero root to the equation $\phi(\theta) = 1$. Let it be θ_0. It has the same sign as $E[X]$.

We apply Wald's identity for $\theta = \theta_0$, so that $\phi(\theta_0) = 1$. Moreover, we make an approximation, assuming the the step size be negligible with respect to the absorption level a, hence $Y_N \approx a$. Then, eq. (8.69) leads to

$$p_a \mathrm{E}[e^{-\theta_0 Y_N} | Y_N \geq a] = 1 \quad \Rightarrow \quad p_a e^{-\theta_0 a} \approx 1 \tag{8.70}$$

Hence, $p_a \approx e^{\theta_0 a}$. This results holds only if $\mathrm{E}[X] < 0$ and hence it is $\theta_0 < 0$.

We apply Wald's identity to our random walk $\{\hat{U}_n, n \geq 0\}$, hence $X \sim \hat{Z}$. Let $x > 0$ be the absorption threshold. Let $\phi_{\hat{Z}}(\theta) = \int_{-\infty}^{\infty} e^{-\theta x} f_{\hat{Z}}(x) \, dx$ be the generating function associated with the PDF of the random walk step $\hat{Z}$. Since the mean and variance of the step random variable $\hat{Z}$ become both negligible as $\rho \to 1$, the approximation used in Wald's identity is asymptotically sharp as $\rho \to 1$. It gives $p_x \sim e^{\theta_0 x}$ for the absorption probability. We can expand the generating function of $\hat{Z}$ in powers of θ as

$$\phi_{\hat{Z}}(\theta) = 1 - \mathrm{E}[\hat{Z}]\theta + \frac{1}{2}\mathrm{E}[\hat{Z}^2]\theta^2 + \dots \tag{8.71}$$

$$= 1 + B(1-\rho)^2\theta + B(1-\rho)^2\theta^2 + o((1-\rho)^2) \tag{8.72}$$

Neglecting the higher powers of $1 - \rho$, thus limiting the expansion of $\phi_{\hat{Z}}(\theta)$ to the first two powers of θ, it is easy to find the root θ_0 of $\phi_{\hat{Z}}(\theta) = 1$ different from 0. It is $\theta_0 = -1$. Then, the absorption probability is $p_x \sim e^{-x}$. Note that the event $\max_{n \geq 0} \hat{U}_n > x$ is equivalent to the absorption event of the random walk U_n on the barrier x. Then, $\mathcal{P}(\hat{W} > x) = \mathcal{P}(\max_{n \geq 0} \hat{U}_n > x) = e^{-x}$, the equality holding asymptotically as $\rho \to 1$. This shows that W tends to a negative exponential random variable with mean $\overline{W}$ as $\rho \to 1$.

The argument can be made rigorous, proving the following [14,127].

Theorem 8.1 Consider a sequence of $G/G/1$ queues indexed by j. For queue j, let T_j and X_j denote inter-arrival and service times random variables. Let $\rho_j < 1$ denote the utilization coefficient and let W_j be the random variable representing the waiting time of the j-th queue. Suppose that $T_j \to T, X_j \to X$ in distribution as $\rho_j \to 1$. Let $\overline{W}_j = (\sigma_X^2 + \sigma_T^2)/[2(\mathrm{E}[T_j] - \mathrm{E}[X_j])]$. Then $W_j/\overline{W}_j \to \hat{W}$, where $\hat{W}$ is a negative exponential random variable with mean 1, provided that: (i) $\sigma_T^2 + \sigma_X^2 > 0$; (ii) $\mathrm{E}[X_j^{2+\delta}]$ and $\mathrm{E}[T_j^{2+\delta}]$ are uniformly bounded for all j for some $\delta > 0$.

Both technical assumptions listed in the theorem statement are required. Two counter-examples illustrate this fact.

Example 8.7 Let us consider a $D/D/1$ queue with $\mathrm{E}[T_j] = 1$ and $\mathrm{E}[X_j] = 1 - \epsilon_j$. It is $\rho_j = 1 - \epsilon_j$ and $W_j = 0$, given that the previous customer has already left the queue, when the new one arrives. We can make ρ_j as close to 1 as we want, still maintaining $W_j = 0$. Hence, it appears that the limit theorem does not hold. In this case, condition (i) of the theorem statement fails to be true.

Example 8.8 Let us assume $T_j = T + A_j$ and $X_j = X + A_j$. Lindley's recursion can be written as $W_j(n) = \max\{0, W_j(n-1) + X_j(n) - T_j(n)\} = \max\{0, W_j(n-1) + X(n) - T(n)\}$. As $n \to \infty$, the sequence $W_j(n)$ converges to a random variable W that is independent of j. Thus, we have $W_j \sim W$ for all j, where W can exhibit a probability distribution different from negative exponential, depending on the probability distributions of X and T. On the other hand, $\rho_j = (E[X] + A_j)/(E[T] + A_j)$ tends to 1 as $A_j \to \infty$. Again, the theorem does not hold, in this case since condition (ii) fails to hold.

The theorem proves that at high loads the mean waiting time is well approximated by the quantity $\overline{W} = \frac{\sigma_X^2 + \sigma_T^2}{2(E[T] - E[X])}$, which is but the upper bound found in Section 8.2.3.

8.5 Approximate Analysis of a Network of $G/G/1$ Queues

The approximate analysis of $G/G/1$ queues, specifically, of the mean system time, can be extended to the case of a network of such queues.

Let us consider a queueing network of infinite buffer queues (no customer loss), independent arrival and service times, memoryless routing, described by means of the probability r_{ij} of joining queue j when leaving queue i. We generalize the Jackson model considered in Chapter 7 by letting inter-arrival times and service times be described by general renewal processes. The approximation consists of neglecting two facts: (i) the output process of a $G/G/1$ queue is not necessarily a renewal process; (ii) the superposition of independent renewal processes does not necessarily result in a renewal process. Even though external inputs to the queueing networks are renewal processes, going through network queues destroys the renewal property of the input inter-arrival processes. However, it can be the case that correlations have small effects on the mean waiting time performance. Moreover, reducing the general arrival process to renewal ones yields to a relatively simple analysis, thus letting us gain insight into the system performance.

The approximation consists of analyzing each queue of the network as if it were in isolation, except that its input process is deemed to be the superposition of the customer flow coming directly from outside of the network plus flows of customers leaving other queues and joining the tagged queue. The analysis of each queue aims at providing the first two moments of the inter-departure times of customers leaving the queue. Those are then used to evaluate the first two moments of the arrival processes at other queues. Along with the first two moments of the service process, that information is enough to evaluate the approximation of the mean waiting time (or the mean system time). Using Little's law enables the evaluation of the mean number of customers in each queue and finally the mean sojourn time of a customer in the network, provided the network is at statistical equilibrium.

The analysis approach overviewed above is known as the "decomposition" approximation. The key ideas are: (i) to approximate every flow process in the system as a renewal process; (ii) to determine formulas to compute the first two moments of the considered processes when they undergo transformations induced by the queueing network. More precisely, the decomposition method consists of three process transformations: superposition of flows, splitting of flows and analysis of flows through a queue. We will tackle each of them in the following sections.

8.5.1 Superposition of Flows

Let us consider a queue fed by the superposition of n renewal arrival processes. The i-th process has mean arrival rate v_i and squared coefficient of variation (SCOV) C_i^2. Its counting function in the interval $(0, t)$ is denoted with $A_i(0, t)$. The counting process associated with the superposition is denoted with $A(0, t)$: it counts the number of customers arrived at the queue in the time interval $[0, t]$. By the very meaning of superposition, we have

$$A(0, t) = A_1(0, t) + \cdots + A_n(0, t) \tag{8.73}$$

Taking expectation and variances we find

$$E[A(0, t)] = \sum_{i=1}^{n} E[A_i(0, t)] = t \cdot \sum_{i=1}^{n} v_i \tag{8.74}$$

$$Var[A(0, t)] = \sum_{i=1}^{n} Var[A_i(0, t)] = t \cdot \sum_{i=1}^{n} v_i C_i^2 \tag{8.75}$$

Working back from these relationships, we can express the mean rate and the SCOV of the superposed process as follows:

$$v = v_1 + \cdots + v_n \tag{8.76}$$

$$C^2 = \sum_{i=1}^{n} \frac{v_i}{v} C_i^2 \tag{8.77}$$

While those expressions are exact, the superposed process is not a renewal one in general, i.e., inter-event times are not necessarily independent random variables.

8.5.2 Flow Through a Queue

Given the first two moments of the inter-arrival time T and of the service time X, we aim at determining the first two moments of the output process of a queue. We derived these results in Section 8.2.2. Denoting the random variable of the inter-departure times as Y, we rewrite eq. (8.22) here:

$$E[Y] = E[T] = 1/\lambda \tag{8.78}$$

$$C_Y^2 = C_T^2 + 2\rho^2 C_X^2 - 2\lambda E[W](1 - \rho) \approx (1 - \rho^2)C_T^2 + \rho^2 C_X^2 \tag{8.79}$$

The last approximation of the SCOV of the inter-departure time has been derived using the RBM approximation of the mean waiting time:

$$E[W] \approx \frac{\rho^2(C_T^2 + C_X^2)}{2\lambda(1 - \rho)} \tag{8.80}$$

Hence, we have

$$C_Y^2 \approx (1 - \rho^2)C_T^2 + \rho^2 C_X^2 \tag{8.81}$$

8.5.3 Bernoulli Splitting of a Flow

Let us consider an arrival process with inter-arrival time Y. We are interested in characterizing the first two moments of the inter-arrival times of the process obtained by sampling the original process according to a Bernoulli pattern. Bernoulli sampling means that each arrival is maintained (sampled) with probability p, independently of all others. In other words, sampling is independent of the original arrival process and it is memoryless.

Let Z denote the inter-arrival time of the sampled process. It is the sum of a random number N or inter-arrival times of the original process, where N has a geometric probability distribution with ratio p. That is to say, $P(N = n) = (1 - p)^{n-1}p$, $n \geq 1$ and $Z = \sum_{k=1}^{N} Y_k$. Given that the Y_k are i.i.d. (renewal process) and are independent of N, it is easy to verify that $E[Z] = E[N]E[Y]$ and

$$E[Z^2] = E[N(N - 1)](E[Y])^2 + E[N]E[Y^2] \tag{8.82}$$

From this equality we derive

$$\sigma_Z^2 = \sigma_N^2(E[Y])^2 + E[N]\sigma_Y^2 \tag{8.83}$$

whence $C_Z^2 = C_N^2 + C_Y^2/E[N]$. Since it is $E[N] = 1/p$ and $\sigma_N^2 = (1 - p)/p^2$, we obtain

$$E[Z] = E[Y]/p \tag{8.84}$$

$$C_Z^2 = 1 - p + pC_Y^2 \tag{8.85}$$

Those are the first two moments of the Bernoulli sampled arrival process as a function of the sampling probability p and of the first two moments of the original arrival process.

Memoryless routing in a queueing network induces Bernoulli sampling of the customer flow leaving a queue. The sub-flows feeding other queues or definitely leaving the network are obtained by sampling the output flow of the tagged queue.

8.5.4 Putting Pieces Together: The Decomposition Method

By now we have reviewed all tools required by the decomposition method for the analysis of a network of $G/G/1$ queues.

We consider J queues, arranged in an open queueing network. The same hypotheses holding for Jackson type open networks are assumed here, except that two of them are relaxed: namely, external arrivals and service processes are allowed to be general renewal processes, rather than only Poisson processes. Other assumptions common to $G/G/1$ queue networks and Jackson-type open networks are: infinite buffer size, memoryless routing, independence of external arrivals and service times. Moreover, in this section, we restrict our attention to single-server queues.

The obtained results are approximate, since we use the expression of the mean waiting time provided by the RBM approximation. Also, we assume that the superposition of independent renewal processes and that the output process of a $G/G/1$ queue are renewal processes, which is not true in general. Yet, accepting those approximations makes it possible to set up a relatively simple computational machinery for the evaluation of metrics of general $G/G/1$ queueing networks, which is a valuable result, at least to provide a first-hand understanding of the system performance.

Let λ_i and $C_{T,i}^2$ be the mean arrival rate and the SCOV of the external arrival process at the i-th queue, $i = 1, \ldots, J$. Let a_i denote the mean arrival rate at the input of the i-th queue, $i = 1, \ldots, J$. Let $E[X_i]$ and $C_{X,i}^2$ denote the mean and the SCOV of the service times of the server of the i-th queue, $i = 1, \ldots, J$. Let p_{ij} be the probability of joining queue j after having departed from queue i.

The mean arrival rate at queue j is obtained by a flow balance, assuming statistical equilibrium:

$$a_j = \lambda_j + \sum_{i=1}^{J} a_i p_{ij} \tag{8.86}$$

The matrix $\mathbf{P}$ is irreducible, since the interconnection graph of the queueing network is connected. A positive solution of this linear equation system always exists, since the matrix $\mathbf{P} = [p_{ij}, i,j = 1, \ldots, J]$ is sub-stochastic.

The queuing network is stable if $a_i E[X_i] = \rho_i < 1$, $\forall i$. The analysis in this section holds under stability, hence we assume $\rho_i < 1$, $\forall i$.

Let C_i^2 denote the SCOV of the input process of queue i. This input process is the superposition of $J + 1$ flows, one coming from outside of the network and all others coming from other queues of the network (possibly including the queue i itself.) The first two moments of the flow out of queue k are a_k and $C_{Y,k}^2$, as given by eqs. (8.78) and (8.79). Only a randomly sampled fraction p_{ki} of that flow moves from queue k to queue i. Then, the first two moments of the sampled subflow are $a_k p_{ki}$ and $1 - p_{ki} + p_{ki} C_{Y,k}^2$, consistent with eqs. (8.84) and (8.85). Finally, the overall arrival flow at the input of queue i is the superposition of the sampled subflows coming from all queues plus the external input. Applying eqs. (8.76) and (8.77), we have:

$$C_i^2 = \frac{\lambda_i}{a_i} C_{T,i}^2 + \sum_{k=1}^{J} \frac{a_k p_{ki}}{a_i} (1 - p_{ki} + p_{ki} C_{Y,k}^2) \tag{8.87}$$

To complete the argument, we need only recall that, according to our approximation of the mean waiting time of a $G/G/1$ queue, we have $C_{Y,k}^2 = (1 - \rho_k^2) C_k^2 + \rho_k^2 C_{X,k}^2$. Then we find

$$C_i^2 = \frac{\lambda_i}{a_i} C_{T,i}^2 + \sum_{k=1}^{J} \frac{a_k p_{ki}}{a_i} (1 - p_{ki} + p_{ki}(1 - \rho_k^2) C_k^2 + p_{ki} \rho_k^2 C_{X,k}^2) \tag{8.88}$$

holding for $i = 1, \ldots, J$. Equation (8.88) defines a linear equation system. The solution yields the SCOVs of the input processes of the network queues, C_i^2, $i = 1, \ldots, J$. Along with the mean input rates a_i, we thus obtain the full two-moment characterization of the input process of each queue.

The mean queue length of the i-th queue can be found by using Little's law and resorting to the RBM approximation of the mean waiting time:

$$E[Q_i] = a_i E[W_i] + a_i E[X_i] \approx a_i \frac{\rho_i^2 (C_i^2 + C_{X,i}^2)}{2 a_i (1 - \rho_i)} + \rho_i = \frac{\rho_i^2 (C_i^2 + C_{X,i}^2)}{2(1 - \rho_i)} + \rho_i \tag{8.89}$$

The mean time through the network $E[D]$ can be found again using Little's law:

$$E[D] = \frac{1}{\sum_{i=1}^{J} \lambda_i} \sum_{i=1}^{J} E[Q_i] = \frac{1}{\sum_{i=1}^{J} \lambda_i} \sum_{i=1}^{J} \left(\frac{\rho_i^2 (C_i^2 + C_{X,i}^2)}{2(1 - \rho_i)} + \rho_i \right) \tag{8.90}$$

with $\rho_i = a_i E[X_i]$.

Summarizing the whole procedure, we identify the following steps.

Given the mean rates and SCOVs of the external arrival processes, the mean rates and SCOVs of the service processes and the routing probability matrix:

1. Calculate the internal mean arrival rates a_j by solving the linear equation system:

$$a_j = \lambda_j + \sum_{i=1}^{J} a_i p_{ij} \tag{8.91}$$

The utilization coefficients of queue servers are then $\rho_j = a_j E[X_j]$. Check that $\rho_j < 1$, $j = 1, \ldots, J$.

2. Calculate the SCOVs of the internal arrival processes by solving the following linear equation system:

$$C_j^2 = \frac{\lambda_j}{a_j} C_{T,j}^2 + \sum_{k=1}^{J} \frac{a_k p_{kj}}{a_j} (1 - p_{kj} + p_{kj}(1 - \rho_k^2) C_k^2 + p_{kj} \rho_k^2 C_{X,k}^2) \qquad (8.92)$$

3. Calculate performance metrics for the j-th queue using approximations for *isolated G/G/1* queues whose input is modeled as a renewal process having mean rate a_j and SCOV C_j^2, for $j = 1, \ldots, J$.

This algorithm can be generalized to networks of $G/G/m$ queues by simply revisiting step 3. Specifically, the mean waiting time is evaluated using, e.g., the approximation (8.50). Let queue i be a multi-server queue with m_i servers. The mean waiting time can be found as

$$E[W_i] = E[X_i] \frac{C(m_i, a_i E[X_i])}{1 - \rho_i} \frac{C_i^2 + C_{X_i}^2}{2m_i} \qquad (8.93)$$

where $\rho_i = a_i E[X_i]/m_i$ and $C(m, A)$ is the Erlang-C function for an $M/M/m$ queue with m servers and mean offered traffic A. The only other modification required to adapt the decomposition method to the case of multi-server queues concerns the expression of the SCOV of the departing process as a function of the SCOVs of the inter-arrival and service times. It is [204]:

$$C_{Y_i}^2 = 1 + \frac{\rho_i^2 (C_{X_i}^2 - 1)}{\sqrt{m_i}} + (1 - \rho_i^2)(C_i^2 - 1) \qquad (8.94)$$

with $\rho_i = a_i E[X_i]/m_i$. This expression reduces to the formula already found for the single server case when $m_i = 1$. It is exact for $M/M/m$ and $M/G/\infty$ queues. In those cases, it is $C_i = 1$ and $C_{X_i} = 1$ or $m_i = \infty$, respectively. In either case, we get $C_{Y_i} = 1$, which is exact, given that those two queueing models have a Poisson output process.

The queueing network analyzer (QNA) [204] provides an algorithm to obtain approximations of the performance measures of a multi-class and multi-server queueing network.

Further generalization involve introducing priorities of different customer classes. Priorities can be global, i.e., a customer maitains its priority level throughout its journey in the queueing network. Alternatively, priority can be local, i.e., a customer is given a priority level independently at each visited queue. Approximation for priority queueing network are discussed, e.g., in Chen and Yao [54].

Example 8.9 *LTE Access Reservation Procedure for Internet of Things* We consider an LTE cell where a large number of objects access periodically the cellular network to upload a small chunk of data. Typical paradigms casting into this framework are smart meters, vehicles, sensors in industrial environments, environmental sensing, notification from smartphones. These application instances are often referred to collectively as the Internet of Things (IoT) paradigm. Massive data collection from a large number of objects is among the targets of 5G cellular networks (massive machine-type communications). This kind of applications is not well supported in LTE, mainly because of the burden of signaling in the access reservation procedure (ARP). Figure 8.7 illustrates the main phases of ARP (additional signaling after message 4, e.g, for security handshake, is neglected).

In this example, we give a simplified model of ARP. For an alternative, more sophisticated model, the reader can refer to [154].

A brief description of the ARP of LTE is in order, before proceeding to state the model. To understand the ARP it suffices to give few details on the multiple access scheme of LTE. Uplink and downlink resource are organized according to a mixed time-frequency divisioni multiple access. Let us consider a frequency channel of 1.4 *MHz* bandwidth. Transmission is based on orthogonal frequency-division multiplexing (OFDM). The time axis is organized in frames, divided into slots. A frame lasts $T_f = 10$ *ms* and has 10 time slots. Each slot is split into two sub-slots. The radio resource made up of 12 OFDM subcarriers (with an overall bandwidth of 180 *kHz*) for one sub-slot is referred to as a resource block (RB). With a channel bandwidth of 1.4 *MHz* it is possible to pack 6 RB per sub-slot. Therefore, one frame contains $6 \cdot 20 = 120$ RBs. The RB is the minimum quantum of assignable radio resource.

We consider the format with six useful OFDM symbols per RB. Since one RB comprises 12 OFDM sub-carriers, overall $12 \cdot 6 = 72$ resource elements are available in one RB. This format maps into a number of data bits carried by one RB that depends on the modulation and coding set. In the following, we assume that the radio channel experienced by users can be either a "bad" one or a "good" one. In

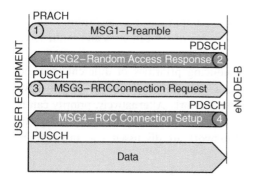

Figure 8.7 Message flow and radio channels of the Access Reservation Procedure in LTE (PRACH = Physical Random Access CHannel; PDSCH: Physical Downlink Shared CHannel; PUSCH: Physical Uplink Shared CHannel).

case of bad channel, QPSK with coding rate $1/2$ is used, resulting in $72 \cdot 2 \cdot 1/2 = 72$ bits per RB, hence a supported bit rate of $r_b = 72$ bit$/0.5$ ms $= 144$ kbit/s. In case of good channel, we assume that 64-QAM with coding rate $2/3$ can be used, so that $72 \cdot 6 \cdot 2/3 = 288$ bits per RB can be transmitted, resulting in a bit rate $r_g = 288$ bit$/0.5$ ms $= 576$ kbit/s. The probability that a user experiences a bad channel in one RB is assumed to be $q = 0.5$.

The RBs are assigned to support logical channels, either devoted to control or reserved for user data. Besides channels devoted to synchronization and other physical layer functions, we consider in this example the following channels:

- PRACH (Physical Random Access Channel): A variable number of slots (pairs of sub-slots) are assigned to it. It is used to start the ARP.
- PUSCH (Physical Uplink Shared Channel): It is the radio channel used to send signaling and data messages from users to the network. Its resources are assigned to different users, according to a dynamic schedule, ruled by the general principle of demand-assignment (no random access).
- PDSCH (Physical Downlink Shared Channel): It is analogous to the PUSCH, except that it goes from the eNodeB, the radio base station of LTE, toward the users. Data and signaling messages destined to users are multiplexed in the PDSCH. An associated control channel signals the RBs where users can find their messages.

A user equipment (UE) starting a new radio session must preliminarily get radio resources to send its traffic. This is the purpose of ARP (see Figure 8.7). At the beginning the UE has no dedicated radio resource. The first step of the ARP consists of selecting at random one of m preambles, defined by the LTE standard, and transmitting it in the PRACH (message 1). The eNodeB detects which of the m preambles have been activated. Then, it sends as many response messages (message 2) in the PDSCH as the number of detected preambles. The response message assigns radio resources for the requesting UE to be able to send an explicit message (message 3), where it can give details on its identity, credentials, what it is requesting and the reason why. If more than one UE has chosen the same preamble, they get the same resources for their messages. Hence, those messages are mixed at the receiver of the eNodeB and cannot be correctly decoded. This event is referred to as a collision. The involved UEs detect the collision event when their timers expire and no reply to their messages 3 has been received. In that case, the UE has to start the whole ARP anew. If instead there was no collision, the reception of message 3 at the eNodeB triggers the sending of the final reply message, concluding the ARP, where the eNodeB specifies the RBs of the PUSCH assigned to the requesting UE (message 4). After receiving this last message from the eNodeB, a UE starts transmitting its data, according to the adopted scheduling policy. Note that PUSCH RBs assigned to UEs are shared dynamically among multiple UEs sending their data.

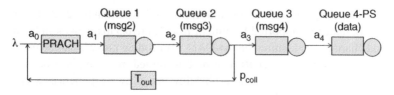

Figure 8.8 Sketch of the queueing network model of LTE access reservation procedure (ARP).

We assume the preambles are always detected, if activated. The collision probability is denoted with p_{coll}. A collision event is detected by an involved UE because it starts a timer on sending message 2. Let T_{out} be the duration of the time-out.

Let us assume that the user terminals generate a Poisson flow of requests of mean rate λ. The overall length of signaling messages, including all overheads, is assumed to be $L_{msg,2} = L_{msg,3} = 36$ bytes, and $L_{msg,4} = 180$ bytes.

The queueing network model of ARP is depicted in Figure 8.8.

Three single-server queues model the transmission of messages 2, 3 and 4 (the queues labeled as 1, 2 and 3, respectively, in Figure 8.8). The block labeled with T_{out} represent the timer mechanism that triggers the restart of ARP upon collision. It corresponds to a fixed delay (a $G/D/\infty$ queue). The last queue represents the multiplexing of user data on the PUSCH for those UEs that have completed ARP successfully (the queue labeled as 4 in Figure 8.8). We assume this last queue uses processor-sharing discipline.

Additionally, we introduce a block representing the PRACH operation. We refer to this block as queue 0. The modeling of this block is explained later.

Let a_i denote the mean rate of arrivals at the input of queue i, $i = 0, 1, \ldots, 4$. At equilibrium, the mean input and output rate at each queue are the same. We can write the following equilibrium equations for the mean rates a_i:

$$a_0 = \lambda + a_2 p_{coll}$$
$$a_2 = a_1 = a_0$$
$$a_3 = a_2(1 - p_{coll})$$
$$a_4 = a_3$$

where we account for the fact that the mean arrival rate a_0 at the input of queue 0 is the sum of fresh arrivals, with mean rate λ, and of arrivals triggered by collisions. Solving the linear system, we find

$$a_0 = a_1 = a_2 = \frac{\lambda}{1 - p_{coll}} \tag{8.95}$$
$$a_3 = a_4 = \lambda$$

The model of the PRACH deserves some attention. This flow of requests at the input of the PRACH is randomly split into m sub-flows, one for each preamble (this is the same as saying that a UE picks one preamble out of m at random). Each preamble works like a slotted ALOHA system (see Section 9.2). Denote with G the overall arrival rate at the slotted ALOHA system. The probability that the "slot" of such slotted ALOHA is busy (i.e., the corresponding preamble is activated) is $1 - e^{-G}$. This expression holds if requests arrive according to a Poisson process, which in our case is only an approximation, since arrivals at PRACH are the superposition of Poisson fresh arrivals and subsequent attempts after collisions. Since m slotted ALOHA systems operate in parallel, for each PRACH opportunity in a frame, we have

$$a_0 = \frac{m}{T_f}(1 - e^{-G})$$
(8.96)

where T_f is the duration of an LTE frame.

Moreover, the probability of collision, conditional on having transmitted, is the ratio of the probabilities that more than a single UE has activated the same preamble to the probability that at least one UE has activated that preamble. Under the Poisson approximation, we have:

$$p_{coll} = \frac{1 - e^{-G} - Ge^{-G}}{1 - e^{-G}} = 1 - \frac{Ge^{-G}}{1 - e^{-G}}$$
(8.97)

Exploiting the two expressions (8.95) and (8.96) of a_0, we obtain an equation for G:

$$\frac{m}{T_f}(1 - e^{-G}) = \frac{\lambda}{1 - p_{coll}} \quad \Rightarrow \quad \frac{\lambda T_f}{m} = Ge^{-G}$$
(8.98)

This result has a simple interpretation. The left-hand side is the mean number or attempts of activating one preamble. The right-hand side is the probability of transmitting on one preamble with no collision. Given all model parameters, G can be found numerically[4] .

Equation (8.98) sets an upper bound to the feasible value of λ:

$$\lambda \leq \frac{m}{T_f e}$$
(8.99)

To apply the decomposition approximation for $G/G/1$ queueing networks, we simplify the model of Figure 8.8 by removing all elements that are analyzed separately (the PRACH block, the time-out block and the data queue). We obtain the reduced queueing network of $G/G/1$ queue depicted in Figure 8.9, where only queues representing transmssion of messages 2, 3 and 4 are accounted for.

---

[4] A very efficient numerical algorithm can be set up by using Newton-Raphson method: the iteration is $x_{k+1} = \frac{be^{x_k} - x_k^2}{1 - x_k}$, for $k \geq 0$, with $x_0 = 0$ and $b = \lambda T_f/m$. This algorithm works fine as long as $b \in (0, 1/e)$.

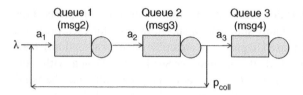

Queue 1 (msg2)   Queue 2 (msg3)   Queue 3 (msg4)

Figure 8.9 Sketch of the reduced queueing network model of LTE ARP. Only the queues modeling the service of message 2, 3, 4, flows are considered.

The first two moments of the service times of the queues in the model of Figure 8.9 are given by:

$$E[X_i] = q\theta_{b,i} + (1-q)\theta_{g,i} \tag{8.100}$$

$$C_{X_i}^2 = \frac{q(1-q)(\theta_{b,i} - \theta_{g,i})^2}{(q\theta_{b,i} + (1-q)\theta_{g,i})^2} \tag{8.101}$$

where $\theta_{b,i} = \frac{L_{msg,i}}{n_i r_b}$ and $\theta_{g,i} = \frac{L_{msg,i}}{n_i r_g}$, for $i = 1, 2, 3$. Here n_i denotes the number of RBs reserved to carry message $i + 1$, for $i = 1, 2, 3$. We recall that r_b and r_g are the bit rates supported by one RB with bad or good radio channel, respectively; q is the probability that the radio channel is bad.

For the stability of these queues, it must be $a_i E[X_i] < 1$ for $i = 1, 2, 3$.

The SCOVs of the arrival processes at the three queues modeling message 2, 3, and 4 service (that is, queue 1, 2, and 3, respectively) are found by using the theory developed in this section. To write down the linear equation system (8.88) we need to specify routing probabilities. By inspection of the model sketch in Figure 8.9, we find easily:

$$\mathbf{P} = \begin{bmatrix} 0 & 1 & 0 \\ p_{coll} & 0 & 1 - p_{coll} \\ 0 & 0 & 0 \end{bmatrix} \tag{8.102}$$

We assume that the 120 RBs managed by the eNodeB are so used:

- $n_{PRACH} = 12 \cdot n_{opp}$ RBs are reserved to the PRACH, where n_{opp} is the number of PRACH opportunities in a frame. We consider only two values: $n_{opp} = 1, 2$. One PRACH opportunity is assigned two sub-slots. The number of preambles is 54 for each PRACH opportunity: hence, $m = 54 n_{opp}$.
- In the uplink, 2 RBs are reserved to physical layer functions in each slot where the PUSCH is allocated. Since the number of sub-slots in a frame is 20, the overall number of RBs consumed by physical layer functions is $n_{PHY} = 2(20 - 2n_{opp})$.
- The remaining $n_{PUSCH} = 120 - n_{PRACH} - n_{PHY}$ RBs of the uplink are assigned to the PUSCH and split into two parts: one reserved to signaling traffic (n_2 for message 3 signaling traffic) and the other one reserved for user data traffic (n_{data}).
- As for the downlink, 12 RBs are assigned to physical layer functions. The remaining $n_{PDSCH} = 120 - 12 = 108$ RBs are split into n_1 and n_3 RBs, reserved for message 2 flow and message 4 flow, respectively.

Table 8.3 Assignment of RBs in the uplink.

	PHY layer	PRACH	PUSCH (n_2)	PUSCH (n_{data})	Total
$n_{opp} = 1$	36	12	4	68	120
$n_{opp} = 2$	32	24	8	56	120

The limits on λ, imposed by the requirement of stability of the queues and of the PRACH, are the following:

$$\lambda < \frac{m}{T_f e} \quad \text{and} \quad \lambda < \frac{1}{E[X_i]} = \frac{n_i}{c_i}, \quad i = 1, 2, 3, \tag{8.103}$$

where c_i are known constants.

We dimension the numbers $n_i, i = 1, 2, 3$, of RBs used to transmit the messages of the ARP so that none of the queues in the network is a bottleneck with respect to PRACH. We set therefore:

$$n_i = \left\lceil \frac{c_i m}{T_f e} \right\rceil, \quad i = 1, 2, 3. \tag{8.104}$$

According to modularization constraints in LTE, we impose that n_2 and n_{data} be multiples of 2. With the numerical values assumed for the model parameters, we obtain $n_1 = 5, n_2 = 4, n_3 = 11$ for $n_{opp} = 1$ and $n_1 = 9, n_2 = 8, n_3 = 21$ for $n_{opp} = 2$. Using the RB allocation described above, we find the number of RBs devoted to UE data in the uplink, i.e., $n_{data} = 68$ for $n_{opp} = 1$ and $n_{data} = 56$ for $n_{opp} = 2$. The overall assignment of RBs in the uplink for the two considered values of n_{opp} is shown in Table 8.3.

Let $E[W_i]$ and $E[S_i]$ be the mean waiting and system time at queue i. We have $E[S_i] = E[W_i] + E[X_i]$ and

$$E[W_i] = E[X_i]\frac{\rho_i(C_i^2 + C_{X_i}^2)}{2(1 - \rho_i)} \tag{8.105}$$

with $\rho_i = a_i E[X_i], i = 1, 2, 3$.

We set $T_{out} = 2T_f$. Since the delay introduced by the timer in the feedback branch of the model is constant, it does not affect the first two moments of the arrival processes. We can therefore preserve the formulas derived for the analysis of the reduced queueing network.

As for UE data, we assume that the RBs reserved for data are shared among all requesting UEs according to a processor sharing policy. Then, the mean delay suffered by a UE to send B bytes of data is

$$E[S_{data}] = \frac{E[X_{data}]}{1 - \rho_{data}} \tag{8.106}$$

where $E[X_{data}] = qB/(n_{data}r_b) + (1-q)B/(n_{data}r_g)$ and $\rho_{data} = \lambda E[X_{data}]$. This holds provided $\rho_{data} < 1$. To provide numerical results, we assume $B = 1000$ bytes. This is consistent with an Internet of Things context, where UEs are smart meters that send periodically small chunks of data to a central server via the cellular network.

We are now ready to derive the overall delay incurred by an UE to complete the ARP and to send its data. The overall mean number of customers in the queueing network, including the constant delay that accounts for the timer, is:

$$E[Q_{ARP}] = \sum_{i=1}^{3} a_i E[S_i] + a_2 p_{coll} T_{out} = \frac{\lambda(E[S_1] + E[S_2] + p_{coll} T_{out})}{1 - p_{coll}} + \lambda E[S_3]$$

By Little's law, the overall mean time required to complete the ARP is:

$$E[S_{ARP}] = \frac{E[Q_{ARP}]}{\lambda} \tag{8.107}$$

The overall mean delay of the ARP $E[S_{ARP}]$ (solid line), the mean delay for sending the data $E[S_{data}]$ (dashed line) and the mean delay of the three queues representing the transmission of ARP messages 2, 3, and 4 (marked lines) are plotted against λ in Figure 8.10. The left plot shows results for $n_{opp} = 1$, the right plot refers to the case $n_{opp} = 2$. The upper bound of λ as determined by the stability of the ARP is given by $m/(eT_f)$. This gives 1.99 req/ms for $n_{opp} = 1$ and 3.97 req/ms for $n_{opp} = 2$. If we account also for the data queue, then we have the further limit $\lambda < 1/E[X_{data}]$. With our numerical values, we find that this requirement translates into $\lambda < 1.90$ req/ms for $n_{opp} = 1$ and $\lambda < 1.56$ req/ms for $n_{opp} = 2$. These limits are apparent by looking at the vertical asymptotes of the curves of $E[S_{ARP}]$ and $E[S_{data}]$ plotted in Figure 8.10.

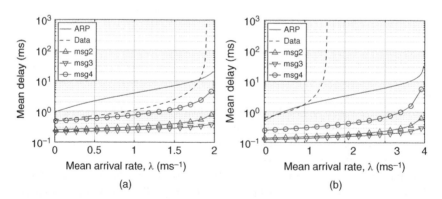

Figure 8.10 Mean queueing delays and overall ARP and data delays of the LTE access model. The left plot refers to $n_{opp} = 1$, the right plot to the case $n_{opp} = 2$.

Some comments emerge from the analysis of the curves in Figure 8.10.

- ARP delay is quite affected by the timeout, as apparent by comparing $E[S_{ARP}]$ with the queueing delays $E[S_i]$ for $i = 1, 2, 3$. On the other hand, the timeout must be set so that it allows enough time for the eNodeB to detect preambles and answer with message 2.
- ARP delay ranges between one and few tens of ms, but it grows rapidly as λ approaches its upper limit. This calls for the deployment of a congestion control algorithm that keeps λ away from its upper bound.
- The most interesting remark comes from a comparison of ARP delay and data delay. For the smaller system with $n_{opp} = 1$, it turns out that $E[S_{data}]$ is way smaller than $E[S_{ARP}]$, that is to say, the overhead required for a UE to be able to send a small chunk of data is overwhelming.
- Since data use the uplink RBs left over by physical layer functions and signaling (ARP), it turns out that, as λ grows, the data queue becomes overloaded eventually. This is apparent by the divergence of the mean data delay for growing λ. In this numerical example the data queue sets the most stringent requirement on λ.

The last comment highlights that the access to LTE can break down because of the overhead implied by ARP. In fact, ARP consumes so many RBs that too few of them are left for data. Most of the potential throughput gain due to the larger PRACH channel is lost because of overload of the data queue. This outcome occurs with a data message size of only 1000 bytes. Obviously, this problem could be circumvented by expanding the bandwidth of the LTE cell, so that more RBs become available.

Several works (e.g., see [201, 166, 114]) highlight that LTE is inadequate to sustain the special kind of traffic offered by a large population of devices sending small chunks of data periodically.

Example 8.10 Let us consider a Voice over IP (VoIP) packet flow. The VoIP source sends one packet of length $L_0 = 160$ bytes (including all overheads) every $T_0 = 20$ ms. The VoIP flow goes through a network path of M links. The path is modeled as a chain of queues, representing the output links of the M crossed routers. Cross traffic sharing each link wuth the VoIP flow is modeled as interfering flows. The k-th interfering flow enters the path at the ingress of queue k and leaves it at the output of the same queue. While this model of cross traffic simplifies the analysis, it can be regarded as reasonable in a large core network, where the probability of flows sharing more than one consecutive link is negligible. Figure 8.11 illustrates an example with $M = 3$.

VoIP
flow

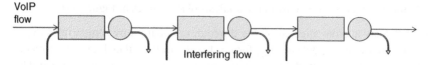

Figure 8.11 Chain of buffers crossed by a VoIP flow. An interfering flow at each buffer models cross traffic.

We assume all links have the same capacity rate, equal to $C = 1$ Mbit/s. Let λ_k and $C^2_{A,k}$ be the mean arrival rate and the SCOV of the k-th interfering traffic flow. The mean rate of the VoIP flow is $\lambda_0 = 1/T_0$, while the SCOV is 0, since the VoIP source is natively a constant bit rate traffic source. Service times are transmission times, given by the ratio of packet length and link capacity rate. As for VoIP, the service time is constant and equal to $Y_0 = L_0/C$. As for the interfering traffic, we choose the first two moments of packet length according to experimental data provided by CAIDA from backbone traffic traces. It is $E[L] = 483.6$ bytes, $\sigma_L = 625.7$ bytes, hence service times for the interfering flow packets are characterized by the first two moments $E[Y] = E[L]/C$ and $E[Y^2] = (\sigma_L^2 + E[L]^2)/C^2$.

The purpose of the analysis is to understand how the cross-traffic affects the VoIP flow, introducing timing jitter on voice packets.

We can directly write an iteration for $k = 1, \ldots, M$, yielding the desired performance measures thanks to the simple feedforward structure of the queueing network in Figure 8.11.

$$a_k = 1/T_0 + \lambda_k$$

$$E[X_k] = \frac{1}{C}\left(\frac{1}{1+\lambda_k T_0}Y_0 + \frac{\lambda_k T_0}{1+\lambda_k T_0}E[Y]\right)$$

$$\rho_k = a_k E[X_k] = \frac{1}{C}\left(\frac{Y_0}{T_0} + \lambda_k E[Y]\right)$$

$$E[X_k^2] = \frac{1}{C^2}\left(\frac{1}{1+\lambda_k T_0}Y_0^2 + \frac{\lambda_k T_0}{1+\lambda_k T_0}E[Y^2]\right)$$

$$C^2_{X,k} = \frac{E[X_k^2]}{E[X_k]^2} - 1$$

$$(8.108)$$

$$C^2_k = \frac{1}{1+\lambda_k T_0}C^2_{VoIP,k-1} + \frac{\lambda_k T_0}{1+\lambda_k T_0}C^2_{A,k}$$

$$C^2_{D,k} = \rho_k^2 C^2_{X,k} + (1-\rho_k^2)C^2_k$$

$$C^2_{VoIP,k} = 1 - \frac{1}{1+\lambda_k T_0} + \frac{1}{1+\lambda_k T_0}C^2_{D,k}$$

to be initialized with $C^2_{VoIP,0} = 0$.

We exploit this model in two ways. First, we consider randomized interfering flows, whose mean arrival rate is drawn uniformly at random in the interval

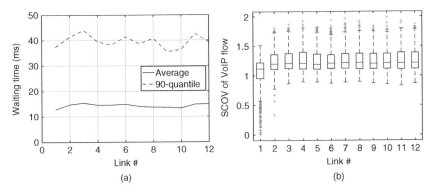

Figure 8.12 Performance measures of the VoIP flow through $M = 12$ buffers with randomized cross traffic load: mean and 90-quantile of the waiting time for each buffer (left plot); box plot of the SCOV of VoIP at the output of each buffer (right plot).

$(0, \lambda_{max})$, where $\lambda_{max} = (C\rho_{max} - Y_0/T_0)/E[Y]$. The SCOV of the interfering flows is drawn uniformly at random from the interval $[C^2_{A,min}, C^2_{A,max}]$.

In this numerical example, we set $\rho_{max} = 0.95$, $C_{A,min} = 0.5$, $C_{A,max} = 2$. Figure 8.12 plots the mean and 90-quantile of the waiting time (left plot) and a box plot of the SCOVs of the VoIP flow out of each buffer (right plot) for $M = 12$ buffers.

The quantile of the waiting time is obtained by exploiting the upper bound of the waiting time probability density function obtained in Section 8.2.3.

The box plot is obtained by generating 1000 samples of interfering flows and collecting the corresponding values of the VoIP flow SCOVs. Each box is delimited by the 25 and the 75 quantiles of the SCOV. The median is marked by a line inside the box, while outliers are shown as plus marks below and above the box.

It is apparent that waiting time performance are essentially the same in each buffer, which is consistent with the fact that the utilization coefficient of the VoIP flow is about 0.06, hence the load of each buffer is dominated by cross interfering traffic. Given the randomized interfering traffic generation, the mean load in each buffer is about 0.5. This explains the relatively low mean waiting time.

As for the SCOV, there is evidently a saturation effect. The SCOV of the VoIP flow, initially equal to 0, grows quickly at the first buffers, then more slowly, tending to about 1.3. The effect of the cross traffic on the VoIP flow is a function of both the mean load and of the variability of the interfering flows.

A second analysis corresponds to assigning a prescribed mean load ρ, hence fixing the mean arrival rate of the cross traffic to $\lambda = (C\rho - Y_0/T_0)/E[Y]$ for all queues. The SCOV of the interfering flows is fixed at 1 for all of them. Figure 8.13 plots the mean of the waiting time (left plot) and the SCOVs of the VoIP flow out of each buffer (right plot) for $M = 12$ buffers and four values of the mean load of each buffer, ranging from 0.1 up to 0.9.

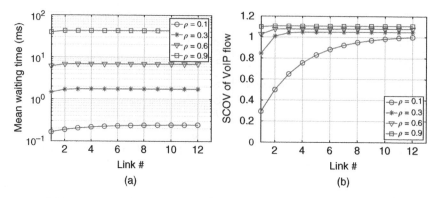

Figure 8.13 Performance measures of the VoIP flow through $M = 12$ buffers for four values of the mean load ρ in each buffer: mean VoIP waiting time for each buffer (left plot); SCOV of VoIP at the output of each buffer (right plot).

The mean waiting time is almost the same at each buffer, apart from a noticeable growth for the first ones. This is due to the growing SCOV of the VoIP flow, which worsens the delay performance exhibited by the queues down the chain.

As for the SCOV of VoIP, it grows as VoIP packets progress through their path, as expected. Each buffer adds a timing jitter on the VoIP packet flow, hence a growing SCOV of the flow. The growth of VoIP SCOV is slower at lower load levels, i.e., more buffers must be crossed for the SCOV of VoIP packets to get close to the SCOV of the cross traffic (1 in this example). At high loads ($\rho = 0.9$), one buffer is enough to make the SCOV of the VoIP flow already as high as the SCOV of the interfering traffic.

This kind of analysis helps dimensioning VoIP delay equalization at the final destination decoder.

8.5.5 Bottleneck Approximation for Closed Queueing Networks

Approximate analysis of closed queueing networks with general service probability distribution cannot be straightforwardly deduced from the $G/G/1$ open network approximation developed so far.

One way to lead the analysis of the closed queueing network back to the approximation for open queueing network is the so called bottleneck approximation [37]. It consists of identifying the most loaded queue and dealing with all others as if they belonged to an open network. This algorithm turns out to provide useful results for large values of the number N of customers circulating in the queueing network.

We use the same notation as in the previous section. The starting point of the analysis is to determine the visit ratios v_j as the nontrivial solution of the following homogeneous linear equation system:

$$v_j = \sum_{i=1}^{J} v_i p_{ij} \tag{8.109}$$

The solution is scaled so that $v_1 + \cdots + v_J = 1$. We then identify the bottleneck node, as the one that maximizes the quantity v_j/μ_j, where $\mu_j = 1/E[X_j]$. Let us label the queues so that $v_1/\mu_1 > v_j/\mu_j$ for $j = 2, \ldots, J$. The queue labeled as 1 is thus the bottleneck one.

We define

$$\rho_j = \frac{v_j/\mu_j}{v_1/\mu_1} \qquad a_j = \rho_j \mu_j = v_j \frac{\mu_1}{v_1} \tag{8.110}$$

for $j = 2, \ldots, J$. This first result covers step 1 of the solution procedure at the end of the previous section. As for step 2, the calculation of the SCOVs of the internal arrival processes, we use the same set of equations as in the open network, except that here there is no external arrival process. Deleting the term *relevant* to external arrivals, we recover a linear equation system whose solution yields the SCOVs C_j^2:

$$C_j^2 = \sum_{k=1}^{J} \frac{a_k p_{kj}}{a_j}(1 - p_{kj} + p_{kj}(1 - \rho_k^2)C_k^2 + p_{kj}\rho_k^2 C_{X,k}^2) \tag{8.111}$$

The mean number of customers at queue j can then be found by using the approximation:

$$E[Q_j] = \rho_j + \frac{\rho_j^2(C_j^2 + C_{X,j}^2)}{2(1 - \rho_j)} \tag{8.112}$$

This expression can be used for all queues, except for queue 1, for which it would blow up to infinity, since $\rho_1 = 1$. At this point, we exploit the fact that the network is closed, hence the overall number of customers is fixed and equal to N. Then, it must be:

$$E[Q_1] = N - \sum_{j=2}^{J} E[Q_j] \tag{8.113}$$

Once the mean number of customers in each queue is found, we can use Little's law to find the mean system time $E[S_j] = E[Q_j]/a_j$ and then the mean waiting time $E[W_j] = E[S_j] - 1/\mu_j$.

An alternative approximate algorithm for closed queueing networks consists of adapting the mean value analysis to the case of general arrival and service times [86].

8.6 Fluid Models

The fluid approximation, leading to fluid models of traffic systems, consists essentially of replacing stochastic phenomena that describe arrivals and service demand by means of their respective averages. To grasp the essence of the fluid model, we first consider the deterministic fluid limit of a stochastic process. Then we extend the fluid modeling to stochastically driven systems. A brief digression is devoted to highlighting that the fluid model is a first-order approximation in a more general framework, where the second-order approximation corresponds to the diffusion model for queues (see Section 8.4.3).

8.6.1 Deterministic Fluid Model

Let us consider a renewal arrival process, described by means of the counting function. We start counting arrivals from $t = 0$. We assume that the first two moments of the inter-arrival time are finite, and that the arrival process is stationary. Then, the counting function depends only on the amount of time t elapsed since the time origin, i.e., we let $A(t)$ denote the number of arrivals in $[0, t)$. We know that for large t the probability distribution of $A(t)$ tends to be Gaussian with mean λt and variance $\lambda t C_T^2$, where $\lambda = 1/E[T]$ is the mean arrival rate, $E[T]$ is the mean inter-arrival time, C_T^2 is the SCOV of the inter-arrival time.

We now introduce a scaling factor n and define the scaled arrival process

$$\overline{A}(t) = \lim_{n \to \infty} \frac{A(nt)}{n} \quad t > 0. \tag{8.114}$$

For a *fixed* t, $A(nt)/n \sim \mathcal{N}(\lambda t, \lambda t C_T^2/n)$. Using Chebychev's inequality, for any given $\epsilon > 0$ we can write

$$P(|\overline{A}(t) - \lambda t| > \epsilon \lambda t) \leq \frac{C_T^2}{\lambda t n \epsilon^2} \tag{8.115}$$

We can bound the right-hand side with an arbitrarily chosen positive number η, by setting $n \geq n(\epsilon, \eta) = C_T^2/(\lambda t \eta \epsilon^2)$. For example, if we let $\epsilon = \eta = 0.1$ (i.e., $\overline{A}(t)$ can differ from λt by at most 10% in 90% of cases), we find $n(\epsilon, \eta) = 10^3 C_T^2/\lambda t$. This is an estimate of the time scale over which a fluid approximation leads to interesting results.

In general, for large values of n, $\overline{A}(t)$ tends to a deterministic process, coinciding with the mean of the original stochastic arrival process, namely λt. Figure 8.14 gives a graphical representation of the effect of scale. Samples of the scaled counting function of a Weibull renewal process with $E[T] = 1$ and $C_T = 2$ are represented as a function of t for the four scales $n = 1, 10, 100, 1000$. The dashed red line plots the mean number of arrivals, λt.

It is clear that the larger the scale the closer the scaled arrival process to the deterministic function λt ($\lambda = 1$ in this numerical example). If we aim to characterize the system behavior on such time scales, we can disregard the details of the discrete steps (the "granularity" of the individual arrivals) and focus on the general "trend" of the counting function.

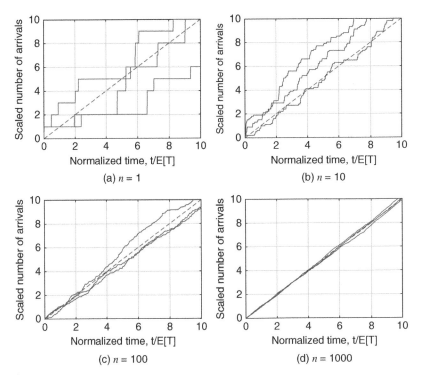

Figure 8.14 Scaled counting function of a Weibull renewal process with $E[T] = 1$ and $SCOV = 4$. Time is normalized with respect to the mean inter-arrival time.

In the case of a stationary arrival process, this leads to a quite uninteresting result, i.e., the linear function of time λt. The reasoning can be carried over to nonstationary processes. Then, at proper time scales, we can restrict our consideration to the time-varying mean of the process, neglecting all the complexity of the full stochastic description.

It is expected that predictions based on such an approximation lead to interesting results when the considered performance metrics depend strongly on the 'trend' (large scale behavior) of the involved processes and only marginally on the 'noise' of the stochastic variability around their respective means.

Let us first apply the fluid model to a trivial case, a stationary $G/G/1$ queue. We re-write eq. (8.56)

$$Q(t) = Q(0) + A(t) - S(B(t)), \quad t \geq 0. \tag{8.116}$$

and scale time by a factor n, thus obtaining:

$$\frac{Q(nt)}{n} = \frac{Q(0)}{n} + \frac{A(nt)}{n} - \frac{S(n \cdot B(nt)/n)}{n} \tag{8.117}$$

For fixed t, $A(nt)/n$ and $B(nt)/n$ tend to λt and $\min\{1, \rho\}t$, respectively, for $n \to \infty$. Moreover, we have also $S(nt)/n \to \mu t$. Putting all together, we can write

$$\overline{Q}(t) = \lim_{n \to \infty} \frac{Q(nt)}{n} = \lambda t - \mu \min\{1, \rho\}t = \begin{cases} 0 & \rho < 1 \\ (\lambda - \mu)t & \rho > 1 \end{cases} \qquad (8.118)$$

This rather disappointing result comes at no surprise if we give it a little thought. At large time scales, the average trend of the queue level of a stationary $G/G/1$ queue is either flat at 0, if the queue is stable, or it grows with time proportionally to the mismatch between the arrival rate and the server capacity rate. If the queue is stable, the mean queue length stays finite on average. Hence, the scaled version (divided by n) of the queue length tends to 0 for a stable queue. In other words, the average queue length of a stable queue amounts to a negligible fraction of the overall number of arrivals over a large time scale. On the contrary, if the queue is unstable, it has a growth drift on the average. Then, the average queue length represents a nonvanishing fraction of the arrivals over a large time scale.

Even if the fluid model of the stationary $G/G/1$ queue is not very useful, it shows that a fluid model can be obtained also for an *unstable* queue. The stochastic description of the transient behavior of a queue does not yield to analytical results except of few cases. Steady-state results only exist if the queue is stable. Hence, the analysis of the transient behavior of a queue is almost unaccessible to a stochastic approach. This is a first hint at the usefulness of the fluid model, i.e., it opens the way to a quantitative understanding of systems where the stochastic analysis is unfeasible.

An enlightening image of the fluid model of a queueing systems is offered by a hydraulic analogy, illustrated in Figure 8.15. Let us consider a tank, where a fluid pours in through an input faucet and drains through an output faucet. The two faucets allow variable flows in and out of the tank. As a consequence, the

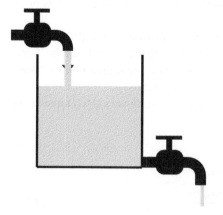

Figure 8.15 Pictorial illustration of the fluid model of a queueing system.

level of the fluid in the tank varies, eventually emptying completely the tank if the drainage out of the tank exceeds the inflow for a sufficiently long time, or, on the opposite, overflowing the limited tank capacity and spilling out, if the inflow exceeds the output capacity long enough.

Let us now consider a nonstationary queueing system. $Q(t)$ denotes the number of customers in the system at time t. We will introduce a fluid approximation $\overline{Q}(t)$ of $Q(t)$, as a result of a scaling technique known as *uniform acceleration*.

Let us assume that the number of arrivals in $(0, t)$, denoted with $A(0, t)$, be a nonstationary Poisson process with mean arrival rate $f_A(t, Q(t))$ that is state dependent. The mean number of arrivals in a time interval (u, v) is $E[A(u, v)] = \int_u^v E[f_A(\tau, Q(\tau))]\, d\tau$. The expectation inside the integral is with respect to the time-dependent probability distribution of $Q(\tau)$. Similarly, the number of departures out of the queue in the interval $(0, t)$ is assumed to be a nonstationary, state-dependent Poisson process, denoted with $D(0, t)$. The mean rate of the departure process is denoted with $f_D(t, Q(t))$.

For any $t \geq 0$ we have

$$Q(t) = Q(0) + A(0, t) - D(0, t) \tag{8.119}$$

We now introduce a scaled stochastic process, depending on an integer index n. The scaled process $Q_n(t)$ is obtained from $Q(t)$ by multiplying the mean arrival and service rates by n and scaling the obtained queue by n. This is called *uniform acceleration*, since it is equivalent to making arrivals and service completions faster by a factor n. We have

$$Q_n(t) = \frac{nQ_n(0) + A_n(0, t) - D_n(0, t)}{n}, \quad n \geq 1. \tag{8.120}$$

where $nQ_n(0) = Q(0)$, $A_n(0, t)$ and $D_n(0, t)$ are the arrival and departure nonstationary, state-dependent Poisson processes with rates $nf_A(t, Q_n(t))$ and $nf_D(t, Q_n(t))$, respectively.

It can be shown that the sequence $Q_n(t)$ converges almost surely to a deterministic process $\overline{Q}(t)$ [138]. Since $\overline{Q}(t)$ is deterministic, we have also $\lim_{n \to \infty} E[Q_n(t)] = \overline{Q}(t)$. For the validity of this asymptotic result we require some technical conditions to hold on the functions $f_A(t, x)$ and $f_D(t, x)$. Specifically, there must exist a finite T such that for any $t \leq T$: (i) it is $|f_A(t, x)| \leq C(1 + x)$ for some constant C; (ii) the function $f_A(t, \cdot)$ satisfies the Lipschitz property, i.e., $|f_A(t, x) - f_A(t, y)| \leq M|x - y|$, $\forall x, y$, for some positive constant M. Similar conditions are required of the function $f_D(t, x)$.

Under these conditions it can be proved that the limit function $\overline{Q}(t)$ is the solution of the following equation (for a sketch of the proof, see, e.g., [86]):

$$\overline{Q}(t) = \overline{Q}(0) + \int_0^t f_A(\tau, \overline{Q}(\tau))\, d\tau - \int_0^t f_D(\tau, \overline{Q}(\tau))\, d\tau \tag{8.121}$$

We can convert the integral equation into a differential equation:

$$\frac{d\overline{Q}(t)}{dt} = f_A(t, \overline{Q}(t)) - f_D(t, \overline{Q}(t)) \tag{8.122}$$

with $\overline{Q}(0)$ equal to an assigned value. To clarify the meaning of the uniform acceleration scaling and of the resulting fluid approximation, we present a simple example.

Example 8.11   Let us consider an $M_t/M/m_t$ queueing system, i.e., a multi-server queue with a nonstationary Poisson input process and a time-varying number of servers $m(t)$. Let $\lambda(t)$ be the time-dependent arrival rate of the input nonstationary Poisson process and let μ be the serving rate of a server. Then, the departure rate of the queue is $f_D(t, Q(t)) = \mu \cdot \min\{m(t), Q(t)\}$. Let $N(a(0, t))$ denote the number of arrivals in the interval $(0, t)$ of the nonstationary Poisson process with mean number of arrivals in $(0, t)$ equal to $a(0, t)$. The number of customers in the queue at time t can be written as the balance between the input and the output flow:

$$Q(t) = Q(0) + N\left(\int_0^t \lambda(u)\,du\right) - N\left(\int_0^t \mu \min\{m(u), Q(u)\}\,du\right) \tag{8.123}$$

To obtain the scaled process, we pick a value of the scaling factor n and define the scaled arrival rate $\overline{\lambda}(t) = \frac{\lambda(t)}{n}$ and the scaled number of servers $\overline{m}(t) = \frac{m(t)}{n}$. Then, we can rewrite eq. (8.123) for the scaled process $Q_n(t) \equiv Q(t)/n$:

$$Q_n(t) = \frac{nQ_n(0) + N\left(\int_0^t n\overline{\lambda}(u)\,du\right) - N\left(\int_0^t \mu n \min\{\overline{m}(u), Q_n(u)\}\,du\right)}{n}$$

$$\tag{8.124}$$

As $n \to \infty$, $Q_n(t)$ converges almost surely to $\overline{Q}(t)$, which is given by:

$$\overline{Q}(t) = \overline{Q}(0) + \int_0^t \overline{\lambda}(u)\,du - \int_0^t \mu\min\{\overline{m}(u), \overline{Q}(u)\}\,du \tag{8.125}$$

Note that the general theoretical result stated above refers to rates f_A and f_D for the arrival and departure processes, which are here replaced by the scaled rates $\lambda(t)/n$ and $\mu\min\{m(t), Q(t)\}/n$. Therefore, we expect that the approximation be accurate when $\lambda(t)/\mu \gg 1$ and $m(t) \gg 1$.

We can use the asymptotic result as follows:

- Given $\lambda(t)$, $m(t)$ and $Q(0)$, pick an arbitrary n and define $\overline{\lambda}(t) = \lambda(t)/n$, $\overline{m}(t) = m(t)/n$ and $\overline{Q}(0) = Q(0)/n$.
- Calculate $\overline{Q}(t)$ by integrating numerically eq. (8.125) with the initial condition $\overline{Q}(0)$.
- Recover the approximation of the original queue state as $Q(t) \approx n \cdot \overline{Q}(t)$.

Table 8.4 Values of the mean arrival rate λ, number of servers m and utilization coefficient ρ as a function of time for the $M_t/M/m_t$ queue of Example 8.11.

t	1	2	3	4	5	6	7	8	9	10	11	12
$\lambda(t)$	400	570	720	850	950	990	990	950	850	720	570	400
$m(t)$	80	100	110	120	120	120	110	100	80	80	80	80
$\rho(t)$	0.5	0.57	0.655	0.708	0.792	0.825	0.9	0.95	1.06	0.9	0.713	0.5

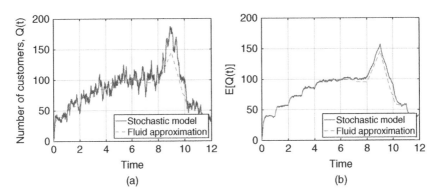

Figure 8.16 Fluid approximation of the $M_t/M/m_t$ queue with nonstationary input and time-varying number of servers. Comparison of the fluid approximation $\overline{Q}(t)/n$ (dashed line) with a sample path of $Q(t)$ (left plot) and with $E[Q(t)]$ (right plot).

As a numerical example, we fix a time unit (e.g., one hour) and define the time-varying arrival rate according to a half-wave sine shape. The number of servers grows linearly up to a maximum level, then it decreases linearly. The numbers used for $\lambda(t)$ and $m(t)$ are shown in Table 8.4. The mean service time of a server is $1/\mu = 1/10$ of the time unit. The resulting server utilization coefficient $\rho(t) = \frac{\lambda(t)}{m(t)\mu}$ as a function of time is shown in the last line of Table 8.4. Note that utilization coefficient values greater than 1 are possible, since we are carrying out a transient analysis.

Figure 8.16(a) compares a sample path of $Q(t)$ (solid line) with the fluid approximation $\overline{Q}(t)/n$ (dashed line), as obtained by integrating eq. (8.125) with $n = 40$. This value of n is used to define the scaled arrival rate and the scaled number of servers. The accuracy of the fluid approximation is remarkable. It turns out that the trend of $Q(t)$ is tracked quite effectively.

The plot in Figure 8.16(b) shows an excellent agreement between the fluid approximation $\overline{Q}(t)/n$ and the mean value of the stochastic process $Q(t)$, estimated by means of 50 sample paths of $Q(t)$. Note that $E[Q(t)]$ is a function of time due to the nonstationarity of the $M_t/M/m_t$ queue.

Note that the fluid approximation allows an easy computation of the transient behavior of the $M_t/M/m_t$ queue. During this transient evolution it can well happen that the *local* coefficient of utilization be greater than 1 (the queue is temporarily unstable). For example, this occurs in the ninth interval, as shown in Table 8.4.

Let us apply the fluid approximation for nonstationary queues to a single server queue. We assume a time-varying, state-independent arrival rate $\lambda(t)$ and serving rate $\mu(t)$. Then, it is $f_A(t, Q(t)) = \lambda(t)$ and

$$
f_D(t, Q(t)) = \begin{cases} \mu(t) & Q(t) > 0 \\ \min\{\lambda(t), \mu(t)\} & Q(t) = 0 \end{cases} \tag{8.126}
$$

The fluid approximation is expressed in differential form as:

$$
\frac{d\overline{Q}}{dt} = f_A(t, \overline{Q}(t)) - f_D(t, \overline{Q}(t)) = \begin{cases} \lambda(t) - \mu(t) & \overline{Q}(t) > 0 \\ \max\{0, \lambda(t) - \mu(t)\} & \overline{Q}(t) = 0 \end{cases} \tag{8.127}
$$

This is the differential equation that governs the evolution of the deterministic fluid model of the original nonstationary queue.

Since $f_D(t, Q(t)) \leq \mu(t)$, given a finite time horizon T, we can write:

$$
\overline{Q}(T) - \overline{Q}(0) \geq T \left(\frac{1}{T} \int_0^T \lambda(t)\, dt - \frac{1}{T} \int_0^T \mu(t)\, dt \right) \tag{8.128}
$$

Let the long-term averages of the arrival rate and of the serving rate, i.e., the limits of the two terms in brackets, be finite, denoted with $\overline{\lambda}$ and $\overline{\mu}$, respectively. Then, the factor within brackets tends to $\overline{\lambda} - \overline{\mu}$ as $T \to \infty$. If this difference is positive, i.e., the long-term mean arrival rate exceeds the long-term average serving rate, it turns out that $Q(T) - Q(0)$ diverges as T grows. In other words, $\overline{\lambda} > \overline{\mu}$, implies the instability of the queue. This simple argument proves that $\overline{\lambda} \leq \overline{\mu}$, is a necessary condition for stability of the fluid queue.

We can informally state that the fluid approximation of the queue length $Q(t)$ consists of replacing the actual arrival and service stochastic processes of the arrivals and departures by their respective averages. The analysis of the system model shifts from equations involving the probability distributions to a dynamical system equation, whose solution gives the fluid queue length $\overline{Q}(t)$. The differential equation that drives the fluid dynamical system can be written down straightforwardly, given its physical meaning. It states that the rate of change of the queue content equals the instantaneous difference between the input rate and the drainage rate. Depending on the structure and on the policy of the system, input and output rates could be functions of the system state or of other environmental parameters. We will see that the fluid approximation provides useful models to describe packet networks under closed-loop flow control (e.g., TCP traffic flows), or multiple access systems. When applying a fluid model one must be aware that effects of stochastic variability around the mean values of the involved processes are neglected, yet the "trend" of system variables is correctly captured. Hence, the deterministic fluid model is extremely useful to gain insight into the dynamics of nonstationary systems, or to model network control mechanisms, or to assess the stability of complex systems, e.g., networks of queues (see Section 7.5).

Example 8.12 Let us consider a queueing system with constant serving rate $\mu = 1$ (in suitable measure units). The arrival rate $\lambda(t)$ has a trapezoidal shape, as illustrated in Figure 8.17. The function $\lambda(t)$ is defined as

$$\lambda(t) = \begin{cases} 0 & 0 \leq t \leq 1 \\ 12(t-1) & 1 \leq t \leq 1.25 \\ 3 & 1.25 \leq t \leq 2.5 \\ 3 - 6(t-2.5) & 2.5 \leq t \leq 3 \\ 0 & 3 \leq t \leq 5 \end{cases} \tag{8.129}$$

The fluid queue length $\overline{Q}(t)$ of the system can be obtained by numerical integration of the differential equation in (8.127). In this simple case, it suffices to use the iteration

$$Q_k = \begin{cases} Q_{k-1} + \delta[\lambda(k\delta) - \mu] & Q_{k-1} > 0 \\ Q_{k-1} + \delta\max\{0, \lambda(k\delta) - \mu\} & Q_{k-1} = 0 \end{cases} \tag{8.130}$$

for $k \geq 1$, initialized with $Q_0 = 0$. Here δ is a time step chosen for numerical integration, small as compared to the time scale of the arrival and servic processes. In this numerical example we set $\delta = 10^{-3}$.

The arrival rate exceeds the serving rate in the interval $[1.08, 2.83]$. This is the *overload* interval. The queue length $Q(t)$ is monotonously increasing during the

Figure 8.17 Time evolution of arrival rate $\lambda(t)$, serving rate $\mu(t)$ and fluid queue length $Q(t)$ of the system in Example 8.12.

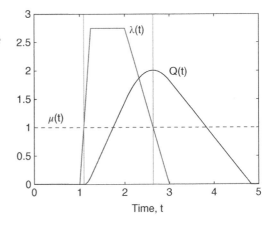

overload interval, since more fluid is offered to the system than the amount that it can dispose of. Therefore, the maximum of the queue length is attained just as the overload interval ends and an underload time starts. During underload, the pressure on the queue is relieved and the excess unfinished work can be drained out, until the queue gets eventually empty, if the underload lasts long enough.

The maximum queue length achieved during the observed time interval depends on how long the overload lasts and how deep it is, i.e., how much the arrival rate exceeds the serving rate.

It is true in general that the worst congestion is to be found as soon as the overload ends, which might seem a bit counterintuitive at first, yet it is clearly highlighted by the fluid model analysis of this example.

8.6.2 From Fluid to Diffusion Model

The deterministic fluid approximation consists essentially of observing the evolution of a stochastic system on a sufficiently gross-grained time scale, so that statistical fluctuations due to the stochastic nature of the involved quantities can be neglected with respect to their average trend.

Part of the stochastic variability can be preserved by using a different scaling. Specifically, let us first consider a sum of i.i.d. random variables, $S_n = Z_1 + \cdots + Z_n, n \geq 1$. Let m and σ^2 denote the mean and variance of $Z \sim Z_j$. It is known by the central limit theorem (CLT) that the random variable $\frac{S_n - nm}{\sigma \sqrt{n}}$ converges in distribution to a standard Gaussian random variable as $n \to \infty$. We can generalize and construct the random process $Y_n(t) = \frac{S_{\lfloor nt \rfloor} - \lfloor nt \rfloor m}{\sigma \sqrt{n}}, t > 0, n \geq 1$, where $\lfloor x \rfloor$ denotes the largest integer not greater than x. By convention, we set $S_0 = 0$. It can be shown that the entire process $Y_n(t)$ converges in distribution to a standard Brownian Motion process[5] $B(t)$ as $n \to \infty$. This is known as the functional central limit theorem (FCLT).

The FCLT can be applied to a counting process $X(t)$ with mean $E[X(t)] = t/m$ and variance $\sigma^2_{X(t)} = t\sigma^2/m^3$, where m and σ^2 are the mean and the variance of inter-event times. It can be shown that the sequence of random processes $Y_n(t)$ defined as follows:

$$Y_n(t) = \frac{X(nt) - nt/m}{(\sigma/m^{3/2})\sqrt{n}} \tag{8.131}$$

converges to a standard Brownian motion process as $n \to \infty$.

The plot of $Y_n(t)$ for the counting process of the departures out of a $G/G/1$ queue with Pareto inter-arrival and service times, is shown in Figure 8.18. It is

---

5 Also known as Wiener process; see the Appendix at the end of the book for its definition and basic properties.

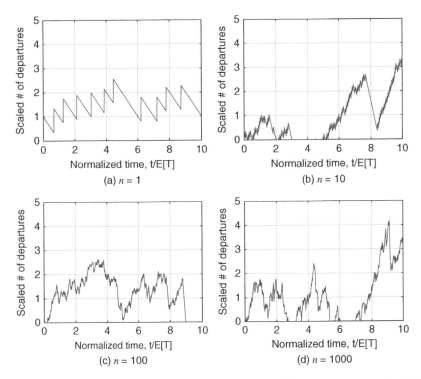

Figure 8.18 Diffusion scaling of the counting function of the output process of a $G/G/1$ queue with Pareto inter-arrival and service times, with $\rho = 0.9$, $E[T] = 1$, $E[X] = E[T]\rho$, $C_T = C_X = 2$. Time is normalized with respect to the mean inter-arrival time.

evident that the sample paths of the scaled process resemble a continuous-time, continuous-state process as n grows. In other words, at large enough time scales, the discrete nature of events can be replaced with a random process having continuous sample paths.

In general, the diffusion approximation of queues consists of finding a scaling of the original process $X(t)$ that converges to some diffusion process as a parameter tends to infinity, e.g., time scale or number of servers. More in depth, let $Z(t)$ be a general stochastic process and let us consider the scaling:

$$Z_n(t) = \frac{Z(nt) - \overline{Z}(nt)}{\sqrt{n}} \tag{8.132}$$

for any integer $n \geq 1$. Here $\overline{Z}(nt)$ is the deterministic fluid approximation of $Z(nt)$, i.e., a deterministic process that approximates $Z(t)$ on suitably large time scales. A typical choice could be $\overline{Z}(t) = E[Z(t)]$.

Given the fluid approximation $\overline{Z}$, the main task is to study the convergence of the scaled stochastic process $\{Z_n(t), t \geq 0\}$ as $n \to \infty$.

A diffusion approximation is obtained when $Z_n(t)$ converges to a diffusion process. A diffusion process is a time-continuous process with almost surely continuous sample paths, that satisfies the Markov property. Brownian motion is an example of diffusion process.

If a diffusion approximation $Z_D(t)$ can be found, i.e., $\lim_{n \to \infty} Z_n(t) = Z_D(t), \forall t \geq 0$, then for suitably large n we can write $Z(nt) \approx \overline{Z}(nt) + Z_D(t)\sqrt{n}$. If a steady state exists for the involved processes, the probability distribution of the steady-state process $Z(\infty)$ can be approximated with that of $\overline{Z}(\infty) + Z_D(\infty)\sqrt{n}$, where $\overline{Z}(\infty)$ is a constant and $Z_D(\infty)$ is a random variable representing a steady-state sample of the diffusion process.

Let us give more details on diffusion process. Let us consider a continuous-time, continuous-state stochastic process $X(t)$. We define the transition probability distribution

$$F(y, t; x, u) = P(X(t) \leq y | X(u) = x), \quad u \leq t, \tag{8.133}$$

that is to say, the CDF of $X(t)$, conditional on the process being in state x at some prior time $u \leq t$. We assume that the associated PDF $f(y, t; x, u) = \frac{\partial F}{\partial y}$ is properly defined, as long as its derivatives with respect to the state variables x and y and with respect to time t.

The process $X(t)$ is Markovian if its evolution from time v is independent of the states visited prior to v, conditional on an assigned state at time v:

$$f(y, t; x, u) = \int_{-\infty}^{\infty} f(w, v; x, u) f(y, t; w, v) \, dw, \quad u < v < t. \tag{8.134}$$

The process is time-homogeneous if $f(y, t; x, u)$ depends only on $t - u$, rather than on t and u separately. In other words, the evolution of the process depends only on the amount of time elapsed and not on which is the initial time instant of the evolution.

The process is space homogeneous, if $f(y, t; x, u)$ depends only on $y - x$, rather than on x and y separately. This means that the evolution of the process depends only on the distance between the initial and final state and not on which they are.

Finally, we define the moment rates as follows:

$$R_n(x, t) = \lim_{\Delta t \to 0} \frac{1}{\Delta t} \int_{-\infty}^{\infty} (y - x)^n f(y, t + \Delta t; x, t) \, dy \tag{8.135}$$

For example, let us focus on $n = 1$. The conditional mean of the process $X(t)$ is defined as

$$M(t; x, u) = E[X(t) | X(u) = x] = \int_{-\infty}^{\infty} y f(y, t; x, u) \, dy \tag{8.136}$$

Since the process is assumed to have continuous sample paths, i.e., it does not make jumps, $\lim_{\Delta t \to 0} M(t + \Delta t; x, t) = M(t, x, t) = x$. Then, it is easy to check that

$$R_1(x, t) = \lim_{\Delta t \to 0} \frac{M(t + \Delta t; x, t) - M(t; x, t)}{\Delta t} = \frac{\partial M(\tau; x, t)}{\partial \tau} \quad (8.137)$$

hence the name of moment rate given to $R_1(x, t)$.

Exploiting the definitions above and the Markovian property of $X(t)$, the following forward Chapman-Kolmogorov equation can be derived for $f(y, t; x, u)$ [131]:

$$\frac{\partial f}{\partial t} = \sum_{n=1}^{\infty} \frac{(-1)^n}{n!} \frac{\partial^n [R_n(y, t) f(y, t; x, u)]}{\partial y^n} \quad (8.138)$$

The differential equation must be solved under boundary conditions depending on specific cases. In general, the solution must be a PDF, i.e., it must be non-negative and such that $\int_{-\infty}^{\infty} f(y, t; x, u) dy = 1$. Moreover, it must be $f(y, u; x, u) = \delta(y - x), \forall u$, where $\delta(\cdot)$ is the delta of Dirac.

If only the first-order term is considered in the series on the right-hand side of eq. (8.138), we write the differential equation satisfied by the fluid approximation of $X(t)$:

$$\frac{\partial f}{\partial t} = -\frac{\partial [R_1(y, t) f]}{\partial y} \quad (8.139)$$

Example 8.13 Let us assume that the first moment rate, also known as drift, is a function of time only, i.e., $R_1(y, t) = m(t)$. Then

$$\frac{\partial f}{\partial t} = -m(t) \frac{\partial f}{\partial y} \quad (8.140)$$

A general solution is $f(y, t; x, u) = \psi \left(y - x - \int_u^t m(\tau) d\tau \right)$ and $\psi(\cdot)$ is a PDF. Applying the initial condition, we have $f(y, u; x, u) = \psi(y - x) = \delta(y - x)$, whence we identify $\psi(\cdot)$ as Dirac's delta function. Summing up, the solution of eq. (8.140) is $f(y, t; x, u) = \delta \left(y - x - \int_u^t m(\tau) d\tau \right)$.

It turns out that the process whose PDF is the solution of eq. (8.140) is a deterministic one, given by $X(t) = X(u) + \int_u^t m(\tau) d\tau$, $t \geq u$. This is the fluid approximation of a stochastic process with time-varying drift and no constraints (e.g., reflecting or absorbing barriers).

The diffusion approximation of $X(t)$ is the stochastic process whose transition PDF solves the differential equation obtained by truncating the sum in the right-hand side of eq. (8.138) to the first two terms:

$$\frac{\partial f}{\partial t} = -\frac{\partial [R_1(y, t) f]}{\partial y} + \frac{1}{2} \frac{\partial^2 [R_2(y, t) f]}{\partial y^2} \quad (8.141)$$

For a time and space homogeneous process, we can let $u = 0$ and $x = 0$, i.e., the initial condition of the process is $X(0) = 0$. Hence, the PDF depends only on y and t and the moments rates are constants. We use the notation $R_1 = \alpha$ for the drift and $R_2 = \beta^2$ for the diffusion coefficient. Then, we get the following simplified differential equation for $f(y, t)$:

$$\frac{\partial f}{\partial t} = -\alpha \frac{\partial f}{\partial y} + \frac{1}{2}\beta^2 \frac{\partial^2 f}{\partial y^2} \tag{8.142}$$

The stochastic process whose PDF is the solution of eq. (8.142) is known as Wiener process or Brownian motion. Not surprisingly, this is the same as eq. (8.63), that is satisfied by the diffusion approximation of the $G/G/1$ queue with stationary arrival and service processes.

8.6.3 Stochastic Fluid Model

A fluid model of a service system substitutes the stochastic processes representing system quantities with their respective average trends. While the small details of the stochastic varibility of the system is lost, the general time evolution is captured, including transient analysis. The system description is based on differential equations governing the time evolution of the deterministic fluid model $\overline{Q}(t)$, rather than equations yielding the probability distributions of the stochastic processes $Q(t)$.

The fluid model approach can be extended to the case where the fluid flow level is a function of random processes modulating some parameter of the service system, e.g., the arrival and service rates. In this setting, the discrete description of the process state is relaxed to a real-valued state variable. The latter is still a random process, due to randomness of the (fluid) arrival and service processes.

Let $Q(t)$ be the number of customers in the system at time t and let $J_A(t), J_S(t)$ be two discrete-state random processes, modulating the arrival and service processes. The mean arrival rate of customers at the system is λ_j when $J_A(t) = j$. Similarly, the serving capacity of the system is c_j when $J_S(t) = j$.

The full stochastic description of the system admits a state space made up of the lattice (k, i, j), where $k = 0, 1, \dots, K, i = 0, \dots, n_A, j = 0, \dots, n_S$. Here K denotes the maximum number of customers that can be admitted into the system (including those under service). The number of states grows with the product Kn_An_S and can easily explode to unfeasible sizes. The stochastic fluid model offers a viable approach, retaining a very good insight, in those cases where the time scale of the stochastic variability of $Q(t)$ is small compared to the time scale of the arrival/service modulating process.

In the stochastic fluid model, we replace $Q(t)$ with its fluid approximation, maintaining the stochastic modulating processes $J_A(t)$ and $J_S(t)$. With a small abuse of notation, we keep using $Q(t)$ to denote the content level of the system. In the fluid setting, however, $Q(t)$ is a real-valued function of time, taking values in the interval $[0, K]$. The time evolution of $Q(t)$ is described by the following equation:

$$\frac{dQ}{dt} = \begin{cases} \max\{0, \lambda_{J_A(t)} - c_{J_S(t)}\} & Q(t) = 0 \\ \lambda_{J_A(t)} - c_{J_S(t)} & 0 < Q(t) < K \\ \min\{0, \lambda_{J_A(t)} - c_{J_S(t)}\} & Q(t) = K \end{cases} \tag{8.143}$$

The interpretation of this equation is quite intuitive: the rate of change of the system content level is given by the difference between the instantaneous rate of the input flow minus the instantaneous rate at which the server drains the system, except at the two boundary states 0 and K. When the system is empty, it remains empty until the input flow rate exceeds the server capacity rate. Symmetrically, when the system is full, it stays full until the input rate gets smaller than the serving rate.

We are interested in characterizing the statistics of $Q(t)$ as $t \to \infty$. A steady state exists if K is finite and the modulating processes admit steady state themselves (e.g., they are irreducible continuous-time Markov chains). Let $J_A(\infty)$ and $J_S(\infty)$ denote the limiting steady-state modulating processes.

If $K = \infty$, we require the following stability condition:

$$\mathrm{E}[\lambda_{J_A(\infty)}] < \mathrm{E}[c_{J_S(\infty)}]. \tag{8.144}$$

In the following we assume that $J_S(t)$ has a constant value, so that we can drop it[6]. We simplify notation by dropping the subscript A as well. Hence, only arrival rate is modulated, by the discrete-state process $J(t)$. The constant serving capacity is denoted with c.

Analytical solutions are available for the following cases of modulating process.

1. $J(t)$ is a continuous-time Markov chain (CTMC). In this case we assume that $J(t)$ is irreducible and positive recurrent. Let also $w_j = P(J(\infty) = j)$ be the limiting state probability and $\mathbf{M}$ be the infinitesimal generator of the modulating CTMC $J(t)$. The mean arrival rate is $\overline{\lambda} = \sum_{j=0}^{n} \lambda_j w_j$.
2. $J(t)$ is an alternating renewal process (see Section 3.4.3). The two states can be characterized as the ON state, where fluid arrives at a rate λ_{ON}, and an OFF

---

6 Let $c_j, j = 1, \ldots, n_S$ be the service rate values corresponding to the states of the CTMC modulating the server. Let $c = \max_j \{c_j\}$. We can redefine the system model as one where there are two traffic sources. One is the original one, modulated by the CTMC $J_A(t)$. The other one is dubbed the *compensating* source and it is defined as one whose fluid arrival rate is $c - c_{J_S(t)}$.

Then, we can rewrite the system content level dynamic equation as $\frac{dQ}{dt} = \lambda_{J_A(t)} + c - c_{J_S(t)}$ $- c = \Lambda_{J(t)} - c$, where $J(t) = (J_A(t), J_S(t))$ is a CTMC and $\Lambda_{J(t)} = \lambda_{J_A(t)} + c - c_{J_S(t)}$. Therefore, we can always restate the model as one where only the input process is modulated by a CTMC, while the server capacity rate is a constant.

state, where fluid arrival rate is λ_{OFF}. To complete the description of the modulating process, we need to assign the PDFs of the ON and OFF times. The mean arrival rate is $\bar{\lambda} = \frac{\tau_{ON}\lambda_{ON}+\tau_{OFF}\lambda_{OFF}}{\tau_{ON}+\tau_{OFF}}$, where τ_{ON} and τ_{OFF} are the mean sojourn times of $J(t)$ in the ON and OFF state, respectively.

3. $J(t)$ is a semi-Markov process (SMP) over the state space $\{0, 1, \ldots, n\}$. Let T_h denote the time of the h-th jump of the process $J(t)$. The sequence $Z_h = J(T_h^+), h \geq 1$, defined by the state attained by $J(t)$ immediately after a jump, is a discrete-time Markov chain (DTMC) over the state space $\{0, 1, \ldots, n\}$. Let $\tau_j = E[T_1 | J(0) = j]$ and $w_j = \lim_{h \to \infty} P(Z_h = j)$ be the mean sojourn time of $J(t)$ in state j and the limiting state probability of state j of the DTMC Z_h, respectively. Then, the mean arrival rate is $\bar{\lambda} = \sum_{j=0}^{n} \lambda_j \tau_j w_j / \sum_{j=0}^{n} \tau_j w_j$.

We are interested in the joint probability distribution

$$F_j(x, t) = P(Q(t) \leq x, J(t) = j), \quad 0 \leq x \leq K; j = 0, 1, \ldots, n. \tag{8.145}$$

or in the corresponding PDF $f_j(x, t) = \frac{\partial F_j(x,t)}{\partial x}$.

In the following, we assume that the modulating process $J(t)$ is an irreducible CTMC on the finite state space $\{0, 1, \ldots, n\}$ (case 1 above). Let q_{ij} be the transition rate from state i to state j. Hence, the limiting state probability distribution is well defined, i.e., there exists the limit

$$\lim_{t \to \infty} P(J(t) = j) = P(J(\infty) = j) = w_j \tag{8.146}$$

with w_j finite and positive for all j and $\sum_{j=0}^{n} w_j = 1$.

Fluid is offered to the system at a mean rate of λ_j when the modulating process is in state j. Fluid is drained out of the system at a constant rate c, whenever there is any fluid in the system. We can define the *drift* of the system in state j as the difference $d_j = \lambda_j - c$, $j = 0, \ldots, n$. The system content $Q(t)$ builds up when $d_j > 0$, whereas the system level decreases when $d_j < 0$. We refer to states with positive drift as *overload* states, while *underload* states designate those states where d_j is negative. The marginal case $d_j = 0$ leads to a temporary equilibrium of the system content level (the fluid amount entering the system balances exactly the amount drained out of the system). We will see soon that those marginal states can be removed in the steady-state analysis.

It is intuitive that this queueing system has an interesting dynamics only if *both* positive and negative drifts are possible. For only positive drifts, the system is permanently overloaded, so that the buffer will be eventually filled up and it will stick to its upper bound K forever. For only negative drifts, the system is always underloaded, so that it drains out eventually and stays empty forever. Therefore, in the following we focus our attention on cases where the drifts d_j are partly positive and partly negative.

Let us now derive the differential equation system whose solution yields the joint CDF $F_j(x, t)$. The key observation is that, to the first order, the variation of the buffer content in a "small" time interval Δt is $\Delta x = (\lambda_j - c)\Delta t = d_j \Delta t$. Considering a vanishingly small time interval Δt, we have

$$F_j(x, t + \Delta t) = \sum_{i \neq j} q_{ij} \Delta t F_i(x - d_i \Delta t, t) + (1 + q_{jj} \Delta t) F_j(x - d_j \Delta t, t) + o(\Delta t) \quad (8.147)$$

then

$$F_j(x, t + \Delta t) - F_j(x, t) + F_j(x, t) - F_j(x - d_j \Delta t, t) = \sum_{i=0}^{n} q_{ij} \Delta t F_i(x - d_i \Delta t, t) + o(\Delta t)$$

$$(8.148)$$

and

$$\frac{F_j(x, t + \Delta t) - F_j(x, t)}{\Delta t} + d_j \frac{F_j(x, t) - F_j(x - d_j \Delta t, t)}{d_j \Delta t} = \sum_{i=0}^{n} q_{ij} F_i(x - d_i \Delta t, t) + o(1)$$

$$(8.149)$$

In the limit for $\Delta t \to 0$, we obtain

$$\frac{\partial F_j}{\partial t} + d_j \frac{\partial F_j}{\partial x} = \sum_{i=0}^{n} q_{ij} F_i(x, t), \qquad j = 0, \dots, n. \quad (8.150)$$

This equations can be written compactly, defining the drift matrix $\mathbf{D} = \text{diag}[d_0, \dots, d_n]$, the transition rate matrix $\mathbf{M}$ of the modulating process, and the row vector $\mathbf{F}(x, t) = [F_0(x, t) : \dots, F_n(x, t)]$:

$$\frac{\partial \mathbf{F}}{\partial t} + \frac{\partial \mathbf{F}}{\partial x} \mathbf{D} = \mathbf{FM} \quad (8.151)$$

Boundary conditions must be assigned for $t = 0$, $0 < x < K$, and for $x = 0$ and $x = K$ for all $t > 0$. We will not pursue this further. Rather, we focus on steady-state analysis.

8.6.4 Steady-State Analysis

If the steady state exists, its CDF $\mathbf{F}(x) = \lim_{t \to \infty} \mathbf{F}(x, t)$ can be obtained solving the ordinary differential equation:

$$\frac{d\mathbf{F}}{dx} \mathbf{D} = \mathbf{FM} \quad (8.152)$$

Let $\mathbf{w} = [w_0, \dots, w_n]$ be the stationary vector of the infinitesimal generator $\mathbf{M}$. We define[7] the set of underload states as $\mathcal{U} = \{j : d_j < 0\}$ and the set of overload

---

7 We assume there is no j with $d_j = 0$. If that were the case, the differential equation system could be rearranged to remove the null drift, since the corresponding equation is just an

states as $\mathcal{O} = \{j : d_j > 0\}$. The boundary conditions are

$$F_j(0^+) = 0, \ j \in \mathcal{O} \qquad F_j(K^-) = w_j, \ j \in \mathcal{V} \tag{8.153}$$

These are exactly $n + 1$ conditions, that suffice to find a unique solution.

The solution of the ordinary differential equation system (8.152) can be written as a linear combination of exponential modes. We define the generalized eigenvalue z and left eigenvector φ of the couple of matrices $(\mathbf{D}, \mathbf{M})$ as a non-null number and non-null vector satisfying the following equality:

$$z\varphi\mathbf{D} = \varphi\mathbf{M} \tag{8.154}$$

The solution in terms of the CDF of the buffer content can be written as

$$\mathbf{F}(x) = \sum_{k=0}^{n} a_k \varphi_k e^{z_k x}, \qquad 0 \le x \le K \tag{8.155}$$

where the coefficients a_k are found from these boundary conditions:

$$\sum_{k=0}^{n} a_k \varphi_k(j) = 0, \ j \in \mathcal{O}, \qquad \sum_{k=0}^{n} a_k \varphi_k(j) e^{z_k K} = w_j, \ j \in \mathcal{V} \tag{8.156}$$

This is a linear equation system of $n + 1$ non-homogeneous equations in the $n + 1$ unknowns a_k, $k = 0, \ldots, n$.

Let us discuss in detail a specific, yet quite interesting example. Let us assume that the input is driven by the superposition of n two-state Markovian ON-OFF traffic sources (see Figure 8.19).

An ON-OFF traffic source is defined as emitting data at rate r when in the ON state, while it is silent in the OFF state. Sojourns in the ON and OFF states are defined by a two-state Markov process. Let α and β be the transition rates out of the OFF and ON states, respectively. The mean duration of the OFF and ON times are $1/\alpha$ and $1/\beta$.

The multiplexer input rate to the system in state j, i.e., when j sources are in the ON state and the other $n - j$ sources are in the OFF state, is $\lambda_j = jr, j = 0, 1, \ldots, n$. The j-th element of the diagonal of the drift matrix $\mathbf{D}$ is $d_j = \lambda_j - c = jr - c$. The

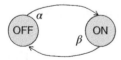

Figure 8.19 Two-state Markov process modeling an ON-OFF traffic source.

---

algebraic one. More in detail, let r_0 be an index such that $d_{r_0} = 0$. The r_0-th equation of the system in eq. (8.152) can be rewritten as $-q_{r_0 r_0} F_{r_0}(x) = \sum_{i \ne r_0} F_i(x) q_{i r_0}$. Feeding this identity back into the other equations, the generic k-th equation can be rewritten as $\frac{dF_k}{dx} = \sum_{i \ne r_0} F_i(x) \tilde{q}_{ik}$, where $\tilde{q}_{ik} = q_{ik} - q_{i r_0} q_{r_0 k}/q_{r_0 r_0}, i = 0, \ldots, n, i \ne r_0$. It is easy to verify that the $\tilde{q}_{ij}$'s still form an infinitesimal generator.

multiplexer analysis is trivial (from a fluid point of view) if $nr \leq c$, i.e., no overload is possible. Hence we assume that $n > c/r$. The underload states form the set $\mathcal{U} = \{0, \dots, \lfloor c/r \rfloor\}$, while the overload states are $\mathcal{O} = \{\lfloor c/r \rfloor + 1, \dots, n\}$. The limiting probabilities of the modulating CTMC are

$$w_j = \binom{n}{j} p^j (1-p)^{n+1-j} , \quad j = 0, \dots, n \tag{8.157}$$

where $p = \alpha/(\alpha + \beta)$ is the probability that a source is ON. Finally, the elements of the matrix $\mathbf{M}$ are given by:

$$\begin{cases} q_{j,j+1} = (n-j)\alpha & j = 0, \dots, n-1 \\ q_{j,j-1} = j\beta & j = 1, \dots, n \\ q_{j,j} = -j\beta - (n-j)\alpha & j = 0, \dots, n \\ q_{i,j} = 0 & \text{otherwise} \end{cases} \tag{8.158}$$

The matrix $\mathbf{M}$ has a tri-diagonal structure. This is the key to derive closed-form results. Thanks to the special structure of the matrices $\mathbf{M}$ and $\mathbf{D}$, it can be shown that all eigenvalues are real and there are exactly n^+ negative eigenvalues and n^- positive eigenvalues. Here n^+ and n^- are the number of positive and negative drift coefficients, respectively [12], hence $n^+ = |\mathcal{O}| = n - \lfloor c/r \rfloor$ and $n^- = |\mathcal{U}| = \lfloor c/r \rfloor + 1$.

It is shown in [12] that the eigenvalues are the solutions of the following $n+1$ quadratic equations:

$$A_k z^2 + 2B_k z + C_k = 0 , \quad k = 0, \dots, n, \tag{8.159}$$

where the coefficients are given by:

$$A_k = U_k - V^2$$
$$B_k = (1 - \alpha/\beta)U_k - (1 + \alpha/\beta)V$$
$$C_k = (1 + \alpha/\beta)^2(U_k - 1)$$

with $U_k = (1 - 2k/n)^2$ and $V = 1 - 2c/(nr)$. We assume, along with [12], that the eigenvalues are indexed so that:

$$z_{n-\lfloor c/r \rfloor -1} < \cdots < z_1 < z_0 < z_n = 0 < z_{n-1} < \cdots < z_{n-\lfloor r/c \rfloor} \tag{8.160}$$

and that the eigenvectors φ_k are normalized so that $\{\varphi_k(n) = 1 \text{ for } k = 0, \dots, n$. With this normalization, it can be shown that

$$\Phi_k(x) \equiv \sum_{j=0}^{n} \varphi_k(j)x^j = (x - \zeta_1)^k (x - \zeta_2)^{n-k} , \quad k = 0, \dots, n, \tag{8.161}$$

where

$$\zeta_1 = \frac{\beta}{2\alpha}[-(z_k + 1 - \alpha/\beta) + \sqrt{(z_k + 1 - \alpha/\beta)^2 + 4\alpha/\beta}]$$

$$\zeta_2 = \frac{\beta}{2\alpha}[-(z_k + 1 - \alpha/\beta) - \sqrt{(z_k + 1 - \alpha/\beta)^2 + 4\alpha/\beta}]$$

The expressions of the eigenvectors turn out to be quite cumbersome and unstable for large size systems. Efficient numerical algorithms can be conveniently used to evaluate the eigenvectors and eigenvalues, given the special tri-diagonal structure of the matrix $\mathbf{M}$.

The left eigenvector associated with the eigenvalue $z_n = 0$ is $\mathbf{w}$, since $\mathbf{wM} = \mathbf{0}$. Then, we can rewrite eq. (8.162) as follows:

$$F(x) = a_n \varphi_n + \sum_{k=0}^{n-1} a_k \varphi_k e^{z_k x} = a_n \mathbf{w} + \sum_{k=0}^{n-1} a_k \varphi_k e^{z_k x} \tag{8.162}$$

The marginal CDF of the buffer content level is obtained by saturating over the state component J, i.e.:

$$G_Q(x) = P(Q > x) = 1 - F(x)\mathbf{e} = 1 - a_n \varphi_n \mathbf{e} - \sum_{k=0}^{n-1} a_k \varphi_k \mathbf{e} \, e^{z_k x} \tag{8.163}$$

It is clear that the probability distribution of the buffer content is given by a linear combination of exponential terms.

8.6.4.1 Infinite Buffer Size ($K = \infty$)

In the special case $K = \infty$, we must have $\rho = nrp/c < 1$, to guarantee the existence of the steady state. The expansion in exponential terms of the CDF $F(x)$ reduces to the $n - \lfloor c/r \rfloor$ nonpositive eigenvalues, since the CDF must be finite in the limit for $x \to \infty$. It is

$$F_j(\infty) = P(Q \leq \infty, J = j) = w_j, \quad j = 0, \ldots, n. \tag{8.164}$$

The expression of the CDF of the buffer content for an infinite buffer can be written as:

$$F(x) = F(\infty) + \sum_{k=0}^{n-\lfloor c/r \rfloor - 1} a_k \varphi_k e^{z_k x}, \quad x \geq 0, \tag{8.165}$$

where the $n - \lfloor c/r \rfloor$ eigenvalues z_k are all negative. The boundary conditions are:

$$F_j(0) = w_j + \sum_{k=0}^{n-\lfloor c/r \rfloor - 1} a_k \varphi_k(j) = 0, \quad j \in \mathcal{O}. \tag{8.166}$$

The solution can be written explicitly as a function of the eigenvalues:

$$a_k = -p^n \prod_{i=0, i \neq k}^{n-\lfloor c/r \rfloor -1} \frac{z_i}{z_i - z_k}, \quad k = 0, \ldots, n - \lfloor c/r \rfloor - 1, \tag{8.167}$$

where $p = \alpha/(\alpha + \beta)$ is the probability that a source is in the ON state.

The CCDF of the buffer content level in the special case $K = \infty$ is

$$G_Q(x) = 1 - \mathbf{F}(x)\mathbf{e} = - \sum_{k=0}^{n-\lfloor c/r \rfloor -1} a_k \varphi_k \mathbf{e} \, e^{z_k x} \tag{8.168}$$

Asymptotically, for large x, $G_Q(x)$ is dominated by the largest negative eigenvalue z_0 (according to the labeling of eq. (8.160)). It can be shown that

$$z_0 = -\frac{1-\rho}{(1-p)\left(1 - \frac{c}{nr}\right)} \tag{8.169}$$

Finally, it can be verified that $\varphi_0 \mathbf{e} = (nr/c)^n$.

Putting all pieces together, we finally get

$$G_Q(x) \sim \rho^n \prod_{i=1}^{n-\lfloor c/r \rfloor -1} \frac{z_i}{z_i + |z_0|} \, e^{-|z_0|x} \qquad (x \to \infty) \tag{8.170}$$

8.6.4.2 Loss Probability

Turning back to the finite buffer model with size K, the loss probability of the stochastic fluid model can be calculated according to the following expression:

$$P_L(K) = \frac{\sum_{j \in \mathcal{O}} d_j [w_j - F_j(K^-)]}{\sum_j w_j \lambda_j} \tag{8.171}$$

Note that $w_j - F_j(K^-)$ is the probability that the buffer content equals K and the phase of the modulating process is j. This probability is 0 when j belongs to the underload states, while it is positive if j is an overload state.

In the simple case $K = 0$, the loss probability is known explicitly: it is the ratio between the average amount of fluid that overflows the server capacity rate c and the average offered fluid rate, i.e.,

$$P_L(0) = \frac{\sum_{j \in \mathcal{O}} w_j (\lambda_j - c)}{\sum_{j=0}^{n} w_j \lambda_j} \tag{8.172}$$

The numerical evaluation of the loss probability, as well as of the CDF of the buffer content, goes through the determination of the coefficients $a_h, h = 0, \ldots, n$. We stress the dependence of those coefficients on the buffer size K by using the notation $a_h(K)$. We let $\mathcal{N}$ denote the set of indices corresponding to real negative eigenvalues. We know that $|\mathcal{N}| = n^- = |\mathcal{U}|$, i.e., the number of negative eigenvalues equals the cardinality of the set of underload states [12]. For ease of notation, we introduce also the diagonal matrix $\mathbf{W} = \text{diag}[w_0 \ldots w_n]$.

In general, the coefficients must by found by solving the linear equation system (8.156) of size $(n+1) \times (n+1)$. There are, however, two special cases where the coefficients can be found exploiting only the knowledge of eigenvalues and eigenvectors. One such case is $K = \infty$, as we have seen in the previous sub-section[8] . The other special case is $K = 0$. Let ψ_h denote the right eigenvector of the matrix $\mathbf{MD}^{-1}$ associated to the eigenvalue z_h, i.e., such that

$$z_h \psi_h = \mathbf{MD}^{-1} \psi_h \tag{8.173}$$

If the modulating CTMC is time-reversible (see the Appendix), it is $w_i q_{ij} = w_j q_{ji}$ for all i and j. In matrix notation, it is $\mathbf{WM} = \mathbf{M}^T \mathbf{W}$.

Multiplying both sides of (8.174) by the diagonal matrix $\mathbf{D}^{-1} \mathbf{W}$, we get

$$z_h \mathbf{D}^{-1} \mathbf{W} \psi_h = \mathbf{D}^{-1} \mathbf{WMD}^{-1} \psi_h = \mathbf{D}^{-1} \mathbf{M}^T \mathbf{WD}^{-1} \psi_h \tag{8.174}$$

Taking the transpose of both sides and multiplying by $\mathbf{D}$ on the right, we obtain finally

$$z_h (\mathbf{D}^{-1} \mathbf{W} \psi_h)^T \mathbf{D} = (\mathbf{D}^{-1} \mathbf{W} \psi_h)^T \mathbf{M} \tag{8.175}$$

We see that $(\mathbf{D}^{-1} \mathbf{W} \psi_h)^T$ is proportional to the left eigenvector φ_h. Choosing the constant of proportionality so that the scalar product of the left and right eigenvectors be 1, i.e., $\varphi_h \psi_h = 1$, we find

$$\psi_h = \frac{1}{b_h} \mathbf{W}^{-1} \mathbf{D} \varphi_h^T \tag{8.176}$$

where the normalization coefficient b_h is defined as

$$b_h = \varphi_h \mathbf{W}^{-1} \mathbf{D} \varphi_h^T = \sum_{k=0}^{n} \frac{d_k}{w_k} \varphi_h^2(k) , \qquad h = 0, \ldots, n \tag{8.177}$$

For $K = 0$, the linear equation system of the boundary conditions can be written as $\mathbf{a}(0)\mathbf{\Phi} = [\mathbf{w}_u \ \mathbf{0}]$, where $\mathbf{a}(0) = [a_1(0), \ldots, a_n(0)]$, $\mathbf{\Phi}$ is the left eigenvector matrix,

---

8 The coefficients $a_h(\infty)$ can be found with a reduced linear equation system also in more general cases than for the superposition of N homogeneous ON-OFF traffic sources. It can be shown that [17]:

$$a_h(\infty) + \sum_{j \in \mathcal{N}} a_j(\infty) u_{jh} = a_h(0) , \quad h \in \mathcal{N}.$$

where

$$u_{ij} = \frac{1}{b_j} \sum_{k \in \mathcal{U}} \varphi_i(k) \varphi_j(k) \frac{|d_k|}{w_k} , \quad i, j \in \mathcal{N}.$$

the h-th row of which is φ_h, and $\mathbf{w}_u$ is a row vector containing the components of $\mathbf{w}$ that correspond to indices of the underload set $\mathcal{U}$. It is easy to verify that $\mathbf{\Phi}^{-1} = \mathbf{\Psi}$, where the h-th column of $\mathbf{\Psi}$ is ψ_h. Then, $\mathbf{a}(0) = [\mathbf{w}_u\ \mathbf{0}]\mathbf{\Phi}^{-1} = [\mathbf{w}_u\ \mathbf{0}]\mathbf{\Psi}$. In scalar form, we have

$$a_h(0) = \frac{1}{b_h} \sum_{k \in \mathcal{U}} d_k \varphi_h(k), \quad h = 0, \dots, n. \tag{8.178}$$

In [17] it is proved that

$$\frac{1}{A} \sum_{h \in \mathcal{N}} a_h^2(\infty) b_h e^{z_h K} \le P_L(K) \le \frac{1}{A} \sum_{h \in \mathcal{N}} a_h(0) a_h(\infty) b_h e^{z_h K} \tag{8.179}$$

where $A = \sum_{k=0}^{n} \lambda_k w_k$ is the mean offered traffic rate.

Those expressions provide tight bounds of the loss probability of the stochastic fluid flow model, whenever the modulating CTMC is time-reversible. This is the case for example, when the input traffic is generated by the superposition of ON-OFF traffic sources, even in the general case where they belong to different classes, each class characterized by its own values of the traffic source parameters α, β and r.

An even simpler (though looser) upper bound can be proved [17]: it is

$$P_L(K) \le \frac{1}{A} \sum_{h \in \mathcal{N}} a_h^2(0) b_h e^{z_h K} \tag{8.180}$$

The strength of the result stated in eq. (8.179) and (8.180) lies both in the numerical efficiency of the evaluation of the bounds, that depend only on eigenvector and eigenvalues of the matrix $\mathbf{MD}^{-1}$, and in the insight provided by the relatively simple expressions of the bounds. Specifically, to the leading term, it is apparent that the loss probability decays exponentially fast for growing K: $P_L(K) \sim \kappa \cdot e^{z_0 K}$, where z_0 denotes the largest negative eigenvalue and κ is a constant.

Example 8.14 Let us consider as an example a two-state source, that emits λ_1 units of fluid in state 1 and λ_2 units in state 2, the mean sojourn times in the two states being $1/\alpha_1$ and $1/\alpha_2$, respectively. The source offers its traffic flow to a transmission link of capacity rate c, equipped with a storage space (buffer) of size K.

We assume that $\lambda_1 < c$ (underload state) and $\lambda_2 > c$ (overload state). Moreover, $\bar{\lambda} \equiv \lambda_1 w_1 + \lambda_2 w_2 < c$, where $\mathbf{w} = [w_1\ w_2] = [\alpha_2\ \alpha_1]/(\alpha_1 + \alpha_2)$ are the limiting state probabilities of the Markov process that modulates the source fluid emission rate.

With simple algebraic manipulations it is easy to find the two eigenvalues $z_1 = 0$ and $z_0 = -\frac{\alpha_1}{\lambda_1 - c} - \frac{\alpha_2}{\lambda_2 - c} = -\frac{\alpha_1 + \alpha_2}{(c - \lambda_1)(\lambda_2 - c)}(c - \bar{\lambda})$. The corresponding left eigenvectors are $\varphi_1 = [w_1\ w_2]$ and $\varphi_2 = [\lambda_2 - c\ \ c - \lambda_1]$. The joint CDF of the buffer content and source state is

$$\begin{aligned} F_1(x) &= a_1 w_1 + a_2(\lambda_2 - c) e^{z_0 x} \\ F_2(x) &= a_1 w_2 + a_2(c - \lambda_1) e^{z_0 x} \end{aligned} \tag{8.181}$$

for $0 < x < K$.

The boundary conditions yield

$$\begin{cases} a_1 w_1 + a_2(\lambda_2 - c)e^{z_0 K} = w_1 & \text{underload state,} \\ a_1 w_2 + a_2(c - \lambda_1) = 0 & \text{overload state.} \end{cases} \tag{8.182}$$

from which we find

$$a_1 = \frac{w_1(c - \lambda_1)}{w_1(c - \lambda_1) + w_2(c - \lambda_2)e^{z_0 K}}$$

$$a_2 = \frac{w_2 w_1}{w_1(c - \lambda_1) + w_2(c - \lambda_2)e^{z_0 K}}$$

Then, finally we have

$$F_1(x) = w_1 \frac{w_1(c - \lambda_1) + w_2(c - \lambda_2)e^{z_0 x}}{w_1(c - \lambda_1) + w_2(c - \lambda_2)e^{z_0 K}}$$

$$F_2(x) = \frac{w_1 w_2(c - \lambda_1)(1 - e^{z_0 x})}{w_1(c - \lambda_1) + w_2(c - \lambda_2)e^{z_0 K}}$$

Note that for $K = \infty$ we recover expected results, i.e., $F_i(\infty) = w_i$ for $i = 1, 2$. The loss probability P_L is:

$$P_L(K) = \frac{c - \overline{\lambda}}{\overline{\lambda}} \frac{w_2(\lambda_2 - c)e^{z_0 K}}{w_1(c - \lambda_1) + w_2(c - \lambda_2)e^{z_0 K}} \tag{8.183}$$

As it is intuitive, the loss probability is larger the more the overload is pronounced (λ_2 larger than c). It decays exponentially with K and reduces to $w_2(\lambda_2 - c)/\overline{\lambda}$ for $K = 0$.

In the following, we address the analysis of first passage times for the stochastic fluid model. Then, we discuss the insight provided by the stochastic fluid model in the analysis of a packet multiplexer loaded by ON-OFF traffic sources.

8.6.5 First Passage Times

With reference to the stochastic fluid model, we define two thresholds, a and b, with $0 \le a \le b \le K$. The system content level $Q(t)$, starting from the initial condition $Q(0) = x$, evolves and eventually hits any of the two thresholds. The time that the process $Q(t)$ takes to attain either level a or b is called *first passage time*. Three cases can be identified:

- $a < x < b$: either a of b can be hit by the process; first passage refers to the first of the two thresholds that is reached by the process.
- $a = b < x$: there is only one lower threshold; the process bounces back when hitting the reflecting barrier at K (upper limit of the system storage), and is eventually absorbed by the lower threshold.

- $x > a = b$: there is only one upper threshold; the process bounces back when hitting the reflecting barrier at 0 and is eventually absorbed by the upper threshold.

The three cases affect the boundary conditions of the differential equation yielding the probability distribution of the first passage time. In the following we focus on the first case, the other two being derived with simple modifications. Let us define the first passage time T as

$$T = \inf\{t > 0 : Q(t) = a \text{ or } Q(t) = b | Q(0) = x\} \tag{8.184}$$

The joint probability distribution of T and $J(t)$ is defined as follows:

$$H_{ij}(x, t) = P(T \le t, J(T) = j | Q(0) = x, J(0) = i) \tag{8.185}$$

The probabilities $H_{ij}(x, t)$ form a $(n + 1) \times (n + 1)$ matrix $\mathbf{H}(x, t)$.

Assuming that the modulating process $J(t)$ is an irreducible CTMC with infinitesimal generator $\mathbf{M}$, and considering a vanishingly small time increment Δt, we can write:

$$H_{ij}(x, t + \Delta t) = (1 + q_{ii}\Delta t)H_{ij}(x + d_i\Delta t, t) + \sum_{k \ne i} q_{ik}\Delta t H_{kj}(x + d_i\Delta t, t) + o(\Delta t)$$

$$\tag{8.186}$$

This identity results from the decomposition of the trajectory of $Q(t)$ into:

1. a first step over the time interval $[0, \Delta t)$, where the buffer content level moves from x to $x + d_i\Delta t$ and the phase switches from i to k;
2. a second step over the time interval $[\Delta t, t + \Delta t)$, where the phase changes from k to j.

Rearranging terms, we obtain

$$\frac{H_{ij}(x, t + \Delta t) - H_{ij}(x, t)}{\Delta t}$$
$$= \frac{H_{ij}(x + d_i\Delta t, t) - H_{ij}(x, t)}{d_i\Delta t}d_i + \sum_k q_{ik}H_{kj}(x + d_i\Delta t, t) + o(1) \tag{8.187}$$

Taking the limit as $\Delta t \to 0$ and using the matrix notation, we find finally

$$\frac{\partial \mathbf{H}}{\partial t} - \mathbf{D}\frac{\partial \mathbf{H}}{\partial x} = \mathbf{MH} \tag{8.188}$$

The boundary conditions state that, if the process is initialized at either absorbing barrier, than the only state transition that is possible is from the initial state $J(0) = i$ to itself:

$$H_{ij}(b, t) = \begin{cases} 1 & i = j, d_i > 0 \\ 0 & i \ne j, d_i > 0 \end{cases} \qquad H_{ij}(a, t) = \begin{cases} 1 & i = j, d_i < 0 \\ 0 & i \ne j, d_i < 0 \end{cases} \tag{8.189}$$

holding for all $t \ge 0$.

The CDF of T can be obtained by saturating the events $J(0) = i$ and $J(T) = j$:

$$F_T(t|x) = \mathcal{P}(T \le t|Q(0) = x) = \mathbf{p}_0 \mathbf{H}(x, t) \mathbf{e} \tag{8.190}$$

where $\mathbf{p}_0$ is a row vector assigning the probability distribution of the initial state $J(0)$, and $\mathbf{e}$ is a column vector of 1's.

8.6.6 Application of the Stochastic Fluid Model to a Multiplexer with ON-OFF Traffic Sources

Let us consider multiplexing n ON-OFF sources. Each source is modeled with a two-state continuous time Markov process (see Figure 8.19).

The superposition of the n sources can be described by an $n + 1$ state birth–death Markov process $J(t)$, with state space $\{0, 1, \dots, n\}$. When $J(t) = k$ the aggregate rate is $\lambda_k = kr$. Birth rate in state k is $(n - k)\alpha$, while death rate is $k\beta$, $k = 0, \dots, n$. This arrival process is offered to a packet multiplexer, served by an output link with capacity c bit/s, equipped with a buffer of size K packets (see Figure 8.20).

Figure 8.21 illustrates a sample path of the buffer content $Q(t)$, in units of packet size, as a function of time, in units of packet transmission times. Packets are assumed to have fixed length. The buffer size of the multiplexer is 50 packets. The data sending rate of a source when in the ON state is 1/15 of the multiplexer output line capacity, the probability of being in the ON state is 0.1 and the mean size of the burst during the ON time is 100 packets. The number of multiplexed sources is $n = 135$. Since the mean rate of an ON-OFF source is 1/150 of the multiplexer output line capacity, the mean offered traffic is $A_o = 135/150 = 0.9$. That is to say, the output capacity of the line serving the multiplexer is busy transmitting packet $(1 - P_L) \cdot 90\%$ of the time on the average.

The sample path of $Q(t)$ is plotted in Figure 8.21 (dark curve). For comparison purposes, we plot the sample path of an $M/D/1/K$ queue with the same mean offered traffic 0.9 and $K = 50$ (light curve). It is apparent that the two queues behave in two qualitatively different ways. The Poisson driven queue exhibits a "smooth" sample path, with limited excursion, never reaching the buffer limit of $K = 50$ packet over the observed time interval. On the contrary, the multiplexer queue driven by ON-OFF sources is dominated by a bi-stable behavior, where low

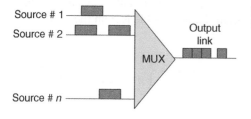

Figure 8.20 Packet multiplexer loaded by ON-OFF traffic sources.

Figure 8.21 Sample path of the buffer content of a packet multiplexer loaded by ON-OFF sources. Time is normalized with respect to the packet transmission time; buffer content is in packet size units.

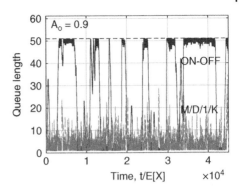

content intervals alternate with essentially full buffer intervals. The former occurs when the input process is in the underload region, i.e., when the number of active sources (in ON state) is such that their aggregate rate does not exceed the output line rate. During overload times, when the aggregate rate of active sources exceeds the output line capacity rate, the buffer content builds up very quickly, hitting its upper limit and essentially sticking there. The transition between the two regimes is very fast, thus making intermediate states of the process $Q(t)$ take very small probability levels.

Looking at the sample path of $Q(t)$, it can be easily realized that neglecting the random fluctuations on short-time scales, due to discrete packets, leads to negligible errors when describing the behavior of the multiplexer. What really matters is to capture the effect of the stochastic process that modulates the number of active sources and hence the aggregate packet rate offered to the multiplexer at any time. Short-term random fluctuations of the buffer content become important only when the buffer size K is very small, so small that even a Poisson input process could saturate the buffer and cause packet loss with non-negligible probability. As a result, we expect that in evaluating the packet loss probability of the multiplexer, P_L, as a function of K, we will find two regimes:

1. *Small buffer size.* K is so small that short-time scale buffer fluctuations due to concurrent packet arrivals cause buffer overflows and packet loss.
2. *Large buffer regime.* Overload dominates buffer overflow and hence packet loss, while the short-time scale pattern of packet arrivals is irrelevant.

The intuition is confirmed by the results plotted in Figure 8.22(a). The packet loss probability of the multiplexer loaded by ON-OFF sources is plotted against the buffer size. The buffer size K is measured in packets, assumed to have fixed length. The number of multiplexed sources is $n = 70$. Time is normalized with respect to the packet transmission time of a source and data is measured in packets. The peak rate of a source is $r = 1$ (1 packet per time unit). The source mean OFF and ON times are $1/\alpha = 900$ and $1/\beta = 100$, so that $p = 0.1$. The capacity of the multiplexer is $c = 12.5$ times the peak bit rate of the source. The mean offered load is $A_o = npr/c = 70 \cdot 0.1 \cdot 1/12.5 = 0.56$.

We compare the loss probability evaluated by means of simulations (shown with 95% confidence intervals) and two models: the fluid model (dashed line) and the $M/D/1/K$ model (dotted line). It is evident that the $M/D/1/K$ model gives accurate predictions for small buffer levels, where the short-time scale effects are dominating, whereas it fails completely when the effect of the overload is dominant. On the contrary, the fluid model captures the effect of the overload, but it cannot account for the effect of discrete packet arrivals that is most relevant at small buffer levels.

As a last numerical example, Figure 8.22(b) shows the loss probability evaluated via the stochastic fluid model for a case where two classes of ON-OFF sources are multiplexed.

The modulating CTMC in this case has $(n_1 + 1) \times (n_2 + 1)$ states, where n_i is the number of ON-OFF sources of class i, $i = 1, 2$.

The three parameters of class S1 sources are: data rate when in the ON state $r_1 = 2/15$ packets per packet transmission time; probability of being in the ON state $p_1 = 0.2$; mean duration of the ON state $1/\beta_1 = 1000$ packet transmission times.

The three parameters of class S2 sources are: data rate when in the ON state $r_2 = 1/15$ packets per packet transmission time; probability of being in the ON state $p_2 = 0.1$; mean duration of the ON state $1/\beta_2 = 100$ packet transmission times.

The mean offered traffic rate is $A_o = (n_1 r_1 p_1 + n_2 r_2 p_2)/c = 0.6$. The three traffic mixes plotted in the figure correspond to the following numbers of sources (n_1, n_2): $(0,90), (11,41), (22,0)$.

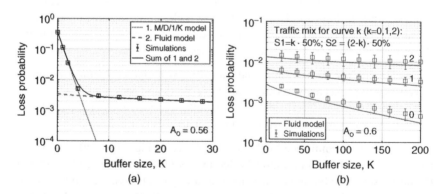

Figure 8.22 Loss probability as a function of the buffer size in packets for a multiplexer loaded by ON-OFF sources. Left plot: $n = 70$ homogeneous ON-OFF sources with mean burst size of 100 packets, activity coefficient 0.1, and peak rate equal to 2/25 of the multiplexer capacity. Right plot: mixes of heterogeneous ON-OFF sources.

Solid lines are obtained with the stochastic fluid model. Simulations at 95% confidence level are shown as well. In the simulation, the discrete packet arrivals and service completions are accounted for.

For the considered buffer sizes, the packet loss is largely dominated by the overloading effect induced by the input traffic rate modulation. Hence, small-scale random fluctuations of the arrival process have a negligible effect and the fluid model yields accurate results. Note that the state space of the considered traffic mixes scales up to 504 states. If a discrete model were considered, with a buffer size $K = 200$ packets, the overall state space would attain a size of 101304. Forgetting about packets and resorting to a fluid model makes the packet loss computation much easier, reducing the computation burden by at least two orders of magnitude, without any appreciable loss of accuracy.

Even simpler calculations are possible losing only a little bit of accuracy, if the approximation (8.179) for the packet loss probability is used.

Summary and Takeaways

This chapter offers an overview of several topics related to bounds and approximations in queueing theory and network traffic engineering applications. Bounds and approximations are most useful and practically used, given that exact solutions are rarely available. Often, we cannot simplify the model of a system to the point where exact analysis is possible, lest it loses practical significance.

In this chapter we have first derived classic bounds and approximations for the $G/G/1$ and $G/G/m$ queues in isolation. We also cover the diffusion approximation of the $G/G/1$ queue leading to the reflected Brownian motion model and to heavy-traffic analysis.

We then extend the analysis to networks of $G/G/1$ queues, both open and closed. The approach to approximation is based on the decomposition method. We hint also to extension to networks of $G/G/m$ queues. Pointers to the technical literature on approximate analysis of more general models of queueing networks are given as well.

Finally, we address the fluid model of service systems. This is a general paradigm, originating from a scaling limit of discrete-event system. We look at discrete-event systems over time scales that are large with respect to the typical times of discrete steps. Then, we can disregard the discrete nature of the sample paths and focus on the "average trend" of the processes. We present the fluid model of a queue, which reduces to a deterministic dynamic system described by means of a differential equation. We discuss at length the significance and expected accuracy of the deterministic fluid model. The key finding is that it is generally quite useless for the steady-state analysis of stable systems. It offers a valuable performance evaluation tool, when considering transient analysis, or nonstationary systems.

We have addressed the stochastic fluid model where arrivals and service are modulated by stochastic processes. The queueing system level is still described by a real-valued variable, i.e., the discrete nature of customers is neglected. The rate at which fluid flow enters the system and is drained out of it is however a function of the state of a stochastic process. This class of models is quite versatile and lends itself to modeling a wide variety of traffic systems. As a relevant example, we discuss at length a packet multiplexer loaded with intermittent ON-OFF traffic sources. It turns out that a remarkable performance prediction accuracy can be attained, at least for some performance metric (the packet loss probability in our example) and in some regime (the large buffer size regime).

Being able to provide performance bounds and approximations to key performance indicators is an extremely useful capability to gain insight into the working of a given system, often at a cost (in terms of resources required to achieve the insight) much lower than simulations or experiments. Crafting a suitable model and an approximate analysis thereof is often more an art than a science. Mathematical techniques come at help once the model has been stated. Under that respect, the array of available mathematical tools and theories is probably wider than one realizes, and getting wider with time and research findings. This is the reason why we strived to give a wide overview of several different techniques, even within the constraint of an introductory chapter.

The really difficult point however is getting to understand the working of the system under study, so as to be able to conceive a simple and parsimonious model, yet a model rich enough to capture the key elements that affect the performance metrics of interest.

Problems

8.1 Observing a single server system, it has been found out that, during equilibrium, the average inter-arrival time is 90 s with standard deviation 30 s. The mean service time is 60 s with standard deviation 45 s. The company that has performed the survey of the service system claims that they have measured a mean number of customers in the system equal to 3.5. Do you believe them? Motivate your answer.

8.2 An on-off traffic source sends a burst of $B = 500$ kbyte every 10 seconds at a peak rate $P = 1$ Mbit/s. The output channel has a capacity C. Excess data is temporarily stored in a buffer waiting to be sent according to a FIFO discipline. Analyze the output channel congestion and the buffer behavior by using a fluid approximation.

a) Find the fluid approximation of the buffer content $Q(t)$. Discuss the behavior of $Q(t)$ as a function of C.

b) Find the maximum and the average time through the buffer when C takes the minimum feasible value for stability, in a fluid approximation.

8.3 Analyze a rate controller interacting with a bottleneck of capacity C by means of a fluid approximation. The rate controller changes the emitted packet rate $\lambda(t)$, aiming at a target rate $\theta(t)$. Let $Q(t)$ denote the content of the bottleneck buffer at time t. Let τ be the feedback delay.

The evolution equation of $\lambda(t)$ is

$$\frac{d\lambda}{dt} = \frac{\theta(t) - \lambda(t)}{2} - \beta\lambda(t)p(t-\tau)\lambda(t-\tau) \tag{8.191}$$

The target rate $\theta(t)$ is set to the current rate whenever the current rate is decreased because of negative feedback, and it increases at a constant rate:

$$\frac{d\theta}{dt} = \alpha - [\theta(t) - \lambda(t)]p(t-\tau)\lambda(t-\tau) \tag{8.192}$$

with $\theta(0) = \lambda(0) = r$ Then $p(t) = I(Q(t) > K)$ for an assigned threshold $K > 0$.

(a) Write the evolution equation of $Q(t)$.

(b) Study the stability of the controller as a function of α assuming $C = 1, r = 4, \beta = 0.5$, and $K = 20$. For this purpose, assume that the feedback delay is equal to the quantization time step when discretizing the differential equations for numerical integration.

8.4 In a CDMA to a base station uplink, the link quality requirement can be expressed as:

$$\frac{W}{R}\frac{SNR_0}{1 + SNR_0(N-1)} \geq \gamma \tag{8.193}$$

where N is the number of simultaneous transmitters, $W = 100$ Mchip/s is the chip rate, R is the bit rate that each user can transmit, $SNR_0 = 30$ dB is the signal-to-noise ratio if there is no interference, and $\gamma = 10$ dB is the SNIR requirement of the link. Users arrive at the system at a mean rate Λ users/sec. On average, a user has to send $Q = 1$ Mbyte of data. Use a fluid approximation.

(a) Write a differential equation governing the number of users $N(t)$ in the system.

(b) Discuss the existence of a stable equilibrium, provided that $\Lambda < \Lambda_{sup}$. Identify Λ_{sup}. For $\lambda = 0.8 \cdot \Lambda_{sup}$, calculate the value N_0 that $N(t)$ tends to as $t \to \infty$.

8.5 Study the scaling of the upper bound of the mean number of customers in the queue, derived from eq. (8.41), as $m \to \infty$, if it is $\lambda = m\beta$ for a fixed positive β and assuming that the SCOVs of the inter-arrival and service times are fixed.

8.6 A traffic source emits bursts of data at a peak rate R, lasting time B. The channel serving the source has a base capacity C_{min}. When a new burst is detected, the capacity is eventually increased to its maximum value C_{max} after a time D since the beginning of the burst, if it is still going on. Once the output capacity is increased to C_{max}, it remains at that value until the buffer is completely drained. Use a fluid approximation in all your calculations.

(a) Assume $C_{max} = R$ and $B > D$. Calculate the amount of time required to clear the backlog.

(b) Do again the calculation of point (a), this time assuming that $R > C_{max}$ and $B > D$.

(c) Repeat point (a), assuming that $C_{max} = R$ and B be a negative exponential random variable with mean $1/\beta$. In this case, calculate the mean time required to clear the backlog.

8.7 Consider a flow of packets originated at a host H and transported first through an access wireless interface (e.g., cellular network), then through a wired network, to a remote server S. We identify two segments: the wireless link between H and the base station BS and the wired network between BS and the server S. The wireless link has a variable capacity rate: we model it with a bimodal random variable, assuming the two values $C_{w1} = 250$ kbit/s and $C_{w2} = 2$ Mbit/s with equal probability. The capacity experienced by packets varies independently from packet to packet. The wired network has a fixed capacity rate $C = 2$ Mbit/s and introduces a fixed delay $RTT_0 = 100$ ms. Packets have fixed length $L = 1000$ bytes. They are sent out by the source H at a rate λ, according to a Poisson process.

There is a probability ϵ_w that packets get lost over the wireless interface. Lost packets are recovered by a data link protocol over the wireless interface. Assume the number of retransmissions is unlimited.

Packets can also be dropped through the wired network, with probability ϵ_d. Dropped packets are detected and recovered by means of end-to-end retransmissions by the transport protocol.

(a) Identify an open queueing network model of the system.

(b) Evaluate the throughput, mean delay, and mean buffer levels at the host H and at the *BS* (uplink) as a function of λ for $\epsilon_w = \epsilon_d = 0.1$.

(c) Study the behavior of the performance indicators as the packet loss and drop probabilities vary.

8.8 Consider a closed-loop packet flow controlled by means of a window-based mechanisms. Let W be the window size. The network path of the packet flow goes through M links, with capacity rates $C_i, i = 1, \dots, M$. The packets have a fixed length $L = 1500$ bytes. The base round trip time of the path is $RTT_0 = 100$ ms.

(a) Identify a closed queueing network model of the system.

(b) Evaluate the mean content levels at the M link buffers, assuming that $W = \beta C, C_1 = \alpha C < C_2 = \dots = C_M = C$, with $\alpha < 1$. Discuss the results as α, β, and C are varied. Can you identify a bottleneck link?

(c) Evaluate the effect of service time variability (e.g., because of underlying subnet protocols that introduce jitter, e.g., random access protocols, ARQ) under the same setting as in the previous point. Assume that the service time of a packet through the wireless access link has a SCOV $C_{X,1}^2 > 1$. How does the bottleneck effect change?

Part III

Networked Systems and Protocols

9

Multiple Access

For a successful technology, reality must take precedence over public relations, for Nature cannot be fooled.

Richard Philips Feynman

9.1   Introduction

Multiple access encompasses both centralized and distributed multiplexing of traffic flows onto a shared communication channel. In the distributed case, a medium access control (MAC) protocol is defined to rule the access to the channel. MAC protocols follow two alternative approaches: (i) deterministic ordered access; (ii) random access. In the first case, the stations sharing the communication channel exchange *signaling messages* to achieve full coordination and thus give rise to an ordered access schedule. Ordered protocols encompass centralized solutions, where a special node is in charge of coordinating the network access through signaling, and distributed approaches, where peer nodes interact through explicit signaling to achieve coordination in sharing the channel. With random access protocols, stations act autonomously, i.e., a station bases its transmission decision only on prior knowledge (protocol state machine) and on events detected by the MAC protocol entity at its interface with the channel. While ordered protocols offer superior performance under heavy traffic load, they usually do not scale nicely as the size of the traffic source population grows, due to the complexity of signaling. Moreover, the overhead implied by signaling can become overwhelming in circumstances where a significant part of the offered traffic is made of isolated, short messages (e.g, sensor networks, app notifications, periodic state updates). Furthermore, random access protocols achieve lower delays under light

Network Traffic Engineering: Stochastic Models and Applications, First Edition. Andrea Baiocchi.
© 2020 John Wiley & Sons, Inc. Published 2020 by John Wiley & Sons, Inc.

traffic load. It is a fact that the most popular MAC technologies have been designed following the random access approach, e.g., WiFi and Ethernet[1] .

In this chapter we focus on random access MAC protocol modeling, analysis and optimization.

The central issue comes out of the multi-access or broadcast communication channels. They are realized by means of a communication medium where a multiplicity of terminals can transmit and receive signals. The broadcast property of such channels implies that *every* terminal receives the signals transmitted onto the channel. Proper information delivery is achieved by means of labels (addresses) added to the transmitted data. As an example, a portion of the radio spectrum corresponding to frequencies ranging from f_1 up to f_2 can be used as a multi-access channel, if each terminal is equipped with a suitable transceiver, able to radiate and sense an electromagnetic field whose frequency content is within the assigned range. Moreover, the electromagnetic field must be the physical support of a signal, whose characteristics (amplitude, phase, frequency) are varied according to the stream of data symbols that must be transmitted. The hardware and software required to transform a sequence of bits into a physical signal that can be impressed over an electromagnetic field and to recover back those bits from the electromagnetic field impinging on the antenna of the receiver is referred to as a transceiver. Transmitter is properly the part that creates and transmits a signal, receiver the part that recovers the bits from the signal. In essence, a communication channel amounts to at least a transmitter, a communication medium, and a receiver. The data source is connected to the transmitter, while the destination is connected to the receiver.

If only a single transmitter can send signals over the given channel, there is no arbitration issue and the data link protocol can focus on data unit delineation, integrity, segmentation and reassembly. If, on the contrary, more than one transmitter can possibly send data over the channel at any time, there arises the problem of ruling the access to the channel.

The MAC *protocol* consists of the specification of data unit format, their meaning and of the state machine executed by each node to perform the access to the shared channel. The piece of hardware/software that realizes the MAC protocol is termed MAC entity. It is interfaced to the upper layer entity and to the physical

---

1 Since the introduction of switches and star topology cabling of ethernet networks, and as their speed has scaled up to the multi-Gbit/s range, the originally CSMA MAC protocol of ethernet network interface cards (NICs) has been substantially made obsolete. In practice, there is no more need of the MAC protocol whenever the physical arrangement of ethernet NICs is such that each transmitter has its own physical medium, like in point-to-point communication systems. Yet, it is at least a historical fact that the technology that won the lead for cabled access networks was originally conceived based on a random access MAC protocol. Random access MAC protocols have also been defined and implemented for other cabled access technologies, eg., cable-modem networks, powerline communication networks, and passive optical networks.

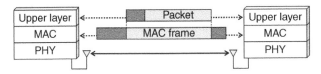

Figure 9.1   Protocol architecture of the wireless access network.

layer entity below (see Figure 9.1). The upper layer entity prompts the MAC entity when it has data to send. Then, the MAC entity commits to transferring the data handed by the upper layer entity. In turn, the MAC layer entity calls for the physical layer entity whenever it has data to transmit and it listens to the physical layer entity to get incoming data. Eventually, the MAC layer entity at the receiving side delivers the received data to the upper layer entity on the receiving side. The arrows in Figure 9.1 recall the direct communication between peer entities. Dark shaded parts of the data unit correspond to overhead used to manage the layer protocol.

With any random access MAC protocol, each node acts on the basis of two inputs:

- local events detected at the interface between the MAC entity and its adjacent entities, the upper layer and the physical layer;
- instructions coded in the MAC protocol algorithm.

A MAC protocol can be defined as a finite-state machine. The state variables depend on the observable events and on the MAC algorithm parameters. The state space is made up of all admissible values of the state variables. An arc is put between states A and B if a transition is possible from A to B. The arc is labeled with two pieces of information: (i) event triggering the transition; and (ii) action taken upon the transition (e.g., send ACK, reset timer, update counters).

In all models considered in this chapter the physical layer is idealized, so that it only introduces a fixed delay and a limited capacity. We assume that the communication link parameters are dimensioned so that any node can communicate with any other node with success, i.e., if a node is *the only one* transmitting in the multi-access channel, any other node can detect the bits sent by the tagged node successfully. Henceforth, a failed reception can only occur if more than a single node is transmitting at a same time[2] .

In the following, nodes are referred to as *stations*, as commonly done in the context of wireless networks. This is motivated by the fact that often such nodes are stationary, as opposed to mobile terminals in a cellular network. To be concrete, we

---

2 We are therefore neglecting what is often referred to as *capture effect* in the technical literature, i.e., the possibility that a message is correctly decoded in spite of interference caused by other overlapping transmissions.

identify the protocol layer above the MAC as the network layer. Then, data units offered to the MAC layer entity from above are named *packets*. Packets are vested with the overhead of the MAC layer to form the MAC protocol data unit (MPDU), also called *frame* (see Figure 9.1). A station having at least one frame waiting to be sent is said to be *backlogged*. We neglect prioritization and hence multiple queues and sophisticated packet scheduling policies.

The rest of this Chapter covers the following topics, with an emphasis on modeling, performance evaluation and optimization: ALOHA (mainly the Slotted ALOHA), CSMA, WiFi random access.

9.2 Slotted ALOHA

In this section we consider the Slotted ALOHA protocol[3]. Each station taking part in the network runs the following algorithm.

1. As long as no packet is standing in the station queue, the station is said to be idle and it takes no action.
2. If a packet arrives at some slot, the station becomes backlogged and moves to the next step.
3. The station will attempt transmitting the packet for the first time in the slot immediately following the one where the packet has arrived.
4. If the transmission attempt fails (the station can detect this event at the end of the slot where the attempt has been carried out), the station moves to step 6.
5. After a successful transmission attempt in the current slot, the station checks whether there are more packets waiting for transmission. If that is the case, it goes to step 6; otherwise, if no more packets are waiting for transmission, it moves back to step 1.
6. The station attempts the transmission of the head-of-line packet in the current slot with probability p; with probability $1 - p$ it backs off for the current slot and repeats step 6 in the subsequent slot.

This protocol statement assumes that the outcome of a transmission attempt is learned by the transmitting station within the end of the same slot where the attempt has been performed. For example, there can be an ACK mechanism in place, so that after completing the transmission attempt, there is room for a short ACK message from the receiver (if it has checked that the received frame is error-free). If the transmitter does not detect the ACK message, i.e., its ACK timer expires and no ACK has been detected, the transmitter deems its attempt to have failed. In the following, we use both the generic term "packet" and the MAC-specific term "frame", since in most implementations there is a one-to-one correspondence.

---

3 ALOHA was originally conceived by Norman Abramson [4] at the University of Hawaii.

9.2.1 Analysis of the Naïve Slotted ALOHA

Let the time axis be divided into fixed length slots, with duration T. A fixed-air bit rate is used in the physical channel, so that L bit long data units can be accommodated into a slot. Let $N(t)$ be the number of stations that are backlogged at the beginning of time slot t.

To simplify the description of the state of the network, we assume that a station can hold at most one frame waiting for transmission. Hence, we can identify the number of stations that are backlogged $N(t)$ with the number of frames contending for the access to the channel.

Let $A(t)$ denote the number of *new* frames that arrive during slot $t - 1$ and are henceforth scheduled for the first transmission attempt in slot t. Besides those new frames, there will be retransmissions of old frames, according to the back-off policy. A station holding an old frame will attempt a transmission with probability p. Then, the number $Y(t)$ of transmissions occurring in the multi-access channel at time slot t is $Y(t) = A(t) + B(t)$, where $B(t)$ is a binomial random variable:

$$P(B(t) = k \mid N(t) = n) = \binom{n}{k} p^k (1 - p)^{n-k}, \qquad k = 0, 1, \dots, n. \qquad (9.1)$$

Our reception model (no capture effect, no channel error) allows only three possible outcomes of the channel usage:

1. $Y(t) = 0$: the slot goes idle, no station attempts transmission.
2. $Y(t) = 1$: a single station attempts transmission, hence the corresponding frame is received with success.
3. $Y(t) > 1$: more than one station attempts transmissions, hence no successful detection is possible.

The last event is usually referred to as a *collision*[4] . Let $I(\mathcal{E})$ be the indicator function of the event $\mathcal{E}$, which is equal to 1 if and only if the event is true, otherwise it is 0. The number of frames that are successfully delivered in slot t is therefore equal to $I(Y(t) = 1)$. The number of backlogged stations found in the system *at the beginning* of slot t evolves according to the following equation

$$N(t + 1) = N(t) + A(t) - I(Y(t) = 1), \qquad t \geq 0, \qquad (9.2)$$

---

4 Note that frames can be unicast or broadcast. In the former case, a frame is addressed to a specific station, hence the frame is lost only if *that* station fails to decode the frame correctly. For a broadcast frame, all stations are addressed. Moreover, if station A sends a frame to station B and station B starts a new transmission whatsoever during A's transmission, B will not be able to receive the frame from A. This is again a sort of collision, i.e., the intended recipient of the frame cannot receive it because it is busy transmitting in turn. The failure of the reception in this case depends on the fact that typically radios are half-duplex, i.e., they cannot receive while transmitting, since self-interference kills the weak signal arriving at the antenna. During this decade more and more experimental evidence has been given that in-band full-duplex is feasible. This opens the way to new interesting possibilities for the design of MAC protocol algorithms and their traffic engineering.

for a given initial state $N(0)$. If the sequence $\{A(t)\}_{t\geq0}$ is made of i.i.d. random variables with the common PDF of the random variable A, then $N(t)$ is a time-homogeneous Markov chain.

To understand how the Markov chain evolves over time and whether it eventually attains a statistical equilibrium regime, we can examine its *drift*, i.e., the average state change at the next transition, given the current state. The drift d_n of the Markov chain in the state $N(t) = n$ is

$$d_n = E[N(t+1) - N(t) \mid N(t) = n] = a - E[I(Y(t) = 1) \mid N(t) = n] =$$

$$= a - P(Y(t) = 1 \mid N(t) = n) = a - a_0 np(1-p)^{n-1} - a_1(1-p)^n$$

where $a_k = P(A = k)$, $k \geq 0$ and $a = E[A] = \sum_{k=0}^{\infty} k a_k$.

Intuitively, the Markov chain attains the statistical equilibrium regime if and only if the drift becomes negative as n grows. From the system point of view, this means that the multi-access channel equipped with the Slotted ALOHA MAC protocol can carry the offered traffic without getting overloaded. If the drift is negative, the mean rate at which the backlog is cleared is bigger than the average arrival rate of *new* frames. We say that the protocol is *stable*.

If A is a Poisson random variable with mean a, the drift for each given n depends only on the two parameters p and a; so the condition $d_n < 0$ (at least definitely for all n bigger than a threshold) leads to

$$a < e^{-a}np(1-p)^{n-1} + ae^{-a}(1-p)^n \tag{9.3}$$

For any fixed p, as n grows, the right hand side tends to 0. We can therefore expect that Slotted ALOHA as defined here cannot be stable for any positive mean load! This is not an approximate finding. It can actually be shown that, if $a > 0$ and $p \in (0,1)$, almost surely there exists a finite time T after which it will always be $Y > 1$ [122, Ch. 5, Prop. 5.3]. In other words, *ALOHA delivers only finitely many packets successfully, then it jams forever.* In terms of Markov chain theory, the result shows that the Markov chain $N(t)$ is not positive recurrent for any $a > 0$. A proof of this result can be obtained also by using the Foster-Lyapunov theorem (see the Appendix at the end of this chapter).

Let us try to understand the "meaning" of this result. First, it can be observed that the explosion of the backlog depends on the fact that the potential population of stations has been assumed to have no limit. Saying that new packets arrive according to a Poisson process *and* that each new packet corresponds to a new, different station implies that the number of stations is not finite. This could be corrected quite easily, by modifying the model. Let M be the maximum number of stations. Assume that an idle station can become backlogged in a slot with probability q. Once it is backlogged, it does not accept any more packets from the network entity, until the standing packet has been successfully delivered. At that point, the station moves back to the idle state and a new packet can be accepted.

Figure 9.2 Sample path of $N(t)$: with Poisson arrivals (dashed line), and Bernoulli arrivals (solid line).

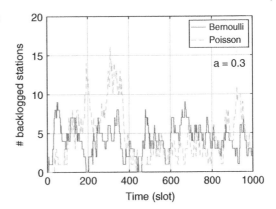

Then, the number of new arrivals $A(t)$ when the system state (number of backlogged stations) is $N(t) = n$ has the following PDF

$$P(A(t) = k \mid N(t) = n) = \binom{M - n}{k} q^k (1 - q)^{M-n-k} , \qquad k = 0, \dots, M - n,$$

$$(9.4)$$

for $n = 0, \dots, M$.

Sample paths of $N(t)$ of the two models of ALOHA are plotted in Figure 9.2. With Poisson arrivals $A(t)$ is a Poisson random variable. The curve labeled "Bernoulli" has been obtained by generating the number of arrivals according to the probability distribution of eq. (9.4) with $M = 10$. The mean number of new arrivals, $a = E[A(t)]$ is the same for the Poisson and for the Bernoulli arrival processes, equal to 0.3.

It is apparent that the ALOHA protocol with Poisson arrivals exhibits wider fluctuations than with Bernoulli arrivals. The latter has an intrinsic limit since the number of stations is finite (10 in the numerical simulation of Figure 9.2). In both cases, it is noted that the number of backlogged stations never gets back to 0 in the case of Poisson arrivals, and only very rarely does with Bernoulli arrivals. This is the mark of a potentially "overloaded" system.

It can be verified that, if M is increased, the process $N(t)$ tends to fluctuate close to its upper bound. It appears that the tendency of ALOHA to boost the number of backlogged stations has deeper reasons than simply an artifact of the Poisson arrival model. A key reason for this behavior is that stations that fail to deliver their frame are *persistent* in their retrials. In other words, the number of retries is unlimited and the probability p that decides whether a station takes a new shot or not is *fixed*. The key lesson learned by the bad performance of the Slotted ALOHA protocol, as we have defined it, is that it is fundamental to limit the number of retries[5].

---

5 Any sane protocol has to guarantee that its execution terminates for any possible situation: hence, a limit must be imposed to the number of attempts for a given frame.

Yet, this limitation alone is not sufficient in general, e.g., if the competing stations are heavily backlogged and keep pushing new frames into the channel. A more structural arrangement is required to stabilize ALOHA, i.e., the probability p of retrying should be modulated according to the amount of backlog.

To grasp the issue, we can state a fluid approximation of the ALOHA protocol. The fluid approximation replaces the stochastic process $N(t)$ with its mean. We denote the approximation with $\overline{N}(t)$, which can assume real values in the interval $[0, M]$, according to the Bernoulli model. The variation with time of $\overline{N}(t)$ is driven by the instantaneous drift, i.e., the difference between the arrival rate $[M - \overline{N}(t)]\lambda$ and the service rate $P_s(\overline{N}(t))/T$, where $P_s(n) = np(1 - p)^{n-1}$ is the probability of a successful packet transmission in a slot and T is the slot duration. Since a station gets ready to transmit a new packet with probability q in each slot, the mean arrival rate of an idle station is $\lambda = q/T$. If we normalize time with respect to the slot duration T, the differential equation that characterizes $\overline{N}(t)$ can be written as:

$$\frac{d\overline{N}(t)}{dt} = [M - \overline{N}(t)]q - \overline{N}(t)p(1 - p)^{\overline{N}(t)-1} = A(\overline{N}(t)) - D(\overline{N}(t)) \qquad (9.5)$$

with the initial condition $\overline{N}(0) = 0$. The two functions $A(x)$ and $D(x)$ appearing in the right hand side of eq. (9.5) are plotted in Figure 9.3 as a function of $x \in [0, M]$ for $q = 1 - a \approx 0.013$, $p = 0.3$. The function $A(x)$ corresponds to the dashed line, while $D(x)$ is the solid curve.

With the chosen numerical values of parameters, there are three points where the two functions coincide and hence the derivative of $\overline{N}(t)$ is zero. The intersections correspond to stationary points of $\overline{N}(t)$, i.e., if $\overline{N}(t)$ is initialized at one of those points, it will stay there forever. This does not mean they are *stable*. Looking at the sign of the right hand side of eq. (9.5), it is easy to recognize that the points marked with a circle in Figure 9.3 are locally stable, i.e., a small perturbation off those points does not kick $\overline{N}(t)$ definitely away. On the contrary, eventually $\overline{N}(t)$

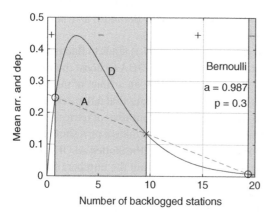

Figure 9.3 Fluid approximation
of $N(t)$. The solid curve is the
mean number of departures,
the dashed curve is the mean
number of arrivals; their
difference is the derivative of
the fluid approximation of $N(t)$
with respect to time. The shaded
regions are those where the
derivative is negative. The
parameters used for this
Bernoulli model of the ALOHA
protocol are: $a = 0.987$, $p = 0.3$.

gets back to x_0 if it starts anywhere around x_0, where x_0 is one of the points marked with a circle. More precisely, the regions shaded in Figure 9.3 denote intervals of the x-axis where the derivative of $\overline{N}(t)$ is negative, hence $\overline{N}(t)$ is pushed to the left (it decreases). On the contrary, white regions are characterized by a positive derivative of $\overline{N}(t)$. Whenever there, $\overline{N}(t)$ tends to move to the right. It is therefore clear that the middle stationary point is unstable, whereas the other two stationary points are locally stable.

Going back from the mathematical model to the system (the ALOHA protocol), the two stable stationary points are expected to be working points of the protocol for long spans of time. Random fluctuations of arrivals and departures can eventually move the state of the system from one locally stable point to the other, yet most of the time we expect to find the state of the system hovering around one of those two stable points. Here comes the pitfall of this naïve version of ALOHA: while the locally stable point on the left is a (relatively) good one, the locally stable point on the right is a terrible one, since the throughput there is close to 0! The system alternates between two working regimes, one with a small backlog and a high throughput, the other one with a heavy backlog and a low throughput.

Before delving into the design of as stabilized version of ALOHA, we develop an extensive analysis of the finite population model of Slotted ALOHA channel. This allows us to gain a deep understanding of the system dynamics.

9.2.2 Finite Population Slotted ALOHA

All times are normalized to the slot duration T. We consider M stations. A station alternates between an idle state, where it has no pending packets to send, and a backlogged state, where it has exactly one outstanding packet. After having sent its packet through the channel successfully, the station goes back into the idle state for a geometrically distributed "think time," with mean value $1/q$.

Once backlogged, a station tries sending a packet with probability p, while it postpones its attempt with probability $1 - p$. Hence p is the transmission probability and $1/p$ represents the mean back-off time before a transmission attempt is carried out. This model is slightly different from the one considered before. In the Slotted ALOHA version of this Section, a new packet arriving at an idle station is immediately considered to be backlogged, hence it is transmitted in the next slot with probability p. In the Slotted ALOHA version of the previous section the transmission of a newly arrived packet is attempted in the next slot with probability 1. Only if a collision occurs will the packet become backlogged and new transmissions be attempted in subsequent slots with probability p. Given that n stations are backlogged at the beginning of a slot, the probability of a successful transmission in that slot is $np(1 - p)^{n-1}$ in the first variant, while it is $(1 - q)^{M-n}np(1 - p)^{n-1} + (M - n)q(1 - q)^{M-n-1}(1 - p)^n$ in the second variant. The latter approach leads to lower delays at light load levels. However, there is little difference between packet delays of the two approaches for the stabilized ALOHA presented in next section. In the following we stick to the first variant, hence assume that all newly arriving packets become immediately backlogged and their transmissions are attempted always with probability p in the next available slot.

According to this model setting, the number $N(t)$ of stations backlogged at the beginning of slot t is a discrete time Markov chain (DTMC) over the state space $[0, \ldots, M]$. It is quite easy to write down the expressions of the one-step transition probabilities $P_{ij} = P(N(t+1) = j \mid N(t) = i)$ for $i, j = 0, \ldots, M$. Given $N(t) = i$, the number of backlogged stations at the beginning of slot $t + 1$ can assume the following values:

1. $N(t + 1) = i - 1$, if and only if no idle station activates and there is a successful transmission in slot t.
2. $N(t + 1) \geq i$, if there is a successful transmission in slot i and $j - i + 1$ idle stations activate *or* no successful transmission takes place in slot t and $j - i$ idle stations activate.
3. $N(t + 1) < i - 1$ is not possible.

Obvious adjustments are required to account for the boundary states 0 and M.

Following the definition of the events listed above, it is easy to find the expressions of the transition probabilities $P_{ij} = P(N(t+1) = j \mid N(t) = i)$:

$$
P_{ij} = \begin{cases} P_s(i)Q(M - i, 0) & j = i - 1, \ i \geq 1 \\ P_s(i)Q(M - i, j - i + 1) + [1 - P_s(i)]Q(M - i, j - i) & j = i, \ldots, M, \ i \geq 0 \\ 0 & \text{otherwise.} \end{cases}
$$

$$(9.6)$$

where

$$
P_s(i) = ip(1 - p)^{i-1}, \qquad i = 0, 1, \ldots, M, \tag{9.7}
$$

and

$$
Q(m, k) = \binom{m}{k} q^k (1 - q)^{m-k}, \qquad k = 0, \ldots, m, \ m = 0, \ldots, M. \tag{9.8}
$$

The one-step transition matrix $\mathbf{P}$ has an Hessenberg structure, just as the matrix of the embedded Markov chain of the $M/G/1$ queue. In fact, this is the common structure of all single-server systems.

The steady-state probability distribution of the number of backlogged stations $\pi_n = \lim_{t \to} P(N(t) = n)$, $0 \leq n \leq M$ can be computed by solving the linear system $\pi\mathbf{P} = \pi$ subject to he congruence condition $\sum_{n=0}^{M} \pi_n = 1$. The limiting probability distribution exists for any value of the model parameters M, q and p, with

$0 < p, q < 1$. In fact, the DTMC $N(t)$ is ergodic since it is irreducible, aperiodic, and lives on a finite state space.

The throughput achieved by the Slotted ALOHA channel is:

$$S = \sum_{n=1}^{M} \pi_n P_s(n) \tag{9.9}$$

The mean delay suffered by a station to send a packet can be evaluated by using Little's law, namely:

$$E[D] = \frac{E[N]}{S} = \frac{\sum_{n=1}^{M} n\pi_n}{\sum_{n=1}^{M} P_s(n)\pi_n} \tag{9.10}$$

The finite population model of the Slotted ALOHA channel is parametrized by only three variables:

1. the population size M;
2. the transmission probability p, or equivalently, the mean back-off time $1/p$;
3. the station activation probability q, i.e., the probability that an idle station becomes backlogged during a slot, or, equivalently, the mean think time of a station $1/q$.

Figure 9.4 displays the throughput-delay trade-off as q is varied for $M = 50$ (left plot) and the trade-off as M is varied, for $q = 0.01$ (right plot).

For small enough values of the back-off probability p, as q is increased the throughput grows steadily as well as the mean delay. The trade-off is not too penalizing, given that the mean delay increases slowly with the throughput. However, the maximum achieved throughput is relatively low, while the mean delay is high, due to the large mean back-off time. If p is increased, bigger levels of throughput can be achieved, still with a relatively low delay. The best trade-off

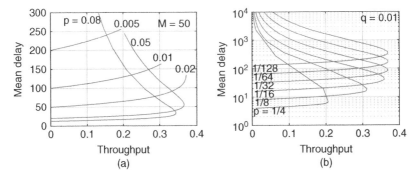

Figure 9.4 Throughput-delay trade-off for several values of p. Left plot: q is varied for $M = 50$ stations. Right plot: M is varied for $q = 0.01$.

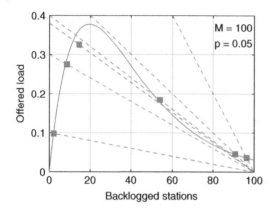

Figure 9.5 Load-lines of Slotted ALOHA for $M = 50$, $p = 0.05$. From right to left, the values of q associated with the load-lines are $q = 0.01, 0.005, 0.004, 0.0038, 0.003, 0.001$. Square markers correspond to steady-state working points.

is obtained for $p = 0.02$ in the figure. After that point, as p grows further, the trade-off exhibits multiple working points. For a same value of the throughput there can be a small and a large mean delay level. This is yet another mark of the bi-stable behavior of Slotted ALOHA.

Similar remarks apply to the trade-off obtained by varying the number of stations M for a fixed level of q (right plot). This last plot highlights that the maximum achievable throughput grows with decreasing levels of p (i.e., as stations become more cautious), though at the cost of increased delay. Moving from high levels of p down, the maximum achievable throughput grows fast at the beginning, then more and more slowly. Correspondingly, the mean delay suffers a little increase at the beginning, then it starts growing more rapidly.

The stability of Slotted ALOHA can be best understood by resorting to a plot like in Figure 9.5. The normalized throughput for a given number of backlogged stations, namely, the success probability $P_s(n)$, is plotted against n (solid line curve). Load-lines defining the mean arrival rate of *new* packets, $(M − n)q$, are plotted for $M = 50, p = 0.05$, and $q = 0.01, 0.005, 0.004, 0.0038, 0.003, 0.001$ (dashed lines from right to left). Square markers show the steady-state working points, i.e., the points with coordinates $(E[N], S)$ for each considered value of q.

Depending on the value of q (for the chosen value of M and p), the load-line intersects the throughput curve in a single point lying on the left of the maximum of the throughput curve, or in three points, or again in a single point, but lying on the right of the maximum of the throughput curve.

In the first case (smallest values of q), the steady-state working point is just the intersection point, as shown by the square markers for the two smallest values of q, $q = 0.001, 0.003$. In the second case, two of the three intersections correspond to locally stable points, while the central one is unstable. Since the actual model is stochastic, there is actually a unique steady state. The system state oscillates between the two locally stable fixed points, so that the steady-state point lies in

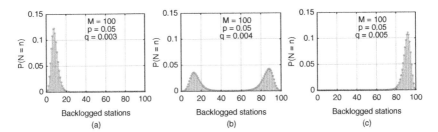

Figure 9.6 Probability distribution of the number of backlogged stations for $M = 100$, $p = 0.05$ and three values of q: $q = 0.003$ (left plot, stable); $q = 0.004$ (middle plot, bi-stable); $q = 0.005$ (right plot, saturated).

between them (see the markers for the two middle load-lines, corresponding to $q = 0.0038$ and $q = 0.004$). Finally, when there is only one intersection to the right of the maximum of the throughput curve (largest values of q), the system moves toward saturation, i.e., all stations are active. The steady-state working point lies close to the right tail of the throughput curve.

The different operation regimes highlighted above can be detected also by inspecting the probability distribution of the number of backlogged stations. Figure 9.6 shows $\pi_n = P(N = n)$ for $M = 100$, $p = 0.05$ and three values of q:

1. $q = 0.003$ (left plot), corresponding to a single intersection, hence a stable system; consistently, the mass of the probability distribution is concentrated at low value of n.
2. $q = 0.004$ (middle plot), corresponding to three intersections, hence a, bi-stable system; this behavior is apparent from the two lobes of the probability distribution, centered on small and large values of n.
3. $q = 0.005$ (right plot), corresponding to one intersection on the right of the maximum of the throughput curve, hence a saturated system; saturation is evident from the concentration of the mass of the probability distribution at large values of the number of backlogged stations.

The analysis of the working regions of slotted ALOHA can be further expanded. Relaxing the system state N to a continuous variable x, the intersections of the load-line with the throughput curve are solutions of the following equations:

$$(M - x)q = xp(1 - p)^{x-1} , \qquad x \in [0, M]. \tag{9.11}$$

We can re-write this equation as

$$f(x) \equiv x + c \, x \, e^{-\beta x} = M \tag{9.12}$$

with $c = \frac{p}{q(1-p)}$ and $\beta = -\log(1 - p)$.

Since $f(0) = 0$ and $f(M) > M$, there is at least one point where $f(x) = M$ for x ranging between 0 and M. The derivative of $f(x)$ is $f'(x) = 1 - c(\beta x - 1)e^{-\beta x}$. This equation has two real solutions if and only if $c \geq e^2$, i.e., for $p \geq \frac{q}{q+e^{-2}}$. Let us denote the two roots as x_1 and x_2 with $x_1 < x_2$. The function $f(x)$ has a relative maximum at x_1 and a relative minimum at x_2. The points x_1 and x_2 are the zeros of $f'(x)$, i.e., they must be such that $e^{\beta x_j} = c(\beta x_j - 1)$, for $j = 1, 2$. Plugging this identity into the expression of $f(x)$, we find

$$f(x_j) = x_j + cx_j e^{-\beta x_j} = \frac{\beta x_j^2}{\beta x_j - 1} = M, \quad j = 1, 2. \tag{9.13}$$

We have

$$x_j = \frac{M}{2}\left(1 \pm \sqrt{1 - \frac{4}{M\beta}}\right), \quad j = 1, 2. \tag{9.14}$$

For these roots to be real, it must be $M\beta \geq 4$, which implies $p \geq 1 - e^{-4/M}$. If this inequality holds, it turns out that the two roots belong to the interval $[0, M]$. Let $M_{min} = f(x_2)$ and $M_{max} = f(x_1)$.

Summarizing this analysis, the existence of the two thresholds M_{min} and M_{max} is guaranteed provided that $p \geq \max\left\{\frac{q}{q+e^{-2}}, 1 - e^{-4/M}\right\}$. Under that condition, for given q and p, the Slotted ALOHA channel operates in the stable regime for $M < M_{min}$. It operates in the bi-stable region, where two locally stable equilibrium points exist, when $M_{min} < M < M_{max}$. Finally, the Slotted ALOHA channel is driven to saturation for $M > M_{max}$. An example plot of $f(x)$ is depicted in Figure 9.7 for $p = 0.08$ and $q = 0.005$.

The left plot in Figure 9.8 illustrates the operating regions of the Slotted ALOHA channel for $q = 0.01$, as a function of the mean back-off time $\overline{B} = 1/p$. The bi-stable region disappears for $\overline{B} > 1 + e^{-2}/q \approx 14.53$ (in slot time units).

For very aggressive system settings, i.e., when the mean back-off time is close to 1 slot, the bi-stable region is substantial and covers most of the range of values of M. In other words, if set to be very aggressive, the Slotted ALOHA channel is easily operated in the bi-stable regime, unless a quite small number of stations are admitted to the channel. As the mean back-off time grows, the number of stations

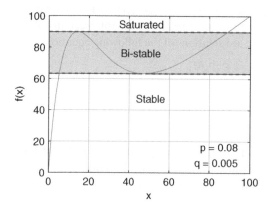

Figure 9.7 Example plot of the function $f(x)$ in eq. (9.12). The three operating regions of Slotted ALOHA are highlighted: $M \leq M_{min}$ (stable), $M_{min} < M < M_{max}$ (bi-stable), and $M \geq M_{max}$ (saturated).

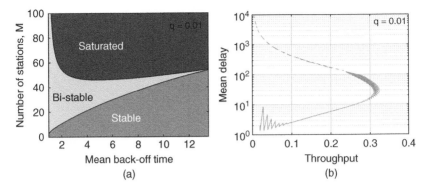

Figure 9.8 Left plot: operating regions of Slotted ALOHA for $q = 0.01$. On the right plot, throughput-delay trade off on the contour separating the stable region from the bi-stable region (solid line curve) and on the contour separating the bi-stable region from the saturation region (dashed-line curve).

that can be accommodated safely in the channel, i.e., under a stable regime, grows as the stable region expands.

The right plot of Figure 9.8 shows the throughput-delay trade-off on the contour separating the stable region from the bi-stable region (solid line curve) and the trade-off on the contour line separating the bi-stable region from the saturation region (dashed line curve). The trade-off is obtained for $q = 0.01$. We move along the trade-off by varying p, i.e, the mean back-off time. It is apparent that, if operated in the saturated region, the Slotted ALOHA offers quite poor performance (high mean delay and low throughput). Better performance is achieved if the Slotted ALOHA channel is operated in the stable region. In that case, the mean delay is always quite low. The throughput is small, if few stations are on the channel. It grows as the number of competing stations increases, still remaining in the stable region.

Finally, we explore the effect of the back-off probability on the steady-state metrics of the Slotted ALOHA channel, i.e., the throughput and the mean delay. Figure 9.9 plots the throughput (left) and the mean delay (right) as a function of the mean back-off time $1/p$, for $M = 100$ and various values of q.

For small mean back-off times the throughput is small and grows with the back-off time, while the mean delay is initally extremely large and drops abruptly as the mean back-off time grows. This part of the curves corresponds to driving the system out of severe congestion, by increasing the mean back-off time.

When the mean back-off time is large, the throughput starts falling, while the mean delay grows fast. This region corresponds to a low load regime, where the channel capacity is wasted due to excessive back-off.

For intermediate values of the mean back-off time the optimal working point is found, where the throughput is maximized and the mean delay is minimized.

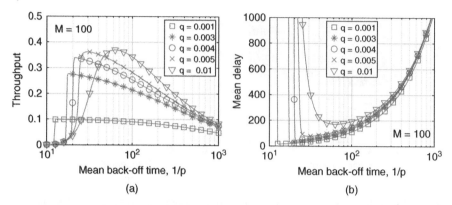

Figure 9.9 Throughput (left plot) and mean delay (right plot) as a function of the mean back-off time $1/p$, for various values of q and $M = 100$.

In the next section we address the core issue of Slotted ALOHA protocol, i.e., adaptation of the retry probability p with the instantaneous load of the system.

9.2.3 Stabilized Slotted ALOHA

The issue with the Slotted ALOHA protocol is that, when the backlog becomes large, the probability of a successful transmission gets small. Since the realized throughput is small, more stations become backlogged while those already backlogged fail to deliver their frames. With Poisson arrivals, the backlog "spiral" leads the system to diverge. With more realistic finite population models or even limiting the maximum number of re-transmissions, things would not be much better under heavy loads, i.e., users would anyway experience low throughput and high delay.

If each station were aware of the number n of backlogged stations (which is not the case in practice, though), it could set the retry probability p to some optimum level. The success probability with n competing stations is $f(p) = np(1 - p)^{n-1}$. It is easy to check that $f(p)$ has a unique maximum for $p \in [0, 1]$, attained for $p^* = 1/n$. Not surprisingly, the optimum retry probability is inversely proportional to the number of backlogged stations.

Stabilization of the Slotted ALOHA protocol can be achieved by using observations of the channel to adjust p *adaptively*, trying to approximate the optimum (unknown) value $p(t) = 1/N(t)$. Define $S(t)$ to be an estimate of the backlog of the system at time t. We aim at making $S(t)$ a proxy of $N(t)$. To that end, it is sensible to increase $S(t)$ whenever a collision event is detected, and to diminish $S(t)$, if slot t turns out to be idle or contains a successfully transmitted frame.

Note that $S(t)$ should be tracked individually by each active station. Hence a station must listen to the shared communication channel. For each slot t, it must log the outcome of that slot. Here success means that the station has decoded the frame correctly. Therefore, the notion of "success" *depends on the point of view of the station*. In many practical cases stations communicate to a central base station (BS), e.g., in the cellular network. In those cases, the BS can observe the channel, estimate $S(t)$, and feed it back to the stations. This avoids any inconsistency that might arise in a fully distributed approach.

We let

$$S(t+1) = \begin{cases} \max\{1, S(t) - \alpha\} & Y(t) = 0, \\ \max\{1, S(t) - \beta\} & Y(t) = 1, \\ S(t) + \gamma & Y(t) > 1 \end{cases} \tag{9.15}$$

i.e., $S(t)$ is decreased by $\alpha > 0$, if an idle slot is observed, it is decremented by $\beta > 0$ if a successful transmission is carried out, it is increased by $\gamma > 0$ for a collision event.[6]

A station that has an estimate $S(t)$, sets its retry probability at $p(t) = 1/S(t)$. It is then clear why imposing that $S(t)$ be no less than 1, whenever it is decreased in eq. (9.15). When $N(t)$ grows, more collisions occur and $S(t)$ is increased, thus relieving the congestion. On the other hand, when $N(t)$ is low, many idle and successful slots appear, $S(t)$ is decreased, hence the stations become more aggressive and the utilization of the channel improves.

A stability analysis can be outlined, by referring to the mean drift of the two processes $N(t)$ and $S(t)$. Given that $N(t) = n$ and $S(t) = s$, we define

$$d_S(n, s) = E[S(t+1) - S(t) \mid N(t) = n, S(t) = s]$$
$$d_N(n, s) = E[N(t+1) - N(t) \mid N(t) = n, S(t) = s]$$

We refer to the Poisson model of new arrivals. The mean number of stations that become backlogged in a slot is denoted with a. For $n, s \geq 1$ we have

$$\begin{cases} d_S(n, s) = \gamma - (\alpha + \gamma)\left(1 - \frac{1}{s}\right)^n - (\beta + \gamma)\frac{n}{s}\left(1 - \frac{1}{s}\right)^{n-1} \\ d_N(n, s) = a - \frac{n}{s}\left(1 - \frac{1}{s}\right)^{n-1} \end{cases} \tag{9.16}$$

---

6 An alternative stabilization law for $S(t)$ is provided by the so called pseudo-Bayesian approach [30, Ch. 4]. Indicating the mean offered load with $a = \lambda T$, the updating rule is:

$$S(t+1) = \begin{cases} \max\{1, S(t) + a - 1\} & \text{in case of idle or successful slot,} \\ S(t) + a + (e - 2)^{-1} & \text{in case of collision.} \end{cases}$$

The idea is that, if an idle or a successful slot, the estimate of the backlog is increased of the new arrivals (a on the average) and reduced by 1. If instead there is a collision, the estimate of the number of backlogged stations is increased by the average number of new arrivals plus an additional term. The choice of the additive corrections, -1 and $(e - 2)^{-1}$ is conceived so as to keep $S(t)$ close to the true backlog when the protocol operates around the optimal working point.

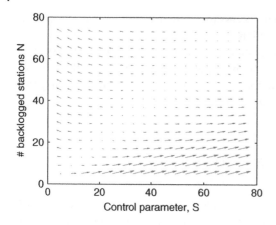

Figure 9.10 Velocity field of the differential equation system (9.17) and (9.16).

In the fluid approximation, the dynamics of the system is described by the differential equation system

$$\begin{cases} \dfrac{ds}{dt} = d_S(n, s) \\ \dfrac{dn}{dt} = d_N(n, s) \end{cases} \tag{9.17}$$

The velocity field of this differential system is plotted in Figure 9.10 for $\gamma = 1$, $\alpha = \beta = e/2 - 1$ and a mean offered load of $a = 0.33$.

Let $x \equiv n/s$ and consider the regime for large n. This is the interesting part of the dynamics as far as stability is concerned, since we should choose the control parameters so as to avoid that $N(t)$ diverges. As $n, s \to \infty$ for a fixed value of x, we have

$$d_S(x) = \gamma - (\alpha + \gamma)e^{-x} - (\beta + \gamma)xe^{-x}$$
$$d_N(x) = a - xe^{-x}$$

For the system to be stable, the mean drift $d_N(x)$ cannot be always positive. Since the function $f(x) = x\,e^{-x}$ peaks at $x = 1$ taking the maximum value $f(1) = 1/e$, we conclude that it must be $a < 1/e$ to guarantee that the drift $d_N(x)$ be negative, when x is around 1. By expanding $d_S(x)$ around 1, we obtain

$$d_S(x) = d_S(1) + \frac{\alpha + \gamma}{e}(x - 1) - \frac{\alpha + \beta + 2\gamma}{2e}(x - 1)^2 + O((x - 1)^3) \tag{9.18}$$

where

$$d_S(1) = \frac{\gamma(e - 2) - \alpha - \beta}{e} \tag{9.19}$$

If we choose the control parameters so that $d_S(1) = 0$, then $x = 1$ becomes an equilibrium point for s (i.e., at $x = 1$ the drift of s becomes 0). Moreover, the drift around $x = 1$ is given by a positive constant multiplying $x - 1$. Then, if x gets bigger than 1, i.e., it is $n > s$, s tends to increase, thus making x decrease. If instead it is $x < 1$, i.e., $n < s$, the drift of s is negative, hence s decreases and correspondingly x grows. Therefore, $x = 1$ is a locally stable stationary point of s (for a given value of n). In other words, the dynamics of s maintains x at 1. When x stays close to 1, the dynamics of n can be approximated as

$$d_N(x) = a - \frac{1}{e} + \frac{1}{2e}(x - 1)^2 + O((x - 1)^3) \tag{9.20}$$

which is negative for x close to 1. Summing it up, by setting $d_S(1) = 0$, the control law of s tends to keep x at 1. Given that, the dynamics of n is negative, as long as x hovers around 1, provided that $a < 1/e$. Overall, the system is stable.

A possible choice of the control parameters to make $d_S(1) = 0$ is

$$\gamma = 1, \qquad \alpha = \beta = \frac{e}{2} - 1 \tag{9.21}$$

With this choice of the parameter values, the adaption of the back-off probability is expressed by $p(t) = 1/S(t)$ with $S(0) = 1$ and

$$S(t + 1) = \begin{cases} \max\{1, S(t) + 1 - e/2\} & \text{if idle or success,} \\ S(t) + 1 & \text{if collision.} \end{cases} \tag{9.22}$$

for $t \geq 1$.

A sample path of $N(t)$ (left plot) and of $p(t)$ (right plot) for $a = 0.36$ is shown in Figure 9.11. The surge of traffic visible in the central part of the $N(t)$ plot is tamed by proper adjustment of $p(t)$ (see the deep valleys in the right plot, matching the peaks in the left plot). When the backlog is light, the back-off probability hovers around 1. Thus, the system experiences a small backlog most of the time, even if the mean load on the system is close to saturation (0.36 is about 98% of the maximum load bearable by the stabilized ALOHA).

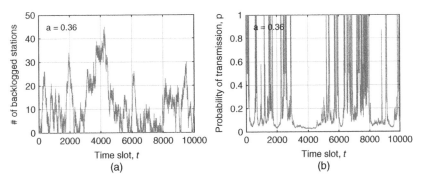

Figure 9.11 Sample paths of $N(t)$ (left plot) and of $p(t)$ (right plot) for the stabilized slotted ALOHA with $\gamma = 1$, $\alpha = \beta = e/2 - 1$ ($a = 0.36$).

We can derive a simple delay analysis for the stabilized Slotted ALOHA. The key remark is that the estimate $S(t)$ of the number of backlogged stations $N(t)$ is close to exact most of the time. We therefore assume that $S(t) = N(t)$ in the delay model. Then, the probability of a successful transmission is $P_s = \frac{n}{s}\left(1 - \frac{1}{s}\right)^{n-1} = \left(1 - \frac{1}{n}\right)^{n-1}$. For $n = 1$ it is $P_s = 1$. For $n > 1$, we approximate the success probability with the value for large n, i.e., $P_s = 1/e$. Further, we consider the Poisson traffic model, i.e., new arrivals occur according to a Poisson process with mean rate a per slot. The Poisson arrival model is consistent with the approximation of the success probability for large n.

Then, we can write

$$N(t + 1) = N(t) + A(t + 1) - U(t + 1) \tag{9.23}$$

where $A(t)$ is a Poisson random variable with mean a and $U(t)$ is the number of successfully transmitted packets per slot, so it takes only values 0 or 1. It is

$$P(U(t + 1) = 1|N(t) = n) = \begin{cases} 0 & n = 0 \\ 1 & n = 1 \\ 1/e & n \geq 2 \end{cases} \tag{9.24}$$

We derive $E[N]$ by using a mean delay analysis. Thanks to stabilization, steady-state exists even with the infinite population Poisson arrivals. Let $\pi_n = P(N(\infty) = n)$. Taking averages on both sides of eq. (9.23) and the limit for $t \to \infty$, we find $E[A] = E[U]$, i.e.,

$$a = \sum_{n=0}^{\infty} \pi_n E[U|N = n] = \sum_{n=0}^{\infty} \pi_n P(U = 1|N = n) = \pi_1 + \frac{1}{e}(1 - \pi_0 - \pi_1) \tag{9.25}$$

Squaring both sides of eq. (9.23), taking averages and then the limit for $t \to \infty$, we find:

$$\begin{aligned} 0 &= E[A^2] + E[U^2] - 2E[NU] + 2E[NA] - 2E[A]E[U] \\ &= a^2 + a + E[U] - 2E[NU] + 2aE[N] - 2a^2 \\ &= 2a - a^2 - 2E[NU] + 2aE[N] \end{aligned} \tag{9.26}$$

The joint moment of N and U can be calculated as follows:

$$\begin{aligned} E[NU] &= \sum_{n=0}^{\infty} \pi_n E[NU|N = n] = \sum_{n=1}^{\infty} n\pi_n P(U = 1|N = n) \\ &= \pi_1 + \frac{1}{e} \sum_{n=2}^{\infty} n\pi_n = \pi_1\left(1 - \frac{1}{e}\right) + \frac{1}{e}E[N] \end{aligned}$$

Figure 9.12 Mean delay of stabilized Slotted ALOHA.

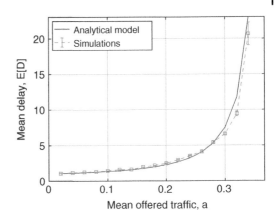

Inserting this result into eq. (9.26), we get

$$2\left(\frac{1}{e} - a\right) E[N] = a(2 - a) - 2\pi_1\left(1 - \frac{1}{e}\right) \tag{9.27}$$

To determine π_1, we need to calculate π_0. We have

$$\pi_0 = P(N + A - U = 0) = P(A = 0)P(N = U) = e^{-a}(\pi_0 + \pi_1) \tag{9.28}$$

Along with (9.25), we can solve the 2×2 linear equation system for π_1 and π_0 to find

$$\pi_1 = \frac{(e^a - 1)(1 - ae)}{1 - (e - 1)(e^a - 1)} \tag{9.29}$$

Inserting this value back into eq. (9.27), we get finally

$$E[N] = \frac{ea(1 - a/2)}{1 - ae} - \frac{(e - 1)(e^a - 1)}{1 - (e - 1)(e^a - 1)} \tag{9.30}$$

Using Little's law, we find the mean delay (expressed in units of slot time):

$$E[D] = \frac{E[N]}{a} = \frac{e(1 - a/2)}{1 - ae} - \frac{(e - 1)(e^a - 1)}{a[1 - (e - 1)(e^a - 1)]} \tag{9.31}$$

As expected, the results holds for $a < 1/e$.

Figure 9.12 plots the mean delay $E[D]$ obtained by using the analytical model (solid line), compared with simulations of the stabilization algorithm of eq. (9.22) (dashed line). Confidence intervals at 95% level are displayed as well.

Despite its simplicity, the analytical model agrees beautifully with simulations.

9.3 Pure ALOHA with Variable Packet Times

Pure ALOHA is the original proposal of Norman Abramson to set up the distributed access via a radio satellite network [4]. It is not much used, due to its limited throughput. It might however be the solution when a very limited traffic is envisaged and slot synchronization is unfeasible or costly. LoRaWAN is an example technology where pure ALOHA is revisited and applied to a sensor network to cover very large distances (in the order of several km) [152].

The typical model of pure ALOHA assumes fixed packet time. In this section we address ALOHA performance with variable size packets. This is interesting both because the analysis is subtle and because it reveals an unfairness issue with pure ALOHA, when packets have different sizes.

An early contribution on the analysis of ALOHA under generally variable packet transmission time appears in [26]. A generalization to multi-packet reception is presented in [19].

We consider a population of transmitting stations sharing an ALOHA wireless channel. New transmissions are offered to the network according to a Poisson process with mean rate v. Transmission attempts (including retransmissions) are modeled as a Poisson process of mean rate $\lambda \geq v$. This is a classic approximation for a large population of sporadically transmitting stations. In addition to v and λ, the following notation is used:

X the random variable representing the packet time of *new packets*. The mean, minimum and maximum packet times are denoted respectively with $E[X]$, X_{min} and X_{max}.

Y the random variable representing the packet time of packets transmitted on the wireless channel, *including retransmissions*. The mean of Y is denoted with $E[Y]$. Maximum and minimum values of Y are the same as for X.

$N(t)$ the number of parallel ongoing transmissions at time t (traffic process).

$A(t, t + \tau)$ the number of packet transmission attempts in the interval $[t, t + \tau]$ (arrival process).

If acknowledgments (ACK) are sent over the same channel as DATA packets, the channel holding time $Y = T_{DATA} + T_a + T_{ACK}$ equals the sum of three components, namely the DATA packet transmission time T_{DATA}, the turn-around time T_a and the ACK transmission time T_{ACK}. This model applies to cases where the same channel is used for DATA and ACK packets. If ACKs are sent on a separate collision-free channel, then $Y = T_{DATA}$. We encompass all cases in one model, by allowing channel holding time to be variable. Moreover, T_{DATA} can vary as well.

Note that the probability density function $f_Y(x)$ of packet time for packets transmitted on the channel is different from the native probability distribution of packet time, $f_X(x)$. In fact, longer packets incur collision with higher probability, hence they are re-transmitted more frequently than shorter packets. Hence, channel operation introduces a bias in the sizes of packets appearing on the channel.

Consider a packet arriving at time t and let x be the corresponding channel holding time. A necessary and sufficient condition for this packet to be successful is that no other transmission is going on at time t, and no new transmission starts during the time interval $(t, t + x)$.

Given the Poisson traffic model, at equilibrium the number of ongoing transmissions $N(t)$ has a Poisson probability distribution with mean $G = \lambda E[Y]$. The number of arrivals in the interval $(t, t + x)$ is a Poisson random variable with mean λx and it is independent of $N(t)$. Therefore the success probability $p_s(x)$ for a packet of size x can be expressed as

$$p_s(x) = P(N(t) = 0, A(t, t + x) = 0) = e^{-\lambda E[Y] - \lambda x} \tag{9.32}$$

The success probability $p_s(x)$ is a decreasing function of the packet time x. Hence, there is an *unfairness* issue among packets of different sizes: larger packets have smaller success probability than smaller ones.

The unconditional success probability p_s can be found as:

$$p_s = \int_{X_{\min}}^{X_{\max}} p_s(x) f_Y(x) dx = \int_{X_{\min}}^{X_{\max}} e^{-\lambda x - \lambda E[Y]} f_Y(x) dx \tag{9.33}$$

and then we derive the normalized throughput

$$S = \lambda p_s E[X] = \lambda E[X] e^{-\lambda E[Y]} \int_{X_{\min}}^{X_{\max}} e^{-\lambda x} f_Y(x) dx \tag{9.34}$$

For variable packet time, the equation linking the statistical characteristics of newly offered packets to those that are transmitted over the channel (including retransmissions) is obtained by equating the mean successful delivery rate to the mean arrival rate of packets of duration x. This balance must hold at equilibrium. Formally:

$$v f_X(x) = \lambda f_Y(x) p_s(x) = \lambda f_Y(x) e^{-\lambda E[Y] - \lambda x}. \tag{9.35}$$

From (9.35) it is easy to derive the expression of the PDF of Y:

$$f_Y(x) = \frac{f_X(x) e^{\lambda x}}{\int_{X_{\min}}^{X_{\max}} f_X(x) e^{\lambda x} dx} \tag{9.36}$$

The mean packet time $E[Y]$ is then calculated as:

$$E[Y] = \int_{X_{\min}}^{X_{\max}} x f_Y(x) \, dx = \frac{\int_{X_{\min}}^{X_{\max}} x f_X(x) e^{\lambda x} \, dx}{\int_{X_{\min}}^{X_{\max}} f_X(x) e^{\lambda x} dx} \tag{9.37}$$

Inserting the expression of the PDF of Y into the throughput equation (9.34), we find

$$S = \frac{\lambda E[X] e^{-\lambda E[Y]}}{\int_{X_{\min}}^{X_{\max}} f_X(x) e^{\lambda x} dx} \tag{9.38}$$

where E[Y] must be computed from eq. (9.37). If the packet time of new packets is fixed, it follows $E[Y] = E[X]$ and we recover the well-known result $S = \lambda E[X]e^{-2\lambda E[X]}$.

Example 9.1 Let us assume the packet time X be distributed as a gamma random variable, with mean E[X] and squared coefficient of variation (SCOV) C_X^2. It is

$$f_X(x) = \frac{\alpha^\beta x^{\beta-1}}{\Gamma(\beta)}e^{-\alpha x}, \quad x \geq 0, \tag{9.39}$$

where $\alpha = \beta/E[X]$ and $\beta = 1/C_X^2$. The extremes of X are $X_{min} = 0$ and $X_{max} = \infty$.

The PDF of Y can be expressed in closed form and it turns out to be still of gamma type:

$$f_Y(x) = \frac{(\alpha - \lambda)^\beta x^{\beta-1}}{\Gamma(\beta)}e^{-(\alpha-\lambda)x}, \quad x \geq 0, \tag{9.40}$$

holding for $\lambda < \alpha$. The mean packet time of packet on the channel is

$$E[Y] = \frac{\beta}{\alpha - \lambda} = \frac{E[X]}{1 - \lambda E[X]/\beta} \tag{9.41}$$

The mean E[Y] is a function of λ, i.e., of the load on the channel. It diverges to infinity as λ approaches its upper limit α. Finally, we can express the throughput in closed form as well:

$$S = \left(1 - \frac{a}{\beta}\right)^\beta a \, \exp\left(-\frac{a}{1 - a/\beta}\right) \tag{9.42}$$

where $a \equiv \lambda E[X]$. The throughput depends only on the two quantities a, namely the mean offered load, and β, that is directly related to the variability of the native packet time. Letting $\beta \to \infty$, which means making the SCOV of the packet size PDF negligible, we recover the well known expression of the throughput of pure ALOHA, namely $S|_{\beta \to \infty} = ae^{-2a}$.

It is possible to find explicitly the optimum load a^* and the maximum throughput S^*, taking the derivative of S with respect to a:

$$a^* = \beta\left(1 - \sqrt{\frac{\beta}{1+\beta}}\right) \tag{9.43}$$

$$S^* = \beta\left(1 - \sqrt{\frac{\beta}{1+\beta}}\right)\left(\frac{\beta}{1+\beta}\right)^{\beta/2} e^{\beta - \sqrt{\beta(1+\beta)}} \tag{9.44}$$

The left plot of Figure 9.13 shows the throughput S as a function of a for various values of the SCOV of packet time C_X^2, assuming $E[X] = 1$ (i.e., the mean packet time is the time unit). The qualitative behavior of the throughput curves

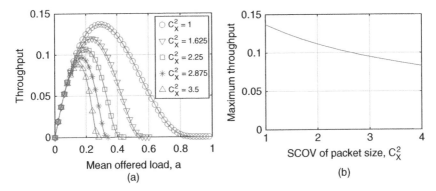

Figure 9.13 Pure ALOHA with gamma distributed offered packet size. Left plot: throughput as a function of the mean offered load for various values of the SCOV C_X^2 of packet size. Right plot: maximum throughput as a function of the SCOV C_X^2 of packet size.

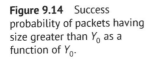

Figure 9.14 Success probability of packets having size greater than Y_0 as a function of Y_0.

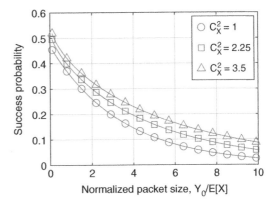

is the same as for Slotted ALOHA, except that the achieved throughput values are smaller and the feasible range of the offered load is limited to $\beta = 1/C_X^2$. As the SCOV of the packet time grows, the peak of the throughput curve narrows and the fall beyond the maximum becomes steeper. This is a sign that engineering the system to work around the optimal throughput level is more and more difficult.

The right plot shows the optimal throughput (9.43) as a function of C_X^2. The maximum achievable throughput of pure ALOHA is a decreasing function of the variability of service times.

To explore the unfairness issue, Figure 9.14 plots the probability of success of packets with size greater than a threshold Y_0 as a function of the normalized threshold $Y_0/E[X]$, for various values of the SCOV of offered packet time.

The success probability for packets longer than Y_0 is found as:

$$p_s(Y_0) = \int_{Y_0}^{\infty} p_s(x) \frac{f_Y(x)}{\int_{Y_0}^{\infty} f_Y(u)\, du}\, dx = p_s \frac{\Gamma(\beta Y_0/E[X], \beta)}{\Gamma((\beta - a)Y_0/E[X], \beta)} \qquad (9.45)$$

where $p_s = S/a$, and $\Gamma(y, \beta) = \int_y^{\infty} \frac{u^{\beta-1}}{\Gamma(\beta)} e^{-u}\, du$. The numerical values displayed in Figure 9.14 are obtained by setting $a = a^*(\beta) = \beta(1 - \sqrt{\beta/(1+\beta)})$.

It can be seen that the success probability decreases as Y_0 grows, i.e., bigger packets are penalized. The effect of the SCOV is marginal. More variable packet sizes entail a slightly better success probability for longer packets.

9.4 Carrier Sense Multiple Access (CSMA)

ALOHA is best suited for a population of terminals that cannot hear each other and communicate to a BS, from which they receive the synchronization signal for Slotted ALOHA.

In many cases, it is possible for a terminal to receive signals coming from other terminals. Then each terminal can assess whether the channel is busy and, if that is the case, abstain from jumping on the channel with its own transmission. This opportunity gives rise to the so called family of Carrier Sense Multiple Access (CSMA) protocols.

We devote this section to analysis and engineering of several variants of CSMA protocols, under different model assumptions. The general framework is based on a population of M stations (possibly infinitely many), sharing a communication channel. A station is said to be *idle* if it has no packets to send. As with ALOHA, an idle station listens anyway to the channel to receive packets addressed to it[7] . When the upper layer entity passes a new packet to the MAC entity of a station, that station is said to be *backlogged*. A backlogged station goes through the following steps to deliver the packet to the destination over the channel.

1. Sense the channel to assess whether it is idle or busy.
2. If the channel is found to be idle, transmit the packet, if it is a new one; if it is a rescheduled packet, adopt a persistence policy.
3. If a feedback is envisaged by the MAC protocol (e.g., an ACK of the successful reception of the transmitted packet), the station waits for the feedback; if the packet turns out to have been successfully delivered, the station is done and can go back to its idle state; otherwise, it reschedules the packet for a later time, selected at random.

---

7 Energy saving algorithms make idle stations doze and only wake up periodically, to check whether packets for them are announced by the access point.

4. If the channel is sensed busy, a persistence policy is adopted, and the packet is re-scheduled for a new attempt, according to a prescribed policy.

Typically, the number of transmission attempts is limited to some maximum number of retries, after which the packet is discarded. According to the high level description of CSMA given above, the time axis is divided into idle times, where no transmission takes place, and *activity* times, when the channel is busy for on going transmissions. The activity time encompasses the actual transmission time, plus all associated overhead, e.g., inter-frame spaces, time to transmit acknowl-edgements (ACKs), turn-around times. Moreover, a station alternates between an idle state, when it has no pending packets, and a backlogged state, when there is at least one packet waiting to be sent over the channel. Once the head-of-line (HOL) packet is transmitted successfully, the station picks another packet from its MAC buffer, if any is waiting; otherwise, it goes back to idle state.

9.4.1 Features of the CSMA Protocol

Key parts of the CSMA protocol are:

- The Clear Channel Assessment (CCA) test, i.e., the procedure to assess whether the channel is idle or busy, typically implemented at the physical layer.
- The persistence policy, to decide when to postpone the packet transmission after having sensed a busy channel.
- The retransmission policy, to schedule a new transmission attempt, after a failed transmission attempt.

9.4.1.1 Clear Channel Assessment

Time is fine-grained slotted in *mini-slots* or *back-off slots* or simply slots, if there is no ambiguity. The mini-slot is the time interval required by the station hard-ware to assess reliably the status of the channel. It is therefore at least equal to the propagation time (e.g., see [128]), but it is often much larger, especially for half-duplex transceivers, to accommodate the time needed to switch the commu-nication hardware from transmitting mode to receiving mode. Typical values in CSMA technology range in the order of microseconds up to tens of microseconds, e.g., the back-off slot lasts 9 μs in most WiFi implementations. Physical chan-nel sensing consists of measuring the energy received at the station during the mini-slot and comparing it with a threshold, set to the background noise level in the given communication channel plus a suitable margin. A typical threshold level in WiFi technology over 20 MHz channels is −90 dBm, i.e., 1 pW.

Besides the *physical* channel assessment, a *logical* channel assessment is possible, in case a preliminary handshake precedes the actual data transmission. The handshake consists of a request-to-send (RTS) control frame and a response clear-to-send (CTS) frame. The initiating station puts the duration θ of the scheduled data transmission (including any additional overhead) into its RTS; θ is the the time that the station is planning to hold the channel to complete its data transmission. The destination station echoes back the channel holding time in its CTS reply. Any other station receiving either RTS or CTS is informed on the forthcoming data transmission and can set a timer to the announced channel holding time. So long as the timer is still running, the channel is deemed to be busy, no matter what physical channel sensing says. This provision is useful whenever physical channel sensing is imperfect (hidden node).

For example, let us consider two stations, A and B, laying on opposite sides of an access point (AP), within the range of the AP, but out of range of each other. A and B cannot hear each other. If A is transmitting a data frame to the AP and at the same time B senses the channel, physical channel sensing at B will assess an idle channel. Hence B will start its own transmission and run over A's ongoing transmission, thus destroying reception of both frames at the AP (collision event). With the RTS/CTS mechanism in place, A would send an RTS, the AP would reply with a CTS, and the latter would be received by B, blocking B for the whole duration of A's data transmission. There remains obviously the possibility that RTS frames collide themselves (nothing is perfect!). This is anyway much more unlikely than with data frames, being an RTS usually much shorter than a data frame.

9.4.1.2 Persistence Policy

Transmission attempts of a station are of two types. The first attempt is dealt with according to the persistence policy. Subsequent retransmissions, if required and allowed, are dealt with according to the retransmission policy. Let us focus here on transmission attempts of the first kind.

After having sensed an idle back-off slot, a backlogged station starts transmitting its packet with probability p. With probability $1 - p$ it defers to the subsequent back-off slot, where the same persistence procedure is repeated. If the station senses a busy channel, it waits for the first idle mini-slot and then applies the persistence policy. Let δ denotes the back-off slot time. If there is a single backlogged station on the channel, the delay induced by the persistence procedure is $K\delta$, where K is a geometrically distributed random variable, with $P(K = k) = (1 - p)^{k-1}p$, $k \geq 1$. The persistence policy here defined is referred to as *slotted nonpersistent* (SNP) policy. The parameter p is called *back-off probability*.

Back-off times can be generated according to a probability distribution other than the geometric one. For example, the most common paradigm is to extract an uniformly distributed integer K in a range $[1, W]$, then to count down K idle back-off slots before starting the transmission. The count-down is frozen whenever the channel is busy. It is decremented only as an idle back-off slot goes by.

In the classic CSMA literature, persistence is defined in a different way. The idea is that the packet is transmitted immediately after its arrival, if the station senses an idle mini-slot. If instead the station senses a busy channel upon packet arrival, it defers according to one of the following three persistence policies:

- 0-persistent[8] : a station sensing a busy channel waits for the channel to become idle again; then it draws a random time Z (e.g., uniformly distributed between a minimum and a maximum level) and postpones a new sensing attempt by Z.
- 1-persistent: a station sensing a busy channel, waits for the channel to become idle again; as soon as an idle mini-slot is sensed, the station starts its transmission.
- P-persistent: a station sensing a busy channel, waits for the channel to become idle; then, it starts transmitting with probability P, while it schedules a new sensing after a random time Z with probability $1 - P$.

The 0-persistent and 1-persistent policies are corner cases of the P-persistence, obtained for $P = 0$ and $P = 1$, respectively.

In the following we consider the SNP policy, unless otherwise stated.

9.4.1.3 Retransmission Policy

Once a station (or possibly, more than one, in case of collision) starts transmitting a packet, a so called activity time starts.

If the transmission attempt fails, the packet is rescheduled for a later transmission attempt, provided that the maximum number of attempts (max-retry limit), has not been exceeded. The re-scheduled packet is transmitted, after having sensed an idle back-off slot, with probability p'. With probability $1 - p'$, the transmission is deferred and the entire procedure is repeated at the next idle back-off slot. In general, the probability p' is different from the probability p used in the persistence procedure (e.g., a more aggressive policy can be used for persistence, whereas retransmissions after a failure must be cautious to avoid more collisions). The probability p' can even change after each failed transmission attempt. A common update law for the retransmission back-off probability is to halve it after each failed transmission. This policy is often called binary exponential back-off (BEB). We will see BEB at work in WiFi CSMA/CA.

The pseudo-code of an example CSMA algorithm is given below. The function channel_state(δ) consists of measuring the channel for a back-off slot of duration δ, at the end of which the station compares the received energy to a threshold to check whether the channel is idle or busy. The execution of this function lasts therefore a time δ.

Many small details play an important role in the actual performance of CSMA networks. We will however consider the essential features common to all CSMA

---

8 Sometimes referred to as nonpersistent, a terminology that we reserve to SNP policy.

Algorithm Pseudo-code of CSMA (sender).

1: Wait for a new packet from upper layer
2: $k \leftarrow 0$
3: done $\leftarrow$ FALSE
4: **while** $(k \leq$ max_retry) **and** (**not** done) **do**
5:    CCA $\leftarrow$ channel_state(δ)
6:    **if** CCA $==$ IDLE **then**
7:      **if** rand $\leq p_k$ **then**
8:       $k \leftarrow k + 1$
9:       send(packet)
10:      set(ACK_timer)
11:      check(ACK)
12:      **if** ACK **then**
13:        done $\leftarrow$ TRUE
14:      **end if**
15:    **end if**
16:   **end if**
17: **end while**
18: GO TO step 1

Table 9.1 Notation used in the analysis of CSMA.

Symbol	Definition
v	mean arrival rate of new packets
λ	mean arrival rate of packets, including rescheduling and retransmissions
δ	back-off slot time
T	mean duration of the activity time
β	ratio of the back-off slot to the mean packet time, $\beta = \delta/T$
p	transmission probability in a mini-slot
M	number of stations on the channel
q	probability that an idle station becomes backlogged during a back-off slot
S	channel throughput, $S = vT$
G	channel offered load, $G = \lambda T$

protocols for a first understanding of the basic trade-offs. This allows us to state simpler models and gain insight, before delving into a specific application of CSMA, namely WiFi networks.

We now turn to modeling CSMA protocols and evaluate their performance. The main notation used in this section is listed in Table 9.1.

Assumptions underpinning the CSMA models stated in this section are as follows.

1. Carrier sense is ideal, i.e., all stations can detect the state of the channel (idle or busy) without errors.
2. Persistence follows the SNP policy defined above, with back-off probability p.
3. Retransmissions use always the same back-off probability p, equal to the probability adopted by the persistence policy.
4. An infinite number of retries is allowed. Thus, no frame is lost, it only gets delayed, until is it eventually delivered with success.
5. The transmission time of frames (including all overhead, ACK, inter-frame spacings) is a constant, T. We will refer to this also as the activity time of a station on the channel.
6. A station detects the outcome of its transmission attempt immediately after having completed the transmission, i.e., by the end of the activity time.
7. If more than a single station transmit on the channel, no receiver can detect the transmitted packets correctly. This is the *collision* event.
8. The only cause of reception failure is a collision event, i.e., transmissions of two or more packets overlapping in time, even partially. Physical channel errors are neglected.
9. There is no queueing at stations. Each station deals with one packet at a time. When it is done with that packet (i.e., the packet has been delivered with success), the station goes back to the idle state.

Those assumptions are common to all models of CSMA considered in this section. We will relax some of them occasionally. Specifically, we will develop models of CSMA that account for variable activity times and allow successful reception of simultaneous transmissions up to some degree, thus relaxing assumptions 5 and 7. When turning to the special CSMA of WiFi, we will relax hypotheses 3, 4 and 5.

9.4.2 Finite Population Model of CSMA

In a finite population model we consider a finite number M of stations, alternating between an idle state, where no packet is waiting to be transmitted, and a backlogged state, where the station has one packet to send.

The strength of such a model is the intrinsic stability (the number of traffic sources is finite), which makes steady-state analysis meaningful. Both throughput and mean delay can be evaluated. The down side is that only numerical solutions can be derived, thus limiting the insight to be gained from the model. In the next section, we will consider infinite population models, an abstraction that leads to simpler results.

We consider the sequence of time points corresponding to the end of idle back-off slots, when the back-off counter of stations is decremented (see Figure 9.15).

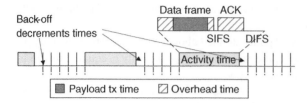

Back-off decrements times

Data frame ACK

SIFS   DIFS

Activity time

☐ Payload tx time   ☑ Overhead time

Figure 9.15 Illustration of the activity times and of the embedded time points of the CSMA Markov chain (decrement times).

We assume that the activity time is an integer multiple of the back-off slot duration. This is a minor approximation, since it is usually the case that $T \gg \delta$, otherwise the efficiency of the CSMA network is low. Let $J = T/\delta$. We denote with $N(t)$ the number of stations backlogged at time point t. Let $Q_{n,m}$ be the probability that the number of backlogged stations is m at the end of a back-off slot, given that n stations were backlogged at the beginning of the back-off slots. It is

$$Q_{n,m} = \begin{cases} \dbinom{M-n}{m-n} q^{m-n}(1-q)^{M-m} & m = n, \dots, M, n = 0, \dots, M, \\ 0 & \text{otherwise.} \end{cases} \tag{9.46}$$

The matrix whose entry (n, m) is $Q_{n,m}$ is denoted with $\mathbf{Q}$.

Under the hypotheses listed at the end of the previous section, $N(t)$ is a DTMC over the state space $\{0, 1, \dots, M\}$. The one-step transition probabilities are (for $n = 0, 1, \dots, M$):

$$P_{n,m} = \begin{cases} P_e(n)Q_{n,m} + P_s(n)Q_{n,m+1}^{(J+1)} + P_c(n)Q_{n,m}^{(J+1)} & m = n, \dots, M-1 \\ P_s(n)Q_{n,n}^{(J+1)} & m = n-1 \\ P_e(n)Q_{n,M} + P_c(n)Q_{n,M}^{(J+1)} & m = M \\ 0 & \text{otherwise.} \end{cases} \tag{9.47}$$

where $Q_{n,m}^{(J+1)}$ is the entry (n, m) of the matrix $\mathbf{Q}^{J+1}$ and for $n \geq 0$ we define:

$$\begin{cases} P_e(n) = (1-p)^n \\ P_s(n) = np(1-p)^{n-1} \\ P_c(n) = 1 - P_e(n) - P_s(n) \end{cases} \tag{9.48}$$

The DTMC $N(t)$ is aperiodic and irreducible, hence it is ergodic. Let π_n denote the limiting probability of the event $N(t) = n$. It is the solution of the linear equation system $\boldsymbol{\pi}\mathbf{P} = \boldsymbol{\pi}$ with the condition $\boldsymbol{\pi}\mathbf{e} = 1$, where $\mathbf{e}$ is a column vector of ones of size $M + 1$.

The probability distribution π_n refers to the embedded process $N(t)$. Since the time intervals between embedded epochs are not constant, the probability distribution of the number of backlogged stations *at any time* is obtained by weighting for the duration of the time intervals between successive embedded epochs. Let T_n be the duration of the time starting at the embedded time t conditional on $N(t) = n$. It is

$$T_n = \begin{cases} \delta & \text{w.p. } (1 - p)^n, \\ \delta + T & \text{otherwise.} \end{cases} \tag{9.49}$$

The mean duration of T_n is $\overline{T}_n = \delta + T - T(1 - p)^n$.

Let x_n denote the probability that $N = n$ at any time (not only embedded points). It is

$$x_n = \frac{\overline{T}_n \pi_n}{\sum_{j=0}^{M} \overline{T}_j \pi_j} \qquad n = 0, 1, \dots, M. \tag{9.50}$$

The throughput of CSMA is

$$S = T \cdot \frac{\sum_{n=0}^{M} P_s(n) \pi_n}{\sum_{n=0}^{M} \overline{T}_n \pi_n} = \frac{\sum_{n=0}^{M} np(1 - p)^{n-1} \pi_n}{\sum_{n=0}^{M} [\beta + 1 - (1 - p)^n] \pi_n} \tag{9.51}$$

where $\beta = \delta/T$. The mean delay (normalized with respect to T) is found by means of Little's law:

$$E[D] = \frac{E[N]}{S} = \frac{1}{S} \sum_{n=0}^{M} n x_n = \frac{\sum_{n=1}^{M} n[\beta + 1 - (1 - p)^n] \pi_n}{\sum_{n=1}^{M} np(1 - p)^{n-1} \pi_n} \tag{9.52}$$

We use the model to explore the performance of CSMA. All times are normalized with respect to the duration of the back-off slot. The mean delay is expressed in units of transmission times T.

Figure 9.16 shows the throughput (left plot) and the mean delay, normalized to T, as a function of M, for $p = 0.01$ and $q = 0.01$.

The throughput curve is qualitatively similar to the Slotted ALOHA throughput curve, except that it peaks to much higher values. The maximum throughput can

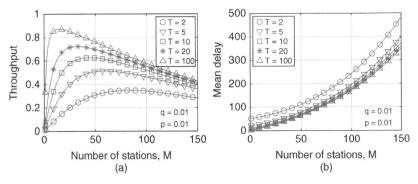

Figure 9.16 Throughput (left plot) and mean delay (right plot) as a function of M for the finite-population slotted-time model of CSMA with SNP policy.

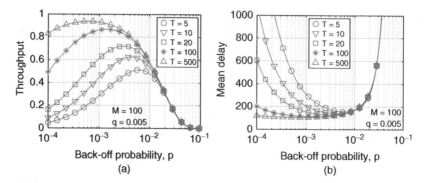

Figure 9.17 Throughput (left plot) and mean delay (right plot) as a function of p for the finite-population slotted-time model of CSMA with SNP policy.

attain values close to 1 when the duration of the transmission time amortizes the back-off slot overhead. On the contrary, for small values of T, the throughput is strongly penalized, even if the achievable throughput performance are much more stable with respect to variations of the number of stations.

The mean delay grows steadily as M increases, which is expected, since q is a constant. The normalized mean delay is weakly dependent on the value of T.

Throughput and mean delay as a function of p are shown in Figure 9.17, on the left and right plots, respectively. The number of stations and the activation probability are fixed at $M = 100$ and $q = 0.005$. Several values of T are considered, ranging between 5 and 500.

It appears that an optimal level of p exists, where throughput is maximized and mean delay is minimized. While the achievable minimum of the mean delay is scarsely sensitive to the value of T, the maximum achievable throughput is strongly affected by the impact of the back-off slot overhead.

The rise and fall of the throughput and the steep increase of the mean delay as p is varied are due to two opposite phenomena, whose balance determines the optimal working point:

- At low p levels, channel capacity is wasted because of too many idle back-off slots.
- At high p levels, stations are too aggressive and a large fraction of time is wasted in collisions.

The throughput delay trade-off as M is varied for $q = 0.05$, $T = 20$ and several values of p is shown in Figure 9.18.

The trade-off has the same appearance as for Slotted ALOHA. The most important finding that is highlighted by the throughput-delay trade-off is that a same level of throughput can be obtained with a low and a high mean delay, i.e., the

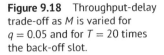

Figure 9.18 Throughput-delay trade-off as *M* is varied for $q = 0.05$ and for $T = 20$ times the back-off slot.

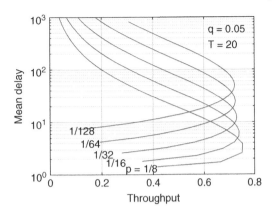

system could settle on a heavily congested or lightly loaded regime. We will see more on this point in the section devoted to the stability of CSMA. Here it suffices to say that, similarly to Slotted ALOHA, the finite population guarantees a finite mean delay. However, depending on the parameter values, the number of active stations can hover close to *M* or it can be small. The latter situation is obviously preferable. It corresponds to the lower branch of the throughput-delay trade-off.

9.4.3 Multi-Packet Reception CSMA

A key, simplifying approximation in the analysis of random access protocols, including CSMA, consists of assuming the traffic made up by backlogged stations behaves as a Poisson flow of arriving packets with a mean rate λ.

Poisson assumption. Transmission attempts, including new packets and already transmitted packets, form a Poisson process with mean rate λ.

The throughput of the channel is $S = \nu T$, where ν is the mean arrival rate of *new* packets. Due to retransmission of packets that have failed previous attempts the actual mean rate λ of *all* packets offered to the channel is greater than ν. We denote the normalized load of the channel as $G = \lambda T$. Under the Poisson model, the channel throughput can be expressed as a function of only two quantities, namely β and G.

The Poisson traffic assumption is only an approximation. More importantly, we will see that the CSMA protocol with Poisson input traffic has zero stable throughput, i.e., the number $N(t)$ of backlogged stations *diverges with probability 1* as $t \to \infty$, for *any positive value* of ν, given that $N(0) = 0$. This apparently odd result calls for some remarks. Some questions arise: why should we care for Poisson traffic models of CSMA? What do those models tell us? Do they provide useful approximations?

On an algorithmic ground, instability can be tamed by introducing adaptation of the back-off probability, i.e., by changing the parameter p according to the outcomes observed on the channel (idle mini-slot, successful transmission, collision).

From a theoretical modeling point of view, the throughput predicted by Poisson models of CSMA turns out to provide an *upper bound of the throughput level achievable by means of stabilized CSMA*. More in depth, let $S(G)$ be the throughput of the Poisson model when the mean number of packet arrivals in a back-off slot is $\alpha = \beta G$. Then, CSMA can be stabilized for any throughput $S < S(G)$ by adjusting the back-off probability according to $p(t) = \alpha/N(t)$. This fact has a nice correspondence with the fact that the equation $S = S(G)$ admits solutions whenever stabilization of CSMA is possible. The statement above is strictly true for nonpersistent CSMA, when newly arrived packets are immediately backlogged. Minor modifications should be made in other cases, but the essence of the statement still holds.

Under a performance evaluation perspective, the mean number of packet arrivals in a back-off slot with a finite population model is $(M - E[N])q/\beta + E[N]p/\beta$. If $p \approx q \ll 1$ and $M \gg 1$, it can be expected that the Poisson approximation yields accurate results. In other words, a very large population of stations, each transmitting sporadically can realize a Poisson stream of packet arrivals on the channel, if the activation probability is close to the back-off probability (i.e., the behavior of backlogged and idle stations is almost the same).

Let us now turn to the development of the throughput model of CSMA under the Poisson offered traffic assumption. To that end, we relax two of the hypotheses listed for CSMA modeling. Specifically, we assume here that:

- The packet size X has a general probability distribution, with CDF $F_X(x)$; for a discrete valued random variable, the packet size takes values in the set $\{X_1, \ldots, X_\ell\}$ with probabilities $q_X(j) = P(X = X_j)$, $j = 1, \ldots, \ell$.
- Multi-packet reception is possible: $k \geq 1$ simultaneous transmissions can be correctly decoded with probability g_k; in the following we assume the all-or-nothing model, in which $g_k = 1$ for $1 \leq k \leq K$ and $g_k = 0$ for $k > K$.

The multi-reception model deserves some attention. In an infra-structured wireless LAN, frames from stations are directed to the AP, while the AP distributes frames to all stations. If k stations transmit simultaneously, including the AP, no frame can be detected by the AP, since a wireless transceiver is normally half-duplex, i.e., transmission and reception are mutually exclusive. As another example, if the AP transmits a frame addressed to station A, while station A transmits in turn, even if two simultaneous transmission are assumed to be decodable, yet the AP and station A miss their respective reception, because of the half-duplex working of their hardware. The simple multi-reception model here defined corresponds to a case where: (i) wireless transceivers are *full-duplex*, i.e., a station can simultaneously transmit and receive; (ii) up to K superposed transmissions can be successfully decoded. Full-duplex is obtained by cancelling the transmitted signal, a quite complex operation that implies both analog and digital cancellation (e.g., see [31]). We are then assuming that multi-packet reception is possible even if the receiving station is simultaneously transmitting, i.e., we assume ideal cancellation.

We assume packet sizes are i.i.d. random variables and independent of the inter-arrival times. The minimum and maximum packet size values are denoted by X_{min} and X_{max}, respectively. As already done with ALOHA, the packet size includes the time required to transmit a data frame plus the time taken by the acknowledgment, inter-frame spacing and any other overhead.

The time axis is split into back-off slots of duration δ and activity times, where transmissions take place. We introduce the notion of "batch" to refer to a group of transmissions with aligned starting times. A batch ends when all concurrent transmissions end *and* an idle back-off slot is detected (the idle back-off slot is just the way that stations can detect the end of the batch).

We shall indicate by $T(j) = \delta + B(j)$ and $L(j)$, respectively, the *j*-th batch *duration* and the *j*-th batch *cardinality*, i.e., the number of packets transmitted during batch *j*. We will occasionally use the term *batch size* as synonymous for batch cardinality.

We develop two models, one for 1-persistence, the other one for nonpersistence.

9.4.3.1 Multi-Packet Reception 1-Persistent CSMA with Poisson Traffic

The considered CSMA protocol version corresponds essentially to the 1-persistent CSMA as introduced in the seminal paper [128]. Therein, only the special case of single-packet reception ($K = 1$) was considered. We extend the analysis to the more general case of multi-packet reception ($K > 1$).

The variable part $B(j)$ of the *j*-th batch, with size $L(j)$, can be expressed as:

$$B(j) = \begin{cases} \max_{i=1,\dots,L(j)} X_i & L(j) > 0 \\ 0 & L(j) = 0 \end{cases} \tag{9.53}$$

where X_i is the activity time associated with the *i*-th packet.

The probability distribution of (the variable part of) the batch duration and batch size can be derived from the following relationship:

$$\begin{cases} \mathcal{P}(L(j) = n \mid B(j-1) = x) = \dfrac{(\lambda x + \lambda\delta)^n}{n!} e^{-\lambda(x+\delta)} \\ \mathcal{P}(B(j) \leq x \mid L(j) = n) \quad = F_X^n(x) \end{cases} \tag{9.54}$$

The first relation holds because $L(j)$ conditional on $B(j-1)$ has a Poisson distribution. The second relation descends from the independence of activity times of different packets. The pair of relations (9.54) means that the two processes $L(j)$ and $B(j)$ are intertwined.

Let $F_B(x) \equiv P(B(j) \leq x)$, for $x \geq 0$, and $\phi_L(n) \equiv P(L(j) = n)$ for $n \geq 0$. We use also the following notation:

$$\Gamma(K, y) = e^{-y} \sum_{k=0}^{K-1} \frac{y^k}{k!} = \int_y^\infty \frac{u^{K-1}}{(K-1)!} e^{-u} \, du \, , \qquad K \geq 1, \, y \geq 0. \qquad (9.55)$$

where K is an integer. From the conditional probabilities (9.54) we derive the following unconditional probabilities:

$$F_B(x) = \sum_n P(B(j) \leq x \mid L(j) = n) \cdot P(L(j) = n) = \sum_{n \geq 0} F_X^n(x) \cdot \phi_L(n) \qquad (9.56)$$

and

$$\phi_L(n) = \int_x P(L(j) = n \mid B(j-1) = x) \, dP(B(j-1) \leq x)$$

$$= \int_0^\infty \frac{(\lambda x + \lambda \delta)^n}{n!} e^{-\lambda(x+\delta)} \, dF_B(x) \qquad (9.57)$$

The pair of functional integral equations (9.56) and (9.57) can be solved for $F_B(x)$ and $\phi_L(n)$, given $F_X(x)$ and λ. Substituting (9.57) into (9.56) we obtain

$$F_B(x) = \sum_{n=0}^\infty F_X^n(x) \cdot \phi_L(n)$$

$$= \int_0^\infty \sum_{n=0}^\infty F_X^n(x) \frac{[\lambda(t+\delta)]^n}{n!} e^{-\lambda(t+\delta)} \, dF_B(t)$$

$$= \int_{X_{min}}^{X_{max}} e^{-\lambda(t+\delta)[1-F_X(x)]} \, dF_B(t) \qquad (9.58)$$

where we have accounted for the bounded range of the packet size X. The CDF $F_B(x)$ can be evaluated numerically, discretizing the integral. This reduces eq. (9.58) to a linear equation system, yielding an approximation of $F_B(x)$.

A batch is successful if the number of transmitted packets is comprised between 1 and K. The success probability P_s, conditional on a station transmitting in that batch, is given by

$$P_s = P(0 \leq L(j) \leq K-1) = \int_{X_{min}}^{X_{max}} \Gamma(K, \lambda y + \lambda \delta) \, dF_B(y) \qquad (9.59)$$

The conditional success probability is independent of the size of the arriving packet. Therefore, the size distribution of transmitted packets (including retransmissions) is the same as that of new packets. This is different from pure ALOHA. The key point is that transmissions start simultaneously in a CSMA batch.

The throughput can be expressed as the ratio between the mean number of successful packets delivered in one batch and the mean duration of the batch:

$$v = \frac{\sum_{k=1}^{K} k\, P(L=k)}{\delta + E[B]} = \lambda \int_{X_{\min}}^{X_{\max}} \frac{\delta + x}{\delta + E[B]} \Gamma(K, \lambda\delta + \lambda x)\, dF_B(x) \qquad (9.60)$$

We define the normalized throughput as $S = vE[X]$. Given K and $F_X(x)$, for each value of λ we can calculate v and $F_B(x)$ using (9.60) and (9.58), respectively.

We can rewrite the final result for a discrete packet size probability distribution. Let the packet size take ℓ values, denoted with $X_j, j = 1, \dots, \ell$, and sorted in ascending order. The random variable B takes values in the same set as X. For notation convenience we define also $X_0 = 0$. Let further $\phi_X(j) \equiv P(X = X_j)$ and $\phi_B(j) \equiv P(B = X_j)$. From

$$P(B \le X_k \mid L = n) = [F_X(X_k)]^n = \left(\sum_{j=1}^{k} \phi_X(j) \right)^n \qquad (9.61)$$

we derive

$$F_B(X_k) = P(B \le X_k) = \sum_{n=0}^{\infty} \left(\sum_{j=1}^{k} \phi_X(j) \right)^n \phi_L(n)$$

$$= \sum_{n=0}^{\infty} \left(\sum_{j=1}^{k} \phi_X(j) \right)^n \int_0^{\infty} \frac{[\lambda(t+\delta)]^n}{n!} e^{-\lambda(t+\delta)}\, dF_B(t)$$

$$= \int_0^{\infty} e^{-\lambda(t+\delta)\left(1 - \sum_{j=1}^{k} \phi_X(j)\right)}\, dF_B(t)$$

Rearranging the sums and accounting for the discrete nature of the probability distribution of X and B, we obtain finally

$$F_B(X_k) = \sum_{j=0}^{k} \phi_B(j) = \sum_{h=0}^{\ell} e^{-\lambda(X_h+\delta)\sum_{j=k+1}^{\ell} \phi_X(j)} \phi_B(h) \qquad (9.62)$$

for $k = 0, 1, \dots, \ell - 1$. Taking differences, we find finally

$$\begin{cases} \phi_B(0) = \sum_{h=0}^{\ell} \phi_B(h) b(h, 1) \\[2mm] \phi_B(k) = \sum_{h=0}^{\ell} \phi_B(h)[b(h, k+1) - b(h, k)], \quad 1 \le k \le \ell - 1 \\[2mm] \phi_B(\ell) = \sum_{h=0}^{\ell} \phi_B(h)[1 - b(h, \ell)] \end{cases} \qquad (9.63)$$

where for $h = 0, 1, \ldots, \ell, k = 1, \ldots, \ell$ we have defined

$$b(h, k) \equiv \exp\left(-\lambda(X_h + \delta)\sum_{j=k}^{\ell}\phi_X(j)\right). \tag{9.64}$$

The probability vector $\mathbf{q}_B = [\phi_B(0) \ \cdots \ \phi_B(\ell)]$ is the nontrivial solution of $\mathbf{q}_B = \mathbf{q}_B\mathbf{P}_B$, normalized so that $\mathbf{q}_B\mathbf{e} = 1$, where $\mathbf{e}$ is a column vector of 1's and $\mathbf{P}_B$ is an $(\ell + 1) \times (\ell + 1)$ matrix, whose entries are for $j = 0, \ldots, \ell$:

$$\begin{cases} p_{j0} = b(j, 1) = e^{-(\lambda\delta + \lambda X_j)} \\ p_{ji} = b(j, i+1) - b(j, i) & i = 1, \ldots, \ell - 1; \\ p_{j\ell} = 1 - b(j, \ell) = 1 - e^{-(\lambda\delta + \lambda X_j)\phi_X(\ell)} . \end{cases} \tag{9.65}$$

The normalized throughput in case of discrete-valued probability distribution of X can be written as

$$S = \lambda E[X]\sum_{k=0}^{\ell}\frac{\delta + X_k}{E[B] + \delta}\Gamma(K, \lambda X_k + \lambda\delta)\phi_B(k) \tag{9.66}$$

Special case: limit for $\delta \to 0$. In the limit for $\delta \to 0$, the linear equation system (9.63) must be modified. For $\delta > 0$ we find

$$\phi_B(0) = \frac{e^{-\lambda\delta}}{1 - e^{-\lambda\delta}}\sum_{h=1}^{\ell}e^{-\lambda X_h}\phi_B(h) \tag{9.67}$$

Plugging this expression into the remaining ℓ equations and letting $\delta \to 0$, we find the new linear equation system

$$\tilde{\phi}_B(k) = \sum_{h=1}^{\ell}(e^{-\lambda X_h\sigma_{k+1}} - e^{-\lambda X_h\sigma_k} + \phi_X(k)e^{-\lambda X_h})\tilde{\phi}_B(h) , \quad k = 1, \ldots, \ell,$$

where $\sigma_k = \sum_{j=k}^{\ell}\phi_X(j)$ for $k = 1, \ldots, \ell$ and $\sigma_{\ell+1} = 0$. The role of the matrix $\mathbf{P}_B$ is here taken by the $\ell \times \ell$ matrix $\tilde{\mathbf{P}}_B$ whose entries are

$$\tilde{p}_{hk} = e^{-\lambda X_h\sigma_{k+1}} - e^{-\lambda X_h\sigma_k} + \phi_X(k)e^{-\lambda X_h} , \qquad h, k = 1, \ldots, \ell. \tag{9.68}$$

In the limiting case for $\delta \to 0$, the normalized throughput simplifies to

$$S|_{\delta \to 0} = \lambda E[X]\frac{\sum_{k=1}^{\ell}\tilde{\phi}_B(k)[e^{-\lambda X_k} + \lambda X_k\Gamma(K, \lambda X_k)]}{\sum_{k=1}^{\ell}\tilde{\phi}_B(k)(e^{-\lambda X_k} + \lambda X_k)} \tag{9.69}$$

Special case: Fixed packet size. In the special case of fixed packet size, equal to T, we have $\ell = 1, X_1 = T$ and $\phi_X(1) = 1$.

The probability distribution of B reduces to two values, $\phi_B(0) = P(B = 0)$ and $\phi_B(1) = P(B = T)$. The linear equation system holding for discrete-valued probability distributions reduces to a 2×2 system, that yields an explicit solution:

$$\phi_B(0) = \frac{e^{-\lambda(\delta+T)}}{1 - e^{-\lambda\delta} + e^{-\lambda(\delta+T)}} \tag{9.70}$$

$$\phi_B(1) = 1 - \phi_B(0) = \frac{1 - e^{-\lambda\delta}}{1 - e^{-\lambda\delta} + e^{-\lambda(\delta+T)}} \tag{9.71}$$

Moreover, we have $E[B] = \phi_B(1)T$ and

$$\phi_L(n) = \phi_B(0)\frac{(\lambda\delta)^n}{n!}e^{-\lambda\delta} + \phi_B(1)\frac{(\lambda\delta+\lambda T)^n}{n!}e^{-\lambda(\delta+T)}, \quad n \geq 0. \tag{9.72}$$

The normalized throughput specializes to

$$
\begin{aligned}
S &= T\frac{\sum_{k=1}^{K} k\, \mathcal{P}(L=k)}{\delta + E[B]} = T\frac{\sum_{k=1}^{K} k\phi_L(k)}{\delta + \phi_B(1)T} \\
&= \frac{\phi_B(0)\,\lambda\delta\,\Gamma(K,\lambda\delta) + \phi_B(1)\,(\lambda\delta+\lambda T)\,\Gamma(K,\lambda\delta+\lambda T)}{\delta/T + \phi_B(1)}
\end{aligned} \tag{9.73}
$$

To compare with the result presented in [128] for fixed packet size and $K = 1$, we set $K = 1$ in the throughput expression, to find out

$$
\begin{aligned}
S|_{K=1} &= \frac{\lambda T\, e^{-(\lambda\delta+\lambda T)}(1 + \delta/T - e^{-\lambda\delta})}{(1 + \delta/T)(1 - e^{-\lambda\delta}) + e^{-(\lambda\delta+\lambda T)}\delta/T} \\
&= \frac{Ge^{-(1+\beta)G}(1 + \beta - e^{-\beta G})}{\beta e^{-(1+\beta)G} + (1 + \beta)(1 - e^{-\beta G})}
\end{aligned}
$$

where we have used the notation $G = \lambda T$ and $\beta = \delta/T$. This is just the result given in eq. (19) of [128].

9.4.3.2 Multi-Packet Reception Nonpersistent CSMA with Poisson Traffic

The probability of n simultaneous transmission attempts is $(\lambda\delta)^n/n!e^{-\lambda\delta}$, since only arrivals in an idle back-off slot of duration δ can possibly transmit, according to the nonpersistence policy. Therefore, the CDF of B can be found explicitly:

$$F_B(x) = \sum_{n=0}^{\infty} \frac{(\lambda\delta)^n}{n!}e^{-\lambda\delta}F_X^n(x) = e^{-\lambda\delta(1-F_X(x))} \tag{9.74}$$

The conditional probability of success is therefore $\Gamma(K, \lambda\delta)$, independently of the frame length. The throughput can be found as the ratio of the mean number of successfully received packets to the mean time it takes to perform the transmission

$$v = \frac{\sum_{k=1}^{K} k\frac{(\lambda\delta)^k}{k!}e^{-\lambda\delta}}{\delta + E[B]} = \frac{\lambda\delta\,\Gamma(K, \lambda\delta)}{\delta + E[B]} \tag{9.75}$$

The normalized throughput is $S = vE[X]$.

For fixed packet size, we have $\ell = 1$ and $X_1 = T$. It is $B = T$ with probability $1 - e^{-\lambda\delta}$ and $B = 0$ with probability $e^{-\lambda\delta}$. The mean batch time is therefore $E[B] = T(1 - e^{-\lambda\delta})$. Letting $G = \lambda T$ and $\beta = \delta/T$, the normalized throughput is given by

$$S = \frac{\beta G\,\Gamma(K, \beta G)}{\beta + 1 - e^{-\beta G}} = \frac{\beta G\, e^{-\beta G}}{\beta + 1 - e^{-\beta G}}\sum_{k=0}^{K-1}\frac{(\beta G)^k}{k!} \tag{9.76}$$

which is a nice generalization of the known expression for $K = 1$ (see eq. (8) of [128]). Looking at S as a function of the variable $z = \beta G$, it can be verifed that it is maximized for z solving the following identity

$$1 - \frac{ze^{-z}}{1 + \beta - e^{-z}} = zB(K - 1, z) \tag{9.77}$$

where $B(\cdot, \cdot)$ is the Erlang B-formula.

Impact of back-off slot overhead for a fixed packet size and $K = 1$. We rewrite here below the formulas for the throughput of nonpersistent and 1-persistent CSMA with Poisson traffic, fixed packet size and $K = 1$:

$$S = \frac{\beta Ge^{-\beta G}}{\beta + 1 - e^{-\beta G}} \quad \text{nonpersistent} \tag{9.78}$$

$$S = \frac{Ge^{-(1+\beta)G}(1 + \beta - e^{-\beta G})}{\beta e^{-(1+\beta)G} + (1 + \beta)(1 - e^{-\beta G})} \quad \text{1-persistent} \tag{9.79}$$

Both nonpersistent and 1-persistent CSMA protocols experience a fast worsening of performance as β grows, definitely vanishing at an exponential rate with β for a fixed level of the mean offered load G.

The impact of the back-off slot duration and in general of contention degrades significantly the efficiency of CSMA, as the activity time shrinks due to increasing transmission speed. The evolution of WiFi is a striking example of how an increasing fraction of capacity is wasted as the transmission rate is boosted by exceptional technology improvements at physical layer, while the MAC protocol lags behind tied to increasingly inadequate concepts. In spite of some new feature, like data unit aggregation and flexible transmission opportunity time, the fraction of wasted capacity in IEEE 802.11n and IEEE 802.11ac is impressive. The deep reconsideration of the MAC design taken over by IEEE 802.11ax has its roots in the aim of regaining a high efficiency, rather than pushing even further the physical layer bit rate.

For small back-off slots as compared to activity times, namely when $\beta \to 0$, we obtain very simple expressions for the throughput of nonpersistent and 1-persistent CSMA protocols:

$$S|_{\beta \to 0} = \frac{G}{G + 1} \quad \text{nonpersistent} \tag{9.80}$$

$$S|_{\beta \to 0} = \frac{Ge^{-G}(G + 1)}{G + e^{-G}} \quad \text{1-persistent} \tag{9.81}$$

It is apparent that nonpersistent CSMA promises to achieve a better throughput, obviously at the price of longer delays, due to the back-offs. On the contrary, 1-persistent CSMA is by far more aggressive, thus achieving smaller delays in the case of light loads, but incurring possibly in heavier congestion and eventually in a throughput collapse outcome.

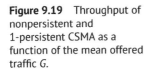

Figure 9.19 Throughput of nonpersistent and 1-persistent CSMA as a function of the mean offered traffic G.

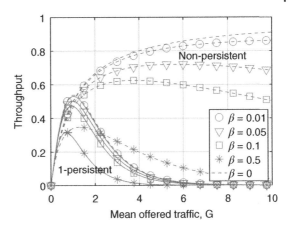

Figure 9.19 shows the throughput of nonpersistent (dashed lines) and 1-persistent (solid lines) CSMA as a function of the mean offered traffic G.

For small levels of offered traffic the more aggressive 1-persistent yields a higher throughput than nonpersistent CSMA. As the offered load grows however, 1-persistent falls down, while the nonpersistent CSMA has much higher throughput, the more the smaller the parameter β.

The engineering of CSMA networks is therefore based on an estimate of typical usage of those networks. If the network is expected to be loaded by few active nodes at any time, being more aggressive pays off. On the other hand, if large networks are to be set up, where a large number of nodes contend for the medium, even if each single node may transmit infrequently, keeping a brake on nodes aggressiveness is shown to be a good idea by the results depicted in Figure 9.19. The picture is even sharper if the cost of collision can be relieved, e.g., because a reliable collision detection mechanism can be put in place. This is just the case of cabled ethernet CSMA used in local area networks. Low offered traffic intensity and collision detection are both features of that technology, which explains why a 1-persistent CSMA has been chosen.

Figure 9.20 compares the throughput predicted by the finite population model of CSMA (solid lines) with that of the Poisson traffic model (star markers), as a function of M, for $T = 1/\beta = 5, 20, 100$ (in units of back-off time).

Since it is $p = q = 0.01$, the accuracy of the Poisson model is high, except at low M values and for high value of β (small T).

Example 9.2 Let us consider the nonpersistent, multi-packet reception CSMA model. Assume the packet time X has a uniform probability distribution between $X_{\min} = 0.25$ and $X_{\max} = 1.75$ (all times are normalized with respect to the average transmission time $E[X] = T$, that is taken as the time unit).

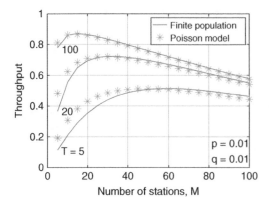

Figure 9.20 Comparison of CSMA throughput with finite population and with Poisson model, in case of single reception, $K = 1$. The throughput is plotted as a function of M for three values of $T = 1/\beta$ and for $p = q = 0.01$.

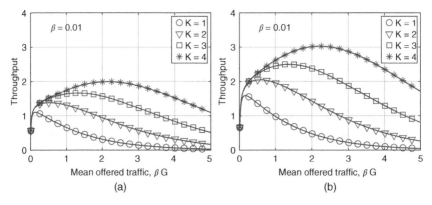

Figure 9.21 Throughput for nonpersistent, multi-packet reception CSMA with Poisson traffic for uniform (left plot) and bimodal (right plot) probability distribution of packet time.

The average batch time is found from the CDF in eq. (9.74):

$$E[B] = \int_{X_{min}}^{X_{max}} [1 - F_B(x)] \, dx = \int_{X_{min}}^{X_{max}} [1 - e^{-\lambda\delta(1 - F_X(x))}] \, dx =$$

$$= (X_{max} - X_{min}) \int_0^1 [1 - e^{-\lambda\delta(1 - u)}] \, du = (X_{max} - X_{min}) \left(1 - \frac{1 - e^{-\lambda\delta}}{\lambda\delta}\right)$$

$$(9.82)$$

The throughput as a function of $\beta G = \lambda\delta$ for $\beta = 0.01$ and several values of K is plotted in Figure 9.21(a). The maximum throughput grows with K, as expected.

There are two reasons the maximum throughput with $K = 4$ is about 2: (i) collisions reduce the achievable throughput; (ii) successful outcomes do not necessarily comprise always 4 packets, rather they can be made up of 1, 2, 3, or 4 packets.

Even better results can be achieved if the packet times have a bimodal probability distribution with two values X_1 and X_2. To maximize the variance of the packet time distribution, we assume the two values are equiprobable, i.e., $w \equiv \mathcal{P}(X = X_1) = 1/2$. Given the desired mean values $E[X] = 1$ and denoting the SCOV of X as c^2, we find $X_1 = 1/2 - c$ and $X_2 = 1/2 + c$. Setting $c^2 = 0.1$, we have $X_1 = 0.1838$ and $X_2 = 0.8167$. The mean value of the batch size is:

$$E[B] = (e^{-\lambda\delta(1-w)} - e^{-\lambda\delta})X_1 + (1 - e^{-\lambda\delta(1-w)})X_2 \tag{9.83}$$

Figure 9.21(b) shows the throughput of the multi-packet reception nonpersistent CSMA with Poisson traffic, as a function of the mean offered traffic G for several values of K and $\beta = 0.01$.

It is apparent that quite higher maximum throughput levels can be achieved with respect to the case of uniformly distributed packet time.

This is a consequence of the higher regularity of the chosen bimodal probability distribution. Specifically, the SCOV of the bimodal distribution is 0.1, while it is 0.1875 for the uniform distribution of the first part of the example.

Apart from this quantitative difference, the qualitative behavior of the throughput curves are similar for the two cases. This is no surprise in view of the fact that the expression of the throughput of the multi-packet reception nonpersistent CSMA highlights that it depends on the packet time probability distribution only through the mean value $E[B]$.

9.4.4 Stability of CSMA

Let us consider CSMA with SNP policy where *new* packet arrivals occur according to a Poisson process of mean rate v and p is the back-off probability. We refer to the usual embedded times corresponding to the end of an idle back-off slot. Note that there is an idle back-off slot after each transmission. There could be more than one, if backlogged stations back off and no new packet arrives.

Let $N(t) = n$ be the number of stations backlogged at the t-th embedded time. We highlight that a same value of p is used in all cases, both for the persistence policy and for retransmissions. The mean number of arrivals during a virtual slot (i.e., the time between two consecutive embedded epochs) is

$$\overline{A}_n = v[\delta(1-p)^n + (\delta + T)(1 - (1-p)^n)] = S[\beta + 1 - (1-p)^n] \tag{9.84}$$

The mean number of packets that are successfully transmitted in the same slot is $\overline{U}_n = np(1-p)^{n-1}$. The resulting drift in state $N(t) = n$ is

$$d_n = \overline{A}_n - \overline{U}_n = S[\beta + 1 - (1-p)^n] - np(1-p)^{n-1} \tag{9.85}$$

The key remark is that, *for fixed* p, d_n becomes definitely positive, for any offered traffic rate $v > 0$. In other words, CSMA with Poisson offered traffic achieves *zero stable throughput*. This is the same phenomenon that we have encountered with Slotted ALOHA. In the Appendix at the end of the chapter it is proved that a

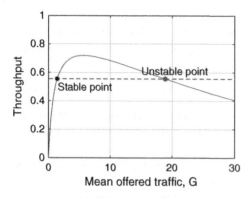

Figure 9.22 Example of load-line with Poisson input traffic: CSMA throughput (solid line), load-line at a level of 75% of the maximum throughput (dashed line). The locally stable and unstable equilibrium points are marked with a dark-shaded and light-shaded dot respectively.

Markov chain of the type here considered cannot be positive recurrent, if the drift d_n tends to a positive limit as $n \to \infty$.

We can visualize the situation by plotting the throughput function of eq. (9.78) as a function of G, along with a load-line corresponding to a given level of throughput S. For example, Figure 9.22 shows the horizontal load-line corresponding to an offered traffic intensity equal to 75% of the maximum throughput achievable by non-persistent CSMA with $\beta = 0.05$.

It is apparent that, provided $S = \nu T < \max_{G \geq 0} S(G)$, the load-line intersects the throughput curve in exactly two points. The left one corresponds to a locally stable equilibrium, while the right one is unstable. The two intersections of the horizontal load-line with the throughput curve are marked as a dark-shaded dot (on the left), corresponding to a locally stable equilibrium, and with a light-shaded dot (on the right), marking the unstable equilibrium point.

Given the statistical fluctuations of the number of backlogged stations, sooner or later $N(t)$ will slide to the right of the unstable point and will then shoot to infinity, for any initial value $N(0)$.

With a finite population of size M (a more realistic model), the system backlog cannot escape to infinity. Even if that dramatic outcome is forbidden and existence of a steady-state is mathematically guaranteed, still from an engineering perspective we can identify the three operation regions already highlighted for Slotted ALOHA:

- *Stable region.* The load-line has a single intersection with the throughput curve, a single stable equilibrium exists and it corresponds to a backlog substantially less than M.
- *Bi-stable region.* The load-line has three intersections with the throughput curve, two of which are locally stable.
- *Saturated region.* The load-line has again a single intersection with the throughput curve, corresponding to a level of backlog close to M.

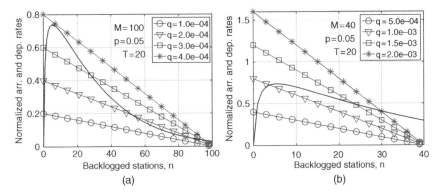

Figure 9.23 Load-lines of CSMA networks. The straight marked lines represent the mean arrival rate of packets, $\hat{\lambda}_n$. The solid line represents the success rate (or departure rate) of packets, $\hat{\mu}_n$. They are plotted as a function of the system backlog $N = n$.

A graphical representation of this situation is shown in Figure 9.23. The plots are obtained by comparing the arrival rate for a given backlog n, namely $\lambda_n = (M - n)q/\delta$ and the mean serving rate, i.e., $\mu_n = np(1 - p)^{n-1}/(\delta + T - T(1 - p)^n)$. Normalizing with respect to T, we obtain

$$\hat{\lambda}_n = \lambda_n T = \frac{(M - n)q}{\beta} \qquad \hat{\mu}_n = \mu_n T = \frac{np(1 - p)^{n-1}}{\beta + 1 - (1 - p)^n} \tag{9.86}$$

In the limit for $M \to \infty$ and $q \to 0$, with $Mq \to \nu\beta$ we find the mean arrival rate and the mean departure rate of the CSMA model with Poisson offered traffic.

The line $\hat{\lambda}_n$ and the curve $\hat{\mu}_n$ are plotted against n in Figure 9.23 for $p = 0.05$, $T = 20$, $M = 100$ (left plot) and $M = 40$ (right plot)

Depending on the shape of the success rate curve $\hat{\mu}_n$, the load-lines $\hat{\lambda}_n$ intersect the curve in one or three points (left plot) or only in one point (right plot). Therefore, there may or may not exist a bi-modal region. A close look at Figure 9.23 and inspection of eq. (9.86) reveals that the intersection between the load-line $\hat{\lambda}_n$ and the success rate curve $\hat{\mu}_n$ shifts toward high backlog levels as M and or q are increased. In other words, the equilibrium of the system moves toward saturation (i.e., most stations are active, contending for the channel, at any given time) as stations become more loaded (higher q) and their overall number grows. As the equilibrium backlog shifts toward M, the CSMA networks tends to settle on a working regime where the high pressure of contending stations induces more and more collisions, thus reducing the efficiency of channel usage.

The transition from lightly loaded systems, where the equilibrium corresponds to few active stations, to heavily loaded system (saturation) may go through an intermediate regime, where bi-stability arises (three intercepts of the load-line

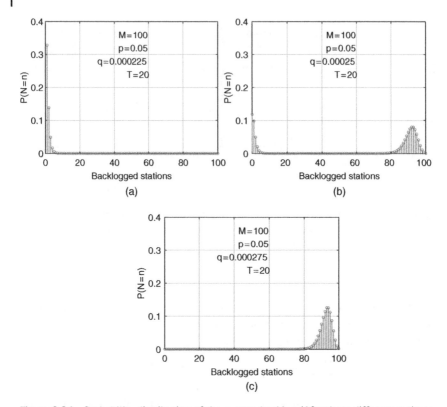

Figure 9.24 Probability distribution of the system backlog *N* for three different regimes with *M* = 100 stations: lightly loaded system (top left plot); bi-stable regime (top right plot); highly loaded system (bottom plot).

with the success rate curve, two of which are locally stable equilibrium points). In that special regime the network backlog oscillates between light and heavy load. The CSMA network state *N* hovers around each of the two locally stable equilibrium points for a very long time (much greater than the time required to send a single packet; see [129]). The characteristic mark of this behavior is a bimodal probability distribution of *N*.

For example, Figure 9.24 and 9.25 show the probability distribution of the backlog *N* for *M* = 100 and *M* = 40 stations, respectively. Three regimes are presented in each figure.

In Figure 9.24 we can see the probability distribution of a lightly loaded network on the top left plot (the mass of the distribution is concentrated on small *n* values), a clear bi-modal behavior of the distribution in the top right plot and the probability distribution of a highly loaded network in the bottom plot. These probability distributions correspond to three of the load-lines of in Figure 9.23(a).

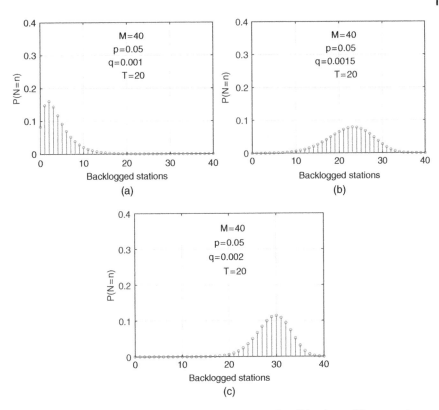

Figure 9.25 Probability distribution of the system backlog N for three different regimes with $M = 40$ stations: lightly loaded system (top left plot); intermediate load regime (top right plot); highly loaded system (bottom plot). No bi-modal behavior appears (compare with the load-line plot in Figure 9.23).

A different behavior emerges in Figure 9.25, where the probability distributions of a CSMA network with $M = 40$ are plotted for three different regimes. In this case there is no bi-modal behavior. As the load generated by the stations grows, the mass of the probability distribution shifts from low to intermediate then to high backlog levels. This corresponds to the load-lines shown in Figure 9.23(b), where only a single intersection of the load-line $\hat{\lambda}_n$ and the success rate curve $\hat{\mu}_n$ is possible.

The Poisson traffic model represents the asymptotic case where the population of stations is very large, each contributing negligibly to the overall generated traffic, i.e., the limit for $M \to \infty$ and $q \to 0$, such that $Mq \to \nu\beta$.

As clear from the load-line analysis, under Poisson offered traffic, the system backlog explodes and there does not exist any steady-state regime. The CSMA network, once started with zero backlog, sooner or later (possibly after a *very* long time: see the analysis in [129]) will escape to a definitely jammed state, where essentially no useful throughput is possible any more.

The cure for instability consists of making p adaptive to the number of backlogged stations. Assume the stations know exactly what is the current value of $N(t)$ at each embedded point t. We set then the back-off probability to $p(n) = \alpha/n$ when the number of backlogged stations is $N(t) = n \geq 1$, for a given $\alpha \in (0, 1)$, to be determined.

With this ideal adaptation of the back-off probability, the drift becomes:

$$d_n = S\left[\beta + 1 - \left(1 - \frac{\alpha}{n}\right)^n\right] - \alpha\left(1 - \frac{\alpha}{n}\right)^{n-1} \tag{9.87}$$

For large n, we have

$$d_\infty = \lim_{n\to\infty} d_n = S(\beta + 1 - e^{-\alpha}) - \alpha e^{-\alpha} \tag{9.88}$$

The limiting drift d_∞ can be made negative, to guarantee stability (see the Appendix at the end of the chapter), provided that we choose the offered traffic rate S that satisfies the following inequality:

$$S < \frac{\alpha e^{-\alpha}}{\beta + 1 - e^{-\alpha}} = S_{\max} \tag{9.89}$$

The right-hand side of eq. (9.89) is just the throughput found for the nonpersistent CSMA under the Poisson model (see eq. (9.78)), once we identify α with βG. In fact, according to our modeling assumptions (packets become backlogged immediately upon their arrival), the mean packet arrival rate in a back-off slot is np. In the Poisson approximation of CSMA, the same quantity is expressed as βG. We can therefore identify βG with $np = \alpha$, the last equality following from the assumed control law of the back-off probability as a function of the backlog.

We have found out that the analysis of CSMA based on the Poisson approximation provides a throughput expression that predicts exactly the *upper bound of the achievable stable CSMA throughput*.

An immediate application of this finding is a guideline to set α, hence to define the (ideal) adaptive control of the back-off probability. *We choose the parameter α equal to the unique positive value α^* maximizing $S_{\max}$*. There remains to turn the ideal adaptation algorithm into a practical one.

We define a Bayesian stabilization algorithm for p. We consider the embedded time points when the back-off counter is decremented. This occurs at the end of a "virtual slot." A virtual slot has three possible outcomes, i.e., it can consist of: (i) an idle back-off slot, when no station transmits; (ii) a successful activity time followed by an idle back-off slot, when a single station transmits; and (iii) a collision activity time, followed by an idle back-off slot, when many stations transmit. For virtual slot t, we let:

$$p(t) \equiv \alpha \min\left\{1, \frac{1}{\hat{n}(t)}\right\} \tag{9.90}$$

where[9]

$$
\hat{n}(t+1) = \begin{cases} \hat{n}(t)(1 - p(t)) + v\delta & \text{if slot } t \text{ is idle,} \\ \hat{n}(t)(1 - p(t)) + v(\delta + T) & \text{if slot } t \text{ is successful,} \\ \hat{n}(t) + \dfrac{[\hat{n}(t)p(t)]^2}{e^{\hat{n}(t)p(t)} - 1 - \hat{n}(t)p(t)} + v(\delta + T) & \text{otherwise.} \end{cases}
$$

(9.91)

The updating equations for the estimate of the number of backlogged stations in slot $t + 1$, $\hat{n}(t + 1)$, are obtained as the expectation of the a posteriori probability distribution of the number of backlogged stations, after having observed the outcome of slot t, given that the a priori distribution is a Poisson one with mean $\hat{n}(t)$.

Let us denote with e, s, and c the three possible outcomes of a virtual slot that can be observed: idle, successful transmission, collision. If the random variable $\hat{N}(t)$ has a Poisson PDF with mean $E[\hat{N}(t)] = \hat{n}(t)$, we have

$$
P(\hat{N} = n|e) = \frac{P(e|\hat{N} = n)P(\hat{N} = n)}{\sum\limits_{k=0}^{\infty} P(e|\hat{N} = k)P(\hat{N} = k)} =
$$

$$
= \frac{(1 - p)^n \frac{\hat{n}^n}{n!} e^{-\hat{n}}}{\sum\limits_{k=0}^{\infty} (1 - p)^k \frac{\hat{n}^k}{k!} e^{-\hat{n}}} = \frac{[\hat{n}(1 - p)]^n}{n!} e^{-\hat{n}(1-p)}, \qquad n \geq 0. \quad (9.92)
$$

The expectation yields $E[\hat{N}|e] = \hat{n}(1 - p)$. Hence the first of (9.91).

Analogously, it can be verified that:

$$
P(\hat{N} = n|s) = \frac{[\hat{n}(1 - p)]^{n-1}}{(n - 1)!} e^{-\hat{n}(1-p)}, \qquad n \geq 1, \quad (9.93)
$$

whence $E[\hat{N}|s] = 1 + \hat{n}(1 - p)$.

The expression in the third line of eq. (9.91) can be dervied following the same approach (see Problem 9.4).

The updating equations are obtained by summing to the expectation of the a posteriori prediction the mean number of new arrivals and subtracting the departure (if any). This leads to a good estimate so long as the a priori probability distribution of the number of backlogged stations is close to the Poisson one. Besides its amenability to calculations, the Poisson distribution is a reasonable choice also in view of the fact that it is the limiting probability distribution of an $M/G/\infty$

---

9 In [30] a pseudo-Bayesian estimate is proposed, where the estimate in case of collision is simplified to $\hat{n}(t + 1) = \hat{n}(t) + 2 + v(\delta + T)$. It can be obtained from (9.91) by using the expansion $e^x \approx 1 + x + x^2/2$.

queue. We will evaluate the mean delay realized by the stabilized CSMA in the next section.

The basic model we are considering assumes new packets become immediately backlogged upon their arrival. Variants can be considered, where new arrivals can be transmitted with probability 1 in their first attempt, if an idle slot is sensed upon their arrival; otherwise, if the channel is busy when the packet arrives, the packet becomes backlogged. The drift d_n, given that the backlog level is n, can be calculated as the difference between the mean number of new arrivals in a virtual slot, minus the mean number of successful transmissions in a virtual slot:

$$d_n = v[\delta + T - T(1 - p)^n e^{-v\delta}] - [np(1 - p)^{n-1}e^{-v\delta} + (1 - p)^n v\delta e^{-v\delta}] \quad (9.94)$$

Let us assume that the back-off probability is adjusted as a function of the backlog level according to $p = \alpha/n$. For large n, we get:

$$d_\infty = S(\beta + 1 - e^{-\alpha-S\beta}) - (\alpha + S\beta)e^{-\alpha-S\beta} \quad (9.95)$$

with $S = vT$, $\beta = \delta/T$. The drift is definitely negative if

$$S < \frac{(\alpha + S\beta)e^{-(\alpha+S\beta)}}{\beta + 1 - e^{-(\alpha+S\beta)}} \quad (9.96)$$

which defines the stability region for the offered load S.

Also in this case we recognize that the right-hand side of (9.96) is the throughput predicted by the CSMA analysis based on the Poisson approximation. In fact, the mean number of offered packets per back-off slot with the considered CSMA variant is $\lambda\delta = np + v\delta$ if n stations are backlogged. Making the back-off probability inversely proportional to the number of backlogged stations, i.e., $p = \alpha/n$, we have $\beta G = \alpha + S\beta$, with $G = \lambda T$.

The maximum achievable stable throughput, under the constraint (9.96), is shown in Figure 9.26 as a function of α for $\beta = 0.01, 0.05, 0.1$. It can be shown that the maximum achievable throughput is the same as with the constraint (9.89) (see Problem 9.4).

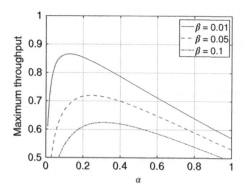

Figure 9.26 Maximum throughput for a stabilized variant of nonpersistent CSMA, where new packets sensing an idle channel upon their arrival are transmitted immediately, while packets finding a busy channel on arrival become backlogged.

There is an optimal choice of α. With $\beta = 0.01$ we find that $\alpha^* \approx 0.13$ attains the best maximum throughput $S^* \approx 0.87$. For $\beta = 0.05$ and $\beta = 0.1$ it is respectively $\alpha^* \approx 0.24$, $\alpha^* \approx 0.31$ and $S^* \approx 0.72$, $S^* \approx 0.62$. The larger the overhead due to the back-off slot, i.e., the value of β, the smaller the optimal achievable stable throughput.

A second variant consists of considering the 1-persistent policy. In that case, we have to be careful with the definition of the stabilized back-off policy. Let n denote the number of backlogged stations. We set $p = \alpha/n$ after an idle slot, so that the mean number of offered packets is $\alpha + S\beta = G\beta$. After a busy slot instead we let $p = \alpha'/n$. Then, the mean number of offered packets is $\alpha' + S(1 + \beta) = G(1 + \beta)$. The two expressions are consistent if we set $\alpha' = \alpha(1 + 1/\beta)$. With this choice, the stability limit is obtained from eq. (9.79), by substituting G with $S + \alpha/\beta$.

9.4.5 Delay Analysis of Stabilized CSMA

The Poisson traffic assumption is useful to derive throughput results in a simple way. It is however deceptive, since the infinite population model is intrinsically unstable, that is to say, it does not admit a steady state. We have seen how to stabilize CSMA, by adjusting the back-off probability in inverse proportion to the number of (estimated) backlogged stations.

We can define a simple model to evaluate the mean delay of a stabilized CSMA protocol with Poisson input traffic. We define an embedded Markov chain at the time points delimiting virtual slot. A virtual slot is an idle slot, if no station transmits. It is an activity time followed by an idle back-off slot, if at least one station transmits. Let $U(t)$ denote the number of packets transmitted successfully in slot t, $A(t)$ the number of new packets arriving during slot t (and immediately backlogged upon their arrival), and $N(t)$ the number of backlogged stations at the end of slot t. Then:

$$N(t + 1) = N(t) + A(t + 1) - U(t + 1) \tag{9.97}$$

Since the protocol is stabilized, a limiting steady-state exists and it is reached whatever the initial state, provided that the normalized input rate S be less than the maximum throughput $S_{max} = \alpha e^{-\alpha}/(\beta + 1 - e^{-\alpha})$. If α is optimally chosen, it can be verified that $S_{max} = 1 - \alpha^*$. We denote the limiting state probabilities as $\pi_n = P(N(\infty) = n)$.

We make a simple model of the stabilized protocol, by assuming that the $(t + 1)$-th virtual slot normalized duration $V(t + 1)$ be a random variable with the following probability distribution, conditional on $N(t) = n$

$$V(t + 1)|_{N(t)=n} = \begin{cases} \beta & \text{w.p. } 1, \text{ if } n = 0, \\ \beta & \text{w.p. } e^{-\alpha}, \text{ if } n \geq 1, \\ \beta + 1 & \text{w.p. } 1 - e^{-\alpha}, \text{ if } n \geq 1. \end{cases} \tag{9.98}$$

Hence, it is $E[V] = \beta + (1 - e^{-\alpha})(1 - \pi_0)$. The number of arrivals in a virtual slot has the following conditional probability distribution:

$$P(A(t+1) = k | N(t) = n) = \begin{cases} P_k(S\beta) & n = 0, \\ e^{-\alpha}P_k(S\beta) + (1 - e^{-\alpha})P_k(S + S\beta) & n \geq 1 \end{cases}$$

(9.99)

where $P_k(a) = \frac{a^k}{k!}e^{-a}$, $k \geq 0$, is the Poisson probability distribution.

Moreover, we define the binary random variable $U(t+1)$, conditional on $N(t) = n$, as follows:

$$P(U(t+1) = 1 | N(t) = n) = \begin{cases} \alpha e^{-\alpha}, & n \geq 1 \\ 0 & n = 0. \end{cases}$$

(9.100)

Taking expectations and limit for $t \to \infty$ on both sides of eq. (9.97), we get $E[U] = E[A]$, i.e.,

$$E[U] = \sum_{n=0}^{\infty} \pi_n P(U = 1 | N = n) = (1 - \pi_0)\alpha e^{-\alpha} = E[A] =$$
$$= SE[V] = S\beta + S(1 - e^{-\alpha})(1 - \pi_0)$$

(9.101)

from which we derive

$$\pi_0 = 1 - \frac{\beta S}{\alpha e^{-\alpha} - S(1 - e^{-\alpha})} \qquad E[V] = \frac{\beta \alpha e^{-\alpha}}{\alpha e^{-\alpha} - S(1 - e^{-\alpha})}$$

(9.102)

Squaring both sides, then taking expectations and limit for $t \to \infty$ on both sides of eq. (9.97), we get

$$0 = E[A^2] + E[U^2] - 2E[AU] + 2E[AN] - 2E[NU] =$$
$$= E[A(A-1)] + 2E[A] - 2E[AU] + 2E[AN] - 2E[NU]$$

(9.103)

After some algebra, it can be found that

$$E[A(A-1)] = S^2\beta^2 + S^2(1 + 2\beta)(E[V] - \beta)$$
$$E[AU] = E[V]S^2(1 + \beta)$$
$$E[AN] = S(\beta + 1 - e^{-\alpha})E[N]$$
$$E[NU] = \alpha e^{-\alpha}E[N]$$

(9.104)

Plugging these expression into (9.103), we get

$$E[N] = \frac{2SE[V] - S^2E[V] - S^2\beta(\beta + 1)}{2[\alpha e^{-\alpha} - S(\beta + 1 - e^{-\alpha})]}$$

(9.105)

The time-averaged mean number of backlogged stations is:

$$\overline{N} = \frac{\sum_{n=0}^{\infty} n\pi_n E[V | N = n]}{E[V]} = \frac{\beta + 1 - e^{-\alpha}}{E[V]}E[N]$$

(9.106)

Figure 9.27 Mean delay of stabilized non-persistent CSMA as a function of the offered traffic S: analytical model (solid line, labeled as "model" in the legend); refined analytical model (dash-dot line, labeled with "model+" in the legend); approximation of eq. (9.108) (dashed line, labeled with 'appx' in the legend); simulations (squared markers).

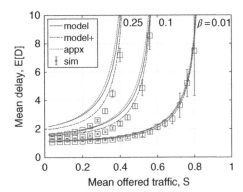

Finally, applying Little's law, we get the desired result:

$$E[D] = \frac{\overline{N}}{S} = \frac{1 - S/2 - S\beta(\beta + 1)/(2E[V])}{S_{max} - S} \tag{9.107}$$

where $S_{max} = \frac{ae^{-a}}{\beta + 1 - e^{-a}}$. It can be verified that

$$E[D] \approx \frac{1 - S/2}{S_{max} - S} - \frac{\beta(\beta + 1)}{2} \frac{S}{S_{max}} \equiv E[D_{appx}] \tag{9.108}$$

Figure 9.27 plots the mean delay of the stabilized CSMA as a function of the offered load S. We compare the mean delay obtained with simulations (square markers), the analytical model delay in eq. (9.107) (solid lines), the mean delay approximation in eq. (9.108) (dashed lines) and the mean delay provided by the refined model introduced at the end of this section (dash-dotted lines).

The model predicts the mean delay very accurately for small values of β, while it overestimates the mean delay, especially at low load levels, for bigger values of β. The approximation $E[D_{appx}]$ in eq. (9.108) is close to the analytical model in all considered cases, especially for $\beta \ll 1$. This is actually the most interesting case in applications, given that CSMA systems are designed so as to minimize the back-off time overhead.

A refined version of the analytical model can be obtained, by redefining the random variable U and A as follows:

$$P(U = 1 | N = n) = \begin{cases} 0 & n = 0; \\ \alpha & n = 1, \\ \alpha e^{-a} & n > 1 \end{cases} \tag{9.109}$$

$$P(A = k | N = n) = \begin{cases} P_k(S\beta) & n = 0, \\ (1 - \alpha)P_k(S\beta) + \alpha P_k(S + S\beta) & n = 1, \\ e^{-\alpha}P_k(S\beta) + (1 - e^{-\alpha})P_k(S + S\beta) & n > 1 \end{cases} \tag{9.110}$$

The accuracy improvement brought about by this refined model is paid at the price of much more cumbersome analytical expressions for the mean backlog level and the mean delay. In spite of its inaccuracy, the model of eq. (9.107) is still useful given that it yields good results for the most interesting (and critical) load levels, and given its extreme analytical simplicity.

9.5 Analysis of the WiFi MAC Protocol

Local wireless communications have become a dominating paradigm, since the end of the 1990s. Except for office and industrial environments, where cabled system are still largely used, most often Internet users access the network via a wireless interface, either cellular or wireless local area network (WLAN). In the latter case, the dominating technology is the IEEE 802.11 standard, also known as WiFi[10] [107,108]. Many variants, technically referred to as amendments, have been defined over the standard lifetime (the first version of the IEEE 802.11 was issued in 1997). Amendments are identified by means of small capital letters appended to the main acronym. The latest versions of the PHY and MAC level protocols as of now are the IEEE 802.11ac [108] and IEEE 802.11ax [7,68].

The IEEE 802.11 standard specifies the physical layer and the MAC protocol. We will focus on the latter in the following. Specifically, we outline models of the standard MAC protocol and some of its variants.

The physical layer has undergone an impressive evolution, from the initial 2 Mbit/s version in the 2.4 GHz bandwidth, up to the multi-Gbit/s IEEE 802.11ac version in the 5 GHz bandwidth. The high speed WiFi exploits wider channel bandwidth (up to 160 MHz against the 20 MHz of the initial version), greater spectral efficiency (up to 256-QAM modulation), Multiple Input Multiple Output (MIMO) communications. On the other hand, the Basic Access (BA) model of the MAC protocol has witnessed little modification since its original design. It is essentially a variant of a nonpersistent CSMA. Before delving into the models, a concise summary of the MAC protocol algorithm is presented, by introducing the so called CSMA/CA (CSMA with Collision Avoidance).

9.5.1 Outline of the IEEE 802.11 DCF Protocol

We outline the so called distributed coordination function (DCF) of the IEEE 802.11 MAC level. The system time evolution from the MAC point of view is

---

10 Wireless Fidelity is the name of the consortium established in 1999, to guarantee the inter-operability of IEEE 802.11 equipment in a multi-vendor environment.

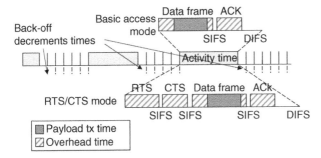

Figure 9.28 Example of activity time and back-off slots with basic access and RTS/CTS mode; back-off decrement times are highlighted as well.

depicted in Figure 9.28. The figure shows both basic access and RTS/CTS mode, to be explained in the rest of this section.

The MAC entity of a station is said to be *idle* if it has no data unit to send (MAC protocol data unit - MPDU). If instead at least one MPDU is waiting to be sent, the station is *backlogged*.

As soon as the upper layer asks for sending a data block, a new MPDU is created, nesting the data block into the MPDU payload, and the access procedure starts.

Let us consider the arrival of a new MPDU at a previously idle MAC entity. First, the channel is sensed, invoking the clear channel assessment (CCA) function of the physical layer. If the channel is sensed to be idle, and if it remains idle for a DCF inter-frame spacing (DIFS) time, then the new frame is transmitted immediately. If instead the channel is sensed to be busy, the station waits for the channel to be clear again. When the channel is back idle, the station checks if the channel remains idle for a DIFS time. It the channel becomes busy again before a DIFS time is elapsed, the procedure starts all over again. Otherwise, once the DIFS time has expired, a number between 0 and $W_0 - 1$ is drawn uniformly at random ($W_0 > 1$). A countdown counter is initialized with the sampled value. W_0 is the width of the basic contention window, i.e., the contention window size used in the first transmission attempt. The station checks whether the channel stays idle for a *back-off slot* time δ. At the end of the idle back-off slot, the counter is checked and, if positive, it is decremented by 1. If the channel becomes busy during the back-off slot time, the counter is frozen. When the counter is checked at 0, the station starts its transmission.

After the MPDU transmission is over, the sending station waits for an acknowledgment (ACK) from the destination[11]. If the MPDU reception was successful, the destination station waits for a short inter-frame spacing (SIFS) time, then starts transmitting the ACK MPDU. Note that no other station can interfere with the

---

11 We refer here to unicast MPDUs. For a broadcast MPDU, no ACK is issued.

ACK transmission. Any other station is forced to wait for an idle channel time of duration DIFS>SIFS, so the only station allowed to use the channel after a SIFS time is the destination one.

A fixed, predefined time-out can be set for the ACK on behalf of the station that transmitted the data MPDU. The time-out is simply the sum of the SIFS plus the constant duration of ACK transmission. If no ACK is received after the timer has expired, the transmitting stations deems a collision has occurred. Therefore, it increments a counter of the number of transmission attempts (initially set at 0), and re-schedules the MPDU. Re-scheduling consists simply in repeating the whole access procedure, i.e., waiting for the channel to stay idle for a DIFS time, then drawing a random integer k, and counting down k idle back-off slots. In general, if the station is starting the i-th transmission attempt, the integer k is drawn uniformly at random between 0 and $W_i - 1$, where W_i is the i-th stage contention window size.

Consistently with the aim of avoiding repeated collision events, it is $W_{i+1} \geq W_i > 1$ for $i \geq 0$. For example, the IEEE 802.11 standard defines the so called binary exponential back-off (BEB) algorithm, setting $W_i = 2 \cdot W_{i-1}$, $i \geq 1$.

Transmission attempts are carried out up to a maximum number of retransmissions, corresponding to the value of the `max_retry` parameter. If a successful transmission is performed (i.e., the transmitting station detects the ACK), the station moves on to the next MPDU to be sent, if any. If no new MPDU is ready, the MAC entity goes back to the idle state, after having drawn a random integer between 0 and $W_0 - 1$ and having completed its countdown (post-back-off).

If instead all `max_retry` retransmissions fail, the MPDU is silently discarded and the station moves on to the next MPDU, if any.

When transmitting, the station selects the modulation and coding set (MCS) according to the physical channel quality that it expects (based on previous measurements and receptions). A data MPDU can be divided into two parts: the MPDU payload, hosting data from upper layer, and the MAC header (see Figure 9.29). The entire data MPDU is transmitted at the bit rate of the selected MCS (referred to as the air bit rate). In turn, a data MPDU is contained into a physical PDU (PPDU). The PPDU adds a so called physical layer convergence protocol (PLCP) header, plus a physical preamble (see Figure 9.29). The PLCP header and the preamble are transmitted at a predefined basic bit rate, corresponding to the lowest bit rate allowed by the available MCSs. Control frames,

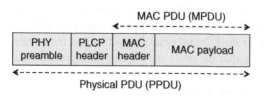

Figure 9.29 Structure of the physical and MAC PDUs.

such as ACK, RTS, CTS, are transmitted at the basic bit rate as well. This choice guarantees that the coverage range of such critical frames is maximum.

A station not involved in a transmission listens to the channel, to collect MPDUs that are addressed to it and to run the access algorithm, if it is backlogged. If it cannot decode successfully an MPDU, after the channel is back idle, it sets a special timer, whose duration is the extended inter-frame spacing (EIFS). The EIFS is typically set as the sum of SIFS plus an ACK time-out plus a DIFS. It is intended to ease the reception of the ACK at the transmitting station, without any interference from other stations not involved.

Current commercial WiFi network interface cards do not support simultaneous transmission and reception. They are said to be half-duplex. Because of nonideal separation between the in and out channels of the transceiver chains, a fraction of the transmitted signal power leaks into the receiving chain and easily overwhelms the feeble received signal. This phenomenon is called self-interference. During the last decade a remarkable progress has been achieved in self-interference suppression, making full-duplex cards a reality, although currently confined mostly to experimental or precommercial setups. [12]

A consequence of the half-duplex mode of communication of WiFi cards is that collision detection is not possible. Collision can only be inferred by means of the ACK time-out at MAC level.

The RTS/CTS mode is alternative to basic access (see Figure 9.28). Request-to-send (RTS) control frames can be sent before the data frame, to seize the channel and possibly enable channel measurements and parameter negotiation with the addressed station. If a collision occurs, since no collision detection is possible with half-duplex equipment, the whole frame transmission time is wasted. To moderate the waste of channel time, a much shorter frame, the RTS, is sent first. A timer is set, upon sending the RTS. If a clear-to-send (CTS) response frame is received from the addressed station within the time-out, then the station starts transmitting the data MPDU, a SIFS time after having received the CTS frame. If instead the CTS time-out expires, the station behaves just as if a collision occurred, rescheduling the RTS. A special maximum retry parameter is defined for the RTS. The default value of such maximum number of retries is 4.

The RTS/CTS mechanism is also useful to enhance carrier sensing. Let us consider two stations, A and B, associated to a same AP C. Let P_{XY} denote the power level received at Y for a signal emitted by X. Since A and B are associated with C it must be the case that P_{CA} and P_{CB} (hence also P_{AC} and P_{BC}) are beyond a threshold, the carrier detect threshold (CDT). It can happen however that A and B are far

---

12 As a matter of fact, the study group of IEEE 802.11ax has postponed introduction and exploitation of full-duplex communication in WiFi to the next standard version. Besides refining the signal processing required to suppress self-interference, the way that full-duplex capability can be exploited at MAC level is an open research topic.

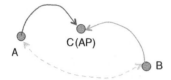

Figure 9.30 Example of hidden nodes. Both *A* and *B* are associated with the AP *C*, yet they cannot hear each other.

away enough that they do not hear each other. Station *A* assesses a busy channel according to physical carrier sensing whenever the signal power level it measures exceeds a threshold, the defer threshold (DT). If $P_{BA} < DT$, then *A* senses an idle channel even if *B* is transmitting a frame to *C*. Then *A* could start its own transmission, thus causing a collision at *C*. This situation is described by saying that *A* and *B* are *hidden* to each other; it is illustrated in Figure 9.30.

To alleviate the hidden node problem, both RTS and CTS frames carry a field announcing the channel holding time of the forthcoming transmission. The initiating station (the one that sends the RTS) knows how much data it has to send and the selected MCS. Hence it can calculate the overall time it needs to use the channel, including any overhead (CTS, preambles, ACK). The announced channel holding time, diminished by the RTS duration, is echoed back by the addressed station, when it issues the CTS. Thus, neighbors of both origin and destination stations hear about the upcoming transmission. They enter this channel holding time into a data structure called network allocation vector (NAV).

When a backlogged station has to assess the channel status, it does usually *both* physical channel sensing (the CCA described above) and *virtual carrier sensing*, i.e., it checks the NAV to verify whether it reports a busy or idle channel.

Another control MAC frame that is worth mentioning is the Beacon frame, periodically issued by any AP, to announce itself and to broadcast a number of parameters related to the WiFi network created by the AP and useful for MAC operations in that network.

Notation and default values of the MAC parameters for the IEEE 802.11g version of the standard are reported in Table 9.2.

9.5.2 Model of CSMA/CA

We focus on an isolated WiFi network made up of *n* stations. In the model we first focus on the back-off processes of the *n* stations.

The model of 802.11 DCF is derived under the following assumptions:

- *Symmetry*: Stations are statistically indistinguishable, i.e., traffic parameters (air bit rate, payload length probability distribution) and multiple access parameters are the same.
- *Proximity*: Every station is within reach of all others, i.e., there are no hidden nodes.
- *Saturation*: Stations are continuously backlogged.

Table 9.2 Notation and parameter values of IEEE 802.11g standard.

Symbol	Description	Value
δ	Back-off slot time	9 µs
C	Air bit rate	$6 \leq C \leq 54$ (Mbit/s)
C_b	Basic bit rate	6 Mbit/s
L	MPDU payload length	variable ($\leq$ 2304 bytes)
U	MPDU payload time	L/C
L_{ACK}	ACK frame length	14 bytes
L_{RTS}	RTS frame length	20 bytes
L_{CTS}	CTS frame length	14 bytes
T_{ACKTO}	ACK Time-Out	$SIFS + L_{ACK}/C_b$
$SIFS$	Short inter-frame spacing	16 µs
$DIFS$	DCF inter-frame spacing	$SIFS + 2\delta$
$EIFS$	Extended inter-frame spacing	$SIFS + L_{ACK}/C_b + DIFS$
m	Maximum number of re-transmissions (`max_retry`)	7
CW_{min}	Minimum contention window	16
CW_{max}	Maximum contention window	1024
W_i	Contention window at stage i for $i = 0, 1, \ldots, m$	$\min\{CW_{max}, CW_{min}2^i\}$

Along with these we introduce a simplifying hypothesis:

- *Independence*: The back-off processes of the stations are independent of one another.

The independence hypothesis is essential to describe the system dynamics by using a low dimensionality Markov chain. Its validity has been discussed from a theoretical viewpoint in [189, 38] and confirmed by extensive simulations in a large number of papers. It turns out that numerical results of models based on the independence hypothesis are extremely accurate not only for evaluating average metrics (throughput, mean delay) but even probabilities and the entire back-off counter probability distribution. A deep insight into the effectiveness and limits of the independence assumption can be gained from [39].

The model is developed from the point of view of a tagged station (any one will do, given the *Symmetry* assumption). We consider embedded times when the tagged station decrements its back-off counter. Let t_k denote the k-th such time point. If $B(t)$ denotes the back-off at time t, by $B(t_k)$ we mean the back-off immediately before the decrement at time t_k. According to the rules of IEEE 802.11 DCF, the back-off counter is decremented after an idle back-off has expired, while the counter is frozen when the channel is sensed to be busy.

A *virtual slot* is the time spanning between two consecutive back-off counter decrements for a tagged station. Slot duration can be described as a random variable X, whose distribution is derived in § 9.5.2.2.

9.5.2.1 The Back-off Process

The slot counting process is completely independent of the duration of the transmission attempt (either successful or not) thanks to the freezing of the back-off counters and to the renewal of back-off counters each time they are reset. Therefore, the description of the counting process is oblivious of details of the transmission mode, of data unit formats, and of the air bit rate. On the contrary, evaluation of the throughput and of the service time probability distribution does depend on the actual transmission attempt duration, hence on the transmission mode, data unit length, bit rate.

Let $I(t)$ denote the retransmission stage and $J(t)$ the back-off count of the tagged station at time t. It is $I(t) \in \{0, 1, \dots, m\}$ and $J(t) \in \{0, 1, \dots, W_{I(t)} - 1\}$. The well-known model of IEEE 802.11 DCF by Bianchi [32,33] consists of a bi-dimensional Markov chain with state $(I(t_k), J(t_k))$, i.e., it is the embedded Markov chain (EMC) sampled at the back-off decrement times t_k of the continuous time process $(I(t), J(t))$.

Let $P(i, j | i', j')$ denote the one-step transition probability from state (i', j') to state (i, j) of the EMC of the back-off process. Let p denote the conditional collision probability, i.e., the probability that the tagged station incurs a collision event upon its transmission. Note that we are assuming that p be the same for all stations and it does not depend on the state of the tagged station. In other words, the independence hypothesis lets us decouple the tagged station with respect to all other stations, that collectively make a *background* stationary process, resulting in the tagged station experiencing a collision upon its transmission with a fixed probability p.

Given the maximum number of retransmissions m and the contention window sizes W_i, $i = 0, \dots, m$, it is easy to verify that:

$$P(i, j | i', j') = \begin{cases} 1 & i = i', j = j' - 1, i' = 0, \dots, m, j' = 1, \dots, W_i - 1 \\ p/W_i & i = i' + 1, j = 0, \dots, W_i - 1, i' = 0, \dots, m - 1, j' = 0 \\ (1 - p)/W_0 & i = 0, j = 0, \dots, W_0 - 1, i' = 0, \dots, m - 1, j' = 0 \\ 1/W_0 & i = 0, j = 0, \dots, W_0 - 1, i' = m, j' = 0 \\ 0 & \text{otherwise.} \end{cases}$$

$$(9.111)$$

Let $\pi_{i,j}$ denote the limiting probability distribution of the EMC. It exists since the EMC is irreducible, finite, and aperiodic, hence ergodic. It is not difficult to find that

$$\pi_{i,j} = \pi_{0,0} p^i \frac{W_i - j}{W_i} = \frac{p^i (W_i - j)/W_i}{\sum_{i=0}^{m} p^i (W_i + 1)/2}, \qquad j = 0, \ldots, W_i - 1, \; i = 0, \ldots, m.$$

$$(9.112)$$

The states $(i, 0)$ correspond to end of countdown, hence to a transmission attempt. Let τ denote the probability of transmission of a station, i.e., the probability of the event: "The tagged station starts transmission at the end of the current time slot." Then, we have

$$\tau = \sum_{i=0}^{m} \pi_{i,0} = \frac{\sum_{k=0}^{m} p^k}{\sum_{k=0}^{m} p^k (W_k + 1)/2} \qquad (9.113)$$

Note that $(W_k + 1)/2$ is just the mean number of back-off slots counted down when in stage k. It is also remarkable that the limiting state probabilities of the back-off process of the tagged station depend only on the contention window size, hence on the *mean* of the back-off time, not on its probability distribution.

The result of eq. (9.113) can be found without resorting to the EMC, by means of an elegant regenerative argument (see [135] for details). Here we summarize the key steps, just to appreciate the elegance and simplicity that leads to (9.113). Let $B_i(k)$ denote the number of back-off slots counted down by the tagged station at the i-th re-transmission attempt of packet k; $i = 0$ corresponds to the initial transmission attempt. Let $M(k)$ denote the number of transmission attempts required to deal with the k-th packet. The overall number of back-off slots counted down for packet k is $Z(k) = \sum_{i=0}^{M(k)} B_i(k)$. By the model assumptions, the sequence $Z(k)$ is a renewal process. We can regard the number of transmission attempts $M(k)$ as the reward associated to the k-th renewal time. By the renewal reward theorem, the mean reward rate R is $R = E[M]/E[Z]$. The probability that h or more attempts are required to deal with a packet is $P(M \geq h) = p^h$, $h = 0, 1, \ldots, m$, thanks to the independence hypothesis. Then, we have $E[M] = \sum_{h=0}^{m} P(M \geq h) = 1 + p + \cdots + p^m$ and

$$E[Z] = \sum_{h=0}^{m} P(M = h) \sum_{i=0}^{h} E[B_i] = \sum_{i=0}^{m} E[B_i] P(M \geq i) = b_0 + b_1 p + \cdots + b_m p^m$$

$$(9.114)$$

where $b_i \equiv E[B_i] = (W_i + 1)/2$, $i = 0, \ldots, m$.

The mean reward rate in our case is the mean fraction of back-off slots at the end of which the tagged station attempts a transmission, i.e., *it is the transmission probability over the embedded times*, $\tau = R$. Using $\tau = E[M]/E[Z]$, we find finally the result already stated in eq. (9.113):

$$\tau = \frac{1 + p + \cdots + p^m}{b_0 + b_1 p + \cdots + b_m p^m} \equiv F(p) \qquad (9.115)$$

The collision probability can be expressed in terms of the transmission probability by observing that it is the probability that at least one of the $n-1$ stations other than the tagged one transmits along with the tagged station. It is therefore:

$$p = 1 - (1 - \tau)^{n-1} \equiv G(\tau) \tag{9.116}$$

thanks to the independence hypothesis. This second equation is the "coupling" between the tagged station and all other stations. It summarizes (in a clever way) the effect that all other stations have on the back-off process of the tagged station.

Equations (9.115) and (9.116) make a nonlinear equation system. We can write formally $\tau = F(G(\tau))$, which is a fixed-point equation. Since both $G(\cdot)$ and $F(\cdot)$ are continuous functions, the map $x \in [0, 1] \mapsto F(G(x)) \in [0, 1]$ is continuous. By Brouwer's theorem we can then say that there exists a fixed point in $[0, 1]$. Uniqueness is more tricky. It can be shown that a sufficient condition to guarantee that a unique fixed point exists in the symmetric scenario here considered is that the sequence W_i, $i = 0, 1, \ldots, m$ be nondecreasing (see Theorem 5.1 in [177]).

Before closing this section, we outline a different back-off process model, which offers an alternative model of the CSMA/CA countdown.

Let us consider the embedded times corresponding to the end of a transmission attempt, when the countdown is resumed or the back-off counter is reinitialized. The latter case applies when the station is involved in the transmission.

The back-off at the k-th embedded time is denoted with B_k. It can take any integer value between 1 and $W \equiv CW_{\max}$. We let $b_i(k) = \mathcal{P}(B_k = i), i = 1, \ldots, W$. When B_k is reset, it is assigned a new value according to the probability distribution q_j, $j = 1, \ldots, W$. In the CSMA/CA of WiFi, the back-off at the h-th retransmission attempt is drawn uniformly at random between 1 and $W_h = 2^h CW_{\min}$. The probability of making h retransmissions is $P(h) = (1 - p)p^h$, $h = 0, \ldots, m - 1$ and $P(m) = p^m$ for $h = m$. Then, we have

$$q_j = \sum_{h=h_j}^{m} \frac{P(h)}{W_h}, \quad h_j = \left\lceil \log_2\left(\frac{j}{CW_{\min}}\right) \right\rceil, \quad j = 1, \ldots, W. \tag{9.117}$$

where $\lceil x \rceil$ is the least integer not less than x.

The sequence B_k is a Markov chain. Let $G_i(k) = \mathcal{P}(B_k \geq i)$ for $i = 1, \ldots, W$. For notational convenience, we define also $G_{W+1}(k) = 0$. It is easy to recognize that the following one-step equations hold

$$b_j(k+1) = \begin{cases} q_j \sum_{i=1}^{W} b_i(k)[G_i(k)]^{n-1} + \\ + \sum_{i=j+1}^{W} b_i(k)([G_{i-j}(k)]^{n-1} - [G_{i+1-j}(k)]^{n-1}) & j = 1, \ldots, W-1, \\ q_W \sum_{i=1}^{W} b_i(k)[G_i(k)]^{n-1} & j = W \end{cases} \tag{9.118}$$

Figure 9.31 Probability distribution of the back-off for the WiFi CSMA/CA and three values of the number n of stations.

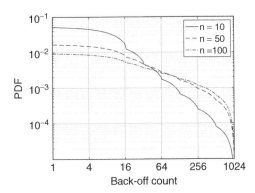

The Markov chain B_k is irreducible, aperiodic, and finite, hence it is ergodic and there exists a well defined limit of $b_i(k)$ as $k \to \infty$. The limiting probability distribution $\{b_i\}_{1 \le i \le W}$ can be computed by solving numerically the fixed point equations:

$$b_j = \begin{cases} q_j \sum_{i=1}^{W} b_i G_i^{n-1} + \sum_{i=j+1}^{W} b_i (G_{i-j}^{n-1} - G_{i+1-j}^{n-1}) & j = 1, \dots, W-1, \\ q_W \sum_{i=1}^{W} b_i G_i^{n-1} & j = W \end{cases} \tag{9.119}$$

where $G_i = \sum_{r=i}^{W} b_r$ and q_j is given in eq. (9.117). The probability that a station transmits is $\tau = \sum_{i=1}^{W} b_i G_i^{n-1}$. The probability that station is successful transmitting its frame is $p_{s1} = \sum_{i=1}^{W} b_i G_{i+1}^{n-1}$. Then, the conditional collision probability is $p = (\tau - p_{s1})/\tau$.

Figure 9.31 plots the probability distribution b_i for $n = 10, 50, 100$. The retransmission stages are visible through "bumps" in the curve profile, especially for the lowest value of n. The bulk of the probability distribution shifts toward higher values of the back-off count as n grows. This is consistent with the higher collision probability with bigger values of n, hence with the higher probability that larger contention windows are used because of multiple retransmissions.

9.5.2.2 Virtual Slot Time

A virtual slot is the time elapsing between two consecutive decrements of the countdown of a tagged station. It is made up of: (i) an idle back-off slot, if no station transmits; (ii) the transmission of a single station, in case of success; or (iii) several simultaneous transmissions in case of collision.

Let U denote the duration of the payload of the MPDU, $T_{\text{oh},s}$, $T_{\text{oh},c}$ the overall duration of overhead in case of successful transmission and collision, respectively.

The overhead time depends on the access mode, on the station role and on the outcome of the transmission attempt. Let us consider first the BA. For any station that decodes the transmitted frame correctly, in case of success, the overhead time is made of preamble and PLCP header transmission time (P_{PHY}), MAC header transmission time (T_{MAChdr}), acknowledgement time (T_{ACK}) and inter-frame spacings (*DIFS, SIFS*). In case of collision, for the transmitting station there is the acknowledgement time-out (T_{ACKTO}). For any other station that does not decode the data frame correctly, a special inter-frame spacing is adopted (*EIFS*).

Summing up, the overhead times for BA are:

$$
\begin{cases}
T_{oh,s}^{(BA)} = T_{PHY} + T_{MAChdr} + SIFS + T_{ACK} + DIFS & \text{(all stations)} \\
T_{oh,c}^{(BA)} = T_{PHY} + T_{MAChdr} + T_{ACKTO} + DIFS & \text{(transmitting station)} \\
T_{oh,c}^{(BA)} = T_{PHY} + T_{MAChdr} + EIFS & \text{(any other station)}
\end{cases}
$$
$$(9.120)$$

In the following, we assume $EIFS = T_{ACKTO} + DIFS = SIFS + T_{ACK} + DIFS$. Therefore, we have $T_{oh} \equiv T_{oh,s}^{(BA)} = T_{oh,c}^{(BA)}$.

For RTS/CTS, the components of the overhead time are the same plus the times required to transmit the RTS and CTS frames:

$$
\begin{cases}
T_{oh,s}^{(RC)} = T_{RTS} + SIFS + T_{CTS} + SIFS + T_D & \text{(all stations)} \\
T_{oh,c}^{(RC)} = T_{RTS} + T_{RTSTO} & \text{(transmitting station)} \quad (9.121) \\
T_{oh,c}^{(RC)} = T_{RTS} + EIFS & \text{(any other station)}
\end{cases}
$$

where $T_D = T_{PHY} + T_{MAChdr} + SIFS + T_{ACK} + DIFS$. The overhead time in case of collision has the same value for all stations, if we set $EIFS = T_{RTSTO}$. Overhead times in case of success and collision are different with RTS/CTS mode. In the following, we drop the superscript (RC) for the sake of a simple notation. We denote the overhead times for success or collision with RTS/CTS as $T_{oh,s}$ and $T_{oh,c}$ respectively.

The duration X of the virtual slot with BA mode, when there are n contending stations, is given by:

$$
X = \begin{cases}
\delta & \text{w.p. } (1 - \tau)^n, \\
\delta + T_{oh} + \max\{U_1, \dots, U_k\} & \text{w.p. } \cdot \binom{n}{k} \tau^k (1 - \tau)^{n-k}, \quad k = 1, \dots, n
\end{cases}
$$
$$(9.122)$$

For RTS/CTS mode the virtual slot duration is:

$$
X = \begin{cases}
\delta & \text{w.p. } (1 - \tau)^n, \\
\delta + T_{oh,s} + U_1 & \text{w.p. } \cdot n\tau(1 - \tau)^{n-1}, \\
\delta + T_{oh,c} & \text{w.p. } \cdot 1 - (1 - \tau)^n - n\tau(1 - \tau)^{n-1}
\end{cases}
$$
$$(9.123)$$

By the symmetry assumption, the payload transmission time U is the same random variable for all stations. If L denotes the length of the payload and C the air bit rate, it is $U = L/C$. U is variable both as a consequence of the variability of the length of the payload and of the selection of the air bit rate among a set of several possible values. In the following we assume a general discrete distribution for U. This is a useful model, since only a finite number of rates are available in each standard specification of IEEE 802.11 physical layer and packet length probability distributions usually have most of their masses on a few values, all the other lengths having a negligible probability.

We denote the probability distribution of U as $q_j = P(U = a_j), j = 1, \ldots, \ell$, where a_j is the j-th payload time. In the following, we assume $a_1 < a_2 < \cdots < a_\ell$. The corresponding CDF is $Q_j = \sum_{i=1}^{j} q_i, j = 1, \ldots, \ell$. For ease of notation we define also $Q_0 = 0$.

By the *Independence* assumption, we have $P(\max\{U_1, \ldots, U_r\} \le a_j) = Q_j^r$ and $P(\max\{U_1, \ldots, U_r\} = a_j) = Q_j^r - Q_{j-1}^r$, for $j = 1, \ldots, \ell$ and $r = 1, \ldots, n$.

In the case of BA mode, according to eq. (9.122), X is a discrete random variable taking $\ell + 1$ values:

$$\begin{cases} P(X = \delta) = (1 - \tau)^n \\ P(X = \delta + T_{\text{oh}} + a_j) = \displaystyle\sum_{r=1}^{n} P(\max\{U_1, \ldots, U_r\} = a_j) \binom{n}{r} \tau^r (1 - \tau)^{n-r} \\ \qquad = (1 - \tau + \tau Q_j)^n - (1 - \tau + \tau Q_{j-1})^n, \quad 1 \le j \le \ell. \end{cases}$$

$$(9.124)$$

In the case of RTS/CTS mode, the probability distribution of X, as given by eq. (9.123), is:

$$\begin{cases} P(X = \delta) = (1 - \tau)^n \\ P(X = \delta + T_{\text{oh},s} + a_j) = n\tau(1 - \tau)^{n-1}q_j \quad 1 \le j \le \ell \\ P(X = \delta + T_{\text{oh},c}) = 1 - (1 - \tau)^n - n\tau(1 - \tau)^{n-1} \end{cases}$$

$$(9.125)$$

To keep notation compact, in the following we use also

$$P_j(n) \equiv (1 - \tau + \tau Q_j)^n \tag{9.126}$$

for $j = 0, 1, \ldots, \ell$ and $n \ge 0$ and

$$\begin{cases} p_e(n) = (1 - \tau)^n, \\ p_s(n) = n\tau(1 - \tau)^{n-1}, \\ p_c(n) = 1 - p_e(n) - p_s(n) \end{cases} \tag{9.127}$$

9.5.2.3 Saturation Throughput

The normalized, long-term saturation throughput ρ can be found as the ratio of the mean time spent to make a successful transmission and the mean virtual slot duration, i.e., $\rho = p_s(n)E[U]/E[X]$. We focus on the BA mode. For n saturated (always backlogged) stations it is:

$$\rho = \frac{n\tau(1-\tau)^{n-1}E[U]}{\delta + T_{oh}[1-(1-\tau)^n] + \sum_{i=1}^{\ell} a_i[P_i(n) - P_{i-1}(n)]} \tag{9.128}$$

In the special case of fixed payload times, $U = a_1$ with probability 1. Equation (9.128) simplifies to:

$$\rho = \frac{n\tau(1-\tau)^{n-1} a_1}{\delta + (T_{oh} + a_1)[1-(1-\tau)^n]} \tag{9.129}$$

The throughput for the RTS/CTS mode is:

$$\rho = \frac{n\tau(1-\tau)^{n-1}E[U]}{\delta + (T_{oh,s} + E[U])p_s(n) + T_{oh,c}p_c(n)} \tag{9.130}$$

For centralized scheduling, neither contention nor back-off are required. Each transmission is successful, hence

$$\rho_{ideal} = \frac{E[U]}{T_{oh} + E[U]} \tag{9.131}$$

We can highlight the close affinity between WiFi CSMA/CA and nonpersistent CSMA as we have seen in previous sections.

As the number n of stations tends to infinity, while the transmission probability tends to 0, so that $n\tau \to \alpha > 0$, we obtain:

$$\rho \to \rho_{ideal} \frac{\alpha e^{-\alpha}}{\beta + 1 - e^{-\alpha}} \tag{9.132}$$

where $\beta = \delta/(T_{oh} + a_1)$. Not surprisingly, this is just the expression of the throughput in case of nonpersistent CSMA with Poisson offered traffic.

To assess the effect of contention on throughput performance we should compare ρ with ρ_{ideal}, not with the theoretical upper bound 1. The ideal throughput ρ_{ideal} accounts for all overhead involved with the MPDU transmission on the channel. The overhead can be substantial with typical WiFi parameter values.

Performance of IEEE 802.11g and IEEE 802.11ac DCF in BA mode are shown in Figure 9.32. Fixed payload lengths are considered, as shown in the plot legend. Figure 9.32(a) and Figure 9.32(b) plot the saturation throughput of IEEE 802.11g and for IEEE 802.11ac, respectively, as a function of the number n of stations. Figure 9.32(c) shows the collision probability, which depends only on the number n of contending stations, the number of retries m, and the contention window sizes. Numerical values of the IEEE 802.11g and IEEE 802.11ac parameters are listed in Table 9.3. The contention window size, according to BEB, is $W_k = \min\{CW_{max}, CW_{min} \cdot 2^k\}$ for $k = 0, \dots, m$, with $CW_{min} = 16$ and $CW_{max} = 1024$.

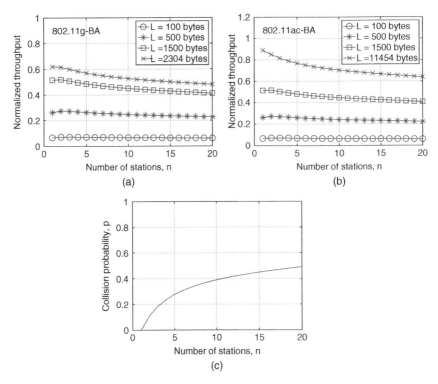

Figure 9.32 Throughput and collision probability as a function of the number n of stations for fixed payload size. Top-left plot: saturation throughput in case of IEEE 802.11g standard BA mode. Top-right plot: saturation throughput in case of IEEE 802.11ac standard BA mode. Bottom plot: collision probability.

Table 9.3 Numerical values of IEEE 802.11g and IEEE 802.11ac DCF parameters.

Parameter	Value	Parameter	Value
Back-off slot, δ	9 μs	max retry, m	7
SIFS	16 μs	DIFS	SIFS+2δ
CW_{min}	16	CW_{max}	1024
MAC header (g)	34 bytes	MAC header (ac)	40 bytes
ACK length (g)	14 bytes	ACK length (ac)	32 bytes
bandwidth (g)	20 MHz	bandwidth (ac)	40 MHz
overhead time, T_{oh} (g)	136.8 μs	overhead time, T_{oh} (ac)	153.9 μs
max payload length (g)	2304 bytes	max payload length (ac)	11454 bytes
basic bit rate (g)	6 Mbps	basic bit rate (ac)	15 Mbps
air bit rate (g)	54 Mbps	air bit rate (ac)	200 Mbps

The normalized throughput is slightly decreasing with the number of stations, especially for longer payloads, where the weight of collision impacts more the achievable performance. On the other hand, for small payload lengths, throughput performance are killed by overhead, even when few stations contend. Improved throughput performance are obtained in case of IEEE 802.11ac, with the considered numerical values of the parameters. This is also a consequence of the bigger maximum payload length allowed by IEEE 802.11ac and of the possibility to aggregate several MPDUs in a single access opportunity (we assume 4 aggregated MPDUs here).

The collision probability is the same for the two considered standard versions. It has quite high values (beyond 0.2 already for $n = 5$ stations), thus showing that collision events are anything but rare in even small WiFi networks.

Throughput and collision probability are shown for a bimodal payload probability distribution in Figure 9.33, in case of IEEE 802.11g BA. Figure 9.33(a) plots the saturation throughput as a function of the number of contending stations, while the collision probability is shown in Figure 9.33(b).

Payload lengths are $a_1 = 52$ bytes and $a_2 = 1500$ bytes, with $q_1 = 1/3$ and $q_2 = 2/3$. This is a model of a case where TCP segments and TCP ACKs are exchanged over the wireless interface. With the delayed ACK mechanism enabled, one TCP ACK is issued any other data TCP segment. Therefore, the probability of long payloads is two times the probability of short payloads.

Solid line curves in Figure 9.33 refer to the standard DCF. Dashed lines correspond to optimized DCF, where BEB is disabled and the contention window is $W^*(n) = 2/\tau^*(n) - 1$, where $\tau^*(n)$ is the transmission probability that maximizes the throughput for each n (see Sec. 9.5.3.1). The dotted line for throughput is the ideal throughput achievable when there is no back-off overhead and no collision.

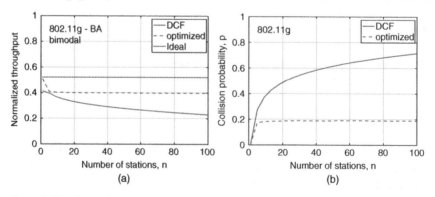

Figure 9.33 Saturation throughput (left plot) and collision probability (right plot) as a function of the number n of stations for bimodal payload size, in case of IEEE 802.11g BA mode (solid line: DCF; dashed line: optimized DCF; dotted line: ideal scheduling with no contention).

It represents the upper bound of throughput, obtained with an ideal scheduler, yet accounting for transmission overhead.

It can be seen that there is a constant performance offset between the optimized DCF and the ideal level, which is due to back-off overhead and collisions. The standard DCF departs more and more from optimized and ideal performance, as the number of stations grows. This is apparent both from the throughput and collision probability performance. This is the effect of the growing number of collisions suffered by the standard DCF, where contention window sizes are fixed. The rough adaptation provided by BEB is inadequate for large WiFi networks. On the other hand, otpimization requires in principle the knowledge of the number of contending stations. We will see in Sec. 9.5.3.1 that a practical smart solution can be set up, by observing the channel and estimating a proxy of the number of contending stations.

9.5.2.4 Service Times of IEEE 802.11 DCF

Let t_j denote the j-th back-off decrement time of the tagged station. Let $t_j^{(s)}$ be the service completion epochs (either with success or failure) as seen by a tagged station. The sequence $\{t_j^{(s)}\}$ is obtained by sampling the full sequence $\{t_j\}$. The j-th service time for the tagged station is denoted as Θ_j. Under the *Saturation* assumption we have $\Theta_j = t_j^{(s)} - t_{j-1}^{(s)}$. At steady state we have $\Theta_j \sim \Theta, \forall j$.

An example of the service time structure, as a sequence of *transmission cycles*, is depicted in Figure 9.34. The top line refers to a successful service time, where the frame is eventually delivered to the destination after k transmission attempts ($k - 1$ failed ones and a successful last one). The bottom line shows a sequence of $m + 1$ failed transmission attempts, after which the frame is discarded. Between two consecutive transmission attempts of the tagged station, countdown takes place. Either idle back-off slots or other stations' transmission attempts go by during the countdown of the tagged station.

A service time Θ is the sum of $K + 1$ attempts, where K is the number of *re-transmissions* ($0 \leq K \leq m$). Given the assumptions of our analysis, we

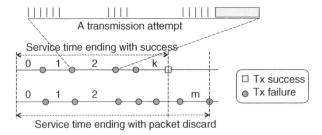

Figure 9.34 Tagged station service time structure: successful service time in $k \leq m$ attempts (top line); unsuccessful service time (bottom line).

have $P(K = k) = (1 - p)p^k$ for $k = 0, \ldots, m - 1$ and $P(K = m) = p^m$, where $p = 1 - (1 - \tau)^{n-1}$ is the collision probability. Each attempt is made up of

- the tagged station countdown time C, during which the tagged station stays idle and some other station possibly attempts transmission;
- a transmission time $Z(U)$ involving the tagged station, where U is the payload time of the tagged station frame.

Let B_j denote the number of back-off slots counted on attempt j. According to the BEB algorithm, it is $P(B_j = r) = 1/W_j$ for $r = 0, \ldots, W_j - 1, j = 0, \ldots, m$. Let $\tilde{X}$ denote the virtual slot time, conditional on the tagged station being on back-off countdown. $\tilde{X}$ has the same probability distribution as X in Sec. 9.5.2.2, except that $n - 1$ stations must be considered. The time required to count down during the j-th attempt is $C_j = \sum_{i=1}^{B_j} \tilde{X}_{ij}$, where $\tilde{X}_{ij} \sim \tilde{X}$.

Let further $T_{\mathrm{oh},j}$ and $Z_j(U)$ denote the overhead time and payload time of the j-th transmission attempt, respectively. The expression of $Z_j(U)$ depends on the outcome of the transmission. If it is successful, it is simply $Z_j(U) = U$. In case of collision, it is $Z_j(U) = \max\{U, U_1, \ldots, U_{Y_j}\}$, where Y_j denotes the number of stations colliding with the tagged one in the j-th transmission attempt ($1 \leq Y_j \leq n - 1$). By construction, $Z_0(U), \ldots, Z_{K-1}(U)$ are payload times of collision events, while $Z_K(U) = U$ is the payload time of the last, successful attempt, if $K \leq m - 1$. For $K = m$, the last attempt can be either a success or yet another collision. Since that is the last possible retry, if it fails, the frame is definitely discarded.

With those definitions, we can write

$$\Theta = \sum_{j=0}^{K} (C_j + T_{\mathrm{oh},j} + Z_j(U)) = \sum_{j=0}^{K} \left(\sum_{i=1}^{B_j} \tilde{X}_{ij} + T_{\mathrm{oh},j} + Z_j(U) \right) \tag{9.133}$$

where $0 \leq K \leq m$,

Let us first address the simpler case where the payload time is fixed, $U = a_1$, and let us restrict our attention to the BA mode. A transmission time lasts $T \equiv T_{oh} + a_1$, independently of the number of transmitting stations. Therefore, $Z_j(U) = a_1$ and $T_{\mathrm{oh},j} = T_{oh}$ for all j, irrespective of the outcome of the transmission (success or collision). From the general expression of Θ, we derive for this special case:

$$\Theta = \sum_{j=0}^{K} C_j + (K + 1)T_{oh} + (K + 1)a_1 = \sum_{j=0}^{K} C_j + (K + 1)T \tag{9.134}$$

where we let $T = T_{oh} + a_1$ for ease of notation.

The Laplace transform of the PDF of $C_j = \sum_{i=1}^{B_j} \tilde{X}_{ij}$ is

$$\varphi_{C_j}(s) = \frac{1}{W_j} \sum_{h=0}^{W_j-1} [\varphi_{\tilde{X}}(s)]^h, \quad j = 0, \ldots, m. \tag{9.135}$$

where, using the notation $p_e(n)$ introduced in eq. (9.127), we have

$$\varphi_{\tilde{X}}(s) = p_e(n-1)e^{-s\delta} + (1 - p_e(n-1))e^{-s(\delta+T)} \tag{9.136}$$

Putting all together, the Laplace transform of the PDF of Θ is

$$\varphi_\Theta(s) = \sum_{k=0}^{m}(1-p)p^k e^{-sT(k+1)}\prod_{j=0}^{k}\varphi_{C_j}(s) + p^{m+1}e^{-sT(m+1)}\prod_{j=0}^{m}\varphi_{C_j}(s) \tag{9.137}$$

More complex expressions hold in case of variable payload size. We introduce the random variable

$$A_k = \begin{cases} U & k = 0, \\ \sum_{j=0}^{k-1} Z_j(U) + U & k = 1, \dots, m \\ \sum_{j=0}^{m} Z_j(U) & k = m+1. \end{cases} \tag{9.138}$$

where $Z_j(U)$ corresponds always to collision events. We can now write the expression of Θ more compactly:

$$\Theta = \sum_{j=0}^{K} C_j + \sum_{j=0}^{K} T_{oh,j} + A_{K+\epsilon_K} \tag{9.139}$$

where $\epsilon_K = 1$ if and only if $K = m$ and the last retransmission fails, otherwise it is $\epsilon_K = 0$. The expression of eq. (9.137) generalizes to:

$$\varphi_\Theta(s) = \sum_{k=0}^{m}(1-p)p^k \Phi_k(s)\prod_{j=0}^{k}\varphi_{C_j}(s) + p^{m+1}\Phi_{m+1}(s)\prod_{j=0}^{m}\varphi_{C_j}(s) \tag{9.140}$$

where

$$\Phi_k(s) = \begin{cases} e^{-sT_{oh}(k+1)}\varphi_{A_k}(s) & \text{BA mode}, \\ e^{-skT_{oh,c}}e^{-sT_{oh,s}}\varphi_{A_1}(s) & \text{RTS/CTS mode}. \end{cases} \tag{9.141}$$

for $k = 0, \dots, m$, and

$$\Phi_{m+1}(s) = \begin{cases} e^{-sT_{oh}(m+1)}\varphi_{A_{m+1}}(s) & \text{BA mode}, \\ e^{-sT_{oh,c}(m+1)} & \text{RTS/CTS mode}. \end{cases} \tag{9.142}$$

The Laplace transform of the PDF of the random variable A_k is

$$\varphi_{A_k}(s) = \begin{cases} \sum_{i=1}^{\ell} q_i e^{-a_i s}[g_i(s)]^k & k = 0, \dots, m, \\ \sum_{i=1}^{\ell} q_i[g_i(s)]^{m+1} & k = m+1. \end{cases} \tag{9.143}$$

with

$$g_i(s) = \sum_{j=1}^{\ell} e^{-s\max\{a_i,a_j\}}\frac{P_j(n-1) - P_{j-1}(n-1)}{p} \tag{9.144}$$

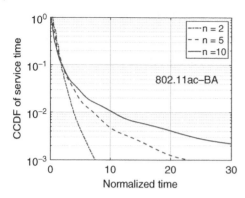

Figure 9.35 CCDF of service time. Time is normalized with respect to the mean service time $E[\Theta]$ (for the three curves it is $E[\Theta] = 0.55$ ms, 1.43 ms, 3.05 ms for $n = 2, 5, 10$, respectively).

where $P_j(n-1) = (1 - \tau + \tau Q_j)^{n-1}$, for $j = 0, \ldots, \ell$. Obviously, the Laplace transform of the PDF of $\tilde{X}$ must be modified as well, i.e., it is:

$$\varphi_{\tilde{X}}(s) = p_e(n-1)e^{-s\delta} + \sum_{i=1}^{\ell} e^{-sa_i}[P_i(n-1) - P_{i-1}(n-1)] \qquad (9.145)$$

Inversion of the Laplace transform can be accomplished numerically (see the Appendix at the end of the book). The CCDF of the service time is plotted in Figure 9.35 for $n = 2, 5, 10$ stations, in the case of IEEE 802.11ac for fixed payload size of 1500 bytes. Time is normalized with respect to the mean service time $E[\Theta]$. It is $E[\Theta] = 0.55$ ms, 1.43 ms, 3.05 ms for $n = 2, 5, 10$, respectively.

As the number of contending stations grows, the tail of the CCDF becomes heavier. For $n = 10$, the service time exceeds about 10 times the mean service time with probability 0.01. In other words, the 99-quantile of the service time is about $10 \cdot E[\Theta] \approx 30$ ms. Such a large delay may affect adversely real-time services, e.g., VoIP and streaming. From this picture of the CCDF of the service time we expect high variability of service time (jitter).

Moments can be found by derivation of the Laplace transform of the PDF of Θ. For a random variable V with Laplace transform of the PDF given by $\varphi_V(s)$, the mean is given by $-\varphi'_V(0)$, while the variance equals $\varphi''_V(0) - [\varphi'_V(0)]^2$

The mean service time is

$$E[\Theta] = \sum_{k=0}^{m}(1-p)p^k\overline{\Theta}(k) + p^{m+1}\overline{\Theta}(m+1) \qquad (9.146)$$

where:

$$\overline{\Theta}(k) = \begin{cases} (k+1)T_{oh} + E[A_k] + E[\tilde{X}]\sum_{j=0}^{k}\dfrac{W_j - 1}{2} & k = 0, \ldots, m \\[2ex] (m+1)T_{oh} + E[A_{m+1}] + E[\tilde{X}]\sum_{j=0}^{m}\dfrac{W_j - 1}{2} & k = m+1 \end{cases} \qquad (9.147)$$

The mean service time is tied to the saturation throughput of the tagged station. Let λ_1 be the successfully delivered frame rate of the tagged station. We have $\lambda_1 = (1 - p^{m+1})/E[\Theta]$, since p^{m+1} is the probability that the frame is ultimately dropped, after $m + 1$ unsuccessful transmission attempts. On the other hand, the mean throughput rate of the tagged station can also be written as $\lambda_1 = (\rho/E[U])/n$. Therefore, we have

$$E[\Theta] = \frac{(1 - p^{m+1})E[U]n}{\rho} \tag{9.148}$$

The variance of service times can be calculated as soon as we know the second moment, since $\sigma_\Theta^2 = E[\Theta^2] - E[\Theta]^2$. As for the second moment of the service time, we have:

$$E[\Theta^2] = \sum_{k=0}^{m}(1 - p)p^k(\overline{\Theta}(k)^2 + \sigma_{A_k}^2) + p^{m+1}(\overline{\Theta}(m + 1)^2 + \sigma_{A_{m+1}}^2) +$$

$$+ \sigma_{\tilde{X}}^2 \sum_{k=0}^{m} p^k \frac{W_k - 1}{2} + E[\tilde{X}]^2 \sum_{k=0}^{m} p^k \frac{W_k^2 - 1}{12} \tag{9.149}$$

Mean and variance of the random variable A_k can be calculated from its Laplace transform, for $k = 0, \dots, m$.

In the special case of fixed payload size, $E[A_k] = (k + 1)a_1$ and $\sigma_{A_k} = 0, \forall k$. Moreover, it is:

$$\begin{cases} E[\tilde{X}] = \delta + (T_{oh} + a_1)[1 - p_e(n - 1)] \\ \sigma_{\tilde{X}}^2 = (T_{oh} + a_1)^2 p_e(n - 1)[1 - p_e(n - 1)] \end{cases} \tag{9.150}$$

It is seen that the source of jitter in the service times lies in repeated transmission attempts, in case of collision, and in the randomized countdown, hence on the contention window sizes. While the former component cannot be removed, the latter one can be greatly reduced by optimizing the contention window sizes. We will see instead that BEB entails a large jitter.

The coefficient of variation (COV) $C_\Theta \equiv \sigma_\Theta/E[\Theta]$ of the service time is plotted against the number of stations in Figure 9.36 in case of IEEE 802.11ac (see Table 9.3 for the numerical values of the parameters). Payloads lengths are uniformly distributed over the set $\{80, 1500, 9000, 11454\}$ bytes. Aggregation of 4 MPDUs is assumed at each transmission opportunity. Payload lengths of the aggregated MPDUs are selected independently of one another.

We select just one example, since it turns out that the COV curves are largely insensitive to assumptions on payload lengths and DCF parameter values, except of the contention window sizes. The dominant term in the service time variance is by far the contribution due to contention windows (the last sum in the right-hand side of eq. (9.149)).

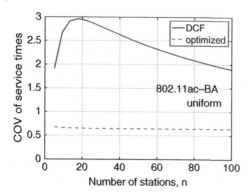

Figure 9.36 COV of service times as a function of the number n of stations for uniformly distributed payload size, in case of IEEE 802.11ac basic access (solid line: DCF; dashed line: optimized DCF).

As seen in Figure 9.36, C_Θ is well beyond 1, denoting a significant jitter of service time induced by BEB. The COV peaks around few tens of contending stations. This is explained as follows. When few stations contend, the collision probability is low and service times consist of one or few attempts. Thus only the smaller contention windows are used. On the contrary, when a large population of stations shares the channel, the variability of the service times is capped by the limited number of retries allowed by DCF.

The dashed line in Figure 9.36 corresponds to optimized DCF, i.e., the transmission probability is chosen so as to maximize the saturation throughput, for each given number of stations. When DCF is optimized, BEB is disabled and the contention window is set to $W^*(n) = 2/\tau^*(n) - 1$, where $\tau^* = \tau^*(n)$ is the optimal transmission probability, which is a function of n. It is apparent that removing BEB reduces the jitter of service time drastically. This brilliant result is achieved provided that the optimal transmission probability is set for each given value of n. In a practical implementation, estimating the number of contending stations is difficult. However, we will see that it is possible to estimate a *proxy* of the number of stations and that is enough to achieve (almost) optimal performance. A practical approach in this sense is offered by Idle Sense (see Sec. 9.5.3.1).

9.5.2.5 Correlation between Service Times

We have assumed service times of a tagged station form a renewal process. In this section we explore the *correlation* among service times, showing that it is generally weak enough that it can be neglected with little harm. To this end, we resort to simulations. We consider the IEEE 802.11ac standard DCF with BA mode (see Table 9.3 for parameter values). The payload size is uniformly distributed across the values obtained by considering different packet lengths and nine code and modulation set values. Assuming a radio channel of 40 MHz and short-prefix OFDM symbols, ten air bit rates are possible: 15, 30, 45, 60, 90, 120, 135, 150, 180, 200 Mbit/s. Frame payload lengths take one of the following values: $80, 1500, 9000, 11454$ bytes. MAC frame aggregation is used: a station winning the channel sends 4 MAC PDUs. Combining payload lengths with air bit rates, overall we find $\ell = 35$ different values of the payload times a_i. We assume a uniform probability distribution for these values, i.e., $q_i = 1/\ell$, $i = 1, \ldots, \ell$.

We estimate Pearson's correlation coefficient between service times via simulations. The coefficient at lag k is denoted with $r(k)$. For a sample of n service times θ_j, $j = 1, \ldots, n$, the correlation coefficient is:

$$r(k) = \frac{1}{n-k-1} \sum_{h=1}^{n-k} \frac{\theta_h - \overline{\theta}}{s_\theta} \frac{\theta_{h+k} - \overline{\theta}}{s_\theta} \tag{9.151}$$

where $\overline{\theta}$ and s_θ are estimates of the mean and of the standard deviation of the service times. At $k = 0$ we have $r(0) = 1$. For all other lags k, it is $|r(k)| \le 1$, the smaller being $|r(k)|$ the weaker the correlation among the samples.

Figure 9.37(a) plots the values of Pearson's coefficient $r(k)$ for k ranging between 0 and 20, for a number of sources ranging from $n = 1$ up to $n = 20$.

Simulation results indicate that service times are essentially uncorrelated. The absolute value of the coefficient $r(k)$ is smaller than 0.0547 for all $k \ge 2$ and for all values of n between 1 and 20. As for lag 1, we have $-0.0243 \le r(1) \le 0.0202$ for n ranging from 1 to 20. We see that we make a minor approximation considering the sequence of the service times of a tagged station as a renewal process.

A sample path of the service counting function is shown in Figure 9.37(b). The high variability of service times is evident from this sample path: sequences of relatively small service times (steep climb of the curve) alternate with very long service times (plateaus of the counting function).

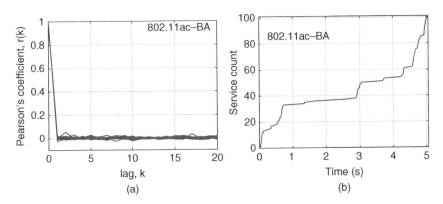

Figure 9.37 Simulation of WiFi DCF for IEEE 802.11ac BA mode with uniformly distributed payload times. Left plot: Pearson's correlation coefficient between service times. Right plot: counting function of service times of a tagged station.

9.5.3 Optimization of Back-off Parameters

We consider two optimization problems:

- Maximization of throughput
- Minimization of service time variance (jitter)

In both cases we adjust the back-off parameter τ (or, equivalently, the contention window size W) to attain the optimization target. In the rest of this section we use a superscript DCF to denote parameter values set according to the standard.

9.5.3.1 Maximization of Throughput

We can write the saturation throughput of a set of n stations, under the hypotheses laid down in previous sections, as:

$$\rho = \rho_{\text{ideal}} \frac{p_s(T_{oh} + \mathrm{E}[U])}{\delta + p_s(T_{oh} + \mathrm{E}[U]) + p_c T_c} \tag{9.152}$$

where T_c is the (average) duration of the collision event, $p_s = n\tau(1 - \tau)^{n-1}$ is the success probability, $p_e = (1 - \tau)^n$ is the probability of an idle back-off slot, and $p_c = 1 - p_e - p_s$ is the collision probability.

Since it is difficult to give an accurate value to the collision duration, which depends on many parameters of the physical and MAC layers, we lower bound the throughput, by substituting T_c with an upper bound $T_{c,\max}$ of the collision duration, e.g., the maximum value of the transmission opportunity, which is a well-defined parameter for each specific 802.11 version. We will maximize the lower bound $\tilde{\rho}$ of the throughput. To that end, we can minimize the following cost function:

$$\frac{\rho_{\text{ideal}}}{\tilde{\rho}} = \frac{1}{p_s} \frac{\delta}{T_{oh} + \mathrm{E}[U]} + 1 + \frac{p_c}{p_s} \frac{T_{c,\max}}{T_{oh} + \mathrm{E}[U]} \tag{9.153}$$

as a function of τ. Removing additive terms that do not depend on τ and scaling the function by a positive constant, we can verify that the function to be minimized depends only on τ, n and on the nondimensional parameter $\eta = 1/(1 + \delta/T_{c,\max})$. Representative values for η are $\eta \approx 0.9697$ for IEEE 802.11g and $\eta \approx 0.9955$ for IEEE 802.11ac.

Setting to zero the first derivative of (9.153) with respect to τ, we find that the optimal value of τ must satisfy:

$$\eta(1 - \tau)^n = 1 - n\tau \tag{9.154}$$

A simple study of the functions of τ appearing on the two sides of (9.154) shows that there exists a unique solution $\tau^* \in (0, 1)$. The quantity τ^* is a function of the number of stations n. Once τ^* is known, the corresponding optimal size of the contention window is $W^* = 2/\tau^* - 1$. This identity follows from the fixed

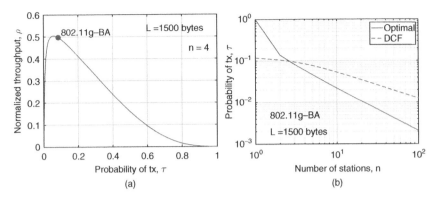

Figure 9.38 Left plot: throughput as a function of τ for $n = 4$ stations and fixed payload size; the dot mark corresponds to the throughput of IEEE 802.11g standard. Right plot: probability of transmission τ as a function of the number of stations n: comparison between optimized (solid line) and standard (dashed line) values.

point equations (9.115), if we use the same window size at any (re)transmission stage, i.e., $b_0 = \cdots = b_m = (W^* + 1)/2$. Note that optimizing the contention window sizes for maximum throughput automatically implies that BEB should be removed. In fact, BEB is but a rough attempt to adapt the contention window size to the contetion level in the network. Note also that the optimization *requires the knowledge of the number n of contending stations.*

To grasp the relationship between the standard DCF mode of operation (BEB) and the optimized contention window size, Figure 9.38(a) shows the throughput for $n = 4$ as a function of τ. The dot marks the throughput level of IEEE 802.11g.

The value of ρ as a function of τ has the bell shape typical of CSMA-like protocols. There is clearly an optimal value of τ. The standard DCF (circle marker) gets quite close to the optimum in this case ($n = 4$ stations, IEEE 802.11g parameters, fixed payload length equal to 1500 bytes).

However, the standard DCF may depart signficantly from the optimal working point as hinted by Figure 9.38(b), where the optimal value τ^* is compared to τ^{DCF} resulting from DCF, as a function of n for IEEE 802.11g and a fixed payload length of 1500 bytes.

It can be seen from Figure 9.38(b) that τ_{DCF} is far from τ^*, except around a specific value of n ($n = 3$ in this case). This is no exception. It can be verified that, as we vary the payload length from few tens of bytes up to few thousands of bytes, the number of stations for which $\tau^{DCF} \approx \tau^*$ lies between $n = 2$ and $n = 4$. The gap between τ^* and τ^{DCF} widens as n grows.

Remarkably, τ^* is inversely proportional to n (Figure 9.38(b) has a log-log scale), i.e., we have $\tau^* \approx \alpha(n)/n$, where $\alpha(n)$ solves the equation $\eta(1 - \alpha/n)^n = 1 - \alpha$. The

Table 9.4 Optimal parameter values in the limit for $n \to \infty$.

Symbol	IEEE 802.11g	IEEE 802.11ac
η	0.9697	0.9955
α^*	0.2278	0.0918
τ^*	$0.2278/n$	$0.0918/n$
W^*	$7.75 \cdot n$	$21.8 \cdot n$
$\overline{N}_{e,\infty}^*$	3.91	10.40

inverse proportion dependence of τ^* on n suggests that $\alpha(n)$ depends weakly on n. In the limit for $n \to \infty$, we obtain $\eta e^{-\alpha} = 1 - \alpha$. Hence, we have $\tau^* \approx \alpha^*/n$, where α^* is the unique root of $1 - \alpha = \eta e^{-\alpha}$ in the interval $(0, 1)$. Solving for the example value of η of IEEE 802.11g, we obtain $\alpha^* \approx 0.2278$. In case of IEEE 802.11ac, we get $\alpha^* \approx 0.0918$. The corresponding optimal window sizes are $W^* = 2n/\alpha^* - 1 \approx 7.75 \cdot n$ for IEEE 802.11g and $W^* \approx 21.8 \cdot n$ for IEEE 802.11ac.

A small implementation detail is that the optimal contention window size is noninteger in general. This can be dealt with by assuming that each time a new value of the back-off counter is drawn, which must be uniformly distributed over $[0, W - 1]$ with $\lceil W \rceil \neq \lfloor W \rfloor$, first the contention window size is rounded randomly, i.e., it is set to $\lfloor W \rfloor$ with probability $\lceil W \rceil - W$, otherwise it is set to $\lceil W \rceil$.

All stations need to do is to adjust their contention window sizes in an optimal way. We could achieve this result, if the number of contending stations on the IEEE 802.11 channel were known. This is by no means a trivial task. However, we do not need really to estimate n, rather we can use a proxy of n. As a matter of fact, to make this theory applicable to a practical network, we need an observable quantity that is directly related to the optimal τ. We now elaborate such a quantity.

An idle back-off slot appears with probability p_e, hence the mean number of consecutive idle back-off slots is $\overline{N}_e = p_e/(1 - p_e) = (1 - \tau)^n/[1 - (1 - \tau)^n]$. At the optimum point, it is $\overline{N}_e^* = (1 - \tau^*)^n/[1 - (1 - \tau^*)^n]$, which is a function of n. It turns out that, for a given value of η, $\overline{N}_e^*$ is weakly dependent on n and converges quickly to the asymptotic value for $n \to \infty$. Since $\tau^* \sim \alpha^*/n$, it is $\overline{N}_e^* \approx \overline{N}_{e,\infty}^* = \frac{1}{e^{\alpha^*} - 1}$. With the numerical example numbers we are using, it is $\overline{N}_{e,\infty}^* \approx 3.9$ for IEEE 802.11g and $\overline{N}_{e,\infty}^* \approx 10.4$ for IEEE 802.11ac. The optimal parameter values in the limit for $n \to \infty$ are summarized in Tab. 9.4.

The key point is that *if the throughput is optimized, the average number of idle back-off slots separating two successive transmission attempts must equal $\overline{N}_e$.*

Algorithm   Pseudo-code of Idle Sense [95, 102].

```
 1: maxtx ← 5;   sum ← 0;   ntx ← 0;
 2: After each transmission on the channel do:
 3: N_e ← number of idle slot preceding tx
 4: sumtx ← sumtx+N_e;
 5: ntx ← ntx+1;
 6: if ntx >= maxtx then
 7:    Ñ_e ← sum/ntx;
 8:    sum ← 0
 9:    ntx ← 0
10: end if
11: if Ñ_e < N_{e,tgt} then
12:    CW ← CW + λ;
13: else
14:    CW ← μ · CW;
15: end if
16: if |Ñ_e − N_{e,tgt}| < Δ then
17:    maxtx ← CW/γ;
18: else
19:    maxtx ← 5;
20: end if
```

In turn, $\overline{N}_e$ depends only weakly on n, hence we can compare the value of $\overline{N}_e$ estimated from wireless channel observations with the target value $\overline{N}^*_{e,\infty}$ for any n. This is the theoretical ground that the Idle Sense algorithm is based on [95, 102].

The idea of the Idle Sense algorithm is to adjust the contention window size as follows:

- If $\overline{N}_e > \overline{N}^*_{e,\infty}$ is observed, then too many idle back-off slots are left over by the contending stations, hence they should become more aggressive, which is obtained by reducing the contention window size.
- If $\overline{N}_e < \overline{N}^*_{e,\infty}$ is observed, stations are being too aggressive, since too few idle back-off slots separate two consecutive transmissions. Therefore, the contention window size should be increased.

This basic idea is implemented according to an additive increase, multiplicative decrease (AIMD) approach. The contention window size is increased by adding a constant increment, while it is decreased multiplying by a factor less than 1. The algorithm is completed by adaptive adjustment of the number of samples used to estimate $\overline{N}_e$.

The algorithm listed above reproduces the algorithm proposed in [95], based on the theory developed in [102] and recalled above. In [102] the authors suggest the following parameter value set for IEEE 802.11g: $N_{e,tgt} = 3.91$, $\lambda = 6$, $\mu = 1/1.0666$, $\Delta = 0.75$, $\gamma = 4$.

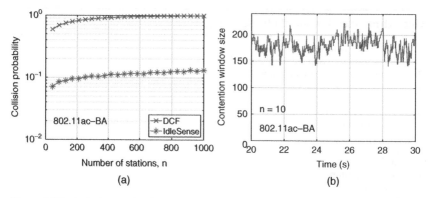

Figure 9.39 Left plot: probability of collision as a function of the number of stations, comparison between Idle Sense and DCF. Right plot: time evolution of the contention window of a tagged station running the Idle Sense algorithm ($n = 10$ stations).

The collision probability resulting from Idle Sense algorithm is compared with DCF in Figure 9.39 against the number of stations (left plot). The time evolution of the contention window size of a tagged station is plotted as well (right plot).

Not only does Idle Sense achieve a collision probability level one order of magnitude smaller than the standard DCF; it also manages to maintain the collision probability essentially insensitive to the growing number of contending stations. The fluctuations of the contention window size of a tagged station are seen to be relatively limited. The optimal contention window size hovers around 200, much bigger than the default base contention window, which is 16, yet much smaller than the maximum contention window according to the standard DCF (1024). It is apparent that the rough contention window size adjustment attempted by BEB falls short of bringing the system to work at an optimal regime. Note however that the adaptive optimization introduced by the Idle Sense algorithm requires setting a number of parameters. A critical one is the threshold on the average number of idle back-off slots between two consecutive transmissions. The threshold that corresponds to the maximum throughput for the given network *depends* on all variables influencing the throughput, namely the probability distribution of payloads and the details of the frame format and control frames of the specific brand of IEEE 802.11.

The throughput achieved by Idle Sense is compared with DCF throughput and ideal throughput as a function of n in Figure 9.40 for the IEEE 802.11ac and in Figure 9.41 for the IEEE 802.11g. Idle Sense guarantees highly stable performance levels in spite of large excursion of the number of contending stations (three orders of magnitude, from 1 to 1000). On the contrary, DCF performance decay as n grows. There is still a gap between the constant level of throughput achieved by Idle Sense and the ideal throughput. The gap is entirely due to collisions. To

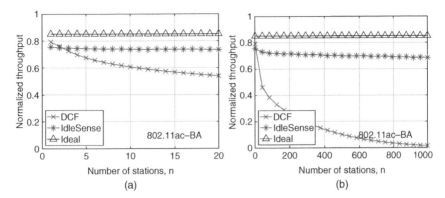

Figure 9.40 Throughput as a function of the number of stations for IEEE 802.11ac: comparison between DCF, Idle Sense and the ideal value.

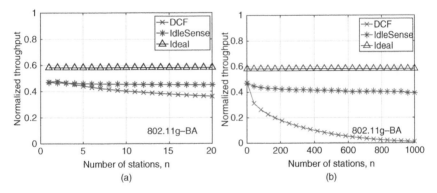

Figure 9.41 Throughput as a function of the number of stations for IEEE 802.11g: comparison between DCF, Idle Sense, and the ideal value.

improve further the multiple access performance calls for a structural change of the contention mechanism (e.g., see [20, 21]).

Besides Idle Sense, several algorithms have been devised to estimate n and hence adapt the random access parameters of CSMA/CA (e.g. see [34, 202, 116]), showing that accurate and fast estimation of n, i.e., of the number of stations actually contending for the channel, is possible in infra-structured WLANs.

9.5.3.2 Minimization of Service Time Jitter

We can search for the contention window sizes that minimize the COV of the service time for a given saturation throughput.

There is an optimal value τ^* that yields the maximum throughput $\rho^* \equiv \rho(\tau^*)$. The function $\rho(\tau)$ is strictly increasing for $\tau \in [0, \tau^*]$ from 0 up to ρ^* and strictly decreasing for $\tau \in [\tau^*, 1]$ from ρ^* down to 0. In the following we consider the two intervals separately, so that we can deal with $\rho(\tau)$ as an invertible function under the restriction $\tau \in [0, \tau^*]$ or $\tau \in [\tau^*, 1]$. With this convention, requiring that ρ takes a given value yields a unique value of τ, hence also of p and $E[\Theta]$, for fixed values of n and $E[U]$. Therefore, minimizing the jitter of the service time for a given throughput, i.e., $\min\{C_\Theta\}$ for $\rho = \rho_0$, is equivalent to minimizing $E[\Theta^2]$ for $\rho = \rho_0$.

Since τ and p are given, eq. (9.113) implies that the contention window sizes must satisfy

$$\sum_{k=0}^{m} p^k W_k = \left(\frac{2}{\tau} - 1\right) \sum_{k=0}^{m} p^k \tag{9.155}$$

We compare the following two approaches.

- Stick to the standard DCF and maintain BEB, i.e., contention window sizes grow geometrically with the retransmission stage.
- Give up to BEB and pursue the minimization of $E[\Theta^2]$ given ρ, with the only constraint that the contention window sizes form a nondecreasing sequence.

In the first case, we develop the DCF-compliant trade-off of the jitter of service time against the throughput. The degrees of freedom to realize the trade-off are the contention window sizes $W_0, \ldots, W_m$, under the constraint (9.155) for a given ρ, hence a given τ. Let $\{W_k^{\mathrm{DCF}}\}_{0 \le k \le m}$ denote the default contention window sizes of the DCF standard. Maintaining BEB implies that we can change window sizes *only by scaling them by a common factor*, i.e., $W_k = \zeta W_k^{\mathrm{DCF}}$ for $k = 0, \ldots, m$. The constraint (9.155) implies that the scale factor ζ must satisfy

$$\zeta = \frac{(2/\tau - 1) \sum_{k=0}^{m} p^k}{\sum_{k=0}^{m} p^k W_k^{\mathrm{DCF}}} = \frac{2/\tau - 1}{2/\tau^{\mathrm{DCF}} - 1} \tag{9.156}$$

where τ^{DCF} is the transmission probability with the standard parameter values of DCF. The trade-off jitter-throughput is obtained with the following steps

1. For a given throughput $\rho_0 \in (0, \rho^*)$ find the unique value of $\tau \in [0, \tau^*]$ such that $\rho(\tau) = \rho_0$.
2. Find the contention windows scaling factor ζ by using (9.156) for the given τ.
3. Evaluate $E[\Theta^2]$ for the given τ, with the contention window sizes $W_k = \zeta W_k^{\mathrm{DCF}}$ for $k = 0, \ldots, m$.

This procedure yields one branch of the throughput-jitter trade-off curve. The entire procedure can be repeated only modifying the first step, by restricting $\tau \in [\tau^*, 1]$. This way, we get the other branch of the trade-off curve.

The second approach aims at minimizing the service time jitter for a given throughput. Since the mean service time is constrained, minimizing the jitter amounts to minimize $E[\Theta^2]$. Recalling (9.149), we reckon that $E[\Theta^2]$ is a quadratic form in the variables W_k, $k = 0, \dots, m$. By accounting for the constraint (9.155), minimizing $E[\Theta^2]$ is equivalent to minimizing the function

$$f(\mathbf{W}) \equiv c + \sum_{k=0}^{m}(1-p)p^k\overline{\Theta}(k)^2 + p^{m+1}\overline{\Theta}(m+1)^2 + E[\tilde{X}]^2 \sum_{k=0}^{m} p^k \frac{W_k^2 - 1}{12}$$

(9.157)

where $\mathbf{W} \equiv [W_0, W_1, \dots, W_m]$, the $\overline{\Theta}(k)$ are given in eq. (9.147), and c does not depend on the contention window sizes.

It is apparent that $W_0 \leq W_1 \leq \cdots \leq W_m$ implies that to minimize $f(\mathbf{W})$ it is necessary and sufficient that $W_k = W_0$, $k = 1, \dots, m$, i.e., all contention window sizes must have the same value. The common size of the contention windows follows from the constraint (9.155): it equals $W_0 = 2/\tau - 1$.

The argument above shows that *the service time jitter is minimized by setting all contention windows to a same size, under the joint constraint that the saturation throughput be assigned a given value and that the contention window sizes form a nondecreasing sequence.* This result sheds some light on the BEB mechanism, pointing out that it hinders smooth service of MAC frames, while gaining nothing from the throughput side.

As a numerical example, let us consider an IEEE 802.11ac network, whose parameters are listed in Table 9.3. The payload length has the same distribution and parameter values as in the example of Figure 9.36.

Figure 9.42 shows the trade-off between the COV of service time $C_\Theta \equiv \sigma_\Theta/E[\Theta]$ and the throughput, for $n = 2, 5, 10, 100$ stations, from top-left to right-bottom. Solid lines refer to the DCF-compliant trade-off, with contention window sizes scaled by the factor ζ. Dashed lines correspond to the selection of contention window sizes that minimizes service time jitter for any given throughput level. To minimize the service time jitter, we set the contention windows all equal to a same value, $2/\tau - 1$, where τ is set according to the desired throughput. The lower branch of the curves is obtained for $\tau \in [0, \tau^*]$. The upper branch is obtained for $\tau \in [\tau^*, 1]$. The two branches join where the throughput is maximum. The black dot marks the trade-off of the standard DCF (nominal values of the contention window sizes).

The best regime lies in the right-bottom corner of the diagram: low jitter, high throughput. The standard DCF achieves the optimal working point only for $n = 2$. If window sizes are not dimensioned properly for the given n, it can depart significantly from the optimal regime, both because of high service time variability, even for as few as five stations, or because of low throughput, e.g., for very crowded networks ($n = 100$). Only for $n = 2$ does DCF attain the best possible working point.

Giving up BEB, and setting the contention window sizes all to the same value yields a much better trade-off, especially from the point of view of service time variability. The resulting COV never exceeds 1. It turns out that the trade-off curve has

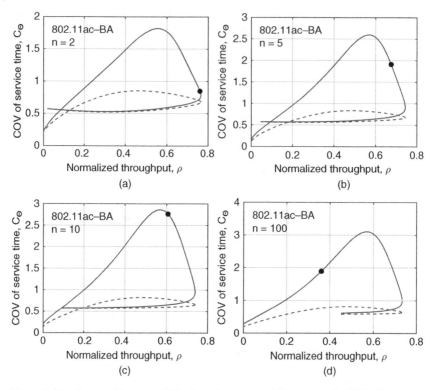

Figure 9.42 Trade-off between COV of service time and throughput for IEEE 802.11ac. Payload lengths are uniformly distributed over values ranging from 80 up to 11454 bytes per MPDU. The black dot marks the working point of standard DCF.

a narrow cusp around the maximum throughput point. As the contention window size moves off the optimal value, the achieved throughput departs sharply from its maximum. This points out that the optimal transmission probability (or, equivalently, the optimal contention window size) must be matched quite accurately to maximize the saturation throughput. The optimal values of the contention window size are $W^* = 25, 80, 170, 1817$, for $n = 2, 5, 10, 100$, respectively. As we move away from W^* the achieved throughput level falls off rapidly, while the service time jitter does not vary sensitively.

Interestingly, we can conclude that BEB entails a less favorable trade-off than setting a unique (suitably chosen for each given number n of stations) contention window size. The jitter of the service time can be minimized *without loss of throughput*, by choosing the contention window size W^* for all retransmission stages. In other words, setting the contention window to W^* achieves the maximum possible throughput and, for that throughput level, minimizes the service time jitter. That W^* is exactly the optimal contention window size that the Idle Sense algorithm tends to.

9.5.4   Fairness of CSMA/CA

We devote this last section on WiFi to fairness. This concept addresses the way stations share the capacity of the channel among them.

A first nontrivial point is to state what we mean precisely with fairness. Given a fixed capacity link, sharing it fairly brings naturally to our minds that each station should get a same share of the overall available capacity. Things become immediately less crisp as soon as we consider individual limitations on the competing stations. For example, with wireless channels, stations may experience different channel quality. Or they could use different standard versions, with different capabilities. Or they could offer traffic flows with different parameters (e.g., a file transfer with large packets, 1500 byte long, versus a VoIP flow with short packets, 80 byte long).

All those examples entail that it is not necessarily possible or wise to impose that each contending station get a same share of the overall capacity. There could even not exist such a concept as "*the* link capacity," i.e., not a unique value for all stations. Therefore, besides equal shares fairness (the so called *output fairness*), other types of fairness have been defined (e.g., max-min fairness, proportional fairness; see Chapter 10).

Three issues need be considered.

- *Long-term fairness*: Do stations get an equal share of the channel capacity over a time horizon much longer than the frame transmission time (ideally, extending to infinity)?
- *Short-term fairness*: Do stations manage to access the channel regularly on time scales comparable with the frame transmission time?
- *Priority handling, scheduling*: How is it possible to assign different shares of capacity to stations?

In the context of WiFi, we will first show that *a CSMA/CA network is long-term fair in the sense of transmission opportunities, under the proximity, symmetry and saturation assumptions*. A transmission opportunity occurs when a station wins the channel for transmitting its own data frames. According to the IEEE 802.11 MAC, a station can send up to some amount of bytes or maintain the use of the channel up to some maximum time. Those limits define the extent of the transmission opportunity. A station heavily loaded with traffic will exploit all of its allowed time and bytes on each opportunity, whereas a lightly loaded station could possibly only send short messages, when it gets the use of the channel.

We can show that WiFi is fair in the sense that it guarantees that a same fraction of the transmission opportunities is granted to each contending station asymptotically, as the observed time horizon tends to infinity. This holds provided that stations are continuously backlogged (saturation), that they can hear each other (proximity) and that they use the same values of the access parameters (symmetry). This result *does not imply at all* that stations get a same throughput (unless their frames have statistically identical lengths, air bit rates are the same, radio channels have the same characteristics).

Let us start by showing the following.

Theorem 9.1 An isolated WiFi network is long-term fair in terms of transmission opportunities under the proximity, saturation and symmetry assumptions.

Proof: Let us consider a tagged station. Let J denote the number of other stations being served between two successful service events of the tagged station. We restrict here our attention only to successful service events.

Other stations can perform successful transmission during the tagged station countdown. The number G of successful transmissions of other stations when the tagged station counts down a back-off B extracted from the interval $[0, W - 1]$ uniformly at random has a probability distribution $P(G = j) = g_j, j = 0, \ldots, W - 1$. Conditional on $B = h$, g_j is a Bernoulli probability distribution. The generating function of G is given by

$$\phi_G(z) = \frac{1}{W} \sum_{h=0}^{W-1} [1 - p_s(n - 1) + p_s(n - 1)z]^h \tag{9.158}$$

where $p_s(n - 1) = (n - 1)\tau(1 - \tau)^{n-2}$ $(n \geq 2)$ is the success probability of the other $n - 1$ stations, when the tagged one is forced idle by the countdown.

Let G_k denote the random variable G when the contention window size is $W = W_k$, $k = 0, \ldots, m$. Let K denote the number of retransmissions of the tagged station, until a successful frame delivery is achieved. Note that there could be frames discarded because the maximum number of retries is exhausted in between two consecutive successful frame deliveries. Thanks to the independence hypothesis, the random variable K is geometrically distributed: $P(K = k) = (1 - p)p^k$, $k \geq 0$.

Taking into account BEB, the generating function of $J = \sum_{k=0}^{K} G_k$ is given by

$$\phi_J(z) = \sum_{k=0}^{\infty} (1 - p)p^k \prod_{j=0}^{k} \phi_{G_{j \bmod(m+1)}}(z) = \frac{\sum_{k=0}^{m} (1 - p)p^k \prod_{j=0}^{k} \phi_{G_j}(z)}{1 - p^{m+1} \prod_{j=0}^{m} \phi_{G_j}(z)} \tag{9.159}$$

The mean of J can be calculated either directly from the definition of J or taking the derivative of $\phi_J(z)$ and setting $z = 1$. We find:

$$E[J] = \phi_J'(1) = \frac{\sum_{k=0}^{m} (1 - p)p^k \sum_{j=0}^{k} \phi_{G_j}'(1) + p^{m+1} \sum_{j=0}^{m} \phi_{G_j}'(1)}{1 - p^{m+1}} \tag{9.160}$$

Since $\phi'_{G_j}(1) = p_s(n-1)\frac{W_j-1}{2}$, rearranging terms in (9.160), we get

$$E[J] = p_s(n-1)\frac{\sum_{k=0}^{m} p^k \frac{W_k-1}{2}}{1-p^{m+1}} = p_s(n-1)\frac{1-\tau}{\tau(1-p)} = n-1 \qquad (9.161)$$

The last result derives from eq. (9.113), the expressions of $p_s(n-1)$ and p as functions of τ and n.

The mean number of service instances to stations other than the tagged one between two consecutive successful transmissions of the tagged station is $n-1$. This means that *on average*, the tagged station *gets one successful transmission opportunity every n*. With n stations contending for access, this is just the fair share of transmission opportunities that the tagged station has a right to. ∎

Some caveats are in order in the face of this result. While in the long-term the fraction of transmission opportunities obtained by each station among the n contending ones is $1/n$, in the short-term access to the channel can be quite unfair. We define *service gap* the number J of frames of stations other than the tagged one transmitted in between two consecutive successful transmissions of the tagged station. We will see that the "service gap" has a large variability, the more the larger the number of contending stations. The main responsible for that is BEB and the freezing of back-off counters. The former tends to put at a disadvantage stations that suffer a collision with respect to other contending stations. The latter introduces a memory that discriminates further stations suffering a collision.

In case of optimization of CSMA/CA for maximum throughput and minimum service time jitter, we know that $W_k = W^* = 2/\tau^* - 1$, $k = 0, \ldots, m$, where τ^* is the transmission probability that maximizes throughput. The CCDF of J simplifies to

$$\phi_J(z) = \frac{(1-p)\phi_{G^*}(z)}{1-p\phi_{G^*}(z)} \qquad (9.162)$$

where

$$\phi_{G^*}(z) = \frac{1}{W^*}\sum_{h=0}^{W^*-1}[1-p_s^*(n-1)+p_s^*(n-1)z]^h \qquad (9.163)$$

with $p_s^*(n-1) = (n-1)\tau^*(1-\tau^*)^{n-2}$.

We illustrate the CCDF of J for various values of n and both standard DCF (solid lines) and optimized CSMA/CA (dashed lines) in Figure 9.43.

The most striking feature of these distributions is the slow decay of the CCDF in case of standard DCF, already for $n = 5$. For example, for $n = 5$ we have $P(J \geq 36) = 0.01$, i.e., in 1% of cases there are at least 36 other frames sent in between two consecutive frames of a tagged station.

The short-term unfairness is completely mitigated for optimized CSMA/CA. In that case, the decay of the CCDF of J is much faster than with the standard DCF.

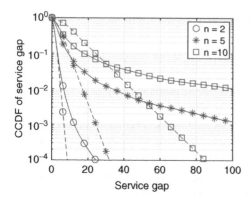

Figure 9.43 CCDF of the number of frames transmitted by stations other than a tagged one between two consecutive transmissions of the tagged station. Solid lines: standard DCF. Dashed lines: optimized CSMA/CA.

Service from the point of view of each station resembles more closely a deterministic round robin, than for standard DCF. With DCF, service from the point of view of a station appears to be highly randomized, with large excursions of the service gap between one service opportunity and the next one.

Finally, we address the analysis of what may happen when stations are not fully equivalent. Specifically, we will disclose the *performance anomaly*, an issue first identified in [101]. We maintain that back-off parameters are the same for all stations, but payload distribution may differ.

Let us consider two groups of stations. Group i consists of n_i stations and sends frames of length L_i at an air bit rate C_i, $i = 1, 2$. Let $a_i = L_i/C_i$ and assume $a_1 < a_2$. Since the back-off parameters are the same for all stations, the transmission probability is the same for all stations. The saturation throughput rate of any station of group i $(i = 1, 2)$ is

$$\mu_i = \frac{L_i\tau(1 - \tau)^{n_1+n_2-1}}{\delta(1 - \tau)^{n_1+n_2} + (n_1 T_1 + n_2 T_2)\tau(1 - \tau)^{n_1+n_2-1} + p_{c1} T_1 + p_{c2} T_2} \quad (9.164)$$

where $T_i = \delta + T_{oh} + a_i$ and

$$\begin{cases} p_{c1} = [1 - (1 - \tau)^{n_1} - n_1\tau(1 - \tau)^{n_1-1}](1 - \tau)^{n_2}, \\ p_{c2} = 1 - (1 - \tau)^{n_1+n_2} - (n_1 + n_2)\tau(1 - \tau)^{n_1+n_2-1} - p_{c1} \\ \quad = 1 - (1 - \tau)^{n_2} - n_2\tau(1 - \tau)^{n_1+n_2-1} \end{cases} \quad (9.165)$$

Note that collision time lasts $T_1 < T_2$ if only stations from group 1 are involved in the collision event. Otherwise, it lasts T_2.

For example, assume that all frames have the same length $(L_1 = L_2)$ and that group 1 stations use a high data rate, while group 2 stations are downgraded to a slower air bit rate, i.e., $C_1 > C_2$. The paradoxical result emerging from eq. (9.164) is that $\mu_1 = \mu_2$, i.e., stations of either group experience the *same* throughput rate, even if they are using possibly very different air bit rates.

To give a numerical example, let us consider IEEE 802.11g, assume $C_1 = 54$ Mbit/s, $C_2 = 6$ Mbit/s (the two extreme bit rates defined by this amendment). Let $L_1 = L_2 = 1500$ bytes.

For $n_1 = n_2 = 5$ stations, we get a throughput rate of 757 kbit/s for any station, irrespective of the group it belongs to. A station that can push its air bit rate up to the maximum rate granted by IEEE 802.11g turns out to achieve the same long-term throughput rate of a station that must transmit as slow as the basic bit rate of the standard. Although it might be conceived as a fair situation in a sense, this is clearly disturbing. To gain more insight, let us make a simple example, with $n_1 = n_2 = 1$ and $L_1 = L_2 = L$. Throughput rates simplify to

$$\mu_i = \mu = \frac{L\tau(1-\tau)}{\delta(1-\tau)^2 + (T_1 + T_2)\tau(1-\tau) + T_2\tau^2} , \quad i = 1, 2. \tag{9.166}$$

If $C_1 = C_2 = 54$ Mbit/s, we have $\mu \approx 13.99$ Mbit/s. If instead $C_1 = 54$ Mbit/s and $C_2 = 6$ Mbit/s, we have $\mu \approx 4.22$ Mbit/s. We see that one station that downgrades its own air bit rate (e.g., because of a bad radio channel to the AP), actually impairs the performance of *other* stations as well. In our example, station 1 uses the same air bit rate (54 Mbit/s) in both cases, yet is suffers a penalty in the order of 70% of its throughput rate just because another station has to downgrade its own air bit rate. The performance loss is so remarkable that it seems justified to call it "anomaly."

The source of the performance anomaly lies in the long-term fairness concept built in the first version of the CSMA/CA as implemented in WiFi, namely transmission opportunity fairness. To break this uncomfortable constraint, we should shift the fairness concept to something suitable for wireless networks. Proportional fairness is known to strike a good balance between throughput performance efficiency and guaranteeing some throughput to all stations. Given n traffic sources, say the r-th source gets an average throughput rate x_r, $r = 1, \ldots, n$. A rate assignment x_r is said to be *proportional fair* if $\sum_{r=1}^{n}(y_r - x_r)/x_r \leq 0$ for any other rate assignment y_r. The rationale of the definition is that changing rate assignment from the proportional fair one may improve the rate obtained by some source relatively, yet summing up all relative changes, losses outweigh gains.

The implementation of the proportional fairness principle into CSMA/CA brings to the so called *airtime fairness*, i.e., providing each station with an equal long-term fraction of transmission *time* on the channel [35]. Airtime fairness can be realized by using the deficit round robin (DRR) scheduler [190], a weighted form of round robin (see Sec. 6.4.5).

A recent work [104] addresses both the performance anomaly and excessive delays at station buffers, due to large buffer backlogs (bufferbloat).

Additional features of recent MAC amendments (IEEE 802.11n and IEEE 802.11ac) to manage airtime are: (i) the transmission opportunity parameter (TxOP), that defines the time a station can hold the channel; (ii) MPDU aggregation, i.e., the possibility of sending more than a single MPDU once a station grabs the channel.

A revolution of MAC is going to be laid down with the forthcoming new version of WiFi, the IEEE 802.11ax. Referring to infra-structured basic service sets, the AP plays a major role in managing radio resources. The random access component will most probably be confined to sporadic, small messages and intermittent traffic, whereas most traffic flows will be granted resources in a demand-assignment fashion. The WLAN access mode will resemble the one currently used in cellular networks much more closely than it does with legacy WLANs, where random access is essentially the only access mode used in practice.

The paradigm shift will bring new opportunities to introduce scheduling algorithms, inspired to the models presented in Chapter 6.

9.6   Further Readings

Random multiple access has been a fruitful research area for more than 40 years. There is still intense activity, especially in connection with wireless communications (which offer naturally a broadcast medium, requiring some form of multiple access coordination) and with new paradigms, such as Internet of Things (IoT), which envisages a high number of appliances, generating more or less heavy traffic both for control and information gathering.

Pointers to the literature are given for the following topics: (i) ALOHA; (ii) CSMA); (iii) WiFi MAC protocol; (iv) random access in LTE.

As for ALOHA, recent work introduced and studied coded random access [171] and specifically coded slotted ALOHA [109, 143, 194]. Paolini et al. [171] address the paradigm of coded random access applied to ALOHA. They discuss several examples of coded random access protocols, pointing at the efficiency gains that can be achieved by coding and doing successive interference cancellation. Ivanov et al. [109] analyze an all-to-all broadcast coded slotted ALOHA (B-CSA) over a packet erasure channel for a finite frame length. They apply their analysis to vehicular communication networks, highlighting the performance improvements of beaconing capacity offered by B-CSA over CSMA, under the same level of reliability. Stefanovic et al. [194] prove a bound on the achievable capacity of coded slotted ALOHA with K-MPR for any given level of a normalized load measure[13] . Lazaro et al. [143] investigate frameless ALOHA under a K-MPR receiver model and show how to optimize throughput as a function of K.

As for CSMA, apart from the classical papers of Kleinrock and Tobagi [128,129], recent lines of research are on utility maximization and CSMA networks. Network utility maximization (NUM) approaches consists of considering a social utility

---

13 MPR stands for multi-packet reception. It is the capability of the receiver to decode successfully up to K packets in a single reception event.

function of the achievable throughput in a constrained capacity system and in devising strategies to achieve the optimal throughput assignment among the competing traffic sources, possibly in a distributed and even autonomous way (i.e., without even message passing among the traffic sources). Application of the NUM principle to CSMA is dealt with in [150, 115],

CSMA schemes implemented in current technologies, such as the 802.11 DCF, expose a limited form of flexibility by enabling the dynamic configuration of contention windows and retry limits, as well as the possibility to activate or not four-way handshake mechanisms. Several research works have been focused on the optimization of these parameters as a function of the network load and topology. For example, in [146], inspired by the throughput-optimal CSMA theory (e.g., see [115][150]), the Authors present the optimal DCF, that implements in off-the-shelf 802.11 devices the principles of adaptation of contention windows and channel holding times as a function of the difference between the bandwidth demand and supply of the node.

As for CSMA networks, a number of works in the last decade have studied various aspects of networks made up of CSMA nodes. A key characteristic of these models is relaxation of the proximity assumption, i.e., nodes do not necessarily hear one another. Rather, a graph model is used to represent proximity, where an arc is laid between two nodes if one can hear the other. A fundamental analysis of the capacity region of such networks is given in [142]. Liew et al. [147] elaborate on the key factors that affect the probability distribution of the node states in CSMA networks. A full understanding of CSMA network dynamics and a fully satisfactory modeling are still open issues.

As for WiFi and its evolution, [124] surveys the development of the currently ongoing efforts to finalize the IEEE 802.11ax standard. IEEE 802.11ax will be a major break with the traditional WLAN approach, based on random multiple access. Resource reservation according to a signaling-based demand-assignment scheme will be a key ingredient of the new WiFi recipe. Overall, the system will look more like cellular networks, which is not entirely surprising, given that cellular networks are more and more devoted to a broad array of data communication applications and that WLAN are been deployed in an increasingly dense pattern, so that some coordination capability among access points is necessary.

Although not included in current standards, another promising pathway to boost wireless network capacity is full-duplex radio, which has become a viable technical solution [125, 191, 105]. For example, in [31] a cancellation capability of up to 110 dB is demonstrated over up to 80 MHz bandwidth with a prototype using a single antenna. How to exploit at best the full-duplex capability of new radio and how to redesign MAC protocol are still open research issues.

Another direction for improving the MAC efficiency is the reduction of control messages' overheads. In [156], control messages like RTS, CTS, and ACK are encoded by using correlatable symbol sequences (CSS). The properties of the CSS allow a substantial reduction of the vulnerability intervals and of the air time wasted to contend for the medium (RTS/CTS) and to send ACKs.

Interaction of CSMA, specifically, the CSMA/CA of WiFi, and TCP has been considered as well. Experimental work points out performance shortcomings that might arise in dense WiFi networks with mixture of long-lived and short-lived TCP connections [155].

Finally, random access is also a component of cellular networks. Since their inception cellular radio access networks make use of a random access channel to allow terminals requesting new radio resource assignment, when they have none. The Random Access CHannel (RACH) is usually managed with some variant of slotted ALOHA. In LTE it consists of a slotted channel where stations can transmit special signal patterns, called preambles. The LTE standard defines 54 orthogonal preambles. This kind of access is modeled in Example 8.9.

Recently, many works have explored whether the LTE RACH can support an IoT paradigm, i.e., massive numbers of terminals, each sending periodically or sporadically, relatively short files [8]. A detailed analytical model of the LTE RACH is presented in [154]. It is applied to an IoT scenario in power grid networks. It is shown how several hundreds of metering terminals lead to a severe congestion of the LTE RACH, with a collapse of the achievable throughput performance. Similar issues and some directions for throughput optimization are discussed in [175, 210, 166], with reference to smart cities and vehicular networking applications. Optimization of the class barring parameter is investigated in [201] as means of controlling the LTE RACH congestions level.

Appendix

Let us consider a Markov chain representing a single server system in discrete time:

$$N(t + 1) = N(t) + A(t + 1) - U(t + 1) \tag{9.167}$$

The number of departure U is either 0 or 1. If $N(t) = A(t + 1) = 0$, we assume it is $U(t + 1) = 0$, hence $N(t + 1) = 0$. It is obviously $N(t + 1) \geq 0$. Therefore, we have always $N(t) + A(t + 1) - U(t + 1) \geq 0$.

Let us define the conditional drift

$$d_n = E[A|N = n] - E[U|N = n] , \qquad n \geq 0. \tag{9.168}$$

We assume that $\sup_{n \geq 0} d_n = d_{\text{sup}} < \infty$ and $\lim_{n \to \infty} d_n < 0$. Applying Foster-Lyapunov theorem (see the Appendix at the end of the book), we will show that the Markov chain $N(t)$ is positive recurrent.

Let $V(x) = |x|$. This is a legitimate Lyapunov function on the state space $S = \mathbb{Z}^+$. Since the drift d_n tends to a negative limit, we can find $\epsilon > 0$ and n_ϵ such that $d_n \leq -\epsilon$ for $n > n_\epsilon$. We can write

$$\Delta_n = E[V(N(t+1)) - V(N(t))|N(t) = n]$$
$$= E[|n + A(t+1) - U(t+1)| - n|N(t) = n]$$
$$= E[A(t+1) - U(t+1)|N(t) = n] = d_n$$

Let $\mathcal{A} = \{n : 0 \leq n \leq n_\epsilon\}$. This is a finite set. It is obviously $\Delta_n \leq d_{\text{sup}}$ for $n \in \mathcal{A}$. Based on the definition of n_ϵ, we have also $\Delta_n \leq -\epsilon$ for $n \in S \backslash \mathcal{A}$. All hypotheses of Foster-Lyapunov theorem are satisfied, then $N(t)$ is positive recurrent.

We can also prove that the Markov chain $N(t)$ is either transient or null recurrent as the drift sequence d_n tends to a positive limit for $n \to \infty$. To that end, we use Theorem A.7, mentioned in the Appendix at the end of the book.

We again pick $V(x) = |x|$ as the Lyapunov function. Since $\lim_{n \to \infty} d_n > 0$, we can find n_0 such that $d_n \geq 0$ for all $n > n_0$. Then, we define the finite set $\mathcal{A} = \{n : 0 \leq n \leq n_0\}$. We have $\Delta_n = d_n \geq 0$ for $n > n_0$ (condition 1 of Theorem A.7). We have obviously $V(n_0 + 1) = n_0 + 1 \geq y = V(y)$, for all $y \leq n_0$ (condition 2 of Theorem A.7). Finally, we have

$$E[|V(N(t+1)) - V(N(t))| \, |N(t) = n]$$
$$= E[|n + A(t+1) - U(t+1)| - n \, |N(t) = n]$$
$$= E[|A(t+1) - U(t+1)| \, |N(t) = n] \leq E[A|N = n] + 1$$

which is finite for any n, if the expected number of arrival is finite.

All hypotheses of Theorem A.7 are satisfied, hence $N(t)$ cannot be positive recurrent.

Summing up, we can say that the Markov chain $N(t)$ is positive recurrent, if the drift d_n tends to a negative limit, it is transient or null recurrent, if the drift tends to a positive limit.

Summary and Takeaways

This chapter is devoted to a special category of multiple access, namely random multiple access. We recall here concisely the most significant highlights emerging from this chapter.

ALOHA is the simplest random multiple access protocol. It is distributed and autonomous, requiring at most synchronization at slot level, in case of Slotted ALOHA. Slotted ALOHA is intrinsically unstable. To work properly, it requires adaptation of the back-off probability. We have discussed how this can be done in detail, pointing at the achievable performance in terms of throughput and delay.

If synchronization is too demanding, pure ALOHA can be used at the price of a reduced throughput. We have shown that pure ALOHA is unfair if stations transmit frames of different sizes. This sheds some light on why ALOHA is always considered with fixed size packets.

The next random multiple access protocol we have studied is CSMA. CSMA entails the capability of performing carrier sensing (certainly physical carrier sensing; possibly also virtual carrier sensing).

We have studied the performance of several variants of CSMA as the fundamental parameters are varied. Major findings are:

- the possibility to optimize the performance by selecting back-off parameters properly;
- the need to stabilize the protocol and the feasibility of a distributed adjustment of the back-off probability that achieves both stability and throughput optimization.

From a modeling point of view, we have gained an understanding of what the Poisson approximation means for packet arrivals in a CSMA protocol. We have seen that the throughput predicted by Poisson based models is the target long-term throughput value that the (optimally) stabilized CSMA converges to.

Finally, we have explored in detail one of most successful application of CSMA to real life, the CSMA/CA protocol defined in WiFi WLAN standard. We have studied a number of useful models that predict performance of WiFi networks at MAC layer, both in terms of throughput and service time. We have discussed at length the suitability of the binary exponential back-off of CSMA/CA. We have raised more than one point *against* the use of BEB. Rather, it is feasible to select the contention window size as a function of the (estimated) number of contending stations, so as to maximize the saturation throughput, while minimizing the jitter of service times. Long-term and short-term fairness of WiFi have been discussed as well.

A number of generalizations, research lines, and new developments are hinted at in the closing section, where references to emerging issues and technologies are given. Performance models that provide insight and assess quantitative trade-offs will be the key to understanding where these new ideas can bring us.

Problems

9.1 **0-persistent CSMA with Poisson offered traffic.** Consider the 0-persistence policy and assume that offered traffic can be modeled as a Poisson process with mean rate λ. The channel time axis can be split into *virtual slots*, defined as the time elapsing between the beginning of an idle back-off slot and the immediately following idle back-off slot.

(a) Show that the mean duration of the virtual slot is

$$E[V] = \delta e^{-\lambda\delta} + (\delta + T)(1 - e^{-\lambda\delta}) = \delta + T - Te^{-\lambda\delta}$$

(b) Find the probability of success P_s.

(c) Show that the throughput S can be written as:

$$S = \frac{TP_s}{E[V]} = \frac{T\lambda\delta e^{-\lambda\delta}}{\delta + T - Te^{-\lambda\delta}} = \frac{\beta Ge^{-\beta G}}{1 + \beta - e^{-\beta G}}$$

9.2 **1-persistent CSMA with Poisson offered traffic.** Consider a cycle time C, defined as the time between the ends of two subsequent activity times. A cycle time is made of a sequence of idle back-off slots followed by an activity time, if no arrival occurs in the previous activity time. Otherwise, the cycle time reduces to the activity time.

(a) Find the mean duration of the cycle time, by showing that:

$$E[C] = e^{-\lambda(\delta+T)}\left(\frac{\delta}{1 - e^{-\lambda\delta}} + T + \delta\right) + (1 - e^{-\lambda(\delta+T)})(T + \delta)$$

where we have accounted for an idle back-off slot necessarily following the activity time (it is the time it takes for stations to assess that the channel has gone idle).

(b) The probability of success P_s is the probability that: (i) only a single arrival takes place in the last back-off slot of the sequence, given that an arrival has taken place (this is the condition that makes the tagged back-off slot the last one of the sequence) and no arrival takes place in the previous activity time; (ii) only one arrival occurs in the previous activity time (and hence there is no sequence of idle back-off slots). Show that:

$$P_s = \frac{\lambda\delta e^{-\lambda\delta}}{1 - e^{-\lambda\delta}}e^{-\lambda(\delta+T)} + \lambda(T + \delta)e^{-\lambda(\delta+T)}$$

(c) Show that the throughput of the 1-persistent CSMA with Poisson traffic is:

$$S = \frac{TP_s}{E[C]} = \frac{\lambda e^{-\lambda(\delta+T)}(T + \delta - Te^{-\lambda\delta})}{\delta e^{-\lambda(\delta+T)} + (T + \delta)(1 - e^{-\lambda\delta})} = \frac{Ge^{-(1+\beta)G}(1 + \beta - e^{-\beta G})}{\beta e^{-(1+\beta)G} + (1 + \beta)(1 - e^{-\beta G})}$$

9.3 **P-persistent CSMA with Poisson offered traffic.** According to P-persistence, packets that arrive during a transmission will be transmitted with probability P as soon as the channel becomes idle again, while they will be postponed by a random time Z much bigger than T with probability $1 - P$. Then, the number of packets arriving during a transmission and scheduled to be transmitted immediately after that transmission is a Poisson random variable with mean $(P + \beta)G$.

(a) Show that the throughput is therefore

$$S = \frac{Ge^{-(P+\beta)G}(P + \beta - P\,e^{-\beta G})}{\beta e^{-(P+\beta)G} + (1 + \beta)(1 - e^{-\beta G})}$$

(b) Check that this throughput reduces to the one for 1-persistent CSMA with $P = 1$ and to the case of 0-persistent CSMA for $P = 0$.

9.4 Consider the stabilization algorithm of Sec. 9.4.3. Specifically, we focus on updating the estimated number of backlogged stations $\hat{n}$ in case a collision is observed.

(a) Show that the updating derived according to the Bayes rule, assuming a prior Poisson probability distribution of the number of backlogged station, is:

$$\hat{n}(t + 1) = \hat{n}(t) + \frac{[\hat{n}(t)p(t)]^2}{e^{\hat{n}(t)p(t)} - 1 - \hat{n}(t)p(t)} + v(\delta + T)$$

(b) If up to K packets can be received simultaneously, show that the update upon collision of the stabilization Bayesian algorithm becomes:

$$\hat{n}(t + 1) = \hat{n}(t) + \frac{\pi(K, \hat{n}(t)p(t))}{\sum_{j=K+1}^{\infty} \pi(j, \hat{n}(t)p(t))} + v(\delta + T)$$

where $\pi(k, a) = \frac{a^k}{k!}e^{-a}$.

9.5 The aim of this exercise is to establish the relationship between the maximum throughput achievable with nonpersistent *stabilized* CSMA where: (i) packets are deemed to be backlogged immediately upon their arrival (basic version); or (ii) packets arriving when the channel is sensed idle are transmitted immediately, otherwise, they become backlogged (alternate version). It can be seen that the maximum throughput of the basic version is $S_1(\alpha) = \alpha e^{-\alpha}/(\beta + 1 - e^{-\alpha}) \equiv f(\alpha)$. We have seen also that the maximum throughput of the alternate version satisfies the following equation: $S_2(\alpha) = f(\beta S_2(\alpha) + \alpha)$. Let α_i^* denote the value of α that maximizes $S_i(\alpha)$, $i = 1, 2$.

(a) Show that the derivative of $S_2(\alpha)$ satisfies

$$S_2'(\alpha) = [1 + \beta S_2'(\alpha)]f'(\beta S_2(\alpha) + \alpha)$$

(b) Show that $\beta S_2(\alpha_2^*) + \alpha_2^* = \alpha_1^*$ and $S_2(\alpha_2^*) = S_1(\alpha_1^*)$, i.e., the maximum achievable throughput of the alternate version is the same as the basic version, while the optimal α for the alternate version is smaller than for the basic version.

(c) Give intuition and insight backing up the analytical result found in point (b).

9.6 **Saturation throughput of WiFi with unequal transmission probabilities**. Let us consider a WiFi network, where transmission probabilities are assigned to stations and held constant (no BEB is used). Let τ_i be the transmission probability of station i, $i = 1, \ldots, n$. Let us assume that packet size is fixed.

(a) Show that the saturation throughput of the i-th station is

$$\rho_i = \rho_{\text{ideal}} \frac{P_e \tau_i / (1 - \tau_i)(T_{oh} + a_1)}{\delta + (T_{oh} + a_1)(1 - P_e)}$$

and find the expression of P_e.

(b) Say the τ_i's are fixed so as to guarantee that the throughput of station i is a fraction ψ_i of throughput of station 1, i.e., such that $\rho_i / \rho_1 = \psi_i$, $i = 2, \ldots, n$. Show that the overall throughput $\rho_1 + \rho_2 + \cdots + \rho_n$ can be maximized and that there exists a unique value of τ_1 that attains such a maximum, for a given sequence of ratios $\psi_1 = 1, \psi_2, \ldots \psi_n \in (0, 1)$.

(c) Set up an algorithm to evaluate numerically the optimal τ_1 and try it for several values of the system parameters.

9.7 In a WiFi network, assume a packet can be received in error due to interference and noise. Let ϵ be the probability that a bit is received in error and assume binary errors are independent of one another. Assume also that only the MAC protocol data unit (MPDU) can be affected by errors, while the data frame preamble and ACK are always received correctly. Write down the expression of the saturation throughput for n homogeneous stations, that have different channel qualities, hence error probabilities.

9.8 The aim of this problem is showing how the parameters DT (defer threshold) and CDT (carrier detect threshold) of the IEEE 802.11 can be set to avoid hidden nodes.

Let us consider an infra-structured WiFi network. For a station to associate with the AP, it must be that the power level received by the station is beyond CDT. To transmit, a station must assess whether the channel is idle. A station deems the channel to be idle, if it measures a power level below DT at its receiver.

Let us assume that the path gain is deterministic, depends only on the distance between transmitter and receiver and it is $G(d) = \kappa/d^\alpha$ at distance d, with $\alpha = 4$.

The background noise level is −91 dBm.

Find *CDT* and *DT* so that there are no hidden nodes, requiring that the channel is deemed busy, if the measured power level is twice the background noise level.

10

Congestion Control

Optimum is hard to achieve; near optimum is usually good enough.

Folk saying

Non faciunt meliorem equum aurei freni.[1]

Latin proverb

Aliud aliis videtur optimum.[2]

Marcus Tullius Cicero

10.1 Introduction

Congestion refers to demand exceeding service capability. Given a service system, let its serving capability be measured by a parameter μ in terms of unit of work per unit of time that the service system can carry out. The parameter μ can be time-varying and dependent on the system state variables. Let λ denote the aggregate service demand on the system. Also λ can be time dependent and a function of system state variables and customer population characteristics. In general, we say the service system is congested when $\lambda > \mu$.

Congestion is characterized by its duration and extent.

The amount by which the demand λ exceeds the serving capability μ determines the extent of congestion. Light congestion can be born for some time without compromising the system functionality, whereas heavy congestion requires immediate action to avoid a system collapse.

The uninterrupted time interval during which it is $\lambda > \mu$ defines the duration of congestion. Usually, a short congestion event can be absorbed by the serving

---

1 Golden harnesses do not make a horse better.
2 Different things appear as optimal to different people (i.e., optimality is subjective).

Network Traffic Engineering: Stochastic Models and Applications, First Edition. Andrea Baiocchi.
© 2020 John Wiley & Sons, Inc. Published 2020 by John Wiley & Sons, Inc.

system with no harm, while persisting congestion requires active modification of the system and/or of the traffic demand, lest the service quality be degraded.

Also, the rate at which congestion events occur matters. If congestion is a rare event, occurring over time scales that are much longer than typical times of customer-system interaction, congestion can be seen as part of the system impairments and could be compatible with an acceptable quality of experience of customers. If congestion arises on a time scale comparable with customer-system interaction times, then it is too frequent not to impose active management on behalf of the system and possibly a back-pressure on customers' demand.

We anticipate that the ultimate defense against congestion consists of turning down the excess demand that cannot be accommodated by the service system with the desired level of quality of service. Given that, if a nonmarginal fraction of the demand is systematically rejected because of congestion, that is a mark that the system is *undersized* with respect to the potential demand. For example, let us consider a cellular network base station (BS) that rejects 20% of offered connection attempts over the busy part of the day. If this rejection rate holds for most days and for several hours per day, this is a sign that the radio resources managed by the BS should be incremented. This is both convenient for cellular users, that would perceive a better access quality, and for the cellular network operator, that could carry more traffic and obtain a revenue increase.[3]

Resource dimensioning goes under the name of network planning. It is the result of network design through optimization procedures, given the available technology, the forecast of the traffic demand, the service and application requirements, the law and normative constraints, the economical opportunities, and feasibility. This is obviously a long-term process, needing investment, design, and deployment.

In the following, we do not consider network planning as means to overcome (systematic) congestion. We refer to situations where the service system resources are given and are not going to be modified on the time scale of the congestion phenomena.

The concept of *fairness* is strictly tied to congestion. We have already encountered this term in previous chapters. In general, let N be the number of customers that contend for the resources of the service system and let λ_i be the demand of the i-th customer. We have $\lambda = \lambda_1 + \cdots + \lambda_N$. As long as $\lambda < \mu$, there is room to accommodate all requests, so it is expected that every customer gets exactly what it is asking for. There is no real competition among customers and no congestion on the system resources. There is no issue, if one customer gets more resource and service (even much more) than another customer. Everyone is getting just what it asks for and is therefore happy with that.

---

3 We are assuming that rejected connections are independent of one another. If a significant repeated-lost attempts phenomenon is in place, many rejection events could be generated by few users making repeated attempts to establish a connection.

Problems arise when $\lambda > \mu$. Unless we can provide more resources (which amounts to increasing μ), the excess demand is not going to be satisfied immediately. It will be either delayed or rejected. The decision of which customer's demand is not accommodated fully concerns the fairness aspect of congestion control. We could select randomly which customer demand is not serviced immediately, or we could resort to priorities or we could try to maximize some collective utility measure. Whatever criteria we follow to decide which part of the offered demand cannot be accommodated immediately, we are implementing a fairness approach. Usually, a trade-off is required between fairness and efficiency of system resource management. In general, pushing an efficient use of system resources means that customers that use system resources less effectively should not be serviced. However, the desire to guarantee some minimum level of service to every customers typically impairs the achievement of top efficiency levels. A broad concept used in this context is *utility*, a quantitative measure of the benefit gained by a customer upon receiving service at a given level of quality. We will see how utility can be used to give a general statement of fairness.

There are several levers than can be geared to face congestion. *Pricing* is a major tool to control traffic demand. Pricing strikes the balance between the benefit achieved thanks to the service obtained by a customer and the cost incurred for that service. It is not surprising that congestion control invokes concepts imported from economics. Game theory, auction theory, and distributed optimization have been used to study congestion control.

Two major approaches can be identified to deal with congestion:

- *Proactive:* Congestion control aims at *avoiding* congestion before it intervenes.
- *Reactive:* Action is taken when congestion is detected, aiming at containing its effects and eventually removing it.

Proactive congestion control is based on a preventive approach encompassing admission control and policing functions. Admission control consists of evaluating service demand against the current system resource availability, to verify whether it can be accommodated with the requested quality of service, while maintaining the quality of service stipulated with customers currently under service. Admission control assumes three facts:

1. Service demand is organized with a connection-oriented mode: before service can start, a negotiation phase takes place, when resources are reserved. This is when admission control takes in.
2. It is possible to describe the traffic demand in a way that is understandable to customers and to service providers and is not too complex to assess (traffic contract or service-level agreement).
3. Quality of service targets can be defined in an unambiguous way and can be guaranteed based on the established traffic contract and the available system resources.

The admission control function should be run at connection setup. At that point, the characteristics of the traffic offered by the customers must be described into the traffic contract (the traffic specifications).

Policing is the second pillar of proactive congestion control. Once a traffic contract has been established between a customer and the service provider, it must be verified that the customer abides by its declared traffic specifications. Quality of service should be guaranteed as long as the customer sticks to its traffic specifications.

Policing consists of controlling the traffic submitted by the customer to the system to verify that it complies with the stipulated traffic contract. Since traffic demand is described usually as a stochastic process, policing is realized as a statistical test of hypothesis, where the null hypothesis is that the traffic entering the system is compliant.

Real-life experience has shown that implementing a proactive congestion control scheme is extremely difficult. It is not obvious that traffic specifications and quality of service targets can be expressed in a simple and easily measurable way. Moreover, the implementation of admission control and policing relies on a number of statistical assumptions that might be violated in practice. A major example of proactive congestion control in telecommunication networks has been the telephone network and is currently the cellular network, at least for some type of services (e.g., phone calls). Trying to implement proactive congestion control in packet networks has turned out to be too challenging. Asynchronous Transfer Mode (ATM) in the nineties has been one of the most significant attempts.

The reactive congestion control approach is the one chosen for the Internet and for several other service systems. With reactive congestion control there is no need to adhere to a connection oriented mode, to describe traffic specifications, to make advance checks for admissibility and to set up continuous monitoring of input traffic to verify if it complies with the stipulated specifications. On the other hand, reactive congestion control requires two factors:

- Congestion can be detected.
- Functions to remove congestion are provided.

The first point entails measuring the level of usage of system resources and the quality of the provided service (e.g., waiting time, buffer content level, amount of lost traffic, link capacity utilization). There must be real-time tests to determine congestion. Once congestion is detected, reaction consists in exerting a back-pressure on the traffic sources until they eventually reduce their offered traffic rate. The traffic control paradigm is akin to a closed-loop feedback control. One nontrivial issue in realizing the reactive congestion control, apart from the stability of the closed-loop control, is that reaction must be directed to individual traffic sources, while resources that can experience congestion see aggregate traffic flows. The point is to realize a congestion control function that scales with the number of traffic sources, i.e., congested network elements should not be required to take actions separately on each individual traffic flow that goes through them.

When service systems are interconnected in a network, or a network itself is the service system, congestion and fairness issues are quite complex. In general, we distinguish between link and node congestion. The former means that the serving capacity of a link connecting two nodes has been exceeded by the demand of traffic addressed to that link. The latter implies that internal node resources, e.g., processing capability and storage memory, are being overwhelmed.

For example, telecommunication networks link congestion arises when more connections than is possible should be routed through a link; or the arrival rate of packets at the link exceeds the packet serving rate of the link. Node congestion arises when the packet processing capability of the node is insufficient to keep up with the rate at which packets arrive or when buffers overflow because there are too many packets to store. As another example, in a transportation road network, link congestion amounts to a significant slow down of vehicles traveling on a road segment. Node congestion manifests with long lines at intersection traffic lights.

Congestion control in a network is often realized by means of distributed functionality introduced in the network nodes. A typical case is what is done in telecommunication networks, specifically in the Internet. Internet congestion control is but one instance of how the problem of resource contention has been addressed at a network level, yet it is most relevant both from a theoretical point of view and because of the importance of the Internet itself.

The next section introduces the architecture of congestion control in the Internet, which is implemented in the Transmission Control Protocol (TCP). Then, several versions of the TCP congestion control are outlined. A fluid model of TCP congestion control is then described and generalization to network-level modeling are addressed.

10.2 Congestion Control Architecture in the Internet

In the Internet protocol architecture the functions aimed at detecting congestion and removing it are implemented in the transport layer.[4] Those functions pertain to the TCP. TCP relies on the IP layer and offers its service to the application layer. An excellent source on TCP, although somewhat dated, is [195].

---

4 This was historically a system design decision consistent with the inter-networking strategy of TCP/IP. It makes full sense to implement congestion and flow controls in other architectural layers, from application layer, down to the data link layer.

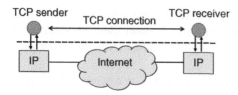

TCP sender   TCP connection   TCP receiver

IP   Internet   IP

Figure 10.1 TCP connection end-points, sender and receiver, and their relationships with IP entites.

TCP offers a connection-oriented service between two parties (see Figure 10.1). Data can be exchanged in both directions, in full-duplex mode. Focusing on a single direction, we distinguish a TCP sender, injecting data packets into the network, and a TCP receiver, who is in charge of acknowledging the received packets, if they are error-free and in sequence. Each TCP endpoint plays both sender and receiver roles, according to the direction of data. The TCP entities (sender and receiver) abstract of the network details. TCP calls for IP in the end system, and hands packet over to the IP entity. Packets make their route through the network, eventually arriving at the receiving end system, where IP delivers the received packets to the TCP receiving entity. Acknowledgment control packets (briefly named ACKs) make the reverse path. The TCP entities do not have any direct communication with routers along the network path of the connection, nor do they know which route their packets are going through. As a consequence, the TCP sending entity does not know what the available capacity on the path is. Moreover, the capacity available to a TCP connection is time-varying, due to other competing TCP connections and other traffic flows sharing the links of the given TCP connection path.

TCP distinguishes two types of congestion: (i) in-network congestion, i.e., congestion at intermediate nodes (routers); and (ii) end-point congestion, i.e, congestion at the destination host. The first type of congestion is targeted by the function called *congestion control*, the second one by the *flow control* function. If we visualize the path of a TCP connection through the network as a pipe and the final host as a bucket, then congestion control provides adaptation to the pipe throughput, while flow control avoids bucket overflow.

The average data rate sustained by a TCP connection is limited according to three different causes:

1. *Source-limited*: The data source (sender application entity) produces data at a maximum rate, intrinsically limited by the application (e.g., streaming).
2. *Destination-limited*: The destination side (receiver application entity) throttles the rate of the connection, to adapt to the capability of the local resources (e.g., a server allocating limited buffer space and processing power to each thread).
3. *Network-limited*: The achievable data throughput is limited by the capacity of data transfer through the network in between the source and the destination.

If the TCP connection throughput is source limited, there is no congestion at the receiver, nor in the network path of the connection. The second case is dealt with by flow control, while congestion control addresses the third case. We will focus on this last case in the following.

The philosophy of TCP is to *probe* the network so as to *discover* what the available capacity is at any time during the connection lifetime. Two key elements allow TCP to perform probing and data rate adjustment: windows and ACKs.

The TCP connection provides a byte stream service to the application. Bytes sent through the TCP pipe are numbered sequentially. Bytes that have been sent through the network, but have not been acknowledged yet, are said to be *in-flight*. The amount of bytes in-flight is the *flightsize*.

The TCP sender paces the emission of new data bytes by means of the *transmission window*. A window is an interval of sequence numbers $[\ell, \ell + W - 1]$, where W is the window size and ℓ is the sequence number of the oldest byte in-flight. Let $W_{tx}(t)$ denote the size of the transmission window at time t. The TCP sender can have up to $W_{tx}(t)$ bytes in-flight at any given time t. The dependence on t is motivated by the fact that the window size can be adjusted adaptively to estimated congestion during the connection lifetime. This is actually what congestion and flow control do.

The transmission window is the minimum between the congestion window and the receiver (or advertised) window, i.e., $W_{tx} = \min\{cwnd, rwnd\}$, where *cwnd* (*rwnd*) denotes the *congestion window* size (*receiver window* size).

The congestion window *cwnd* is updated according to an algorithm which is internal to the TCP sender. This enables the possibility to change the congestion control algorithm at the sender side, still keeping full inter-operability with any other TCP receiver. On the contrary, the receiver window *rwnd* is updated based on TCP segments exchange between the TCP sender and TCP receiver, exploiting the receiver window field of the TCP segment header. This implies that the receiver window updating algorithm is frozen into the TCP standard and any modification entails modifying *both* the TCP sender and receiver, thus impacting heavily the interoperability issue.

Let us focus on the congestion window *cwnd*. The updating principle of the congestion window is the additive increase multiplicative decrease (AIMD) paradigm. Increments of the congestion window are additive, whereas, when time comes to decrease the congestion window, this is done by multiplying it by a factor less than 1. The purpose is to grow the window cautiously, to avoid incurring into deep congestion, and to shrink the window quickly, to wipe congestion out as soon as possible.

When are increments and decrements applied? Here comes into the picture the second key element, ACKs. They represent the *feedback* that the TCP sender entity gets from the TCP receiver through the network. Whenever a TCP segment is sent out by the TCP sender, a timer is started. When the corresponding ACK is received, the timer is stopped and the TCP sender can gauge the round-trip time (RTT) associated to the segment. Then, the TCP sender collects a sequence of RTTs as packets are delivered through the connection. Seeing a growing trend in the RTT sequence is an indication of congestion onset along the network path of the TCP connection (if the TCP receiver were congested, it would signal so, by means of the ACK receiver window field). On the contrary, diminishing RTTs highlight congestion relief. A more dramatic sign of congestion is detecting packet loss. In fact, TCP segments may get lost for a number of reasons, the most prominent of which, at least in wired networks, is router buffer overflow. Then, when the TCP sender detects a packet loss, besides triggering the retransmit function, it shall reduce its congestion window size. Summing up, the congestion window size shall be decremented whenever the TCP sender has evidence (increase of RTTs, packet loss) that congestion might have set on along the network path. This evidence is gained by analyzing ACKs (or the lack thereof).

As for increments, whenever an ACK is received, a new packet can be sent. At that point, the TCP sender can decide to probe the network for more capacity, by incrementing the congestion window. Since congestion window increments are triggered by ACK reception, the TCP algorithm is said to be *self-clocking*. An important consequence of the self-clocking is that a TCP flow can be heavily affected by time-varying capacity of the forward and reverse path of the TCP connection, since the sending rate is based on a delayed version of the time-varying bottleneck capacity.

Figure 10.2 illustrates the self-clocking concept. The sender starts injecting data segments at the highest rate allowed by its access link. Along the network path, the data segments go through a bottleneck, thus they get spaced out. Once they are spaced out by the time T required to transmit a data segment on the slowest link along the path, they maintain that average spacing (packets can be delayed, but they cannot be anticipated). As a consequence, ACKs are spaced out as well, by the same amount. New data segments are released by the TCP sender when ACKs come in. Therefore, new segments are sent out at the pace dictated by the arrival of ACKs, ultimately determined by the path bottleneck.

Let W_k denote the congestion window size immediately after the k-th update time t_k, $k \geq 0$. The initial value W_0 is named initial window (*IW*). The congestion control algorithm defines

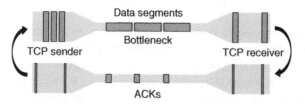

Figure 10.2
Self-clocking mechanism
of the TCP connection.

1. what are the updating times t_k;
2. how W_{k+1} is calculated, given W_k and the event occurring at time t_{k+1} (and possibly other state variables).

Although TCP windows are measured in bytes, in most congestion control algorithms and in some implementations the congestion window is measured in units of maximum segment size (MSS), i.e., the maximum length of a TCP segment payload. We will follow this convention. Thus the congestion window size W denotes the maximum number of MSSs that can be in flight. The MSS parameter is negotiated at connection setup, a typical value being 1460 bytes. The MSS value is limited by the Maximum Transfer Unit (MTU), the maximum length of an IP packet allowed by an IP sub-network. The limiting value is the minimum of MTUs of the sub-networks crossed by the TCP connection path.

The precise algorithm according to which the congestion window is updated is the heart of what goes under the name of TCP congestion control. The seminal work that laid down the foundations of TCP congestion control is due to Van Jacobson [112]. Since then, several flavors of TCP congestion control have been proposed over the years. In the following we will review some of them.

10.3 Evolution of Congestion Control in the Internet

Since the 1980s, the congestion control has been recognized as a key function for the proper working of the Internet, after major performance degradation was experienced during that decade because of too aggressive packet retransmission policies (congestion collapse).

Since the first releases of TCP congestion control at the end of the 1980s, a continuing evolution has taken place, driven both by technological advances, calling for new functionality or revised algorithms, and by a deeper understanding of congestion control theory in packet networks. This evolution process is still ongoing and it is not expected that a "period" will eventually be marked.

In the following subsections we review briefly the congestion control algorithms defined by some major versions of TCP, covering about three decades of the evolution (from 1988 up to 2017). We consider a selection of TCP congestion control algorithms: the "classic" TCP Reno and the more recent TCP CUBIC, as representatives of the loss-based class; TCP Vegas, as representative of the delay-based class; DCTCP, as representative of network-assisted congestion control (ECN based); BBR, as representative of a new class, neither loss- nor delay-based.

A basic knowledge of TCP main features is assumed, specifically, error recovery and timeout management. The reader can consult [195] as an excellent reference for a review of TCP, specifically Chapter 14 for TCP timeout and retransmission mechanisms.

10.3.1 TCP Reno

There is no such thing as *the* standard TCP congestion control algorithm. The algorithm closest to a standard is the "classic" TCP congestion control, dubbed Reno[5]. TCP Reno probes the network to discover the best size for the congestion window. We will see that the optimal value of the congestion window size is the *bandwidth delay product* (BDP), i.e., the product of the connection bottleneck capacity by the base RTT of the connection[6].

At connection start, the TCP sender knows neither the bottleneck capacity nor the base RTT. The sender has to accommodate two conflicting requirements: (i) bring the congestion window size quickly at the optimal level; (ii) avoid causing deep congestion, by sending a number of packets largely exceeding the capacity of the network. The empirical solution of TCP Reno consists of defining *two* algorithms: slow start and congestion avoidance[7].

The slow start algorithm is a fast probing mode, the purpose of which is to take the congestion window size rapidly close to a value that allows full use of the end-to-end capacity available to the TCP connection. To that end, slow start increments of the congestion window size are gross-grained.

The congestion avoidance algorithm increases the congestion window size much slower than slow start, with fine-grained increments, thus guaranteeing that only one or few packets get lost normally, when congestion arises.

Two other algorithms are invoked to improve the TCP congestion control: fast retransmit and fast recovery. The fast retransmit algorithm consists of detecting packet loss by means of duplicate ACKs, hence reacting more quickly than if retransmission were based only on timer expiry. The fast recovery algorithm consists of maintaining the segments in-flight by sending one new segment for each duplicate ACK received by the TCP sender, until a "good" ACK is received, i.e., a nonduplicate ACK. The rationale is that a packet has been actually delivered to the TCP receiver for each duplicate ACK, even if the gap in the segment sequence has not been recovered yet. Then, to maintain the number of packets in-flight it is necessary to inject a new packet into the connection pipe.

---

5 In the rest of this chapter we refer to classic TCP as the "standard" TCP.

6 The base RTT is the time elapsing since a TCP segment is issued until the corresponding ACK is received, assuming no loss and *no queueing delay in the network buffers*.

7 At first glance, these seem to be kind of misnomers. The increase of the congestion window size is exponentially fast in the slow start phase. Actually, "slow" refers to the fact that the congestion window size starts from a very low value. The steady increase of the congestion window size in the congestion avoidance phase is bound to provoke congestion eventually. However, "avoidance" here refers to the fact that the increment of the congestion window size is minimal. It aims at probing the network path capacity, avoiding the loss of more data than strictly necessary.

10.3.1.1 TCP Congestion Control Operations

Let us follow the sequence of operations for the classic TCP congestion control step-by-step.

The congestion control state machine is initialized to slow start at connection setup. The congestion window *cwnd* is initially set to *IW*. The updating rule in slow start is

$$cwnd \leftarrow cwnd + 1 \qquad \text{on ACK reception} \tag{10.1}$$

Since the TCP receiver issues by default an ACK for each TCP segment received correctly, the ACK rate is the same as the packet rate. The default behavior of the TCP receiver can be modified, to reduce the overhead due to ACKs, by adopting the *delayed ACK* option. According to delayed ACK, the TCP receiver issues a cumulative ACK for any other segment[8]. The TCP receiver also manages a timer. Whenever a correct segment is received, the TCP receiver starts the timer. Either a second correct segment in sequence is received, and then the ACK is issued, or the timer expires. In the latter case, the ACK for the first segment is issued anyway. The motivation of the timer is to avoid delaying the ACK too much at the risk of triggering a timeout event at the TCP sender. The typical value of the delayed ACK timer is 200 ms. The ACK is issued immediately in case it is duplicated, i.e., a data segment is received out of sequence.

Let us see the effect of the slow start updating rule under a generalized delayed ACK option (one ACK every *r* segments), assuming that the TCP sender always has packets to send. Initially *IW* segments are sent and *IW*/*r* ACKs are received after one RTT (we assume no packet gets lost). Then, *cwnd* is incremented to *IW* + *IW*/*r* and so many segments are sent, causing the reception of $(IW + IW/r)/r$ ACKs after one more RTT. If W_k denotes the *cwnd* at the end of the *k*-th RTT, the slow start update rule results in[9] $W_{k+1} = W_k + W_k/r$ for $k \geq 0$, with $W_0 = IW$. The solution of the difference equation is $W_k = IW(1 + 1/r)^k$, $k \geq 0$. The diagram on the left in Figure 10.3 illustrates an example of slow start evolution with $IW = 1$ and $r = 1$.

---

8 Delayed ACK can be generalized to issuing one cumulative ACK for *r* consecutive data segments.

9 This is only an approximation, since we are neglecting the fact that the ratio W_k/r is not an integer in general. The exact recursion is:

$$
\begin{cases}
W_{k+1} = W_k + \left\lfloor \dfrac{W_k + \eta_k}{r} \right\rfloor \\[2mm]
\eta_{k+1} = W_k + \eta_k - r \left\lfloor \dfrac{W_k + \eta_k}{r} \right\rfloor
\end{cases}
$$

for $k \geq 0$, initialized with $W_0 = IW$ and $\eta_0 = 0$. $\lfloor x \rfloor$ denotes the largest integer not greater than *x*.

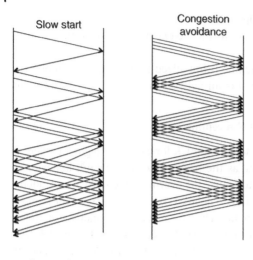

Slow start   Congestion avoidance

Figure 10.3 Examples of TCP congestion control. Left plot: Slow start with initial window size of 1 and no delayed ACK. Right plot: Congestion avoidance starting with $cwnd = 4$.

The marking trait of slow start is the exponential growth of the congestion window with the RTTs. Such a rapid growth promises to reach the right level of the congestion window fast, but, on the other side, can cause a major overshoot, bringing to massive packet loss. To mitigate that risk, slow start is stopped once the congestion window hits the slow start threshold *ssthresh*. The quantity *ssthresh* is itself a time-varying, adaptive parameter of TCP congestion control. It is initalized at connection setup, to some large value. A typical choice for the initial value of *ssthresh* is 64 MSSs.

Once the congestion window size exceeds *ssthresh*, the TCP congestion control switches to congestion avoidance. In this state the *cwnd* update rule is:

$$cwnd \leftarrow cwnd + 1/cwnd \qquad \text{on ACK reception} \qquad (10.2)$$

In words, *cwnd* increments by 1 for each window-worth of data that is acknowledged.[10] Assuming that a window-worth of data is delivered in one RTT, congestion avoidance produces a linear growth of the congestion window size, namely *cwnd* grows by 1 every RTT.

The TCP sender is still probing the network to check whether more data can be pushed in. Congestion avoidance does it more gently than slow start, but it still does it steadily. If no packet loss is detected, the congestion window size of TCP Reno keeps growing, until eventually the connection is finished or the maximum transmission window size is hit.

---

10 The updating rule in (10.2) does not account for the constraint that *cwnd* be an integer. A more precise statement is based on an auxiliary variable *xcwnd*, initialized to *cwnd* on congestion avoidance start. Then, the update rule is as follows: $xcwnd \leftarrow xcwnd + 1/cwnd$, on ACK reception, and $cwnd \leftarrow \lfloor xcwnd \rfloor$. The result is that *cwnd* increments by 1 after exactly *cwnd* ACKs have been received.

The probing phase (be either slow start or congestion avoidance) stops when congestion is experienced. In the TCP Reno algorithm, congestion is recognized only upon packet loss detection. A packet loss event is detected by the TCP sender upon a retransmission time out (RTO) expiry. If the fast retransmit algorithm is active, packet loss is declared also upon reception of *DupThresh* duplicate ACKs (DUPACKs). The default value of *DupThresh* is 3. Whatever the loss detection algorithm, when a retransmission takes place, *ssthresh* is set to

$$ssthresh \leftarrow \max\{2, flightsize/2\} \tag{10.3}$$

where the *flightsize* is the amount of data in-flight into the TCP pipe. Note that *flightsize* $\leq W_{tx} = \min\{cwnd, rwnd\}$, i.e., the *flightsize* is always no more than the congestion window size.

If an RTO occurs, the state of congestion control is brought back to slow start, *cwnd* is reset to *IW*, *ssthresh* is updated according to eq. (10.3) and the congestion control process starts all over again. The rationale is that timeout is a dramatic event in the life of the TCP connection, announcing a major discontinuity in the connection experience. Then it is worth restarting the probing anew.

If instead *DupThresh* DUPACKs are received before an RTO occurs, a packet loss is declared, but the received DUPACKs give evidence of the fact that the connection path is still operational and carrying packets. Hence, we only need to adjust the *cwnd*. This is done by invoking the fast recovery function. After having recovered the lost data, fast recovery sets *cwnd* to the value assigned to *ssthresh* by fast retransmit (see eq. (10.3)). Fast recovery entails a multiplicative decrease of *cwnd*:

$$cwnd \leftarrow cwnd - \beta \cdot cwnd \tag{10.4}$$

The default value of β for TCP Reno is 1/2. In fact, the implementation of (10.4) consists of setting *cwnd* $\leftarrow$ *ssthresh*, where *ssthresh* is updated according to eq. (10.3).

When the packet loss is detected, TCP enters a recovery phase that differs according to different TCP versions (NewReno TCP [11] and SACK [36]). In particular, NewReno TCP recovery phase is based only on the cumulative ACK information whereas SACK TCP receiver exploits the TCP selective acknowledge option [159] to advertise the sender about out-of-order received data. This information is employed by the sender to recover from multiple losses more efficiently than with NewReno TCP. Once the fast recovery phase is completed, congestion avoidance is resumed, with the reduced congestion window size given in (10.4).

The rules for updating the congestion window size and the slow start threshold are summarized as follows. Upon receiving a good ACK, the congestion window size is updated according to:

$$cwnd \leftarrow \begin{cases} cwnd + 1 & cwnd < ssthresh, \\ cwnd + 1/cwnd & cwnd \geq ssthresh \end{cases} \tag{10.5}$$

If an RTO is detected, *ssthresh* and *cwnd* are updated as follows:

$$ssthresh \leftarrow \max\{2, flightsize/2\} \tag{10.6}$$

$$cwnd \leftarrow IW \tag{10.7}$$

If instead fast retransmit is invoked because of reception of *DupThresh* duplicate ACK, the following actions are performed:

1. The *ssthresh* is set as

$$ssthresh \leftarrow \max\{2, flightsize/2\} \tag{10.8}$$

2. The congestion window size is set as

$$cwnd \leftarrow ssthresh + 3 \tag{10.9}$$

3. *cwnd* is increased by 1 for each duplicate ACK received.
4. When a good ACK is received, the *cwnd* is reset to *ssthresh* and congestion avoidance is resumed.

Let us explain the motivation behind these glib rules. Detection of packet loss via timeout entails a long delay, since the RTO is easily much longer than the average RTT. The fast retransmit algorithm has been introduced to overcome this issue. According to fast retransmit, after three duplicate ACKs a packet is deemed to be lost. Hence it is retransmitted and the congestion window is halved. A key point is the following general rule, that the TCP sender entity always complies with, throughout the lifetime of the TCP connection: *a new packet can be sent only if the flighsize is less than the cwnd*[11]. Let W denote the *cwnd* when the three DUPACKs have been received and let F denote the *flightsize*. If the TCP sender is greedy, it is $F = W$. By setting the new *cwnd* value to $W/2$, the TCP sender could not send any new packet until it receives the ACK of the lost packet, after its retransmission. In the meantime, the TCP sender receives DUPACKs corresponding to all other flying packets (we assume a single packet is lost). Therefore, the pipe of the TCP connection empties completely. Once the packet loss is recovered, i.e., as soon as the TCP source receives the ACK for the retransmitted packet, the *flightsize* drops to 0 and hence the TCP sender can send $W/2$ packets.

The fast recovery procedure is conceived to accelerate the recovery of the *flightsize* to the level $W/2$. According to this procedure, the *cwnd* is set to $W/2 + 3$ (since three DUPACKs have already been received); then it is inflated by one each time a new DUPACK arrives. Overall, $W - 1$ DUPACKs arrive, if only a single packet of the window has been lost. Three of them are used to detect the loss, therefore $W - 4$ more DUPACKs arrive before the retransmitted packet is acknowledged.

---

11 Here we assume that both *flightsize* and *cwnd* are measured in MSSs.

Figure 10.4 Example of fast recovery evolution from when the packet loss is detected (three DUPACKs have been received) until when it is recovered (reception of ACK of the retransmitted packet). Crosses mark reception events of DUPACKs, the circle marks the reception of the good ACK that closes the recovery phase.

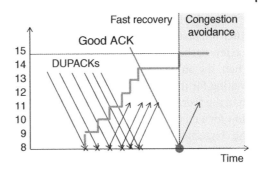

Then, the *cwnd* is inflated up to $W/2 + 3 + W - 4 = W/2 + W - 1$. Since it is $F = W$, it follows that the TCP sender is enabled to send up to $W/2 + W - 1 - F = W/2 - 1$ *new* packets. One more new packet can be sent upon reception of the ACK of the retransmitted packet. Overall, the TCP sender has *already* sent up to $W/2$ *new* packets, at the time when the congestion avoidance is resumed. Recovery is attained immediately this way, which is why this procedure is called fast recovery.

An example of the evolution of the *cwnd* is shown in Figure 10.4. In the figure, we assume that $W = 10$. The plot of the *cwnd* (the thick step-wise line) starts from the time that the packet loss event has been detected by means of the fast retransmit function. At that time the *cwnd* is set at $W/2 + 3 = 10/2 + 3 = 8$. It increments by one upon each received DUPACK. After six DUPACKs, the ACK of the retransmitted packet finally arrives.

The *flightsize* equals 10 when loss is detected. As *new* packets are sent (upward arrows in the figure), the *flightsize* increments by one for each transmitted packet, attaining 14 immediately before the good ACK is received. As the ACK of the retransmitted packets arrives at the TCP sender, one more new packet is sent (hence the *flightsize* is incremented to 15), and 10 old packets are cumulatively ACKed (hence the *flightsize* is decremented to 5). A neat explanation of fast recovery implementation alternatives is given in [195].

A number of nuances have been added to the classic algorithm over the years. We review some of them, given their relevance (the devil is in the details, a proverb that suits TCP perfectly).

10.3.1.2 NewReno

When a packet loss is detected by means of the fast retransmit algorithm (duplicate ACKs), the lost packet is retransmitted and the congestion window is temporarily inflated by incrementing it for any duplicate ACK received, until a good ACK is received. When a good ACK comes in, the congestion window is reset to *ssthresh*, wiping out the inflation.

For multiple packet losses this can affect adversely the connection throughput. After the first loss is recovered and normal operation is resumed, there might be too few packets in-flight to trigger again the fast retransmit for other lost packets. Then, the only way the TCP sender can realize that more packets are lost is by waiting for the RTO to expire. The RTO is typically set to a much higher value than the average RTT[12]. Waiting for the RTO to expire freezes the connection data flow for a relatively long time. Moreover, it causes the *cwnd* to roll back to *IW*, and the congestion control state of the connection to be set back to slow start.

To avoid problems with multiple losses, fast recovery has been modified. The modification is active provided the SACK option is off (see next subsection for details on the SACK option). The classic Reno congestion control with the modified fast recovery is dubbed NewReno.

The modification consists in defining a new variable, the *recovery point*. The recovery point is the highest sequence number from the last transmitted window of data. The congestion control keeps track of the recovery point. As the fast retransmit is invoked, the congestion window is managed according to fast recovery, with temporary inflation, until an ACK with a sequence number at least as high as the recovery point is received. Only at that point is the inflation removed and normal operation in congestion avoidance is resumed.

This simple modification allows TCP to recover from multiple packet losses without exiting prematurely from the fast recovery procedure.

10.3.1.3 TCP Congestion Control with SACK

Selective ACK defines an option of the TCP header, implementing a selective ACK mechanism. It specifies not only the sequence number of the first expected byte, but also holes in the received byte sequence as seen by the TCP receiver.

With SACK, a TCP sender is informed of multiple missing chunks of data, so that it can recover all of them without unneeded retransmissions. Upon reception of a SACK, the TCP sender could theoretically send all missing segments, given that they fall inside the permitted window. However, for large window sizes and multiple losses, this could cause the injection of too many packets at high speed in the TCP connection path, thus causing more congestion.

To avoid this effect, SACK calls for separation of the retransmission and the congestion control logic. While SACK helps identifying *which* segments must be retransmitted, congestion control dictates *when* they shall be retransmitted.

A new variable is defined, called *pipe*. The variable *pipe* accounts for the bytes in-flight, that are not deemed to be lost. The TCP sender is permitted to send a segment provided that $cwnd * MSS - pipe \geq MSS$, i.e., provided the congestion

---

12 Typical implementation set the RTO to the currently estimated average RTT, the smoothed RTT, plus 4 times the estimated deviation of the RTT from the average value.

window has room to accommodate an MSS in addition to the effective number of bytes still in flight.

TCP with SACK can perform better than NewReno when multiple packets get lost in a window of data.

10.3.1.4 Congestion Window Validation

The congestion window size is an estimate of the amount of data that a TCP sender should be able to send into the network without causing significant congestion. This estimate is continuously updated on the basis of the feedback collected from the network. The feedback is essentially provided by the ACK flow arriving at the TCP sender. In turn, ACKs are triggered by data segments pushed into the connection pipe by the TCP sender. If the TCP connection is idle for some time or it is otherwise stuck for any reason, no data segment is sent and no ACK is received. That way, the information condensed into the current value of *cwnd* becomes stale. When operation is resumed, it might be that the value of *cwnd* is no more appropriate. For example, if *cwnd* grew up to a relatively high value, when operation is resumed, a large burst of segments could be launched into the network at high speed, possibly causing congestion and multiple packet losses.

The algorithm called congestion window validation (CWV) has been introduced to handle those situations. First we have to distinguish between two circumstances. A TCP connection can be *idle* if the TCP sender has no new segment to send, and all sent segments have been successfully acknowledged. The connection is therefore truly quiescent; it will become active again as soon as a new segment is ready to be sent at the TCP sender. Alternatively, a connection is *blocked* if the TCP sender has received ACKs for all previously sent segments, it has data ready to be sent, but it is unable to send segments for some reason. For example, the sending host could be busy in other tasks, so that it cannot assign CPU time and resources to the TCP sender thread. Or, the network interface is blocked by some lower-layer protocol issue. Also, in this case, the pipe is empty. As soon as the TCP sender resumes sending segments, the connection exits this blocked state.

In both cases the TCP sender condition is referred to as *application-limited.*

The CWV algorithm works as follows. Let t_s denote the last time that a segment has been sent out. Let a new segment, ready to be sent out, arrive at an *idle* TCP sender at time t_a. The sender checks whether $t_a - t_s > RTO$. If that is the case, the following steps are performed.

1. $ssthresh \leftarrow \max\{ssthresh, (3/4) \cdot cwnd\}$, i.e., *ssthresh* is assigned a value that maintains a "memory" of the current value of the congestion window size.
2. $cwnd \leftarrow \max\{1, cwnd \cdot 2^{-(t_a - t_s)/RTT}\}$, i.e., the congestion window size is halved for each RTT elapsed during the idle time.

For a *blocked* TCP sender, the following steps are carried out:

1. The amount of window actually used is saved into the variable *w_used*.
2. *ssthresh* ← max{*ssthresh*, (3/4) · *cwnd*}.
3. *cwnd* ← (*cwnd* + *w_used*)/2.

In both cases, the *cwnd* decays with respect to the last registered value, while *ssthresh* keeps a memory of the last attained *cwnd* value. In the first case, the *cwnd* can be affected substantially. Moreover, reducing *cwnd* and keeping *ssthresh* close to the former value of the congestion window size results easily in the congestion control state move to slow start. This is consistent with the fact that, after a long pause, the TCP sender has to start its probing almost from scratch.

CWV is enabled by default in Linux TCP implementations.

10.3.2 TCP CUBIC

The CUBIC algorithm [181, 96] is an evolution of the binary increase congestion control algorithm. Initialization, slow start and handling of congestion window reductions are done as in standard TCP, including variants, e.g., SACK. The only difference lies in the amount that the congestion window is reduced upon packet loss. The reduction of *cwnd* is a fraction β of the current value of the congestion window, i.e., *cwnd* ← *cwnd* − β · *cwnd*. The default value of β is 0.3.

The key original idea of the CUBIC algorithm affects the congestion avoidance phase. If *cwnd* ≥ *ssthresh*, congestion control is in congestion avoidance. Then, *cwnd* growth is governed by a cubic equation:

$$W(t) = W_{max} + C(t - t_0 - K)^3 \tag{10.10}$$

where t_0 is the last time before the current time t that the congestion window has been reduced and $W_{max} = W(t_0^-)$ stores the value of the congestion window immediately before time t_0, i.e., before the reduction. Note that CUBIC uses the true time variable t. It is therefore not simply self-clocking as classic TCP.

The constant K can be found, by imposing that $W(t_0^+) = W_{max}(1 - \beta)$. It is

$$K = \left(\frac{\beta W_{max}}{C}\right)^{1/3} \tag{10.11}$$

The default value chosen by the Authors for C is 0.4. The CUBIC rule to increase the congestion window is made of a first part that is concave, from time t_0 until time $t_0 + K$. The congestion window grows fast at the beginning, to regain the level it attained before the reduction. As the congestion window gets close to the previous level, the curve flattens and the increase becomes slower and slower. If no packet is lost and the growth can continue, the convex part of the curve is involved. Then the congestion window grows faster and faster, eventually causing a packet loss. The fast growth is useful to reap high values of throughput, when there is a large bandwidth that has become available.

CUBIC implements a TCP-friendly mechanism. To be fair with respect to the standard TCP (Reno). As long as *cwnd* is less than the congestion window size that a standard TCP congestion control would set, *cwnd* is updated following the rule of the standard TCP, instead of the CUBIC law in eq. (10.10). To find the time evolution of the standard TCP *cwnd*, we consider that in congestion avoidance, the standard TCP grows linearly with the RTT. Accounting for the different reduction factor ($\beta = 0.3$ for CUBIC, instead of the factor $1/2$ of the standard TCP), the congestion window size of the standard TCP can be put in the form

$$W_{\mathrm{TCP}}(t) = (1 - \beta)W_{\max} + \frac{3\beta}{2 - \beta}\frac{t - t_0}{RTT}, \qquad t > t_0. \tag{10.12}$$

During congestion avoidance, upon the arrival of an ACK at the TCP sender at time t_a, if it is $cwnd < W_{\mathrm{TCP}}(t_a)$, then CUBIC sets $cwnd \leftarrow W_{\mathrm{TCP}}(t_a)$. Otherwise, if $cwnd \geq W_{\mathrm{TCP}}(t_a)$, CUBIC sets $cwnd$ to $W(t_a)$, as given by eq. (10.10).

Apart from congestion avoidance, TCP CUBIC behaves as the standard TCP on packet loss, except that the congestion window is reduced by a factor β, rather than being halved. On timeout, TCP CUBIC resets all its variables and rolls back the congestion window size to the initial window level.

The detailed pseudo-code of TCP CUBIC algorithm is listed below. Note that the clause "On congestion detection" refers to congestion events detected by means of: (i) packet loss detected with three DUPACKs; (ii) ECN-Echo flag set in the header of ACKs received in the last RTT

The core function runs on each ACK; it is listed in a separate pseudo-code. In this second algorithm, *t_tcp* indicates the TCP timestamp (current time at the TCP sender). The variable *SRTT* denotes the average (smoothed) round-trip time, measured on ACK reception[13].

The boolean *fast_convergence* enables the mechanism that facilitates new TCP connections to gain their own share of the bottleneck capacity (fairness to other CUBIC connections). According to fast convergence, if packet loss is detected so early that the *cwnd* is still below the switching point between concave growth and convex growth, i.e., if it is $cwnd < W_{\max}$, the value of $W_{\max}$ for the next congestion avoidance phase is set to a level lower than the last attained value of *cwnd*. With the suggested value of $\beta = 0.3$, the reduction factor is $1 - \beta/2 = 0.85$. The rationale of this mechanism is to make room for new connections, whose onset is

---

13 On each good ACK reception (except of an ACK of a retransmitted segment) the variable *SRTT* is updated according to $SRTT = (1 - \gamma)SRTT + \gamma T_k$, where T_k is the sample of RTT measured on the k-th good ACK reception. This recursion is initialized at connection set-up time with $SRTT = T_0$, where T_0 is the sample of RTT collected during the three-way handshake. A typical default value of γ is $\gamma = 0.125$.

Algorithm Pseudo-code of TCP CUBIC congestion control algorithms.

Initialization:

  1: *fast_convergence* = **true**
  2: *C* = 0.4
  3: *β* = 0.3
  4: *IW* = 3
  5: s_0 = 64
  6: *reset_tcp_cubic_cc()*

On each ACK:

  1: *cwnd_update_on_ACK()*

On congestion detection:

  1: *wcnt* = 0
  2: t_0 = 0
  3: **if** *cwnd* < W_{max} **and** *fast_convergence* **then**
  4:     W_{max} = *cwnd* · (2 − *β*)/2
  5: **else**
  6:     W_{max} = *cwnd*
  7: **end if**
  8: *ssthresh* = max{2, ⌊*cwnd* · (1 − *β*)⌋}
  9: *cwnd* = *ssthresh*

On timeout:

  1: *reset_tcp_cubic_cc()*

reset_tcp_cubic_cc()

  1: *wcnt* = 0
  2: t_0 = 0
  3: W_{max} = 0
  4: *cwnd* = *IW*
  5: *ssthresh* = s_0

---

likely the cause of early packet loss. The unit of measure used for all windows and window-related quantities is the MSS.

CUBIC is set as the default TCP congestion control in Linux kernels since version 2.6.18. The full TCP CUBIC implementation has more details. We do not pursue those detail further here. The interested reader can find them in [96] and ultimately in the code of the Linux implementation.

10.3.3 TCP Vegas

TCP Vegas [43] is based on the usual five fundamental mechanisms: slow start, congestion avoidance, retransmission timeout, fast retransmit, and fast recovery.

Vegas is the first example of delay-based (rather than packet loss–based) congestion control. The key idea of Vegas congestion control is comparing the amount of data it expects to be able to transfer in one RTT and the actual amount

---

Algorithm Pseudo-code of function *cwnd_update_on_ACK()*.

1: **if** *cwnd* < *ssthresh* **then**
2:    *cwnd* = *cwnd* + 1
3: **else**
4:    **if** $t_0 \leq 0$ **then**
5:       $t_0 = t\_tcp$
6:       **if** *cwnd* < W_{max} **then**
7:          $K = \left(\frac{W_{max} - cwnd}{C} \right)^{1/3}$
8:          $W_0 = W_{max}$
9:       **else**
10:         $K = 0$
11:         $W_0 = cwnd$
12:       **end if**
13:       $W_{tcp} = cwnd$
14:    **end if**
15:    $t = t\_tcp + SRTT - t_0$
16:    $target = W_0 + C(t - K)^3$
17:    **if** *target* > *cwnd* **then**
18:       $w = \frac{cwnd}{target - cwnd}$
19:    **else**
20:       $w = 100 \cdot cwnd$
21:    **end if**
22:    $W_{tcp} = W_{tcp} + \frac{3\beta}{2-\beta} \frac{1}{cwnd}$
23:    **if** W_{tcp} > *cwnd* **then**
24:       $w' = \frac{cwnd}{W_{tcp} - cwnd}$
25:       $w = \min\{w, w'\}$
26:    **end if**
27:    **if** *wcnt* > *w* **then**
28:       *cwnd* = *cwnd* + 1
29:       *wcnt* = 0
30:    **else**
31:       *wcnt* = *wcnt* + 1
32:    **end if**
33: **end if**

---

of transferred data. If the expected throughput level is not attained, there is likely some slowdown due to queue build-up in some intermediate router. If this symptom of congestion persists, Vegas slows down, by reducing the congestion window. This is in contrast to the standard TCP approach, which forces a packet drop in order to determine the point at which the network is congested. On the contrary, if data transfer proceeds smoothly with no sign of queue build-up, Vegas probes for further increase of *cwnd*. Differently from standard TCP, Vegas can also maintain the *cwnd* unchanged, if it cannot infer any definite drift from the RTT analysis.

More in depth, TCP Vegas differs from TCP Reno by the way that slow start, congestion avoidance, and fast retransmit are implemented.

The TCP Vegas congestion control is based on two parameters representing, respectively, the *expected* and the *actual* rate, calculated on ACK receipt as follows:

$$Expected = \frac{cwnd}{BaseRTT} \qquad Actual = \frac{flightsize}{RTT} \tag{10.13}$$

where *BaseRTT* is the minimum round-trip time experienced by the connection, and *flightsize* is the number of segments already sent and not acknowledged yet.

Every RTT, TCP Vegas computes the normalized difference Δ between *Expected* and *Actual*:

$$\Delta = (Expected - Actual) \cdot BaseRTT \tag{10.14}$$

During congestion avoidance, TCP Vegas compares Δ with two thresholds α and β. Suggested values are $\alpha = 1$ and $\beta = 3$. If Δ is less than α, the congestion window grows linearly in the next RTT; if Δ is larger than β, TCP Vegas decreases linearly *cwnd* in the next RTT; otherwise, if Δ is between α and β, *cwnd* is not changed:

$$cwnd = \begin{cases} cwnd + 1 & \Delta \leq \alpha \\ cwnd - 1 & \Delta > \beta \\ cwnd & \text{otherwise} \end{cases} \tag{10.15}$$

The slow start mechanism is based on the same concepts of the congestion avoidance phase: TCP Vegas computes Δ and compares it with a unique threshold γ. The suggested value is $\gamma = 1$. As long as Δ is less than γ and *cwnd* is less than *ssthresh*, the TCP Vegas congestion window is doubled every other round trip time. This can be obtained by freezing the *cwnd* in one RTT and allowing *cwnd* to grow by 1 per received ACK in the next RTT. Alternatively, it can be obtained approximately by incrementing the *cwnd* by 1 any other received ACK. Switching between slow start and congestion avoidance is decided by comparing *cwnd* with the *ssthresh* as usual. If three duplicate ACKs are received, the Vegas sender performs fast retransmit and fast recovery as in TCP Reno. Actually, Vegas develops a more refined fast retransmit mechanism based on a fine-grained clock, whose details are described in [43]. After fast recovery, TCP Vegas sets the congestion window to 3/4 of the previous maximum attained congestion window and performs the congestion avoidance algorithm.

The parameters α and β can be thought of in terms of buffer utilization at the bottleneck link [43]. The reasoning behind these values is as follows. At least one packet should be buffered in the network path, at the queue of the bottleneck link router along the TCP connection path, to maintain full utilization of the available network capacity. If extra bandwidth becomes available, buffering additional packets (up to 3, according to the set value of β) obviates the need to wait an extra RTT in order to inject more packets, which would be required if Vegas tried to stick to just a single extra packet buffered in the network path. The reason to have $\beta > \alpha$ is to introduce dumping in the congestion window closed-loop, so as to avoid oscillations triggered by minor changes in the available network capacity.

Vegas is supported by Linux, but not enabled by default. Though it has been shown that Vegas offers good throughout and fairness performance, when operated in a homogeneous Vegas environment, it has been found that Vegas is starved when confronted with loss-based TCP congestion control algorithms. Those algorithms are much more aggressive than Vegas, since they keep pumping new packets into the network, until a loss is detected by the sender. This increases the bottleneck queue, and Vegas sources sense the increased delay, thus shrinking their windows. In a world of aggressiveness, there is no room for the gentle Vegas.

Furthermore, under certain circumstances, Vegas can be "fooled" into believing that the forward-direction delay is higher than it really is. This happens when there is significant congestion in the reverse direction (recall that the paths in the two directions of a TCP connection may be different and have different congestion levels). In such cases, ACKs returning to the TCP sender are delayed, even though the sender is not really contributing to the congestion of the forward path. This causes Vegas to reduce the congestion window even though such an adjustment is not really necessary. This is a potential pitfall for most techniques based on RTT measurement as a basis for congestion control decisions.

Vegas is fair relative to other Vegas connections sharing the same bottleneck link. However, Vegas and standard TCP flows do not share the network capacity equally. A standard TCP sender tends to fill queues in the network, whereas Vegas tends to keep them nearly empty. Consequently, as the standard sender injects more packets, the Vegas sender sees increased delay and slows down. Ultimately, this leads to an unfair bias in favor of the standard TCP, which can result in Vegas connection starvation.

10.3.4 Data Center TCP (DCTCP)

The data center TCP (DCTCP) [9] has been proposed as an improvement over traditional TCP congestion control for the specific environment of data center networks. Data centers are the heart of cloud computing. They comprise from several thousands to hundreds of thousands servers, interconnected by a high performance network, so as to enable parallel computing paradigms that require intensive communications among servers. The most prominent special features of a data center network are: (i) the typically small base end-to-end delay in the absence of queueing, in the order of few tens of microseconds up to several hundreds of microseconds; (ii) the very large link capacity (often 10 Gbit/s, or even more, if optical technology is used); (iii) the burstiness and volatility of the traffic, that can cause significant queue build-up in switches; and (iv) TCP incast. The

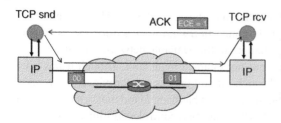

Figure 10.5 Illustration of the ECN logic. A router sets the ECN flag on a packet sent on a congested link and the TCP receiver echoes back the flag into the ACK fed back to the TCP sender.

last issue is triggered by distributed computing applications (see Section 10.9.4). A large number of TCP connections are opened to different servers simultaneously when a computation task is spread among them. Completion of the distributed computation tasks and return of results may synchronize, so that there is a major surge of traffic at the switch attached to the server that launched the distributed computation.

To improve the performance offered by standard TCP in these environments, the DCTCP has been conceived. It leverages upon a standard cooperation mechanism between network nodes and traffic sources, the Explicit Congestion Notification (ECN) [178]. According to ECN, each router defines a link congestion criterion (e.g., exceeding a level threshold of the output link buffer). When congestion is detected on the link, the outgoing packet is marked. Marking amounts to setting a bit (CE flag) in the IP packet header.[14] The CE flag is echoed by the TCP receiver, using the ECN-Echo (ECE) flag in the TCP ACK header. A scheme of the ECN principle is illustrated in Figure 10.5.

The DCTCP algorithm exploits ECN. Although not activated consistently in the Internet, ECN is supported by most off-the-shelf routers and switches. It is sensible to define a TCP congestion control algorithm that relies upon the cooperation of network switches in the special environment of data center networks, since this is a controlled einvornment, where the data center operator can guarantee consistency of configuration of all switches.

DCTCP has three main components: (i) marking at the switch; (ii) ECN-Echo at the receiver; and (iii) controller at the sender.

10.3.4.1 Marking at the Switch

A simple active queue management scheme is used at the switches to support DCTCP. A single parameter is assigned, the marking threshold, K. A packet arriving at the output buffer of a switch port is marked if it finds a buffer queue length greater than K. Otherwise, it is not marked. Looking at the output of the buffer, the sequence of CE flags forms a 1-0 signal that is high (code 1) as long as the buffer occupancy overshoots the threshold, low (code 0) otherwise.

---

14 Two bits of the ToS field of IPv4 header are devoted to ECN: one is for signaling the ECN capability, the other is the congestion experienced (CE) flag.

10.3.4.2 ECN-Echo at the Receiver

The packets traveling the forward connection path collect the congestion notification flags. It is a task of the receiver to feed this notification back to the TCP sender via the TCP ACKs. ECN as specified in the standard RFC 3168 states that a receiver sets the ECN-Echo flag in a series of ACK packets, until it receives confirmation from the sender that the congestion notification has been received. Given the way that CE flags are set through the switches according to DCTCP, the procedure stated by RFC 3168 introduces a distorsion of the congestion signal. The simplest way for the DCTCP receiver to convey an accurate feedback signal is to ACK every packet, setting the ECN-Echo flag if and only if the acknowledged data packet has a CE flag set to 1.

There remains to deal with the delayed ACK mechanism. When this option is selected, the TCP receiver can send a single ACK for up to r TCP data segments. The DCTCP receiver behavior is represented by a two state machine. State S_0 is maintained as long as the incoming TCP data segments carry a CE flag of 0. While in state S_0, the TCP receiver sends a single ACK with ECN-Echo flag set to 0 every r data segments with CE=0. The arrival of a data segment with a CE flag of 1 triggers the transition to state S_1. On the transition, the TCP receiver issues immediately an ACK with ECE=1. While in state S_1, the TCP receiver sends a single ACK with ECN-Echo flag set to 1 every r data segments with CE=1. The arrival of a data segment with a CE flag of 0 triggers the transition to state S_0. On the transition, the TCP receiver issues immediately an ACK with ECE=0. The state machine at the TCP receiver to handle CE and ECE flags with delayed ACKs is shown in Figure 10.6. Looking at the sequence numbers of the received ACKs, the TCP sender is able to recover the correct congestion notification signal.

10.3.4.3 Controller at the Sender

The sender maintains an estimate of the average fraction α of packets that are marked. α is updated once for every window of data (about once every RTT), according to:

$$\alpha \leftarrow (1 - g)\alpha + gF \tag{10.16}$$

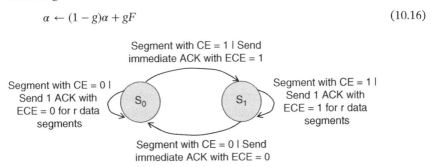

Figure 10.6 State-machine for handling CE and ECE flags at the TCP receiver when the delayed ACK mechanism is used.

where F is the fraction of packets marked in the last window and $g \in (0, 1)$ is the weight given to new samples against the value of α accumulated in the past. At steady state, α converges to E[F]. This is the probability that the bottleneck buffer on the forward path of the TCP connection is beyond the threshold K. The heavier the congestion, the larger α.

The parameter α is used to reduce the congestion window. Here lies the main difference between the DCTCP sender and the standard TCP sender. The DCTCP sender reduces *cwnd* when it receives at least one marked ACK in the last window of segments. The DCTCP sender reacts by reducing the window by a factor that depends on the average fraction of marked packets: the larger α, the smaller the decrease factor. Formally:

$$cwnd \leftarrow cwnd \left(1 - \frac{\alpha}{2}\right) \tag{10.17}$$

When α is near 0 (low congestion), the window is only slightly reduced. When congestion is high ($\alpha \sim 1$), DCTCP cuts *cwnd* by a factor of $1/2$, just like the standard TCP.

Other features of DCTCP are the same as the standard TCP. In slow start the *cwnd* grows by 1 for each received ACK. In congestion avoidance, the applied rule is *cwnd* $\leftarrow$ *cwnd* + 1 after *cwnd* received ACKs. Fast retransmit and fast recovery are also left unchanged, except that the congestion signal is not conveyed by packet loss detection, rather it consists of marked ACKs.

10.3.5 Bottleneck Bandwidth and RTT (BBR)

The bottleneck bandwidth and RTT (BBR) variant of TCP congestion control was first published at the end of 2016 [49,50]. The Authors claim that development and experimenting with BBR started at Google around 2014. It is a major variant of TCP congestion control. Besides the weight of Google on the Internet, the qualification "major" is due also to innovations that break longstanding "tradition" in TCP congestion control, even if the Authors interestingly root their main ideas in fundamental statements of Kleinrock's papers dating back to the mid-1970s.

BBR is not a loss-based nor a delay-based congestion control. It is still based on the closed-loop reactive paradigm of TCP. The sender takes decisions on the amount of data it can send and when it can send, based on information derived from ACKs.

In a nutshell, analysis of ACK sequence numbers and arrival times enables the sender to estimate the minimum path RTT and the bottleneck bandwidth, hence the BDP. The sender maintains a sending rate consistent with the estimated bottleneck bandwidth, capping the amount of data in-flight so as to make sure that the pipe is full, yet there is no (or little) queueing at the bottleneck. This is obtained by limiting the *flightsize* according to the estimated BDP value. Sent packets are paced according to the sending rate. Periodic probing is carried out to assess whether more bandwidth is available.

Even from these few lines, it emerges that BBR gives up to TCP self-clocking (it uses packet pacing instead), it does not use the congestion window, it is not responsive to packet loss (segments detected to be lost are obviously retransmitted, but no congestion window update occurs; the sending rate is adjusted according to the estimated bottleneck bandwidth). Therefore, BBR entails a full decoupling of congestion control and error control.

A starting motivation of the designers of BBR comes from the remark that loss-based TCP congestion control tends to make the TCP connection work on the brink of a full buffer at the bottleneck link. In fact, loss-based congestion controllers react only on packet loss, hence when buffer overflow occurs at the bottleneck. On the contrary, BBR tries to maintain the TCP connection to work at maximum throughput, yet with an essentially empty buffer.

Figure 10.7 illustrates the concept of BBR versus loss-based TCP. The top plot shows a stylized behavior of the RTT as a function of the amount W of data in flight for a single greedy connection with a single bottleneck. Let T be the base RTT, C the bottleneck link capacity and Q the queue length at the bottleneck. As long as $W \leq C \cdot T \equiv BDP$, we have $Q = 0$ and the RTT sticks to its minimum T. As the amount of data in-flight exceeds the BDP, the queue at the bottleneck starts building up, until when buffer overflow occurs eventually, as $W > BDP + B$, where B is the buffer size.

Figure 10.7 Concept of BBR versus loss-based TCP.

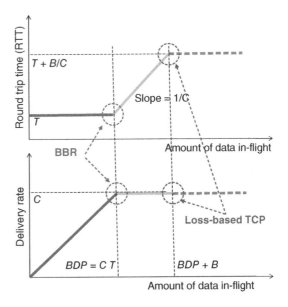

Correspondingly, the delivery rate grows linearly from 0 up to C. Once the bottleneck capacity is saturated, no further growth is possible. A loss-based TCP keeps inflating the amount of data in-flight, until a loss is detected. Hence, it drives TCP to work around the points marked with circles labeled "Loss-based TCP" in the figure. That is not a convenient working point, especially if the buffer size is large.

The best working points would be the ones marked by circles labeled "BBR", where BBR tries to drive TCP. That corresponds to having no delay at the bottleneck buffer, yet achieving the maximum possible delivery rate.

However, it is difficult to strike that ideal working point. The TCP sender cannot measure the minimum RTT, until it knows positively that the bottleneck buffer is empty. On the other hand, it cannot assess what the bottleneck capacity is until it has started filling the bottleneck buffer. In other words, observations made during time intervals when the system is operating on the initial part of the curve (empty buffer, minimum RTT) do not bring information on the bottleneck capacity, whereas observations collected when the system runs on the intermediate part of the curve do not reveal the minimum RTT.

To operate near the point with maximum throughput and minimum delay [84,132], the system needs to maintain two conditions:

- *Rate balance*: The bottleneck packet arrival rate equals the bottleneck bandwidth available to the TCP flow.
- *Full pipe*: The total data in-flight along the path is equal to the BDP.

To achieve these goals, BBR is based on a model of the network path consisting of two key parameters:

- C_{BBR} (denoted with `BBR.BtlBw` in [51]): The estimated bottleneck bandwidth available to the TCP connection.
- T_{BBR} (denoted with `BBR.RTprop` in [51]): The estimated base RTT of the TCP connection path.

Both estimates result from sliding window filters. Specifically, C_{BBR} is the maximum of a sliding window of *delivery rate* samples, while T_{BBR} is the minimum of a sliding window of RTT samples. BBR defines which samples are included in the filter windows, how they are calculated and how the windows are delimited. Given the two quantities C_{BBR} and T_{BBR}, the principles mentioned above are met by ensuring that the sending rate matches C_{BBR}, which is done by adjusting the *pacing rate* parameter (rate balance), and by maintaining the amount of data in-flight at $C_{\text{BBR}} \cdot T_{\text{BBR}}$ (full pipe).

The BBR state machine is shown in Figure 10.8. After a StartUp state, that plays a role similar to slow start, and has the task to drive the estimate of the bottleneck link bandwidth close to its target, there is the Drain state, shortly visited after

Figure 10.8 BBR state diagram.

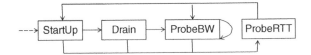

completing StartUp, to drain the bottleneck buffer of excess packets due to the StartUp overshoot. At this point BBR enters the ProbeBW state, which is intended to be the state where a BBR connection spends most of its time. In that state the pacing gain of the BBR algorithm cycles through values that allow probing for checking whether more bandwidth has become available. The ProbeRTT state is visited when the estimate of the RTT is deemed to need a refresh. For that purpose, the pipe is drained until the bottleneck buffer is deemed to be empty. Then the base RTT can be reliably estimated.

The BBR state machine runs periodic, sequential experiments, sending faster to check for C_{BBR} increases or sending slower to check for T_{BBR} decreases. The frequency, magnitude, duration, and structure of these experiments differ depending on what is already known (startup or steady-state) and sending application behavior (intermittent or continuous).

10.3.5.1 Delivery Rate Estimate

A key operation of BBR is estimation of the delivery rate, i.e., the current rate at which data is being effectively delivered to the final destination. The delivery rate sample taken upon receipt of the ACK of packet P is the ratio of the amount of data that was in-flight when P was sent to the maximum between the time interval it took to send all of that data and the time interval over which ACKs of that data arrived at the sender. The algorithm used by BBR is detailed in [57]. We give a concise account in the following.

Formally, we define the following connection state variables of the TCP connection at time t.

$C_D(t)$ the total amount of data delivered up to time t.
$C_T(t)$ the wall clock time when $C_D(t)$ was last updated.
$C_S(t)$ the send time of the packet that was most recently marked as delivered.

When a packet is sent, at time s, the following quantities are stored in a per-packet data structure: $[s, C_D(s), C_T(s), C_S(s)]$.

Let s_k and a_k be the times when the k-th packet delivered over the connection has been sent and ACKed, respectively. Then, the sample R_k of the delivery rate calculated upon reception of the ACK at time a_k is

$$R_k = \frac{\Delta d_k}{\Delta t_k} = \frac{C_D(a_k) - C_D(s_k)}{\max\{C_T(a_k) - C_T(s_k), s_k - C_S(s_k)\}} \tag{10.18}$$

Note that $C_T(a_k) - C_T(s_k)$ is the time required to collect ACKs of the packets that were in-flight when packet k was sent, while $s_k - C_S(s_k)$ is the duration of the time interval in which packets that were in-flight when packet k was sent have been transmitted. The connection state variables are updated as follows upon reception of the ACK of packet k at time a_k: $C_D(a_k^+) = C_D(a_k^-) + L_k$, $C_T(a_k) = a_k$, $C_S(a_k) = s_k$, where L_k is the length of packet k. A boolean variable marks each delivery rate sample, to highlight whether it is the result of limitations induced by the application rather than by the network bottleneck capacity.

10.3.5.2 StartUp and Drain

In StartUp, the RTT estimator T_{BBR} and the bottleneck bandwidth estimator C_{BBR} are initialized to the first available value of the Smoothed RTT (SRTT), after connection set-up, and IW/T_{BBR}, where IW is the initial congestion window size. During StartUp packets are sent at a pacing rate that is a multiple of the currently estimated delivery rate. Therefore, the sending rate is steadily increased by a multiplicative factor (as well as the congestion window size), aiming at a fast probing of the available bandwidth. The pacing gain during this phase is chosen to be $\alpha = 2/\log(2)$. We can realize where such a number comes from, by writing the expression of the estimated delivery rate $R(t)$ for a packet whose ACK is received at time t during StartUp:

$$R(t) = \frac{NL}{\frac{L}{C(t_1)} + \frac{L}{C(t_2)} + \cdots + \frac{L}{C(t_N)}} \tag{10.19}$$

where T denotes the RTT, N is the number of packets delivered in the time interval between $t - 2T$ and $t - T$, L is the packet size, t_k is the sending time of the k-th packet, $k = 1, \ldots, N$, with $t - 2T \le t_1 < \cdots < t_N \le t - T$, and $C(t)$ is the sending rate at time t. In writing eq. (10.19) we assume that all round-trip times have a fixed duration T (the base RTT) during StartUp. This is true until when the probing rate is less than or equal to the bottleneck link capacity.

The sending rate is set to $C(t) = \alpha R(t)$, where α is the pacing gain. We can write:

$$C(t) = \alpha \left[\frac{1}{N} \left(\frac{1}{C(t_1)} + \frac{1}{C(t_2)} + \cdots + \frac{1}{C(t_N)} \right) \right]^{-1} \tag{10.20}$$

Let us now make a fluid approximation, that can be expected to yield a reasonable result when $N \gg 1$. Reminding that N is the number of packets sent in an RTT and that RTTs are assumed to have a constant duration T, we get:

$$C(t) = \alpha \left[\frac{1}{T} \int_{t-2T}^{t-T} \frac{1}{C(u)} \, du \right]^{-1} \tag{10.21}$$

Taking reciprocals of both sides and deriving with respect to t, we find:

$$\frac{C'(t)}{C(t)^2} = \frac{1}{\alpha} \frac{1/C(t-2T) - 1/C(t-T)}{T} \tag{10.22}$$

Since $C(t)$ grows multiplicatively, we assume an exponential form, i.e., $C(t) = C_0 b^{t/T}$. Using this expression in eq. (10.22), we derive an equation for the nondimensional growth rate b:

$$\log b = \frac{b(b-1)}{\alpha} \quad \Rightarrow \quad \alpha = \frac{b(b-1)}{\log b} \tag{10.23}$$

If we require that the sending rate doubles each RTT, hence $b = 2$, it turns out that we should set $\alpha = 2/\log(2)$.

During the StartUp phase, the pacing rate and the congestion window are inflated by using the pacing gain α. This goes on until the sender estimates that the pipe is full. This event is detected when the estimated delivery rate is found to be almost constant (to within a given percentage of the previous level; 25% in the default implementation) for a preset number of times (3 in the default implementation).

At that point the Drain state is visited, where the pacing gain is set to $1/\alpha$ for one RTT. The purpose is to drain the bottleneck queue, which might have built up due to the overshoot of the StartUp phase. The state moves from Drain to Probe BW when the number of packets in-flight matches the estimated BDP. BBR checks the this condition upon every ACK.

10.3.5.3 ProbeBW

ProbeBW is actually the typical state where the BBR state machine should be found most of the time (the Authors claim 98% of the time in normal usage conditions). ProbeBW consists of scanning periodically a sequence of pacing gain values. In the default implementation, ProbeBW uses an eight-phase cycle with the following pacing gain values: 5/4, 3/4, 1, 1, 1, 1, 1, 1. The initial phase is chosen at random (except that it cannot be pacing gain = 3/4). The purpose of this modulation of the pacing gain is to probe for more available bandwidth at the bottleneck (pacing gain > 1), and to drain the possibly accumulated queue immediately afterward (pacing gain < 1). Each of the eight phases lasts a time equal to the current estimate of the RTT. Checking for phase advancement is performed on each ACK reception. The whole round is cycled through in about eight RTTs.

At any time, the pacing rate C is set at the estimated bottleneck capacity C_{BBR} multiplied by the pacing gain. The pacing rate is used to determine the next packet send time. After having sent a packet at time t_0, the next packet of length L can be sent no earlier than time $t_0 + L/C(t_0)$, where $C(t_0)$ is the pacing rate at time t_0. BBR ensures also that at any time there cannot be more than *cwnd* packets in-flight. In ProbeBW, the *cwnd* is maintained at twice the estimated BDP.

In turn, the bottleneck capacity is the result of a max-filter of the samples of delivery rate collected over the last K RTTs. In the default implementation, it is $K = 10$. The max-filter is simply a sliding window of delivery rate values, the maximum of which is picked at any time as the estimated bottleneck capacity. The idea behind choosing the maximum is that BBR should hover on the working point shown in the lower graph of Figure 10.7. Then, the bottleneck bandwidth could be easily underestimated. The underestimation error is reduced by taking the max of K samples of delivery rate. The reason of using a sliding window is to track a time-varying environment.

10.3.5.4 ProbeRTT

In any state other than ProbeRTT itself, if the estimate of T_{BBR} has not been updated (i.e., by getting a lower RTT measurement) for more than `ProbeRT-TInterval` = 10 seconds, then BBR enters ProbeRTT and reduces the *cwnd* to a minimal value, `BBRMinPipeCwnd` (four packets). After maintaining `BBRMin-PipeCwnd` or fewer packets in-flight for at least `ProbeRTTDuration` (200 ms) or one RTT, whichever is larger, BBR leaves ProbeRTT and moves to either ProbeBW or StartUp, depending on whether it estimates the pipe was already filled.

ProbeRTT lasts long enough (at least `ProbeRTTDuration` = 200 ms) to allow flows with different RTTs to have overlapping ProbeRTT states, while still being short enough to bound the throughput penalty of ProbeRTT, due to the reduction of *cwnd*, to roughly 2% (200 ms/10 seconds).

Samples of the RTT, collected upon every ACK, are inserted in the RTT Min-Filter data structure, a sliding window that must contain a number of samples corresponding to the last 10 seconds of TCP connection lifetime.

An `RTpropFilterLen` of 10 seconds is short enough to allow quick convergence, if traffic levels or routes change, but long enough so that we exploit interactive applications low-rate periods or pauses that drain the bottleneck queue. Then the `BBR.RTprop` filter opportunistically picks up these RTT measurements, and T_{BBR} refreshes without requiring a visit to the ProbeRTT state. Summing up, BBR connections typically need only pay 2% throughput penalty, if there are multiple bulk flows busy sending over the entire `ProbeRTTInterval` window.

10.3.5.5 Pseudo-code of BBR Algorithm

A simplified pseudo-code that summarizes the key BBR operations is as follows. Upon ACK reception at the sender the following code is executed.

```
rtt = now - packet.sendtime
update_min_filter(RTpropFilter, rtt)
delivered += packet.size
delivered_time = now
data = delivered - packet.delivered
time = delivered_time - packet.delivered_time
deliveryRate = data / time
```

```
if ((deliveryRate > BtlBwFilter.currentMax) ||
    (! packet.app_limited))
    update_max_filter(BtlBwFilter, deliveryRate)
if (app_limited_until > 0)
    app_limited_until = app_limited_until - packet.size
```

Note that the Max-filter for bottleneck bandwidth estimation is updated in case the sample is not marked as application-limited, or if it is greater than the current bottleneck bandwidth capacity. In the latter case, the filter is updated, irrespective of whether the estimated delivery rate sample is marked as application-limited.

On packet send time, the following function is executed.

```
bdp = BtlBwFilter.currentMax * RTpropFilter.currentMin
if (inflight >= cwnd_gain * bdp){
    // wait for ack or retransmission timeout
    return}
if (now >= nextSendTime){
    packet = nextPacketToSend()
    if (! packet){
        app_limited_until = inflight
        return}
    packet.app_limited = (app_limited_until > 0)
    packet.sendtime = now
    packet.delivered = delivered
    packet.delivered_time = delivered_time
    ship(packet)
    rate = pacing_gain * BtlBwFilter.currentMax
    nextSendTime = now + packet.size / rate}
```

Widespread use of BBR is controversial. It has been found experimentally that BBR can cause severe throughput fluctuations [163]. Fairness to other congestion control algorithms is an issue as well. It has been verified that the BBR algorithm becomes too aggressive, hence unfair, to loss-based TCP algorithms when bottleneck buffers are shallow [103]. Performance of BBR in cellular network environments have raised further issues [213] and stirred the study of a new version of the protocol, still under way as of the time of writing.

10.4   Traffic Engineering with TCP

In this section we establish simple relationships between the steady-state throughput, congestion window size and packet loss of a long-lived, greedy TCP connection. By long-lived we mean a TCP connection whose duration is theoretically

TCP senders

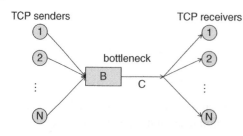

TCP receivers

Figure 10.9 Dumb-bell topology model of N TCP connections sharing a bottleneck link.

unlimited. Practically, it means that the TCP connection lifetime is much bigger than the average RTT. By greedy we mean that the TCP source always has new packets to send whenever possible.

We refer to the classic dumbbell topology model, where N TCP flows share a bottleneck link (see Figure 10.9). Let T be the common value of the base RTT of the N TCP connections. Let B denote the buffer size at the bottleneck. Let C denote the bottleneck capacity.

In the simple analysis of this section we assume that the RTT is constant, equal to T. Hence the sending rate of the TCP source is $x(t) = W(t)/T$, where $W(t)$ denotes the *cwnd* at time t. We assume that a single packet loss per connection occurs as *cwnd* attains its maximum value, given by $(C\dot{T} + B)/N$, where CT is the BDP of the TCP connections path. Hence the *cwnd* grows from its initial value W_0 up to $W_f = (CT + B)/N$. Then, upon packet loss detection, *cwnd* is reduced to $W_f(1 - \beta)$ and a new congestion avoidance phase starts. Therefore, $W(t)$ behaves as a periodic function of time, increasing steadily from $W_0 = W_f(1 - \beta)$ up to W_f. Let us consider one period, between 0 and t_f, where $W(0) = (1 - \beta)W_f$ and $W(t_f) = W_f$.

The mean sending rate of the TCP connection is

$$\bar{x} = \frac{1}{t_f} \int_0^{t_f} \frac{W(t)}{T} dt \qquad (10.24)$$

Since a single packet is dropped over the entire time window of duration t_f, the packet loss fraction can be written as:

$$p = \frac{1}{\int_0^{t_f} \frac{W(t)}{T} dt} \qquad (10.25)$$

The throughput of the TCP connection is

$$\Lambda = \bar{x}(1 - p) = \frac{1 - p}{t_f p} \approx \frac{1}{t_f p} \qquad (10.26)$$

where the last expression holds if $p \ll 1$.

We consider two congestion control paradigms: a general AIMD model and CUBIC.

The general AIMD model is characterized by two parameters, α and β [79, 181]. During congestion avoidance, the *cwnd* is incremented by $\alpha/cwnd$ upon each received good ACK. When a packet loss is detected via duplicated ACKs, *cwnd* is reduced by $\beta \cdot cwnd$. Then, we have

$$W_{\mathrm{AIMD}}(t) = (1 - \beta)W_f + \alpha \frac{t}{T}, \qquad 0 \le t \le t_f. \tag{10.27}$$

Imposing $W_{\mathrm{AIMD}}(t_f) = W_f$, we find $t_f = T\beta W_f / \alpha$. Introducing the expression of $W_{\mathrm{AIMD}}(t)$ into eq. (10.25), we get $p = \frac{2\alpha}{\beta(2-\beta)W_f^2}$, whence $W_f = \sqrt{\frac{2\alpha}{p\beta(2-\beta)}}$. Using this result and eq. (10.27) into eq. (10.24), we get

$$\bar{x}_{\mathrm{AIMD}} = \left(1 - \frac{\beta}{2}\right)\frac{W_f}{T} = \frac{1}{T\sqrt{p}}\sqrt{\frac{\alpha(2-\beta)}{2\beta}} \tag{10.28}$$

With the parameter setting of classic TCP, $\alpha = 1$ and $\beta = 1/2$, we have

$$\bar{x}_{\mathrm{classic}} = \frac{\sqrt{1.5}}{T\sqrt{p}} \tag{10.29}$$

For CUBIC, we have[15] $W(t) = W_f + \Gamma(t - K)^3$, with $K = (\beta W_f / \Gamma)^{1/3}$, for $0 \le t \le t_f = K$. Following the same steps as with AIMD, we get $p = \frac{T/K}{(1-\beta/4)W_f}$, whence $W_f = \left(\frac{\Gamma}{\beta}\right)^{1/4}\left(\frac{T}{p(1-\beta/4)}\right)^{3/4}$. Substituting the expression of the *cwnd* for CUBIC into eq. (10.24), we get

$$\bar{x}_{\mathrm{CUBIC}} = \left(1 - \frac{\beta}{4}\right)\frac{W_f}{T} = \frac{1}{T^{1/4}p^{3/4}}\left[\frac{\Gamma(4-\beta)}{4\beta}\right]^{1/4} \tag{10.30}$$

Using the default values of CUBIC, $\beta = 0.3$ and $\Gamma = 0.4$, we find

$$\bar{x}_{\mathrm{CUBIC}} = \frac{1.054}{T^{1/4}p^{3/4}} \tag{10.31}$$

where T is in seconds.

It is interesting to exploit this simple relationships to assess the level of packet integrity required to make the congestion window reach the average value corresponding to full capacity utilization. To attain the maximum throughput with no buffer delay, at equilibrium the average *cwnd* size must be equal to the BDP, i.e., $\overline{W} = C \cdot T$, where C is measured in packets per second. For example, choosing a packet size of 1500 bytes and RTT equal to $T = 100$ ms, the average *cwnd* at equilibrium for a 1 Gbit/s link is $\overline{W} = 8333.3$ pkts.

For the general AIMD it is $\overline{W} = (1 - \beta/2)W_f$. Using eqs. (10.25) and (10.27), we find $p = \frac{\alpha(2-\beta)}{2\beta\overline{W}^2}$, with $\overline{W} = C \cdot T$. For a throughput of 1 Gbit/s to be achieved by a standard TCP AIMD mechanism with $\alpha = 1$ and $\beta = 1/2$ in a pipe with 100 ms base RTT, it must be $p \approx 2.2 \cdot 10^{-8}$. The requirement on the packet loss probability

---

15 We replace the conventional CUBIC parameter notation C with Γ, to avoid confusing with the bottleneck link capacity.

becomes two orders of magnitude smaller for each order of magnitude of increase of the bottleneck capacity. Even such a simple model points out that exploiting fully the capacity of fat, long pipes is critical for standard TCP.

As for CUBIC, the average congestion window size for full link exploitation is $\overline{W} = (1 - \beta/4)W_f$. Then, the packet loss probability can be expressed as $p = T\left[\frac{\Gamma(4-\beta)}{4\beta}\right]^{1/3}\frac{1}{\overline{W}^{4/3}}$, with $\overline{W} = C \cdot T$. Using the default values for the CUBIC parameters, with $T = 100$ ms and a bottleneck link capacity of 1 Gbit/s we find a requirement of $p \approx 6.35 \cdot 10^{-7}$. In general, the packet loss requirement is improved by an order of magnitude by CUBIC with respect to the standard TCP congestion control, as highlighted in Table 3 of [181].

Note that a more realistic packet loss level of 10^{-4} (10^{-5}) over a connection path with a base RTT of 100 ms, with MSS size of 1500 bytes, would make the achievable throughput about 14.7 Mbit/s (46.5 Mbit/s) with the standard TCP congestion control and about 22.5 Mbit/s (126.5 Mbit/s) for CUBIC, no matter how much the bottleneck link capacity is bigger than that level. For lower values of the packet loss, the behavior of TCP CUBIC falls back to the standard TCP, thanks to the TCP friendliness function.

10.5 Fluid Model of a Single TCP Connection Congestion Control

We present here a classic analysis of a TCP connection with a single bottleneck link, using a fluid approximation.

Figure 10.10 depicts the TCP connection model abstraction for a single bottleneck scenario. Let $C(t)$, T, and B denote the capacity of the bottleneck link, the base RTT of the TCP connection and the size of the bottleneck buffer. The model abstraction assumes that the bottleneck lies in the forward path of the TCP connection, while the reverse path suffers no congestion, i.e., the delay of ACKs from the receiver to the sender is constant.

We denote with $W(t)$ and $Q(t)$ the congestion window and the occupancy level of the bottleneck buffer at time t, respectively. Let also $\lambda(t)$ and $\mu(t)$ denote the instantaneous rate in and out of the bottleneck buffer. We assume all quantities are measured in consistent units. Thanks to our assumptions, we can measure W and B in units of MSS (or packets) and C in MSS/s (or packets/s).

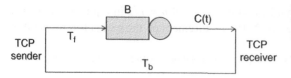

Figure 10.10 Model of a TCP connection with a single bottleneck.

In the model statement and analysis we assume that:

1. After being established, the considered TCP connection lasts forever (long-lived TCP connection).
2. The TCP sender is greedy, i.e., it has always new packets ready to send.
3. The internal TCP sender buffer is large enough not to be a bottleneck.
4. The receiver window is always bigger than *cwnd* (virtually it is infinite), i.e., the TCP sender is only constrained by the congestion window (network-limited TCP connection).
5. There is no loss of packets except because of buffer overflow at the bottleneck buffer.
6. Packet loss is always detected thanks to fast retransmit (i.e., reception of *DupThresh* duplicate ACKs).
7. The duration of fast retransmit and fast recovery reduces to one RTT.
8. Upon packet loss detection, the congestion window is scaled down by a factor $1 - \beta$.
9. Fixed length data segments are sent on the connection, with length equal to the negotiated MSS.[16]

One consequence of assumptions 5 and 6 is that slow start is never performed, except at connection opening. Given that the connection is long-lived, slow start has a negligible impact on the steady-state connection throughput, hence we disregard it. As for assumption 7, it is a minor approximation, if some mechanism is employed to recover multiple packet losses within one or few RTTs, e.g., NewReno or SACK.

We start with the simpler case of fixed capacity bottleneck link, then we generalize to the case where the bottleneck link capacity is time-varying.

10.5.1 Classic TCP with Fixed Capacity Bottleneck Link

Let us consider a link of capacity C equipped with a buffer of size B, which is the bottleneck link of a TCP connection. To the assumptions listed above, we add the following:

> The bottleneck link capacity is constant, i.e., $C(t) = C, \ \forall t$.

In the following we develop a fluid model of the TCP connection evolution. Hence, all quantities are represented by real numbers, and data flow in and out of the bottleneck buffer is regarded as continuous. The state of the model is completely described by assigning the instantaneous values of the congestion window $W(t)$ and of the bottleneck buffer occupancy level $Q(t)$.

---

16 Packets in the network include also overhead. We assume that the buffer size is set so as to accommodate also that overhead.

Under our assumptions, the evolution of the congestion window of the TCP connection is deterministic and periodic, with a steady increase from an initial value W_0 at time t_0, up to a maximum value, when finally packet loss is detected at the TCP sender, at time t_f and the congestion window has ramped up to the value W_f. At that point the congestion window is cut by the factor $1 - \beta$ and a new window growth cycle starts.

Depending on values of C and B, hence of W_0, the interval $[t_0, t_f]$ can be divided up into two subintervals. The first one, possibly void, is $[t_0, t_b]$, during which the bottleneck buffer is empty and the output rate $\mu(t)$ is equal to the input rate $\lambda(t)$. In the remaining interval $[t_b, t_f]$, $\mu(t) = C$ and the buffer content builds up.

A simple analysis of the TCP connection goes as follows. As long as the bottleneck buffer is empty, the *cwnd* grows linearly by 1 MSS every RTT and RTTs have a constant duration T. Hence

$$W(t) = W_0 + \frac{1}{T}(t - t_0), \qquad t_0 \leq t \leq t_b \tag{10.32}$$

This behavior stops as soon as the pipe is full, i.e., $W(t_b) = CT$. This condition yields:

$$t_b = t_0 + T(CT - W_0) \tag{10.33}$$

After time t_b, ACKs arrive at a rate C at the TCP sender. Then, the congestion windows grows according to the law:

$$\frac{dW}{dt} = \frac{C}{W(t)}, \qquad t_b \leq t \leq t_f. \tag{10.34}$$

This is but the Reno *cwnd* update rule, according to which *cwnd* is increased by $1/cwnd$ upon each good ACK reception in congestion avoidance. The initial condition is $W(t_b) = CT$, whence

$$W(t) = \sqrt{W(t_b)^2 + 2C(t - t_b)}, \qquad t_b \leq t \leq t_f. \tag{10.35}$$

The buffer overflow time t_f is found by imposing that $W(t_f) = B + CT + 1 \equiv W_f$. We have

$$t_f = t_b + \frac{W_f^2 - W(t_b)^2}{2C} = t_0 + T(CT - W_0) + \frac{W_f^2 - (CT)^2}{2C} =$$
$$= t_0 + \frac{W_f^2 + (CT)^2 - 2CTW_0}{2C} \tag{10.36}$$

The average throughput Λ is found as the ratio of the amount of data delivered in one cycle and the duration of the cycle:

$$\Lambda = \frac{\int_{t_0}^{t_f} \mu(\tau)d\tau}{t_f - t_0} \tag{10.37}$$

Since it is $\mu(t) = W(t)/T$ for $t \in [t_0, t_b]$, and $\mu(t) = C$ for $t \in [t_b, t_f]$, and recalling that $W(t_b) = CT$, we can write

$$\rho = \frac{\Lambda}{C} = \frac{\int_{t_0}^{t_b} \frac{W(\tau)}{T} d\tau + C(t_f - t_b)}{C(t_f - t_0)} = \frac{\frac{(CT)^2 - W_0^2}{2} + \frac{W_f^2 - (CT)^2}{2}}{\frac{W_f^2 + (CT)^2 - 2CTW_0}{2}} =$$

$$= \frac{W_f^2[1 - (1 - \beta)^2]}{W_f^2 + (CT)^2 - 2CTW_f(1 - \beta)} \tag{10.38}$$

where $\rho = \Lambda/C$ is the normalized throughput and we have accounted for the continuity condition $W_0 = (1 - \beta)W_f$. This last condition comes from the congestion window reduction at the end of the fast recovery. The new congestion avoidance cycle is started with an initial window W_0, which is a fraction of the congestion window attained in the previous cycle, W_f.

Substituting $W_f = B + CT + 1 \approx B + CT$ and using the simplifying notation $a = CT$, we get

$$\rho = \frac{(B + a)^2[1 - (1 - \beta)^2]}{B^2 + 2\beta aB + 2\beta a^2} \tag{10.39}$$

As expected, the ratio $b = B/a$ plays a key role in the throughput formula. This is the ratio of the buffer size to the BDP of the TCP connection. In terms of b, reminding that the throughput Λ can never exceed the bottleneck link capacity C, we have:

$$\rho = \min\left\{1, \frac{(b + 1)^2[1 - (1 - \beta)^2]}{b^2 + 2\beta b + 2\beta}\right\} \tag{10.40}$$

In the standard TCP it is $\beta = 1/2$, hence we find $\rho = \min\{1, \frac{3}{4}\frac{(b+1)^2}{b^2+b+1}\}$. The maximum is achieved for $b \geq 1$. Hence, it is seen that the minimum amount of buffer space at the bottleneck node for the considered connection to attain the full throughput allowed by the bottleneck link capacity is $B = CT$, i.e., *a buffer size equal to the BDP*. This is the famous "rule of thumb" of standard TCP buffer sizing.

10.5.2 Classic TCP with Variable Capacity Bottleneck Link

In this section we generalize the model of the previous section to the case of a bottleneck link with time-varying capacity $C(t)$. Assumptions listed at the beginning of Section 10.5 hold also for this case, except of the bottleneck link capacity. We assume here that the bottleneck link capacity is described by a function of time $C(t)$ such that $0 \leq C(t) \leq C$.

We denote with $W(t)$ the congestion window at time t. The bottleneck buffer content at time t is denoted by $Q(t)$. The base RTT T is split into the forward base

RTT, T_f, from the TCP sender to the bottleneck buffer input, and the backward base RTT, T_b, from the output of the bottleneck buffer to the TCP receiver and then back to the TCP sender[17].

Let $S(t)$ be the sending rate of the TCP sender at time t. The input rate at the bottleneck buffer is $\lambda(t) = S(t - T_f)$. It must always be $0 \leq Q(t) \leq B$, where B is the buffer size. For $0 < Q(t) < B$ the buffer content satisfies the differential equation $\frac{dQ}{dt} = \lambda(t) - C(t)$. The buffer content fluctuates between two reflecting barriers, namely 0 and B. Once the buffer gets empty at time t, which can only occur if $\lambda(t) < C(t)$, it will stay empty for all $\tau \geq t$ such that $\lambda(\tau) \leq C(\tau)$ and in the meanwhile, the output rate is $\lambda(\tau)$. As long as the buffer is nonempty, the output rate is $C(t)$. When the buffer gets full at time t, which implies that $\lambda(t) > C(t)$, it will stick to the upper limit B for all $\tau \geq t$ such that $\lambda(\tau) \geq C(\tau)$. For all those times, the output rate will be $C(\tau)$. Summing up, we have:

$$\frac{dQ}{dt} = \begin{cases} \lambda(t) - C(t) & 0 < Q(t) < B, \\ \max\{0, \lambda(t) - C(t)\} & Q(t) = 0, \\ \min\{0, \lambda(t) - C(t)\} & Q(t) = B. \end{cases} \tag{10.41}$$

The output rate of the buffer is denoted by $\mu(t)$. It is

$$\mu(t) = \begin{cases} \min\{\lambda(t), C(t)\} & Q(t) = 0, \\ C(t) & 0 < Q(t) \leq B \end{cases} \tag{10.42}$$

The TCP sender receives ACKs at rate $A(t) = \mu(t - T_b)$. The congestion window is updated as a function of $A(t)$. For TCP Reno, the congestion window is increased by $1/W(t)$ for each received good ACK. The differential equation is $\frac{dW}{dt} = \frac{\mu(t - T_b)}{W(t)}$. This equation holds in the congestion avoidance phase, i.e., when $W(t) \geq ssthresh$. If $W(t) < ssthresh$, $W(t)$ is incremented by 1 per received ACK, hence the rate of increase of $W(t)$ is equal to the ACK rate $A(t)$. Then, the evolution equation is $\frac{dW}{dt} = \mu(t - T_b)$. Putting all together, we can write:

$$\frac{dW}{dt} = \begin{cases} \frac{\mu(t - T_b)}{W(t)} & W(t) \geq ssthresh, \\ \mu(t - T_b) & W(t) < ssthresh. \end{cases} \tag{10.43}$$

The connection between the sending rate $S(t)$ and the congestion window size can be derived by expressing the number of packets in-flight at time t and matching this number with $W(t)$ (greedy TCP sender). The amount of data in-flight at time t corresponds to all packets sent since a piece of data is emitted by the TCP sender, say at time t_0, until the relevant ACK is received back at the sender at time t. The

---

17 We are assuming no bottleneck is in the way of the TCP ACKs; this is not the case with half-duplex links, e.g. IEEE 802.11 WLAN.

difference $t - t_0$ is the RTT measured by the TCP sender when it receives the ACK at time t. The starting time t_0 is a function of the RTT end time t. We denote this function with $\theta(t)$. It is implicitly defined by the following equation:

$$\theta(t) + T_f + D(\theta(t) + T_f) + T_b = t \tag{10.44}$$

where $D(t)$ denotes the waiting time of a piece of data arriving at the bottleneck buffer at time t. The RTT $R(t)$ of a piece of data whose ACK arrives at the sender at time t can be in turn written as follows: $R(t) = t - \theta(t) = T_f + D(t - R(t) + T_f) + T_b$, which yields another implicit equation.

In case of FIFO queueing at the bottleneck buffer, $D(t)$ can be found as the minimum non-negative solution to

$$Q(t) = \int_t^{t+D(t)} C(\tau)d\tau \tag{10.45}$$

This comes from the assumption of a work-conserving server (buffer output link) and the FIFO discipline. Taking derivatives of both sides of eq. (10.45), we find

$$\frac{dQ}{dt} = C(t + D(t))\left[1 + \frac{dD}{dt}\right] - C(t) \quad \Rightarrow \quad \frac{dD}{dt} = \frac{\frac{dQ}{dt} + C(t)}{C(t + D(t))} - 1 \tag{10.46}$$

For later use, we calculate the derivative of $D(t)$ at time $\theta(t) + T_f$. Since $\frac{dQ}{dt} = \lambda(t) - C(t)$ for $0 < Q(t) < B$ and $\lambda(t) = S(t - T_f)$, we have

$$\left.\frac{dD}{dt}\right|_{\theta(t)+T_f} = \frac{\lambda(\theta(t) + T_f)}{C(\theta(t) + T_f + D(\theta(t) + T_f))} - 1 = \frac{S(\theta(t))}{C(t - T_b)} - 1 \tag{10.47}$$

The derivative of $D(t)$ is used to obtain the relation between the congestion window and the sending rate. Since the TCP source is greedy, the congestion window size $W(t)$ equals the flightsize:

$$W(t) = \int_{\theta(t)}^t S(\tau)d\tau$$

From this we get

$$\frac{dW}{dt} = S(t) - S(\theta(t))\frac{d\theta}{dt} \tag{10.48}$$

By taking derivatives of both sides of eq. (10.44), we get

$$\frac{d\theta}{dt} = \begin{cases} 1 & Q(t - T_b) = 0, \\ \dfrac{1}{1 + \frac{dD}{dt}\big|_{\theta(t)+T_f}} = \dfrac{C(t - T_b)}{S(\theta(t))} & 0 < Q(t - T_b) < B. \end{cases} \tag{10.49}$$

where we used eq. (10.47).

If $Q(t - T_b) = 0$, we have simply $\theta(t) = t - T$. Then, from eqs. (10.48) and (10.49) we get

$$\frac{dW}{dt} = \begin{cases} S(t) - S(t - T) & Q(t - T_b) = 0, \\ S(t) - C(t - T_b) & 0 < Q(t - T_b) < B. \end{cases} \tag{10.50}$$

Looking at eq. (10.42), we can rewrite (10.50) as follows:

$$\frac{dW}{dt} = S(t) - \mu(t - T_b) \tag{10.51}$$

This last identity gives an expression for the rate $S(t)$ as a function of buffer output rate and the congestion window:

$$S(t) = \frac{dW}{dt} + \mu(t - T_b) \tag{10.52}$$

Equation (10.52) has an intuitive interpretation. The sending rate of the greedy TCP source equals the ACK rate (one new segment is injected into the TCP pipe for each received good ACK) plus the increment of the congestion window, which accounts for the additional amount of traffic that the TCP source sends into the pipe to probe it.

Since we assume a greedy TCP source, there is constantly a balance between the congestion window size and the amount of data in-flight in the pipe, as long as there is no loss. From (10.41) and (10.52) we have for any t such that $Q(t) < B$:

$$\frac{dQ}{dt} = \lambda(t) - \mu(t) = S(t - T_f) - \mu(t) = \frac{dW(t - T_f)}{dt} + \mu(t - T) - \mu(t) \tag{10.53}$$

Integrating both sides over the interval $[t_0, t]$, where we assume no loss takes place, we get

$$Q(t) - W(t - T_f) = Q(t_0) - W(t_0 - T_f) + \int_{t_0}^{t} [\mu(\tau - T) - \mu(\tau)]d\tau$$

$$= Q(t_0) + \int_{t_0 - T}^{t_0} \mu(\tau)d\tau - W(t_0 - T_f) - \int_{t - T}^{t} \mu(\tau)d\tau \tag{10.54}$$

Rearranging terms, it follows:

$$Q(t) + \int_{t-T}^{t} \mu(\tau)d\tau - W(t - T_f) = constant \tag{10.55}$$

The constant must be 0 to be consistent with initialization at TCP connection startup or with re-initialization of the congestion window after a fast retransmit/fast recovery or after a timeout. For a greedy source, in all these cases the congestion window size equals the *flightsize*. Therefore, we have

$$W(t - T_f) = Q(t) + \int_{t-T}^{t} \mu(\tau)d\tau \, , \qquad t \in [a, b] \tag{10.56}$$

where [a.b] is any interval during which there is no packet loss detection.

Summing up, the evolution of $Q(t)$ and $W(t)$ starting from an initial time t_0 is given by eqs. (10.41) and (10.43), respectively. These equations hold for $t \geq t_0$, until a buffer overflow occurs, t_0 being the initial time of a congestion avoidance phase. The time-varying bottleneck link capacity $C(t)$ appears indirectly through the output rate of the buffer, $\mu(t)$. To start the equations, we need an initial value $W(t_0)$ along with the specification of $\mu(\tau)$ for $\tau \in (t_0 - T, t_0]$.

As for performance, the average TCP throughput Λ can be obtained as

$$\Lambda = \lim_{\Delta t \to \infty} \frac{1}{\Delta t} \int_{t_0}^{t_0 + \Delta t} \mu(\tau) d\tau$$

The flightsize $Y(t)$, i.e., the amount of fluid in the pipe at time t, can be evaluated as

$$Y(t) = \int_{t-T_f}^{t} S(\tau) d\tau + Q(t) + \int_{t-T_b}^{t} \mu(\tau) d\tau \tag{10.57}$$

Using eq. (10.52), we find $Y(t) = W(t) - W(t - T_f) + Q(t) + \int_{t-T}^{t} \mu(\tau) d\tau$. This reduces to $Y(t) = W(t)$ as long as there is no loss in the bottleneck buffer; instead, it is $Y(t) < W(t)$ after loss has occurred, until full recovery is completed. From the time average $\overline{Y}$ of $Y(t)$ we can compute the average packet delay of the TCP connection as $\overline{Y}/\Lambda$, according to Little's law.

To start the evolution of the fluid model equations at $t = 0$ we assign the following initial conditions: the congestion window is set to IW packets (e.g., $IW = 1$, according to the old default value of TCP Reno), $S(0)$ is set to a Dirac pulse of area IW packets, the buffer is empty and the pipe is empty too, which means $\mu(\tau) = 0$ for $\tau \in (-T, 0]$.

There remains to define what happens as soon as a packet loss occurs. Assume a congestion avoidance phase starts at time t_0. Let t_B be the first time after t_0 such that $Q(t) = B$. Formally, $t_B = \inf_{t \geq t_0} \{ t : Q(t) = B \}$. The TCP sender becomes aware of the packet loss after the whole content of the buffer has been drained and up to *DupThresh* duplicate ACKs requesting the lost packet have been received, i.e., at time $t_1 \equiv t_B + D_e + T_b$, where D_e is given by

$$B + DupThresh = \int_{t_B}^{t_B + D_e} C(\tau)\, d\tau \tag{10.58}$$

The additional term *DupThresh* in the left hand side accounts for the fast retransmit mechanism.

Let $W_f \equiv W(t_1)$. At time t_1 the sender is aware of the packet loss event. First, the sender retransmits the packet it has detected as lost at time t_1. Immediately after the retransmission, the sending rate is set to zero, until an amount βW_f of in-flight data is delivered, say at time t_2. This event is detected at the sender by means of duplicate ACKs. From time t_2 the sending rate is set to the ACK rate, i.e., so much data is pushed into the pipe as it leaves the pipe. This recovery phase goes on from time t_2, until time t_3, when the ACK of the lost packet is received. This event occurs when an additional amount $(1 - \beta)W_f$ of data has left the pipe and has been replaced by as much data, according to the Fast Recovery procedure. At time t_3 congestion avoidance operation is resumed, starting from a congestion window of size $(1 - \beta)W_f$. Also the *ssthresh* is updated and set equal to the same value as the *cwnd*, namely $(1 - \beta)W_f$.

The details of the fluid model of the recovery phase are as follows. Packet loss is detected at time t_1 at the TCP sender. Then $S(t) = 0$ for $t \in [t_1, t_2)$, where t_2 is such that

$$\beta W_f = \int_{t_1}^{t_2} \mu(\tau - T_b) \, d\tau \tag{10.59}$$

After that, we set $S(t) = \mu(t - T_b)$ for $t \in [t_2, t_3)$, where t_3 is such that

$$(1 - \beta)W_f = \int_{t_2}^{t_3} \mu(\tau - T_b) \, d\tau \tag{10.60}$$

At t_3, congestion avoidance operation is resumed with a congestion window $(1 - \beta)W_f$.

The recovery model works provided there is enough "fluid" in the pipe as the packet loss is detected at the sender at time t_1. In fact, the sender waits for an amount of fluid βW_f to be delivered, while no new fluid is sent into the pipe. If so many packets have been lost that the overall fluid in the pipe at t_1 is less than βW_f, the recovery procedure will be stalled waiting for the never occurring event that up to an amount βW_f of data will be delivered. Real TCP is not trapped this way thanks to the retransmission time-out (RTO) mechanism.

The model described above applies to fast retransmit/fast recovery. There is, however, the possibility that timeout is triggered if fast retransmit or fast recovery fail. For the fast recovery procedure to work in our model, two conditions must be satisfied:

1. Lost data is less than $(1 - \beta)W_f$.
2. The bottleneck link capacity is such that recovery be completed before the time-out is triggered.

The first condition ensures that there exists a finite time t_2 such that the pipe is drained by an amount of data βW_f, while the sending rate is 0. The second condition could be violated, e.g., if the bottleneck link is shut off long enough for the TCP RTO to expire.

To keep things simple, we do not model the adaptive RTO estimate. Rather, we assume that RTO is set to a fixed value equal to $5T$, i.e., five times the base RTT. This approximation replaces the true RTO, equal to the current estimate of the smoother RTT plus four times the RTT deviation. In this replacement, we assume that the smoothed RTT and the RTT deviation both are equal to T.

When loss is detected at time t_1, the lost packet has been sent at time $\theta(t_1)$. If it turns out that $t_3 > \theta(t_1) + 5T$, then it is assumed that a time-out is triggered. Therefore, the pipe is drained completely, the congestion window is reinitialized to IW, and the *ssthresh* to $(1 - \beta)W_f$.

Example 10.1 *Analysis for periodic capacity*   The equations governing the congestion window during congestion avoidance can be given a rather simple analytical form, useful to gain insight, if we assume the buffer stays not empty and the capacity function has an analytically convenient form. For that purpose, we consider a sinusoidal behavior for the bottleneck link capacity in this example. We let

$$C(t) = C_0 + C_1 + C_1 \cos(2\pi Ft + \phi_0),\qquad(10.61)$$

with $C_0, C_1 > 0$. This is useful to gain insight into the effect of the time scale of variation of the capacity on TCP throughput performance. With a sinusoidal function, the time scale over which $C(t)$ has significant variations is of the order of half the period $1/F$.

Let us assume the bottleneck buffer is non empty in the interval $[t_0 - T, t_B]$. Then, the congestion window is determined by the following dynamical equation:

$$\frac{dW}{dt} = \frac{C(t)}{W(t)}\qquad(10.62)$$

Given the initial condition $W(t_0) = W_0$, the solution for $t \in [t_0, t_B]$ is easily found to be

$$W(t) = \sqrt{W_0^2 + 2\frac{t - t_0}{T}f(t)}\qquad(10.63)$$

with

$$f(t) = (C_0 + C_1)T + C_1 T \frac{\sin[\pi F(t - t_0)]}{\pi F(t - t_0)}\cos[\pi F(t + t_0) + \phi_0]\qquad(10.64)$$

If there is no loss in the interval $[t_0, t_B]$, the flightsize equals the congestion window size. Since the buffer is never empty, the flightsize is the integral of the capacity function over a base RTT, i.e.

$$W(t) = Q(t) + \int_{t-T}^{t} C(\tau)\, d\tau = Q(t) + g(t)\qquad(10.65)$$

where

$$g(t) = (C_0 + C_1)T + C_1 T\frac{\sin(\pi FT)}{\pi FT}\cos(\pi F(2t - T) + \phi_0)\qquad(10.66)$$

Note that $g(t)$ is the time-varying BDP.

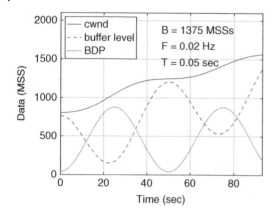

Figure 10.11 Time evolution of the congestion window, the queue length at the bottleneck and the time-varying BDP for the TCP fluid model with sinusoidal bottleneck link capacity.

Under the double assumption that the buffer never empties and that it does not overflow, we can find the buffer content evolution from eq. (10.65): $Q(t) = W(t) - g(t)$. The evolution of $W(t)$ and $Q(t)$ according to the time functions given above stops when the buffer overflows, i.e., at time t_B such that $Q(t_B) = B$. After loss recovery, a new congestion avoidance phase starts, with initial congestion window size equal to $(1 - \beta)W(t_B)$.

The behavior of $W(t)$, $Q(t)$ and $g(t)$ is plotted in Figure 10.11 for $B = 1375$ MSSs, $T = 50$ ms, $C_0 = 10$ Mbit/s, $C_1 = 100$ Mbit/s, $F = 0.02$ Hz, $t_0 = 0$, $\phi_0 = \pi$. The MSS is 1500 bytes. The initial congestion window at $t = 0$ is set to 800 MSSs.

In the situation depicted in Figure 10.11 the buffer size can cope with the BDP fluctuations, for the given set of parameter values.

In general, as $F \to \infty$, we have $f(t) \to (C_0 + C_1)T$, hence the congestion window and the buffer content evolve as if the capacity were constant and equal to the average value of the actually available link capacity. In other words, for very fast time variation of the link capacity ($FT \gg 1$) the TCP congestion control averages out the variation.

In the limit $F \to 0$, we have $f(t) \to (C_0 + C_1)T + C_1 T \cos \phi_0$, i.e., TCP sees a constant capacity, equal to the initial value of the link capacity. This too matches with intuition, since for $FT \ll 1$, it is almost as if the link capacity were constant from the point of view of the TCP sender.

The intermediate range of F is where the most interesting interaction between the TCP control loop and the time-varying bottleneck link capacity takes place. The analysis is not straightforward. We resort to fluid simulations, based on the discretization of the differential equation model developed in this section.

10.5.2.1 Discretization of the Evolution Equations

To solve numerically the differential equations governing the congestion window and the buffer content, we sample time with a step δ, i.e., at times $t_k = t_0 + k\delta$, $k \geq 0$.

In the discretization, we distinguish between quantities denoting amount of data (window size, buffer content) and data rates (rate of arrivals, departures, lost data). As for the first ones, the discrete sample of the continuous time function $H(t)$ is denoted with $H(k) \equiv H(t_k)$. As for rates, we let $H(k) \equiv \int_{t_k}^{t_{k+1}} H(\tau)d\tau$. Note that rates are converted into amount of data, i.e., the amount of data sent over a time interval of duration δ starting at t_k.

Let us define the normalized fixed delays $K \equiv T/\delta$, $K_f \equiv T_f/\delta$ and $K_b \equiv T_b/\delta$. We assume these numbers are integer valued.

Let $X(k)$, $U(k)$ and $L(k)$ denote the arrivals at the bottleneck buffer, the departures from the bottleneck buffer, and the lost data, respectively. Then

$$Q(k+1) = Q(k) + X(k) - U(k) - L(k)$$

$$U(k) = \min\{C(k), Q(k) + X(k)\}$$

$$L(k) = \max\{0, Q(k) + X(k) - C(k) - B\}$$

For TCP Reno the congestion window discrete evolution equation, during the congestion avoidance phase, is

$$W(k+1) = W(k) + \frac{U(k - K_b)}{W(k)} \tag{10.67}$$

The sending rate is

$$S(k) = U(k - K_b) + \frac{U(k - K_b)}{W(k)} \tag{10.68}$$

Since the input rate at the bottleneck buffer is a delayed copy of the sending rate, we have $X(k) = S(k - K_f)$ and therefore $X(k) = U(k - K) + U(k - K)/W(k - K_f)$.

Equations (10.67) and (10.68) change to $W(k+1) = W(k) + U(k - K_b)$ and $S(k) = 2U(k - K_b)$, respectively, when the TCP connection congestion control is in the slow start state.

To simplify the notation, in the following we align the time axis to the ingress of the bottleneck buffer, i.e., we shift the time origin by T_f. To implement the change of the time origin, it suffices to rewrite all equations, letting $K_f = 0$ and $K_b = K$.

We can summarized the discretized equations as follows.

Initialization. At the initial step $k = 0$ we let $W(0) = IW$, $Q(0) = 0$, and $U(-K) = \cdot = U(-1) = 0$.

The first RTT. For $k = 1, \ldots, K$, we let[18]

$$X(k) = IW/K$$

$$U(k) = \min\{C(k), Q(k) + X(k)\}$$

$$L(k) = \max\{0, Q(k) + X(k) - C(k) - B\}$$

$$W(k + 1) = IW$$

$$Q(k + 1) = Q(k) + X(k) - U(k) - L(k)$$

The first round trip time is a special case, since the connection is operated open-loop during that time. As soon as the first feedback arrives at the sender, we move to the next phase.

Normal operations. For $k \geq K + 1$ the evolution of the TCP connection is described by the following equations in slow start:

$$X(k) = 2U(k - K)$$

$$U(k) = \min\{C(k), Q(k) + X(k)\}$$

$$L(k) = \max\{0, Q(k) + X(k) - C(k) - B\}$$

$$W(k + 1) = W(k) + U(k - K)$$

$$Q(k + 1) = Q(k) + X(k) - U(k) - L(k)$$

and the following equations in congestion avoidance:

$$X(k) = U(k - K) + \frac{U(k - K)}{W(k)}$$

$$U(k) = \min\{C(k), Q(k) + X(k)\}$$

$$L(k) = \max\{0, Q(k) + X(k) - C(k) - B\}$$

$$W(k + 1) = W(k) + \frac{U(k - K)}{W(k)}$$

$$Q(k + 1) = Q(k) + X(k) - U(k) - L(k)$$

The state of the congestion control is switched from slow start to congestion avoidance as soon as it is $W(k) \geq ssthresh$. The evolution equations listed above are used until the connection is finished or loss is detected. A loss event is detected at step k_L, if $L(k_L) > 0$ occurs for the first time, after the previous loss event or since the connection start. At that time the modeling of the recovery phase starts.

The discrete version of the loss recovery phase is quite straightforward. It is worth mentioning that the buffer delay can be computed by discretizing

---

18 The sending rate is paced so that IW segments are sent out during the whole initial RTT. Alternatively, we could set $X(1) = IW$ and $X(2) = \cdots = X(K) = 0$.

the differential equation governing $D(t)$ for a nonempty buffer, i.e., $\frac{dD}{dt} = \lambda(t)/C(t + D(t)) - 1$. Therefore, we have

$$D(k + 1) = \begin{cases} 0 & Q(k + 1) = 0, \\ D(k) + \delta \left[\dfrac{X(k)}{C(k + D(k))} - 1 \right] & Q(k + 1) > 0. \end{cases}$$

Given this recurrence, we are able to compute the discretization of D_e and of the loss recovery times t_1, t_2, t_3.

10.5.2.2 Accuracy of the Fluid Approximation of TCP

We assess the accuracy of the fluid model of TCP connection as compared to ns-2 based simulations, where all details of TCP and packetized flows are considered. Simulated bottleneck link capacity variation is obtained by means of higher priority interfering traffic flows, not by an instantaneously varying function $C(t)$. This is a realistic scenario where TCP would see a bottleneck link capacity varying independently of its closed-loop control. By modulating the arrival rate of the high-priority traffic flow, we can obtain any desired shape of the residual link capacity for TCP connection packets.

Figure 10.12 depicts the ns-2 simulation model: it is composed of a greedy TCP sender (SND) and a TCP receiver (RCV), connected to SND through two tandem routers, R1 and R2.

The bottleneck is the link between R1 and R2 and it is provided with a capacity C_{max}, whereas the links between SND and R1, between R2 and RCV, as well as the reverse path from RCV to SND are provided with a capacity much greater than C_{max}. Capacity variations are simulated by a Constant Bit Rate (CBR) source with a time-varying sending rate $R_{CBR}(k\Delta)$ that is kept constant during a sampling period of duration Δ and depends on the discretized version of the target time-varying capacity function $C(t)$. During the k-th sampling interval, the CBR transmission rate is $R_{CBR}(k\Delta) = C_{max} - C(k\Delta)$ bps. If L_{CBR} is the CBR packet size in bits (in all simulations L_{CBR} is 400 bits), the CBR inter-arrival time is L_{CBR}/R_{CBR}. The sampling time Δ has been chosen much smaller than the time scale of the capacity function $C(t)$. The scheduler at the bottleneck link always serves CBR data packets with priority. As regards the TCP settings, we used the NewReno TCP variant, but the same results can be obtained with the SACK version of TCP. The TCP receiver

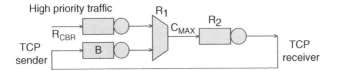

Figure 10.12 ns-2 simulation model for the variable capacity bottleneck link.

does not pose any limitations to the TCP sender, i.e., the TCP receiver window is always supposed to be larger than *cwnd*. TCP packet size is constant and equal to 1500 bytes. All simulations last 1000 *s*.

In the following, we compare the performance predicted by the fluid model with those obtained from ns-2 simulations for a periodic capacity profile at the bottleneck. We therefore use a sinusoidal function, i.e.,

$$C(t) = \frac{C_{max} + C_{min}}{2} + \frac{C_{max} - C_{min}}{2} \sin(2\pi Ft) \tag{10.69}$$

In the numerical example, $C_{min} = 10$ Mbps and $C_{max} = 100$ Mbps. The mean bottleneck link capacity is $(C_{max} + C_{min})/2 = 55$ Mbps.

Figure 10.13 depicts the mean normalized utilization of the bottleneck link capacity, obtained by means of the fluid model (left plot) and of ns-2 simulations (right plot), when T_f is 10 ms and T_b is 90 ms (hence it is $T = 100$ ms). Figure 10.14 reports the scenario where $T_f = 30$ ms and $T_b = 270$ ms (hence $T = 300$ ms).

All figures show the link utilization as a function of the frequency of the sinusoid F, for different values of the bottleneck buffer size B. The buffer size B is set to a fraction of the maximum BDP (e.g. $B = 0.1 \cdot C_{max} \cdot T$).

Focusing our attention on fluid model results obtained with $T = 100$ ms in Figure 10.13, we note that the link utilization achieved by the TCP connection strongly depends on the frequency of the bottleneck capacity variation. When the rate of variation is very low, the link utilization is high and TCP congestion window is able to follow the capacity variation. Obtained results reflect the ones achievable for constant capacity. Similar results are obtained when the rate of variation is very high. In this case, TCP behaves as if it saw the average capacity.

For intermediate values of F (between 0.1 Hz and 10 Hz), the TCP performance are highly degraded. The smaller is the buffer size the stronger is the degradation.

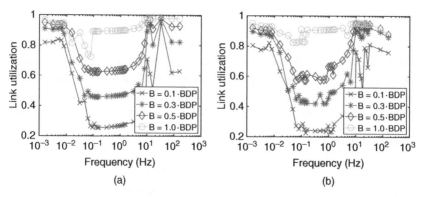

(a)   (b)

Figure 10.13 Comparison between fluid model (left plot) and ns-2 simulations (right plot): sinusoidal capacity, $T = 100$ ms.

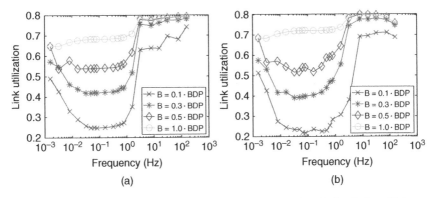

Figure 10.14 Comparison between fluid model (left plot) and ns-2 simulations (right plot): sinusoidal capacity, $T = 300$ ms.

For example, when $B = 0.1 \cdot C_{max}T = 84$ packets, the utilization falls below 30% of the available average capacity, whereas when the buffer is big, the effect is marginal since the buffer absorbs the capacity variation.

Comparing simulation results with fluid model results, we notice that results obtained for low and high values of F are the same as the ones obtained with the fluid model, whereas for intermediate values of F the degradation region is enclosed in a smaller range of frequencies.

Figure 10.14 depicts results obtained with T equal to 300 ms. Also in this scenario, the general behavior is similar, however, the degradation zone is shifted on the left side of the frequency scale since the ratio between the period of the sinusoid, $1/F$, and T is lower.

Overall, the fluid model turns out to be quite accurate.

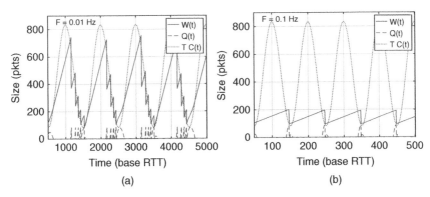

Figure 10.15 Evolution of $W(t)$, $Q(t)$, and $C(t) \cdot T$ for $T = 100$ ms and $B = 0.1 \cdot C_{max}T$. Sinusoidal capacity with $F = 0.01$ Hz (left plot) and $F = 0.1$ Hz (right plot).

Further insight into the fluid simulation can be gained looking at Figure 10.15, where the congestion window $W(t)$, the instantaneous BDP $C(t) \cdot T$ and the buffer content $Q(t)$ are plotted over time in the case of the sinusoidal capacity function in eq. (10.69), with $F = 0.01$ Hz (left plot) and $F = 0.1$ Hz (right plot). Other relevant parameters are $T = 100$ ms and $B = 0.1 \cdot C_{max}T = 84$ packets.

When the period of the capacity function is small with respect to the base RTT (Figure 10.15(a)), the TCP congestion control is able to follow the slow capacity variation and to grow the congestion window toward high values, leading to a high value of the link utilization (about 0.73 in this case).

When F is higher (Figure 10.15(b)), TCP congestion window synchronizes with capacity period and packet losses occur at small values of the congestion window. Observing the figure, it is possible to note that the congestion window oscillates approximately between $(C_{min}T + B)/2$ and $C_{min}T + B$, proving that TCP is not able to exploit the average capacity but just the minimum one. The result is an average throughput of about 0.29.

In both cases, the *cwnd* fails to follows the rising front of the bottleneck link capacity curve. The growth rate of the *cwnd* is too small, especially for the larger value of the RTT. This is the major source of inefficiency of TCP with this kind of time-varying link.

10.5.3 Application to Wireless Links

The fluid model can be applied to the analysis of TCP performance on radio links. This is a typical case where the link capacity varies for reasons that are not dependent on the TCP congestion control, e.g., interference, quality of the radio link. Another reason for having a variable capacity link, is occasional link blockage due to intervening obstacles or to handoff. Moreover, it is typically the case that the bottleneck of the TCP connection path is on the radio link, when such a link is involved. In this section we consider different capacity variation patterns: random stationary process and LTE case study.

10.5.3.1 Random Capacity

We assume that $C(t)$ is the realization of a stationary stochastic process. This is a simple model of the behavior of wireless channels.

We aim at highlighting the effect of the capacity variation time scale with a given fixed capacity marginal distribution. This way, the average capacity left over for TCP and its probability distribution are the same while the time scale over which a given variation is seen can be adjusted with respect to the base RTT T of the TCP connection. To keep things simple, we choose the stationary random process as a first-order autoregressive model with innovation uniformly distributed in the interval $[C_{min}, C_{max}]$.

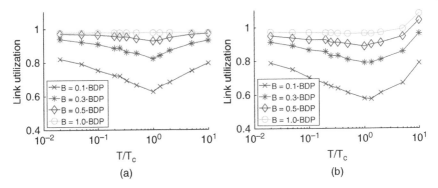

Figure 10.16 Comparison between fluid model (left plot) and ns-2 simulations (right plot): random capacity, $T = 100$ ms.

The discrete time model of the capacity is stated as follows:

$$C(k + 1) = \alpha C(k) + (1 - \alpha)I(k)$$

where $I(k)$ is a sequence of identically distributed, independent random variables with uniform distribution between C_{min} and C_{max}. The weighting coefficient α is chosen such that the autocorrelation coefficient of the sequence $C(k)$ at a time lag of T_c equals a conveniently low value. We require $\alpha^{T_c/\delta} = 0.1$. The meaning of T_c is a coherence time of the time-varying capacity function, i.e., a time span over which the capacity function de-correlates. In the numerical results, we use the ratio $x = T/T_c$ as an independent variable. Given the RTT T, for each value of x, the corresponding α is obtained as $\alpha = 0.1^{\frac{x\delta}{T}}$. The initial value of the capacity is set to the average capacity, i.e. $C(0) = (C_{max} + C_{min})/2$. In the numerical examples we set $C_{max} = 100$ Mbps and $C_{min} = 10$ Mbps.

In Figures 10.16 the bottleneck link utilization obtained with the fluid model (left) and with ns-2 simulations (right) is plotted against the ratio $x = T/T_c$ for $T = 100$ ms and for four values of the buffer size (normalized as before with respect to $C_{max}T$).

Two main facts are apparent as compared to the results for deterministic periodic capacity function. First, the qualitative effect of the capacity function time scale is analogous to the case of deterministic sinusoidal capacity function: values of T_c much larger and much smaller than T yield better performance as compared to intermediate values ($\frac{T}{T_c}$ in the range $0.1 \div 10$ in both figures.) Second, the quantitative effect is quite reduced with respect to the deterministic case, since the range of values of each utilization curve for a given buffer size value never entails a throughput penalty larger than 33% with respect to the largest achieved throughput. Also, the curve behavior is much smoother than in the case of sinusoidal capacity, due to the randomization of the variable capacity.

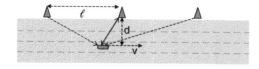

Figure 10.17 Road layout for the TCP connection over LTE case study.

10.5.3.2 TCP over Cellular Link

We consider a wireless link with adaptive modulation and coding, and a link adaptation mechanism that adjust the transmission configuration to follow the time-varying radio channel.

We consider a user on board a vehicle, moving along a road where radio coverage is offered through evenly spaced base stations (BSs). We let $\ell = 2$ km be the distance between two consecutive BSs. The user moves along an ideal line. BSs are aligned at a distance $d = 20$ m off the trajectory of the vehicle. Figure 10.17 illustrates the layout of the road where the vehicle moves. The triangles stand for BSs along the road side.

The LTE radio access standard is assumed. Time is divided in slots of duration $T_s = 0.5$ ms. Time slots are organized in frames. The downlink LTE frame is made up of 20 slots, so its duration is $T_f = 20 \cdot T_s = 10$ ms. The whole channel is split into frequency sub-channels of bandwidth $W_{RB} = 180$ kHz. One sub-band for one time slot is a resource block (RB) and is the unit of assignable radio resource.

The user terminal is assigned a number n_{RB} of RBs in the downlink channel out of the overall set of $n_{RB,max}$ RBs available in the downlink for one slot time. Capacity is assigned to the user terminal in one time slot per frame.

We now introduce the user mobility and radio link models. The time evolution of the considered models is discretized, with time steps equal to the frame time T_f. Let $t_k = kT_f + t_0$, $k \in \mathbb{Z}^+$, for some initial time t_0.

The user mobility is described by a linear accelerated motion, where the acceleration is sampled once per LTE frame, according to $a(t_k) = g\, a(t_{k-1}) + (1-g)u_k$, with $u_k \sim \mathcal{N}(0, \sigma^2)$ and $g = 0.99$. The standard deviation σ is set to 3 m/s^2. The speed of the user is given by $v(t_k) = \min\{v_{max}, \max\{0, v(t_{k-1}) + T_f a(t_k)\}\}$, where $v_{max} = 40$ m/s. In turn, the position is $x(t_k) = x(t_{k-1}) + T_f v(t_k)$. This is initialized with $x(t_0) = x_0$.

The radio link is modeled as an AWGN channel with time-varying path gain. The capacity sustained by a single RB is given by:

$$C_{RB} = \min\left\{ C_{RB,max},\, W_{RB}\log_2\left(1 + \frac{SNIR}{\Gamma}\right)\right\}, \qquad (10.70)$$

where we let $C_{RB,max} = 720$ kbit/s. The *SNIR* is calculated as

$$SNIR = \frac{G_i P_{RB}}{P_N + \sum_{j\neq i} G_j P_{RB}}, \qquad (10.71)$$

where $P_{RB} = P_{tx}/n_{RB,max}$ is the transmission power on one RB, $P_{tx} = 10\ W$ is the overall power budget of each BS, P_N is the background noise power level, and Γ is the gap factor. This last quantity is calculated as $\Gamma = (2/3)\log(2/P_e)$, where P_e is the target BER of the physical link. We let $P_e = 10^{-6}$. In the *SNIR* expression, we denote with i the serving station, while the signal received by all other BSs is accounted for as interference.

The path gain between the user terminal and BS k is a function of their distance $d(t_k)$. We let $G(t_k) = G_{det}(d(t_k))S(t_k)$. The distance dependent deterministic path gain $G_{det}(\cdot)$ is defined according to the two-ray interference model presented in [192]. The shadowing component $S(\cdot)$ is assumed to be log-normal and spatially correlated. The sequence of shadowing gains is generated as an auto-regressive process of order 1, i.e., $S(t_k) = \gamma_k\ S(t_{k-1}) + (1 - \gamma_k)U_k$, where $U \sim \mathcal{N}(0, \sigma_S^2)$, with $\sigma_S = 10\ dB$. The coefficient γ_k is given by $\gamma_k = 0.1^{\Delta x_k/D_S}$. The quantity Δx_k is the distance traveled by the user terminal between two sampling times, i.e., $\Delta x_k = x(t_k) - x(t_{k-1})$. The parameter D_S is the spatial coherence distance of the shadowing gain, set to $D_S = 50$ m.

The wireless bottleneck link capacity of a user connected to BS i is given by

$$C(t) = n_{RB}\min\left\{ C_{RB,max}, W_{RB}\log_2\left(1 + \frac{1}{\Gamma}\frac{G_i(t)P_{tx}}{n_{RB,max}P_N + \sum_{j\neq i}G_j(t)P_{tx}}\right)\right\}$$

(10.72)

where we have emphasized the dependence of the path gains on time, due to user mobility. The capacity varies frame by frame, so in the fluid model we hold it constant over time intervals of duration T_f.

In the ensuing numerical example we set $n_{RB} = 40$ (hence the maximum bottleneck link bit rate is 230.4 Mbit/s) and we set the buffer size B to a fraction ξ of the maximum BDP of the connection, i.e., $B = \xi\ T\ n_{RB}\ C_{RB,max}/(8 \cdot L_0)$, where $L_0 = 1500$ bytes is the packet size and T is the base RTT. We consider a connection lifetime of 300 s. Therefore, the user travels an average distance between 900 m and 9 km as the average speed varies between 3 m/s and 30 m/s. Since BSs are spaced out by 2000 m, the user experiences a very different channel as the speed changes. The numerical data are obtained by averaging over 10 different fluid simulations obtained by generating random path gains, random user acceleration values and setting the initial position of the vehicle at random with respect to the BS pattern.

Figure 10.18 depicts the link utilization, i.e., the mean throughput of the TCP connection divided by the mean of $C(t)$ during the connection lifetime, as a function of the mobile user speed for three values of the bottleneck link buffer size (100%, 50%, and 25% of the maximum BDP). The plot on the left refers to $T = 100$ ms, the one on the right to the case $T = 300$ ms. Correspondingly, the maximum BDP is 240 packets in the first case, 720 packets in the second case.

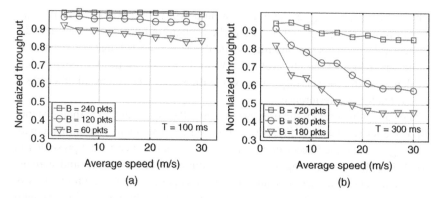

Figure 10.18 Normalized throughput of a mobile user in an LTE network as a function of the average user speed. The base RTT is 100 ms (left plot) and 300 ms (right plot). The corresponding maximum BDPs are 240 and 720 packets. The bottleneck link buffer size is set to a fraction of the maximum BDP (100%, 50% and 25%).

The results shown in Figure 10.18 point out that the throughput performance degrades as the user moves faster, i.e., as the time scale over which the link capacity has a sensitive variation becomes smaller. The degradation is much worse for the larger RTT value. This is yet another point in favor of placing contents that mobile users wish to download close to the edge of the mobile network, so as to reduce as much as possible the RTT.

The buffer has a beneficial effect: the bigger it is, the more it absorbs the bottleneck link capacity variation. However, bigger buffers entail higher delays. This is highlighted by Figure 10.19, that plots the throughput-delay trade-off for the same

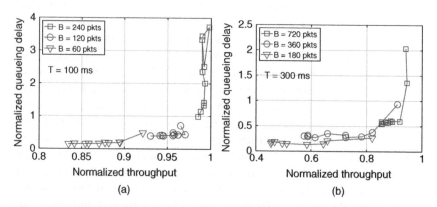

Figure 10.19 Throughput-delay trade-off of a mobile user in an LTE network as the average user speed varies: The base RTT is 100 ms (left plot) and 300 ms (right plot). The delay is the queueing delay in the bottleneck buffer, normalized with respect to the base RTT.

cases and numerical values as in Figure 10.18. The throughput is measured as in Figure 10.18. The delay is the queueing delay at the bottleneck buffer, normalized with respect to the base RTT. The trade-off is obtained by varying the average user speed (lower values correspond to points exhibiting larger throughput values).

It appears that large buffer sizes can achieve a high throughput at the cost of large delays (up to almost four times the base RTT if $T = 100$ ms and more than twice the base RTT for $T = 300$ ms). The best trade-off is offered by the inter-mediate buffer size (50% of the maximum BDP), which brings the connection to work around the "knee" of the trade-off curve, i.e., high throughput and low mean queueing delay.

10.6 Fluid Model of Multiple TCP Connections Congestion Control

We refer to the dumbbell topology in Figure 10.9 and to the notation introduced in Section 10.5. With respect to the assumptions made in Section 10.5, we add the following ones.

1. N TCP connections share the fixed capacity bottleneck link.
2. The *cwnd* is incremented by $\alpha/cwnd$ per received ACK during congestion avoidance.
3. All N TCP connections have the same base RTT T and start at the same time.

This last assumption can be relaxed, considering asynchronous TCP connections with different base RTT values, at the cost of added complexity to the fluid model.

10.6.1 Negligible Buffering at the Bottleneck

The strongest assumption we make in this section is that the base RTT is dominant with respect to queueing delay at the bottleneck buffer, so that we can neglect the latter. We can therefore approximate the RTT as a constant T.

The instantaneous variation of the *cwnd* of any of the N TCP sources is the sum of two contributions: (i) a positive term accounting for increments of the congestion window size on ACK reception; (ii) a negative term, accounting for congestion window reduction upon detection of packet loss. Formally, the increment of the congestion window $W(t)$ at time t per received ACK is $\alpha/W(t)$, the decrement upon loss detection is $\beta W(t)$.

Let $p(t)$ denote the probability that the bottleneck buffer is full at time t. In the fluid simulation, it is replaced by the indicator function of the event $Q(t) = B$. The net rate of ACKs arriving at time t at a TCP source is then $[1 - p(t - T)]x(t - T)$, where $x(t)$ is the TCP source sending rate at time t. We can state the following differential equation for the evolution of the congestion window of each TCP connection:

$$\frac{dW}{dt} = \frac{\alpha[1 - p(t - T)]\, x(t - T)}{W(t)} - \beta W(t)p(t - T)\, x(t - T) \tag{10.73}$$

The amount of data sent by a source in an RTT equals the congestion window size. Then, we can write $x(t) = W(t)/T$. Substituting into (10.73), we get:

$$\frac{dx}{dt} = \frac{\alpha[1 - p(t - T)]x(t - T)}{T^2 x(t)} - \beta x(t)p(t - T)x(t - T) \tag{10.74}$$

If a steady-state is achieved by this dynamical system as $t \to \infty$, let $\bar{x}$ denote the steady-state mean sending rate of the TCP connection and $\bar{p}$ the steady-state packet dropping probability at the bottleneck buffer (compare with the analysis in Section 10.4). The following relationship holds:

$$0 = \frac{\alpha\, [1 - \bar{p}]\, \bar{x}}{T^2\, \bar{x}} - \beta\, \bar{p}\, \bar{x}^2 \quad \Rightarrow \quad \bar{x} = \frac{1}{T}\sqrt{\frac{\alpha(1 - \bar{p})}{\beta\bar{p}}} \tag{10.75}$$

The achieved throughput is $\Lambda = \bar{x}(1 - \bar{p})$. The prediction of this simple model is that the average throughput of a long-lived TCP connection is inversely proportional to the RTT and to the square root of the loss probability at the bottleneck buffer. This result has been actually confirmed by more sophisticated models, that account for the discrete nature of packets, and by accurate, packet-level simulations. Recall that the model has been derived by assuming that the bottleneck buffer is small as compared to the BDP, so that the queueing delay can be neglected with respect to the base RTT.

We can use this simple model by expressing the packet-dropping probability at the bottleneck buffer. With N TCP traffic sources, the load on the bottleneck link is $y(t) = \sum_{j=1}^{N} x_j(t)$. If we can approximate the aggregated packet flow arriving at the bottleneck as a Poisson arrival process (which we know is a good approximation with a high degree of multiplexing, i.e., if $N \gg 1$ and no connection has a sending rate much bigger than all others), then the packet dropping probability can be expressed by using the approximation derived in Section 4.3.2 for the $M/G/1/B$ queue, where B is the bottleneck buffer size.

The buffer size is a fraction of the BDP, i.e., $B = \gamma C \cdot T/L_0$, where L_0 denotes the packet length.

To keep things simple, let us assume a negative exponential probability distribution for the packet length. So, the packet dropping probability at steady-state is:

$$\bar{p} = \frac{1 - \bar{y}/C}{1 - (\bar{y}/C)^{B+1}}\left(\frac{\bar{y}}{C}\right)^B \tag{10.76}$$

with $\bar{y} = N\bar{x}$.

Let us define the normalized load $\bar{u} = \bar{y}/C$. Putting all pieces together, we find:

$$\bar{u} = \frac{N\bar{x}}{C} = \frac{N}{a}\sqrt{\frac{\alpha}{\beta}\left(\frac{1}{\bar{p}}-1\right)} = b\sqrt{\frac{1-\bar{u}^B}{\bar{u}^B(1-\bar{u})}} \tag{10.77}$$

where $a = CT$ is the BDP, and $b = \frac{N}{a}\sqrt{\frac{\alpha}{\beta}}$. Finally, we derive the following fixed point equation

$$\bar{u} = \left(b\sqrt{\frac{1-\bar{u}^B}{1-\bar{u}}}\right)^{\frac{2}{B+2}} \quad\Rightarrow\quad \bar{u}^{B+2} = b^2\sum_{j=0}^{B-1}\bar{u}^j \tag{10.78}$$

Once the fixed point $\bar{u}$ is found, the steady-state average throughput of one TCP connection can be calculated as $\Lambda = \bar{x}(1-\bar{p}) = \frac{\bar{u}}{N}\frac{1-\bar{u}^B}{1-\bar{u}^{B+1}}$.

The simple model outlined here holds approximately only for $N \gg 1$ and $\gamma \ll 1$, i.e., a small buffer size at the bottleneck.

10.6.2 Classic TCP with Drop Tail Buffer at the Bottleneck

To explore the dynamics of the congestion window and buffer size, let us assume that the buffer size B cannot be neglected with respect to the BDP.

We assume a drop-tail buffer policy, i.e., packets are dropped at the bottleneck only if the buffer is full. If all sources experience the same base RTT, each one gets a same fraction of the bottleneck capacity. The ACK rate at any TCP sender equals C/N, whenever the buffer is not empty, $W(t)/T$ when it is empty. The buffer is empty if and only if $NW(t) \leq CT$. Therefore, we can write the ACK rate as $A(t) = \min\left\{\frac{W(t)}{T}, \frac{C}{N}\right\}$. When the buffer is not empty, the pipe balance implies that $NW(t) = Q(t) + CT$, where $Q(t)$ is the queue length at the bottleneck at time t. Then, it is easy to verify that the ACK rate can be expressed as $A(t) = W(t)/R(t)$, where $R(t) = T + Q(t)/C$ is the RTT at time t (including the queueing delay at the bottleneck).

As for the packet loss rate, applying the fluid approximation, the mean rate of lost fluid can be expressed as $L(t) = \max\{0, W(t)/R(t) - C/N\}I(Q(t) = B)$, where $I(E)$ is the indicator function of the event E.

Since the congestion window grows by $\alpha/W(t)$ on each received ACK and is reduced by $\beta W(t)$ on each detected packet loss, the rate of change of the congestion window can be written as:

$$\frac{dW}{dt} = \frac{\alpha}{W(t)}A(t) - \beta W(t)L(t - T^*) =$$

$$= \frac{\alpha}{R(t)} - \beta W(t)\max\left\{0, \frac{W(t-T^*)}{R(t-T^*)} - \frac{C}{N}\right\}I(Q(t-T^*) = B) \tag{10.79}$$

with $R(t) = T + Q(t)/C$. T^* is the *reaction time*, i.e., the time between the packet loss event at the buffer and the time when packet loss is detected at the TCP sender. It is therefore $T^* = T + B/C$, since packet loss occurs only when the buffer is full.

The evolution equation of the buffer size is:

$$
\frac{dQ}{dt} = \begin{cases} \max\left\{0, \dfrac{NW(t)}{R(t)} - C\right\} & Q(t) = 0, \\[2mm] \dfrac{NW(t)}{R(t)} - C & 0 < Q(t) < B, \\[2mm] \min\left\{0, \dfrac{NW(t)}{R(t)} - C\right\} & Q(t) = B. \end{cases} \tag{10.80}
$$

The initial conditions for numerical integration can be set as $Q(t) = 0$ for $t \le 0$, $W(0) = \min\{ssthresh, a/N\}$, where a is the BDP and $ssthresh$ is the Slow Start threshold. The initial $ssthresh$ value is typically set at 64 MSSs.

Under the assumptions of this model, timeout is never triggered, so the slow start is only executed once, at the beginning of the TCP connection. This is consistent with the assumption that the bottleneck capacity is constant. Note that the model studied in section 10.5.1 refers to the analysis of the steady-state average throughput, while the analysis presented here aims at characterizing the dynamical behavior of N TCP connections sharing a bottleneck link.

10.6.3 Classic TCP with AQM at the Bottleneck

Let us assume that the bottleneck buffer is managed according to an active queue management (AQM) policy, specifically the random early detection (RED) algorithm (see Section 10.9.3 for a brief note on AQM). According to RED, a packet arriving at the bottleneck buffer at time t is dropped with a probability depending on the queue occupancy at time t. In RED, the packet dropping probability is calculated based on a smoothed version of the buffer occupancy level. If t_k denotes the time of the k-th packet arrival at the bottleneck buffer, the smoothed buffer size is updated as follows: $S(t_k) = (1 - g)S(t_{k-1}) + gQ(t_k)$, where $Q(t)$ denotes the buffer occupancy level at time t. The probability that the packet arriving at time t_k is dropped is calculated as $p(t_k) = p_{max}[S(t_k) - B_{th}]/(B - B_{th})$ when $B_{th} < S(t_k) < B$. All incoming packets are dropped as long as $Q(t) = B$. The quantities p_{max} and B_{th} are parameters of the AQM algorithm. They define, respectively, the maximum rate of packet dropping as the buffer is not full and the buffer threshold below which no packet is dropped.

When the bottleneck buffer is congested, we can assume that sampling is done at a rate equal to the bottleneck packet rate, i.e., $t_k - t_{k-1} = 1/\lambda(t_{k-1})$, with $\lambda(t) = \frac{NW(t)}{R(t)}$. Then, the smoothed buffer updating equation can be converted into a continuous time fluid approximation by replacing $S(t_k) - S(t_{k-1})$ with $\frac{1}{\lambda(t)}\frac{dS}{dt}$.

The differential equation that drives the evolution of the congestion windows becomes:

$$\frac{dW}{dt} = \frac{\alpha}{R(t)} - \beta W(t)\frac{W(t - \tau(t))}{R(t - \tau(t))}p(t - \tau(t)) \tag{10.81}$$

$$\frac{dS}{dt} = g\,\frac{NW(t)}{R(t)}\,[Q(t) - S(t)] \tag{10.82}$$

with $p(t)$ given by:

$$p(t) = \begin{cases} 0 & 0 \le S(t) \le B_{th}, \\ p_{max}\dfrac{S(t) - B_{th}}{B - B_{th}} & B_{th} < S(t) < B, \\ 1 & S(t) = B. \end{cases} \tag{10.83}$$

The evolution differential equation of the buffer occupancy level $Q(t)$ is the same as in eq. (10.80).

The reaction delay $\tau(t)$ can be found as follows. The delay from when the packet is dropped until when the effect of the congestion window reduction is visible at the bottleneck buffer is $\tau(t) = \frac{Q(t - \tau(t))}{C} + T$. By deriving both sides and rearranging, we obtain:

$$\frac{d\tau}{dt} = \frac{Q'(t - \tau(t))}{Q'(t - \tau(t)) + C} \tag{10.84}$$

where $Q'(u)$ denotes the derivative of $Q(t)$ calculated at time u. This equation can be solved with the initial condition $\tau(0) = T$.

A simpler approximation can be obtained by replacing $\tau(t)$ with the constant value $T^* = T + \overline{B}/C$, where $\overline{B}$ is the average buffer level sampled by packets that are dropped. It can be verified that $\overline{B} = (2B + B_{th})/3$.

10.6.4 Data Center TCP with FIFO Buffer at the Bottleneck

In this section we abandon the classic TCP to consider the DCTCP, described in Section 10.3.4. The statement of the fluid model is derived from [10]. It is:

$$\frac{dW}{dt} = \frac{1}{R(t)} - \frac{W(t)\alpha(t)}{2R(t)}p(t - T^*) \tag{10.85}$$

$$\frac{d\alpha}{dt} = \frac{g}{R(t)}[p(t - T^*) - \alpha(t)] \tag{10.86}$$

where $p(t) = I(Q(t) > K)$, $R(t) = T + Q(t)/C$, $T^* = T + K/C$ and $g \in (0, 1)$ is a smoothing coefficient. The evolution differential equation of the buffer occupancy level $Q(t)$ is the same as in eq. (10.80).

Starting from the initial state, the DCTCP dynamical system tends to settle on a limit cycle for $t \to \infty$. If we average out the state of the model over one orbit through the limit cycle, we obtain the following relationship among the steady-state averages of the system quantities, denoted with a bar: $\bar{p} = \bar{\alpha}$ and $1 = \overline{W}\,\bar{\alpha}\,\bar{p}/2$. Hence, it is $\bar{\alpha} = \sqrt{2/\overline{W}}$.

A refined approximation can be found by the following argument. The buffer level $Q(t)$ follows a periodic pattern in time: it grows steadily, until it hits the level K. In the subsequent RTT all packets going through the bottleneck buffer are marked. Getting back to the source they cause the congestion window to be reduced. Then, the buffer content falls back to its minimum level and a new growth cycle starts. Let $W^* = (CT + K)/N$ denote the critical value of the congestion window where the buffer level hits the threshold K for the first time. Since one more RTT is required for the marking feedback to be effective, the congestion window grows up to $W^* + 1$ before being reduced by a factor $1 - \alpha/2$. Let $A(u, v)$ be the number of packets sent when the congestion window grows from u up to v. It is $A(u, v) = \sum_{k=u}^{v} k \approx \int_u^v z\,dz = (v^2 - u^2)/2$. The marking rate can be expressed as the ratio of the packets that get marked in one period and the overall number of packets sent in the period, i.e.,

$$\alpha = \frac{A(W^*, W^* + 1)}{A((1 - \alpha/2)(W^* + 1), W^* + 1)} = \frac{2W^* + 1}{\alpha(1 - \alpha/4)(W^* + 1)^2} \tag{10.87}$$

where we have used the result $A(u, v) = (v^2 - u^2)/2$. It turns out that α is the solution of the cubic equation

$$\alpha^2 \left(1 - \frac{\alpha}{4}\right) = \frac{2W^* + 1}{(W^* + 1)^2} \tag{10.88}$$

in the interval $(0, 1)$. For large BDP networks, it is $W^* \gg 1$, hence it is $\alpha \ll 1$. Therefore, eq. (10.88) simplifies to $\alpha^2 \approx 2/W^*$ and we recover the result found by averaging over one limit cycle.

The period of the buffer level evolution (which is also the period of the steady-state limit cycle) is

$$T_c = T\left[W^* + 1 - (W^* + 1)\left(1 - \frac{\alpha}{2}\right)\right] \approx \frac{TW^*\alpha}{2} = \frac{T}{2}\sqrt{2W^*} = T\sqrt{\frac{CT + K}{2N}} \tag{10.89}$$

The maximum and minimum buffer levels are:

$$Q_{max} = N(W^* + 1) - CT = N + K$$

$$Q_{min} = N(W^* + 1)(1 - \alpha/2) - CT \approx N + K - N\sqrt{\frac{W^*}{2}}$$

$$= N + K - \sqrt{\frac{N(K + CT)}{2}}$$

To avoid packet loss, it must be $N + K \leq B$. To avoid buffer depletion, hence under-utilization of the bottleneck capacity, it suffices that $Q_{\min} > 0$. The expression of $Q_{\min}$ as a function of N has a minimum for $N = (K + CT)/8$. Imposing that $Q_{\min}$ be positive even for this value of N, we find that it must be $K > CT/7$. This is a much more favorable condition than the one holding for classic TCP. The classic TCP requires that it be $B = CT$ to achieve full utilization of the bottleneck link, whereas with DCTCP a buffer size in the order of $CT/7$ is enough. This is especially interesting for large BDP pipes.

The gain g must be chosen small enough to ensure the exponential moving average covers at least one congestion event. In our model, there is exactly one congestion event every period T_c. Hence we can require that $(1 - g)^{T_c/T} \geq \eta$, with $T_c/T = \sqrt{(K + CT)/(2N)}$. The worst case (longest) period occurs for $N = 1$. Then we can set $g \leq 1 - \eta^{\sqrt{2/(K+CT)}}$ for a suitably large η, e.g., $\eta = 1/2$.

To give numerical examples, we assume a packet length of 1250 bytes and a bottleneck link capacity of 100 Mbit/s. Hence it is $C = 10^4$ pks/s. We set the base RTT at $T = 100$ ms. The resulting BDP is 1000 pkts. The buffer size is set at 50% of the BDP, i.e., $B = 500$ pkts. The number of flows is chosen as $N = 10$. Specific parameters for RED are chosen as $p_{\max} = 0.25$, $B_{th} = 0.1 \cdot B$. As for DCTCP, we let $K = 0.1 \cdot B$ and $g = 0.0227$.

Figure 10.20(a) compares the congestion window sizes as a function of time for classic TCP with a drop-tail buffer at the bottleneck (top plot), classic TCP with RED at the bottleneck (middle plot) and DCTCP (bottom plot).

The large excursion of the *cwnd* of classic TCP is clearly visible as well as the slow climbing up of *cwnd* to fill the pipe and the buffer. When RED is used, random

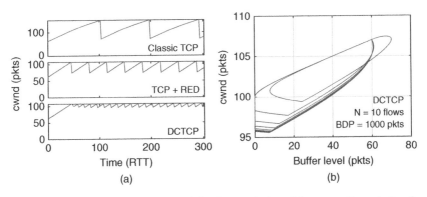

Figure 10.20 Left plot: comparison of the time evolution of the congestion window for classic TCP with drop-tail buffer (top), classic TCP with RED at the bottleneck (middle), DCTCP (bottom). Right plot: limit cycle of the DCTCP orbit in the buffer level-*cwnd* plane. Time is measured in RTTs, *cwnd* and queue length in packets.

early discard makes the sawtooth of the *cwnd* span a smaller range and react faster. A beneficial effect of RED that cannot be appreciated directly through the fluid model is the de-synchronization of packet losses of different TCP flows. On the contrary, drop-tail buffers tend to correlate packet losses as seen by different TCP flows, hence TCP sources tend to synchronize *cwnd* updates (as long as the base RTT of all TCP flows is the same).

Figure 10.20(b) depicts the limit cycle of DCTCP orbit in the buffer level-*cwnd* plane. DCTCP offers the best behavior in terms of oscillations of the *cwnd* size. This behavior reflects in a higher and more stable throughput, with smaller delays at the bottleneck. These good results are achieved at the price of introducing ECN, i.e., making modifications both at the TCP recevier and at the intermediate nodes. On the contrary, RED does not require anything of the TCP receiver, but is network-assisted, i.e., RED algorithm must be deployed in the intermediate nodes.

10.7 Fairness and Congestion Control

Let us consider a generic resource and let C denote the amount of available resource. It can be memory space, processing capability, link bandwidth, transportation capacity, energy, money. Consider N sources requesting the resource and let R_j be the amount of demand for the resource submitted by source j. The demand can be time-dependent.

As long as $R_1 + \cdots + R_N \leq C$ no congestion arises, all demands can be accommodated without conflicts. There is no *fairness* issue.

Problems start whenever $R_1 + \cdots + R_N > C$, so that part of the demand will not be satisfied, at least for some time, if not definitely (that depends on the overflow handling, either delay or loss based). Here we have a fairness problem, i.e., we must give some *fair* criterion according to which we determine what part of the demand will not be met.

Clearly, fairness is tightly coupled with congestion control. With proactive congestion control, a typical approach is to deal with demands as they arrive and reject the excess demand. This basic paradigm can be altered as a function of priorities, possibly time-dependent. Yet, the typical setting envisages that it is fair to preserve already ongoing services, rejecting new demand as long as the available resource has no room to accommodate it.

Reactive congestion control requires a more sophisticated approach. Assume sources are greedy, i.e., they could possibly use up to the whole available resource, if granted. When a new demand arrives at the system, it is necessary to rearrange the resource sharing to make room for the new demand. This is especially fit if the sources cope with an elastic quality of service, i.e., they can live with a variable quantity of resource. For example, when a new TCP connection is set up, its bottelneck link capacity could be entirely used by already ongoing TCP connections. Some mechanism is required to induce the ongoing connections to yield to the new one and achieve a new sharing equilibrium that is fair in some sense.

There is no general consensus on what might be deemed as "fair," even in the restricted realm of network traffic engineering. If there is no other constraint and all sources can use up to the entire resource, it seems reasonable to state that it is fair to assign an equal share of the available resource to each source, i.e., C/N for N sources. This is what Processor Sharing achieves. More generally, a weighted sharing can be defined, leading to Generalized processor sharing (GPS) and to its practical implementation (e.g., weighted fair queueing (WFQ); see Section 6.4.3). However, things could become much more complicated, specifically when we move to a networked resource.

Let us introduce quantitative approaches to fair sharing of a resource with reference to communication link capacity shared among traffic sources.

Assume N traffic sources share a link of capacity C. Let λ_j be the mean offered flow rate of source j. As long as $\sum_{j=1}^{N} \lambda_j \leq C$, no issue arises. Each sources gets exactly the average rate it requests. There is no congestion as well as no fairness problem in sharing the capacity, even if the realized throughputs λ_j might be quite different one from another.

Problems start whenever $\sum_{j=1}^{N} \lambda_j > C$. The traffic demand offered by the traffic sources cannot be accommodated in that case. Congestion control will enforce some lower level of throughput for the involved traffic sources. It is clear how much the excess offered traffic is. What is not evident at all is *which* sources should be penalized, and assigned a throughput rate lower than their demand. This is the key question to which a *fairness* principle should answer[19].

A simple answer seems to be: assign a same share of the resource to all requesting sources. This appears immediately to be too simplistic. What if the equal share is more than a source request? Again, the intuition suggests: limit the assignment to at most what is requested and share the exceeding portion among the sources that request more. We could make the picture a bit more complicated by assuming that each source has an individual constraint on the achievable amount of resource.

To be concrete, let us refer to the example of link capacity sharing. Assume it must be $\lambda_j \leq R_j, j = 1, \dots, N$, with $\sum_{j=1}^{N} R_j > C$ (otherwise there is no congestion). We define a fairness principle informally as follows: let each source get a same share, unless it is limited by its individual constraint.

Assume sources are labeled so that the constraint rates are ordered in increasing progression: $R_1 \leq R_2 \leq \cdots \leq R_N$. Let j^* be defined as the smallest integer $j \in$

---

19 It has been even questioned that fairness should apply to flow rates at all [45].

$\{1, \ldots, N\}$ such that the inequality $\sum_{i=1}^{j-1} R_i + (N + 1 - j)R_j > C$ holds (note that it holds by hypothesis for $j = N$; moreover, we let $\sum_{i=1}^{0} \equiv 0$ by definition). The "fair" rate assignment is as follows:

$$R_j^* = \begin{cases} \dfrac{C}{N} & j^* = 1 \\[2ex] \min\left\{ R_j, R_{j^*-1} + \dfrac{C - \sum_{i=1}^{j^*-1} R_i}{N - j^* + 1} \right\} & \text{otherwise.} \end{cases} \tag{10.90}$$

for $j = 1, \ldots, N$.

Note that, if we try to shift any amount of rate, the only feasible outcome is that we subtract some rate from a source and assign it to sources that already are assigned a bigger rate than the penalized source. Said informally, we steal from the poor to favor the rich.

The resource assignment described above is the *max-min* fairness principle. We can state this fairness principle formally as follows. Let x_j, $j \in S$ be the max-min fair rate assignment for a set of traffic sources S, and let y_r be any other feasible rate assignment. Feasible means that the rate assignment fulfills all constraints. Then the following property holds: $\min_{j \in S} x_j \geq \min_{j \in S} y_j$. This means that the "poorest" source is still better off with the fair assignment than with any other feasible rate assignment. We are not claiming that all sources get the same flow rate. In fact, it may well be the case that the obtained flow rates are quite different from one another, depending on the constraints. This is typically the case when we apply the max-min rate assignment in a network and the constraints on the flow rates apply on each link of the routing path assigned to each flow.

We discuss one more example of "fair" rate assignment, namely *proportional fairness* (PF). This shows that a completely different rate assignment principle can still be considered sensibly a "fair" one.

Let x_k, $k \in S$ denote a rate allocation. We define x_k to be proportional fair if, for any other feasible rate allocation y_k it is $\sum_k (y_k - x_k)/x_k \leq 0$. In other words, proportional fairness implies that any modification to the rate allocation causes a loss in relative terms: sources that improve their rates are outbalanced by sources that lose a fraction of their rates. Note that proportional fairness guarantees that no source can ever get a zero average rate.

We can show that the property defining proportional fairness is equivalent to maximizing the sum of logarithms of the rates, under the same constraints that define the feasible rate allocations.

Theorem 10.1 A rate allocation x_k is proportional fair if and only if it maximizes $\sum_k \log x_k$ in the set of feasible rate allocations.

Proof: The result is a direct application of the property stated in point 10.5 of Theorem 10.5 in the Appendix of this chapter. That property states that, for a concave differentiable function $f(x)$ defined for $x \in \mathcal{A}$, where $\mathcal{A}$ is a compact set, a point $x^* \in \mathcal{A}$ is a maximum if and only if $\nabla f(x^*) \cdot (x - x^*) \leq 0$ for any $x \in \mathcal{A}$.

We consider the function $f(x) \equiv \sum_k \log x_k$ for x belonging to the feasible set. This is a strictly concave function in the feasible set defined by the constraints on the rate allocation x_k. Since the k-th component of the gradient of $f(x)$ is $\nabla f(x)_k = 1/x_k$, a rate allocation x_k maximizes the function $\sum_k \log x_k$ if and only if $\nabla f(x) \cdot (y - x) = \sum_k \frac{1}{x_k}(y_k - x_k) \leq 0$ for any feasible rate allocation y_k. This proves the result. ∎

Example 10.2 Let us consider a cascade of n links and $n + 1$ flows, one of which is routed through all n links, say it is x_0, while flow i is routed only on link i for $i = 1, \dots, n$. The proportional fairness assignment is found as the one maximizing the sum $\log x_0 + \log x_1 + \cdots + \log x_n$ under the constraint $x_0 + x_i \leq c_i$ for $i = 1, \dots, n$.

Let us consider the balanced case, where $c_i = c$ for all i. It is clear that the optimum rate assignment will satisfy all constraints with the equality sign (if we left some unused capacity on a link, it would be possible to improve the sum of $\log x$ by increasing the rate assigned to the flow that uses only that link). Therefore, we have $x_i = c - x_0$. Substituting into the target function, the optimization reduces to the unidimensional problem

$$\max_{0 < x_0 < c} \log x_0 + n \log(c - x_0) \tag{10.91}$$

It is easy to find that the optimum is $x_0^* = c/(n+1)$ and $x_i^* = nc/(n+1)$. The proportional fairness assignment has an intuitive interpretation in this case: since flow 0 engages n links in its routes, it "consumes" n times more resource than each of the other flows. Therefore, it is sensible to ask that the ratio between x_0 and any ot the x_i must be 1 to n.

A framework for fairness, encompassing the max-min and proportional fairness concepts is sometimes deinfed as the so called α-fairness. Rather than lingering on this specific definition, we introduce a much more general framework in the next section, which can be considered as the current state-of-the-art of fair resource sharing in a network where end-to-end flows share the capacity of the network links. It is a model applicable to different domains, e.g., telecommunications, transportation systems, energy grid network, water distribution.

10.8 Network Utility Maximization (NUM)

Let us consider a network modeled as an undirected graph. Links correspond to arcs in the graph, traffic sources, and sinks correspond to graph nodes. Let $\mathcal{L}$

denote the set of links of the network graph. We associate a positive quantity c_i to link $i \in \mathcal{L}$, c_i being the capacity of link i. End-to-end flows are established between nodes. Let $\mathcal{F}$ denote the set of end-to-end flows. Let x_j denote the average bit rate of flow $j \in \mathcal{F}$.

Defining routing of flows inside the network is equivalent to assigning a *routing matrix* $\mathbf{A}$, whose entry $(i, j) \in \mathcal{L} \times \mathcal{F}$ is defined as follows:

$$a_{ij} = \begin{cases} 1 & \text{flow } j \text{ is routed through link } i, \\ 0 & \text{otherwise.} \end{cases} \tag{10.92}$$

This definition is consistent with single-path routing, e.g., obtained by choosing the least cost path (shortest path). It can happen that multiple minimum cost paths exist in a network for given source and sink nodes. Multi-path routing is also possible more generally, by considering the k least cost paths among all possible paths between a given source and sink. The definition of the routing matrix can be easily adapted to encompass also the general multi-path scenario. It suffices to let nonzero entries of the matrix assume fractional values, i.e., a_{ij} is either 0, if flow j does not use link i, otherwise it equals the fraction of flow rate x_j that is allocated on link i.

Given the routing matrix $\mathbf{A}$ and the link capacities $\mathbf{c} = [c_1, \ldots, c_\ell]$, where $\ell = |\mathcal{L}|$, the flow rates x_j must satisfy the following set of linear constraints:

$$\sum_{j \in \mathcal{F}} a_{ij} x_j \leq c_i , \qquad i \in \mathcal{L}. \tag{10.93}$$

The set of constraints can be stated compactly as $\mathbf{Ax} \leq \mathbf{c}$, where $\mathbf{x} = [x_1, \ldots, x_n]$, with $n = |\mathcal{F}|$, and the inequality is entry-wise. In addition, the obvious set of inequalities $\mathbf{x} \geq \mathbf{0}$ must hold for the flow rates.

Any assignment of non-negative values to the vector $\mathbf{x}$ that complies with the inequalities (10.93) is a *feasible* assignment. The set of feasible assignments is not empty, since inequalities (10.93) can always be satisfied, provided we take small enough positive values of the flow rates (link capacities are positive).

A fairness criterion consists of a way of identifying one assignment of the flow rate vector in the set of feasible assignments. To that end, we define for each flow a *utility* function $U_j(z)$. It represents the degree of "usefulness" for the j-th source of being granted flow rate $z = x_j$. It is quite plain that a sensible utility function should be monotonous non-decreasing. It is often the case that it is also concave. Getting an increment Δx of its rate can be highly useful for flow j that has an assigned rate of Δx, but is less useful for a flow that has an already assigned rate of $10 \cdot \Delta x$. In other words, marginal increments are less and less useful as the already available rate grows, which is a typical mark of concavity. In the following we assume $U_j(z)$ is a differentiable, monotonously increasing and strictly concave function.

A fair target is to let the utility functions of all flows grow as much as possible, without preferring any of them. It is therefore natural to pursue the target of maximizing the *sum* of all utility functions (social utility). We define a flow rate assignment $x_j, j \in \mathcal{F}$, as fair if it solves the following optimization problem (see the Appendix of this chapter for an essential introduction to definitions and properties of convex optimization):

$$\max_{\mathbf{x}} f(\mathbf{x}) \equiv \sum_{j \in \mathcal{F}} U_j(x_j) \tag{10.94}$$

subject to

$$\sum_{j \in \mathcal{F}} a_{ij} x_j \le c_i , \qquad i \in \mathcal{L} \tag{10.95}$$

$$x_j \ge 0, \qquad j \in \mathcal{F} \tag{10.96}$$

Let D denote the bounded and closed (hence compact) subset of $\mathbb{R}^n$ defined by the constraints.

Since the target function is strictly concave, the constraints are linear and the feasible set is nonempty, compact and convex, this is a convex optimization problem. It has therefore a unique solution, the maximizer rate allocation $\mathbf{x}^*$. Hence, the fairness definition is well grounded.

As an example, by setting $U_j(z) = \log z$, we recover the proportional fairness criterion.

In general, to find the fair rate allocation $\mathbf{x}^*$, we can introduce the Lagrange multipliers p_i and define the new function

$$g(\mathbf{x}, \mathbf{p}) = \sum_{j \in \mathcal{F}} U_j(x_j) - \sum_{i \in \mathcal{L}} p_i \left(\sum_{j \in \mathcal{F}} a_{ij} x_j - c_i \right) \tag{10.97}$$

to be maximized for $\mathbf{x} \ge \mathbf{0}$ and $\mathbf{p} \ge \mathbf{0}$. This is the *primal problem* of flow rate optimization.

We can define a dual function as follows:

$$d(\mathbf{p}) = \max_{\mathbf{x} \in D} g(\mathbf{x} . \mathbf{p}) \tag{10.98}$$

It can be shown that $d(\mathbf{p})$ is convex.

Since Slater's conditions hold for the optimization problem stated above, it is

$$\min_{\mathbf{p} \ge 0} d(\mathbf{p}) = \max_{\mathbf{x} \in D} f(\mathbf{x}) \tag{10.99}$$

Moreover, the unique maximizer of the problem, $\mathbf{x}^*$, lies in the interior of the feasible domain D and it satisfies the Karush-Kuhn-Tucker (KKT) conditions:

$$\frac{\partial g}{\partial x_k} = 0 \quad \Rightarrow \quad U'_k(x_k) - \sum_{i \in \mathcal{L}} p_i a_{ik} = 0 , \qquad k \in \mathcal{F} \tag{10.100}$$

$$p_i \left(\sum_{j \in \mathcal{F}} a_{ij}x_j - c_i \right) = 0 \qquad i \in \mathcal{L}. \tag{10.101}$$

The first set of equations establishes a relationship between a function of the flow rate of each flow, namely $U'_k(x_k)$ for the k-th flow, and the quantity $q_k = \sum_{i \in \mathcal{L}} p_i a_{ik}$ for the k-th flow. The quantity q_k is the sum of multipliers p_i associated to the links belonging to the route of flow k. We can interpret the multiplier p_i as the *cost* of link i capacity and q_k as the *price* to be paid for the route of flow k. The KKT condition (10.100) states that the optimum rate allocation corresponds to equating the source k marginal utility for an increase of its flow rate x_k and the price it has to pay for network resources.

The second set of conditions in eq. (10.101) says that, either a link capacity is saturated, or the relevant multiplier must be 0. Only bottlenecks (i.e., links where the whole available capacity is assigned) have a non-null cost and hence contribute to the sum of the path price for each flow.

To gain more insight into the application of the optimization framework to the fairness concept, we modify slightly the primal problem. We pursue the maximization of the following target function for $\mathbf{x} \geq \mathbf{0}$:

$$F(\mathbf{x}) = \sum_{j \in \mathcal{F}} U_j(x_j) - \sum_{i \in \mathcal{L}} \int_0^{y_i} h_i(u)du \tag{10.102}$$

where $h_i(u)$ is a positive, continuous, and increasing function of u for $u \geq 0$ and $y_i = \sum_{j \in \mathcal{F}} a_{ij}x_j$ is the load on link i. The quantity $H_i(y) = \int_0^y h_i(u)du$ represent the "cost" of link i when its load is y. Given the properties assumed for $h_i(u)$, the cost is convex and differentiable, hence the function $F(\mathbf{x})$ is differentiable and strictly concave. Then $F(\mathbf{x})$ has a unique maximum in the set $\mathbf{x} \geq \mathbf{0}$, provided that $\lim_{\|\mathbf{x}\| \to \infty} F(\mathbf{x}) = -\infty$ (intuitively: for very high rate values, the cost overwhelms the utility).

A suitable cost function could be the amount of lost flow, as a consequence of bottleneck link buffer overflows. As y_i gets close to c_i, the amount of lost flow grows sharply and it diverges linearly with y_i (hence, faster than a strictly concave function).

By letting the gradient of F to be zero, we find the conditions that must be satisfied by the maximizer $\mathbf{x}^*$ of $F(\mathbf{x})$:

$$U'_k(x_k) - \sum_{i \mathcal{L}} h_i(y_i)a_{ik} = 0 , \qquad k \in \mathcal{F}. \tag{10.103}$$

Inspired by the iterative ascent gradient method for the numerical computation of the maximizer, we consider the following dynamical system to control the flow rate at the source of flow k:

$$\frac{dx_k}{dt} = \zeta_k(x_k)[U'_k(x_k) - q_k] , \qquad k \in \mathcal{F}, \tag{10.104}$$

where $\zeta_k(\cdot)$ is a positive function and q_k is the path price of flow k, defined as the sum of prices of links belonging to the route of flow k:

$$q_k = \sum_{i \in \mathcal{L}} h_i(y_i) a_{ik} , \qquad k \in \mathcal{F}. \tag{10.105}$$

The q_k's depend on the flow rates through $y_i = \sum_{j \in \mathcal{F}} a_{ij} x_j$.

The interesting points on (10.104) are as follows:

- The right-hand side, i.e., the update of the flow rate, can be computed at the source of the flow as soon as the path price is known to the source.
- Only the path price matters to the source of flow k; the price of any other link not belonging to the routing of flow k does not affect directly the updating rule of the rate of flow k.
- The path cost can be collected by packets of flow k, sent through the path from source to destination, and then fed back to the source, if the data transfer envisages acknowledgement packets flowing back from the destination to the source; this is just the case, if we identify flows with TCP connections.
- The set of controllers (10.104) defines an asymptotically globally stable dynamical system, converging to the maximizer $\mathbf{x}^*$ for any initial value $\mathbf{x}_0$ belonging to the feasible set.

The last statement requires some justification (for essential definitions on stability of dynamical systems, the reader might consult the Appendix at the end of this chapter). A Lyapunov function for the system of controllers defined by (10.104) is $V(\mathbf{x}) = F(\mathbf{x}^*) - F(\mathbf{x})$, where $\mathbf{x}^*$ is the flow assignment that maximizes $F(\mathbf{x})$. It is $V \geq 0$ with equality if and only if $\mathbf{x} = \mathbf{x}^*$ (the maximizer is unique, since $F(\mathbf{x})$ is strictly concave). Moreover, it is

$$\frac{dV}{dt} = -\nabla F \cdot \frac{d\mathbf{x}}{dt} = -\sum_{j \in \mathcal{F}} \zeta_j(x_j) \left[\frac{\partial F}{\partial x_j} \right]^2 \leq 0 \tag{10.106}$$

with equality holding if and only if the gradient of F is null, i.e., $\mathbf{x} = \mathbf{x}^*$, since $F(\mathbf{x})$ is strictly concave. Finally, thanks to the limiting property of $F(\mathbf{x})$ as $\| \mathbf{x} \| \to \infty$, the function $V(\mathbf{x})$ is unbounded as the norm of $\mathbf{x}$ diverges. Since all conditions stated in case 3 of Theorem 10.9 in the Appendix to this chapter apply, we conclude that the dynamical system (10.104) is asymptotically globally stable, i.e., for any initial state $\mathbf{x}_0 \in D$, it is $\mathbf{x}(t) \to \mathbf{x}^*$ for $t \to \infty$.

We have found that the optimal flow rate allocation can be achieved by having the traffic sources control their respective sending rate according to the dynamical system (10.104). The interesting feature of these equations is that the k-th source can compute the adjustment of its rate by using local information (the variation of the utility function around the current rate level) and the "cost" of its own network path, q_k. To implement such a control, we need a protocol where data packets probe the path, "collect" information on the path cost (which is seen from (10.105) to be a sum of terms, each one related to a link of the network path). We need also a feedback to convey the collected path cost back to the source of the flow. This can be implemented by having acknowledgment packets sent by the receiver back to the data sender. Acknowledgments can carry the information on the path cost. The scheme here described fits perfectly into the TCP paradigm, except that we have to define: (i) what we mean for path cost; and (ii) what is the utility function that sums into the social utility maximized by the dynamic controllers defined in eq. (10.104).

The dynamical system (10.104) that converges asymptotically to the maximizer of the target function in eq. (10.102) can be read through TCP lenses as follows. For small bottleneck buffer size, the classic TCP congestion window evolution equation in congestion avoidance can be written as (see eq. (10.73)):

$$\frac{dW}{dt} = \frac{\alpha x(t-T)[1 - p(t-T)]}{W(t)} - \beta W(t)x(t-T)p(t-T) \tag{10.107}$$

where T is the base RTT, $p(t)$ is the loss probability at the bottleneck and $x(t)$ is the sending rate. Since it is $x(t) = W(t)/T$, we have

$$\frac{dx}{dt} = \frac{\alpha x(t-T)[1 - p(t-T)]}{T^2 x(t)} - \beta x(t)x(t-T)p(t-T) \approx \frac{\alpha}{T^2} - \beta x^2(t)p(t) \tag{10.108}$$

where we have assumed $p(t) \ll 1$ and we have neglected the delay in the right-hand side of the differential equation, i.e., we assume a slowly varying flow rate: $x(t) \approx x(t-T)$. The dynamical system (10.108) can be rewritten as:

$$\frac{dx}{dt} = \beta x^2 \left[\frac{\alpha}{\beta T^2 x^2} - p \right] \tag{10.109}$$

The similarity with eq. (10.104) is evident, once we interpret the loss probability p as the path price, the multiplicative term βx^2 as the coefficient $\zeta(x)$ and the term $\alpha/(\beta T^2 x^2)$ as the derivative of the utility function. Then, it is $U(x) = -\alpha/(\beta T^2 x)$. Maximizing the sum of such utility functions is equivalent to minimizing the sum $\sum_{j \in F} 1/x_j$. Since $1/x$ is proportional to the time required to transfer a given amount of bytes at an average rate x, we recognize that the optimization target is minimization of a potential delay. Under the stated assumptions *classic TCP aims at allocating average connection throughput so as to minimize the potential data transfer delay.*

In our revisiting of classic TCP, we have assumed that all flows have a same RTT T. If RTTs are different, the utility functions are $U_j(x) = -\alpha/(\beta T_j^2 x)$. The function that is minimized is $\sum_{j \in F} \frac{1}{T_j^2 x_j}$. It is clear that source with larger RTTs get smaller flow rates.

Going back to the original constrained optimization problem (10.94), we can solve it by resorting to the dual problem, i.e., minimization of the dual function $d(\mathbf{p})$ defined in (10.98), under the constraint $\mathbf{p} \geq \mathbf{0}$. For given $\mathbf{p}$, let $\hat{\mathbf{x}}$ be the value of the source rates that maximizes the Lagrangian $g(\mathbf{x}, \mathbf{p})$. The components of $\hat{\mathbf{x}}$ satisfy $U_k'(\hat{x}_k) = q_k = \sum_{i \in \mathcal{L}} p_i a_{ik}$, $k \in \mathcal{F}$. We can write $d(\mathbf{p}) = g(\hat{\mathbf{x}}, \mathbf{p})$, where $g(\cdot, \cdot)$ is given in eq. (10.97). Since $d(\mathbf{p})$ is convex, the minimizer is found by equating the gradient to 0. We have

$$
\begin{aligned}
\frac{\partial d}{\partial p_h} &= \sum_{j \in \mathcal{F}} U_j'(x_j) \frac{\partial x_j}{\partial p_h} - (y_h - c_h) - \sum_{i \in \mathcal{L}} p_i \sum_{j \in \mathcal{F}} a_{ij} \frac{\partial x_j}{\partial p_h} \\
&= \sum_{j \in \mathcal{F}} q_j \frac{\partial x_j}{\partial p_h} - (y_h - c_h) - \sum_{j \in \mathcal{F}} \frac{\partial x_j}{\partial p_h} \sum_{i \in \mathcal{L}} p_i a_{ij} \\
&= c_h - y_h , \qquad h \in \mathcal{L},
\end{aligned}
$$

where we have used the definitions of the path price q_j of source j and the notation y_h for the load of link h.

The expression of the gradient of $d(\mathbf{p})$ suggests the following dynamical system, inspired by the gradient descent method (we search a minimizer for $d(\mathbf{p})$):

$$
\frac{dp_h}{dt} =
\begin{cases}
\eta_h(p_h)(y_h - c_h) & p_h > 0, \\
\eta_h(p_h)\max\{0, y_h - c_h\} & p_h = 0,
\end{cases}
\tag{10.110}
$$

where $\eta_h(\cdot)$ is a positive function and we have taken into account the fact that the multipliers p_h must be non-negative. In the dynamical system (10.110) we set

$$
y_h = \sum_{j \in \mathcal{F}} a_{hj} x_j , \quad h \in \mathcal{L},
\tag{10.111}
$$

$$
U_j'(x_j) = \sum_{i \in \mathcal{L}} p_i a_{ij} , \quad j \in \mathcal{F}.
\tag{10.112}
$$

Given the p_i's, we compute the x_j's using eq. (10.112). Then we can compute the link loads y_i's, using eq. (10.111). Finally, we update the p_i's, according to the dynamical equations (10.110).

We can interpret p_h as the queue size at the buffer of link h, if we set $\eta_h \equiv 1$. Hence, p_h is proportional to the queueing delay at the h-th link, provided the link capacity is constant, as we have assumed here. The dynamical system (10.110) can be translated into a distributed algorithm, where sources adapt their flow rates according to eq. (10.112), by collecting the path cost as the overall queueing delay through the path.

The global asymptotic stability of the dynamical system (10.110) can be established by considering the Lyapunov function $V(\mathbf{p}) = d(\mathbf{p}) - d^*$, where d^* is the global minimum of $d(\mathbf{p})$ for $\mathbf{p} \geq \mathbf{0}$. Note that

$$
\frac{dV}{dt} = \sum_{h \in \mathcal{L}} \frac{\partial d}{\partial p_h} \frac{dp_h}{dt} = -\sum_{h \in \mathcal{L}} \eta_h(p_h)(y_h - c_h)^2 \leq 0
\tag{10.113}
$$

All other conditions for $V(\mathbf{p})$ to be a Lyapunov function of the dynamical system (10.110) hold as well (see the conditions enumerated in Theorem 10.9 at the end of the Appendix to this chapter). Then, we have $\mathbf{p}(t) \rightarrow \mathbf{p}^*$ as $t \rightarrow \infty$, where $\mathbf{p}^*$ is the minimizer of the dual function $d(\mathbf{p})$ in the region $\mathbf{p} \geq \mathbf{0}$.

The algorithms derived from the utility optimization problems do not account for path delays, i.e., they assume that costs (whatever their meaning) are immediately available at the source. This is not the case in a real network, e.g., with TCP congestion control algorithms. The round trip delay between source actions to control the sending rate and the reaction of the network as perceived by the source is non null. As a consequence, stability problems arise for the dynamical systems defining the distributed adaptation algorithm of the source flow rate. A linear analysis shows that stabilization entails essentially picking suitably small values of the loop gains ζ or η.

10.9 Challenges to TCP

TCP has been conceived during the 1980s. At the time the Internet was still in its infancy. It spanned a limited extension and penetration, mostly limited to public institutions and to the academic environment. The dominant applications were remote shell, e-mail, file transfer. The link capacity ranged between kbit/s up to several Mbit/s.

If compared to the Internet today, it must be recognized that the design of TCP mechanisms, and above all, of the flow and congestion controls, has been amazingly successful and robust. The classic TCP still works decently in most networking situations, even if a number of key improvements have amended the original version of TCP over the decades. A comprehensive survey of TCP evolution is offered in [6]. A more recent survey, focused on TCP issues in modern networks is given in [176].

Notwithstanding this excellent future-proof longevity of TCP, it is increasingly evident that a deep revision of end-to-end congestion control would be beneficial to reap the potential capacity and performance gain offered by modern networking technologies. It is not by chance that proposals for new congestion control design have been put forward since the second half of the 1990s.

The end-to-end approach of TCP abstracts from details of the network and links, nor is it envisaged that TCP entities interact explicitly with network entities (e.g., routers). The view of a TCP connection consists of a communication pipe whose capacity is determined by a bottleneck link, characterized by a given RTT. The details of the links making the path of the TCP connections, the buffer sizes, the link layer protocol features, the volatility of the link layer capacity do affect however the performance of TCP. This has motivated extensive investigation of alternative congestion controls and even spurred criticism on the overall TCP approach and suitability as a one-size-fits-all solution. Conceptually, congestion control is not limited to the end-to-end (transport) layer. It could be implemented in the network layer. It has actually been implemented in the application layer, e.g., quick UDP Internet connections (QUIC) [110] and congestion control for real-time traffic [65, 215].

Evolution of networking technologies and opportunities to develop new applications are the main drive forces that call for new designs of end-to-end congestion control algorithms. The pace of that evolution suggests that the interest of the research community will hardly fade away in the future. Two directions that congestion control (in general, design of transport protocols) appear to aim at are ways to overcome the ossification of the Internet and multi-path. The former addresses the proliferation of middle-boxes that filter Internet traffic and de facto impair the introduction of new transport layer protocols (besides TCP and UDP) or major modifications to the logic of TCP. Multi-path stems from the increasingly available ability of user terminals to support multiple network interfaces (e.g., smartphones can manage simulateneous connections to the cellular network and to WiFi). Running TCP on top of multiple paths provides robustness against link failure, diversity, the opportunity to overcome head-of-line blocking (see Section 10.9.4). In spite of various proposals, design of a reliable and efficient multi-path TCP, specifically for the wireless environment, is still an open problem.

In the following, we review a number of issues affecting classic TCP performance and calling for new design.

10.9.1 Fat-Long Pipes

To fill the TCP connection pipe and fully utilize the bottleneck link capacity, the sender transmission window should scale up to the order of the BDP of the connection path. A path having BDP $\gg 1$ (e.g., hundreds of MSSs or more) is said to be a fat-long pipe. The name stems from the fact that a connection path with RTT in the order of hundreds of ms is typically a geographic link (long pipe) and to make a large BDP the bottleneck capacity should be in the order of hundreds of Mbit/s or more (fat pipe).

The issue of large BDP came up as the progress of communication technologies started providing geographical links with capacity in the order of Gbit/s, as well as wireless cellular links with capacities in the range of tens of Mbit/s. Coupled with RTTs typical of those environments, the effect was evidence that NewReno TCP could not harvest the promise of large sustained throughput on such fat-long pipes.

Given that TCP congestion control is loss-based, at least for most of currently deployed TCP versions, and hence packet loss is inherent to TCP operations, at least in long-lived connections, it follows that the congestion window is often reduced during the lifetime of a connection. With NewReno TCP, the *cwnd* is cut by half, then increased by one for each RTT as long as all packets are successfully delivered. We have seen that, in spite of this sawtooth behavior of the *cwnd*, 100% efficiency of the TCP connection throughput is possible, provided that the buffer space available at the bottleneck is equal to the BDP. Here comes the pitfall; with connection path exhibiting BDPs in the order of thousands of packets, typical router buffer sizes fall short of the BDP, thus reducing the achievable throughput of long-lived TCP connections. Let us make a simple numerical example. Let $RTT = 100$ ms and $C = 1$ Gbit/s. With $MSS = 1460$ bytes it is BDP ≈ 8560 MSSs. Let the bottleneck buffer size be $B = 10$ Mbytes ≈ 6850 MSSs. The maximum value reached by the *cwnd* before packet loss occurs is $B + BDP \approx 15410$ MSSs. According to classic TCP, the *cwnd* is cut to half upon packet loss detection, therefore dropping to about 7705 MSSs. Since this value is less than the *BDP*, the pipe is drained partially. To fill it again, the *cwnd* must be incremented by $8560 - 7705 = 855$ MSSs. Since the *cwnd* is incremented by 1 *MSS* per RTT, it takes about 85.5 s to fill the pipe. As the bottleneck capacity grows, the time to fill the pipe grows proportionally. The fill-up time grows also in case of smaller buffer sizes. For example, for a buffer size of 1 Mbyte, the fill-up time jumps to about 394 s, more than 6 minutes and a half.

The example is extreme and one could correctly object that a single long-lived connection using a full 1 Gbit/s link in a wide area network is quite unrealistic. While not typical, still the scenario outlined above is not unrealistic. Probably the first community that became aware of the poor performance of classic TCP over fat-long pipes was composed of experimental physics scientists, who often need to move huge amounts of data, produced by experiment sensors, between sites located in different continents. Researchers from this community started claiming data transfer limitations due to TCP in the early 2000s. With fast expansion of the network access capacity, Gbit/s-level connections are going to be commonplace.

Even if not so extreme, an increasing number of networking settings exhibit a high BDP with respect to the bottleneck buffer size. Hence the classic AIMD mechanism has become inadequate. Many new proposals have been advanced, most of them still in the mainstream of loss-based kind of control. CUBIC is one of them. Common features to most proposals are: (i) more aggressive *cwnd* increases with respect to classic TCP, when in congestion avoidance; and (ii) attempt to estimate the bottleneck bandwidth, so as to adjust the multiplicative *cwnd* reduction on packet loss detection to avoid pipe depletion.

The challenge to tame fat-long pipes is still open, as network link capacity scales to higher levels. New trends that strongly influence the quest for a suitable TCP congestion control are the pervasiveness of wireless links and the edge computing/information centric/content delivery paradigm. The latter pushes contents close to the user access network, so as to cut delays. A side effect is reducing RTT for TCP connections, hence helping reduce the BDP. As for the impact of wireless links on TCP, it deserves an ad-hoc discussion, provided in the next section.

10.9.2 Wireless Channels

Wireless links differ from wired ones in several respects: packet loss is more probable; capacity can fluctuate significantly, especially in mobile networks; links might experience interruptions and volatility. Therefore, the end-to-end model abstraction that classic TCP bases its congestion control upon can be inadequate. For example, packet loss due to link issues cause undue reductions of the congestion window. This is a consequence of classic TCP mingling error recovery with congestion control. Volatility and wide fluctuations of link capacity are other issues that classic TCP hardly copes with.

Comprehensive reviews of existing works on improving TCP performance over wireless and mobile networks are provided in [22, 180, 187]. Approaches to solve those issues follow different lines: (i) improving channel quality and stability of wireless links; (ii) re-designing TCP mechanisms, including proxy TCP approaches; and (iii) setting-up cross-layer algorithms, by exploiting lower layer knowledge of the link status.

The first approach consists of providing retransmissions, diversity, and other protection mechanisms over the wireless link. The side effect is increasing the mean delay and the delay variability.

A common redesign approach of TCP mechanisms is based on the introduction of a proxy within the TCP connection path. Recognizing that wired links should be dealt with differently than the wireless ones, those solutions envisage breaking down the path into two segments: the wireless one and the rest of the path. The congestion control loop and error recovery can be managed in different ways over the two segments, by introducing a TCP-aware gateway in between. Proxy solutions implement protocol optimizations in an intermediate network device to maintain compatibility with the TCP sender, the TCP receiver, or both. These solutions can be further divided into two sub-types: non-split TCP proxy, and split-TCP proxy. In non-split TCP proxies, the proxy will generate (or forward) an ACK packet to the TCP sender only when the ACK is received from the TCP receiver, thereby preserving the end-to-end semantic of TCP. In split-TCP proxies, the proxy effectively terminates the TCP connection from the Internet sender and implements different TCP algorithms for the wireless/mobile leg. This approach breaks the end-to-end semantic of TCP, as an ACK from the proxy no longer guarantees that the acknowledged packet has been delivered to the intended receiver. For work on split TCP solutions over cellular network links, see, e.g., [126].

It is also possible to design a cross-layer congestion control mechanism over the wireless link, assuming that information can be passed to the intermediate TCP entity by the link layer management. An interesting redesign proposal of TCP along these lines is presented in [151].

Also TCP performance over WiFi has been studied extensively. A work, highlighting throughput instability and unfairness with TCP in WiFi networks, is reported in [155].

A new twist in the study of TCP over wireless links comes from the adoption of mmWave communication channels in 5G cellular networks. The propagation characteristics of radio waves with frequency beyond 6 GHz are quite different from traditional cellular networking radio links. Blockage of the radio link due to transitions from line-of-sight (LOS) to non-LOS conditions of the radio path cause extreme capacity fluctuations, e.g., from Gbit/s to almost zero. The volatility of the mmWave links coupled with the very high BDP makes these links highly challenging for TCP. Moreover, it has been shown that classic TCP makes a suboptimal use of the very high mmWave data rates, since it takes a long time to reach full capacity after a loss or at the beginning of a connection. A discussion on this topic is offered in [212].

10.9.3 Bufferbloat

The currently dominant paradigm of TCP congestion control is loss-based. Hence, packet loss is inherent to TCP working. On the other hand, packet loss exposes traffic to the need of recovering lost data by means of coding or retransmissions, therefore introducing additional and possibly variable delays, or even unrecovered data. The desire to remove packet loss as far as possible has pushed manifacturers to equip routers with very large buffers. A large buffer at a bottleneck link may translate into large delays, especially in the access network, where congestion is more likely to arise. For example, a 10 Mbit buffer in front of a link with capacity 100 Mbit/s can introduce up to 100 ms delay. Much larger buffer sizes can be found in the nowadays Internet, bloating delays in the range of seconds in worst cases. Since classic TCP detects congestion only upon packet loss, buffers must be saturated to make TCP sources back off and relieve congestion.

The bufferbloat issue [88] came up around the end of the first decade of the 2000s as a consequence of the interaction of loss-based TCP congestion control and increasingly large buffers deployed in routers.

Bufferbloat degrades the QoS of applications, in particular when real-time or web browsing flows share the same bottleneck with file transfer TCP connections. Bufferbloat has worsened over the years, mainly due to loss-preventing design strategies that place large buffers in front of low capacity access links. Moreover, large delay fluctuations make the estimate of the RTT less stable and impair the possibility of using variations of end-to-end delays as a congestion signal.

The most adopted fix consists of AQM, run locally at routers. AQM is a buffer management policy where packets are dropped or marked as a function of the congestion level observed in the buffer. The basic principle is relatively simple: since traffic sources probe the network for capacity, increasing their sending rate until packet loss is detected, AQM provides packet loss evidence to sources, *before* the buffer gets completely filled up.

AQM algorithms define a dropping probability p as a function of the status of the queue in the bottleneck buffer (e.g., the queue occupancy level, the trend of the queue, or quantities obtained by filtering those variables). Upon arrival, a packet is dropped with probability p. Since p is typically an increasing function of the buffer occupancy level, the more the buffer is congested the more likely a packet is dropped, hence causing a TCP source to back-off its sending rate. Randomly selecting the packet to be dropped is motivated by fairness considerations. Since the sending rate of a TCP connection is an increasing function of its congestion window size, TCP connection having larger *cwnd* sizes are more likely to be hit by the random packet drop. For example, if two TCP connections send at rates of 10 and 100 packets per second respectively, the probability that a randomly selected arriving packet belongs to the more demanding connection is $100/(10 + 100) \approx 0.9$. Consistently, those connections are more responsible for congestion than those having smaller *cwnd* sizes. If TCP sources are designed to react also on marking signals carried back to the source on ACKs (e.g., ECN marking), then the AQM can be designed to mark rather than drop packets with probability p.

Simple as it sounds, crafting a good AQM algorithm has turned out to be a very hard task. Experience with AQM has shown that performance results are sensitive to parameter values. Since many different variants of TCP congestion control are coexistent in the Internet ecosystem, the effect of dropping packets is not easily predictable, which makes design of AQM algorithm difficult and proposed solutions often quite complex. On the other hand, the decoupling between end-to-end congestion management algorithms and network layer buffering and scheduling of packets inside the routers is a cornerstone of the Internet architecture. Introducing explicit cross-layer cooperation between end-to-end transport (or application) entities and network layer entities inside routers would induce undesirable dependencies and strongly affect the flexibility and capability to evolve of the Internet.

AQM is not a recent idea: numerous techniques were proposed in the 1990s, such as RED. Pointers to the literature are provided in [93] and in the extensive survey presented in [5]. Despite these many proposals, they have encountered limited adoption, partly due to the issues we mention above with parameter tuning and partly because of the computational cost of the algorithms. More recently, new and easier to tune and deploy AQM schemes, such as controlled delay management (CoDel) [170], have been adopted in several commercial products. CoDel is an AQM algorithm that limits the buffering latency by monitoring the queueing delay.

Nonetheless, the bufferbloat issue remains relevant in the wireless domain, where typically base stations are equipped with large per-flow buffers.

10.9.4 Interaction with Applications

We outline two examples where new application paradigms pose challenges to TCP.

Data centers are the technological platform for implementing cloud computing. They provide computational power and storage to support the execution of applications, including big data. A data center is composed of a large number of servers (from thousands to hundreds of thousands), interconnected with ethernet switches. The whole system is connected to the Internet via border routers and load balancers. Current data centers are based on TCP/IP networking protocols. Most traffic generated by servers is confined to the data centers and essentially all of it is carried on TCP. A survey of transport protocol issues in the data center environment can be found in [211].

Typical application paradigms used for heavy computational tasks in data centers are distributed, i.e., the task execution is split on a large number of servers that work in parallel and partial results are reassembled at other servers (partition/aggregation design pattern, e.g., as done with PageRank). Such a paradigm (e.g., map/reduce) entails intense communications among servers.

A typical traffic pattern consists of a cluster-head server A launching subtasks on n other servers $B_1, \ldots, B_n$. Let S denote the switch that A is connected to through port p (see Figure 10.21). A opens TCP connections to each of the n servers. Packets back from servers $B_1, B_2, \ldots, B_n$ to A arrive almost simultaneously, causing a major surge of traffic in the buffer of port p of switch S. Ethernet switch buffers are rather small, so the parallelism degree n can be large enough to cause buffer overflow. If this occurs at the start of the TCP connection, it is highly likely that lost packets are recovered by means of timeout, thus causing large delays.

This issue is referred to as *incast*. It arises because of the interaction among many-to-one traffic patterns induced by the application level, small buffers at the switch port and TCP loss recovery mechanisms.

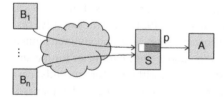

Figure 10.21 Networking layout of TCP incast. The end-system A initiates n TCP connections with servers $B_1, \ldots, B_n$. The n servers reply sending data to A almost simultaneously, through the output port p of switch S.

Many attempts have been made to analyze and solve this problem. Proposed solutions may be classified into four categories [56]: (i) system parameters adjustment, like disabling slow start to avoid retransmission timeout; (ii) enhanced in-network and client-side algorithm design, to minimize the number of packet losses, and to improve the quick recovery of lost packets; (iii) replacement of loss-based TCP algorithm with delay-based ones; and (iv) design of completely new algorithms for this particular environment, like DCTCP [9], a variant of TCP congestion control that uses packet marking to adjust the congestion window, or IATCP [106], a rate-based congestion control that counts the total number of packets injected to constantly meet the path BDP.

A second example of an issue arising from the interaction of TCP with application is the so called head-of-line (HOL) blocking of web traffic. A web page is composed of several distinct objects. When a client requests a page to the server, all these objects are downloaded with a single HTTP GET request. HTTP/1.1 did not allow multiplexing, so the client was forced to open one TCP connection for every object. To reduce the overhead implied by many TCP connections, version HTTP/2 was shipped in 2015. In this new version, a single TCP connection may be used to download all objects of the web page.

The potential performance advantage of HTTP/2, expected to reduce download times with respect to the previous version, turned out to be significantly impaired by packet losses, e.g., in cellular networks. When a packet is lost, out-of-order packets are accepted and SACKed by the TCP receiver (assuming selective acknowledgments are used), but their payloads cannot be delivered to the application. Data delivery takes place only when the complete sequence of packets is recovered at the receiver, by means of retransmissions. As a result, objects whose data have been completely received cannot be passed to the application level and displayed, until the lost packets are recovered, because of the in-sequence delivery constraint of TCP.

The solution of the HOL blocking issues can be obtained by using multi-path transport protocols. A multi-path transport protocol uses different parallel paths. The in-sequence delivery constraint is restricted to each single path.

An alternative approach consists of delegating error recovery and congestion control to the application layer (as done with QUIC over UDP).

Appendix

We review here basic definitions and results. For a systematic introduction to convex optimization the reader may consult, e.g., [41].

Definition 10.1 *(Convex set)*   A set $S \subset \mathbb{R}^n$ is convex if, for any $x, y \in S$ and any $\alpha \in [0, 1]$, it is $\alpha x + (1 - \alpha)y \in S$.

In a convex set, the entire line segment with extremes x and y lies within the set. For example, a circular ring in not convex, a circle is. A C-shaped set is not convex either.

Definition 10.2 *(Convex function)*   Given a set $S \subset \mathbb{R}^n$ and a function $f : S \mapsto \mathbb{R}$, f is said to be convex if S is convex and if, for any $x, y \in S$ and $\alpha \in [0, 1]$ it is

$$f(\alpha x + (1 - \alpha)y) \le \alpha f(x) + (1 - \alpha)f(y) \tag{10.114}$$

f is strictly convex, if the inequality holds strictly for any $x \ne y$ and $\alpha \in (0, 1)$.

Definition 10.3 *(Concave function)*   Given a set $S \subset \mathbb{R}^n$ and a function $f : S \mapsto \mathbb{R}$, f is said to be concave (strictly concave) if $-f$ is convex (strictly convex).

Checking whether a function is convex or concave, according to the definition, can be hard. We give properties that are useful to check convexity.

Theorem 10.2 *(First-Order condition for uni-dimensional x domain)*   If a function $f(x)$ of a uni-dimensional variable x is differentiable in an interval S and its derivative is non-decreasing (increasing) in S, then $f(x)$ is convex (strictly convex) in S.

Theorem 10.3 *(First-Order condition for multi-dimensional x domain)*   A function $f(x) : S \subset \mathbb{R}^n \mapsto \mathbb{R}$, differentiable in the convex set S, is convex if $f(y) \ge f(x) + \nabla f(x)(y - x)$, $\forall x, y \in S$, where $\nabla f(x)$ is the gradient of f calculated at x. The function f is strictly convex if the inequality is strict for any $x \ne y$.

Theorem 10.4 *(Second-Order condition)*   A function $f(x) : S \subset \mathbb{R}^n \mapsto \mathbb{R}$, twice differentiable in the convex set S, is convex (strictly convex) if the associated Hessian matrix $\mathbf{H}(x)$ is positive semi-definite (positive definite) for $x \in S$. The entry (i, j) of the Hessian matrix is given by $H_{i,j}(x) = \frac{\partial^2 f}{\partial x_i \partial x_j}(x)$ for $i, j = 1, \ldots, n$.

We also introduce affine functions, i.e., functions that are composed of a linear term plus a constant.

Definition 10.4 *(Affine function)*   A function $f : \mathbb{R}^n \mapsto \mathbb{R}^m$ is said to be affine if $f(x) = Ax + b$, where A is an $m \times n$ matrix and b is a column vector of size m.

Let us now consider an unconstrained optimization problem for a function defined over a set S, namely

$$\max_{x \in S} f(x) \qquad (10.115)$$

We define x^* as a *local maximizer* of f in S if there exists $\varepsilon > 0$ such that $f(x + \delta x) \leq f(x^*)$ for any δx with $\| \delta x \| \leq \varepsilon$ and $x + \delta x \in S$, where $\| \cdot \|$ is any norm over $\mathbb{R}^n$. The point x^* is said to be a *global maximizer* if $f(x) \leq f(x^*)$ for any $x \in S$. Properties that help calculate the maximizer x^* are summarized in the following theorem.

Theorem 10.5 Given a function $f : S \subset \mathbb{R}^n \mapsto \mathbb{R}$, the following results hold.

1. If f is continuous in S and S is compact (closed and bounded), there exists a global maximizer $x^* \in S$ of f.
2. If f is differentiable in S, any local maximizer x^* in the interior of S must satisfy the condition $\nabla f(x^*) = 0$. If f is concave in S, that condition is sufficient for x^* to be a local maximizer of f.
3. If f is concave, a local maximizer is also a global maximizer. If f is strictly concave, there is a unique global maximizer in S.
4. All properties stated above hold for a convex function f if the optimization problem becomes $\min_{x \in S} f(x)$, "concave" is replaced with "convex" and "maximizer" is replaced with "minimizer."
5. If f is concave and differentiable in S, x^* is a maximizer of f in S if and only if $\nabla f(x^*) \delta x \leq 0$ for any δx such that $x^* + \delta x \in S$.

Let us now consider the constrained optimization problem defined by:

$$\max_{x \in S} f(x) \qquad (10.116)$$

under the constraints:

$$h_i(x) \leq 0 \qquad i = 1, \dots, I, \qquad (10.117)$$

$$g_j(x) = 0 \qquad j = 1, \dots, J, \qquad (10.118)$$

A point x is said to be feasible if it belongs to S and it satisfies all constraints. Let us denote the feasible set with $\mathcal{F}$. We assume it is nonempty. The *Lagrangian* associated to this optimization problem is defined as:

$$L(x, \lambda, \mu) = f(x) - \sum_{i=1}^{I} \lambda_i h_i(x) + \sum_{i=1}^{J} \mu_j g_j(x) \qquad (10.119)$$

where $\lambda_i \geq 0$ and μ_j are called *Lagrange multipliers*. We define the *Lagrange dual function* as follows:

$$D(\lambda, \mu) = \sup_{x \in S} L(x, \lambda, \mu) \tag{10.120}$$

Let x^* denote a feasible maximizer of the optimization problem (10.116) and let $f^* \equiv f(x^*)$. We can prove the following.

Theorem 10.6 The function $D(\lambda, \mu)$ is convex and it is $D(\lambda, \mu) \geq f^*$ for any $\lambda \geq 0$ and any μ.

Proof: The function $D(\lambda, \mu)$ is convex as a consequence of the fact that it is the point-wise supremum of affine functions, parametrized by x.

As for the second part of the theorem statement, since $\lambda \geq 0$, $h_i(x) \leq 0$, and $g_j(x) = 0$ for all $x \in \mathcal{F}$, we have $L(x, \lambda, \mu) \geq f(x)$ for any $x \in \mathcal{F}$. Hence

$$D(\lambda, \mu) = \sup_{x \in S} L(x, \lambda, \mu) \geq \sup_{x \in \mathcal{F}} L(x, \lambda, \mu) \geq \max_{x \in \mathcal{F}} f(x) = f^* \tag{10.121}$$

∎

The best bound on f^* provided by the dual function is obtained by solving the following *dual optimization problem*:

$$D^* = \inf_{\lambda \geq 0, \, \mu} D(\lambda, \mu) \tag{10.122}$$

For contrast, the optimization problem on $f(x)$ is called the *primal optimization problem*.

According to the theorem above, it is $D^* \geq f^*$. The difference between D^* and f^* is called the *duality gap*. If the duality gap is 0 we say that *strong duality* holds. Strong duality is interesting since it allows solving either the primal or the dual problem equivalently. Sometimes, the dual problem turns out to be easier than the primal one.

Slater's condition provides a sufficient condition for strong duality to hold. Before stating it, we define the relative interior of a convex set S. A point x belongs to the relative interior of S if, for any $y \in S$, there exist $z \in S$ and $\xi \in (0, 1)$ such that $x = \xi y + (1 - \xi)z$.

Theorem 10.7 (*Slater's condition*) Strong duality for the constrained optimization problem (10.116)–(10.118) holds if the following conditions are true.

1. $f(x)$ is a concave function.
2. $h_i(x)$ are convex functions.
3. $g_j(x)$ are affine functions.

4. there exists x belonging to the relative interior of S such that $h_i(x) < 0$ for all i and $g_j(x) = 0$ for all j.

Next, we give the famous Karush-Kuhn-Tucker (KKT) conditions for the solution of convex optimization problems.

Theorem 10.8 *(Karush-Kuhn-Tucker conditions)* Consider the constrained optimization problem (10.116)–(10.118). Assume f, h_i, $i = 1, \dots, I$, and g_j, $j = 1, \dots, J$ are differentiable functions in S and that Slater's condition holds. A feasible point x^* is a global maximizer for the constrained optimization problem if and only if there exist $\lambda^* \geq 0$ and μ^* such that

$$\frac{\partial f}{\partial x_k}(x^*) - \sum_{i=1}^{I} \lambda_i^* \frac{\partial h_i}{\partial x_k}(x^*) + \sum_{i=1}^{J} \mu_j^* \frac{\partial g_j}{\partial x_k}(x^*) = 0 \qquad k = 1, \dots, n, \qquad (10.123)$$

$$\lambda_i^* h_i(x^*) = 0 \qquad i = 1, \dots, I. \tag{10.124}$$

If f is strictly concave, x^* is the unique global maximizer of the constrained optimization problem. Moreover, eqs. (10.123) and (10.124) are also necessary and sufficient conditions for (λ^*, μ^*) to be a global minimizer of the dual problem.

We close this short summary of mathematical facts by giving some definitions of stability of dynamical systems and the concept of Lyapunov function.

Let us consider a continuous mapping $f : \mathbb{R}^n \mapsto \mathbb{R}^n$ and trajectories $x(t)$ obtained by solving the differential equation system

$$\dot{x} = f(x) \tag{10.125}$$

for a given initial point $x(0) = x_0$. The dot stands for derivation with respect to time t. We assume that f satisfies suitable conditions so as to guarantee that the solution $x(t)$, $t \geq 0$, for the differential equation (10.125) exists and is unique for a given initial condition.

We say that a point x_e is an *equilibrium* point of the dyanmical system if $f(x_e) = 0$. If the system is placed into the position x_e, no "force" acts on it, so that it stays in x_e forever.

We give the following notions of stability for an equilibrium point.

Definition 10.5 *(Local stability)* The point x_e is said to be locally stable if, for any given $\epsilon > 0$ there exists $\delta > 0$ such that $\| x(t) - x_e \| \leq \epsilon$, $\forall t \geq 0$, provided that $\| x_0 - x_e \| \leq \delta$.

Definition 10.6 *(Asymptotic local stability)* The point x_e is said to be asymptotically locally stable if there exists $\delta > 0$ such that $\lim_{t \to \infty} \| x(t) - x_e \| = 0$, provided that $\| x_0 - x_e \| \leq \delta$.

Definition 10.7 *(Asymptotic global stability)*   The point x_e is said to be asymptotically globally stable if $\lim_{t \to \infty} \| x(t) - x_e \| = 0$, for any x_0.

We assume in the following that the dynamical system (10.125) has a unique equilibrium point at $x_e = 0$. We give sufficient conditions to assess the stability of the equilibrium point $x_e = 0$.

Theorem 10.9 *(Lyapunov stability)*   Consider a continuous and differentiable function $V : \mathbb{R}^n \mapsto \mathbb{R}$ such that

$$V(x) > 0, \ \forall x \neq 0, \qquad V(0) = 0. \tag{10.126}$$

The following conditions are sufficient for the different types of stability of the point $x = 0$.

1. If $\dot{V}(x) \leq 0$, $\forall x$, then the point 0 is locally stable.
2. If additionally $\dot{V}(x) < 0$ for all $x \neq 0$, then 0 is asymptotically locally stable.
3. If the conditions in the two points above hold and additionally the function $V(x)$ is radially unbounded, i.e., $V(x) \to \infty$ as $\| x \| \to \infty$, then the point 0 is asymptotically globally stable.

The crucial point with the Lyapunov approach to stability analysis is to be able to define a suitable Lyapunov function, given the dynamical system.

Summary and Takeaways

This chapter addresses congestion control, i.e., protection of network resources from overload. After a general taxonomy of congestion control approaches, the chapter focuses on how congestion control is realized in the Internet, highlighting the basic principles and their implications in protocol deployment. The role played by the Transmission Control Protocol (TCP) is discussed in detail. We present the main congestion control algorithms of TCP, the classic one and several variants.

Fluid models of TCP long-lived connections are introduced. We consider first a single long-lived TCP connection, running the classic TCP Reno congestion control, with a fixed bottleneck link capacity. Those models allows us to highlight the saw-tooth behavior of TCP under constant bottleneck link capacity. We also discover the relationship between the average throughput, the bandwidth-delay product and the bottleneck buffer size. We derive the well known result on the dimensioning of the bottleneck buffer size to achieve the maximum bottleneck link utilization. We then generalize the fluid model to the case of variable bottleneck link capacity. This brings major added complexity. The resulting model can be used to investigate the performance of TCP on wireless links, which can be naturally described by time-varying capacity functions. We highlight a "resonance" phenomenon, i.e., worst-case throughput performance of TCP when the time scale of bottleneck link capacity variations is comparable to the TCP connection RTT.

Next we consider models for a multiplicity of TCP connections sharing a same bottleneck link. We state these models in terms of differential equations that describe the congestion state evolution (the *cwnd*, the buffer occupancy level and other state variables, depending on the specific TCP congestion control algorithm). We provide easy-to-use fluid models for the classic TCP congestion control (Reno) with a drop-tail bottleneck buffer, for an AQM (RED) buffer, for DCTCP.

We introduce and discuss the concept of fairness and its tight connection with congestion control. After defining some popular fairness paradigms, we give a general framework for fairness, the network utility maximization. This theory casts a new light on TCP congestion control as a distributed algorithm for the achievement of an optimized flow rate allocation to TCP connections. It provides also a general framework for a quantitative definition of fairness.

Finally, we review hot topics on TCP evolution. Technology evolution, new services and networking paradigms are challenging TCP congestion control and in general call for rethinking the entire approach to congestion control. It can therefore be expected that research interest will still focus on congestion control as a key component of future networks and, in general, shared resource systems (not only in the context of telecommunication networks).

Problems

10.1 A TCP connection is routed through a big fat pipe, i.e., a path with a large BDP. The MSS is 1500 bytes long. The slow start threshold equals 64 MSS. The RTT is essentially constant, i.e., the queuing delay of the path can be neglected with respect to the base RTT. Also packet loss is negligible. The path capacity and base RTT are 1 Gbit/s and 20 ms respectively. The initial window is equal to 1. The connection is set up to transfer a file of size $F = 25$ Mbyte.

(a) Calculate the number of RTTs that it takes to complete the file transfer through the connection (assume all TCP procedures are successful). Compare the result to the time strictly required to transfer the file at full link speed and comment on the result.

(b) Repeat the calculation of the time required to transfer the file by using a fluid approximation of the evolution of the window $W(t)$ and of the corresponding instantaneous throughput $x(t) = W(t)/T$.

10.2 A TCP is set up between a client and a very restrictive host that imposes a receiver window of 4 MSSs. The MSS length is 1500 bytes. The client is downloading a file of 600 kbytes.

(a) Assuming that the queuing delay and the packet loss of the path can be neglected, find how many RTTs it takes for the client to complete the download (consider two RTTs overall for SYN and FIN handshakes).

(b) Now assume there is a packet loss probability p, with loss events independent of one another. What is the probability that no timeout occurs in one RTT?

10.3 Let us consider the same setting as in Problem 10.2, but generalizing the size of the receiver window and the amount of data to transfer. Let W_{rx} be the size of the receiver window in packets (we assume all packets have the same size) and let M denote the overall number of packets to be transferred through the connection. Assume no packet loss takes place at the bottleneck and that buffering delay is negligible.

Plot the number of RTTs required to complete the data transfer (including one RTT for opening the TCP connection) as a function of M for several values of W_{rx}. Plot also the mean utilization of the bottleneck link capacity C as a function of M and comment on the significance of this performance indicator as M varies. Assume $C = W_{rx}/T$, where T is the RTT.

10.4 Consider the fluid model of a long-lived TCP Reno connection with a single bottleneck of capacity C. Let B denote the buffer size at the bottleneck, T the base RTT. Calculate the mean and the maximum delay through the buffer during backlogged time, i.e., the time when the buffer is nonempty.

10.5 Consider the fluid model of a TCP connection with a single bottleneck of capacity C. Let B denote the buffer size at the bottleneck, T the base RTT. Assume that M packets must be transferred, that the initial handshake takes one RTT and that the receiver window is large enough not to impose any restriction. Assume also that the buffer size is equal to the connection BDP.

Plot the number of RTTs required to complete the data transfer as a function of M for $IW = 1$, $ssthresh = 64$, for several values of the BDP (consider values larger than the $ssthresh$).

10.6 Consider a TCP connection set up to transfer M packets. Assume all packets have equal length L (except of the SYN and SYN-ACK initial signaling messages). Let IW be the initial value of the congestion window and let *ssthresh* = 64 MSS. Assume no congestion arises at the bottleneck link and no packet loss occurs.

Plot the number of RTTs required to complete the data transfer as a function of M for $IW = 1$, 3, and 10. Comment on the impact of IW as M grows.

10.7 Write a code script to generate the *cwnd* time evolution of a TCP CUBIC connection according to the algorithms in Section 10.3.2. Assume no packet loss occurs. Plot *cwnd* as a function of time:

(a) since the start of the connection (assume $IW = 3$);

(b) immediately after the recovery of a packet loss, occurred when the *cwnd* had attained the value $W_f = 100$ pkts.

11

Quality-of-Service Guarantees

Delay is preferable to error.

Thomas Jefferson

11.1   Introduction

Network traffic engineering is all about modeling service system to provide performance evaluation metrics and hence assess the quality of the service. Service systems can be set up in different ways, following different approaches. In the last two chapters we have seen networked service systems defined to share a common channel in a multiple access scenario (Chapter 9) or to achieve a fair sharing of network resources among end-to-end traffic flows that are flow-controlled via a closed-loop algorithm (Chapter 10).

In this chapter we consider service systems that give *a priori* guarantees on the performance target that the system offers to customers. A priori means that some level of performance (e.g., throughput, delay bound, bound on loss) is stipulated at the time the traffic flow starts, through a negotiation between the traffic source and the network. This approach couples naturally with connection-oriented communication mode, where data transfer is preceded by a negotiation phase to set-up an end-to-end relationship (the connection).

We present both deterministic and stochastic approaches for supporting quality of service (QoS) guarantees. The first one is based on the deterministic traffic theory. It provides a worst-case bound to achievable performance, given a deterministic bound on the offered traffic and on the service capability of the network elements. The stochastic approach provides probabilistic guarantees, thus relaxing the worst-case approach of deterministic traffic theory. Along these lines, we will address the approach based on the effective bandwidth concept.

Network Traffic Engineering: Stochastic Models and Applications, First Edition. Andrea Baiocchi.
© 2020 John Wiley & Sons, Inc. Published 2020 by John Wiley & Sons, Inc.

11.2 Deterministic Service Guarantees

Let us consider a service system, observed starting at an arbitrary time origin, labeled with 0. The cumulated amount of data arrived at the system input up to time t is denoted with $A(t)$. The function $A(t)$ is either continuous or it has jumps. In the latter case, we assume in the following that $A(t)$ is right-continuous, i.e., $A(t) = \lim_{s \downarrow t} A(s)$, $\forall t \geq 0$. Moreover, we let $A(t) = 0$ for $t < 0$.

Example 11.1   Let us observe a flow of packets. Let L_k be the length of the k-th packet, arriving at time t_k, $k \geq 1$, with $t_1 \geq 0$. Then, the arrival function is defined by

$$A(t) = \begin{cases} 0 & 0 \leq t < t_1, \\ \sum_{j=1}^{n} L_j & t_n \leq t < t_{n+1}, \ n \geq 1. \end{cases} \tag{11.1}$$

The service system has a capacity $C(t)$ for $t \geq 0$, where $C(t)$ is the cumulated amount of data that the service system can process up to time t, provided it is continuously backlogged in $[0, t]$. We let $C(t) = 0$ for $t < 0$. We assume that the system does not reject any arriving customer (lossless system).

Note that both $A(t)$ and $C(t)$ are monotonously nondecreasing functions of time.

The generalized Reich's formula (see Section 4.9) states that the content $Q(t)$ of the system buffer at time t is given by

$$Q(t) = \sup_{0_- \leq s \leq t_+} \{A(t) - A(s) - (C(t) - C(s))\}, \quad t \geq 0, \tag{11.2}$$

given that $Q(0_-) = 0$ (the buffer is initially empty). The limits of the interval where the sup is taken account for possible jumps of the arrival and service functions at the interval boundaries.

Proof: For any $s \in [0_-, t_+]$, the amount of data arrived in $[s, t]$ at the system equals $A(t) - A(s)$. The maximum amount of data that can be processed in the same interval is $C(t) - C(s)$. Therefore

$$Q(t) \geq A(t) - A(s) - (C(t) - C(s)) \tag{11.3}$$

since it might be the case that the system is idle for some time in the interval $[s, t]$ and hence it cannot exploit all its potential processing capability. As the inequality (11.3) holds for any $s \leq t$, it follows that

$$Q(t) \geq \sup_{0_- \leq s \leq t_+} \{A(t) - A(s) - (C(t) - C(s))\} \tag{11.4}$$

To prove the reverse inequality, let us define:

$$u = \sup\{s : 0_- \le s \le t_+, Q(s) = 0\} \tag{11.5}$$

In other words, u is the last time in the interval $[0_-, t_+]$ that the buffer is empty. The definition is well posed, since by hypothesis it is $Q(0_-) = 0$. Therefore, it is $u \in [0, t]$.

Given the definition of u, the buffer is always nonempty in the interval $[u, t]$. Then, the full potential processing capacity of the service system is used in that interval and we can write:

$$\begin{aligned}
Q(t) &= Q(u_-) + A(t) - A(u_-) - [C(t) - C(u_-)] \\
&= A(t) - A(u_-) - [C(t) - C(u_-)]
\end{aligned} \tag{11.6}$$

since $Q(u_-) = 0$ by the definition of u.

The point u belongs to the interval $[0_-, t_+]$, hence we can write

$$Q(t) = A(t) - A(u_-) - [C(t) - C(u_-)] \le \sup_{0_- \le s \le t_+} A(t) - A(s) - (C(t) - C(s)) \tag{11.7}$$

Putting together eqs. (11.4) and (11.7), we obtain finally the result in eq. (11.2). ∎

For constant capacity rate C, it is $C(t) = C \cdot \max\{0, t\}$.

Let $D(t)$ denote the cumulative amount of data output by the service system up to time t. We can prove the following formula:

$$D(t) = \inf_{0_- \le s \le t_+} \{A(s) + C(t) - C(s)\}, \quad t \ge 0. \tag{11.8}$$

Proof: Since the amount of data out of the system from time s up to time t cannot exceed $C(t) - C(s)$ and the amount of data out of the system up to time s cannot exceed the amount of data arrived up to that time, we have

$$D(t) = D(s) + D(t) - D(s) \le A(s) + C(t) - C(s), \forall s \in [0_-, t_+]. \tag{11.9}$$

Since the inequality holds for any s in the interval $[0_-, t_+]$, we have proved that:

$$D(t) \le \inf_{0_- \le s \le t_+} \{A(s) + C(t) - C(s)\} \tag{11.10}$$

Let us now consider the time u defined in eq. (11.5). Since $Q(u_-) = 0$, all arrivals up to u_- have been served. Moreover, the system is never idle in the interval $[u, t]$, hence it uses all its potential serving capability $C(t) - C(u_-)$. Then, we have

$$D(t) = A(u_-) + C(t) - C(u_-) \ge \inf_{0_- \le s \le t_+} \{A(s) + C(t) - C(s)\} \tag{11.11}$$

Putting together eqs. (11.10) and (11.11), we obtain the equality in eq. (11.8). ∎

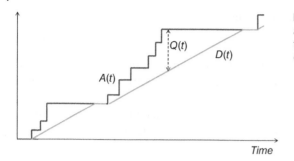

Figure 11.1 Example of arrival function $A(t)$, output function $D(t)$ and queue level $Q(t)$.

Note the consistency of the two results we have proved. Collecting them together, we find $Q(t) = A(t) - D(t)$, which is the obvious input-output balance of the system, given that $Q(0) = 0$. It is also obviously $D(t) \leq A(t)$ (only data that has arrived at the system can leave the system).

Note that the expressions of $Q(t)$ and $D(t)$ are independent of the scheduling algorithm implemented by the service system, provided that it behaves as a work-conserving server.

Thanks to the definition of the arrival and service functions (null for negative times, monotonously nondecreasing), we can rewrite eq. (11.2) as follows:

$$Q(t) = \sup_{s \in \mathbb{R}} \{A(t) - A(s) - (C(t) - C(s))^+\} \tag{11.12}$$

where $(y)^+ \equiv \max\{0, y\}$. For the constant capacity rate case we have:

$$Q(t) = \sup_{s \in \mathbb{R}} \{A(t) - A(s) - C \cdot (t - s)^+\} \tag{11.13}$$

Moreover, we have

$$D(t) = \inf_{s \in \mathbb{R}} \{A(s) + C \cdot (t - s)^+\} \tag{11.14}$$

An example of the behavior of $A(t)$ and $D(t)$ is plotted in Figure 11.1. The queue content $Q(t)$ is shown as well. It can be found as the instantaneous difference between the input function $A(t)$ and the output function $D(t)$.

Equation (11.14) reminds us of the min-plus convolution (see the Appendix to this chapter). To define the convolution, we first define the set of function we will work with. We say a function $f(t)$ is causal if $f(t) = 0$ for $t < 0$. We denote with $\mathcal{F}$ the set of non-negative, nondecreasing, right-continuous, causal functions $f(t)$: $\mathcal{D} \mapsto \{\mathbb{R}^+\} \cup \{\infty\}$, where $\mathcal{D}$ can be either $\mathbb{R}$ or $\mathbb{Z}$.

A function is sub-additive if $f(t + s) \leq f(t) + f(s)$ for all t, s.

Definition 11.1 Given two functions $A, B \in \mathcal{F}$, we define the convolution between them as

$$(A \otimes B)(t) = \inf_{s \in \mathbb{R}} \{A(s) + B(t - s)\} \tag{11.15}$$

For example, let

$$A(t) = \begin{cases} \rho t + \sigma & \text{if } t \geq 0, \\ 0 & \text{otherwise.} \end{cases} \tag{11.16}$$

and $B(t) = C \cdot \max\{0, t\}$, with $C > \rho$. It is easy to check that $A \otimes B = \min\{A, B\}$, i.e.,

$$(A \otimes B)(t) = \begin{cases} Ct & \text{if } 0 \leq t \leq \sigma/(C - \rho), \\ \rho t + \sigma & \text{if } t \geq \sigma/(C - \rho) \\ 0 & \text{otherwise.} \end{cases} \tag{11.17}$$

Let us define the delayed impulse function for $T \geq 0$:

$$\delta_T(t) = \begin{cases} \infty & \text{if } t \geq T, \\ 0 & \text{otherwise.} \end{cases} \tag{11.18}$$

Applying the definition of convolution, it is easy to see that $A \otimes \delta_0 = A$, i.e., the impulse function δ_0 is the neutral element of the convolution operation. Since $B \leq \delta_0$ for any $B \in \mathcal{F}$, we see that $A \otimes B \leq A \otimes \delta_0 = A$. Similarly, we see that $A \otimes B \leq B$. Then, it is $A \otimes B \leq \min\{A, B\}$. As a special case, $A \otimes A \leq A$. If A is sub-additive, we derive $A \leq A \otimes A$ from the definition of sub-additivity. If A is sub-additive and causal, putting together the inequalities, we find $A = A \otimes A$. More definitions and properties related to convolution are listed in the Appendix to this chapter.

Let us consider a service system with arrival function $A(t)$ and departure function $D(t)$. Those functions represent the input-output description of the system.

We can define performance metrics based only on the input-output description of the system. The *backlog* at time t is given by $Q(t) = A(t) - D(t)$. We can define the delay introduced by a service system in this framework. In a lossless and work-conserving system, all data arrived up to time t, that is to say, $A(t)$, will be served by the time $t + \tau$ such that $D(t + \tau) \geq A(t)$. Hence, we define the *virtual delay* at time t as $d(t) = \inf\{\tau \geq 0 : D(t + \tau) \geq A(t)\}$. The maximum delay is defined as $d^* = \sup_{t \geq 0} d(t)$. We call it virtual, since this would be the delay experienced by a customer arriving at time t for FCFS service.

To set bounds on these performance metrics, we need two elements:

1. setting a bound on the arrival function;
2. guaranteeing some minimum amount of service.

We tackle these two points in the next two sub-sections.

11.2.1 Arrival Curves

Any function in $\mathcal{F}$ can be an arrival function, i.e., the amount $A(t)$ of data offered by an arrival flow up to time t.

All definitions and properties we state in this section hold if applied to continuous functions in $\mathcal{F}$ (for which it is necessarily $f(0) = 0$). They hold also for general right-continuous functions, provided that we take inf and sup on intervals of the kind $[a_-, b_+]$.

Given a function $E(t) \in \mathcal{F}$, we say that an arrival function $A(t)$ is constrained by $E(t)$ if for all $s \leq t$ we have

$$A(t) - A(s) \leq E(t - s) \tag{11.19}$$

Note that the inequality can be applied to overlapping time intervals.

The term *envelope curve* or simply envelope is also used to designate a function $E(t)$ satisfying eq. (11.19) for a given arrival function $A(t)$.

We define two special arrival curves we will use in the following:

- *Rate-latency curve*: given two positive parameters r and T, we let

$$\beta_{r,T}(t) = r \cdot \max\{0, t - T\} \tag{11.20}$$

This functions sets a limit r on the rate of the arrival stream, with an initial delay T, during which no data can be sent.

- *Affine arrival curve*. Let us consider a traffic source that emits at a maximum rate of r data units per unit time. We could say that the amount of data emitted in any interval $(\tau, t]$ is upper bounded by $r \cdot (t - \tau)$. However, this envisages a smooth arrival profile. If instead the source can send bursts of data, we can account for that by allowing an initial offset b. This leads to the definition of the *affine arrival curve* defined as

$$\gamma_{r,b}(t) = \begin{cases} rt + b & \text{if } t \geq 0, \\ 0 & \text{otherwise.} \end{cases} \tag{11.21}$$

with r and b positive constants.

More complex functions can be obtained by combining elementary arrival curves with the min operator. For example, given n affine arrival curves $\gamma_{r_i,b_i}(t)$, $i = 1, \ldots, n$, we define a new arrival function as $\gamma(t) = \min\{\gamma_{r_1,b_1}(t), \ldots, \gamma_{r_n,b_n}(t)\}$. Playing with the parameters r_i, b_i, $i = 1, \ldots, n$ we can obtain any piecewise linear function belonging to the set $\mathcal{F}$. Another example of composite arrival curve is $\beta(t) = \inf_{x \geq 0}\{\beta_{r,xT}(t) + xa\} = \min\{a, rT\}\frac{t}{T}$, $t \geq 0$, where a is a constant.

The arrival curve associated with a data flow, i.e., such that the inequality (11.19) is satisfied, is not unique. One could wonder what the "best" arrival curve could be. By "best" we mean here the tightest possible envelope for the given flow $A(t)$.

To make this concept quantitative, we recall the definition of sub-additive closure of a function $A \in \mathcal{F}$.[1] Given a function $A \in \mathcal{F}$, we have $A \leq \delta_0$, hence

---

[1] The definitions and properties from min-plus algebra required for the rest of this section are summarized in the Appendix to this chapter.

$A \otimes A \leq A \otimes \delta_0 = A$. Let $A^{(n)}$ denote the convolution of A with itself $n - 1$ times. The sequence $\{A^{(n)}(t), n \geq 1\}$ is non-negative and monotonously nonincreasing for each t. Hence it admits a proper non-negative limit. We are therefore justified in letting the sub-additive closure of A be defined by $\overline{A} = \inf_{n \geq 1}\{A^{(n)}\}$.

Definition 11.2   We say that a function $E \in \mathcal{F}$ is a "good" arrival function if one of the following equivalent statements holds:

(a) E is sub-additive.
(b) $E = E \otimes E$.
(c) $E = \overline{E}$, where $\overline{E}$ is the sub-additive closure of E.

   The main result on arrival curves is that any arrival curve can be replaced with its sub-additive closure, which is a "good" arrival curve. To obtain this result, we need some preliminary lemmas.

Lemma 11.1   A flow $A(t)$ is constrained by an arrival curve $E(t)$ if and only if $A \leq A \otimes E$.

Proof: The inequality $A \leq A \otimes E$ is equivalent to $A(t) \leq A(s) + E(t - s)$, $0 \leq s \leq t$, by the very definition of the convolution operation. Then, the inequality in the Lemma statement is equivalent to (11.19). ∎

Lemma 11.2   If X and Y are arrival curves for a flow $A(t)$, then so is $X \otimes Y$.

Proof: By the closure property of the $\otimes$ operation, $X, Y \in \mathcal{F}$ implies that $X \otimes Y \in \mathcal{F}$. By the associativity property of $\otimes$ and Lemma 11.1, we have $A \leq A \otimes Y \leq (A \otimes X) \otimes Y = A \otimes (X \otimes Y)$, which proves the Lemma. ∎

Theorem 11.1   If $E(t)$ is an arrival curve for a flow $A(t)$ then so is $\overline{E}(t)$.

Proof: By Lemma 11.2, since E is an arrival curve for A, so is $E \otimes E$. Iterating this reasoning, we find that $E^{(n)}$ is also an arrival curve for A for any $n \geq 1$. We can conclude that $\overline{E}$ is an arrival curve for A. Conversely, since $\overline{E} \leq E$, if $\overline{E}$ is an arrival curve for A, so is E. ∎

   The following property is useful in the calculation of "good" arrival functions.

Lemma 11.3   If A and B are two "good" arrival functions, then the sub-additive closure of $\min\{A, B\} = A \oplus B$ is given by $A \otimes B$. Formally, if $A = \overline{A}$ and $B = \overline{B}$, it is $\overline{A \oplus B} = A \otimes B$,

Proof: We know that $A = A \otimes A$ and $B = B \otimes B$, since A and B are good functions. We have

$$(A \oplus B) \otimes (A \otimes B) = [(A \oplus B) \otimes A] \otimes B = [(A \otimes A) \oplus (B \otimes A)] \otimes B$$
$$= [A \oplus (A \otimes B)] \otimes B = (A \otimes B) \oplus (A \otimes B \otimes B)$$
$$= (A \otimes B) \oplus (A \otimes B) = A \otimes B \qquad (11.22)$$

Moreover, we have

$$(A \oplus B) \otimes (A \oplus B) = (A \otimes A) \oplus (A \otimes B) \oplus (B \otimes A) \oplus (B \otimes B)$$
$$= A \oplus (A \otimes B) \oplus (A \otimes B) \oplus B = A \otimes B \qquad (11.23)$$

For ease of notation, let $X = A \oplus B$ and $Y = A \otimes B$. We have shown that $X \otimes Y = Y$ and $X \otimes X = Y$. It is then easy to show by induction that $X^{(n)} = Y$ for all $n \geq 2$. Since it is further $X \geq Y$, it follows that $\overline{X} = Y$, which concludes the proof. ∎

The following theorem reveals the structure of the minimal arrival curve for a given flow.

Theorem 11.2 (Minimal Arrival Curve) Given a flow $A(t)$ we have

(a) the function $A \oslash A$ is an arrival curve for A.
(b) for any arrival curve E that constrains the flow A it is $A \oslash A \leq E$.
(c) $A \oslash A$ is a "good" arrival curve.

The function $E_A = A \oslash A$ is called the minimal arrival curve that constrains the given flow A.

Proof: We recall that the deconvolution is defined as

$$C(t) = (A \oslash B)(t) = \sup_{u \geq 0} A(t + u) - B(u) \qquad (11.24)$$

If A and B belong to the set $\mathcal{F}$, also their deconvolution does.

For any $s \leq t$ we have $A(t) - A(s) \leq A(t - s + s) - A(s) \leq \sup_{u \geq 0}\{A(t - s + u) - A(u)\} = E_A(t - s)$, which shows the first statement of the theorem.

If E is an arrival curve for the flow A, by Lemma 11.1 we have $A \leq A \otimes E$, that is to say $A(t) \leq A(s) + E(t - s)$, $0 \leq s \leq t$. Letting $v = t - s \geq 0$, we have $A(v + s) - A(s) \leq E(v)$, $\forall v, s \geq 0$. Then, $\sup_{s \geq 0} A(v + s) - A(s) \leq E(v)$. This proves the second statement of the theorem.

For the third statement it suffices to prove that the function $g(t) = (A \oslash A)(t)$ is sub-additive. We have for any $u \geq 0$ and for any s, t:

$$A(t + s + u) - A(u) = A(t + s + u) - A(s + u) + A(s + u) - A(u)$$
$$\leq \sup_{s + u \geq 0}\{A(t + s + u) - A(s + u)\} + \sup_{u \geq 0}\{A(s + u) - A(u)\}$$
$$= g(t) + g(s) \qquad (11.25)$$

Figure 11.2 Comparison of the arrival function, minimum arrival curve and affine arrival curve for the sequence produced by an MPEG coding of the *Star Wars* movie.

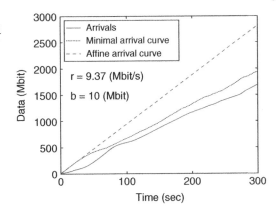

Hence $g(t + s) = \sup_{u \geq 0}\{A(t + s + u) - A(u)\} \leq g(t) + g(s)$, which proves that $g(\cdot)$ is sub-additive and then a "good" function by virtue of Definition 11.2. ∎

In practice, the usefulness of this theorem is limited by the amount of parameters required to describe the minimal arrival curve for a general flow. While the minimal arrival curve can always be computed numerically for a given flow, it does not yield necessarily a parsimonious model. It is however useful to check the tightness of bounds provided by other arrival curves.

Example 11.2 We apply the minimal arrival curve theory to the data stream produced by the MPEG audio/video encoding of the *Star Wars* movie. The MPEG coded sequence of the *Star Wars* movie lasts about 114 minutes: it consists of 171,000 video frames, of duration $T_f = 40$ ms. When packetized with a payload length $L_0 = 1000$ bytes and header length $H_0 = 40$ bytes, its mean bit rate is $\overline{R} = 5.95$ Mbit/s and its peak bit rate is $R_{max} = 16.7$ Mbit/s.

The arrival function $A(t)$ gives the number of bits at the output of the coder at the end of each frame. Figure 11.2 plots the arrival function $A(t)$ and the associated minimal arrival curve for the first 300 s of the sequence.

For comparison purposes, we plot also an affine arrival curve for the given data stream. The parameters are found by setting the burstiness b to 10 Mbit and finding r as the minimum rate that guarantess that $A(t) - A(s) \leq r(t - s) + b$ for all $s \leq t$.

The price for the simplicity of the affine arrival curve is that it gives a significantly looser description of the actual traffic $A(t)$, with respect to the minimal arrival curve.

11.2.2 Service Curves

In this section we aim at setting bounds on the minimum and maximum amount of service provided by a system to its input arrival flow.

Let us consider a generalized processor sharing (GPS) server. We have seen in Section 6.4.2 that GPS guarantees a minimum rate to any multiplexed flow. Say $A(t)$ is the arrival function of a flow that is continuously backlogged between t_0 and t. If r denotes the guaranteed minimum rate of the flow, then $D(t) - D(t_0) \geq r(t - t_0)$. Let s be the largest time in $[0, t]$ such that the flow is not backlogged (this is true at least for $t = 0_-$. It is $D(s) = A(s)$. Then, we can write $D(t) - A(s) \geq r(t - s)$; hence $D(t) \geq \inf_{s \in \mathbb{R}} \{A(s) + r \cdot \max\{0, t - s\}\} = (A \otimes \beta_{r,0})(t)$.

As another example, let us consider a lossless input-output system that imposes a maximum delay T to the served customers. Then, it must be $D(t + T) \geq A(t)$, $\forall t \geq 0$ or $D(s) \geq A(s - T) = (A \otimes \delta_T)(s)$.

In both cases, we find that the output of the system D is lower bounded by an expression of type $A \otimes S$, where $S(t)$ is a function characteristic of the considered service system. This brings us to the following definition.

Definition 11.3 *Service curve*   Let us consider a service system with input and output given by the functions $A(t)$ and $D(t)$, belonging to the set $\mathcal{F}$. Let $S, \overline{S}, \underline{S} \in \mathcal{F}$. We say that $\underline{S}(t)$ is a lower service curve for the system, if $D \geq A \otimes \underline{S}$, $\overline{S}(t)$ is an upper service curve for the system, if $D \leq A \otimes \overline{S}$, $S(t)$ is a service curve for the system if both inequalities hold simultaneously for S, i.e., $D = A \otimes S$.

Service curves give bounds on the amount of data that can be processed by the service system and transferred to the output. They give a description of the "transformation" imposed on the arrival function by the service system, adopting an input/output, black-box point of view. We will see that this is a key building block of a system theory of service systems.

Example 11.3 *Two-class priority system*   Let us consider a service system dealing with two classes of customers. The system has a serving capacity C, i.e., it can work out up to C units of work per unit of time (e.g., if it is a communication link, it can transmit C bits/s). Service is offered according to a head-of-line (HOL) priority discipline (see Section 6.3.2). The arrival, output and backlog functions are denoted with $A(t)$, $D(t)$, and $Q(t)$, respectively. We use a subscript H (L) for the high (low) priority class. To guarantee some minimum service to the low priority class, we assume that the high-priority class admits the arrival curve $E_H(t)$.

Let $t_0 > 0$ be the time when a backlog period of the high-priority class starts. This implies that $Q_H(t_0) = 0$ and hence $D_H(t_0) = A_H(t_0)$. For any time t belonging to the same backlog period it must be $D_H(t) - D_H(t_0) \geq C(t - t_0) - L_{max}$, where L_{max} is the maximum length of a packet of the low-priority class. The inequality follows from the fact that the service capacity of the system is fully devoted to the high-priority packet flow, as soon as the high-priority class becomes backlogged, except possibly of the amount of capacity required to clear a low-priority packet under transmission when the high-priority backlog period starts. Hence

$$D_H(t) \geq A_H(t_0) + C(t - t_0) - L_{max}, \quad t \geq t_0 \tag{11.26}$$

Since it is $D_H(t) - A_H(t_0) = D_H(t) - D_H(t_0) \geq 0$, we have

$$D_H(t) \geq A_H(t_0) + \max\{0, C(t - t_0) - L_{max}\}, \quad t \geq t_0 \tag{11.27}$$

Following the same steps that lead to eq. (11.8), we can prove that $D_H \geq A_H \otimes S_H$, where $S_H(t) = \max\{0, Ct - L_{max}\}$. So, $S_H(t)$ is a lower service curve offered by the system to the high priority class.

As for the low-priority class, let t_b denote the latest time before t_0, when a busy period of the server starts. It is $Q_L(t_b) = Q_H(t_b) = 0$, and hence $D_L(t_b) = A_L(t_b)$ and analogously for the high-priority class. For any time $t \geq t_b$ during the busy period we have

$$D_L(t) - D_L(t_b) + D_H(t) - D_H(t_b) =$$
$$= D_L(t) - A_L(t_b) + D_H(t) - A_H(t_b) \geq C \cdot (t - t_b), \tag{11.28}$$

since the server is work-conserving. Moreover, thanks to the constraint posed by the envelope $E_H(t)$ on the high-priority class, we can write

$$D_H(t) - A_H(t_b) \leq A_H(t) - A_H(t_b) \leq E_H(t - t_b) \tag{11.29}$$

Substituting back into eq. (11.28), we get:

$$D_L(t) \geq A_L(t_b) + C \cdot (t - t_b) - E_H(t - t_b) \tag{11.30}$$

Thanks to $D_L(t) - A_L(t_b) = D_L(t) - D_L(t_b) \geq 0$ and following the same steps as in the proof of eq. (11.8), we find finally that $D_L \geq A_L \otimes S_L$, with $S_L(t) = \max\{0, Ct - E_H(t)\}$. $S_L(t)$ is a lower service curve offered by the system to the low-priority class.

Example 11.4 *Packetized stream of data*   Let us consider a stream of data produced by a coder, e.g., an audio/video coder. To be transferred over a network, the data at the output of the coder are packetized. Let L_{max} denote the maximum length of a packet and R the minimum bit rate at which data is produced by the coder. Then, the maximum delay suffered by coded data when packetized is $\tau_{max} = L_{max}/R$. We can think of the data produced by the coder as an arrival flow $A(t)$ offered to a packetizer. Let $D_p(t)$ be the output function of the packetizer. Since the maximum delay imposed by the packetizer is τ_{max}, we have $D(t) \geq A(t - \tau_{max}) = (A \otimes \delta_{\tau_{max}})(t)$. This shows that $\delta_{\tau_{max}}(t)$ is a lower service curve for the packetizer.

Example 11.5 *Constant rate link*   Let us consider a communication link transmitting packets with maximum length L_{max} at rate C. Assume the data at the output of the link cannot be used other than by integral packets, i.e., a chunk of data emerging out of the link can be processed by downstream service systems only if it consists of an integral number of packets (at least one). This is the case for example with router processors receiving packets on their input lines and analyzing them to perform forwarding actions. Then, a lower service curve for the link is $\max\{0, C \cdot t - L_{max}\} = C \cdot \max\{0, t - L_{max}/C\}$. This is the so called *rate-latency* service curve.

Example 11.6 Let us now consider the cascade of two systems. Let a coder feed a packetizer and then packets be sent over a fluid link of capacity C. By fluid link we mean one where there are no packets. In terms of service curves, a fluid link is characterized by $\underline{S}(t) = \overline{S}(t) = S_l(t) = C \cdot \max\{0, t\}$

The input $A_l(t)$ to the fluid link is the output $D_p(t)$ of the packetizer, that is $A_l(t) = D_p(t) \geq A \otimes \delta_{\frac{L_{max}}{R}}$. Hence

$$D_l = A_l \otimes S_l \geq \left(A \otimes \delta_{\frac{L_{max}}{R}} \right) \otimes S_l = A \otimes \left(\delta_{\frac{L_{max}}{R}} \otimes S_l \right) \tag{11.31}$$

where we have used the associativity of the min-plus convolution. We see that *the cascade of the two systems is characterized by a lower service curve that is the convolution of the lower service curves of the two systems*, namely:

$$S(t) = \left(\delta_{\frac{L_{max}}{R}} \otimes S_l \right)(t) = C \cdot \max\left\{ 0, t - \frac{L_{max}}{R} \right\} \tag{11.32}$$

The last example suggests a powerful property of service curves. Let us consider n service systems in tandem and an arrival flow $A(t)$ that goes through all of them. The output function of the k-th service system $D_k(t)$ is the input function to the $(k+1)$-th system. From the definition of lower service curves and the properties of min-plus convolution, it follows that:

$$D = D_n \geq D_{n-1} \otimes \underline{S}_n \geq D_{n-2} \otimes \underline{S}_{n-1} \otimes \underline{S}_n = \cdots = A \otimes (\underline{S}_1 \otimes \cdots \otimes \underline{S}_n) \tag{11.33}$$

This proves that the lower service curve of the tandem of n service systems is the min-plus convolution of the lower service curves of the systems. This can be obviously proved also for upper service curves.

This property matches perfectly what holds for the impulse response of a cascade of linear, time-invariant systems. The ordinary convolution operation between real-valued integrable functions is replaced by min-plus convolution of functions belonging to the set $\mathcal{F}$.

Example 11.7 Let us consider n systems, each one characterized by a rate-latency curve of its own. Let $S_k(t) = r_k \max\{0, t - d_k\}$ be the service curve of the k-th system. Note that $S_k = F_{r_k} \otimes \delta_{d_k}$, where $F_C(t) = C \cdot \max\{0, t\}$ is the service curve of a fluid link of capacity C.

Then, we have

$$S(t) = \otimes_{k=1}^n S_k = (\otimes_{k=1}^n F_{r_k}) \otimes (\otimes_{k=1}^n \delta_{d_k}) = F_r \otimes \delta_d \qquad (11.34)$$

where $r = \min\{r_1, \ldots, r_n\}$ and $d = \sum_{k=1}^n d_k$. That is to say, the service curve of the tandem composition of n latency-rate servers is still a latency rate server, with rate equal to the minimum among the rates of the component systems and delay equal to the sum of the individual delays.

11.2.3 Performance Bounds

We have learned how we can set deterministic bounds on the arrival process (the arrival curve) and on the minimal amount of service guaranteed by a service system (the service curve).

Let us consider a service system offering a lower service curve $\underline{S}(t)$ and having an input flow that is constrained by the envelope $E(t)$. We assume that $E(t)$ is causal, i.e., it is 0 for $t < 0$. Let $Q(t)$ denote the backlog of the service system at time t and $D(t)$ denote the cumulative output flow of the system at time t. We assume that $Q(0_-) = 0$.

We can prove the following performance bounds.

Theorem 11.3 (Backlog bound) The maximum level achieved by $Q(t)$ is $Q^* = \sup_{u \geq 0}\{E(u) - \underline{S}(u)\}$.

Proof: Since $D \geq A \otimes \underline{S}(t)$, it is

$$Q(t) = A(t) - D(t) \leq A(t) - A(\tau) - \underline{S}(t - \tau) \leq E(t - \tau) - \underline{S}(t - \tau) \qquad (11.35)$$

holding for some τ such that $0 \leq \tau \leq t$. Then, we have $Q(t) \leq \sup_{0 \leq \tau \leq t}\{E(t - \tau) - \underline{S}(t - \tau)\} = \sup_{0 \leq u \leq t}\{E(u) - \underline{S}(u)\}$. Taking the supremum over all $t \geq 0$, we complete the proof of the result. ∎

We can give a deterministic bound also for the output $D(t)$ in terms of the arrival and service curves.

Theorem 11.4 The output flow $D(t)$ is constrained by the envelope $E_D(t) = (E \oslash \underline{S})(t) = \sup_{u \geq 0}\{E(t + u) - \underline{S}(u)\}$.

Proof: Since $D \geq A \otimes \underline{S}(t)$, we have $D(s) \geq A(u) + \underline{S}(s - u)$ for some u such that $0 \leq u \leq s$. Then, for $s \leq t$, we have:

$$D(t) - D(s) \leq D(t) - A(u) - \underline{S}(s - u)$$
$$\leq A(t) - A(u) - \underline{S}(s - u)$$
$$\leq E(t - s + s - u) - \underline{S}(s - u) , \quad \forall u : 0 \leq u \leq s$$

Then, replacing $s - u$ with τ, we have

$$D(t) - D(s) \leq \sup_{0 \leq \tau \leq s} \{E((t - s + \tau) - \underline{S}(\tau)\} \leq (E \oslash \underline{S})(t - s) \tag{11.36}$$

which proves the result by the definition of the ø operation, since the inequality holds for all $s \leq t$. ∎

Finally, we define as delay of a customer arriving at time t the time it takes to empty the buffer of the content $Q(t)$. This is the quantity $d(t)$ defined as

$$d(t) = \inf_{\tau \geq 0} \{\tau : D(t + \tau) \geq A(t)\} \tag{11.37}$$

and referred to as virtual delay. The maximum delay is the supremum over all times: $d^* = \sup_{t \geq 0} d(t)$.

Theorem 11.5 The maximum delay is bounded by the horizontal distance between the service and arrival curves, i.e.,

$$d^* \leq h(E, \underline{S}) \equiv \sup_{t \geq 0} \inf_{\tau \geq 0} \{\tau : E(t) \leq \underline{S}(t + \tau)\} \tag{11.38}$$

Proof: By the definition of lower service curve, we have $D(t + \tau) \geq \underline{S}(t + \tau)$ for all $t, \tau \geq 0$. By the definition of envelope, we have also $E(t) \geq A(t)$ for all $t \geq 0$. Then, the inequality $\underline{S}(t + \tau) \geq E(t)$ implies $D(t + \tau) \geq A(t)$. This means that the set $\{\tau : \underline{S}(t + \tau) \geq E(t)\}$ is a subset of $\{\tau : D(t + \tau) \geq A(t)\}$. The infimum of the first set is therefore not less than the infimum of the second set, which is but the virtual delay at time t. Since this reasoning holds for all $t \geq 0$, we see that we have shown that $h(E, \underline{S}) \geq d^*$. ∎

The main advantage of those bounds is that they apply deterministically (i.e., they bound the worst-case performance). On the down side, they may be loose, depending on the tightness of the bounds set by the arrival curve on the input flow and by the service curve on the minimum amount of guaranteed service offered by the service system.

Figure 11.3 Scheme of a leaky bucket with parameters ρ (token rate) and σ (token bucket size).

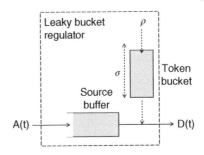

11.2.4 Regulators

How can we enforce a given arrival function or check the conformance of an arrival process to a given arrival function? Let $A(t)$ be an arrival flow, $E(t)$ the envelope we desire to cast over the arrivals, and $D(t)$ the output of the system we build to achieve this purpose.

We say that the system is a regulator with envelope E if for *any* arrival flow A, the output D of the system admits E as an envelope.

From the properties of envelopes, we know that, if E is an envelope for D, then $D \leq D \otimes E$ or $D(t) - D(s) \leq E(t - s), \forall s \leq t$. The question is now: can we define a system that behaves like a regulator for a given envelope?

The answer to this question is relatively simple if we choose as envelope the affine arrival curve. The system that we define is known as the *leaky bucket*. Let then

$$E_{LB}(t) = \begin{cases} \sigma + \rho t & t \geq 0, \\ 0 & t < 0. \end{cases} \tag{11.39}$$

The leaky bucket regulator (also known as the leaky bucket shaper) is also referred to as (σ, ρ)-regulator. A scheme of a leaky bucket regulator is depicted in Figure 11.3. The flow buffer, where the arrival flow data is stored, is assumed to be infinite, i.e., no loss of arriving data is considered. On the contrary, at most σ tokens can be stored in the token bucket.

The leaky bucket generates "tokens" at rate ρ. The tokens accumulate in a buffer of size σ. If the token bucket is full, excess tokens are dropped. The leaky bucket is defined as a service system that receives a data flow at its input and allows the amount of flow that matches the available tokens. A unit of data can move to the output of the leaky bucket, if it finds at least one token in the bucket. In that case, the token is consumed.

The result of this algorithm is that the output rate can exhibit a burst of maximum size σ. Besides that, data leaves the leaky bucket at a sustained rate of at most ρ data units per unit time.

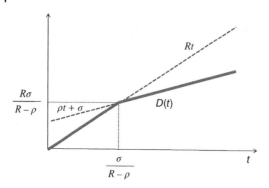

Figure 11.4 Output flow $D(t)$ of a leaky bucket shaper with parameters (σ, ρ), regulating an input flow constrained by the arrival curve $R\max\{0, t\}$, with $R > \rho$.

For example, let us consider a flow $A(t) = R \cdot \max\{0, t\}$, with $R > \rho$, passing through a leaky bucket shaper with parameters (σ, ρ) (see Figure 11.4). This arrival curve is an upper bound for a flow arriving from a link with bit rate R.

At time 0, when the flow starts, the bucket is full. So, initially data can flow out of the shaper at the peak rate R. This goes on until the token bucket is depleted and all newly arrived tokens have been matched by arriving data, i.e., up to time $t_0 = R\sigma/(R - \rho)$. After that time, the arriving data is let out of the shaper with a rate ρ and the token bucket remains empty. The obtained output flow is upper bounded by the arrival curve $\min\{Rt, \rho t + \sigma\}$ for $t \geq 0$.

Let $Q_s(t)$ be the amount of data of the input traffic flow that is stored in the flow buffer of the leaky-bucket at time t, and $Q_b(t)$ be the amount of tokens stored in the token bucket at time t. We prove three key properties of the leaky bucket in the following.

Theorem 11.6 The output flow of the leaky bucket is constrained by the envelope $E_{LB}(t)$.

Proof: In any given time interval $[\tau, t]$ the output of a leaky bucket cannot be more than the amount of tokens found in the leaky bucket at time τ plus the new tokens arrived during the interval. Formally, $D(t) - D(\tau) \leq Q_b(\tau) + \rho(t - \tau) \leq \sigma + \rho(t - \tau) = E_{LB}(t - \tau)$. The second inequality is a consequence of the limited token bucket size. ∎

This following property means that E_{LB} is the service curve of the leaky bucket regulator as a service system.

Theorem 11.7 If A is an arrival flow into a leaky bucket shaper with envelope $E_{LB}(t)$, the output flow $D(t)$ satisfies $D = A \otimes E_{LB}$.

Proof: For any $t \geq 0$ and $s \leq t$, we have:

$$D(t) = D(t) - D(s) + D(s) \leq E_{LB}(t - s) + A(s), \ \forall s : 0_- \leq s \leq t_+ \qquad (11.40)$$

This implies $D \leq A \otimes E_{LB}$.

Conversely, let us consider again a time $t \geq 0$. If the flow buffer is empty at time t, then it is

$$D(t) = A(t) \geq \inf_{0_- \leq s \leq t_+} \{A(s) + E_{LB}(t - s)\} \tag{11.41}$$

where the inequality holds since $E_{LB}(t - t_+) = E_{LB}(0_-) = 0$ and $A(t_+) = A(t)$. Assume the flow buffer is not empty at time t, i.e., it is $Q_s(t) > 0$. There exists a time u such that $0_- \leq u \leq t_+$ and the token bucket is full at time u. Formally, $u = \sup_{\tau \leq t}\{Q_b(\tau_-) = \sigma\}$. Since the bucket starts out full, i.e., $Q_b(0_-) = \sigma$, it must be $u \geq 0$. The flow buffer must be empty at time u_- as a consequence of the bucket being full at time u. On the other hand, $Q_s(t) > 0$ implies that the token bucket must be empty at time t (otherwise there would not be data waiting at the flow buffer). In general, we can write the state of the token bucket at a given time as the sum of the state at a previous time, plus the new arrived tokens, minus the tokens that have been used, i.e.,

$$Q_b(v) = Q_b(s) + \rho(v - s) - [D(v) - D(s)] \tag{11.42}$$

We write this identity for $v = t$ and $s = u_-$, when it holds that $Q_b(t) = 0$ and $Q_b(u_-) = \sigma$. Then, it is

$$\begin{aligned}0 = Q_b(t) &= Q_b(u_-) + \rho(t - u) - [D(t) - D(u_-)] \\ &= \sigma + \rho(t - u) + A(u_-) - D(t)\end{aligned} \tag{11.43}$$

since the flow buffer is empty at time u_-, and hence $D(u_-) = A(u_-)$. Rearranging the terms of eq. (11.43), we find

$$\begin{aligned}D(t) = \sigma + \rho(t - u) + A(u_-) &\geq \inf_{0_- \leq s \leq t_+} \{A(s) + \sigma + \rho(t - s)\} \\ &= \inf_{s \in \mathbb{R}} \{A(s) + E_{LB}(t - s)\}\end{aligned} \tag{11.44}$$

hence $D \geq A \otimes E_{LB}$. Together with the other inequality we have already established, we find finally $D = A \otimes E_{LB}$. ∎

Theorem 11.8 If the input $A(t)$ of the leaky bucket is constrained by the envelope $E_{LB}(t)$, then it is $D(t) = A(t)$, i.e., the input flow goes through the leaky bucket unchanged.

Proof: For any system it is $A \geq D$. For the leaky bucket, we can write $A \geq D = A \otimes E_{LB} \geq A$, where the last inequality is due to the property of envelopes. Therefore, it is $A = D$. ∎

The leaky bucket shaper is a key element of the deterministic traffic control architecture. Let us derive an equivalent fluid queueing model of the shaper fed by a given packet arrival flow. Let L_k be the length of the k-th packet and t_k its arrival time for $k \geq 1$. We assume that $t_k > 0$ for all $k \geq 1$. Let $T_k = t_k - t_{k-1}$, $k \geq 1$ be the k-th inter-arrival time, with $t_0 = 0$ for ease of notation.

We denote the buffer content level of the token bucket and of the flow buffer at time t as $Q_b(t)$ and $Q_s(t)$, respectively. Let $Q_{b,k} = Q_b(t_{k-})$ and $Q_{s,k} = Q_s(t_{k-})$ be the buffer content levels just before the k-th packet arrival.

If $Q_{s,k} > 0$, it must necessarily be $Q_{b,k} = 0$. Then, we have:

$$Q_{s,k+1} = \begin{cases} \max\{0, Q_{s,k} + L_k - \rho T_{k+1}\} & \text{if } Q_{s,k} > 0 \Rightarrow Q_{b,k} = 0, \\ \max\{0, L_k - \rho T_{k+1} - Q_{b,k}\} & \text{if } Q_{s,k} = 0. \end{cases} \quad (11.45)$$

The two expressions can be unified in the following equation, holding for whatever value of $Q_{s,k}$:

$$Q_{s,k+1} = \max\{0, Q_{s,k} + L_k - \rho T_{k+1} - Q_{b,k}\} \quad (11.46)$$

As for the token bucket buffer, we have

$$Q_{b,k+1} = \begin{cases} \min\{\sigma, \max\{0, Q_{b,k} + \rho T_{k+1} - L_k\}\} & \text{if } Q_{b,k} > 0 \Rightarrow Q_{s,k} = 0, \\ \min\{\sigma, \max\{0, \rho T_{k+1} - L_k - Q_{s,k}\}\} & \text{if } Q_{b,k} = 0. \end{cases}$$

where we account for the upper limit σ of the token bucket level. Here too we can write a single equation holding for any value of $Q_{b,k}$ is:

$$Q_{b,k+1} = \min\{\sigma, \max\{0, Q_{b,k} + \rho T_{k+1} - L_k - Q_{s,k}\}\} \quad (11.47)$$

We can prove that the flow buffer content $Q_s(t)$ can be studied by means of an equivalent single server queue.

Theorem 11.9 The process $Q(t) = Q_s(t) + \sigma - Q_b(t)$ is the queue length of a single server system having the same input process as the leaky bucket flow buffer and a constant rate server at rate ρ.

Proof: Let $B_k \equiv Q_{s,k} + L_k - \rho T_{k+1} - Q_{b,k}$. We have established that $Q_{s,k+1} = \max\{0, B_k\}$ and $Q_{b,k+1} = \min\{\sigma, \max\{0, -B_k\}\}$. Let

$$Q_k \equiv Q_{s,k} + \sigma - Q_{b,k} \quad (11.48)$$

Using the recursions that we have established, we find

$$\begin{aligned} Q_{k+1} &= Q_{s,k+1} + \sigma - Q_{b,k+1} = \max\{0, B_k\} + \sigma - \min\{\sigma, \max\{0, -B_k\}\} \\ &= \max\{0, \sigma + B_k\} = \max\{0, \sigma + Q_{s,k} + L_k - \rho T_{k+1} - Q_{b,k}\} \\ &= \max\{0, Q_k + L_k - \rho T_{k+1}\} \end{aligned} \quad (11.49)$$

The recursion obtained for Q_k is nothing but a form of Lindley's recursion for a $G/G/1$ queue, with a constant rate server of capacity ρ. In other words, we can study the leaky bucket flow buffer by looking at an equivalent queue fed by the same arrivals as the flow buffer and served by a constant rate server of capacity ρ. ∎

Since $0 \leq Q_{b,k} \leq \sigma$, we have $Q_{s,k} \leq Q_k \leq Q_{s,k} + \sigma$. Therefore, the event $Q_{s,k} > x$ implies $Q_k > x$ and we can use the probability $\mathcal{P}(Q_k > x)$ to upper bound the probability $\mathcal{P}(Q_{s,k} > x)$.

The delay introduced by a leaky bucket shaping an arrival stream of data can be calculated by observing that a packet of length L bit, arriving at time t, will be released by the leaky bucket after a time d such that

$$Q_b(t) + \rho \cdot d \geq Q_s(t) + L \quad \Rightarrow \quad d(t) = \max\left\{ 0, \frac{Q_s(t) + L - Q_b(t)}{\rho} \right\}$$

$$(11.50)$$

In terms of the equivalent queueing length $Q(t)$ introduced in Theorem 11.9, we have

$$d(t) = \max\left\{ 0, \frac{Q(t) + L - \sigma}{\rho} \right\} \tag{11.51}$$

Imposing $d(t) \leq d_{max}$ reduces to $Q(t) \leq \rho d_{max} + \sigma - L$.

Example 11.8 Let us apply the leaky bucket theory to the *Star Wars* sequence. The main characteristics of this measured traffic trace are described in Example 11.2. We assume that the *Star Wars* data flow is packetized using a maximum payload size of 1000 bytes and a header length of 40 bytes.

Since the sequence has finite length (we are given a specific realization of traffic over a finite time horizon), it makes sense to evaluate the maximum delay, i.e., $\max_{0 \leq t \leq T} d(t)$, where T is the duration of the sequence and $d(t)$ is calculated as in eq. (11.50).

Figure 11.5 plots the minimum required leaky bucket rate of the packetized *Star Wars* movie coded sequence as a function of the delay requirement, for three values of the token bucket size σ.

For each given delay requirement d we find the minimum leaky bucket rate ρ required to meet the requirement. Since the sequence is finite, when the delay requirement falls below a threshold (that depends on σ), the required rate is so high that the achieved delay drops to 0. Therefore, the required rate ρ is kept constant to the maximum value found for nonzero delay.

As expected, the rate is a decreasing function of the delay requirement. The rate appears to be weakly dependent on the delay requirement. The larger the bucket size, the less rate we need. It can be noted also that, the larger the bucket size, the less sensitive ρ is to d_{max}.

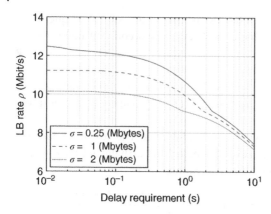

Figure 11.5 *Star Wars* sequence: minimum leaky bucket rate as a function of the delay requirement for three values of the token bucket size σ.

Finally, to appreciate the range of values allowed for the delay requirement, we should keep in mind that the delay through the shaper contributes to the end-to-end delay of the streaming flow. Setting a large value of the delay requirement implies that we need a large delay equalization buffer at the receiver, as highlighted in Chapter 1.

11.2.5 Network Calculus

Network calculus is a theory developed to study network performance under deterministic worst-case bounds. The aim is to characterize performance bounds, given constraints on the network input flows and on the guaranteed service offered by network elements (routers, switches, processing middle-boxes).

The essence of the network calculus approach is to replace the description of the arrival flow as a stochastic process with a deterministic upper bound (in the sense of the envelope). We will see soon that the stochastic nature of the arrival flow does not disappear. The network calculus approach confines the effect of the random nature of traffic to network edges, where the input arrival flow is shaped and imposed a given envelope. The constraints on arrival flows and guarantees on the minimum amount of service devoted to each flow in the network elements, i.e., envelopes and service curves, allow a simple network dimensioning to meet local and global delay constraints.

We will see first how to apply the theory developed so far to define arrival and service curves, for the purpose of giving bounds on the delay through a single service system and to dimension its capacity as a function of the delay requirement. Then, we will extend the analysis to establish a bound on the end-to-end delay through a network .

11.2.5.1 Single Node Analysis

We start with the analysis and dimensioning of a single network element, characterized by a service curve $S(t)$. Let $A(t)$ be the arrival flow at the input of the system and assume that the arrival flow is constrained by an envelope $E(t)$. Let $S(t) = C\max\{0, t - d\}$ be the service curve of the service system (rate-latency server, with capacity $C > 0$ and delay $d \geq 0$).

The delay $d(t)$ through the system is defined in eq. (11.37). Let $d_{max} = \sup_{t \geq 0} d(t)$ be the maximum delay through the service system. Theorem 11.5 gives an upper bound on d_{max} in eq. (11.38). As a consequence of the delay bound, we have

$$E(t - d_{max}) \leq C\max\{0, t - d\} \quad \Rightarrow \quad E(t) \leq C(t + d_{max} - d) , \ \forall t \geq 0.$$
$$(11.52)$$

for the considered rate-latency server.

We can derive a more explicit expression of d_{max} as

$$d_{max} = d + \sup_{t \geq 0} \left\{ \frac{E(t) - Ct}{C} \right\} \tag{11.53}$$

For d_{max} to be finite, it must be $E(t)/t \leq C$ definitely as t grows.

If we impose that $d_{max} \leq d_0$ for a given requirement on the delay d_0, we can find the minimum capacity C that meets that requirement. From $d_{max} \leq d_0$ and eq. (11.53), we find

$$\frac{E(t) - Ct}{C} \leq d_0 - d \quad \Rightarrow \quad C \geq \frac{E(t)}{t + d_0 - d} , \ \forall t \geq 0. \tag{11.54}$$

Therefore, we get

$$C_{min} = \sup_{t \geq 0} \left\{ \frac{E(t)}{t + d_0 - d} \right\} \tag{11.55}$$

Let us apply the capacity dimensioning result to the important case where the envelope is obtained with a leaky bucket with parameters (σ, ρ) applied to an arrival flow with peak rate R and maximum packet size L. Figure 11.6 illustrates the block scheme of the network calculus model. The leaky bucket shaper is the first block on the left of the chain. The offered traffic flow $A(t)$ enters the shaper. At the output of the shaper, the traffic flow is bounded by the envelope $E_{LB}(t)$.

The service system is modeled with a rate-latency service curve, i.e., it guarantees a minimum rate C and introduces a delay d (second and third blocks in Figure 11.6). We assume also $\rho \leq C \leq R$. The inequality $C \geq \rho$ is required to guarantee that the service system is stable under the regulated input flow. We require the inequality $C \leq R$, otherwise the problem becomes trivial (there is no congestion in the service system and no need of a shaper, if the service system guarantees to the source flow a bandwidth larger than the flow peak rate).

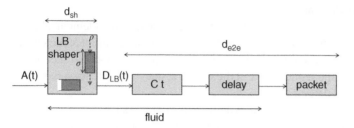

Figure 11.6 Block scheme of the network calculus model of a leaky bucket-shaped traffic flow offered to a rate-latency link.

The first three blocks in Figure 11.6 are based on a fluid model of the traffic flow. The last block introduces the packetization delay, to account for the fact that the link models a router output port and routers use the store-and-forward principle, i.e., a packet can be forwarded only after it has been completely received.

The arrival flow generated by the source has an envelope $A(t) = Rt + L$ for $t \geq 0$. This follows from the fact that the source emits at a peak rate R and that it can release a packet of length up to L. We shape the source traffic through a leaky bucket with parameters $\sigma \geq L$ and $\rho \leq R$. Passing through the leaky bucket we obtain a flow bounded by $D_{LB} = A \otimes E_{LB} = \min\{A, E_{LB}\}$, since both A and E are sub-additive functions belonging to $\mathcal{F}$. The input of the service system is therefore described by the arrival curve $D_{LB}(t) = \min\{Rt + L, \rho t + \sigma\}$, $t \geq 0$.

Using the result in eq. (11.53) to the service system with service curve $S(t) = C\max\{0.t - d\}$, with an input flow constrained by the curve D_{LB}, we get

$$d_{\max} = d + \frac{R - C}{R - \rho}\frac{\sigma - L}{C} + \frac{L}{C} \tag{11.56}$$

This function ranges from $d + L/R$ to $d + \sigma/\rho$ as the capacity C is decreased from R to ρ. Note that it must be $\rho \leq R$ (otherwise the shaper flow buffer is not stable). Moreover, we impose that $\sigma \geq L$. Then, it is $L/R \leq \sigma/\rho$. We can impose a delay requirement d_0 only if $d_0 \in [d + L/R, d + \sigma/\rho]$. Requiring $d_{\max} \leq d_0$ and inverting eq. (11.56) we get

$$C_{\min} = \frac{R\sigma - \rho L}{\sigma - L + (R - \rho)(d_0 - d)} \tag{11.57}$$

This is a monotonously increasing function of σ and monotonously decreasing function of ρ. Setting the service system capacity C to its minimum, i.e., ρ, we find that it must be $\rho(d_0 - d) = \sigma$. Since it is $\sigma \geq L$, we have $\rho \geq L/(d_0 - d)$.

Example 11.9 Assume that the rate-latency server represents a network crossed by a VoIP traffic flow, shaped at the ingress of the network by a leaky bucket. The delay d represents the base end-to-end delay (often referred to as propagation delay). For a Voice over IP (VoIP) flow we have $L = 80$ bytes. If VoIP packets are sent every 10 *ms*, the peak bit rate is $R = 64$ kbit/s. The minimum delay (apart from the network delay d) is $L/R = 10$ ms.

If we assign $\rho = 16$ kbit/s and $\sigma = rL$, the maximum delay (apart from the network delay d) is $\sigma/\rho = r \cdot 40$ ms. Assuming $d = 100$ ms, the delay bound ranges between 110 ms and $100 + r \cdot 40$ ms. Let $r = 2.5$, hence $\sigma = 200$ bytes, so that the maximum delay becomes 200 ms.

We set a limit of $d_0 = 150$ ms to the end-to-end delay, including the base delay d and the network delay. Applying eq. (11.57), we find that it must be $C \geq 27.43$ kbit/s.

As expected, this is an intermediate value between the minimum rate $\rho = 16$ kbit/s and the peak rate $R = 64$ kbit/s. We are guaranteeing deterministically that the end-to-end delay will never exceed d_0; nonetheless we provide a multiplexing gain with respect to peak rate assignment. We shall remember however that we have not accounted for the shaping delay in our calculations.

There remains to assess the delay suffered by input traffic in the shaper flow buffer. This is no more characterized by deterministic bounds, since the input of the leaky bucket is ultimately the traffic process of the source, which is described as a stochastic process. We will address this point in the next section.

Before moving to the application of network calculus to a full network, we establish a useful property of a stable server loaded by traffic sources controlled by leaky bucket shapers.

Let C be the constant service rate of the link, N the number of traffic sources loading the link, and let (σ_i, ρ_i) be the leaky bucket parameters for the flow of traffic source i, $i = 1, \ldots, N$.

Theorem 11.10 Given a constant rate server of capacity C loaded by N leaky bucket controlled sources, assume the server is stable, i.e., it is $\sum_{i=1}^{N} \rho_i < C$. Then, the busy period of the server is upper bounded by $\sum_{i=1}^{N} \sigma / \left(C - \sum_{i=1}^{N} \rho_i \right)$.

Proof: Let u and v be the start and the end times of a busy period. We have $Q(u) = Q(v) = 0$. If $D(t)$ denotes the output flow up to time t, we must have $A(v) - A(u) = D(v) - D(u) = C \cdot (v - u)$, the last equality being a consequence of the fact that we assume a work-conserving server. The duration of the busy period is $y = v - u$. Thank to the envelope bound on the input, we have

$$C \cdot (v - u) = A(v) - A(u) \leq \rho(v - u) + \sigma \quad \Rightarrow \quad y = v - u \leq \frac{\sigma}{C - \rho} \quad (11.58)$$

where $\rho = \sum_{i=1}^{N} \rho_i$ and $\sigma = \sum_{i=1}^{N} \sigma_i$. ∎

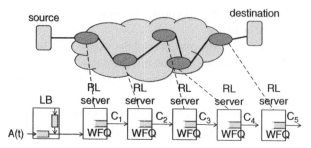

Figure 11.7 Example of end-to-end path with the corresponding model based on deterministic network calculus.

11.2.5.2 End-to-End Analysis

Let us now apply the network calculus approach to the traffic management in a packet network. The key point is to set deterministic bounds on packet arrivals at network edge nodes and service offered by the nodes of the network. The scheme in Figure 11.7 illustrates an example of an end-to-end path between a source and a destination through five routers.

The corresponding model based on deterministic traffic theory is shown in the lower part of the figure. The source generates an arrival flow $A(t)$ that goes through a leaky bucket shaper. The ouput of the shaper is offered to the path through the network routers. Each router is modeled by a rate-latency server, thanks to the weighted fair queueing (WFQ) scheduling algorithm used in the output line of the router. In the following we explain the scheme and the relevant models in detail.

Packets offered to the network are generated by traffic sources. A traffic source can be a user equipment, where applications run, or it can be the peer link with another network, from which external traffic enters the tagged network. In both cases, the traffic entering the tagged network is subject to rules as a consequence of agreements between the network provider and the external source. The approach taken here to constrain the input source traffic into a deterministic bound envelope is to assume that it passes through a leaky bucket shaper before entering the tagged network. Then, the description of the traffic can be reduced to its original peak rate R, the maximum length of the flow packets L, and to the two parameters of the leaky bucket filter, the burstiness σ and the token rate ρ.

The nodes of the network are routers. They provide switching to packets. Packets arriving at a router from any input port of the router are processed and sent to the output port designated by the router forwarding table. The capacity available at an output link of a router is shared by all flows that are concurrently ongoing on that link. To set deterministic bounds on the minimum amount of service provided by the output port of the router to each flow we must guarantee a minimum rate to each flow, *independent* of the backlog of any other flow on the same link. This requires a multiplexing algorithm that guarantees non-null minimum rate to each flow and isolation among flows. We have encountered a scheduling algorithm that is suitable for this purpose in Section 6.4.3, namely WFQ. This scheduling algorithm yields to feasible implementation and realizes a good approximation of the ideal traffic handling of processor sharing in a packet network.

Let us consider a tagged flow. The flow path through the network is made up of N links and crosses $N - 1$ routers (the two end-points are the traffic source and the destination).

Let us consider link h along the path. Let C_h denote the capacity of link h. The capacity of the link is shared by a multiplicity of flows, including the tagged one. Let S_h denote the set of active flows through link h.

We assume that capacity sharing is handled with WFQ, with weight $\phi_i^{(h)}$ for flow i on link h. The minimum rate guaranteed to flow i is $r_i^{(h)} = C_h \frac{\phi_i^{(h)}}{\sum_{j \in S_h} \phi_j^{(h)}}$. Since packets are dealt with as a whole and cannot be transmitted partially, there can be a delay in getting the capacity of the link, if a packet is being transmitted. Let $L_{\max,h}$ denote the maximum length of packets sent through link h (the maximum transfer unit (MTU) of the link). The delay is bounded by $L_{\max,h}/C_h$.

Finally, we account for the fact that the next router can process a packet only when it receives it completely (store-and-forward paradigm). There is therefore an additional delay equal to the time it takes to transmit (and receive) a packet of the flow, i.e., $L_i/r_i^{(h)}$. We conclude that the tagged link behaves as a rate-latency service system for the tagged flow, with service curve given by $r_i^{(h)}\max\{0, t - (L_{\max,h}/C_h + L_i/r_i^{(h)})\}$.

For the sake of a simpler notation, we drop the index i of the flow. We refer to a specific flow and evaluate a bound for its delay end-to-end through the network.

The tagged flow is characterized by a peak rate R and maximum packet length L. It is shaped by a (σ, ρ)-regulator at the ingress to the network. The flow is routed through N links, each link corresponding to the output port of an intermediate router. Link h is modeled as a rate-latency service system with service curve $r^{(h)}\max\{0, t - (L_{\max,h}/C_h + L/r^{(h)})\}$. There is no point in assigning different values of the rates $r^{(h)}$ of the tagged flow on links of its path. The end-to-end throughput achieved by the flow is constrained by the bottleneck link, i.e., it equals $r = \min_{1 \leq h \leq N} r^{(h)}$. In the following we assume that $r^{(h)} = r, \forall h$.

The power and elegance of the network calculus stems from the observation that a chain of N rate-latency service systems is equivalent to a single rate-latency service system having a rate equal to the minimum of the rates and a delay equal to the sum of the delays. Then, the entire network, i.e., the path of N links that the flow travels through the network, as chosen by the routing function, can be represented by a *single* rate-latency server with service curve $r\max\{0, t - d_{\text{link}}\}$, where $d_{\text{link}} = NL/r + \sum_{h=1}^{N} L_{\max,h}/C_h$.

We are now in a position to apply the result found in eq. (11.56) with C replaced by r. To account also for propagation delay on links, we add a component τ_h for link h and let $\tau_{\text{link}} = \sum_{h=1}^{N} \tau_h$. We obtain the following bound on the end-to-end delay of the tagged flow through the network:

$$d_{\text{e2e}} = \underbrace{\sum_{h=1}^{N} \left(\tau_h + \frac{L_{\max,h}}{C_h} \right) + \frac{NL}{r}}_{\text{Link base delay}} + \underbrace{\frac{R-r}{R-\rho} \frac{\sigma-L}{r} + \frac{L}{r}}_{\text{Buffering delay}} \tag{11.59}$$

Setting a requirement on the end-to-end delay, we can find the minimum capacity to be granted on each link to the tagged flow so that the delay requirement is met.

The application of this approach requires the knowledge of:

- the routing of the tagged flow;
- the capacity and the MTU of each crossed link in the network;
- the base delay (sum of propagation delays) of the network path;
- the leaky bucket parameters.

Two approaches can be considered. In a centralized approach, the traffic source contacts a central network manager, that knows the network topology, the load level of the network and can find a suitable routing where the required capacity is available. If there is at least one such routing, the central manager sets the new state of each involved router and grants the requested resources to the traffic source.

In a distributed approach, a control protocol must be defined. The traffic source requests the set up of a new flow, giving its traffic description (e.g., (R, σ, ρ) in this case) to its local router (an edge router of the network). Routers talk among them via the control protocol to find a path through the network toward the designated destination, such that the delay requirement is met. For a given path, it is possible to collect all the information needed to compute the required capacity r for a given requirement d_0 on the end-to-end delay. The collection of this data can be realized by launching a control message that hops from one router to the next one. The data relevant to each crossed link is appended to the message payload, so that the final node receives all the information and can check whether the delay requirement is met or not.

Note that the minimum required capacity $r_{\min}$ on the path links can be calculated as in eq. (11.57), given the traffic flow description (R, σ, ρ) and the path characteristics. In general the set up of a path that maintains the required delay prescription is not necessarily feasible, i.e., there could be less than the required minimum capacity $r_{\min}$ available on the explored path, or there could even not be any feasible path between the source and the destination at the time the flow set up request is issued.

Figure 11.8 Model of the fluid leaky bucket followed by a packetizer.

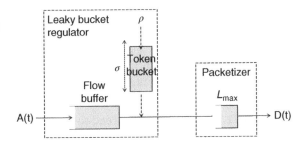

The network function that takes care of checking whether a new flow can be set up is referred to as *admission control*.

There remains one open point. Inspecting Figure 11.7, we see that the flow suffers a delay in each crossed router *and* in the flow buffer of the initial leaky bucket. Up to now we have not considered this delay, since it escapes the deterministic framework that we have pursued in this section. As a matter of fact, the input to the flow buffer of the leaky bucket is a random process, the traffic flow generated by the traffic source.

Let us consider a leaky bucket shaper fed by a packet flow. Figure 11.8 shows a *fluid* model of the leaky bucket. To complete the model, we add a packetizer in tandem with the fluid leaky bucket.

Let $L_{\max}$ be the maximum length of the packets out of the packetizer. Based on the analysis in Section 11.2.4, and specifically on the result of eq. (11.51), the delay through the flow buffer of the leaky bucket satisfies $d \leq d_0$ provided that $Q \leq \rho d_0 + \sigma - L_{\max}$, where Q is the queue length of a fluid queue with constant serving rate equal to ρ and input process equal to the data flow offered to the leaky bucket input. Therefore, the dimensioning of the leaky bucket under a flow buffer delay requirement d_0 reduces to imposing that:

$$P(Q > \rho d_0 + \sigma - L_{\max}) \leq \epsilon \tag{11.60}$$

A metric that seems more appropriate than the delay bound is the time-average fraction of the input data that arrives at the flow buffer and finds a backlog larger than the threshold $Q_{th} = \rho d_0 + \sigma - L_{\max}$:

$$\lim_{T \to \infty} \frac{\frac{1}{T} \int_0^T I(Q(t) > Q_{th}) a(u) \, du}{\frac{1}{T} \int_0^T a(u) \, du} \leq \lim_{T \to \infty} \frac{\frac{1}{T} \int_0^T I(Q(t) > Q_{th}) R \, du}{\frac{1}{T} \int_0^T a(u) \, du}$$

$$= \frac{R}{\lambda} P(Q > Q_{th}) \tag{11.61}$$

where $I(E)$ is the indicator function of the event E, $a(t)$ is the arrival rate (derivative of the arrival function $A(t)$) and λ is the time average rate of the arrival flow. This relationship establishes a connection between the tail of the buffer content probability distribution and the metric defined based on time averages. This explains why we focus on evaluating the tail $P(Q_s > x)$ (or the upper bound $P(Q > x)$, as discussed with reference to eq. (11.60)). This leads however to a stochastic approach, that is addressed in the second part of this chapter.

Example 11.10 Let us apply the end-to-end resource dimensioning to the *Star Wars* traffic sequence. The main characteristics of the *Star Wars* traffic sequence are described in Example 11.2. We fix an end-to-end delay requirement and apply the network calculus approach to dimension network resources to meet the requirement.

We evaluate the maximum delay through the leaky bucket flow buffer for the entire sequence, for given (σ, ρ) parameters, using eq. (11.51). Let it be $d_{sh,max}$, where

$$d_{sh,max} = \max_{k \geq 0} \left\{ 0, \frac{Q_k - (\sigma - L)}{\rho} \right\} \tag{11.62}$$

where $Q_{k+1} = \max\{0, Q_k + L_k - \rho T_{k+1}\}$ for $k \geq 0$, with $Q_0 = 0$. L_k and T_{k+1} are the length of the k-th packet of the arrival flow and the inter-arrival time between packet k and $k + 1$. These two sequences describe the traffic flow offered to the shaper.

In this example, we fix $\rho = 9$ Mbit/s and we let σ vary from 0.1 Mbytes up to 10 Mbytes. As for the network parameters, we assign four different values to the guaranteed bandwidth r. Note that it must be $r \geq \rho$, to guarantee a finite delay bound. Moreover, it must obviously be $\overline{R} \leq r \leq R_{max}$, where $\overline{R}$ is the average bit rate of the *Star Wars* traffic flow and R_{max} is the peak rate. The other parameters appearing in eq. (11.59) are: number of network links $N = 10$; propagation delay $\tau_h = 500$ µs (corresponding to an optical fiber span of 100 km), link capacity $C_h = 10$ Gbit/s, and maximum packet length $L_{max,h} = 1500$ bytes.

Figure 11.9(a) plots the end-to-end delay bound for the *Star Wars* sequence as a function of σ for four values of r. The end-to-end delay bound is computed as $D = d_{sh,max} + d_{e2e}$, where d_{e2e} is given by eq. (11.59).

There is an optimal value of σ, that minimizes the delay bound. While the component d_{e2e} is monotonously increasing with σ, the bound on the shaper delay is, on the contrary, monotonously decreasing with σ.

As an example of stochastic dimensioning of the leaky bucket parameter, let us consider and ON-OFF traffic source, with $R_{max} = 100$ Mbit/s, $\overline{R} = 5$ Mbit/s. During the ON time the source emits packets of length $L = 1500$ bytes at the peak bit rate. The packet burst has a geometric size with mean 200 packets. The idle time has a negative exponential probability distribution. The mean idle time is given by $E[T_{OFF}] = T_0 b(1 - a)/a$, where $T_0 = L/R_{max}$ is the packet transmission time,

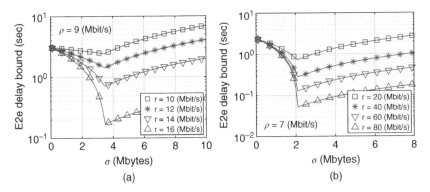

Figure 11.9 Examples of application of the end-to-end delay bound.

$a = \overline{R}/R_{max}$ is the activity coefficient of the source and $b = 200$ packets is the mean burst size.

The delay through the shaper is now a random variable. Through simulation, we can estimate the 90% quantile of this random variable, $D_{sh,90}$. Then, the stochastic delay bound can be computed as $D_{90} = D_{sh,90} + d_{e2e}$. We guarantee that at least 90% of the packets will suffer a delay not larger than D_{90}.

Figure 11.9(b) plots D_{90} for the ON-OFF traffic source as a function of σ, for $\rho = 7$ Mbit/s and four values of r. Here too an optimum value of σ can be found. As in the case of *Star Wars* sequence, the curve is cuspid-shaped around the minimum, so that even small deviations from the optimal value of σ quickly degrade the delay bound.

A comment is in order on the description of a traffic flow, as the input of the shaper. In application cases, it is possible that the traffic flow is already fully generated and ready to be transferred at connection set-up time. For example, audio/video streaming service consists of downloading in real-time a predefined coded sequence of data stored in a server. The coded sequence has been prepared in advance and is entirely stored in a server. In that case, it is conceivable that we prepare a *deterministic* arrival curve associated with a traffic flow. The arrival curve is used at connection set-up time, to evaluate the end-to-end delay bound, specifically, the component due to the shaping delay.

Alternatively, the traffic flow could be generated in real time, during the evolution of the connection (this is, for example, the case of a telephone call or a live show streaming). In this second case, we can only provide a stochastic description of the traffic flow, i.e., we can identify a class of stochastic processes that specific instances of the considered traffic flows belong to. This is the case where the delay bound in the flow buffer can only be probabilistic.

We conclude this section with a traffic engineering consideration on the dimensioning of bandwidth assignment in the network nodes.

The traffic control architecture we have so far described rests on two key assumptions:

1. Each traffic source is constrained by a leaky bucket shaper at the entrance point into the network.
2. Each network link is shared according to a GPS policy (implemented by means of a WFQ algorithm) among the flows that are routed through that link[2]

We say that the flow i is locally stable at link h if $\rho_i \leq r_i^{(h)}$, where $r_i^{(h)} = C_h \phi_i^{(h)} / \sum_{j \in S_h} \phi_j^{(h)}$ and S_h is the set of flows going through link h. In words, flow i is assigned a minimum guaranteed bandwidth that is no less than the bound on its long-term average rate.

Let us assume that flow i is constrained by an arrival curve $E_i^{(h)}(t) = \rho_i t + \sigma_i^{(h)}$ at the input of link h. Since the ideal GPS is equivalent to a rate-latency server with rate $r_i^{(h)}$ and zero delay, the service curve of GPS is given by $S_i^{(h)} = r_i^{(h)} \max\{0, t\}$ for flow i. Then, flow i at the output of link h is bounded by a curve $D_i^{(h)}$ such that $D_i^{(h)} \leq E_i^{(h)} \otimes S_i^{(h)} = E_i^{(h)}$, if $\rho_i \leq r_i^{(h)}$. This conditions means that flow i is locally stable at link h. We can summarize this result saying that, for a locally stable flow, the input constraint of the leaky bucket envelope carries over to the output of the link. Therefore, we see that a flow that is locally stable at each link of its route, can be bounded uniformly with its initial envelope, the one imposed by the leaky bucket shaper at the input of the network.

The numerical examples discussed above show however that, to achieve small delay bounds, we might need to assign a bandwidth which is significantly larger than the leaky bucket average rate ρ.

We can do that for all flows routed on the link, if we have enough bandwidth on that link. Otherwise, we can assign larger bandwidth to delay-sensitive flows, especially if they have limited burstiness. We will then guarantee that these delay-sensitive flows are locally stable. Flows that are not delay-sensitive (elastic flows) can receive less bandwidth, even such that they are not locally stable at some node, provided that at each node we guarantee that the *global stability* is met, i.e., $\sum_{j \in S_h} \rho_j < C_h$ and that all traffic source are shaped according to a leaky bucket envelope at their respective entrance points into the network.

This is particularly interesting from a network traffic engineering point of view, since it gives us the flexibility of assigning a large bandwidth provision to delay-sensitive flows, so that, once they are backlogged, their data can be cleared

---

2 In the analysis of the end-to-end delay bound we have assumed that routers implement WFQ on each link. WFQ is the practical implementation in a packet network of the theoretical fluid GPS policy. There is a constant (small) offset between the guarantees provided by GPS and those realized by WFQ (see Section 6.4.3).

out of the buffer with small delay. The price to pay is an increase of the delay of elastic flows, but still guaranteeing that their delay is bounded *deterministically* (except of the delay incurred by each flow in its leaky bucket flow buffer, which is characterized in general by a random variable). The deterministic guarantee on the delays comes from the leaky bucket envelope that we impose to all flows. By playing with the weights $\phi_i^{(h)}$ at nodes, we can shift more bandwidth to those flows that need tight delay bounding, even to the point that some other flow becomes locally unstable. Note that local instability does not mean that there is not enough bandwidth at the link for all flows going through that link. Since global stability is always met, there is in fact enough bandwidth for all. The large bandwidth guaranteed to a delay-sensitive flow is not used continuously, but only when the flow is backlogged. In the long run, only a fraction of that guaranteed bandwidth will be actully used, i.e., the fraction $\rho_i / r_i^{(h)}$ at link h. The unused part will be "recycled" by flows that are locally unstable at the link, and hence need using more than their reserved minimum bandwidth. The balance on the link is kept by the guarantee that the overall sum of leaky buckets average rates of all flows through the link is less than the capacity of the link.

We make this point more precise in the following theorem. Before stating the theorem, we establish two lemmas required for the main result.

Lemma 11.4 Consider a flow constrained by a leaky bucket shaper passing through a constant rate server. If the backlog of the server is bounded, the flow at the output of the server is still constrained by a leaky bucket envelope.

Proof: Let us consider a flow constrained by the envelope $\rho t + \sigma$, going through a stable server, i.e., a server where the sum of the average rates of the input flow is less than the capacity of the server. Since the server is stable, we can place a bound Q^* on the backlog of the buffer of this link. Then, we can write for the output of the tagged flow:

$$D(t) - D(s) \le A(t) - A(s) + A(s) - D(s)$$
$$\le \rho(t-s) + \sigma + Q(s) \le \rho(t-s) + \sigma + Q^* \tag{11.63}$$

That is to say the output of the flow is constrained by a leaky bucket envelope, with burstiness parameter $\sigma' = \sigma + Q^*$. ∎

Note that we have derived this result, without assuming anything on the service offered to the tagged flow (except that the link must be globally stable, otherwise no bound on the backlog could be set). Therefore, if we can set a deterministic bound on the backlog of the buffer link, we show that a leaky bucket constrained input flow remains leaky bucket constrained also at the output, *irrespective of whether the flow be locally stable.*

Lemma 11.5   The backlog of a constant rate stable server of capacity C loaded by leaky bucket constrained flows is bounded.

Proof: Let us consider a link with capacity C, loaded by N flows, the i-th of which is constrained by an envelope $\rho_i t + \sigma_i$. The overall arrival flow is therefore constrained by $A(t) - A(\tau) \leq \sum_{i=1}^{N} \rho_i \cdot (t - \tau) + \sum_{i=1}^{N} \sigma_i, \ \forall \tau \leq t$. The buffer content at time t is expressed by:

$$Q(t) = \sup_{0 \leq \tau \leq t} \{A(t) - A(\tau) - C(t - \tau)\}$$

$$\leq \sum_{i=1}^{N} \sigma_i + \sup_{0 \leq \tau \leq t} \left\{ (t - \tau) \cdot \left(\sum_{i=1}^{N} \rho_i - C \right) \right\} = \sum_{i=1}^{N} \sigma_i \equiv Q^*$$

since we assume that the link is stable, i.e.,

$$\sum_{i=1}^{N} \rho_i < C \tag{11.64}$$

We see that the backlog of the buffer of a stable link, loaded by sources constrained by leaky-bucket envelopes, is bounded by the sum of the burstiness parameters of the envelopes.   ∎

Let us introduce some notation.

We consider a network composed of ℓ links loaded by n traffic sources. Let $\mathcal{F}$ denote the set of flows of the network. Assume the flow of each source is routed through the network along a single path (e.g., shortest path routing). The routing is defined by the coefficients a_{hi}, where $a_{hi} = 1$, if link h belongs to the route of flow i, and $a_{hi} = 0$ otherwise.

We are now ready to state the main result, which first appeared in [174].

Theorem 11.11   Let us consider a network loaded by n traffic sources constrained by leaky bucket envelopes at network edge. Flow i is constrained by the arrival curve $E_i(t) = \rho_i t + \sigma_i$ at the input to the network. The capacity of the links is shared among the flows loading them by means of GPS, with weight $\phi_i^{(h)}$ for flow i on link h. We assume that all network links are stable, i.e., $C_h > \sum_{i=1}^{n} a_{hi} \rho_i$. We assume further that there exists a unique flow ordering such that if $i < j$ then

$$\frac{\phi_i^{(h)}}{\rho_i} \geq \frac{\phi_j^{(h)}}{\rho_j}, \quad \forall h : a_{hi} a_{hj} = 1 \tag{11.65}$$

Under these assumptions, the backlog and delay at each link (and hence through the network) are bounded for all flows, no matter whether they are locally stable or not.

Proof: We say that flow j affects flow i on link h if $\frac{\phi_i^{(h)}}{\rho_i} < \frac{\phi_j^{(h)}}{\rho_j}$. Note that this inequality can be rewritten as $\frac{r_i^{(h)}}{\rho_i} < \frac{r_j^{(h)}}{\rho_j}$, i.e., the ratio of the guaranteed bandwidth to the leaky bucket average rate of flow j is bigger than the same quantity of flow i.

Let $\mathcal{H}_1$ be the set of all flows such that

$$i \in \mathcal{H}_1 \quad \Rightarrow \quad \frac{\phi_i^{(h)}}{\rho_i} \geq \frac{\phi_j^{(h)}}{\rho_j} , \; \forall j \in \mathcal{F}, \; \forall h : a_{hi} = 1. \tag{11.66}$$

In general, let us define

$$i \in \mathcal{H}_k \quad \Rightarrow \quad \frac{\phi_i^{(h)}}{\rho_i} \geq \frac{\phi_j^{(h)}}{\rho_j} , \; \forall j \in \mathcal{F} \backslash \bigcup_{r=1}^{k-1} \mathcal{H}_r, \; \forall h : a_{hi} = 1, \tag{11.67}$$

for $k = 2, \dots, K$ with $K \leq n$. We see that $\mathcal{H}_1$ is the set of all flows that are not affected by any other flow in any link of their routes. The set $\mathcal{H}_k$ is the set of all flows that are affected only by flows belonging to $\mathcal{H}_1, \dots, \mathcal{H}_{k-1}$ at some link of their routes, but are not affected by any flow belonging to $\mathcal{H}_r$ with $r > k$.

We first show that flows belonging to $\mathcal{H}_1$ are locally stable at all links of their routes. Let us consider a flow $i \in \mathcal{H}_1$ and let h be the first link on its route. Since the link is stable, it is $\sum_{i=1}^{n} a_{hi} \rho_i < C_h$. There must exist at least one flow, say s, such that

$$\rho_s \leq \frac{\phi_s^{(h)}}{\sum_{i=1}^{n} a_{hi} \phi_i^{(h)}} C_h = r_s^{(h)} \tag{11.68}$$

Since no flow can affect i, it must be

$$\frac{r_i^{(h)}}{\rho_i} \geq \frac{r_s^{(h)}}{\rho_s} \geq 1 \tag{11.69}$$

which implies that flow i is locally stable at link h. Since flow i is constrained by the leaky bucket envelope at the input of link h, the local stability at h implies that flow i is still constrained by a suitable leaky bucket envelope at the output of link h, i.e., at the input of the next link on the route of flow i. We can repeat the reasoning we have just done, to conclude that flow i is locally stable at each link on its route. This conclusion holds for all flows belonging to class $\mathcal{H}_1$.

It is easy to verify that the buffer content $Q_h(t)$ of link h is upper bounded by the buffer content of the reduced capacity link, i.e., a link loaded with only sources that do not belong to $\mathcal{H}_1$ and with capacity $C_h - \sum_{i \in \mathcal{H}_1} a_{hi}\rho_i$. It is

$$Q_h(t) = \sup_{0 \le \tau \le t} \left\{ \sum_{i=1}^{n} a_{hi}[A_i(t) - A_i(\tau)] - C_h(t - \tau) \right\}$$

$$= \sup_{0 \le \tau \le t} \left\{ \sum_{i \notin \mathcal{H}_1} a_{hi}[A_i(t) - A_i(\tau)] - C_h(t - \tau) + \sum_{i \in \mathcal{H}_1} a_{hi}[A_i(t) - A_i(\tau)] \right\}$$

$$\le \sup_{0 \le \tau \le t} \left\{ \sum_{i \notin \mathcal{H}_1} a_{hi}[A_i(t) - A_i(\tau)] - C_h(t - \tau) + \sum_{i \in \mathcal{H}_1} a_{hi}[\rho_i(t - \tau) + \sigma_i^{(h)}] \right\}$$

$$= \sup_{0 \le \tau \le t} \left\{ \sum_{i \notin \mathcal{H}_1} a_{hi}[A_i(t) - A_i(\tau)] - \left(C_h - \sum_{i \in \mathcal{H}_1} a_{hi}\rho_i \right)(t - \tau) \right\}$$

$$+ \sum_{i \in \mathcal{H}_1} a_{hi}\sigma_i^{(h)}$$

$$= Q_h^{(1)}(t) + \sum_{i \in \mathcal{H}_1} a_{hi}\sigma_i^{(h)}$$

We see that the buffer content $Q_h(t)$ is upper bounded by $Q_h^{(1)}(t)$, apart from a constant offset. $Q_h^{(1)}(t)$ is the buffer content of link h once we remove all flows belonging to $\mathcal{H}_1$ and reduce the capacity by the average leaky bucket rates of those flows.

Let us remove then all flows belonging to the sets $\mathcal{H}_r$, $r = 1, \ldots, k-1$, and let us consider a flow $i \in \mathcal{H}_k$, for $k \ge 2$. Since there is no flow in the reduced network that can affect flow i, we can retrace all steps of the reasoning above for a flow belonging to $\mathcal{H}_1$ and conclude that:

$$\rho_i \le \frac{\phi_i^{(h)}}{\sum_{j \in \mathcal{H}_k \cup \cdots \cup \mathcal{H}_K} a_{hj}\phi_j^{(h)}} \left(C_h - \sum_{j \in \mathcal{H}_1 \cup \cdots \cup \mathcal{H}_{k-1}} a_{hj}\rho_j \right) \tag{11.70}$$

This inequality states that flow $i \in \mathcal{H}_k$ is locally stable at the reduced link h. We can then show that it is locally stable on each link along its route. We conclude that all flows belonging to $\mathcal{H}_k$ are locally stable, hence have a bounded delay. In turn, the bounded delay guarantees that a flow at the output of the link is still bounded by a leaky bucket envelope.

Note, however, that the local stability is with respect to only flows belonging to $\mathcal{H}_k \cup \cdots \cup \mathcal{H}_K$ and with respect to the reduced link capacity.

Repeating the bounding of link h buffer content for each class, we can prove finally that

$$Q_h(t) \le \sum_{i \in \mathcal{H}_1} a_{hi}\sigma_i^{(h)} + \cdots + \sum_{i \in \mathcal{H}_K} a_{hi}\sigma_i^{(h)} \tag{11.71}$$

that is, link h buffer level is bounded. ∎

The dimensioning criterion of GPS weights established with this theorem is defined in [174] and called there consistent relative session treatment (in this chapter we use the word "flow" instead of "session"). Its meaning is to treat flows consistently at each link where they contend for the link capacity, i.e., the ratio of the guaranteed rate to the leaky bucket mean rate of a flow is always not less (or not greater) than the same ratio for another flow on all links common to the two flows. Mixed situations are not allowed.

11.3 Stochastic Service Guarantees

The force of the deterministic traffic theory resides in its capability of providing worst-case bounds on delay performance to delay-sensitive traffic flows. The approach allows end-to-end network analysis and dimensioning, given that we introduce leaky-bucket shapers at the edge of the network and use WFQ policy on each network link to share the link capacity among the flows.

The price to pay is that bounds might be loose in general, hence the achieved dimensioning can be quite conservative. This motivates the introduction of probabilistic traffic control, where we exploit the statistical variability of traffic to achieve statistical multiplexing gain. We try, however, to maintain the simplicity of the circuit-switching approach, where we are able to associate a bandwidth to each flow that is offered to the network. Flow admission control in that case amounts to checking that there is at least that bandwidth available along the chosen network path. To reproduce this approach also in the realm of stochastic traffic control, we will define the notion of *effective bandwidth*.

Following [137], we consider a discrete time model. The embedded time points of the discrete system are $t_k = k\Delta$, $k \in \mathbb{Z}$, for a given constant time slot duration $\Delta > 0$. We study multiplexing of traffic sources in a link of capacity C. The multiplexer is equipped with a buffer of size K.

Let n be the number of traffic sources. The number of bits produced by the source in the time interval $(t_{k-1}, t_k]$ is denoted with $B_{j,k}$ for source j. These bits are assumed to arrive at the multiplexer input at time t_k.

All quantities are measured in consistent units, e.g., buffer size and arrivals in bits, rates in bits/s, time in seconds.

We consider first the special case $K = 0$ (marginal buffer). It is representative of multiplexers with buffer size much smaller than the average burst size of the sources.

11.3.1 Multiplexing with Marginal Buffer Size

With $K = 0$ there is no wait. The only relevant performance metric is therefore bit loss due to overload of the multiplexer output link capacity. Let us consider a finite time horizon, up to time slot t. The average fraction of time slots where loss occurs is given by

$$Z_L(T) = \frac{1}{T} \sum_{k=1}^{T} I\left(\sum_{j=1}^{n} B_{j,k} > C \right) \tag{11.72}$$

where $I(E)$ is the indicator function of the event E. If the arrival process is stationary, we can take the limit for $T \to \infty$ of $Z_L(T)$ and define the long-term average fraction Z_L of time slots where loss occurs.

The loss probability can be found as

$$P_L(T) = \frac{\sum_{k=1}^{T} \max\left\{ 0, \sum_{j=1}^{n} B_{j,k} - C \right\}}{\sum_{k=1}^{T} \sum_{j=1}^{n} B_{j,k}} \tag{11.73}$$

In the limit for $T \to \infty$ we obtain the steady-state loss probability P_L. Note that the loss probability is defined as in fluid models, without taking packets into account (a packet is either transmitted or lost as a whole).

We can establish a relationship between Z_L and P_L, if the amount of bits offered by source j in a time slot is limited by a peak value R_j, i.e., $B_{jk} \le R_j$, $\forall k$, for $j = 1, \ldots, n$. To simplify notation, let $S_k = \sum_{j=1}^{n} B_{j,k}$. Then

$$P_L = \lim_{T \to \infty} \frac{\frac{1}{T} \sum_{k=1}^{T} \max\{0, S_k - C\}}{\frac{1}{T} \sum_{k=1}^{T} S_k} = \lim_{T \to \infty} \frac{\frac{1}{T} \sum_{k=1}^{T} I(S_k > C)\max\{0, S_k - C\}}{\frac{1}{T} \sum_{k=1}^{T} S_k}$$

$$\le \frac{\max\{0, R - C\}}{E[S]} \lim_{T \to \infty} \frac{1}{T} \sum_{k=1}^{T} I(S_k > C) = \frac{\max\{0, R - C\}}{E[S]} Z_L$$

where $R = \sum_{j=1}^{n} R_j$. Note that $R \le C$ implies that there is no loss. The multiplexer problem is therefore interesting only if $R > C$.

The inequality between P_L and Z_L shows that we can focus on the metric Z_L. Setting a requirement of the type $Z_L \le \epsilon$ implies in turn that $P_L \le \epsilon'$, for a suitable ϵ'.

If the steady-state limit exists, we can write:

$$Z_L = E\left[I\left(\sum_{j=1}^{n} B_j > C \right) \right] = P\left(\sum_{j=1}^{n} B_j > C \right) \tag{11.74}$$

The key point of the analysis and dimensioning of the multiplexer with marginal buffering consists then of studying the tail of the probability distribution of the random variable $S = \sum_{j=1}^{n} B_j$, the sum of n random variables. In the following we assume that the B_j's are i.i.d. random variables.

We use the Chernov bound[3] to give the following upper bound of the tail of the probability distribution of a random variable X:

$$P(X > x) \leq e^{-\theta x} \phi_X(\theta), \quad \forall \theta \geq 0, \tag{11.75}$$

where

$$\phi_X(\theta) = \mathrm{E}[e^{\theta X}] = \int_{-\infty}^{\infty} e^{\theta u} f_X(u) \, du \tag{11.76}$$

is the moment-generating function (MGF) of the random variable X. The MGF is defined for any $\theta \leq 0$, since $f_X(u)$ is a PDF and hence it is integrable. It may exist also for $0 < \theta < \theta_0$ for some positive (possibly infinite) θ_0. If there is a positive θ_0 such that $e^{\theta x} f_X(x)$ is integrable for all $\theta \in (0, \theta_0)$, then the PDF $f_X(x)$ must decay to 0 exponentially fast as $x \to \infty$ and all moments of X are finite. If the tail of $f_X(x)$ tends to 0 slower than an exponential (e.g., it follows a power-law, as is the case of the Pareto PDF), then there is no positive θ_0 and hence the Chernov bound is not applicable.

We introduce the capacity per source c, namely we write $C = nc$. Let $f_B(x)$ be the PDF of the amount of bits B offered by a traffic source in a time slot, and let $\phi_B(\theta)$ be the corresponding MGF. We assume that $f_B(x)$ is light-tailed, i.e., it decays exponentially fast as $x \to \infty$. The PDF of the sum $S = \sum_{j=1}^{n} B_j$ of bits emitted by all sources, $f_S(x)$, is the n-fold convolution of $f_B(x)$. The corresponding MGF is $\phi_S(\theta) = \phi_B^n(\theta)$. Therefore, we have

$$Z_L = P\left(\sum_{j=1}^{n} B_j > nc \right) \leq \inf_{\theta \geq 0} \{e^{-\theta nc} \phi_B(\theta)^n\} = e^{-n \sup_{\theta \geq 0} \{c\theta - \log \phi_B(\theta)\}} \tag{11.77}$$

This bound says that Z_L decays at least exponentially fast with n, since the supremum in the exponent is positive, as we will see.

The bound on Z_L, our performance metric, depends on the function:

$$J(y) = \sup_{\theta \geq 0} \{y\theta - \log \phi_B(\theta)\} = \sup_{\theta \geq 0} \{y\theta - \log \mathrm{E}[e^{\theta B}]\} \tag{11.78}$$

This is called the rate function.

Some properties of $J(y)$ can be readily proved. For ease of notation, let us define $h(y, \theta) = y\theta - \log \phi_B(\theta)$:

1. $J(y) \geq 0$. Since $h(y, 0) = 0$ for any y, it follows that the supremum defining $J(y)$ is non-negative.
2. $J(y)$ is increasing. Given $y_1 \leq y_2$, we have $h(y_1, \theta) \leq h(y_2, \theta) \leq J(y_2)$, holding for any $\theta \geq 0$. Then, it is also $J(y_1) \leq J(y_2)$.

---

3 See the Appendix at the end of the book.

3. $J(y)$ is convex. The proof is similar to the previous point, this time given $y_1 < y_2$ and any $\alpha \in [0, 1]$. We have $h(\alpha y_1 + (1 - \alpha)y_2, \theta) = \alpha h(y_1, \theta) + (1 - \alpha)h(y_2, \theta) \leq \alpha J(y_1) + (1 - \alpha)J(y_2)$.

4. If $y > E[B]$, $J(y)$ does not change if we extend the supremum to $\theta \in \mathbb{R}$. Since the exponential function is convex, Jensen's inequality yields $E[e^{\theta B}] \geq e^{\theta E[B]}$. Then, $h(y, \theta) = y\theta - \log \phi_B(\theta) \leq \theta(y - E[B])$. If $y > E[B]$, the supremum of $h(y, \theta)$ for $\theta \leq 0$ is attained at $\theta = 0$.

5. If $y > E[B]$, the supremum is achieved at a point $\theta^* > 0$ such that $\phi'_B(\theta^*) = y \phi_B(\theta^*)$. The derivative with respect to θ of $h(y, \theta)$ is $h'_\theta(y, \theta) = y - \phi'_B(\theta)/\phi_B(\theta)$. It is $h'_\theta(y, 0) = y - E[B] > 0$. Using Schwartz's inequality, it can be readily verified that $h'_\theta(y, \theta)$ is a monotonously decreasing function of θ^4. From the definition of MGF, it can be shown also that, for any given $y > 0$, it is definitely $\phi'_B(\theta)/\phi_B(\theta) > y$ for large θ. Hence, $h'_\theta(y, \theta) < 0$. Then, there is a unique positive θ^* where $h'_\theta(y, \theta^*) = 0$. This is where the supremum is attained.

Example 11.11 Let us consider n ON-OFF traffic sources. An ON-OFF traffic source offers data to the multiplexer at peak rate R or it is silent. The probability that the source emits data is p (i.e., p is the average fraction of time the source is active). For this source the random variable B has only two values, R and 0, with probabilities p and $1 - p$, respectively. We have therefore:

$$\phi_B(\theta) = 1 - p + pe^{\theta R} \tag{11.79}$$

The rate function can be found after some calculations:

$$J(y) = \frac{y}{R} \log \left(\frac{y/R}{p} \right) + \left(1 - \frac{y}{R} - \right) \log \left(\frac{1 - y/R}{1 - p} \right) \tag{11.80}$$

for $0 < y < R$. The probability of the event $B_1 + \cdots + B_n > ny$ is 1 for $y \leq 0$, while it is 0 for $y > R$, since sources have a peak rate of R. The bounding provided by $J(y)$ is meaningful only for $0 < y < R$, which is in fact the range of values of y where the supremum is finite. The function $J(y)$ is monotonously increasing for $y \in [pR, R]$, rising from 0 to $\log(1/p)$.

Since $E[B] = pR$, we consider values of c belonging to (pR, R). This corresponds to assigning each source with a capacity level intermediate between its mean and peak rates. For example, let $p = 0.1$ (highly intermittent source) and assign $c = 0.5 \cdot R$. It is $J(0.5 \cdot R) \approx 0.511$. For $n = 10$ sources, the probability of overflow is bounded by $Z_L \leq e^{-10 \cdot 0.511} = 6 \cdot 10^{-3}$. A bound for the loss probability is obtained by multiplying the bound of the overflow probability Z_L by $(nR - nc)/(nE[B]) =$

---

4 It suffices to show that the derivative of $\phi'_B(\theta)/\phi_B(\theta)$ is positive. The derivative is $(\phi''_B(\theta)\phi_B(\theta) - [\phi'_B(\theta)]^2)/[\phi_B(\theta)]^2$. This is positive if $[\phi'_B(\theta)]^2 < \phi''_B(\theta)\phi_B(\theta)$. Letting $f(x) = x\sqrt{e^{\theta x}f_B(x)}$ and $g(x) = \sqrt{e^{\theta x}f_B(x)}$, we apply $\left(\int f(x)g(x)dx \right)^2 \leq \int f^2(x)dx \int g^2(x)dx$ and we are done.

$(1 - c/R)/p = 5$. Then, we have $P_L \leq 5 \cdot Z_L \leq 0.03$. The bounds for $n = 20$ become $Z_L \leq 3.66 \cdot 10^{-5}$ and $P_L \leq 1.83 \cdot 10^{-4}$, respectively.

This is yet another manifestation of the fact that large-scale systems offer good performance. In this case, reserving an overall capacity equal to nR guarantees no loss deterministically. We can save half of that capacity, if we are willing to accept a loss probability that is no more than 0.000183, for $n = 20$ sources. Larger number of sources allow even larger efficiency gains for the same requirement on the loss probability.

We can rewrite the key result of this section, eq. (11.77), as follows:

$$\frac{1}{n} \log P \left(\sum_{j=1}^{n} B_j > nc \right) \leq -J(c) \tag{11.81}$$

The question is how tight the bound is. To explore this point, we refer to Cramer's theorem.

Theorem 11.12 (Cramer's theorem)   Let c be a positive constant and $B_1, \ldots, B_n$ be independent and identically distributed random variables with $E[B] < c$ and $P(B > c) > 0$. Assume that the MGF $\phi_B(\theta) = E[e^{\theta B}]$ is finite for $\theta < \theta_0$, with $\theta_0 > 0$. Then

$$\lim_{n \to} \frac{1}{n} \log P \left(\sum_{j=1}^{n} B_j > nc \right) = \inf_{\theta \geq 0} \{ \log \phi_B(\theta) - c\theta \} = -J(c) \tag{11.82}$$

Proof: We have already seen that $-J(c)$ is an upper bound of the left-hand side in eq. (11.82). We now prove that it is a lower bound, asymptotically as $n \to \infty$.

Let θ^* be the point where the supremum of $J(c)$ is attained, so that $\phi'_B(\theta^*)/\phi_B(\theta^*) = c$. We define the exponentially tilted random variable V whose PDF is given by:

$$f_V(x) = \frac{e^{\theta^* x} f_B(x)}{\phi_B(\theta^*)} \tag{11.83}$$

This is a non-negative function, and it sums to 1 when integrated over x, because of the definition of the MGF. We also have $E[V] = c$, since $E[V] = E[Be^{\theta^* B}]/\phi_B(\theta^*) = \phi'_B(\theta^*)/\phi_B(\theta^*) = c$.

By the central limit theorem, given any $\epsilon > 0$, we have:

$$P \left(0 < \frac{1}{n} \sum_{j=1}^{n} (V_j - c) < \epsilon \right) \to \frac{1}{2} \quad \text{as } n \to \infty \tag{11.84}$$

We are now ready to show that $-J(c)$ is a lower bound for the left-hand side limit in eq. (11.82). We have

$$P(B_1 + \cdots + B_n > nc) > P(nc < B_1 + \cdots + B_n < n(c+\epsilon))$$

$$= \int_{D_\epsilon} f_{B_1}(x_1) \ldots f_{B_n}(x_n) \, dx_1 \ldots dx_n$$

$$= \phi_B(\theta^*)^n \int_{D_\epsilon} e^{-\theta^*(x_1 + \cdots + x_n)} f_{V_1}(x_1) \ldots f_{V_n}(x_n) \, dx_1 \ldots dx_n$$

$$> e^{n \log \phi_B(\theta^*)} e^{-\theta^* n(c+\epsilon)} \int_{D_\epsilon} f_{V_1}(x_1) \ldots f_{V_n}(x_n) \, dx_1 \ldots dx_n$$

$$= e^{-n(c\theta^* - \log \phi_B(\theta^*))} e^{-\theta^* n\epsilon}.$$

$$P\left(0 < \sum_{j=1}^n V_j - nc < n\epsilon \right)$$

$$= e^{-n \, J(c)} e^{-n\theta^* \epsilon} \left(\frac{1}{2} + o(1) \right)$$

where $D_\epsilon = \{(x_1, \ldots, x_n) \in \mathbb{R}^n : nc < x_1 + \cdots + x_n < n(c+\epsilon)\}$.

Taking the logarithm of both sides, dividing by n and letting $n \to \infty$ and $\epsilon \to 0$ we complete the proof. ∎

Let us apply this result to the dimensioning of the multiplexer for a prescribed requirement ϵ on the loss probability. The problem statement can be as follows: given the output capacity of the multiplexer, C, and the requirement $P_L \le \epsilon$, find the value of c that maximizes the utilization of the capacity C.

For n homogeneous traffic sources, we have $P_L \le \frac{nR-C}{nE[B]} Z_L$, where R and $E[B]$ are the peak and mean number of bits per slot emitted by a source. The requirement on Z_L shall be $Z_L \le \epsilon' = \frac{nE[B]}{nR-C} \epsilon$. Using the theory we have developed in this section for marginal buffering, a sufficient condition for this requirement to hold is

$$Z_L \le e^{-nJ(C/n)} \le \epsilon' = \frac{nE[B]}{nR-C} \epsilon \tag{11.85}$$

Introducing the link capacity per source $c = C/n$ and the source activity factor $p = E[B]/R$, we obtain:

$$g(c) \equiv \frac{C}{c} J(c) + \log \left(\frac{pR}{R-c} \right) \ge -\log \epsilon \tag{11.86}$$

The range of c is $pR < c < R$. The lower limit comes from the requirement that c be more than the mean bandwidth of the source, as discussed in the development of the rate function theory. The upper limit is obvious, since there is no point in setting a requirement on the loss probability once we assign a bandwidth per source equal or greater than its peak rate.

The function $g(c)$ defined in eq. (11.86) is monotonously increasing. The optimal value of c is the smallest $c > E[B]$ such that the inequality is satisfied. Let it be

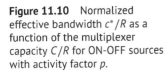

Figure 11.10 Normalized effective bandwidth c^*/R as a function of the multiplexer capacity C/R for ON-OFF sources with activity factor p.

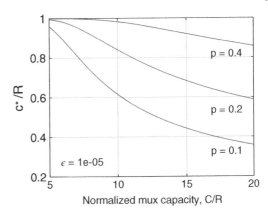

denoted with c^*. This is the portion of multiplexer link bandwidth that must be reserved for one traffic source, under the loss probability constraint. We can admit new traffic flows, as long as their number n is such that $nc^* \le C$. Since a bandwidth c^* is set aside for each on-going flow, as long as $n = n^* = \lfloor C/c^* \rfloor$ we must reject new flow set-up requests, not to impair the quality of service guarantee for the on-going flows.

We see that c^* is the *effective bandwidth* of the traffic source. We can reduce the admission rule to a very simple check: verify that the number of admitted flows does never exceed the threshold n^*.

Example 11.12 Let us consider again the ON-OFF source that alternates between a peak rate R and 0. We recall that p is the probability that the source is emitting, hence pR is the mean rate.

Figure 11.10 plots c^*/R as a function of C/R for three values of p. The requirement on the loss probability P_L is set to $\epsilon = 10^{-5}$.

As the multiplexer capacity increases, a bigger statistical gain is reaped. This is evident from the decreasing behavior of the effective bandwidth. The statistical gain, i.e., the number of sources that can be multiplexed with respect to peak rate assignment, is bigger for lower p. The more the source is intermittent, the more it makes sense to exploit statistics.

We can generalize this result to a heterogeneous environment, where there are r classes of sources. Let $B_j^{(i)}$ be the amount of bits emitted in a time slot by source j of class i. Let n_i be the number of sources of class i. For a compact notation, let $S^{(i)}(n_i) = \sum_{j=1}^{n_i} B_j^{(i)}$ for $i = 1, \ldots, r$ and $S = S^{(1)}(n_1) + \cdots + S^{(r)}(n_r)$. Using the Chernov bound, since the random variables $S^{(i)}(n_i)$ are independent of one another, we have:

$$Z_L = P(S > C) \le e^{-\theta C} \phi_S(\theta) = e^{-\theta C} \prod_{i=1}^{r} \phi_{S^{(i)}}(\theta) \tag{11.87}$$

Hence

$$\log Z_L \le \sum_{i=1}^{r} \log \phi_{S^{(i)}}(\theta) - \theta C = \sum_{i=1}^{r} n_i \log \phi_{B^{(i)}}(\theta) - \theta C \tag{11.88}$$

where $B^{(i)}$ is the random variable defined as the amount of bits emitted by a source of class i in a time slot.

We see that $Z_L \le e^{-\gamma}$ if

$$C\theta - \sum_{i=1}^{r} n_i \log \phi_{B^{(i)}}(\theta) \ge \gamma \tag{11.89}$$

For a given value of θ, let us define the admission region $\mathcal{A}(\theta)$ as follows:

$$\mathcal{A}(\theta) = \left\{ (n_1, \ldots, n_r) \in (\mathbb{Z}^+)^r : \theta C - \sum_{i=1}^{r} n_i \log \phi_{B^{(i)}}(\theta) \ge \gamma \right\} \tag{11.90}$$

We have shown that

$$(n_1, \ldots, n_r) \in \mathcal{A} \quad \Rightarrow \quad Z_L \le e^{-\gamma} \tag{11.91}$$

Invoking Cramer's theorem, we can also show that:

$$\lim_{N \to \infty} \frac{1}{N} \log P\left(\sum_{i=1}^{r} \sum_{j=1}^{n_i N} B_j^{(i)} > NC \right) = \sup_{\theta \ge 0} \left\{ C\theta - \sum_{i=1}^{r} n_i \log \phi_{B_i}(\theta) \right\} \tag{11.92}$$

which points out that the bound we are using is asymptotically tight as the number of multiplexed sources grows.

The whole admission region can be obtained as the union of all the regions $\mathcal{A}(\theta)$ as θ varies over $\mathbb{R}^+$.

In practice, to set up an admission control algorithm, we should fix a value of θ and define

$$e_i(\theta) = \frac{\log \phi_{B^{(i)}}(\theta)}{\theta} \tag{11.93}$$

to be the *effective bandwidth* of a source of type i. The admission rule would then resemble perfectly the one that we use in a circuit-switched network, namely, we maintain that

$$\sum_{i=1}^{r} n_i e_i(\theta) < C - \frac{\gamma}{\theta} = C + \frac{\log \epsilon}{\theta} \tag{11.94}$$

We will resume the generalization of the effective bandwidth with multiple source classes in the next section, devoted to the analysis of multiplexing with non-negligible buffer size.

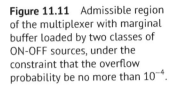

Figure 11.11 Admissible region of the multiplexer with marginal buffer loaded by two classes of ON-OFF sources, under the constraint that the overflow probability be no more than 10^{-4}.

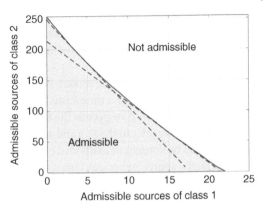

Example 11.13 Let us consider a multiplexer loaded by two classes of ON-OFF sources. The capacity of the multiplexer output link is $C = 1$. The parameters of the ON-OFF sources are $p_1 = 0.1, R_1 = C/10$ for the first class and $p_2 = 0.2 R_2 = C/80$ for the second class.

Figure 11.11 shows the admissible region, i.e., the region of the plane that covers couples (n_1, n_2) that satisfy $n_1 e_1(\theta) + n_2 e_2(\theta) < C - \gamma/\theta$, for some θ. n_1 and n_2 denote the number of sources of class 1 and class 2, respectively. The performance target is $\epsilon = e^{-\gamma} = 10^{-4}$. The solid curve in Figure 11.11 represents the boundary of the admissible region. It can be found as the envelope curve of the set of lines $n_1 e_1(\theta) + n_2 e_2(\theta) = C - \gamma/\theta$ as θ is varied. The infimum of allowed θ values is such that $C - \gamma/\theta > 0$, that is $\theta_{\inf} = \gamma/C \approx 9.22$. Two of those lines are shown in the figure.

It can be seen that the admissible region is delimited by a concave boundary curve. The line obtained for a fixed value of θ is tangent to the boundary curve, therefore it lies entirely below the boundary curve. Managing admission control for a fixed value of θ causes some inefficiency, since part of the admissible region remains out of the full region delimited by the line.

11.3.2 Multiplexing with Non-Negligible Buffer Size

We have seen that statistical gain is possible even if no buffer is provided at the multiplexer. This should come at no surprise, thinking for example to the analysis of ON-OFF sources multiplexing discussed at the end of Chapter 8.

In this section we consider multiplexing of traffic sources modeled with stochastic processes, where the multiplexer has a non-negligible buffer size. The term of comparison to assess the size of the buffer is the burst size of the source, e.g., the ON time for an ON-OFF traffic source. We set a requirement on the probability of buffer overflow and establish a rule for admission control of traffic sources under that requirement. We will see that we can reduce admission control in a packet network to a simple check, akin to what is done in circuit-switched networks, or with peak rate assignment. The role of the peak rate is played by a suitably defined effective bandwidth.

We stick to the discrete-time model. Let $B_k^{(j)}$ denote the amount of bits emitted by traffic source j during slot time k (assumed to arrive at the end of the interval). Time is normalized with respect to the duration of the fixed time slot. Therefore, in the following we refer to the B_k's and any other data related quantities as "rates", since they can be thought of as divided by the fixed unit time. In the following, we drop the traffic source index j, as long as we deal with a single source.

We assume that the traffic generation process is modulated by an ergodic, discrete-time, discrete-state process Z_k. The state space of Z_k is the set $S_Z = \{0, 1, \ldots, m\}$. Let $\pi_i = P(Z_\infty = i), i \in S_Z$ be the steady-state probability distribution of the process Z_k. It is $B_k = V(Z_k)$, where V is a random variable whose PDF depends on the state of Z_k. Let $f_i(x)$ denote the PDF of B_k and $\phi_i(\theta) = \mathrm{E}[e^{\theta V} | Z_k = i]$ be the MGF of B_k, conditional on $Z_k = i$, for $i \in S_Z$. To keep notation simple, we denote the random variable V conditional on being $Z_k = i$ as V_i. The mean rate of the traffic source is $\bar{r} = \sum_{i=0}^{m} \pi_i \mathrm{E}[V_i]$. If the V_i's all have finite support, say $[0, r_{\max}]$, then we can define also the peak rate as $r_{\max}$. A special case is obtained by assuming $V(Z_k) = r(Z_k)$, where $r : S_Z \mapsto \mathbb{R}^+$ is a deterministic rate function. In that case, we denote the state-dependent rates as $r_i, i \in S_Z$. The mean rate is $\bar{r} = \sum_{i=0}^{m} \pi_i r_i$ and the peak rate is $r_{\max} = \max_{i \in S_Z} \{r_i\}$.

As usual, C denotes the amount of bits that the multiplexer output can send out during one time slot. Q_k represents the amount of bits found in the multiplexer buffer at the end of the k-th time slot. Then

$$Q_k = \max\{0, Q_{k-1} + B_k - C\}, \quad k \geq 1, \tag{11.95}$$

with the initial condition $Q_0 = 0$.

Following the same line of reasoning used to derive Reich's formula in continuous time, we find

$$Q_n = \max_{0 \leq k \leq n} \{A_n - A_k - (n - k)C\} \tag{11.96}$$

where $A_n = \sum_{k=1}^{n} B_k$ is the cumulative arrival process. We let $A_0 = 0$.

We assume that the sequence B_k is stationary. Hence the cumulative arrival process A_n has stationary increments.

Reich's formula gives the evolution of the buffer starting from a given time. We have to modify it, to be able to study the steady state. We consider still a range on $n + 1$ time slots, but starting from time $-n$ instead of 0. Then

$$Q_0 = \max_{-n \leq k \leq 0} \{A_0 - A_k - C(-k)\} = \max_{0 \leq k \leq n} \{A_0 - A_{-k} - Ck\} \tag{11.97}$$

The sequence appearing in the rightmost-hand side of eq. (11.97) is nondecreasing, since the set $\{A_0 - A_{-k} - Ck\}_{0 \le k \le n}$ is included in the set $\{A_0 - A_{-k} - Ck\}_{0 \le k \le n+1}$ for a given sample path $\{A_k\}_{k \in \mathbb{Z}}$. Hence, the limit for $n \to \infty$ exists (path-wise). If it is finite, then we can define the limiting random variable Q as

$$Q = \sup_{k \ge 0}\{A_0 - A_{-k} - Ck\} \tag{11.98}$$

We say that the multiplexer is stable if the sequence of probability distributions of the random variables

$$Q(n) \equiv \max_{0 \le k \le n}\{A_0 - A_{-k} - Ck\} \tag{11.99}$$

converges to a finite probability distribution with probability 1. If[5] $\lim_{k \to \infty} \frac{A_{k+h} - A_h}{k} = \Lambda < C$, the sequence $A_0 - A_{-k} - Ck$ behaves asymptotically as $k(\Lambda - C)$ and hence it tends to $-\infty$. This ensures that the supremum in eq. (11.98) is finite.

The target performance requirement is the buffer overflow probability, i.e., $P(K) = P(Q > K)$. To find a useful expression of this metric, we consider the sequence of overflow probabilities defined as:

$$P_n(x) = P(Q(n) > x) = P(\max_{0 \le k \le n}\{A_0 - A_{-k} - Ck\} > x)$$

$$= P\left(\bigcup_{k=0}^{n}\{A_0 - A_{-k} - Ck > x\}\right)$$

$$\le \sum_{k=0}^{n} P(A_0 - A_{-k} - Ck > x) \tag{11.100}$$

where the last inequality derives from the union bound. Using the Chernov bound, we find:

$$P_n(x) \le \sum_{k=0}^{n} P(A_0 - A_{-k} - Ck > x) \le e^{-\theta x}\sum_{k=0}^{n} E[e^{\theta(A_0 - A_{-k} - Ck)}]$$

$$= e^{-\theta x}\sum_{k=0}^{n} e^{-k\theta C}E[e^{\theta(A_0 - A_{-k})}] \tag{11.101}$$

We would like to find an upper bound on P_n in steady state, i.e., taking the limit for $n \to \infty$ of the eq. (11.101). To that end, we introduce the log-moment generating function (LMGF) of the source and we will end up defining the effective bandwidth of the traffic source. This is the topic of next section.

---

5 Notice that the probability distribution of the increment $A_{k+h} - A_h = B_{h+1} + \cdots + B_{h+k}$ does not depend on h, because of the stationarity of the sequence B_k.

11.3.3 Effective Bandwidth

11.3.3.1 Definition of the Effective Bandwidth

We define the LMGF, assuming it exists finite in a neighborhood of the origin:

$$\Gamma(\theta) = \lim_{n \to \infty} \frac{1}{n} \log \mathrm{E}[e^{\theta(A_n - A_0)}] \tag{11.102}$$

Thanks to the stationarity of the increments of A_n, we get $\mathrm{E}[e^{\theta(A_0 - A_{-k})}] \sim e^{k\Gamma(\theta)}$. The k-th term of the sum appearing in the right-hand side of eq. (11.101) behaves asymptotically as $e^{-kC\theta} \mathrm{E}[e^{\theta(A_0 - A_{-k})}] \sim e^{-kC\theta} e^{k\Gamma(\theta)}$ for $k \to \infty$. If $\Gamma(\theta) - C\theta < 0$, the k-th term of the sum tends exponentially to 0 as $k \to \infty$. Therefore, the sum converges to a finite constant κ. Under the assumption $\Gamma(\theta) < C\theta$, it follows that

$$P(x) = \lim_{n \to \infty} P_n(x) \leq \kappa(\theta) e^{-\theta x} \tag{11.103}$$

that is to say, the buffer overflow probability is bounded by an exponentially decreasing function of the threshold x.

Let us state some properties of the LMGF $\Gamma(\theta)$, holding for a cumulative arrival process with stationary increments with a light-tailed PDF.

1. $\Gamma(\theta)$ is monotonously increasing and convex for $\theta \geq 0$.
2. For small θ we have

$$\log \mathrm{E}[e^{\theta(A_n - A_0)}] = \log \mathrm{E}[1 + \theta(A_n - A_0) + o(\theta)]$$
$$= \log(1 + \theta n \mathrm{E}[B] + o(\theta)) = n \mathrm{E}[B]\theta + o(\theta)$$

   Dividing by n and taking the limit for $n \to \infty$, we find $\Gamma(\theta) = \theta \mathrm{E}[B] + o(\theta)$ for $\theta \to 0$. Specifically, we have $\Gamma(0) = 0$ and $\Gamma'(0) = \mathrm{E}[B]$.
3. Using Jensen's inequality, it can be seen that $\Gamma(\theta) \geq \theta \mathrm{E}[B]$.
4. If $B_k \leq R, \forall k$ (i.e., R is the peak rate), then $\Gamma(\theta) \leq \theta R$.
5. If the arrival process is the superposition of ℓ independent processes, i.e., $A = A^{(1)} + \cdots + A^{(\ell)}$, it is $\Gamma(\theta) = \Gamma_1(\theta) + \cdots + \Gamma_\ell(\theta)$, where $\Gamma_i(\theta)$ is the LMGF of the process $A^{(i)}$.

The multiplexer analysis is interesting only if $\mathrm{E}[B] < C < R$. In that case, the properties of the LMGF imply that the curve of $\Gamma(\theta)$ is initially below the line θC for $\theta > 0$. As θ grows, $\Gamma(\theta)$ crosses the line $C\theta$, if $\Gamma(\theta)$ is strictly convex. Say the crossing occurs at η. Then, we have $\Gamma(\theta) < C\theta$ for $0 < \theta < \eta$.

The condition for the bound in eq. (11.103) to hold is that there exist values of θ such that $\Gamma(\theta) < C\theta$. In that case we can give an exponentially decaying bound of the buffer overflow probability. The decay rate can be maximized by choosing the supremum of the set of values of θ for which $\Gamma(\theta) < C\theta$. The bound holds for any $\theta < \eta$, where

$$\eta = \sup\{\theta \geq 0 : \Gamma(\theta) < C\theta\} \tag{11.104}$$

We have thus proved that $P(Q > K) \sim e^{-\theta K}$, with $\theta < \eta$. In fact, a stronger result holds. It can be shown that, at least for Markovian traffic sources, the tail of the buffer content decays exponentially with a decay rate η that is the unique positive solution of $\Gamma(\eta) = C\eta$. Asymptotically we have $P(Q > x) \sim e^{-\eta x}$ (see references in Section 11.4 and [86, Ch. 10]).

The last property listed above means that, for a superposition of ℓ independent traffic sources at the input of the multiplexer, the overflow probability can be bounded by an exponentially decaying function of the buffer size, provided that

$$\frac{\Gamma_1(\theta)}{\theta} + \cdots + \frac{\Gamma_\ell(\theta)}{\theta} < C \tag{11.105}$$

We find out that the ratio $\Gamma(\theta)/\theta$, for a suitable value of θ, behaves as the "bandwidth" of each multiplexed traffic source. We define the *effective bandwidth* of a stationary traffic source with associated LMGF $\Gamma(\theta)$ as

$$e(\theta) = \frac{\Gamma(\theta)}{\theta} \tag{11.106}$$

11.3.3.2 Properties of the Effective Bandwidth

Let us consider a traffic flow B_k with mean $\bar{r} = E[B]$ and peak rate R. Since $A_n \le nR$, it is easy to see that $\Gamma(\theta) \le R\theta$, hence $e(\theta) \le R$. Using Jensen's inequality we have $E[e^{\theta A_n}] \ge e^{\theta E[A_n]} = e^{n\theta E[B]}$, hence $\Gamma(\theta) \ge \bar{r}\theta$, and $e(\theta) \ge \bar{r}$.

The effective bandwidth is monotonously increasing with θ, since $\Gamma(\theta)$ is a convex function. It is $\lim_{\theta \to 0} e(\theta) = \bar{r}$. To prove this, we write for small θ:

$$e(\theta) = \lim_{n \to \infty} \frac{1}{n\theta} \log E[e^{A_n \theta}]$$

$$= \lim_{n \to \infty} \frac{1}{n\theta} \log(1 + E[A_n]\theta + o(\theta))$$

$$= \lim_{n \to \infty} \frac{1}{n\theta}(nE[B]\theta + o(\theta)) = \bar{r}$$

It is also $\lim_{\theta \to \infty} e(\theta) = R$. If $A_n = nR, \forall n$, the result is obvious. Otherwise, let n_1 be the smallest value of n such that $A_{n_1} < n_1 R$. Then, for $n > n_1$ and a sufficiently large θ, we have

$$\log E[e^{\theta A_n}] = \log(e^{nR\theta} E[e^{\theta(A_n - nR)}]) = nR\theta + \log E[e^{\theta(A_n - nR)}] \sim nR\theta \tag{11.107}$$

since $A_n < nR$ with positive probability. Dividing by n and θ and taking the limit for $n \to \infty$, it follows that $\lim_{\theta \to \infty} e(\theta) = R$.

If the traffic flow is described by a sequence of i.i.d. random variables $B_k \sim B$, we have $E[e^{\theta(A_n - A_0)}] = E[e^{\theta \sum_{k=1}^n B_k}] = (E[e^{\theta B}])^n = (\phi_B(\theta))^n$. Therefore, it is $\Gamma(\theta) = \log \phi_B(\theta)$. We see that the definition (11.106) of the effective bandwidth reduces to $e(\theta) = \log \phi_B(\theta)/\theta$ in case the traffic flow is described by a sequence of i.i.d. random variables.

Note that, in the special case where the traffic flow is made up of i.i.d. random variables $B_k \sim B$, we can connect the effective bandwidth theory to the upper bound of the waiting time CCDF given in Section 8.2.4. We proved there that $P(W > x) \leq e^{-s_0 x}$, where W is the random variable to which the sequence $W_{k+1} = \max\{0, W_k + X_k - T_{k+1}\}$ tends with probability 1 as $k \to \infty$. The coefficient s_0 is the unique root of the equation $\varphi_Z(-s) = 1$ different from $s = 0$. The random variable Z is defined by $Z = X - T$, where $X \sim X_k$ and $T \sim T_k$. It is assumed that service times X_k and inter-arrival times T_k are i.i.d. sequences (renewal processes).

The backlog of the buffer we consider in this section obeys the recursion $Q_{k+1} = \max\{0, Q_k + B_k - C\}$. We can identify the inter-arrival time variable with the constant C and the service time variable with B; hence $Z = B - C$ (rememeber that time is slotted, the slot time being the unit time). Passing to the MGF rather than the Laplace transform, we have $\phi_Z(\theta) = \phi_B(\theta)e^{-\theta C}$. The parameter s_0 becomes the unique positive root of the equation $\phi_Z(\theta) = 1$; that is to say, $\log \phi_B(\theta) = \theta C$. In terms of effective bandwidth, this equation becomes $e(\theta) = C$. We see then that we can identify s_0 with the decay rate η defined in eq. (11.104) for the special case of i.i.d. traffic flow sequence B_k. We have therefore $P(Q > x) \leq e^{-\eta x}$.

The QoS requirement set on the buffer overflow probability can be met, provided the sum of the effective bandwidths of the multiplexed traffic sources does not exceed the multiplexer capacity C.

The admission rule is especially simple, in this framework. It is enough to check that we do not exceed the multiplexer capacity C, if we add the effective bandwidth of the requesting new traffic source to the sum of the effective bandwidths of the traffic sources that are already set up through the multiplexer.

11.3.3.3 Effective Bandwidth of a Markov Source

We assume that the traffic source modulating process Z_k is an irreducible, aperiodic, finite-state Markov chain. Let p_{ij} be the one-step transition probability of making a transition from state i to state j. Let $\mathbf{P}$ denote the one-step transition matrix, collecting the probabilities p_{ij}. The steady-state probability of state i of the Markov chain is denoted with π_i and the corresponding row vector is π.

To find the LMGF we need to evaluate the expectation of $e^{\theta(A_n - A_0)}$, where $A_n - A_0 = \sum_{k=1}^{n} B_k$. Conditioning on the first step of the modulating Markov chain, we have for $n \geq 1$:

$$h_i(n) \equiv E[e^{\theta \sum_{k=1}^{n} B_k}|Z_0 = i] = \sum_{j=0}^{m} E[e^{\theta \sum_{k=1}^{n} B_k}|Z_0 = i, Z_1 = j]P(Z_1 = j|Z_0 = i)$$

$$= \sum_{j=0}^{m} p_{ij}E[e^{\theta V_i}]E[e^{\theta \sum_{k=2}^{n} B_k}|Z_1 = j] = \sum_{j=0}^{m} \phi_i(\theta)p_{ij}h_j(n-1), \quad n \geq 1,$$

with $h_i(0) = 1, i = 0, \ldots, m$. Using a compact matrix notation, we can write:

$$\mathbf{h}(n) = \mathbf{D}(\theta)\mathbf{P}\,\mathbf{h}(n-1), \quad n \geq 1, \tag{11.108}$$

where $\mathbf{h}(n)$ is a column vector with i-th entry equal to $h_i(n)$, and $\mathbf{D}(\theta)$ is a diagonal matrix with i-th diagonal entry equal to $\phi_i(\theta)$.

The difference equation (11.108) is solved with the initial condition $\mathbf{h}(0) = \mathbf{e}$, where $\mathbf{e}$ denotes a column vector of 1's. The solution is $\mathbf{h}(n) = (\mathbf{D}(\theta)\mathbf{P})^n\mathbf{e}$. We remove the conditioning on the event $Z_0 = i$ pre-multiplying $\mathbf{h}(n)$ by $\boldsymbol{\pi}$:

$$\mathrm{E}[e^{\theta(A_n - A_0)}] = \sum_{i=0}^{m} \pi_i h_i(n) = \boldsymbol{\pi}(\mathbf{D}(\theta)\mathbf{P})^n\mathbf{e} \tag{11.109}$$

The matrix $\mathbf{M}(\theta) = \mathbf{D}(\theta)\mathbf{P}$ is positive and element-wise increasing with θ. The Perron-Frobenius theorem states that the maximum modulus eigenvalue of $\mathbf{M}(\theta)$ is real, simple and has positive left and right eigenvectors, denoted with $\mathbf{u}(\theta)$ and $\mathbf{v}(\theta)$, respectively. Let $\eta(\theta)$ denote the maximum modulus eigenvalue of $\mathbf{M}(\theta)$. Since $\mathbf{M}(0) = \mathbf{P}$ is a stochastic matrix, we have $\eta(0) = 1$. The function $\eta(\theta)$ is monotonously increasing with θ. Therefore, we have $\eta(\theta) > 1$ for $\theta > 0$.

Since $\eta(\theta)$ is the maximum modulus eigenvalue of $\mathbf{M}(\theta)$, we have $[\mathbf{M}(\theta)]^n \sim \mathbf{u}(\theta)\mathbf{v}(\theta)[\eta(\theta)]^n$ as $n \to \infty$. We can now find the LMGF of the Markov source:

$$\Gamma(\theta) = \lim_{n \to \infty} \frac{1}{n} \log \mathrm{E}[e^{\theta(A_n - A_0)}] = \lim_{n \to \infty} \frac{1}{n} \log \boldsymbol{\pi}(\mathbf{D}(\theta)\mathbf{P})^n\mathbf{e}$$

$$\sim \lim_{n \to \infty} \frac{1}{n} \log \boldsymbol{\pi}\mathbf{u}(\theta)\mathbf{v}(\theta)\mathbf{e}[\eta(\theta)]^n = \lim_{n \to \infty} \frac{1}{n}[\log(b) + n\log\eta(\theta)] = \log\eta(\theta)$$

where $b = \boldsymbol{\pi}\mathbf{u}(\theta)\mathbf{v}(\theta)\mathbf{e}$ is a positive scalar.

Example 11.14 *Effective bandwdith of a two-state Markov traffic source*
Consider a two-state Markov chain ($m = 2$). The matrix $\mathbf{M}(\theta)$ can be written explicitly as

$$\mathbf{M} = \mathbf{D}(\theta)\mathbf{P} = \begin{bmatrix} \phi_1(\theta)p_{11} & \phi_1(\theta)(1 - p_{11}) \\ \phi_2(\theta)(1 - p_{22}) & \phi_2(\theta)p_{22} \end{bmatrix} \tag{11.110}$$

The eigenvalues are the solutions of the quadratic

$$z^2 - z[\phi_1(\theta)p_{11} + \phi_2(\theta)p_{22}] + \phi_1(\theta)\phi_2(\theta)(p_{11} + p_{22} - 1) \tag{11.111}$$

The maximum modulus eigenvalue is easily found to be

$$\eta(\theta) = \frac{A_1 + A_2 + \sqrt{(A_1 - A_2)^2 + 4B_1 B_2}}{2} \tag{11.112}$$

where

$$A_i = \phi_i(\theta)p_{ii}, \quad i = 1, 2,$$

$$B_i = \phi_i(\theta)(1 - p_{ii}), \quad i = 1, 2.$$

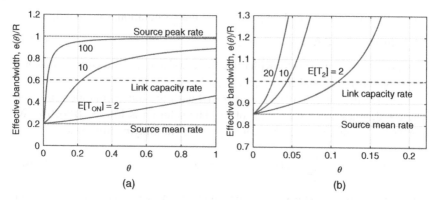

Figure 11.12 LMGF of a two-state Markov source. Left plot: ON-OFF source. Right plot: source with Gamma-distributed rate in either state.

The effective bandwidth is $e(\theta) = \log \eta(\theta)/\theta$. Given the multiplexer capacity C, we know that the tail of the buffer content PDF can be bounded as follows:

$$P(Q > K) \leq \kappa(\theta)e^{-\theta K} \tag{11.113}$$

With the Markov modulated traffic source, this upper bound holds for any positive θ such that $\theta < \eta$, where η is the unique positive solution of $e(\eta) = C$. Moreover, we have also a closed-form expression of the coefficient $\kappa(\theta)$:

$$\kappa(\theta) = \pi(\mathbf{I} - e^{-\theta C}\mathbf{D}(\theta)\mathbf{P})^{-1}\mathbf{e} \tag{11.114}$$

where $\mathbf{I}$ denotes the identity matrix.

As a special case, the ON-OFF source is modeled by letting $\phi_1(\theta) = 1$ (OFF state: $B = 0$ with probability 1) and $\phi_2(\theta) = e^{\theta R}$ (ON state: $B = R$ with probability 1). Figure 11.12(a) plots the effective bandwidth $e(\theta)$ as a function of θ, for this source for three values of the mean sojourn time in the ON state $E[T_{ON}]$ (2, 10 and 100 time slots) and for $\pi_1 = p = 0.2$. The multiplexer output link capacity is $C = 0.6$ (rates are normalized with respect to the source peak rate).

The effective bandwidth grows monotonously from $\bar{r} = pR = 0.2$ to the peak rate $R = 1$ when θ goes from 0 to ∞.

The intercept of the effective bandwidth with the horizontal line at level C marks the critical value η of θ that gives the upper bound of the decay rate of the CCDF of the multiplexer buffer content. Table 11.1 lists the values of η.

As a second example we consider a traffic source with bit rate modulated by a two-state Markov chain. In either state the bit rate is a gamma random variable. The parameters of the gamma PDF depend on the modulating Markov chain state.

Let ρ denote the coefficient of utilization of the capacity, i.e., $\rho = E[B]/C = (\pi_1 E[B_1] + \pi_2 E[B_2])/C$, where C is the the capacity of the multiplexer. B_1 and B_2 denote the bit rates random variables associated with states 1 and 2, respectively.

Table 11.1 Critical decay rate η of the CCDF of the multiplexer buffer queue length with ON-OFF source.

	ON-OFF source			Gamma source		
$E[T_{ON}]$	2	10	100	2	10	100
η	1.5510	0.2220	0.0200	0.1078	0.0431	0.0249

$\pi_1 = 1 - p$ and $\pi_2 = p$ are the state probabilities of the two-state modulating Markov chain, where p is the activity factor of the source.

The activity factor of the source is $p = 0.1$. The standard deviation for state 1 is $\sigma_{B|Z=1} = 0.5$. The mean rate $\bar{r}_1$ in state 1 is set so that the overall mean rate $\bar{r}$ of the source equals a fraction $\rho = 0.85$ of the output capacity C. This is obtained by setting $\bar{r}_1$ so that $\pi_1 \bar{r}_1 + \pi_2 \bar{r}_2 = (1 - p)\bar{r}_1 + p\bar{r}_2 = \rho C$. The mean rate and standard deviation of the bit rate in state 2 are $\bar{r}_2 = E[B|Z = 2] = 2 \cdot C$ and $\sigma_{B|Z=2} = 3$, respectively. The multiplexer link capacity is set to $C = 1$.

Figure 11.12(b) shows the effective bandwidth of this two-state source as a function of θ for three values of the mean sojourn time in state 2. The properties of the effective bandwidth can be still verified, except that in this case there is no peak rate. Therefore, $e(\theta)$ is unbounded as θ grows.

Example 11.15 In this example we elaborate on the admissible region for a single multiplexer loaded by two classes of ON-OFF sources.

We refer to discrete time model and let the time slot be the unit of time. The multiplexer output link capacity C is the unit of capacity. The parameters of the first class of ON-OFF sources are as follows: peak rate $R_1 = 0.25 \cdot C$, activity factor $p_1 = 0.2$, mean ON time $T_{ON,1} = 20$. As for the second class, we let: peak rate $R_2 = 0.05 \cdot C$, activity factor $p_2 = 0.4$, mean ON time $T_{ON,2} = 150$. The mean rates of the two classes are $\bar{r}_1 = 0.05$ and $\bar{r}_2 = 0.02$.

The delay requirement is $d_{max} = 100$. This is interpreted as the 99.9% delay quantile, i.e., the probability that the delay through the multiplexer exceeds d_{max} is no more than $\epsilon_D = 10^{-3}$: $P(W > d_{max}) \leq e^{-\theta d_{max} C} \leq \epsilon_D$. We have converted the bound on the tail of the queue length Q to the bound on the delay. The connection is established by the relationship $W = Q/C$. Therefore, $\theta = -\log \epsilon_D/(d_{max}C)$. In our case we find $\theta = 69.0776$.

The effective bandwidth of the sources of the two considered classes can be calculated with the same approach as in Example 11.14. Numerical values in our example are $e_1 = 0.0546$ and $e_2 = 0.0215$.

Figure 11.13(a) shows the admissible region corresponding to the requirement that $n_1 e_1 + n_2 e_2 \leq C$, where n_i is the number of sources of class i, $i = 1, 2$ (shaded

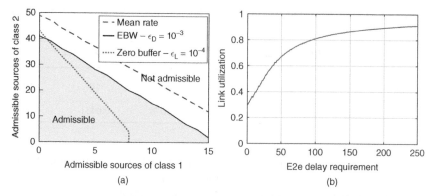

Figure 11.13 Left plot: admissible region for a multiplexer loaded by two classes of ON-OFF sources; the shaded area refers to a non-negligible buffer and effective bandwidth under a delay constraint; the dotted line is the boundary of the admissible region with effective bandwidth under a loss requirement with marginal buffer size; the dashed line is the boundary of the theoretical admissible region based on mean rate assignment. Right plot: multiplexer link utilization as a function of the delay requirement.

region bounded by the solid curve). For comparison purposes we show also the boundary line corresponding to mean rate assignment, i.e., the line such that $n_1 \bar{r}_1 + n_2 \bar{r}_2 = C$ (dashed line). Finally, the dotted line corresponds to the acceptance region boundary if marginal buffer and loss requirement equal to $\epsilon_L = 10^{-4}$. For this requirement, we search for the optimum value of θ, i.e., the one that maximizes the multiplexer link utilization. It turns out that $\theta^* = 9.2564$. Then, the effective bandwidths for the loss requirement under marginal buffer size are $e_1^{(L)} = 0.1121$ and $e_2^{(L)} = 0.0228$.

This plot shows clearly the penalty for guaranteeing (with probability ϵ_D) a delay bound with respect to pure mean rate assignment. The admissible region restricts, with the boundary line dropping by about 20% off the mean rate region boundary line. A much greater penalty is paid if the marginal buffer size is provided and hence a loss requirement is set on the multiplexer. If the mix of traffic sources is composed essentially only of sources of type 2 (those with smaller peak rate), multiplexing under marginal buffer size is even more efficient than providing substantial buffer space. Otherwise, it appears that a significant drop of the efficiency is incurred with marginal buffer size.

Going back to a non-negligible buffer size and to the delay requirement, Figure 11.13(b) plots the multiplexer link utilization $\rho = (n_1 \bar{r}_1 + n_2 \bar{r}_2)/C$ as a function of the delay requirement $d_{\max}$. For each value of $d_{\max}$, we find the mean link utilization level achieved when the maximum number of allowable traffic sources are admitted.

As expected, the link utilization is monotonously increasing with d_{max}. At first, it grows fast, then it saturates and tends to settle on a level slightly higher than 0.91.

11.3.4 Network Analysis and Dimensioning

The theory of effective bandwidth has been developed for a single link (one multiplexer). Moving to a network-wide scope, we consider an end-to-end path, composed of a number of cascaded links. A possible approach is to split the end-to-end delay requirement in delay requirements on each link of the path. The delay requirement on a link can be translated into a requirement for the buffer overflow probability. Let d be the delay target on a tagged link. We interpret d as a requirement on a quantile of the delay, i.e., we require that $P(W > d) \leq \epsilon$. The delay W is tied to the backlog by $W = Q/C$, where C is the capacity of the link and we assume that the buffer is managed according to a FIFO discipline. Then, we translate the requirement in terms of backlog and use the effective bandwidth approach bound, i.e., we have $P(Q > d \cdot C) \approx e^{-\theta d C} \leq \epsilon$. We derive the value of θ that guarantees the prescribed delay quantile requirement. The exponential approximation of the tail of the backlog probability distribution holds under the constraint on the sum of effective bandwidths of the traffic flows loading the tagged link does not exceed the link capacity, i.e., $\sum_i e_i(\theta) < C$.

A critical point in carrying over the theory of effective bandwidth from a single node to multiple nodes is that the source traffic is modified by the passage through a multiplexing stage (where it interacts with other concurrent traffic flows). As a consequence, the effective bandwidth changes from node to node.

We observe that the output process of the n-th multiplexer D_n is upper bounded by the arrival process at the input of the multiplexer, $D_n \leq A_n, \forall n$. Then, it is easy to derive that the LMGF of D_n is upper bounded by the LMGF of A_n, that is $\Gamma_D(\theta) \leq \Gamma_A(\theta)$. Consequently, the effective bandwidth of the traffic flow D_n is bounded by the effective bandwidth of the same flow, at the input of the multiplexer, namely A_n. Therefore, $e_D(\theta) \leq e_A(\theta)$.

A stronger result is found in [86, Ch. 10] in a continuous-time setting, assuming that a capacity c is *reserved* for a traffic flow $A(t)$. In that case, it is

$$\Gamma_D(\theta) = \begin{cases} \Gamma_A(\theta) & \theta < \theta^*, \\ \Gamma_A(\theta^*) - c\theta^* + c\theta & \theta \geq \theta^*. \end{cases} \tag{11.115}$$

where θ^* is the unique positive solution of the equation $\Gamma'_A(\theta) = c$.

Even in the general case where there is no reserved bandwidth (e.g., FIFO multiplexing), the effective bandwidth of a traffic flow does not increase when passing

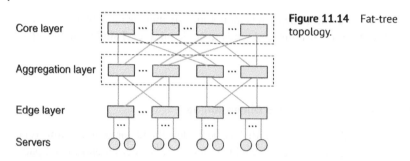

Core layer

Aggregation layer

Edge layer

Servers

Figure 11.14 Fat-tree topology.

through a multiplexer. Therefore, we can use the *same* quantity $e(\theta)$ for a flow on all links of its network path.

Summing up, we can set a delay requirement on link i, say d_i. Given the capacity c_i of link i, the exponential bound on the link queue backlog provides us with a bound for the tail of the time W_i through the buffer of link i, namely, $\mathcal{P}(W_i > d_i) = e^{-\theta_i c_i d_i}$. The requirement $\mathcal{P}(W_i > d_i) \le \epsilon_D$ is met if we set

$$\theta_i = \frac{-\log \epsilon_D}{d_i c_i} \tag{11.116}$$

We assign the effective bandwidth $e_k(\theta_i)$ to the k-th traffic flow on link i. The admission control rule reduces to verifying that at any given time the following inequalities are satisfied:

$$\sum_{k \in S} r_{ik} e_k(\theta_i) \le c_i, \quad i \in \mathcal{L}, \tag{11.117}$$

where S is the set of active flows, $\mathcal{L}$ is the set of network links, r_{ik} equals 1 if and only if the route of flow k uses link i, it is 0 otherwise.

The end-to-end delay of flow k is given by

$$d_{e2e,k} = \sum_{i \in \mathcal{L}} d_i r_{ik} \le d_{\max} \tag{11.118}$$

where $d_{\max}$ is the end-to-end delay requirement.

Example 11.16 In this example we dimension the amount of traffic that can be sustained by a data center with a fat-tree topology under an all-to-all traffic matrix.

The fat-tree topology for data center network depicted in Figure 11.14 is among the most widespread interconnection solutions.

The lower layer consists of servers, where processing is carried out. The data center network purpose is to enable servers to cooperate. The interconnection network is based on switches, represented by the rectangular boxes in Figure 11.14. Let n denote the number of ports of a switch and let us assume that all switches are equipped with the same number of ports. The switches are organized in three layers. The *edge layer* comprises switches placed on top of racks where servers are located. Half of the ports of an edge switch are connected to $n/2$ servers. The other $n/2$ ports are connected to $n/2$ switches of the upper layer. The second layer is referred to as *aggregation layer*. Half of the ports of an aggregation layer switch are connected to as many edge switches, while the other half of the ports are connected to the top layer switches. A set of $n/2$ aggregation switches, $n/2$ edge switches and the relevant $n/2 \cdot n/2$ servers are referred to as *pod*. The top layer of the topology is called *core layer*. Core switches guarantee the connectivity among pods. The full topology is made up of n pods. Hence there are $n^2/4$ core switches overall. Across all pods, there are $n^3/4$ servers and hence $n^3/4$ links interconnecting servers to edge switches, $n^3/4$ links to connect edge switches to aggregation switches, $n^3/4$ links to connect aggregation switches to core switches.

The traffic is described at flow level and packet level. Flow level refers to arrivals of end-to-end (server-to-server) connection requests, while packet-level description applies to the statistics of packets sent within a connection.

We assume that offered traffic at flow level is Poisson. The mean offered traffic intensity of server s is denoted with $A_{o,s}$, $s \in S$, where S is the set of traffic sources (servers).

In this example, we assume the traffic pattern defined as all-to-all, i.e., each server addresses its flows uniformly to all other servers. We have therefore $A_{o,s} = A_o \equiv a(n_{2h} + n_{4h} + n_{6h})$, where a is a scale parameter, and

- $n_{2h} = n/2 - 1$ are servers that belong to the same rack as the tagged one and are therefore reached with a two-hop route;
- $n_{4h} = n^2/4 - n/2$ are the servers that belong to the same pod as the tagged one, but not the same rack; they are reached by means of four-hop routes;
- $n_{6h} = n^3/4 - n^2/4$ are the remaining servers, that can be reached using six-hop routes.

We define the traffic matrix of the network. The entry (i, j) of the traffic matrix represents the amount of traffic directed from node i to node j. With our flow level model, the traffic matrix entries are all equal to A_o, except of diagonal entries, that are equal to 0.

Packet level traffic is modeled as an ON-OFF process with peak rate R, activity factor p, mean rate $\bar{r} = pR$. The mean duration of the ON time is denoted with T_{ON}.

Let d_{e2e} denote the delay requirement. Since the longest route used to connect any two servers is made of six hops, we have $W_{e2e} = W_{\ell_1} + \cdots + W_{\ell_6}$, where W_i is the delay through link i. We can guarantee that the end-to-end delay requirement d_{e2e} is exceeded with probability not larger than ϵ_D, provided that each summand W_{ℓ_j} does not exceed $d_{e2e}/6$ with probability larger than ϵ_D. According to the effective bandwidth bound, we have $\mathcal{P}(W > d_{max}) \leq e^{-\theta d_{max} C}$, where C is the value of the link capacity (same for all links), and d_{max} is the requirement on the quantile of the delay through a single buffer. We impose therefore that $e^{-\theta C d_{e2e}/6} \leq \epsilon_D$,

Table 11.2 Parameter definition and values for Example 11.16.

Symbol	Meaning	Value
C	Link capacity	10 Gbit/s
R	Flow peak rate	1 Gbit/s
p	Flow activity factor	0.1
T_{ON}	Flow mean burst duration	200 μs
ϵ_D	Probability of exceeding delay requirement	10^{-2}
ϵ_L	Requirement on flow set-up rejection probability	10^{-2}
n	Number of ports of a switch	64
n_{2h}	Number of servers reachable in 2 hops from a tagged server	31
n_{4h}	Number of servers reachable in 4 hops from a tagged server	992
n_{6h}	Number of servers reachable in 6 hops from a tagged server	64512

from which we obtain the value of $\theta = -6 \log \epsilon_D/(d_{e2e}C)$. The effective bandwidth $e(\theta)$ is calculated as in Example 11.14. The number of admissible flows per link is therefore $m = \lfloor C/e(\theta) \rfloor$.

Table 11.2 lists the numerical values of the parameters used in the ensuing numerical evaluation.

The effective bandwidth as a function of the end-to-end delay requirement d_{e2e} is plotted in Figure 11.15(a). As the delay requirement is relaxed from a strict 100 *μs* to a looser 5 *ms*, the effective bandwidth decreases monotonously from around the peak rate to close to the mean rate. It is expected that the statistical multiplexing gain provided by the effective bandwidth approach grows accordingly.

The flow level of the network can be modeled by means of a loss network (see Section 7.4). Nodes of the loss network correspond to links of the fat-tree topology. Due to the symmetry of the topology and of the traffic matrix, there are only three types of links: links between servers and edge switches (type 1), between edge and aggregation switches (type 2), between aggregation and core switches (type 3).

Type 1 links carry flows that use routes 2, 4, and 6 hops long. These are the links connecting servers to edge switches.

Type 2 links carry only flows that travel routes of 4 and 6 hops. These are the links between the edge and the aggregation layer.

Type 3 routes carry only flows that go through the longest routes, comprising 6 hops. These are links connecting the aggregation switches to the core switches.

We denote the loss probability of link of type i with E_i, for $i = 1, 2, 3$.

The probability of rejection of flow set-up has different values for the three types of links. Let L_{xh} denote the rejection probability for routes comprising x hops, $x =$

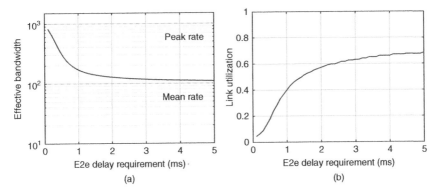

Figure 11.15 Fat-tree topology dimensioning with end-to-end delay requirement d_{e2e}. Left plot: effective bandwidth of ON-OFF sources as a function of d_{e2e}. Right plot: average network link utilization as a function of d_{e2e}.

2, 4, 6 We have

$$L_{2h} = 1 - (1 - E_1)^2$$
$$L_{4h} = 1 - (1 - E_1)^2(1 - E_2)^2$$
$$L_{6h} = 1 - (1 - E_1)^2(1 - E_2)^2(1 - E_3)^2$$

The number of routes of the three types, originating from any server, is denoted with n_{xh}, for $x = 2, 4, 6$, as noted above. The average rejection probability is

$$L = \frac{n_{2h}L_{2h} + n_{4h}L_{4h} + n_{6h}L_{6h}}{n_{2h} + n_{4h} + n_{6h}} \tag{11.119}$$

since the traffic matrix is uniform (all-to-all traffic pattern).

The Erlang fixed point equation whose solution provides the (approximate) values of the E_i's are:

$$
\begin{aligned}
E_1 &= B(m, a \ [n_{2h}(1 - E_1) + n_{4h}(1 - E_1)(1 - E_2)^2 \\
&\quad + n_{6h}(1 - E_1)(1 - E_2)^2(1 - E_3)^2]) \\
E_2 &= B(m, a \ [n_{4h}(1 - E_1)^2(1 - E_2) + n_{6h}(1 - E_1)^2(1 - E_2)(1 - E_3)^2]) \\
E_3 &= B(m, a \ n_{6h}(1 - E_1)^2(1 - E_2)^2(1 - E_3))
\end{aligned}
$$

where $B(m, A)$ is the Erlang-B formula for a system with m servers and mean offered traffic A. We recall that $m = \lfloor C/e(\theta) \rfloor$ is the number of ON-OFF sources that can be multiplexed on a link of capacity C under the requirement on the delay quantile.

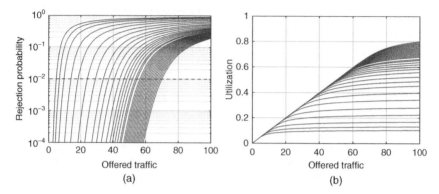

Figure 11.16 Fat-tree topology dimensioning with end-to-end delay requirement d_{e2e}. Left plot: probability of rejection of a connection set-up as a function of the mean offered traffic per server (the delay requirement grows moving from curves on the left to curves on the right). Right plot: average link utilization as a function of the mean offered traffic per server (the delay requirement grows moving from lower to upper curves).

The utilization factors of the three types of links are

$$\rho_1 = \frac{\bar{r}\,a}{C}[n_{2h}(1 - L_{2h}) + n_{4h}(1 - L_{4h}) + n_{6h}(1 - L_{6h})]$$

$$\rho_2 = \frac{\bar{r}\,a}{C}[n_{4h}(1 - L_{4h}) + n_{6h}(1 - L_{6h})]$$

$$\rho_3 = \frac{\bar{r}\,a}{C}n_{6h}(1 - L_{6h})$$

The overall mean link utilization of the network is

$$\rho = \frac{\ell_1\rho_1 + \ell_2\rho_2 + \ell_3\rho_3}{\ell_1 + \ell_2 + \ell_3} = \frac{1}{3}(\rho_1 + \rho_2 + \rho_3) \tag{11.120}$$

since the number of links of type 1, 2 and 3 are the same: $\ell_1 = \ell_2 = \ell_3 = n^3/4$.

The average link utilization of the fat-tree topology under all-to-all traffic is shown in Figure 11.15(b) as a function of the delay requirement. We see that for very strict delay requirements a very low utilization can be achieved. Much higher values of utilization can be attained as the delay requirement is relaxed. Still, we are off full utilization by more than about 30%. Part of this capacity waste is due to the loose bounds we are using to assess the delay requirement end-to-end.

The flow rejection probability and the average link utilization as a function of the mean offered traffic A_o are shown in Figure 11.16(a) and Figure 11.16(b), respectively. Each curve refers to one value of the delay requirement d_{e2e}. In Figure 11.16(a) we go from low to high delay requirement levels moving from left to right. In Figure 11.16(b) we explore delay requirements from the lowest to the highest one moving from the curve at the bottom upward.

It is apparent that the rejection probability grows steeply as A_o is increased, while the average utilization grows much slower. The level of the loss requirement $\epsilon_L = 10^{-2}$ is highlighted in Figure 11.16(a), by means of a dashed horizontal line.

If results seem disappointing, we should consider instead that we are able to *guarantee* a 99% quantile of delays of, e.g., no more than 5 ms, still achieving around 70% average link utilization. This traffic management approach applies to delay-sensitive (inelastic) traffic. The residual capacity can be exploited by elastic traffic, e.g., managed by mean of the reactive congestion control of TCP, thus achieving very high link utilization, yet preserving delay performance of inelastic applications.

11.4 Further Readings

Among the early papers that deal with deterministic traffic theory, applied to packet networks, we mention [61, 62, 173, 174]. An early systematic treatment can be found in [52]. The application of the min-plus algebra to network traffic modeling and analysis is discussed extensively in [144]. A more recent monograph on application of max-plus algebra to the modeling of networked service system is [100].

The key points of network calculus are arrival and service curves and the min-plus convolution, which provides the relationship between the input flow, described by an arrival curve, and the output flow of a service system, described by means of its service curve. As in system theory, the strength of the min-plus convolution is the ability to concatenate tandem systems along a network path. As a consequence, we are able to characterize the whole network path by a single transfer function.

The deterministic network calculus has been extended to relax the deterministic guarantees to stochastic guarantees, trading-off simplicity of the approach with potential efficiency of the resulting dimensioning. The aim of the stochastic network calculus is to account for statistical multiplexing and scheduling of traffic sources in a framework for end-to-end analysis of multi-node networks, as we have discussed for the deterministic traffic theory.

A systematic review of deterministic and stochastic network calculus is provided in [78]. It discusses the concept of service curves, its use in the network calculus, and the relation to systems theory under the min-plus algebra. The modeling of service curves and its derivation from measurements are discussed as well. The paper also reviews stochastic service curve theory, which allows importing the statistical multiplexing gain in a network calculus framework

In the later work [77] the Authors aim to obtain fundamental results of stochastic network calculus, by using a method based on moment generating functions, known from the theory of effective bandwidths, to characterize traffic arrivals and network service. Affine envelope functions with an exponentially decaying overflow profile are derived to compute statistical end-to-end backlog and delay bounds for networks.

The effective bandwidth theory has been treated in a large number of papers and books. The theory of effective bandwidth for Markovian traffic sources is developed in the works of Elwalid et al. [70–72]. Other relevant sources are [123,134].

An extensive account on effective bandwidth with several examples is given in [86].

Appendix

The deterministic traffic theory is based on a special mathematical structure known as the min-plus algebra. Let $\mathbb{R}_{min}$ denote the set or the real numbers extended with ∞, i.e., $\mathbb{R}_{min} = \mathbb{R} \cup \{\infty\}$. We define two operations on this set, denoted with $\oplus$ and $\odot$. The operations are defined as follows:

$$a \oplus b \equiv \min\{a, b\}, \quad \forall a, b \in \mathbb{R}_{min} \tag{11.121}$$

$$a \odot b \equiv a + b, \quad \forall a, b \in \mathbb{R}_{min} \tag{11.122}$$

As an example, $3 \oplus (-1) = -1$ and $3 \odot -1 = 2$. Note also that $a \oplus \infty = a$ and $a \odot 0 = a$ for any $a \in \mathbb{R}_{min}$. The two elements ∞ and 0 play therefore the role of neutral element for the $\oplus$ and $\odot$ operations, respectively. We denote them with $\epsilon = \infty$ and $e = 0$.

The tuple $\mathcal{M} = \{\mathbb{R}_{min}, \oplus, \odot, \epsilon, e\}$ is called a min-plus algebra. The operations of the min-plus algebra have a number of properties, listed below.

- **Associativity**: $\forall x, y, z \in \mathbb{R}_{min}$ it is

$$x \oplus (y \oplus z) = (x \oplus y) \oplus z, \qquad x \odot (y \odot z) = (x \odot y) \odot z \tag{11.123}$$

- **Commutativity**: $\forall x, y \in \mathbb{R}_{min}$ it is

$$x \oplus y = y \oplus x, \qquad x \odot y = y \odot x \tag{11.124}$$

- **Distributivity of $\odot$ over $\oplus$**: $\forall x, y, z \in \mathbb{R}_{min}$ it is

$$x \odot (y \oplus z) = (x \odot y) \oplus (x \odot z) \tag{11.125}$$

- **Existence of a zero element**: $\forall x \in \mathbb{R}_{min}$ it is

$$x \oplus \epsilon = \epsilon \oplus x = x \tag{11.126}$$

- **Existence of a unit element**: $\forall x \in \mathbb{R}_{min}$ it is

$$x \odot e = e \odot x = x \tag{11.127}$$

- **The zero is absorbing for** $\odot$: $\forall x \in \mathbb{R}_{\min}$ it is

$$x \odot \epsilon = \epsilon \odot x = \epsilon \qquad (11.128)$$

- **Idempotency of** $\oplus$: $\forall x \in \mathbb{R}_{\min}$ it is

$$x \oplus x = x \qquad (11.129)$$

The min-plus algebra belongs of a class of algebraic structures known as semi-rings. It is actually an example of commutative and idempotent semi-ring. Another example is the max-plus algebra, obtained by re-defining the operation $\oplus$ with the max operator rather than min and replacing $\epsilon = \infty$ with $\epsilon = -\infty$. The monograph [100] is devoted to the presentation of the max-plus algebra, its properties and an application to transportation services scheduling.

Note that applying the $\odot$ operation $n-1$ times to the same operand x used n times we get $x^{(n)} \equiv x \odot \cdots \odot x = x + \cdots + x = n \cdot x$. We can use this identity to generalize the notation $x^{(n)}$ to non integer exponents, i.e., $x^{(w)} = w \cdot x$ for any real w.

When taking the minimum operator over infinite sets, we will replace it with the infimum operator, which is always well defined.

In the following, we consider wide-sense increasing function of a real variable, i.e., functions $f : \mathbb{R} \mapsto \mathbb{R}$, such that $f(x) \leq f(y)$ for all $x \leq y$. We assume right-continuous functions. Moreover, we let $f(t) = 0$ for $t < 0$. Let $\mathcal{F}$ denote the set of such functions.

Definition 11.4 *Min-plus convolution* Let f, g be two functions belonging to the set $\mathcal{F}$. We define the min-plus convolution of f and g as $h = f \otimes g$, where $h(t)$ is given by:

$$h(t) = (f \otimes g)(t) = \inf_{0 \leq s \leq t} \{f(s) + g(t - s)\} \qquad (11.130)$$

A number of properties can be shown for the min-plus convolution of functions in the set $\mathcal{F}$ (e.g., see [144, Ch. 3]).

1. **Closure:** $f \otimes g \in \mathcal{F}$ if $f, g \in \mathcal{F}$.
2. **Associativity:** $\forall f, g, h \in \mathcal{F}$ it is $f \otimes (g \otimes h) = (f \otimes g) \otimes h$.
3. **Commutativity:** $\forall f, g \in \mathcal{F}$ it is $f \otimes g = g \otimes f$.
4. **Unit element:** If we define $\delta_0(t) = \epsilon = \infty$ for $t \geq 0$ and $\delta_0(t) = 0$ for $t < 0$, we have[6] $f \otimes \delta_0 = f$, i.e., the function $\delta_0(t)$ plays the same role for min-plus convolution as the Dirac delta function does for ordinary convolution.

---

6 We remind that the inf operator in the definition of the min-plus convolution between functions belonging to $\mathcal{F}$ must be considered on the interval $0_- \leq s \leq t_+$.

5. **Distributivity:** $\forall f, g, h \in \mathcal{F}$ it is $(f \oplus g) \otimes h = (f \otimes h) \oplus (g \otimes h)$. We recall that $m = f \oplus g$ is a function defined by $m(t) = \min\{f(t), g(t)\}, \forall t$. It is clear that $m \in \mathcal{F}$ if $f, g \in \mathcal{F}$.

6. **Addition of a constant:** Let K be a real non-negative constant. Then $(f + K) \otimes g = f \otimes g + K$.

7. **Bound:** It is $f \otimes g \leq f \oplus g = \min\{f, g\}$.

8. **Convexity:** If f, g are convex functions belonging to $\mathcal{F}$, then $f \otimes g$ is convex as well.

9. **Isotonicity:** If $f_1 \leq f_2$ and $g_1 \leq g_2$, it is $f_1 \otimes g_1 \leq f_2 \otimes g_2$.

For a proof of these properties the reader is referred to, e.g., [144, Ch. 3]. Some caution is in order since in that text a function in $\mathcal{F}$ is assumed to be left-continuous rather than right-continuous as here.

Definition 11.5 A function f is said to be sub-additive if $f(s + t) \leq f(s) + f(t)$ for all s and t.

Examples of sub-additive functions are given by concave or piece-wise linear functions. Here is a list of some properties of sub-additivity:

1. The sum of two sub-additive functions is itself sub-additive, i.e., if f and g are sub-additive, so is $f + g$.

2. If f and g are sub-additive functions, so is their min-plus convolution $f \otimes g$.

3. Given a function $f \in \mathcal{F}$, it is $f \geq f \otimes f \geq 0$, by the very definition of the min-plus convolution.

4. Given a function $f \in \mathcal{F}$ that is sub-additive, it is $f = f \otimes f$. The preceding property says that $f \geq f \otimes f$. Sub-additivity implies that it is also $f \leq f \otimes f$.

By repeating the convolution operation, we generate a sequence $f^{(n)}$ given by $f^{(n+1)} = f^{(n)} \otimes f \leq f^{(n)} \oplus f \leq f^{(n)}$ (thanks to property 6), with $f^{(1)} = f$. Since the sequence $f^{(n)}$ is point-wise monotonously nonincreasing, it converges to some non-negative limit as $n \to \infty$. The limit function is the largest sub-additive function smaller than f. It is called the sub-additive closure of f.

Definition 11.6 *Sub-additive closure* Let $f \in \mathcal{F}$ and $f^{(n)}$ be the function obtained by convolution of f with itself $n - 1$ times. We let also $f^{(0)} = \delta_0$. The sub-additive closure of f, denoted with $\bar{f}$, is defined by $\bar{f}(t) = \inf\{f^{(n)}(t), n \geq 0\}, \forall t$.

The following theorem establishes a key result on the sub-additive closure of f, namely that it is the upper bound of all sub-additive functions that are not bigger than f.

Theorem 11.13   Let $f \in \mathcal{F}$ and let $\bar{f}$ be its sub-additive closure. Then, $\bar{f} \in \mathcal{F}$ and $\bar{f} \leq f$. Moreover, for any sub-additive function $g \in \mathcal{F}$ such that $g \leq f$, it is $g \leq \bar{f}$.

As a consequence of this theorem, the following three statements are equivalent for $f \in \mathcal{F}$.

1. f is sub-additive.
2. $f = f \otimes f$.
3. $f = \bar{f}$.

The following properties of the sub-additive closure hold as well.

- **Isotonocity:** $f \leq g$ implies that $\bar{f} \leq \bar{g}$.
- **Sub-additive closure of a minimum:** $\overline{\min\{f, g\}} = \bar{f} \otimes \bar{g}$.
- **Sub-additive closure of a convolution:** $\overline{f \otimes g} \geq \bar{f} \otimes \bar{g}$. If $f(0) = g(0) = 0$, then $\overline{f \otimes g} = \bar{f} \otimes \bar{g}$.

The following operation we define, the de-convolution, is strictly tied to the min-plus convolution.

Definition 11.7 *Min-plus de-convolution*   Let f, g be two functions belonging to the set $\mathcal{F}$. We define the min-plus de-convolution of f and g, as $h = f \oslash g$, where $h(t)$ is given by:

$$h(t) = (f \oslash g)(t) = \sup_{u \geq 0}\{f(t + u) - g(u)\} \tag{11.131}$$

If both $f(t)$ and $g(t)$ are infinite for some t, the definition of the de-convolution fails. Contrary to min-plus convolution, the function $f \oslash g$ does not necessarily belong to $\mathcal{F}$, since it can take non-null values for $t < 0$.

We report the following properties of the de-convolution operation.

1. **Isotonicity**: if $f \leq g$, then $f \oslash h \leq g \oslash h$ and $h \oslash f \geq h \oslash g$.
2. **Composition of $\oslash$**: $(f \oslash g) \oslash h = f \oslash (g \otimes h)$.
3. **Duality between $\oslash$ and $\otimes$**: $f \oslash g \leq h$ if and only if $f \leq g \otimes h$.
4. **Self-de-convolution**: let $h = f \oslash f$. It is $h \in \mathcal{F}$.

Finally, we give a definition of two quantities that are related to performance metrics (backlog and delay).

Definition 11.8 *Vertical and horizontal deviations*   For $f, g \in \mathcal{F}$ the vertical deviation $v(f, g)$ and horizontal deviation $h(f, g)$ are defined as:

$$v(f, g) = \sup_{t \geq 0}\{f(t) - g(t)\} \tag{11.132}$$

$$h(f,g) = \sup_{t \geq 0} \{\inf\{d \geq 0 : f(t) \leq g(t+d)\}\} \qquad (11.133)$$

Using the de-convolution operation, the vertical and horizontal deviations can be re-stated as follows:

$$v(f,g) = (f \oslash g)(0) \qquad (11.134)$$

$$h(f,g) = \inf\{d \geq 0 : (f \oslash g)(-d) \leq 0\} \qquad (11.135)$$

Summary and Takeaways

This chapter introduces models for network traffic control functions that provide performance guarantees. It explores two approaches.

Deterministic traffic theory is based on a deterministic description of the traffic flows offered to a network. The deterministic function that bounds the profile of the offered traffic flow is derived by the intrinsic properties of the traffic source. It can also be modified (shaped) by a network element located at the network edge, the traffic regulator or shaper. The leaky bucket is a major example of this kind of element. In either case, we have an arrival curve called an envelope that bounds the offered traffic flow.

A second cornerstone of the deterministic traffic theory consists of a bound on the minimum amount of service provided by a network element, the lower service curve. The key property of the lower service curve is concatenation: the lower service curve of a cascade of N service elements is the min-plus convolution of the lower service curve of the N elements. This is reminiscent of system theory input-output function composition through the usual convolution.

Thanks to the deterministic bounds on the arrival flow and on the minimum amount of service of the network elements (routers) it is possible to find deterministic bounds of the end-to-end delay of each flow. It is therefore possible to define simple admission control rules that guarantee the end-to-end delay requirement, in case of successful flow set-up.

The practical implementation of this approach requires leaky buckets at network entrance and WFQ buffers on network links.

The second approach investigated in this chapter is based on a stochastic description of the traffic flows offered to the network. Focusing on a single link, if the buffer size is negligible, it is possible to set a bound on the loss probability of the link. Through the bound we arrive at defining a simple rule to meet a loss requirement, namely that the sum of effective bandwidths of the traffic sources be less than the link capacity. The effective bandwidth is defined in terms of the MGF of the probability distribution function of the offered traffic flow. The interest of this approach lies in the fact that it allows dealing with packet traffic flows as in circuit-switching networks. Admission control translates into a simple sum of effective bandwidths.

This approach can be carried over to the case where the link is equipped with a non-negligible buffer size. Using Chernov bound, it is possible to bound the tail of a multiplexer buffer backlog probability distribution. Again, the performance requirement on the delay through a buffer is met, provided the sum of effective bandwidths is less than the link capacity. In this case, the definition of the effective bandwidth is more involved, but it depends still on the probabilistic description of the offered traffic flow.

The approach can be extended to a network, setting bounds on end-to-end performance.

Both approaches can incur in a low utilization of the link capacity, due to possibly loose upper bounds involved in the development of the theories. Those approaches make sense when dealing with traffic flows having strict delay and/or loss requirements. Both approaches trade-off manageability of the traffic control functions (policing, admission control), with efficiency. Elastic traffic can raise the utilization of the link, by exploiting the capacity not reserved for inelastic flows. Elastic flow use the leftover network capacity according to a reactive congestion control paradigm, that adapts to the available capacity on each link (see Chapter 10).

Problems

11.1 Calculate the min-plus convolution of the arrival curves $\gamma_{r,b}(t) = rt + b$, $t \geq 0$ and $\beta_{R,T}(t) = R\max\{0, t - T\}$. Discuss all cases with respect to the values of r and R. Remember that arrival curves are assumed to be 0 for $t < 0$.

11.2 Consider an arrival flow $A(t)$ constrained by a leaky bucket shaper with parameters (σ, ρ). The arrival flow pass through a service system having a rate-latency service curve $S(t) = C \cdot \max\{0, t - d\}$, with $C \geq \rho$. Prove that the output flow $D(t)$ is constrained by the arrival curve $\rho\max\{0, t - d\} + \sigma$. Apart from the delay d, the output is therefore constrained by a leaky bucket with the same parameters as the input flow.

11.3 Use the equivalent model Q of the flow buffer queue of the leaky bucket shaper introduced in Theorem 11.9 to derive the CCDF of the system time through the buffer for a two-state Markovian ON-OFF traffic source. The two-state Markovian source is described by a two-state Markov process with transition rates α from state 0 to state 1, and β from state 1 to state 0. The rate in state 0 is 0, in state 1 it is R (the peak rate).

[Hint: Make a fluid model of the flow buffer of the shaper, inspired to Example 8.14 of Chapter 8.]

11.4 Consider a Markovian traffic source in continuous time. The flow rate is given by $X(t) = r(Z(t))$, where $Z(t)$ is a Markov process over the state space $\{1, \ldots, m\}$, where $\mathbf{Q}$ is the infinitesimal generator of $Z(t)$ and $r(1) < \cdots < r(m)$ are m real, positive rate values. Let $\mathbf{R}$ denote the diagonal matrix with diagonal elements equal to the rates $r(j)$, $j = 1, \ldots, m$.

Show that the effective bandwidth of this traffic source is $e(\theta) = \eta(\theta)/\theta$, where $\eta(\theta)$ is the largest real part eigenvalue of the matrix $\mathbf{Q} + \theta\mathbf{R}$.

11.5 Consider the marginal buffer effective bandwidth of an ON-OFF source with peak rate R and activity factor p.

We multiplex homogeneous sources of that kind on a link of capacity C with a loss constraint of ϵ_L.

Find the optimal value of θ, i.e., the one that maximizes the number of traffic sources that can be multiplexed on the link under the loss constraint. Plot the number of sources that can be multiplexed as a function of θ to highlight that there is actually a maximum. For numerical values use $C = 1$, $R = C/10$, $p = 0.1$, $\epsilon_L = 10^{-3}$.

11.6 Consider a multiplexer with capacity C. Using the marginal buffer effective bandwidth definition, write a dimensioning algorithm that guarantees that ON-OFF flows with given peak rate R and mean rate $\bar{r}$ do not lose more than a fraction ϵ_L of their data and that the set-up rejection probability of a flow be less than a prescribed level ϵ. Assume connection requests arrive according to a Poisson process.

A

Refresher of Probability, Random Variables, and Stochastic Processes

The true logic of this world is the calculus of probabilities.

James Clerk Maxwell

This Appendix offers a concise refresher of the concepts of probability space, random variable and stochastic process. It is merely a collection of definitions and key properties that are used in this book. For more extended introductory treatments, specifically thought for non specialists, the textbook of Stewart [196] gives an excellent account of all basic concepts, from probability to Markov chains. Classic specialized textbooks on stochastic processes are [85, 184].

A.1 Probability

We refer to an experiment where some system runs and an outcome is observed. If it is not possible to anticipate the outcome, the experiment can profitably be modeled as a *stochastic* one by introducing the concept of probability. The mathematical model of probability is given by a triplet $(\Omega, \mathcal{E}, P)$, where:

- Ω is a set of elements (e.g., numbers, strings) called *samples*; they are outcomes of the stochastic experiment;
- $\mathcal{E}$ is a set of *events*, where an event is a set of samples; hence $\mathcal{E}$ is nothing but a subset of the power set of Ω;
- P is a *probability measure*, i.e., a map that assigns uniquely real numbers to events.

Network Traffic Engineering: Stochastic Models and Applications, First Edition. Andrea Baiocchi.
© 2020 John Wiley & Sons, Inc. Published 2020 by John Wiley & Sons, Inc.

In the axiomatic theory of probability, any probability measure is acceptable as long as it satisfies three axioms, namely:

1. For any event A, it is $0 \leq P(A) \leq 1$, i.e., probabilities are real number lying in the interval $[0, 1]$.
2. $P(\Omega) = 1$, i.e., the universal or certain event is assigned probability 1.
3. For any countable collection of events $\{A_i\}_{i \geq 1}$ that are mutually exclusive (i.e., $A_j \cap A_k = \emptyset$, $\forall j \neq k$), it must be $P(\bigcup_{i \geq 1} A_i) = \sum_{i \geq 1} P(A_i)$.

As long as the sample space is finite or denumerable, no special difficulty arises, and it is always possible to define a probability value for any event, i.e., $\mathcal{E}$ can coincide with the power set of Ω. This is no longer true in general when Ω is infinite and nondenumerable (e.g., the points belonging to a segment of the real line). In that case we must restrict $\mathcal{E}$ to those events for which we can assign probabilities satisfying all three axioms, i.e., events must be sets to which it is possible to assign a "measure." Indeed, the mathematical theory of probability is intimately connected to the theory of measure. We also insist that the set of events $\mathcal{E}$ be closed under the operations of countable union and intersection. A collection of subsets of a given set Ω that has those properties is referred to as a σ-field or also σ-algebra.

Summing up, a probability space is a triplet $(\Omega, \mathcal{E}, P)$, where Ω is a set, $\mathcal{E}$ is a σ-field of Ω that includes Ω itself, and P is a probability measure on $\mathcal{E}$ that satisfies the three axions given above.

The conditional probability of the event A given the event B is the probability measure that the first event occurs, once we are informed that the second event has taken place. In other words, in evaluating the probability measure of the conditional event, we account for those outcomes that are favorable to *both* A and B out of those that are favorable to B, since this last event has occurred. In mathematical terms, it is

$$P(A|B) = \frac{P(A \cap B)}{P(B)} \tag{A.1}$$

The joint probability of two events A and B is the probability measure that both of them occur, i.e., it is the probability measure of the set of outcomes that belong simultaneously to A and B.

Two events are said to be independent of each other if their joint probability factors in the product of their individual probability measures:

$$P(A \cap B) = P(A)P(B) \tag{A.2}$$

In general, we have $P(A \cap B) = P(A|B)P(B)$.

Given a set of events $\{B_i\}_{1 \leq i \leq n}$, that makes up a partition of the sample space Ω, the law of total probability can be stated as:

$$P(A) = \sum_{i=1}^{n} P(A \cap B_i) = \sum_{i=1}^{n} P(A|B_i)P(B_i) \tag{A.3}$$

The Bayes' rule can be also derived from the definitions above. Given any event A and any set of events $\{B_i\}_{i\geq 1}$, we have

$$P(B_i|A) = \frac{P(A|B_i)P(B_i)}{\sum_{j\geq 1} P(A|B_j)P(B_j)} \tag{A.4}$$

A.2 Random Variables

A random variable is actually a function. It maps elements belonging to the sample space Ω to a subset of the real numbers (we confine ourselves to real random variables), i.e., $X : \Omega \mapsto V \subseteq \mathbb{R}$.

A random variable induces an event space. Given $x \in V$, we can define the event $A_x \equiv \{\omega \in \Omega | X(\omega) = x\}$. The collection of the events A_x as x spans V is the event space associated to the random variable X. If V is a finite or denumerable set, then X is said to be a discrete random variable. Otherwise (e.g., if $V = \mathbb{R}$) we say that X is a continuous random variable.

Now we can define naturally probability measures on the random variable. For a discrete random variable we define the mass probability function as:

$$p_X(x) = P(A_x) = \sum_{\omega \in A_x} P(\omega), \quad x \in V \tag{A.5}$$

It is

$$\sum_{x \in V} p_X(x) = P(\cup_{x \in V} A_x) = P(\Omega) = 1 \tag{A.6}$$

In applications, one forgets about the event space and deals with the probability mass function as a function that assign probability values to each of the possible elements of V.

As for continuous random variable, we can define the cumulative distribution function, CDF, as

$$F_X(x) = P(X \leq x) = P(\{\omega | X(\omega) \leq x\}), \quad x \in \mathbb{R} \tag{A.7}$$

Since $\{\omega | X(\omega) \leq x_1\} \subseteq \{\omega | X(\omega) \leq x_2\}$ for $x_1 \leq x_2$, it is easy to verify that $F_X(x)$ is a monotonously nondecreasing function of x, with $F_X(\infty) = 1$ (since the set $\{\omega | X(\omega) \leq \infty\} \equiv \Omega$) and $F_X(-\infty) = 0$ (since the set $\{\omega | X(\omega) \leq -\infty\}$ is empty and $P(\emptyset) = 0$). It can also be shown that $F_X(x)$ is right-continuous. For any $a < b$, it is also

$$P(a < X \leq b) = P(X \leq b) - P(X \leq a) = F_X(b) - F_X(a) \tag{A.8}$$

If $a \to b$, then we get $P(X = b) = F_X(b) - F_X(b^-)$. Hence, the probability of a given point is always 0 except if the CDF has a jump at that point.

If the CDF has a derivative, we define the probability density function (PDF) of X as $f_X(x) = F'_X(x)$. Since

$$f_X(x) = \lim_{\Delta x \to 0} \frac{F_X(x + \Delta x) - F_X(x)}{\Delta x} = \lim_{\Delta x \to 0} \frac{P(x < X \leq x + \Delta x)}{\Delta x} \tag{A.9}$$

it follows that

$$f_X(x)\Delta x \approx P(x < X \leq x + \Delta x) \tag{A.10}$$

The CDF can be recovered from the PDF by integration:

$$F_X(x) = \int_{-\infty}^{x} f_X(t)\, dt \tag{A.11}$$

Any integrable, nonnegative function that sums to 1, i.e., such that $\int_{-\infty}^{\infty} f_X(t)\, dt = 1$, can be a PDF. In applications, e.g., measurements and computer simulations, it is preferable to use the CDF (or its complement, the survivor function $G_X(x) = 1 - F_X(x)$), rather than the PDF. This is because: (i) the PDF is dimensional, hence its values need be expressed in given measure units; on the contrary, the CDF is non dimensional; (ii) the PDF is not bounded in general, whereas the CDF is limited to the range $[0, 1]$; this simplifies numerical computations.

Functions of a random variable can be defined. In the special case where $Y = g(X)$ and $g(\cdot)$ is monotonous (increasing or decreasing) function, then

$$f_Y(y) = f_X(g^{-1}(y)) \left| \frac{dg^{-1}(y)}{dy} \right| \tag{A.12}$$

This can be shown by exploiting the following equality (we assume that $g(\cdot)$ is increasing):

$$F_Y(y) = P(Y \leq y) = P(g(X) \leq g(x)) = P(X \leq x)|_{x=g^{-1}(y)} = F_X(g^{-1}(y)) \tag{A.13}$$

Conditioned random variables can be defined. Let B be an event. The CDF of the random variable X conditional on the event B is given by

$$F_{X|B}(x) = \frac{P(X \leq x, B)}{P(B)} \tag{A.14}$$

Joint random variables can be defined. The joint CDF of the two random variables X and Y is

$$F_{X,Y}(x, y) = P(X \leq x, Y \leq y) \tag{A.15}$$

The joint density is found by partial derivation with respect to x and y: $f_{X,Y}(x, y) = \frac{\partial^2 F_{X,Y}(x,y)}{\partial x \partial y}$. The CDF of a single random variable is called the marginal CDF. For example, the marginal CDF of X is found as $F_X(x) = F_{X,Y}(x, \infty)$. If the random variables X and Y are independent, the joint CDF factors in the product of the marginal CDFs: $F_{X,Y}(x, y) = P(X \leq x)P(Y \leq y) = F_X(x)F_Y(y)$.

Finally, let us consider a random variable $Z = X + Y$. The CDF of Z can be expressed as a function of the joint PDF of X and Y:

$$F_Z(z) = P(X + Y \leq z) = \int_{-\infty}^{\infty} \int_{-\infty}^{z-x} f_{X,Y}(x,y) \, dy dx \tag{A.16}$$

By deriving with respect to z, we find

$$f_Z(z) = \frac{d}{dz} \int_{-\infty}^{\infty} \int_{-\infty}^{z-x} f_{X,Y}(x,y) \, dy dx = \int_{-\infty}^{\infty} f_{X,Y}(x, z - x) dx \tag{A.17}$$

If the random variables X and Y are independent of each other, eq. (A.18) yields

$$f_Z(z) = \int_{-\infty}^{\infty} f_X(x) f_Y(z - x) dx \tag{A.18}$$

i.e., the PDF of Z is obtained as the convolution of the PDFs of the summands X and Y. For a non-negative random variable, i.e., a random variable with $F_X(x) = 0$ for $x < 0$, the convolution integral is $f_Z(z) = \int_0^z f_X(x) f_Y(z - x) dx$.

Expectations of a random variable are defined as

$$E[g(X)] = \int_{-\infty}^{\infty} g(t) f_X(t) \, dt \tag{A.19}$$

for a continuous random variable and

$$E[g(X)] = \sum_{x \in V} g(x) f_X(x) \tag{A.20}$$

for a discrete random variable. If $g(x) = x^n$ the resulting expectation is called moment of order n.

The minimum and the maximum of n independent random variables $X_1, X_2, \ldots, X_n$ deserve a special mention. Let us assume that the n random variables have the same probability distribution. Then, letting $Y = \min\{X_1, X_2, \ldots, X_n\}$, we get

$$P(Y > y) = P(X_1 > y, X_2 > y, \ldots, X_n > y) = \prod_{i=1}^{n} P(X_i > y) = P(X_1 > y)^n \tag{A.21}$$

Analogously, in case of the maximum, i.e., $Z = \max\{X_1, X_2, \ldots, X_n\}$:

$$P(Z \leq z) = P(X_1 \leq z, X_2 \leq z, \ldots, X_n \leq z) = \prod_{i=1}^{n} P(X_i \leq z) = P(X_1 \leq z)^n \tag{A.22}$$

A.3 Transforms of Probability Distribution Functions

It is common to resort to transforms when dealing with probability distribution functions. Let X be a continuous, non-negative random variable, with PDF $f_X(x)$.

Its Laplace transform is defined as

$$\varphi_X(s) = \int_0^\infty f_X(x)e^{-sx}\, dx \tag{A.23}$$

for s such that $\Re[s] \geq -\beta$, β being a non-negative constant, that depends on the convergence properties of $f_X(x)$ as $x \to \infty$. If $f_X(x)$ decays exponentially with rate η, the abscissa of convergence $-\beta$ equals $-\eta$ and $\varphi_X(s)$ has a singularity (simple pole) at $s = -\eta$. This can be seen by using the final value theorem. The theorem states that $\lim_{t\to\infty} f_X(t) = \lim_{s\to 0} s\, \varphi_X(s)$. Moreover, the Laplace transform of $e^{\eta t}f_X(t)$ is $\varphi_X(s - \eta)$. Then, if $f_X(t)$ decays exponentially with rate η, we have

$$\lim_{s\to-\eta}(s+\eta)\varphi_X(s) = \lim_{s\to 0} s\,\varphi_X(s-\eta) = \lim_{t\to\infty} e^{\eta t}f_X(t) = c \tag{A.24}$$

where c is a positive constant.

Note that $\varphi_X(0) = 1$ since $f_X(x)$ is a PDF. This transform can be often calculated more easily than the density $f_X(x)$. Then, the density can be recovered by numerical inversion. If only moments are required, they can be obtained by deriving $\varphi_X(s)$:

$$E[X^n] = \int_0^\infty x^n f_X(x)\, dx = (-1)^n \varphi_X^{(n)}(0), \qquad n \geq 1, \tag{A.25}$$

where $\varphi_X^{(n)}(s)$ denotes the n-th derivative of $\varphi_X(s)$.

A very useful way of writing the Laplace transform of a PDF is formally $\varphi_X(s) = E[e^{-sX}]$. As a matter of example, this way it is trivial to derive that the Laplace transform of the PDF of the sum of independent random variables is the product of the respective Laplace transforms, i.e., if $Y = X_1 + \cdots + X_k$, then

$$\varphi_Y(s) = E[e^{-sY}] = E[e^{-s(X_1+\cdots+X_k)}] = \prod_{j=1}^k E[e^{-sX_j}] = \prod_{j=1}^k \varphi_{X_j}(s) \tag{A.26}$$

In case of real random variables that take also negative values, it is customary to define the moment generating function (MGF) as

$$\phi_X(\theta) = E[e^{\theta X}] = \int_{-\infty}^\infty f_X(x)e^{\theta x}\, dx \tag{A.27}$$

The integral converges for any negative θ. If the PDF of X decays exponentially with rate η, the integral converges also for $\theta < \eta$.

Analogously, the MGF of a discrete probability distribution $p_k = P(X = k), k \geq 0$ is defined as

$$\phi_X(z) = E[z^X] = \sum_{k=0}^\infty p_k z^k \tag{A.28}$$

for $|z| \leq 1$. Also for discrete random variables it can be verified that the MGF of the sum of independent random variables is obtained as the product of the MGFs of the random variables.

Derivatives of $\phi_X(z)$ yield factorial moments, namely: $\phi_X^{(n)}(1) = E[X(X-1)\cdots(X-n+1)]$. As a matter of example, the first two central moments of X are found as $E[X] = \phi_X'(1)$ and $\sigma_X^2 = \phi_X''(1) + \phi_X'(1) - [\phi_X'(1)]^2$.

Handy as it might be to derive moments, the Laplace transform of a probability distribution would be of limited use, if it were not possible to get back to the probability distribution, by means of numerical inversion. The topic has been extensively studied, given the extensive applications of transforms in various scientific and engineering fields. We confine ourselves to the works of Abate and Whitt, and specifically to two inversion methods: (i) the Fourier transform inversion; and (ii) the Euler method. Here we outline the essential points to grasp the numerical recipes and to be able to implement them. The theory backing numerical methods, estimate of approximation errors, a discussion on numerical stability, hints as to a proper choice of the algorithm parameter values, numerical examples can be found in [1–3].

As for the first method, we recall that the Fourier transform of a function $f(t)$ integrable over the real axis is defined by:

$$\phi(u) = \int_{-\infty}^{\infty} e^{iut} f(t) \, dt \tag{A.29}$$

Under mild conditions (e.g., it is sufficient that $f(t)$ be continuous, besides being integrable over the real axis) the inversion formula is:

$$f(t) = \frac{1}{2\pi} \int_{-\infty}^{\infty} e^{-iut} \phi(u) \, du \tag{A.30}$$

We are interested in the case where $f(t)$ is the PDF of a random variable X, hence a real function. Moreover, it is $\phi(u) = \mathrm{E}[e^{iuX}]$. Focusing our attention on random variables recurring in network traffic engineering, we further restrict ourselves to the consideration of a non-negative X. Hence the PDF and the CDF are non-null only for non-negative argument values. Then, the connection between the Laplace transform of the PDF of X and $\phi(u)$ is easily established:

$$\phi(u) = \int_{0}^{\infty} e^{iut} f(t) \, dt = \int_{0}^{\infty} e^{-st} f(t) \, dt\big|_{s=-iu} = \varphi(s)\big|_{s=-iu} = \varphi(-iu) \tag{A.31}$$

It is shown, e.g., in [1, eq. (3.6)], that the CDF $F(t) = P(X \le t)$ can be obtained as follows:

$$F(t) = \frac{2}{\pi} \int_{0}^{\infty} \mathrm{Re}[\phi(u)] \frac{\sin(ut)}{u} \, du \tag{A.32}$$

By discretizing the integral with step h, we get

$$F(t) \approx \frac{ht}{\pi} + \frac{2}{\pi} \sum_{k=1}^{N} \mathrm{Re}[\phi(kh)] \frac{\sin(kht)}{k} \tag{A.33}$$

There are two sources of error: (i) discretization (depends on the step size h); (ii) truncation (depends on the upper limit N of the summation).

If the function $F(t)$ has to be computed in a number of different points in the range $[0, T]$, it can be convenient to make some changes in (A.33), by letting $t = jT/n$ and $h = \pi/(mT)$ for some integers n and m. Then

$$F(jT/n) \approx \frac{j}{mn} + \frac{2}{\pi} \sum_{k=1}^{2m\,n-1} \alpha_k \sin\left(\frac{\pi j k}{m n}\right), \qquad j = 0, \dots, n. \tag{A.34}$$

where

$$\alpha_k = \sum_{l=0}^{N/(2mn)} \frac{\mathrm{Re}[\phi((2mnl+k)h)]}{2mnl+k} , \qquad k = 1, \dots, 2mn-1. \qquad (A.35)$$

An implementation in Matlab code of this algorithm is shown next[1] . In that code we consider a gamma random variable X. Hence the PDF of X is $f(t) = a\frac{(at)^{b-1}}{\Gamma(b)}e^{-bt}$, $t \geq 0$, and $f(t) = 0$, $t < 0$, with $\mathrm{E}[X] = b/a$, $\sigma_X^2 = b/a^2$; $\Gamma(b)$ is the Euler gamma function defined by $\Gamma(b) = \int_0^\infty u^{b-1}e^{-u}\,du$. The Laplace transform of this PDF is simply $\varphi(s) = \left(\frac{a}{s+a}\right)^b$.

```
% first two moments of the rv X
EX = 1;
sigmaX = 3;
% parameters of gamma PDF of X
aa = EX/sigmaX/sigmaX;
bb = (EX/sigmaX)^2;
tmax = 20*EX;
deltat = EX/10;
np = ceil(tmax/deltat);
tv = tmax*[0:np]/np;
% points for LT inversion
lmax = 7;    % N = lmax*2*np*mp
mp = 11;
hh = pi/mp/tmax;
sv = i*[1:2*mp*np*lmax+2*mp*np-1]*hh;
nums = length(sv);
phiXsv = (aa./(aa+sv)).^bb;      % LT of the gamma PDF
for kk=1:2*mp*np-1
   indvet = [kk:2*mp*np:nums];
   alfav(kk) = sum(real(phiXsv(indvet))./indvet);
end
for jj=0:np
   sinv = sin([1:2*mp*np-1]*jj*pi/mp/np);
   CDFXv(jj+1) = (jj/mp/np)+(2/pi)*sum(alfav.*sinv);
end
CCDFXv = 1-CDFXv;
Gv = 1-gammainc(aa*tv,bb);    % CCDF of X
```

---

[1] This one and all other pieces of Matlab code provided in this text are only examples and they are not intended to be production-level software.

The Euler algorithm is an implementation of the Fourier-series method, using Euler summation to accelerate convergence of the infinite series; e.g., see Abate and Whitt [3]. The approximation of the PDF at a point t is computed according to the expression:

$$f(t) \approx \frac{10^{M/3}}{t} \sum_{k=0}^{2M} \eta_k \mathrm{Re}\left[\varphi\left(\frac{\beta_k}{t}\right)\right] \tag{A.36}$$

where M is an integer, that should be set as $\lceil 1.7n \rceil$, n being the number of significant digits desired, and the following definitions are used:

$$\beta_k = \frac{M \log(10)}{3} + i\pi k \qquad \eta_k = (-1)^k \xi_k \tag{A.37}$$

$$\xi_k = \begin{cases} \frac{1}{2} & k = 0 \\ 1 & k = 1, \dots, M \\ \frac{1}{2^M} \sum_{j=0}^{2M-k} \binom{M}{j} & k = M+1, \dots, 2M \end{cases} \tag{A.38}$$

A Matlab code to implement the Euler algorithm for the CCDF of X is reported below.

```
% the first lines of code are the same as in the code
% for the Fourier series method
EX = 1;
sigmaX = 3;
% parameters of gamma PDF of X
aa = EX/sigmaX/sigmaX;
bb = (EX/sigmaX)^2;
tmax = 20*EX;
deltat = EX/10;
np = ceil(tmax/deltat);
tv = tmax*[0:np]/np;
Fv = gammainc(aa*tv,bb);   % CDF of X
MM = 20;
Mbino(1) = 1;
for kk=1:MM
  Mbino(kk+1) = Mbino(kk)*(MM-kk+1)/kk;
end
csiv = ones(1,2*MM);
sumcsiv = cumsum(Mbino);
csiv(MM:2*MM) = sumcsiv(MM+1:-1:1)/(2^MM);
csiv = [.5 csiv];
```

```
etav = cos(pi*[0:2*MM]).*csiv;
betav = MM*log(10)/3+i*pi*[0:2*MM];
CCDFXv(1) = 1;
for nn=1:np
  tt = tv(nn);
  phiXsv = (aa./(aa+betav/tt)).^bb;
  c = exp(MM*log(10)/3);
  CCDFXv(nn+1) = c*sum(etav.*real((1-phiXsv)./betav));
end
```

A.4 Inequalities and Limit Theorems

Exact results are rarely applicable, mainly since they are derived under too restrictive hypotheses. Often, approximations are required. Sometimes bounding limits may also be useful. This section is devoted to the presentation of some basic inequalities, namely, Markov inequality, Chebychev inequality, Jensen inequality, Chernov bound, and the union bound. A hint is given to the central limit theorem as well.

A.4.1 Markov Inequality

Let X be a random variable with PDF $f_X(x)$. Let $h(x)$ be a non-decreasing non-negative function of x that has finite expectation

$$E[h(X)] = \int_{-\infty}^{\infty} h(u)f_X(u)\, du \tag{A.39}$$

For any t we can write

$$E[h(X)] = \int_{-\infty}^{t} h(u)f_X(u)\, du + \int_{t}^{\infty} h(u)f_X(u)\, du$$

$$\geq h(t) \int_{t}^{\infty} f_X(u)\, du = h(t)P(X > t)$$

Hence, if $h(t) > 0$, we get

$$P(X > t) \leq \frac{E[h(X)]}{h(t)} \tag{A.40}$$

As a matter of example, in case of a non-negative random variable X and $h(x) = x$, we obtain the following upper bound of the tail of the PDF of X:

$$P(X > t) \leq \frac{E[X]}{t}, \qquad t > 0. \tag{A.41}$$

As simple as it is, the Markov inequality turns out often to provide a quite loose bound. This limits its practical usefulness. For instance, if X is a negative exponential random variable with mean 1, it is $P(X > t) = e^{-t}$ and the Markov inequality states that $e^{-t} \leq 1/t$.

A.4.2 Chebychev Inequality

Let us assume that X has finite variance σ_X^2. We can form the new random variable $Y = (X - E[X])^2$. By applying the Markov inequality to Y with $h(x) = x$, we find

$$P(Y > t^2) \leq \frac{E[Y]}{t^2} = \frac{E[(X - E[X])^2]}{t^2} = \frac{\sigma_X^2}{t^2} \tag{A.42}$$

that is to say, for $t > 0$,

$$P(|X - E[X]| > t) \leq \frac{\sigma_X^2}{t^2} \tag{A.43}$$

This can be rewritten by setting $t = a\sigma_X$, with a a positive constant:

$$P\left(\frac{|X - E[X]|}{\sigma_X} > a\right) \leq \frac{1}{a^2} \tag{A.44}$$

The Chebychev bound yields tighter results with respect to Markov inequality, yet it is still often too loose to give practical results in many applications. Nevertheless, these inequalities are sometimes useful in proving properties of random variables.

As an application of the Chebychev bound, let us consider the sum of n i.i.d. random variables, having the same PDF as X, namely, $S_n = X_1 + \cdots + X_n$. We have $E[S_n] = nE[X]$ and $\sigma_{S_n}^2 = n\sigma_X^2$. By applying the inequality to S we get

$$P(|S_n - nE[X]| > t) \leq \frac{n\sigma_X^2}{t^2} \tag{A.45}$$

By setting $t = n\epsilon$, we obtain

$$P\left(\left|\frac{S_n}{n} - E[X]\right| > \epsilon\right) \leq \frac{\sigma_X^2}{n\epsilon^2} \tag{A.46}$$

holding for all positive integer n. Taking the limit for $n \to \infty$, we have

$$\lim_{n \to \infty} P\left(\left|\frac{S_n}{n} - E[X]\right| > \epsilon\right) = 0 \tag{A.47}$$

This is the *weak law of large numbers*. What it means is that the statistical average obtained from repeated experiments (S_n/n) converges in probability to the mean of the random variable as the number of experiments grows. It provides a solid ground to the study of random phenomenon through repeated experiments, e.g., as done in stochastic simulations of service systems.

A.4.3 Jensen Inequality

A function is said to be convex in an interval $[a, b]$ if, for any two points $x, y \in [a.b]$ and any $\lambda \in [0, 1]$ it is

$$g(\lambda x + (1 - \lambda)y) \leq \lambda g(x) + (1 - \lambda)g(y) \tag{A.48}$$

that is to say, the plot of the function $g(\cdot)$ is under any secant of the curve.

Jensen's inequality states that for a random variable X with state space contained in the convexity domain of $g(x)$ it is

$$E[g(X)] \geq g(E[X]) \tag{A.49}$$

For a simple proof, assume $g(x)$ is differentiable. The curve $g(x)$ is above any tangent to the curve, i.e., $g(x) \geq g(x_0) + g'(x_0)(x - x_0)$. Let us consider the tangent at the point $x_0 = E[X]$. Then, it is $g(x) \geq g(E[X]) + g'(E[X])(x - E[X])$. Multiplying both sides of the inequality by the PDF of X and integrating over the domain D of X, we find

$$\int_D f_X(x)g(x)\, dx \geq \int_D f_X(x)g(E[X])\, dx + g'(E[X]) \int_D f_X(x)(x - E[X])\, dx$$

$$= \int_D f_X(x)g(E[X])\, dx = g(E[X])$$

If X is a discrete random variable over a finite set of values, Jensen inequality is a direct consequence of the definition of convexity. Let $p_i = P(X = x_i)$, for $i = 1, \ldots, m$. Since $\sum_{i=1}^m p_i = 1$, we have

$$E[g(X)] = \sum_{i=1}^m p_i g(x_i) \geq g\left(\sum_{i=1}^m p_i x_i\right) = g(E[X]) \tag{A.50}$$

A.4.4 Chernov Bound

It can be obtained as another application of the Markov inequality, by choosing $h(x) = e^{\theta x}$ for a positive parameter θ. Then

$$P(X > t) \leq \frac{E[e^{\theta X}]}{e^{\theta t}} = e^{-\theta t} \phi_X(\theta), \qquad \theta > 0. \tag{A.51}$$

where $\phi_X(\theta) = E[e^{\theta X}]$ is the MGF of the random variable X. Since the upper bound holds for all positive θ for which $\phi_X(\theta)$ is finite, we have

$$P(X > t) \leq \inf_{\theta \geq 0} \{e^{-\theta t} \phi_X(\theta)\} \tag{A.52}$$

As a matter of example, let us consider a Gaussian random variable with mean μ and variance σ^2. It is $\phi_X(\theta) = e^{\mu\theta + \sigma^2\theta^2/2}$; then

$$P(X > t) \leq \inf_{\theta \geq 0} \{e^{-\theta t} e^{\mu\theta + \sigma^2\theta^2/2}\} = e^{\inf_{\theta \geq 0} \{(\mu - t)\theta + \sigma^2\theta^2/2\}} \tag{A.53}$$

For $t > \mu$ the quadratic is minimized for $\theta^* = (t - \mu)/\sigma^2$. Therefore

$$P(X > t) \le e^{-\frac{(t-\mu)^2}{2\sigma^2}} \tag{A.54}$$

The Chernov bound is clearly useful whenever it is difficult to calculate the CCDF of a random variable, but the Laplace transform of its PDF can be obtained. A typical case is when $X = V_1 + \cdots + V_n$, i.e., X is the sum of n independent random variables. If the V's are also identically distributed, then $\phi_X(\theta) = [\phi_V(\theta)]^n$.

A.4.5 Union Bound

This is a simple, but often useful bound. It refers to the probability of an event $\mathcal{A}$ that is the union of several events $\mathcal{B}_i$, for $i = 1, \ldots, n$, where n can also be infinite.

If an outcome x belongs to any of the $\mathcal{B}_i$ for some i, then it belongs also to $\mathcal{A}$. The events $\mathcal{B}_i$ need not be mutually disjoint. Therefore, the sum of their measures is no less than the measure of their union, i.e., the event $\mathcal{A}$. This brings to the following inequality:

$$P\left(\bigcup_{i=1}^{n} \mathcal{B}_i\right) \le \sum_{i=1}^{n} P(\mathcal{B}_i) \tag{A.55}$$

A.4.6 Central Limit Theorem (CLT)

Given n i.i.d. random variables X_i, with mean μ and variance σ^2, the celebrated CLT states that

$$\lim_{n \to \infty} P\left(\sum_{i=1}^{n} \frac{X_i - \mu}{\sigma\sqrt{n}} > x\right) = \frac{1}{\sqrt{2\pi}} \int_x^\infty e^{-u^2/2} \, du \tag{A.56}$$

In words, we can say that $Y_n = \frac{1}{\sqrt{n}} \sum_{i=1}^{n}(X_i - \mu)$ tends to a zero-mean Gaussian random variable with variance σ^2.

We give an informal proof. Let $\varphi_X(s)$ be the Laplace transform of the PDF of $X \sim X_i - \mu$, which is a zero-mean random variable. We have

$$
\begin{aligned}
E[e^{-sY_n}] &= E\left[\exp\left(-\frac{s}{\sqrt{n}} \sum_{i=1}^{n}(X_i - \mu)\right)\right] = \prod_{i=1}^{n} E\left[\exp\left(-s\frac{X_i - \mu}{\sqrt{n}}\right)\right] \\
&= \left[\varphi_X\left(\frac{s}{\sqrt{n}}\right)\right]^n = \left(1 + s\frac{E[X_i - \mu]}{\sqrt{n}} + \frac{1}{2}s^2\frac{E[(X_i - \mu)^2]}{n} + o\left(\frac{1}{n}\right)\right)^n \\
&= \left(1 + \frac{\sigma^2 s^2}{2n} + o\left(\frac{1}{n}\right)\right)^n \to e^{\sigma^2 s^2/2} \quad (n \to \infty)
\end{aligned}
$$

This proves that $\varphi_{Y_n}(s) \to \varphi_Y(s) = e^{\sigma^2 s^2/2}$ as $n \to \infty$. Inverting the Laplace transform of the limiting random variable Y, it is verified that Y is a zero-mean Gaussian random ariable with variance σ^2.

A.5 Stochastic Processes

Let us assign a probability space $(\Omega, \mathcal{E}, P)$. Let us also define a set of "times" $\mathcal{T}$. A stochastic process can be defined as follows. For each sample $\omega \in \Omega$, we assign a time function $X(t, \omega)$, $t \in \mathcal{T}$. In other words, a stochastic process is a collection of random variables $\{X(t, \omega)\}$ indexed by $t \in \mathcal{T}$. For ease of notation, a stochastic process is usually denoted with $X(t)$. The statistical description of the stochastic process can be done by defining the CDF as a function of the parameter t:

$$F_X(x; t) = P(X(t) \leq x) \tag{A.57}$$

This is also known as a first-order description of the process. It can be generalized to any integer order n as

$$F_{X_1 X_2 \dots X_n}(x_1, x_2, \dots, x_n; t_1, t_2, \dots, t_n) = P(X(t_1) \leq x_1, X(t_2) \leq x_2, \dots, X(t_n) \leq x_n) \tag{A.58}$$

A stochastic process $X(t)$ is said to be stationary if a time shift does not affect its probability distributions. Formally

$$P(X(t_1) \leq x_1, \dots, X(t_k) \leq x_k) = P(X(t_1 + h) \leq x_1, \dots, X(t_k + h) \leq x_k) \tag{A.59}$$

for any integer $k \geq 1$, any h, and any set of k times $t_1, \dots, t_k$ and any set of k values $x_1, \dots, x_k$. As a consequence of stationarity, all moments of the first-order probability distribution are independent of time. Specifically we define

$$m_X \equiv E[X(t)]$$
$$\sigma_X^2 \equiv E[(X(t) - m_X)^2] = E[X(t)^2] - E[X(t)]^2$$
$$\gamma_X(h) \equiv E[X(t)X(t + h)] - m_X^2$$

where $\gamma_X(h)$ is the auto-covariance of the process $X(t)$. If only the equalities above hold, we define the process $X(t)$ to be wide-sense stationary.

We are also interested in the *ergodicity* of stochastic processes. Let us confine ourselves to discrete state processes. Intuitively, ergodicity is often stated by saying that each realization of the process is "typical." To be more explicit, the idea of ergodicity is that a realization of the process $x(t)$, during its evolution over time, "touches" all states. As a matter of example, let us assume that we collect a trace of a wireless link path loss over time. If the collected trace has a duration comparable or even less than the typical time over which fluctuations of the path loss take place, e.g., because of user mobility or environmental changes that modify the pattern of radio wave propagation, then the values that we have registered do not tell us "the whole story," i.e., they are affected by the specific conditions under which we have observed the wireless link (e.g., we might be lucky and register a good channel or, conversely, we could run into a deep fading because of a temporary obstruction). On the opposite, if our collected trace lasts for much longer than the time constants that affect the radio link path loss fluctuations, we can deem our measurements to be "representative" of anything a real user could experience. This same example points out that ergodicity is a useful property, if we are interested in metrics collected over a sufficiently long observation time.

To give a more precise statement, suffice it to say here that, given the state space S, the limit $\lim_{t\to\infty} P(X(t) = s) = \pi_s$ must exist for each state $s \in S$, independent of the initial state probability distribution $P(X(0) = s)$. The set of numbers $\{\pi_s\}_{s\in S}$ must be a proper probability distribution, i.e., $\pi_s \geq 0$ for each $s \in S$ and $\sum_{s\in S} \pi_s = 1$.

A.6 Markov Chains

Extensive accounts on Markov chains and Markov processes can be found in any stochastic process textbook. Just to mention a few good references, the interested reader could find excellent material in [60, 117, 186]. Here we confine ourselves to basic definitions and key properties.

Let S denote a countable set representing a state space. In the following we think of S as the set of non-negative integers $\mathbb{N}$, unless stated otherwise.

Definition A.1 A stochastic process $X = \{X_n, \ n \geq 0\}$ over the state space S is a Markov chain if [158]

$$P(X_n = i_n | X_0 = i_0, X_1 = i_1, \dots, X_{n-1} = i_{n-1}) = P(X_n = i_n | X_{n-1} = i_{n-1})$$

(A.60)

for all n and all $i_0, i_1, \dots, i_n \in S$. The Markov chain is *homogeneous* if for all $n, k \in \mathbb{N}$ and all $i, j \in S$ we have

$$P(X_{n+k} = j | X_k = i) = P(X_n = j | X_0 = i)$$

(A.61)

By definition, a Markov chain is a discrete-time, discrete-state stochastic process where all the past history up to any time n can be summarized by the current state of the process. Once that is known, all previous states are irrelevant for the future evolution of the process after n. A time-homogeneous Markov chain is insensitive to time shifts, i.e., the stochastic mechanism that makes the chain evolve is time-invariant. In the following we refer to time-homogeneous Markov chains only, unless stated otherwise.

A key role is played by the so called one-step transition probability matrix **P**, whose entry (i, j) is defined as:

$$P_{i,j} = P(X_{n+1} = j | X_n = i), \quad i, j \in S$$

(A.62)

Since the Markov chain is time-homogeneous, $\mathbf{P}$ does not depend on the time index n. The elements of $\mathbf{P}$ are non-negative and such that $\sum_{j \in S} P_{i,j} = 1$ for all $i \in S$. The last equality states simply that the process moves for sure to some state when leaving i. A matrix having non-negative entries and row sums equal to 1 is called *stochastic*. It can be shown that a necessary and sufficient condition for a discrete process X to be a Markov chain is that

$$P(X_n = i_n, \ldots, X_0 = i_0) = \alpha_{i_0} P_{i_0, i_1} P_{i_1, i_2} \cdots P_{i_{n-1}, i_n} \tag{A.63}$$

for all $n \le 0$ and all $i_0, \ldots, i_n \in S$, where α is a vector of the initial state probabilities $\alpha_i = P(X_0 = i)$, $i \in S$. Hence, the matrix $\mathbf{P}$ is a complete description of the process from a probabilistic point of view.

It can be shown from eq. (A.63) that $P(X_n = i)$ is the i-th entry of the row vector $\alpha \mathbf{P}^n$ and $P(X_n = j | X_0 = i)$ is the entry (i, j) of the matrix $\mathbf{P}^n$. It is easy to show also that for all $n \ge 0$ the matrix $\mathbf{P}^n$ is stochastic.

A.6.1 Classification of States

A state i is *accessible* from j if there exists $n \ge 0$ such that $P_{i,j}^n > 0$, that is, it is possible to move from i to j in a finite number of steps with positive probability. If i is accessible from j and j is accessible from i, we say that i and j *communicate*. Communication is an equivalence relationship. It is easy to show that it is reflexive and symmetric, and it admits transitivity. Then, we can partition the states of S in classes of equivalence. States belonging to a same class C_k communicate to one another. It is also possible that some states in a class are accessible by states in another class, though states belonging to the two classes cannot communicate, otherwise the two classes could be merged into a single one. An example matrix $\mathbf{P}$ of a Markov chain with two classes is as follows:

$$\mathbf{P} = \begin{bmatrix} \mathbf{P}_{1,1} & \mathbf{0} \\ \mathbf{P}_{2,1} & \mathbf{P}_{2,2} \end{bmatrix} \tag{A.64}$$

where the block matrix $\mathbf{P}_{2,1}$ could have all entries equal to 0, while the diagonal block $\mathbf{P}_{1,1}$ cannot be completely equal to 0 (rows must sum to 1).

A Markov chain that has only a single class, i.e., where all states communicate, is said to be *irreducible*. Formally:

Definition A.2 (Irreducibility) A Markov chain is irreducible if for any couple of states i and j there exist a finite $n \ge 1$ such that $P(X_n = j | X_0 = i) > 0$. A Markov process is irreducible if for any couple of states i and j there exist a finite $t > 0$ such that $P(X(t) = j | X(0) = i) > 0$.

From an algebraic point of view, there does not exist any permutation of the states such that the one-step transition probability matrix can be put in the form of eq. (A.64).

The notion of *period* of a state i, denoted with $d(i)$, can be defined as the greatest common divisor (gcd) of all integers $n \geq 1$ such that $P_{i,i}^n > 0$ (if $P_{i,i}^n = 0$ for all n, we set $d(i) = 0$). A state with period $d(i)$ can be visited only at times that are multiples of $d(i)$, given that the chain is initialized in i at time 0. If $P_{i,i} > 0$, state i has period 1. In general, if $d(i) = 1$, we say that i is aperiodic. As an example, the m states of the Markov chain defined by the $m \times m$ matrix

$$\mathbf{P} = \begin{bmatrix} 0 & 1 & 0 & 0 & \cdots & 0 \\ 0 & 0 & 1 & 0 & \cdots & 0 \\ 0 & 0 & 0 & 1 & \cdots & 0 \\ \cdots & \cdots & \cdots & \cdots & \cdots & \cdots \\ 1 & 0 & 0 & 0 & \cdots & 0 \end{bmatrix} \tag{A.65}$$

have all period m (this is a circular chain, where the state moves from i to $1 + i \bmod m$.)

It can be shown that if two states communicate, they must have the same period.

Markov chains are often represented as graphs. The correspondence between a Markov chain and a directed labeled graph can be established as follows. States are mapped to nodes of the graphs. An arc is defined between nodes i and j if $P_{i,j} > 0$. The label of the arc is $P_{i,j}$. The notions of accessibility and communication can be translated into reachability and connectivity. Node j is accessible from i if there exists a path in the graph from i to j. An irreducible Markov chain corresponds to a strongly connected graph and vice versa.

The most interesting aspect of a Markov chain is its dynamics, i.e., the evolution over time of the state probability distribution, given the initial probability distribution at time 0 (i.e., the initialization of the Markov chain). To understand the asymptotic behavior of a Markov chain we have to introduce the notion of *recurrence*.

A.6.2 Recurrence

Give a state $i \in S$, we define the first return probability for any non-negative n

$$f_{i,i}^{(n)} = P(X_n = i, X_k \neq i, \; k = 1, \ldots, n-1 \mid X_0 = i), \quad n \geq 2. \tag{A.66}$$

For $n = 0$ we let $f_{i,i}^{(0)} = 0$ and for $n = 1$ we have obviously $f_{i,i}^{(1)} = P_{i,i}$. $f_{i,i}^{(n)}$ is the probability that state i is reached again for the first time only at the n-th step, given that the chain leaves from i at time 0. This is an example of *first-passage probability*. For a time-homogeneous Markov chain, this probability depends only on the number of steps since departure from state i, not on the absolute time when the journey starts. We can also define

$$f_{i,j}^{(n)} = P(X_n = j, X_k \neq j, \; k = 1, \ldots, n-1 \mid X_0 = i), \quad n \geq 2, \tag{A.67}$$

i.e., the first-passage probability to state j, given that the chain starts out from state i. Note that $f_{i,j}^{(0)} = 0$. Moreover, $f_{i,j}^{(1)} = P_{i,j}$. It can be shown that

$$P_{i,j}^{(n)} = \sum_{k=1}^{n} f_{i,j}^{(k)} P_{j,j}^{(n-k)}, \qquad n \geq 1, \ \forall i, j \in S, \tag{A.68}$$

where $P_{i,j}^{(n)} = P(X_n = j | X_0 = i)$. Note that $P_{i,j}^{(0)} = \delta_{i,j}$, with $\delta_{i,j}$ denoting the delta of Kronecker, equal to 1 if $i = j$, to 0 otherwise.

Equation (A.68) states simply that the probability of moving from i to j in n steps is the sum of the probabilities of the disjoint and exhaustive events that state j is first touched at the k-th step, for $k = 1, \ldots, n$. First-passage arguments can be used also to find a recursion for the computation of the probabilities $f_{i,j}^{(n)}$. The probability of touching state j for the first time after n steps, starting from state i, can be decomposed based on the state $\ell \neq j$ visited at the first step. It can be shown that

$$f_{i,j}^{(n)} = \begin{cases} P_{i,j} & n = 1 \\ \displaystyle\sum_{\ell \in S \setminus \{j\}} P_{i,\ell} f_{\ell,j}^{(n-1)} & n \geq 2 \end{cases} \tag{A.69}$$

We define also

$$f_{i,j} = \sum_{n=1}^{\infty} f_{i,j}^{(n)} \tag{A.70}$$

that is to say the probability of ever getting to state j when leaving from state i. Note that j being accessible from i is a necessary condition for $f_{i,j} = 1$, but it is not sufficient. It could happen that the state "escapes" from i to a set of states that do not communicate with j and remains locked there forever. Summing eq. (A.69) over n, we find:

$$f_{i,j} = P_{i,j} + \sum_{\ell \in S \setminus \{j\}} P_{i,\ell} f_{\ell,j} \tag{A.71}$$

Definition A.3   A state i is said to be *recurrent* if $f_{i,i} = 1$, *transient* [2] if $f_{i,i} < 1$. A Markov chain is said to be recurrent or transient if all of its states are recurrent or transient, respectively. A state is said to be *absorbing* if $P_{i,i} = 1$.

If i is absorbing, it must be $f_{i,i}^{(n)} = 0$ for $n \geq 2$, since, once the Markov chain visits i, it cannot leave i anymore. It is also $f_{i,i}^{(1)} = 1$ and hence $f_{i,i} = 1$, i.e., an absorbing state is recurrent.

---

2 A state is called *ephemeral* if it can be visited at most once. A transient state can be visited more times in general.

We can define the generating function of $f_{i,j}^{(n)}$ as $F_{i,j}(z) = \sum_{n=1}^{\infty} z^n f_{i,j}^{(n)}$. Analogously, we let $P_{i,j}(z) = \sum_{n=1}^{\infty} z^n P_{i,j}^{(n)}$. Both functions are defined for $|z| < 1$. By using eq. (A.68), we obtain

$$P_{i,j}(z) = F_{i,j}(z)[P_{j,j}(z) + 1] \tag{A.72}$$

In case $i = j$, we obtain $P_{i,i}(z) = F_{i,i}(z)[P_{i,i}(z) + 1]$, hence $P_{i,i}(z) = F_{i,i}(z)/[1 - F_{i,i}(z)]$. Then from eq. (A.72) we find for all $i, j \in S$

$$P_{i,j}(z) = \frac{F_{i,j}(z)}{1 - F_{j,j}(z)} \tag{A.73}$$

Inverting this relationship for $i = j$, we have

$$F_{i,i}(z) = \frac{P_{i,i}(z)}{1 + P_{i,i}(z)} \tag{A.74}$$

As $z \to 1$, the series of non-negative terms $P_{i,i}(z)$ can either converge or diverge. In the first case, $P_{i,i}(1)$ is finite and hence $f_{i,i} = \lim_{z \to 1} F_{i,i}(z) < 1$. In words, the state i is transient. If the series is divergent, then it is apparent that $f_{i,i} = \lim_{z \to 1} F_{i,i}(z) = 1$ and the state i is recurrent. From eq. (A.73) it can be seen that the reverse is also true, i.e., if the state i is transient (resp., recurrent) then the series $P_{i,i}(z)$ must be convergent (resp., divergent) as $z \to 1$.

We have thus proved the following.

Theorem A.1 A state i is recurrent if and only if $\sum_{n \geq 1} P_{i,i}^{(n)}$ is divergent; it is transient if and only if the series is convergent.

It can be shown that, if states i and j communicate, they must be *both* either recurrent or transient. In other words, the recurrence property of a state i spreads over all states that communicate with i. The same holds for the period: if state i and j communicate, then $d(i) = d(j)$. More technically, transience, recurrence, and periodicity are properties shared by all states of a same communication equivalence class. Therefore, if the Markov chain is irreducible (there is a single communication class), all of its states are either transient or recurrent. A Markov chain is called *aperiodic* if the period of all states is 1. This is by far the most used kind of Markov chain model.

If a state i is transient, the series $\sum_{n \geq 1} P_{i,i}^{(n)}$ must converge. From eq. (A.72) we deduce that

$$\lim_{z \to 1} P_{i,j}(z) = F_{i,j}(1) \left[1 + \sum_{n \geq 1} P_{i,i}^{(n)} \right] < \infty \tag{A.75}$$

i.e., $\sum_{n \geq 1} P_{i,j}^{(n)}$ converges as well. It is then necessary that $\lim_{n \to \infty} P_{i,j}^{(n)} = 0$, $\forall i, j \in S$. If the Markov chain is finite, i.e., there is a finite number of states, those limits are inconsistent with the fact that $\mathbf{P}^n$ is a stochastic matrix for all n. So, we can exclude that a finite Markov chain is transient (i.e., all states are transient). If the finite Markov chain is irreducible, it must therefore be recurrent.

In general, the following decomposition theorem can be proved.

Theorem A.2 The states of a Markov chain can be divided into two sets (one of which can be empty), one comprising all transient states, the other all recurrent states. This last set can be decomposed in a unique way in subsets, each one representing an equivalence recurrence class. States in the same class communicate, hence they are of same type and period. No communication is possible among states of two different classes.

As a consequence, the one-step transition matrix can be decomposed according to the following structure:

$$
\mathbf{P} = \begin{bmatrix}
\mathbf{P}_1 & \mathbf{0} & \mathbf{0} & \cdots & \mathbf{0} & \mathbf{0} \\
\mathbf{0} & \mathbf{P}_2 & \mathbf{0} & \cdots & \mathbf{0} & \mathbf{0} \\
\mathbf{0} & \mathbf{0} & \mathbf{P}_3 & \cdots & \mathbf{0} & \mathbf{0} \\
\cdots & \cdots & \cdots & \cdots & \cdots & \cdots \\
\mathbf{0} & \mathbf{0} & \mathbf{0} & \cdots & \mathbf{P}_{m-1} & \mathbf{0} \\
\mathbf{T}_1 & \mathbf{T}_2 & \mathbf{T}_3 & \cdots & \mathbf{T}_{m-1} & \mathbf{T}_m
\end{bmatrix}
\tag{A.76}
$$

where each of the matrices $\mathbf{P}_j$ is stochastic.

A.6.3 Visits to a State

Given a state i, we define N_i to be the number of visits that the Markov chain makes to state i excluding the initial state:

$$
N_i = \sum_{n=1}^{\infty} I(X_n = i)
\tag{A.77}
$$

where $I(E)$ is the indicator function of the event E. The state i is ever visited only if its first passage time is finite, i.e., $N_i > 0 \Leftrightarrow \tau(i) < \infty$, where $\tau(i)$ denotes the number of steps for touching i for the first time after leaving from i, i.e., the first passage time of i.

It can be shown that $P(N_j > k \mid X_0 = i) = f_{i,j}(f_{j,j})^k$ for all $k \geq 0$. There are only three possibilities:

1. If $f_{i,j} = 0$, it is $P(N_j > k \mid X_0 = i) = 0$, $\forall k \geq 0$, i.e., no visit at j can take place starting from i.

2. If $f_{i,j} > 0$ and $f_{j,j} = 1$ (j is recurrent), then $P(N_j > k \mid X_0 = i) = f_{i,j} > 0$ for all k and hence $E[N_j \mid X_0 = i] = \infty$; that is, provided j is reachable from i, the Markov chain will visit state j infinitely often (whence the name "recurrent").

3. If $f_{i,j} > 0$ and $f_{j,j} < 1$ (j is transient), then $E[N_j \mid X_0 = i] = f_{i,j}/(1 - f_{j,j}) < \infty$; that is, provided j is reachable from i, the Markov chain will visit state j only a finite number of times with probability 1 (whence the name "transient").

For a recurrent state, we have $P(N_i > k \mid X_0 = i) = f_{i,i} = 1$, hence $P(N_i = \infty \mid X_0 = i) = 1$. If instead i is transient, it is $P(N_i < \infty \mid X_0 = i) = 1$.

For a Markov chain starting from some state at time 0, we can define the visit time of state i as

$$\tau(i) = \inf\{n \geq 1 \ : \ X_n = i\} \tag{A.78}$$

that is the first-passage time by state i. We adopt the convention that $\inf\{\emptyset\} = \infty$. We can consider successive visits at state i by defining the sequence of times

$$\tau_h(i) = \inf\{n > \tau_{h-1}(i) \ : \ X_n = i\}, \qquad h \geq 1, \tag{A.79}$$

with $\tau_0(i) = 0$. The time elapsing between two successive visits defines a sequence of discrete, positive random variables $S_h(i) \equiv \tau_h(i) - \tau_{h-1}(i)$ for $h \geq 1$, provided $\tau_{h-1}(i) < \infty$[3]. It can be shown that $S_h(i)$ is independent of the past evolution of the Markov chain up to time $\tau_{h-1}(i)$, provided it is $\tau_{h-1}(i) < \infty$. We have

$$P(S_h(i) = n \mid \tau_{h-1}(i) < \infty) = P(\tau(i) = n \mid X_0 = i) = f_{i,i}^{(n)} \tag{A.80}$$

If the Markov chain is irreducible, the random variables $\{S_h(i)\}_{h \geq 1}$ are i.i.d. Let $m_i = E[S_1(i)]$. In general, we can define the mean recurrence time, for a recurrent state i:

$$m_i = E[S_1(i)] = E[\tau(i) \mid X_0 = i] = \sum_{n=1}^{\infty} n f_{i,i}^{(n)} \tag{A.81}$$

Let

$$\gamma_j(i) = E\left[\sum_{n=1}^{\tau(i)} I(X_n = j \mid X_0 = i)\right] \tag{A.82}$$

be the mean number of visits at state j during the first passage time of state i. It is $\gamma_i(i) = 1$. It can be shown that

$$\gamma_j(i) = \sum_{\ell \in S} \gamma_\ell(i) P_{\ell j} \tag{A.83}$$

i.e., in matrix terms $\gamma(i) = \gamma(i)\mathbf{P}$. It can be shown that $m_i = \sum_{j \in S} \gamma_j(i)$.

---

3 If $\tau_{h-1}(i) = \infty$ we just set $S_h = 0$.

A.6.4 Asymptotic Behavior and Steady State

Let

$$\pi_{j|i}(n) = P(X_n = j \mid X_0 = i) \tag{A.84}$$

be the state probability of the Markov chain at time n, given that is has started from state i. If we denote the initial probability $\alpha_i = P(X_0 = i)$, we have:

$$\pi_j(n) = \sum_{i \in S} \alpha_i \pi_{j|i}(n) = P(X_n = j) \tag{A.85}$$

A key question on Markov chains is to understand the behavior of the state probabilities $\pi_j(n)$ as n grows. Many problems modeled by Markov chains address issues that live theoretically on the whole time axis, i.e., those are *steady-state* issues. Hence the importance of understanding if and how the transient dies out and to what limit the probability distribution tends eventually, if any.

The fundamental limit theorem of Markov chain states the following.

Theorem A.3 Consider an irreducible, aperiodic Markov chain. Then

$$\lim_{n \to \infty} P_{j,i}^{(n)} = \lim_{n \to \infty} P_{i,i}^{(n)} = \frac{1}{m_i} \equiv \pi_i \tag{A.86}$$

where $m_i = \sum_{n \geq 1} n f_{i,i}^{(n)}$ is the mean recurrence time of state i. If m_i is finite, the quantities $\{\pi_i\}_{i \in S}$ form a probability distribution and they are the unique solution of the equations

$$\pi_j = \sum_{i \in S} \pi_i P_{i,j}, \quad \sum_{i \in S} \pi_i = 1, \quad \pi_i \geq 0. \tag{A.87}$$

We can define two kinds of recurrent states:

1. A state i is *positive recurrent* if $m_i < \infty$.
2. A state i is *null recurrent* is $m_i = \infty$.

It can be shown that positive or null recurrence property is shared by all states of a communication equivalence class. Recurrence of state i guarantees that the process gets back to i for sure, i.e., with probability $f_{i,i} = 1$. Positive recurrence of state i guarantees that the return to that state will be accomplished in a finte time on the average, i.e., a finite time with probability 1.

A consequence of Theorem A.3 for irreducible, aperiodic, positive recurrent Markov chains is that

$$\lim_{n \to \infty} P(X_n = i) = \lim_{n \to \infty} \sum_{j \in S} P(X_0 = j) P_{j,i}^{(n)}$$

$$= \sum_{j \in S} P(X_0 = j) \lim_{n \to \infty} P_{j,i}^{(n)} = \sum_{j \in S} P(X_0 = j) \pi_i = \pi_i$$

In words, the state probability approaches a limiting distribution as time grows. Thus, the Markov chain evolves toward a *statistical equilibrium*, where the probability distribution of the states of the random process becomes independent of time. Under these conditions, we can identify a clear path in the dynamics of the Markov chain, once it is triggered from some initial state. First there is a transient phase, during which the probability that the process is found in a given state i does depend on the observation time n. As n grows (mathematically: as it tends to infinity), the transient behavior dies out and the process settles in an equilibrium. This does not mean that the state does not change anymore at all. The Markov chain is anyway a stochastic process. Statistical equilibrium means that the probability distribution of the process does not change with time, i.e., the probability to find the process in state i is expressed by a number π_i that does not depend on the time of observation. Whence the great interest in assessing whether this sort of "steady state" can set on. Some systems are intrinsically transient. These include the characterization of breakdowns, the process of retransmission to deliver a given amount of data through an unreliable channel, the number of collisions of a given population of stations working according to a random MAC protocol, and the initial evolution of the congestion control of a TCP connection. However, most performance evaluation models address equilibrium situations. This is a strong simplification of the analysis (PDFs become independent of time).

For a finite Markov chain we already know that states cannot be all transient. Now we see that they cannot be either all null recurrent. In fact, $\sum_{j \in S} P_{i,j}^{(n)} = 1$ for all n. If $|S| < \infty$, we can invert limits with the finite sum and obtain

$$1 = \lim_{n \to \infty} \sum_{j \in S} P_{i,j}^{(n)} = \sum_{j \in S} \lim_{n \to \infty} P_{i,j}^{(n)} = \sum_{j \in S} \pi_j \qquad (A.88)$$

Then, it cannot be that all π_j's are null. We conclude that a finite irreducible, aperiodic Markov chain is necessarily positive recurrent. In case of infinite state space, an irreducible aperiodic Markov chain can be positive, null recurrent, or transient. An asymptotic statistical equilibrium does exist only in the first case.

A key aspect of the result of Theorem A.3 is that the limits are independent of the initial state of the Markov chain. Markov chains, whose limiting probability distribution exists and is independent of the probability distribution of the initial state, are referred to as *ergodic*. Ergodicity refers to the coincidence of time averages with probabilistic measures. The ergodic theorem for Markov chains states the following.

Theorem A.4 For an irreducible, aperiodic, positive recurrent Markov chain it is

$$\lim_{n \to \infty} \frac{1}{n} \sum_{k=0}^{n-1} I(X_k = i) = \pi_i \qquad w.p.\ 1 \qquad (A.89)$$

Given a function $f : S \mapsto \mathbb{R}$, it is also

$$\lim_{n\to\infty} \frac{1}{n} \sum_{k=0}^{n-1} f(X_k) = \sum_{i\in S} f(i)\pi_i \qquad w.p.\ 1 \tag{A.90}$$

In general, we define *invariant vector* of a Markov chain any nontrivial vector $\mathbf{v}$ such that $\mathbf{v} = \mathbf{v}\mathbf{P}$. The elements of $\mathbf{v}$ can be positive, null, or negative and their sum over the states of Markov chain need not be finite.

A *steady-state probability distribution* of the Markov chain is a non-negative vector $\mathbf{p}$ such that $\mathbf{p} = \mathbf{p}\mathbf{P}$ and $\mathbf{p}\mathbf{e} = 1$, where $\mathbf{e}$ is a column vector of 1's. In words, $\mathbf{p}$ is a probability distribution. When initialized according to the PDF $\mathbf{p}$, the Markov chain maintains that same PDF forever. $\mathbf{p}$ is a stationary probability distribution for the states of the Markov chain. If a stationary PDF $\mathbf{p}$ exists, then it is obviously also an invariant vector of the Markov chain; in general, however, the opposite does not hold.

The *limiting state probability distribution* is a probability distribution π_i such that

$$\lim_{n\to\infty} P(X_n = i) = \pi_i , \qquad \forall i \in S \tag{A.91}$$

Theorem A.3 tells us that, when a limiting PDF exists, it is also a stationary PDF, i.e., it satisfies $\pi = \pi\mathbf{P}$. The reverse is not necessarily true.

The stationary probability vector of a Markov chain can be found by solving the linear equation system $\pi = \pi\mathbf{P}$, one equation of which must be substituted with $\pi\mathbf{e} = 1$. The equations of this systems can be interpreted as *probability flow balance* equations by looking at the transition diagram of the Markov chain. Let $\pi_i p_{ij}$ be the probability flow from state i to state j. The overall flow out of state i is $\sum_j \pi_i p_{ij} = \pi_i(1 - p_{ii})$. The overall probability flow directed to state i is $\sum_j \pi_j p_{ji}$. It is readily seen that equating these two flows yields the i-th equation of the linear system $\pi = \pi\mathbf{P}$. This is the reason why those equations are sometimes referred to as *full balance equations*.

Based on the decomposition theorem for Markov chains, we see that a reducible Markov chain may have many stationary probability distributions. If at least one class is positive recurrent, there also exists a limiting probability distribution, but it depends on the initialization of the Markov chain.

As an example, let us consider a finite Markov chain made up of two aperiodic, recurrence classes and a transient class. The one-step transition probability matrix can be written as follows:

$$\mathbf{P} = \begin{bmatrix} \mathbf{A} & \mathbf{0} & \mathbf{0} \\ \mathbf{0} & \mathbf{B} & \mathbf{0} \\ \mathbf{C}_1 & \mathbf{C}_2 & \mathbf{C}_3 \end{bmatrix} \tag{A.92}$$

Let α and β be the left eigenvectors corresponding to the eigenvalue 1 of the stochastic matrices $\mathbf{A}$ and $\mathbf{B}$, respectively. We assume the two vectors are normalized so that the sum of their elements be 1. By Theorem A.3, we know that such vectors do exist. Then, $[\alpha \ \mathbf{0} \ \mathbf{0}]$ and $[\mathbf{0} \ \beta \ \mathbf{0}]$ are two stationary vectors for the Markov chain. The limiting probability distribution depends on the initial state. If the Markov chain is initialized in a state of recurrence class $\mathbf{A}$ or $\mathbf{B}$, it will never get out of that class, so that the limiting probabilities of all other states are necessarily zero.

We mention one further algebraic property of one-step transition matrices of finite Markov chains. The following can be shown:

- For each recurrence class comprising m states, the matrix $\mathbf{P}$ has exactly one simple eigenvalue equal to 1, and all other $m-1$ eigenvalues have modulus less than 1.
- For each periodic class of period d the matrix $\mathbf{P}$ has d eigenvalues with modulus 1 (those are the order d roots of unity, $e^{i2\pi k/d}$ for $k = 0, \dots, d-1$) and all other eigenvalues pertaining to that class have modulus less than 1.
- All other eigenvalues, corresponding to transient states, have modulus less than 1.

Reminding the spectral theorem of matrices, it is now possible to cast a new light on the limiting behavior of the probabilities $P_{i,j}^{(n)}$ when n tends to infinity, since

$$\mathbf{P}^n = \sum_{k=1}^{|S|} \lambda_k^n \mathbf{u}_k \mathbf{v}_k \tag{A.93}$$

where λ_k is the k-th eigenvalue, $\mathbf{u}_k$ is the k-th right eigenvector and $\mathbf{v}_k$ is the k-th left eigenvector corresponding to λ_k. If the Markov chain is aperiodic, irreducible and positive recurrent, the maximum modulus eigenvalue is simple and equals 1. All other eigenvalues have modulus strictly less than 1. Letting $\lambda_1 = 1$, we have $\mathbf{u}_1 = \mathbf{e}$ and $\mathbf{v}_1 = \pi$, where π is the stationary vector of the Markov chain. Hence

$$\mathbf{P}^n = \mathbf{e}\pi + \sum_{k \ : \ |\lambda_k|<1} \lambda_k^n \mathbf{u}_k \mathbf{v}_k \tag{A.94}$$

If the Markov chain is irreducible and periodic with period d, the eigenvalues are the d-th root of unity.

We summarize the characteristics of states of a Markov chain in Table A.1.

We conclude this section with two useful theorems to assess the existence of the stationary probability vector of a discrete-time Markov chain (DTMC) and its recurrence.

Theorem A.5 Let $\mathbf{P}$ be the one-step transition probability matrix of a time-homogeneous DTMC. Assume the DTMC is irreducible and aperiodic.

Table A.1 Classification of states of a Markov chain.

Types of state	Description
Periodic	Return to state possible only at times kd, $k \in \mathbb{Z}$, $d > 1$.
Aperiodic	Not periodic
Ephemeral	Visited at most once
Transient	Eventual return to state is uncertain (probability < 1)
Recurrent	Eventual return to state is certain (probability $= 1$)
Positive recurrent	Recurrent, with finite mean recurrence time
Null recurrent	Recurrent, with infinite mean recurrence time
Absorbing	State i with $P_{i,i} = 1$ (once in i, forever in i)
Ergodic	Aperiodic, positive recurrent.

If there exist a positive vector π such that $\pi_i p_{ij} = \pi_j p_{ji}$, $\forall i, j$, and $\sum_i \pi_i = 1$, that vector must be the stationary probability vector of the DTMC.

Proof: Summing the identities $\pi_i p_{ij} = \pi_j p_{ji}$ over i, we find $\sum_i \pi_i p_{ij} = \pi_j$, since **P** is a stochastic matrix. We see that $\pi \mathbf{P} = \pi$, which proves that π is the stationary probability vector of the DTMC with one-step transition probability matrix **P**. ∎

Theorem A.6 (*Foster-Lyapunov*) Let X_k be an irreducible Markov chain on the state space S. Assume there exists a function $V : S \mapsto \mathbb{R}^+$ and a finite set $A \subseteq S$ such that the following conditions hold:

$$\begin{cases} E[V(X_{k+1}) - V(X_k)|X_k = x] \leq -\epsilon & \text{if } x \in S \setminus A, \\ E[V(X_{k+1}) - V(X_k)|X_k = x] \leq b & \text{otherwise.} \end{cases} \tag{A.95}$$

for a given $\epsilon > 0$ and a finite constant b. Then the Markov chain X_k is positive recurrent.

Proof: The ensuing proof applies to aperiodic Markov chains, though the theorem does not require this condition.

Let $\overline{A} = S \setminus A$. The conditions in the statement of the theorem can be summarized in the following equation

$$E[V(X_{k+1}) - V(X_k)|X_k = x] \leq -\epsilon I(x \in \overline{A}) + bI(x \in A) \tag{A.96}$$

where $I(E)$ is the indicator function of the event E. Multiplying both sides by $P(X_k = x)$ and summing over x, we get

$$E[V(X_{k+1})] - E[V(X_k)] \leq -(\epsilon + b)P(X_k \in \overline{A}) + b \tag{A.97}$$

holding for all $k \geq 0$. Summing up for $k = 0, \dots, N$ for some finite N, we have

$$\mathrm{E}[V(X_{N+1})] - \mathrm{E}[V(X_0)] \leq -(\epsilon + b) \sum_{k=0}^{N} P(X_k \in \overline{\mathcal{A}}) + (N+1)b \qquad \text{(A.98)}$$

and, rearranging terms

$$\frac{\epsilon + b}{N + 1} \sum_{k=0}^{N} P(X_k \in \overline{\mathcal{A}}) \leq b + \frac{\mathrm{E}[V(X_0)] - \mathrm{E}[V(X_{N+1})]}{N + 1} \leq b + \frac{\mathrm{E}[V(X_0)]}{N + 1} \qquad \text{(A.99)}$$

We have also

$$\frac{1}{N + 1} \sum_{k=0}^{N} P(X_k \in \mathcal{A}) = 1 - \frac{1}{N + 1} \sum_{k=0}^{N} P(X_k \in \overline{\mathcal{A}}) \qquad \text{(A.100)}$$

Hence, the inequality above yields

$$\frac{1}{N + 1} \sum_{k=0}^{N} P(X_k \in \mathcal{A}) \geq 1 - \frac{b}{\epsilon + b} - \frac{\mathrm{E}[V(X_0)]}{(N + 1)(\epsilon + b)} \qquad \text{(A.101)}$$

From the inequality above it follows that[4]

$$\liminf_{N \to \infty} \frac{1}{N + 1} \sum_{k=0}^{N} P(X_k \in \mathcal{A}) \geq \frac{\epsilon}{\epsilon + b} > 0 \qquad \text{(A.102)}$$

Assume now that there exists a nonrecurrent state. Since the Markov chain is irreducible, all states are then nonrecurrent and hence it should be

$$\lim_{k \to \infty} P(X_k \in \mathcal{A}) = 0 \qquad \text{(A.103)}$$

but this limit contradicts eq. (A.102). This completes the proof. ∎

The following theorem is also useful.

Theorem A.7 Let X_k be an irreducible Markov chain on the state space S. Assume there exists a function $V : S \mapsto \mathbb{R}^+$ and a finite set $\mathcal{A} \subseteq S$ such that the following conditions hold:

1. $\mathrm{E}[V(X_{k+1}) - V(X_k)|X_k = x] \geq 0, \ \forall x \in \overline{\mathcal{A}}$.
2. There exists some $x \in \overline{\mathcal{A}}$ such that $V(x) > V(y), \ \forall y \in \mathcal{A}$.
3. $\mathrm{E}[|V(X_{k+1}) - V(X_k)| \ |X_k = x] \leq b$ for some $b < \infty$ and $\forall x \in S$.

where $\overline{\mathcal{A}} = S \smallsetminus \mathcal{A}$. Then the Markov chain X_k is either transient or null recurrent.

---

4 The liminf is defined as follows. Given a sequence x_n, $n \geq 0$, it is $\liminf_{n \to \infty} x_n \equiv \lim_{n \to \infty} \inf_{k \geq n} x_k$. The liminf of a sequence always exists. It is also $\liminf_{n \to \infty} x_n \leq \lim_{n \to \infty} x_n$.

A.6.5 Absorbing Markov Chains

We focus on Markov chain comprising only transient and recurrent absorbing states. The set comprising the m absorbing states is called A, the remaining states (transient states) are collected in set B. The one-step transition probability matrix is

$$P = \begin{bmatrix} I & 0 \\ P_{BA} & P_{BB} \end{bmatrix} \tag{A.104}$$

where I is the $m \times m$ identity matrix. Since the set B is made up of transient states, all eigenvalues of P_{BB} have modulus less than 1. Since the spectral radius of P_{BB} is less than 1, then $I - P_{BB}$ is invertible and

$$\sum_{n=0}^{\infty} P_{BB}^n = (I - P_{BB})^{-1} \tag{A.105}$$

Moreover

$$P^n = \begin{bmatrix} I & 0 \\ \sum_{k=0}^{n-1} P_{BB}^k P_{BA} & P_{BB}^n \end{bmatrix} \tag{A.106}$$

Let $\alpha = [\alpha_A \ \alpha_B]$ be the initial state PDF and $p(n) = [p_A(n) \ p_B(n)]$ the Markov chain state PDF at time n. It is $p(n) = \alpha P^n$. Specifically, $p_B(n) = \alpha_B P_{BB}^n$

Let also T denote the time to absorption, i.e., $T = \inf\{n \geq 0 \ : \ X_n \in A\}$. It is evident that $P(T > n) = P(X_n \in B) = p_B(n)e$, where e is a column vector of 1's. Then

$$P(T > n) = \alpha_B P_{BB}^n e, \qquad n \geq 0. \tag{A.107}$$

It is also $P(T = 0) = \alpha_A e$, i.e., immediate absorption occurs, if the chain is initialized directly into an absorbing state. The PDF of the absorption time is

$$P(T = n) = \begin{cases} 1 - \alpha_B e & n = 0 \\ \alpha_B P_{BB}^{n-1}(I - P_{BB})e & n \geq 1. \end{cases} \tag{A.108}$$

and the mean is $E[T] = \alpha_B(I - P_{BB})^{-1}e$.

It is also possible to find the joint probability distribution of the time to absorption and of the absorbing state. For $j \in A$ we have

$$P(T = n, X_T = j) = P(X_{n-1} \in B, X_n = j)$$
$$= \sum_{h \in B} P(X_{n-1} = h)P(X_n = j|X_{n-1} = h)$$
$$= \sum_{h \in B} (\alpha_B P_{BB}^{n-1})_h P_{hj} = \alpha_B P_{BB}^{n-1} c_j, \qquad n \geq 1,$$

where c_j is the j-th column of P_{BA} for $j = 1, \ldots, m$. For $n = 0$ it is $P(T = 0, X_T = j) = \alpha_j$. The probability of absorption in state j can be found as

$$P(X_T = j) = \sum_{n=0}^{\infty} P(T = n, X_T = j) = \alpha_j + \alpha_B(I - P_{BB})^{-1} c_j \tag{A.109}$$

The mean time to absorption in state $j \in A$ is

$$E[TI(X_T = j)] = \sum_{n=0}^{\infty} nP(T = n, X_T = j) = \alpha_B(\mathbf{I} - \mathbf{P}_{BB})^{-2}\mathbf{c}_j \qquad (A.110)$$

Finally, in a general case where there are r recurrent classes $A_1, \ldots, A_r$, besides the transient states, we define $\pi_i(A_q)$ to be the probability of absorption into class q, given that the process is initialized in state $i \in B$. We have

$$\pi_i(A_q) = \sum_{j \in A_q} P_{i,j} + \sum_{j \in B} P_{i,j}\pi_j(A_q), \qquad i \in B, \qquad (A.111)$$

for $q = 1, \ldots, r$. In matrix form, we can write $\pi(A_q)^T(\mathbf{I} - \mathbf{P}_{BB}^T) = (\mathbf{P}_{BA_q}\mathbf{e})^T$, whence $\pi(A_q) = (\mathbf{I} - \mathbf{P}_{BB})^{-1}\mathbf{P}_{BA_q}\mathbf{e}$, where $\mathbf{e}$ is a column vector of 1's.

A.6.6 Continuous-Time Markov Processes

We devote a brief section to continuous-time Markov processes, since most definitions are similar to the discrete time case. This kind of process is often referred to as continuous-time Markov chain (CTMC). The major new fact is that transitions are not "clocked" by a slotted time. Thus periodicity disappears, while new issues come to surface. Specifically, the sojourn time into a state must have a special PDF for the Markov property to hold, namely it must be negative exponential (see Section A.6.7).

A continuous-time process $X(t)$ with a countable state space is a Markov process if

$$\mathcal{P}(X(t_n) = i_n | X(t_0) = i_0, \ldots, X(t_{n-1}) = i_{n-1}) = \mathcal{P}(X(t_n) = i_n | X(t_{n-1}) = i_{n-1}) \qquad (A.112)$$

for all n, all $i_0, \ldots, i_n$, and all $t_0 < \cdots < t_n$. We consider time-homogeneous Markov processes, whose probability distribution depend only on the differences between the involved times. Let $H_{i,j}(t) = \mathcal{P}(X(u + t) = j | X(u) = i)$, for $t \geq 0$. The matrix $\mathbf{H}(t)$, with entry (i,j) equal to $H_{i,j}(t)$, is stochastic for any given t. If $t > u > v$, thanks to the Markov property we have

$$H_{i,j}(t - v) = \mathcal{P}(X(t) = j | X(v) = i) = \sum_{k \in S} \mathcal{P}(X(t) = j, X(u) = k | X(v) = i)$$

$$= \sum_{k \in S} \mathcal{P}(X(u) = k | X(v) = i)\mathcal{P}(X(t) = j | X(u) = k)$$

$$= \sum_{k \in S} H_{i,k}(u - v)H_{k,j}(t - u)$$

More compactly, in matrix form we can write $\mathbf{H}(t - v) = \mathbf{H}(u - v)\mathbf{H}(t - u)$ for all $v \leq u \leq t$. It is obviously $\mathbf{H}(0) = \mathbf{I}$. We can choose the three times v, $v + t$, and

$v + t + \Delta t$ to obtain

$$H(t + \Delta t) = H(t)H(\Delta t) \quad \Rightarrow \quad \frac{H(t + \Delta t) - H(t)}{\Delta t} = H(t)\frac{H(\Delta t) - I}{\Delta t}$$

$$(A.113)$$

Let us assume that the limit

$$\lim_{\Delta t \to 0} \frac{H(\Delta t) - I}{\Delta t} = Q \tag{A.114}$$

exists. Since $H(\Delta t)$ is stochastic, we have $Qe = 0$. The matrix Q is called *rate transition* matrix or *infinitesimal generator* of the process. It is formed by non-negative elements of the diagonal, while the entries in the diagonal are such that the sum of each row is 0. From eq. (A.113) we get the following differential equation

$$H'(t) = H(t)Q \tag{A.115}$$

with the initial condition $H(0) = I$. The unique solution is $H(t) = \exp(Qt)$. With a similar reasoning, it can be verified that

$$p(t) = p(0)H(t) = p(0)\exp(Qt), \quad t \geq 0, \tag{A.116}$$

where $p(t)$ is a row vector whose i-th entry is $P(X(t) = i)$. For an irreducible, positive recurrent Markov process there exists a unique positive vector π such that $\pi = \lim_{t \to \infty} p(t)$. The vector π is the unique positive solution of $\pi Q = 0$ that can be normalized so that $\pi e = 1$. π is the left eigenvector of Q corresponding to the eigenvalue 0. Since the sum of each row of Q is zero, 0 is always an eigenvalue of Q. For an irreducible, positive recurrent Markov process all other eigenvalues of Q have strictly negative real part.

The **limiting** probability distribution of the Markov process is $\pi = \lim_{t \to \infty} p(t)$, if the limit exists.

A **stationary** probability distribution is a non-negative vector s such that $se = 1$ and such that $P(X(t_0) = i) = s_i \Rightarrow P(X(t) = i) = s_i$ for all $t > t_0$ and all $i \in S$.

The Markov process is **ergodic**, if $\frac{1}{T}\int_0^T I(X(t) = i)\, dt \to \pi_i$ as $T \to \infty$ for all $i \in S$. For an ergodic Markov process, the limiting probability π_i can be interpreted as the average fraction of time spent by the process in state i. This is a key feature to enable computer simulations and measurements of Markov processes metrics, based on time averages.

A continuous Markov process can be simulated by constructing the associated jump process, i.e., the sequence of visited states. Let $\mu_i = -q_{i,i} = \sum_{j \neq i} q_{i,j}$, where $q_{i,j}$ is the entry (i,j) of the matrix Q. We will see in Section A.6.7 that the time spent in a visit to state i (sojourn time) has a negative exponential PDF with mean $1/\mu_i$. When leaving state i, the process moves to state j with probability $q_{i,j}/\mu_i$. Given that the process enters state i_k, a sample of the sojourn time is drawn as

$T_{i_k} = -\log(\texttt{rand})/\mu_{i_k}$, where $\texttt{rand}$ is a random number generator uniformly distributed in $[0, 1]$. When time T_{i_k} expires, the process moves to state i_{k+1} with probability $q_{i_k,i_{k+1}}/\mu_{i_k}$ for all possible $i_{k+1} \in S \setminus \{i_k\}$. Once the initial state i_0 is give, the whole sequence of the jump process $\{i_k\}_{k \geq 1}$ can be generated.

A.6.7 Sojourn Times in Process States

It can be shown that also the sojourn time in a state of a Markov process has a negative exponential PDF. More generally, let us consider an irreducible Markov process $X(t)$ with infinitesimal generator $\mathbf{Q}$. Let us split the process state space in two regions, denoted as S_1 and S_2, and let

$$\mathbf{Q} = \begin{bmatrix} \mathbf{Q}_{11} & \mathbf{Q}_{12} \\ \mathbf{Q}_{21} & \mathbf{Q}_{22} \end{bmatrix} \tag{A.117}$$

Let Θ_1 be the sojourn time of the process in the region S_1. The probability distribution function of Θ_1 can be found by considering an associated Markov process $\tilde{X}(t)$ with $\mathbf{Q}_{21} = \mathbf{0}$, which is composed of a transient class of states (S_1) and a positive recurrent, absorbing class of states (S_2). The state probability vector at time t, starting from an initial vector $\mathbf{q} = [\mathbf{q}_1 \ \mathbf{0}]$, satisfies

$$\frac{d\mathbf{p}}{dt} = [\mathbf{p}_1(t) \ \mathbf{p}_2(t)] \begin{bmatrix} \mathbf{Q}_{11} & \mathbf{Q}_{12} \\ \mathbf{0} & \mathbf{0} \end{bmatrix} \tag{A.118}$$

with initial condition $\mathbf{p}(0) = [\mathbf{q}_1 \ \mathbf{0}]$. This yields $\mathbf{p}_1(t) = \mathbf{q}_1 \exp(\mathbf{Q}_{11}t)$ and $\mathbf{p}_2(t) = \mathbf{q}_1 \exp(\mathbf{Q}_{11}t)\mathbf{Q}_{12}$. The latter is the probability of ultimate abosrption into a state of region S_2 at time t. The probability of absorption at some time is

$$\mathbf{p}_{2a} = \int_0^\infty \mathbf{p}_2(u) \, du = \mathbf{q}_1(-\mathbf{Q}_{11}^{-1})\mathbf{Q}_{12} \tag{A.119}$$

Note that $-\mathbf{Q}_{11}$ is an M-matrix[5] , hence it is invertible and the inverse is positive. The integral of the matrix exponential can be carried out by integrating over a finite interval $[0, T]$ the series expansion term by term, thus obtaining

$$\int_0^T \exp(\mathbf{Q}_{11}t) \, dt = [\exp(\mathbf{Q}_{11}T) - \mathbf{I}]\mathbf{Q}_{11}^{-1} \tag{A.120}$$

where $\mathbf{I}$ is the identity matrix. Letting $T \to \infty$, the matrix $\exp(\mathbf{Q}_{11}T)$ tends to $\mathbf{0}$, since all eigenvalues of $\mathbf{Q}_{11}$ have negative real part.

The vector $\mathbf{p}_{2a}$ is also the initial probability vector of another Markov process, where now region 1 is recurrent and region 2 is transient. Repeating the reasoning above, with the initial probability vector $[\mathbf{0} \ \mathbf{q}_2]$, we find the probability of

---

[5] A matrix with positive diagonal elements, non positive off-diagonal elements, such that the row sums are non negative and at least one row sum is strictly positive is called an M-matrix.

absorption in states of the region 1, namely: $\mathbf{p}_{1a} = \mathbf{q}_2(-\mathbf{Q}_{22}^{-1})\mathbf{Q}_{21}$ with $\mathbf{q}_2 = \mathbf{p}_{2a}$. This leads to

$$\mathbf{q}_1 = \mathbf{p}_{1a} = \mathbf{q}_2(-\mathbf{Q}_{22}^{-1})\mathbf{Q}_{21} = \mathbf{q}_1(-\mathbf{Q}_{11}^{-1})\mathbf{Q}_{12}(-\mathbf{Q}_{22}^{-1})\mathbf{Q}_{21} \tag{A.121}$$

with the additional condition that $\mathbf{q}_1\mathbf{e} = 1$, where $\mathbf{e}$ is a column vector of 1's. It is easy to verify that the vector $\mathbf{q}_1 = \mathbf{p}_1\mathbf{Q}_{11}/\mathbf{p}_1\mathbf{Q}_{11}\mathbf{e}$ satisfies these equations, where $\mathbf{p} = [\mathbf{p}_1 \ \mathbf{p}_2]$ is the stationary vector of the Markov process, i.e., the solution of

$$[\mathbf{p}_1 \ \mathbf{p}_2] \begin{bmatrix} \mathbf{Q}_{11} & \mathbf{Q}_{12} \\ \mathbf{Q}_{21} & \mathbf{Q}_{22} \end{bmatrix} = [\mathbf{0} \ \mathbf{0}] \tag{A.122}$$

with $\mathbf{pe} = 1$. So, the initial vector of any visit to the region S_1 is $\mathbf{q}_1 = \mathbf{p}_1\mathbf{Q}_{11}/\mathbf{p}_1\mathbf{Q}_{11}\mathbf{e}$.

By referring to the modified Markov process where states of region 1 are transient and states of region 2 are recurrent and absorbing, it is easy to see that $\mathcal{P}(\Theta_1 > t) = \mathbf{p}_1(t)\mathbf{e} = \mathbf{q}_1\exp(\mathbf{Q}_{11}t)\mathbf{e}$. Then, it can be found that $E[\Theta_1] = \mathbf{q}_1(-\mathbf{Q}_{11}^{-1})\mathbf{e} = -\mathbf{p}_1\mathbf{e}/\mathbf{p}_1\mathbf{Q}_{11}\mathbf{e}$.

This result can be interpreted as follows. The mean sojourn time in a region of a Markov process can be obtained as the ratio of the limiting probability of the states making up that region divided by the probability flow out of that region. Remember that the probability flow from state i to state j is defined as p_iq_{ij}.

Application of this general result to the special case where the region S_1 is made up of the state j brings to $G_{\Theta_j}(t) = \mathcal{P}(\Theta_j > t) = e^{q_{jj}t}$, i.e., the sojourn time into a state of a Markov process has negative exponential PDF with a mean value of $E[\Theta_j] = -1/q_{jj} = 1/\sum_{i\neq j}q_{ji}$. This is a revealing result, since it is intimately tied to the Markov property of the process. Giving the state at time t is sufficient to summarize the entire past history of the process prior to time t *just because* the time the process has already spent into the given state has a memoryless PDF, so that we need not know how long it has been, since the visit to the state has started, to predict the future process evolution.

An analogous argument can be developed for Markov chains and geometric probability distributions. Specifically, it can be found that the time spent in a region of a Markov chain has CDF given by $f_{\Theta_1}(n) \equiv P(\Theta_1 = n) = \mathbf{q}_1\mathbf{P}_{11}^{n-1}\mathbf{P}_{12}\mathbf{e}$ for $n \geq 1$, with $\mathbf{q}_1 = \pi_1(\mathbf{I} - \mathbf{P}_{11})/\pi_1(\mathbf{I} - \mathbf{P}_{11})\mathbf{e}$, $\pi = [\pi_1 \ \pi_2]$ being the stationary vector of the original recurrent Markov chain.

A.6.8 Reversibility

Reversibility is an important concept that has many useful applications in the study of CTMCs. Consider a stationary CTMC $X(t)$. Stationarity implies that there exists a probability vector π such that, if for $k \in S$ it is $p_k(0) = \pi_k$, than $p_k(t) = P(X(t) = k) = \pi_k$, $\forall t > 0$, where S is the state space of the CTMC

We define the reversed process associated with $X(t)$ as

$$Y(t) = X(-t). \tag{A.123}$$

Note that $Y(t)$ is stationary because $X(t)$ is, and it has the same stationary probability distribution.

We can prove that $Y(t)$ is a CTMC. We shall prove that

$$P_Y \equiv \mathcal{P}(Y(t_{n+1}) = i_{n+1} | Y(t_1) = i_1, \ldots, Y(t_n) = i_n) = \mathcal{P}(Y(t_{n+1}) = i_{n+1} | Y(t_n) = i_n) \tag{A.124}$$

for any $n \geq 1$ and any set of times $t_1 < \cdots < t_n < t_{n+1}$ and any set of states $i_1, \ldots, i_{n+1}$.

We have

$$P_Y = \mathcal{P}(X(-t_{n+1}) = i_{n+1} | X(-t_1) = i_1, \ldots, X(-t_n) = i_n)$$

$$= \frac{\mathcal{P}(X(-t_{n+1}) = i_{n+1}, X(-t_n) = i_n, \ldots, X(-t_1) = i_1)}{\mathcal{P}(X(-t_n) = i_n, \ldots, X(-t_1) = i_1)}$$

$$= \frac{\mathcal{P}(X(-t_{n+1}) = i_{n+1}) \prod_{k=1}^{n} \mathcal{P}(X(-t_k) = i_k | X(-t_{k+1}) = i_{k+1})}{\mathcal{P}(X(-t_n) = i_n) \prod_{k=1}^{n-1} \mathcal{P}(X(-t_k) = i_k | X(-t_{k+1}) = i_{k+1})}$$

$$= \frac{\mathcal{P}(X(-t_{n+1}) = i_{n+1}) \mathcal{P}(X(-t_n) = i_n | X(-t_{n+1}) = i_{n+1})}{\mathcal{P}(X(-t_n) = i_n)}$$

$$= \mathcal{P}(X(-t_{n+1}) = i_{n+1} | X(-t_n) = i_n) = \mathcal{P}(Y(t_{n+1}) = i_{n+1} | Y(t_n) = i_n)$$

which proves the thesis.

We can also express the rate transition matrix of Y as a function of the rate-transition matrix of X. Let $\mathbf{Q}^{(X)}$ and $\mathbf{Q}^{(Y)}$ denote the rate transition matrices of the CTMCs X and Y, respectively. We have

$$q_{ij}^{(Y)} = \lim_{h \to 0} \frac{\mathcal{P}(Y(t+h) = j | Y(t) = i)}{h} = \lim_{h \to 0} \frac{\mathcal{P}(X(-t-h) = j | X(-t) = i)}{h}$$

$$= \lim_{h \to 0} \frac{\mathcal{P}(X(-t) = i | X(-t-h) = j) \mathcal{P}(X(-t-h) = j)}{h \, \mathcal{P}(X(-t) = i)} = q_{ji}^{(X)} \frac{\pi_j}{\pi_i}$$

where the last passage is due to the stationarity of the process $X(t)$.

We have thus found the simple relationship between the rate transition matrix of a CTMC and its reversed process. As a corollary of this property we deduce that, given a CMTC X with rate transition matrix $\mathbf{Q}$, if we can construct a rate transition matrix $\mathbf{Q}'$ and a probability vector π such that $\pi_j q_{ji} = \pi_i q'_{ij}$, $\forall i, j$, then π is the stationary probability distribution of the CTMC X.

We define also a reversible CTMC.

Definition A.4 (*Reversibility*) A stationary CTMC $X(t)$ is time reversible if it is statistically indistinguishable from its reversed chain $X(-t)$, i.e., if the rate transition matrix $\mathbf{Q}$ of X and that of its reversed chain, i.e., $\mathbf{Q}'$, are the same.

It is easy to show that a stationary CTMC $X(t)$ with rate transition matrix entries q_{ij} is time reversible if and only if

$$\pi_i q_{ij} = \pi_j q_{ji} \tag{A.125}$$

Equations (A.125) are called *partial balance* (or, local balance) equations, in contrast with the usual equations $\sum_i \pi_i q_{ij} = 0$, which are called *full balance* equations.

A.6.9 Uniformization

Let us consider a continuous-time Markov process with infinitesimal generator $\mathbf{Q}$. Let $\lambda_i = -\sum_{j \in S} q_{ij}$ the mean rate out of state i. Let also λ be a positive constant such that $\lambda \geq \{\lambda_i, \ i \in S\}$. We let

$$\mathbf{U} = \mathbf{I} + \frac{1}{\lambda}\mathbf{Q} \tag{A.126}$$

It is easy to verify that $\mathbf{U}$ is stochastic, i.e., all its entries are non-negative and each row sums to 1.

We consider then a Poisson process $N(t)$ with mean rate λ and a discrete Markov process with one-step transition probability matrix $\mathbf{U}$, denoted with Z_n. We initialize Z_n with the same PDF as the original Markov process, i.e., $P(Z_0 = i) = P(X_0 = i)$. The two processes Z_n and $N(t)$ are constructed independently of each other.

The random process $\{Z_{N(t)}, \ t \geq 0\}$ is called the *uniformed process* of X with respect to the uniformization rate λ. We have

$$\mathbf{H}(t) = \exp(\mathbf{Q}t) = \exp(-\lambda t(\mathbf{I} - \mathbf{U})) = e^{-\lambda t} \exp(\lambda t \mathbf{U}) = \sum_{k=0}^{\infty} e^{-\lambda t} \frac{(\lambda t)^k}{k!} \mathbf{U}^k \tag{A.127}$$

Then, it is possible to show that for every $k \geq 1$, times $0 < t_1 < \cdots < t_k$ and states $j_1, \dots, j_k \in S$, we have

$$P(X(t_1) = j_1, \dots, X(t_k) = j_k) = P(Z_{N(t_1)} = j_1, \dots, Z_{N(t_k)} = j_k) \tag{A.128}$$

That is to say, the uniformed process is stochastically equivalent to the original Markov process.

Uniformization of continuous-time Markov processes is a very powerful technique, both for theoretical arguments (e.g., see the elegant proofs of limiting properties of Markov processes in [186, Ch. 3]) and for applications (e.g., see the algorithms for the computation of transient behavior of a Markov process in [196, Ch. 10]).

A.7 Wiener Process (Brownian Motion)

The Brownian motion or Wiener process is a continuous-time, continuous state-space Markov process.

Let $X(t)$ be a random process and let

$$F(y, t; x, u) = \mathcal{P}(X(t) \le y | X(u) = x) \tag{A.129}$$

be the kernel of the process, i.e., the conditional CDF that rules the evolution of $X(t)$. We assume that the corresponding PDF exists. It is $f(y, t; x, u) = \frac{\partial F(y, t; x, u)}{\partial y}$.

The process $X(t)$ is said to be Markovian if

$$f(y, t; x, u) = \int_{-\infty}^{\infty} f(z, v; x, u) f(y, t; z, v) \, dz, \qquad u < v < t. \tag{A.130}$$

Definition A.5 A standard Wiener process (or Brownian motion) is a stochastic process $W(t), t \ge 0$, with $W(0) = 0$, and such that:

1. with probability 1, the function $t \mapsto W(t)$ is continuous;
2. the process $W(t)$ has stationary, independent increments;
3. the increment $W(u + t) - W(u)$ is distributed as a normal random variable with zero mean and variance t, i.e., $W(u + t) - W(u) \sim \mathcal{N}(0, t)$.

The definition refers to a normalized Wiener process. We can introduce a *drift* μ and a variance coefficient σ^2 by slightly changing the probability distribution of the increments, i.e., by letting it be a Gaussian random variable with mean μt and variance $\sigma^2 t$: $W(u + t) - W(u) \sim \mathcal{N}(\mu t, \sigma^2 t)$

The sample paths of a Wiener process are continuous functions of time, by definition, yet they exhibit infinite "spikes" in each given interval, given that any increment, however small, is a Gaussian random variable. Then, sample paths are nondifferentiable.

We can conceive Wiener process as a limit of a simple random walk. Let us consider a random walk $X(t)$ starting from the origin and making jumps at times $t_n = n\tau$, where τ is a given time quantum. Jumps are independent and identically distributed according to the binary random variable Z with $\mathcal{P}(Z = \Delta) = p$ and $\mathcal{P}(Z = -\Delta) = 1 - p \equiv q$. We have $X_{n+1} = X(t_{n+1}) = X(t_n) + Z_{n+1}, n \ge 0$, with $X(0) = 0$. It follows that $X_n = \sum_{k=1}^{n} Z_k$. Let $t = t_n = n\tau$. The Laplace transform of the PDF of $X(t) = X(t_n) = X_n$ is

$$E[e^{-sX(t)}] = (pe^{-s\Delta} + qe^{s\Delta})^{t/\tau} \tag{A.131}$$

The mean and the variance of $X(t)$ are:

$$E[X(t)] = (p - q)\frac{\Delta}{\tau}t \qquad \sigma_{X(t)}^2 = 4pq\frac{\Delta^2}{\tau}t \tag{A.132}$$

We now wish to shrink the time quantum τ and the step size in a time quantum, Δ, both to 0, so as to obtain a meaningful limit, *for a fixed t*, i.e., so that

$$(p - q)\frac{\Delta}{\tau} \to \mu \qquad 4pq\frac{\Delta^2}{\tau} \to \sigma^2 \qquad (\Delta, \tau \to 0) \tag{A.133}$$

To that purpose, we note that it must be $\Delta \sim \sqrt{\tau}$ (see the limit for the variance per unit time). If that is the case, then it must be also $p - q \sim \sqrt{\tau}$. Since it must be $p + q = 1$, we can let $\Delta = a\sqrt{\tau}, p = 1/2 + b\sqrt{\tau}$, and $q = 1/2 - b\sqrt{\tau}$, where a and b are suitable constants.

Inserting those expressions into the formulas of the mean and variance per unit time and carrying out the limit, we find $a = \sigma$ and $2ab = \mu$, hence $b = \mu/(2\sigma)$. Using the new expressions of $\Delta = \sigma\sqrt{\tau}, p = (1 + \mu\sqrt{\tau}/\sigma)/2$, and $q = (1 - \mu\sqrt{\tau}/\sigma)/2$ into the Laplace transform of the PDF of $X(t)$ we get

$$E[e^{-sX(t)}] = \left[\left(\frac{1}{2} + \frac{\mu\sqrt{\tau}}{2\sigma} \right) e^{-s\sigma\sqrt{\tau}} + \left(\frac{1}{2} - \frac{\mu\sqrt{\tau}}{2\sigma} \right) e^{s\sigma\sqrt{\tau}} \right]^{t/\tau} \tag{A.134}$$

Taking the limit for $\tau \to 0$ with fixed t, we find

$$E[e^{-sX(t)}] \to \exp\left(-\mu ts + \frac{1}{2}\sigma^2 ts^2 \right) \tag{A.135}$$

This is the transform of a Gaussian PDF with mean μt and variance $\sigma^2 t$. We see that in the limit, the increment $X(t) - X(0)$ is normally distributed with a drift μ per unit time and a variance coefficient σ^2 per unit time. This limiting process is just the Wiener process.

Formally, we can write a differential stochastic equation for Wiener process with drift μ and variance coefficient σ^2. Considering a finite and "small" time interval Δt, we can write:

$$X(t + \Delta t) = X(t) + \mu\Delta t + \sigma Z\sqrt{\Delta t} \tag{A.136}$$

where Z is a standard Gaussian random variable, $Z \sim \mathcal{N}(0, 1)$.

This relationship offers a simple way to generate a sequence of points belonging to a Brownian Motion. Let δ be the desired time resolution. Then $X_{k+1} = X(k\delta + \delta) = X_k + \mu\delta + \sigma Z_{k+1}\sqrt{\delta}$, for $k \geq 0$, with $X_0 = 0$.

The definition of Wiener process allows a direct derivation of its kernel PDF. Given $X(u) = x$, we have $X(t) = X(t) - X(u) + X(u) = x + X(t) - X(u)$. The increment $X(t) - X(u)$ is independent of $X(u)$, normally distributed with mean $\mu(t - u)$ and variance $\sigma^2(t - u)$. Hence

$$f(y, t; x, u) = \frac{1}{\sqrt{2\pi\sigma^2(t - u)}} e^{-\frac{[y - x - \mu(t-u)]^2}{2\sigma^2(t-u)}} \tag{A.137}$$

for $t > u$.

It is easy to verify that $f(y, t; x, u)$ satisfies the following differential equation:

$$\frac{\partial f}{\partial t} = -\mu \frac{\partial f}{\partial y} + \frac{1}{2} \sigma^2 \frac{\partial^2 f}{\partial y^2} \tag{A.138}$$

with the boundary condition $f(y, u; x, u) = \delta(y - x)$, $\delta(\cdot)$ being the Dirac delta function. Moreover, in solving (A.138), we impose that $f(y, t; x, u)$ be summable over y and its integral over the domain of y be 1.

Equation (A.138) is the diffusion equation of Wiener process. It is also called *forward* equation. A similar diffusion equation, referred to as *backward* equation, can be derived, by considering spatial derivation with respect to the initial position x of the process:

$$\frac{\partial f}{\partial t} = \mu \frac{\partial f}{\partial x} + \frac{1}{2} \sigma^2 \frac{\partial^2 f}{\partial x^2} \tag{A.139}$$

In the following, we let $u = x = 0$ to simplify notation, i.e., we fix the time origin at the initial time u and the origin of the x-axis at x. Consistently, we simplify the notation of the kernel PDF as $f(y, t)$. To recover general expression it suffices to replace y with $y - x$ and t with $t - u$.

A.7.1 Wiener Process with an Absorbing Barrier

Let as consider a Wiener process $X(t)$ with an absorption barrier at $a > 0$. The process starts from $x = 0$ at time 0. As soon as hitting the barrier a for the first time, the process is stuck there forever.

We can find the probability distribution of $X(t)$ at time t by solving the diffusion equation (A.138) of Wiener process with suitable boundary conditions. Besides the initial condition $f(y, t) = \delta(y)$, the absorbing barrier implies that

$$f(a, t) = 0 \qquad t > 0. \tag{A.140}$$

The explicit form of the solution is as follows (see, e.g., [60, Ch. 5]):

$$f(y, t) = \frac{1}{\sigma \sqrt{2\pi t}} \left[\exp\left(-\frac{(y - \mu t)^2}{2\sigma^2 t} \right) - \exp\left(\frac{2\mu a}{\sigma^2} - \frac{(y - 2a - \mu t)^2}{2\sigma^2 t} \right) \right] \tag{A.141}$$

Note that the event $X(t) < a$, given that $X(0) = 0$, implies that the time to absorption T be greater than t, i.e., if the process is still located to the left of the absorbing barrier, absorption cannot have occurred yet. Formally

$$G_T(t) = \mathcal{P}(T > t) = \int_{-\infty}^{a} f(y, t)\, dy \tag{A.142}$$

Integrating the expression of $f(y, t)$ in eq. (A.141), we have:

$$F_T(t) = 1 - G_T(t) = 1 - \Phi\left(\frac{a - \mu t}{\sigma\sqrt{t}}\right) + \exp\left(\frac{2\mu a}{\sigma^2}\right)\Phi\left(\frac{-a - \mu t}{\sigma\sqrt{t}}\right)$$

(A.143)

where $\Phi(z) = \frac{1}{\sqrt{2\pi}}\int_{-\infty}^{z} e^{-u^2/2}\,du$ is the CDF of the standard Gaussian random variable. A closed form expression can be deduced for the PDF of T, as well as for its Laplace transform, in case $\mu \geq 0$. Namely, it is

$$f_T(t) = \frac{a}{\sigma\sqrt{2\pi t^3}}\exp\left(-\frac{(a - \mu t)^2}{2\sigma^2 t}\right), \quad t > 0,$$

(A.144)

and

$$\varphi_T(s) = \exp\left(\frac{a}{\sigma^2}[\mu - \sqrt{\mu^2 + 2s\sigma^2}]\right)$$

(A.145)

More interestingly, we can observe that, for $\mu \geq 0$, the limit of $F_T(t) = P(T \leq t)$ for $t \to \infty$ is 1, i.e., absorption is certain. In that case the random variable has a proper PDF, summing up to 1, even though it can be verified that the mean absorption time is finite only if $\mu > 0$. In that case, it is $E[T] = a/\mu$ and $\sigma_T^2 = a\sigma^2/(2\mu^3)$.

If instead it is $\mu < 0$, the process can drift away from the barrier and never touch it. Taking the limit for $t \to \infty$, we find $P(T < \infty) = \exp(-2|\mu|a/\sigma^2)$. This is the absorption probability of Wiener process, starting from $X(0) = 0$. For a general starting point x_0, to the left of the absorbing barrier a, the absorption probability is $\exp(-2|\mu|(a - x_0)/\sigma^2)$.

The time to absorption can also be interpreted as the first passage time of $X(t)$ by the point located at a.

Moreover, the event $\max_{0 \leq \tau \leq t} X(\tau) < a$ is equivalent to the event $T > t$. As a matter of fact, if the process position up to time t has always been to the left of a, it cannot have touched a, i.e., the first passage time by a must be greater than t. Let us define the random variable $Y(t) = \max_{0 \leq \tau \leq t} X(\tau)$ for a given t. The argument above shows that $F_{Y(t)}(a) = P(Y(t) \leq a) = P(T > t) = 1 - F_T(t)$, where $F_T(t)$ is given in eq. (A.143). In case of negative drift, letting $t \to \infty$, we find the CDF of the random variable $Y = \max_{\tau \geq 0} X(\tau)$, i.e., $F_Y(a) = 1 - \exp(-2|\mu|a/\sigma^2)$. We can summarize this result by stating that the supremum of a Wiener process with negative drift, starting at $X(0) = 0$, is a negative exponential random variable with mean $\sigma^2/(2|\mu|)$.

A.7.2 Wiener Process with a Reflecting Barrier

Let us again consider a Wiener process starting at $X(0) = x_0$ and having a reflecting barrier at $a < x_0$. A reflecting barrier is defined as follows. Whenever the process exceeds a in the negative direction, the state is held at a, until it jumps again in the region $x > a$.

It can be shown that the boundary condition for a reflecting barrier at a is

$$\left[\frac{1}{2}\sigma^2 \frac{\partial f}{\partial y} - \mu f(y, t) \right]_{y=a} = 0 \tag{A.146}$$

Consider a Wiener diffusion process starting at x_0, evolving over the x-axis, with a reflecting barrier at $x = a < x_0$. Due to the barrier, the process is confined to the interval (a, ∞). hence, it must be:

$$\int_a^\infty f(x, t)\, dx = 1, \qquad \forall t \geq 0. \tag{A.147}$$

Deriving with respect to time, we find

$$\frac{\partial}{\partial t} \int_a^\infty f(x, t)\, dx = \int_a^\infty \frac{\partial f}{\partial t}\, dx = 0, \qquad \forall t \geq 0. \tag{A.148}$$

Using the forward diffusion equation, we have

$$0 = \int_a^\infty \frac{\partial f}{\partial t}\, dx = \int_a^\infty \left(\frac{1}{2}\sigma^2 \frac{\partial^2 f}{\partial x^2} - \mu \frac{\partial f}{\partial x} \right) dx$$

$$= \int_a^\infty \frac{\partial}{\partial x} \left(\frac{1}{2}\sigma^2 \frac{\partial f}{\partial x} - \mu f(x, t) \right) dx = - \left[\frac{1}{2}\sigma^2 \frac{\partial f}{\partial x} - \mu f(x, t) \right]_{x=a}$$

Solving the diffusion equation for a reflecting barrier at $a = 0$ with this boundary condition, along with the initial condition $f(y, 0) = \delta(y - x_0)$, $x_0 > 0$, it is possible to see that [60, Ch. 5]:

$$f(y, t) = \frac{1}{\sigma\sqrt{2\pi t}} \left[\exp\left(-\frac{(y - x_0 - \mu t)^2}{2\sigma^2 t} \right) + \exp\left(-\frac{4x_0\mu t + (y + x_0 - \mu t)^2}{2\sigma^2 t} \right) \right]$$

$$- \frac{2\mu}{\sigma^2} \exp\left(\frac{2\mu y}{\sigma^2} \right) \left[1 - \Phi\left(\frac{y + x_0 + \mu t}{\sigma\sqrt{t}} \right) \right] \tag{A.149}$$

It is intuitive that the process $X(t)$ drifts away without reaching an equilibrium, if the drift is positive or null. If instead the drift is negative, the process tends to move toward the origin, where it bounces back, due to reflecting barrier.

By exploiting the mathematical expression of $f(y, t)$ in (A.149), we can find that there exists a limit for $t \to \infty$ in case $\mu < 0$:

$$f(y, \infty) = \frac{2|\mu|}{\sigma^2} \exp\left(-\frac{2|\mu|y}{\sigma^2} \right) \qquad y \geq 0. \tag{A.150}$$

This is a negative exponential PDF with mean $\frac{\sigma^2}{2|\mu|}$. This is the stationary probability distribution of the so called reflected Brownian motion, a Wiener process living on the positive real line, with a reflecting barrier at 0, negative drfit μ and variance coefficient σ^2.

For an in-depth analysis of the Brownian motion the reader can refer to the monograph by Harrison [99].

References

1 Abate, J. and Whitt, W. (1992). The Fourier-series method for inverting transforms of probability distributions. *Queueing Systems* 10 (1): 5–87.

2 Abate, J. and Whitt, W. (1995). Numerical inversion of Laplace transforms of probability distributions. *ORSA Journal on Computing* 7 (1): 36–43.

3 Abate, J. and Whitt, W. (2006). A unified framework for numerically inverting Laplace transforms. *INFORMS Journal on Computing* 18 (4): 408–421.

4 Abramson, N. (1970). The ALOHA system: another alternative for computer communications. In: *Proceedings of the fall joint computer conference (AFIPS '70 - Fall)* (17–19 November 1970), 281–285. New York, NY: ACM.

5 Adams, R. (2013). Active queue management: a survey. *IEEE Communications Surveys & Tutorials* 15 (3): 1425–1476.

6 Afanasyev, A., Tilley, N., Reiher, P., and Kleinrock, L. (2010). Host-to-host congestion control for TCP. *IEEE Communications Surveys & Tutorials* 12 (3): 304–342.

7 Afaqui, M.S., Garcia-Villegas, E., and Lopez-Aguilera, E. (2017). IEEE 802.11ax: challenges and requirements for future high efficiency WiFi. *IEEE Wireless Communications* 24 (3): 130–137.

8 Ali, M.S., Hossain, E., and Kim, D.I. (2017). LTE/LTE-A Random Access for Massive Machine-Type Communications in Smart Cities. *IEEE Communications Magazine* 55 (1): 76–83.

9 Alizadeh, M., Greenberg, A., Maltz, D.A. et al. (2010). Data center TCP (DCTCP). *SIGCOMM Computer Communication Review* 40 (4): 63–74.

10 Alizadeh, M., Javanmard, A., and Prabhakar, B. (2011). Analysis of DCTCP: stability, convergence, and fairness. In: *Proceedings of the ACM SIGMETRICS Joint International Conference on Measurement and Modeling of Computer Systems (SIGMETRICS '11)* (7–11 June 2011), 73–84. San Jose, CA: ACM.

11 Allman, M., Paxson, V., and Stevens, W. (April 1999). TCP congestion control, IETF RFC 2581.

Network Traffic Engineering: Stochastic Models and Applications, First Edition. Andrea Baiocchi.
© 2020 John Wiley & Sons, Inc. Published 2020 by John Wiley & Sons, Inc.

12 Anick, D., Mitra, D., and Sondhi, M.M. (1982). Stochastic theory of a data-handling system with multiple sources. *The Bell System Technical Journal* 61 (8): 1871–1894.

13 Asadi, A. and Mancuso, V. (2013). A survey on opportunistic scheduling in wireless communications. *IEEE Communications Surveys & Tutorials* 15 (4): 1671–1688.

14 Asmussen, S. (2003). *Applied Probability and Queues*, 2e. New York, NY: Springer.

15 Baiocchi, A. (1992). Asymptotic behaviour of the loss probability of the M/G/1/K and G/M/1/K queues. *Queueing Systems* 10 (3): 235–248.

16 Baiocchi, A. (1993). Accurate formulae for the loss probability of a large class of queueing systems. *Performance Evaluation* 18 (2): 125–132.

17 Baiocchi, A. and Blefari-Melazzi, N. (1993). An error-controlled approximate analysis of a stochastic fluid flow model applied to an ATM multiplexer with heterogeneous on-off sources. *IEEE/ACM Transactions on Networking* 1 (6): 628–637.

18 Baiocchi, A. (1994). Analysis of the loss probability of the MAP/G/1/K queue. Part I: asymptotic theory. *Stochastic Models (Marcel Dekker Ed.)* 10 (4): 867–893.

19 Baiocchi, A. and Ricciato, F. (2018). Analysis of pure and slotted ALOHA with multi-packet reception and variable packet size. *IEEE Communications Letters* 22 (7): 1482–1485.

20 Baiocchi, A., Tinnirello, I., Garlisi, D., and Lo Valvo, A. (2017). Random access with repeated contentions for emerging wireless technologies. In: *IEEE INFOCOM 2017 - IEEE Conference on Computer Communications*, (1–4 May), 1–9. Atlanta, GA: IEEE.

21 Baiocchi, A., Garlisi, D., Lo Valvo, A. et al. (2020). 'Good to repeat': Making random access near-optimal with repeated contentions. *IEEE Transactions on Wireless Communications* 19 (1): 712–726.

22 Balakrishnan, H., Padmanabhan, V.N., Seshan, S., and Katz, R.H. (1997). A comparison of mechanisms for improving TCP performance over wireless links. *IEEE/ACM Transactions on Networking* 5 (6): 756–769.

23 Barabási, A.L. (2005). The origin of bursts and heavy tails in human dynamics. *Nature* 435: 207–211.

24 Barabási, A.L. (2016). *Network Science*. Cambridge, UK: Cambridge University Press.

25 Baskett, F., Chandy, K.M., Muntz, R.R., and Palacios, F. (1975). Open, closed and mixed network of queues with different classes of customers. *Journal of the ACM* 22: 248–260.

26 Bellini, S. and Borgonovo, F. (1980). On the throughput of an ALOHA Channel with variable length packets. *IEEE Transactions on Communications* 28 (11): 1932–1935.

27 Beneš, V.E. (1956). On queues with Poisson arrivals. *Annals of Mathematical Statistics* 28: 670–677.

28 Bengio, Y. (2009). Learning deep architectures for AI. *Foundations and Trends in Machine Learning* 2 (1): 1–127.

29 Bensaou, B., Tsang, D.H.K., and Chan, K.T. (2001). Credit-based fair queueing (CBFQ): a simple service-scheduling algorithm for packet-switched networks. *IEEE/ACM Transactions on Networking* 9 (5): 591–604.

30 Bertsekas, D. and Gallager, R. (1992). *Data Network*, 2e. Englewood Cliffs, NJ: Prentice Hall.

31 Bharadia, D., McMilin, E., and Katti, S. (2013). Full duplex radios. In: *Proceedings of the ACM SIGCOMM 2013 Conference on SIGCOMM (SIGCOMM'13)*, 375–386. New York, NY: ACM.

32 Bianchi, G. (1998). IEEE 802.11 Saturation throughput analysis. *IEEE Communications Letters* 2 (12): 318–320.

33 Bianchi, G. (2000). Performance analysis of the IEEE 802.11 distributed function. *IEEE Journal of Selected Areas in Communications* 18 (3): 535–547.

34 Bianchi, G. and Tinnirello, I. (2003). Kalman filter estimation of the number of competing terminals in an IEEE 802.11 network. In: *IEEE INFOCOM'03* (30 March– 3 April 2003), 844–852. San Francisco, CA: IEEE.

35 Jiang, L.B. and Liew, S.C. (2005). Proportional fairness in wireless LANs and ad hoc networks. In: *IEEE Wireless Communications and Networking Conference, 2005*, vol. 3, 1551–1556. New Orleans, LA: IEEE.

36 Blanton, E., Allman, M., Fall, K., and Wang, L. (April 2003). A conservative selective acknowledgment (SACK)-based loss recovery algorithm for TCP, RFC 3517.

37 Bolch, G., Greiner, S., de Meer, H., and Trivedi, K.S. (2006). *Queueing Networks and Markov Chains—Modeling and Performance Evaluation with Computer Science Applications*, 2e. Wiley.

38 Bordenave, C., McDonald, D., and Proutiere, A. (2007). Random multi-access algorithms in networks with partial interaction: a mean field analysis. In: *Managing Traffic Performance in Converged Networks. ITC 2007*, Lecture Notes in Computer Science, vol. 4516 (eds. L. Mason, T. Drwiega and J. Yan), 779–790. Berlin, Heidelberg: Springer.

39 Bordenave, C., McDonald, D., and Proutiere, A. (2009). A particle system in interaction with a rapidly varying environment: Mean field limits and applications. arXiv:math/0701363v3 [math.PR], last revised 16 February 2009.

40 Borovkov, A.A. (1976). *Stochastic Processes in Queueing Theory*. New York, NY: Springer-Verlag.

41 Boyd, S. and Vandenberghe, L. (2004). *Convex Optimization*. Cambridge University Press.

42 Braess, D. (1968). Uber ein paradoxen der verkehrsplanung. *Unternehmensforschung* 12: 258–268.

43 Brakmo, L.S. and Peterson, L.L. (1995). TCP Vegas: end to end congestion avoidance on a global Internet. *IEEE Journal on Selected Areas in Communications* 13 (8): 1465–1480.

44 Bramson, M. (2008). Stability of queueing networks. *Probability Surveys* 5: 169–345.

45 Briscoe, B. (2007). Flow rate fairness: dismantling a religion. *SIGCOMM Computer Communication Review* 37 (2): 63–74.

46 Brown, L.D. and Zhao, L.H. (2002). A test for the Poisson distribution. *The Indian Journal of Statistics, Series A* 64 (3, Part 1): 611–625.

47 Brumelle, S.L. (1971). Some inequalities for parallel-server queues. *Operations Research* 19: 402–413.

48 Burke, P.J. (1956). The output of a queueing system. *Operations Research* 4 (6): 699–704.

49 Cardwell, N., Cheng, Y., Stephen Gunn, C. et al. (2016). BBR: congestion-based congestion control. *Queue* 14 (5): 50–83.

50 Cardwell, N., Cheng, Y., Stephen Gunn, C. et al. (2017). BBR: congestion-based congestion control. *Communications of the ACM* 60 (2): 58–66.

51 Cardwell, N., Cheng, Y., Hassas Yeganeh, S., and Jacobson, V. (July 2017). BBR Congestion Control. draftcardwell- iccrg-bbr-congestion-control-00, Internet Congestion Control Research Group, Internet-Draft.

52 Chang, C.-S. (2000). *Performance Guarantees in Communications Networks*. Springer.

53 Chao, X., Miyazawa, M., and Pinedo, M. (1999). *Queueing Networks: Customers, Signals and Product Form Solutions*. New York, NY: Wiley.

54 Chen, H. and Yao, D.D. (2001). *Fundamentals of Queueing Networks*. New York, NY: Springer-Verlag.

55 Chen, H. and Ye, H.-Q. (2012). Asymptotic optimality of balanced routing. *Operations Research* 60 (1): 163–179.

56 Chen, W., Ren, F., Xie, J. et al. (2015). Comprehensive understanding of TCP incast problem. In: *IEEE Conference on Computer Communications (INFOCOM)*, Hong Kong, China (26 April–1 May 2015), 1688–1696.

57 Cheng, Y., Cardwell, N., Hassas Yeganeh, S., and Jacobson, V. (2017). Delivery Rate Estimation. draft-cheng-iccrg-delivery-rate-estimation-00, Internet Congestion Control Research Group, Internet-Draft.

58 Chiang, M. (2012). *Networked Life: 20 Questions and Answers*. Cambridge, UK: Cambridge University Press.

59 Cox, D.R. (1962). *Renewal Theory*. London: Methuen.

60 Cox, D.R. and Miller, H.D. (1965). *Stochastic Processes*. London: Chapman and Hall.

61 Cruz, R.L. (1991). A calculus for network delay. Part I: network elements in isolation. *IEEE Transactions on Information Theory* 37 (1): 114–131.

62 Cruz, R.L. (1991). A calculus for network delay. Part II: network analysis. *IEEE Transactions on Information Theory* 37 (1): 132–141.

63 Dai, J.G. (1999). *Stability of Fluid and Stochastic Processing Networks*, Miscellanea Publication, vol. 9. Denmark: Center for Mathematical Physics and Stochastics (http:/www.maphysto.dk/).

64 Davin, J.R. and Heybey, A.T. (1990). A simulation study of fair queueing and policy enforcement. *ACM SIGCOMM Computer Communication Review* 20 (5): 23–29.

65 De Cicco, L., Carlucci, G., and Mascolo, S. (2013). Experimental investigation of the google congestion control for real-time flows. In: *Proceedings of the 2013 ACM SIGCOMM Workshop on Future Human-Centric Multimedia Networking (FhMN '13)* (12–16 August 2013), 21–26. Hong Kong, China: ACM.

66 Demers, A., Keshav, S., and Shenker, S. (1989). Analysis and simulation of fair queueing algorithm. In: *Proceedings of ACM SIGCOMM '89* (19–22 September 1989), 21–26. Austin, TX: ACM.

67 Deng, L. and Yu, D. (2013). Deep learning: methods and applications. *Foundations and Trends in Machine Learning* 7 (3–4): 197–387.

68 Deng, D., Lin, Y.P., Yang, X. et al. (2017). IEEE 802.11ax: highly efficient WLANs for intelligent information infrastructure. *IEEE Communications Magazine* (December) 55 (12): 52–59.

69 De Smit, J.H.A. (1973). Some general results for many server queues. *Advances in Applied Probability* 5: 153–169.

70 Elwalid, A.I. and Mitra, D. (1993). Effective bandwidth of general Markovian traffic sources and admission control of high speed networks. In: *Proceedings of IEEE INFOCOM'93*, vol. 1, 256–265. San Francisco, CA: IEEE.

71 Elwalid, A., Heyman, D., Lakshman, T.V. et al. (1995). Fundamental bounds and approximations for ATM multiplexers with applications to video teleconferencing. *IEEE Journal on Selected Areas in Communications* 13 (6): 1004–1016.

72 Elwalid, A. and Mitra, D. (1995). Analysis, approximations and admission control of a multi-service multiplexing system with priorities. In: *Proceedings of IEEE INFOCOM'95*, vol. 2, 463–472. Boston, MA: IEEE.

73 Erlang, A.K. (1917). Solution of some problems in the theory of probabilities of significance in automatic telephone exchanges. *Elektrotkeknikeren* 13.

74 ETSI. 2014. ETSI standard EN 302 637-2 V1. 3.1-Intelligent Transport Systems (ITS); Vehicular Communications; Basic Set of Applications; Part 2: Specification of Cooperative Awareness Basic Service. ETSI, May.

75 Fawzi, A., Fawzi, O., and Frossard, P. (2018). Analysis of classifiers' robustness to adversarial perturbations. *Machine Learning* 107 (3): 481–508.

76 Feller, W. (1968). *An Introduction to Probability Theory and Its Applications*, vol. 1. New York, NY: Wiley.

77 Fidler, M. and Rizk, A. (2015). A guide to the stochastic network calculus. *IEEE Communications Surveys & Tutorials* 17 (1): 92–105.

78 Fidler, M. (2010). Survey of deterministic and stochastic service curve models in the network calculus. *IEEE Communications Surveys & Tutorials* 12, 1: 59–86.

79 Floyd, S., Handley, M., and Padhye, J. (May 2000). A comparison of equation-based and AIMD congestion control. http://www.aciri.org/tfrc/.

80 Foschini, G. and Salz, J. (1978). A basic dynamic routing problem and diffusion. *IEEE Transactions on Communications* 26 (3): 320–327.

81 Fratta, L., Gerla, M., and Kleinrock, L. (1973). The flow deviation method: An approach to store-and-forward communication network design. *Networks* 3 (2): 97–133.

82 Fratta, L., Gerla, M., and Kleinrock, L. (2014). Flow deviation: 40 years of incremental flows for packets, waves, cars, and tunnels. *Computer Networks* 66: 18–31.

83 Fredericks, A.A. (1980). Congestion in blocking systems. A simple approximation technique. *Bell System Technical Journal* 59: 805–827.

84 Gail, R. and Kleinrock, L. (1981). An invariant property of computer network power. In: *Proceedings of the International Conference on Communications (ICC)*, 63.1.1–63.1.5.

85 Gallager, R.G. (2014). *Stochastic Processes: Theory for Applications*. Cambridge University Press.

86 Gautam, N. (2012). *Analysis of Queues—Methods and Applications*. Boca Raton, FL: CRC Press.

87 Georgii, H.-O. (2008). *Stochastics: Introduction to Probability and Statistics*. Berlin, Germany: Walter de Gruyter.

88 Gettys, J. and Nichols, K. (2011). Bufferbloat: dark buffers in the internet. *Queue* 9 (11): 40–54.

89 Gibbens, R.J., Kelly, F.P., and Key, P.B. (1995). Dynamic alternative routing. In: *Routing in Communications Networks* (ed. M. Steenstrup), 13–47. Englewood Cliffs, NJ: Prentice Hall.

90 Gnedenko, B.V. (1997). *Theory of Probability*, 6e. Amsterdam: Gordon and Breach.

91 Golestani, S.J. (1994). A self-clocked fair queueing scheme for broadband applications. In: *Proceedings of the IEEE INFOCOM '94. Networking for Global Communications* (12–16 June), vol. 2, 636–646. Toronto, Ontario: IEEE.

92 Gordon, W.J. and Newell, G.F. (1967). Closed queueing systems with exponential servers. *Operations Research* 15: 145–155.

93 Grazia, C.A., Patriciello, N., Klapez, M., and Casoni, M. (2017). A cross-comparison between TCP and AQM algorithms: which is the best couple for congestion control? In: *14th International Conference on Telecommunications (ConTEL)* (June 2017), 75–82. Zagreb, Croatia: IEEE.

94 Gross, D., Shortle, J.F., Thompson, J.M., and Harris, C.M. (2008). *Fundamentals of Queueing Theory*, 4e. Hoboken, NJ: Wiley.

95 Grunenberger, Y., Heusse, M., Rousseau, F., and Duda, A. (2007). Experience with an implementation of the Idle Sense wireless access method. In: *Proceedings of the 2007 ACM CoNEXT Conference (CoNEXT '07)*, 12. New York, NY, Article 24: ACM.

96 Ha, S., Rhee, I., and Xu, L. (2008). CUBIC: a new TCP-friendly high-speed TCP variant. *ACM SIGOPS Operating Systems Review* 42 (5): 64–74.

97 Haenggi, M. (2013). *Stochastic Geometry for Wireless Networks*. Cambridge, UK: Cambridge University Press.

98 Harchol-Balter, M. (2013). *Performance Modeling and Design of Computer Systems—Queuing Theory in Action*. Cambridge, UK: Cambridge University Press.

99 Harrison, J.M. (1985). *Brownian Motion and Stochastic Flow Systems*. New York, NY: Wiley.

100 Heidergott, B., Olsder, G.J., and van der Woude, J. (2006). *Max Plus at Work—Modeling and Analysis of Synchronized Systems: A Course on Max-Plus Algebra and Its Applications*. Princeton, NJ: Princeton University Press.

101 Heusse, M., Rousseau, F., Berger-Sabbatel, G., and Duda, A. (2003). Performance anomaly of 802.11b. In: *IEEE INFOCOM 2003. Twenty-second Annual Joint Conference of the IEEE Computer and Communications Societies (IEEE Cat. No.03CH37428)*, vol. 2, 836–843. San Francisco, CA: IEEE.

102 Heusse, M., Rousseau, F., Guillier, R., and Duda, A. (2005). Idle Sense: an optimal access method for high throughput and fairness in rate diverse wireless LANs. *SIGCOMM Computer Communication Review* 35 (4): 121–132.

103 Hock, M., Bless, R., and Zitterbart, M. (2017). Experimental evaluation of BBR congestion control. In: *IEEE 25th International Conference on Network Protocols (ICNP)*, 1–10. Toronto, ON: IEEE.

104 Høiland-Jørgensen, T., Kazior, M., Täht, D. et al. (2017). Ending the anomaly: achieving low latency and airtime fairness in WiFi. In: *Proceedings of 2017 – USENIX Annual Technical Conference* (12–14 July 2017), 139–151. Santa Clara, CA: USENIX ATC 17.

105 Hu, J., Di, B., Liao, Y. et al. (2018). Hybrid MAC protocol design and optimization for full duplex Wi-Fi networks. *IEEE Transactions on Wireless Communications* 17 (6): 3615–3630.

106 Hwang, J., Yoo, J., and Choi, N. (2012). IA-TCP: a rate-based incast-avoidance algorithm for TCP in data center networks. In: *2012 IEEE International Conference on Communications*, 1292–1296. Ottawa, ON: ICC.

107 IEEE Computer Society LAN/MAN Standards Committee. 2007. IEEE Standard for Information technology – Telecommunications and information exchange between systems. Local and metropolitan area networks. Specific requirements Part 11: Wireless LAN Medium Access Control (MAC) and Physical Layer (PHY) Specifications. IEEE Std 802.11.

108 IEEE Computer Society LAN/MAN Standards Committee. 2013. IEEE Standard for Information technology—Telecommunications and information exchange between systems Local and metropolitan area networks. Part 11: Wireless LAN Medium Access Control (MAC) and Physical Layer (PHY) Specifications. Amendment 4: Enhancements for Very High Throughput for Operation in Bands below 6 GHz. IEEE Std. 802.11ac-2013.

109 Ivanov, M., Brännström, F., Graell i Amat, A., and Popovski, P. (2017). Broadcast coded slotted ALOHA: A finite frame length analysis. *IEEE Transactions on Communications* 65 (2): 651–662.

110 Iyengar, J. and Thomson, M. (April 2019). QUIC: A UDP-Based Multiplexed and Secure Transport. IETF Internet-Draft, draft-ietf-quic-transport-20.

111 Jackson, J.R. (1957). Networks of waiting lines. *Operations Research* 5: 518–521.

112 Jacobson, V. (1988). Congestion avoidance and control. *ACM SIGCOMM Computer Communications Review* 18 (4): 314–329.

113 Jain, R., Chiu, D., and Hawe, W. (September 1998). A quantitative measure of fairness and discrimination for resource allocation in shared computer systems. arXiv:cs/9809099 [cs.NI], https://arxiv.org/abs/cs/9809099.

114 Jayawickrama, B.A., He, Y., Dutkiewicz, E., and Mueck, M.D. (2018). Scalable spectrum access system for massive machine type communication. *IEEE Network* 32 (3): 154–160.

115 Jiang, L. and Walrand, J. (2010). A distributed CSMA algorithm for throughput and utility maximization in wireless networks. *IEEE/ACM Transactions on Networking* 18 (3): 960–972.

116 Kadota, I., Baiocchi, A., and Anzaloni, A. (2014). Kalman filtering: estimate of the numbers of active queues in an 802.11e EDCA WLAN. *Computer Communications* 39: 54–64.

117 Karlin, S. (1966). *A First Course in Stochastic Processes*. New York, NY/London: Academic Press.

118 Kaufman, J.S. (1981). Blocking in a Shared Resource Environment. *IEEE Transactions on Communications* 29 (10): 1474–1481.

119 Kaul, S., Yates, R., and Gruteser, M. (2012). Real-time status: how often should one update? In: *Proceedings of IEEE INFOCOM* (25–30 March), 2731–2735. Orlando, FL: IEEE.

120 Kelly, F.P. and Laws, C.N. (1993). Dynamic routing in open queueing networks: Brownian models, cut constraints and resource pooling. *Queueing Systems* 13 (1–3): 47–86.

121 Kelly, F.P. (2011). *Reversibility and Stochastic Networks*. Cambridge, UK: Cambridge University Press.

122 Kelly, F. and Yudovina, E. (2014). *Stochastic Networks*. Cambridge, UK: Cambridge University Press.

123 Kesidis, G., Walrand, J., and Chang, C. (1993). Effective bandwidths for multiclass Markov fluids and other ATM sources. *IEEE/ACM Transactions on Networking* 1 (4): 424–428.

124 Khorov, E., Kiryanov, A., Lyakhov, A., and Bianchi, G. (2019). A tutorial on IEEE 802.11ax high efficiency WLANs. *IEEE Communications Surveys & Tutorials* 21 (1): 197–216.

125 Kim, D., Lee, H., and Hong, D. (2015). A survey of in-band full-duplex transmission: from the perspective of PHY and MAC layers. *IEEE Communications Surveys & Tutorials* 17 (4): 2017–2046.

126 Kim, B.H. and Calin, D. (2017). On the split-TCP performance over real 4G LTE and 3G wireless networks. *IEEE Communications Magazine* 55 (4): 124–131.

127 Kingman, J.F.C. (1962). On queues in heavy traffic. *Journal of the Royal Statistical Society. Series B (Methodological)* 24 (2): 383–392.

128 Kleinrock, L. and Tobagi, F. (1975). Packet switching in radio channels: Part I – Carrier sense multiple-access modes and their throughput-delay characteristics. *IEEE Transactions on Communications* 23 (12): 1400–1416.

129 Kleinrock, L. and Tobagi, F. (1977). Packet switching in radio channels: Part IV – Stability considerations and dynamic control in carrier sense multiple access. *IEEE Transactions on Communications* 25 (10): 1103–1119.

130 Kleinrock, L. (1975). *Queueing Systems*, Theory, vol. 1. New York, NY: Wiley.

131 Kleinrock, L. (1976). *Queueing Systems*, Computer Applications, vol. 2. New York, NY: Wiley.

132 Kleinrock, L. (1979). Power and deterministic rules of thumb for probabilistic problems in computer communications. In: *Proceedings of the International Conference on Communications (ICC)* (10–14 June 1979), 43.1.1–43.1.10. Boston, MA: ICC.

133 Korilis, Y.A., Lazar, A.A., and Orda, A. (1999). Avoiding the Braess paradox in non-cooperative networks. *Journal of Applied Probability* 36 (1): 211–222.

134 Kulkarni, V.G. (1996). Effective bandwidth for Markov regenerative sources. *Queueing Systems* 24: 137–153.

135 Kumar, A., Altman, E., Miorandi, D., and Goyal, M. (2007). New insights from a fixed-point analysis of single cell IEEE 802.11 WLANs. *IEEE/ACM Transactions on Networking* 15 (3): 588–601.

136 Kumar, P.R. and Seidman, T.I. (1990). Dynamic instabilities and stabilization methods in distributed real-time scheduling of manufacturing systems. *IEEE Transactions on Automatic Control* 35 (3): 289–298.

137 Kumar, A., Manjunath, D., and Kuri, J. (2004). *Communication Networking: An Analytical Approach*. Morgan Kaufmann Publishers (Elsevier).

138 Kurtz, T.G. (1978). Strong approximation theorems for density dependent Markov chains. *Stochastic Processes and Their Applications* 6 (3): 223–240.

139 Lakatos, L., Szeidl, L., and Telek, M. (2013). *Introduction to Queueing Systems with Telecommunication Applications*. Springer.

140 Latouche, G. and Ramaswami, Y. (1994). A logarithmic reduction algorithm for Quasi Birth and Death processes. *Journal of Applied Probability* 30: 650–674.

141 Lau, V.K.N. (2005). Proportional fair space-time scheduling for wireless communications. *IEEE Transactions on Communications* 53 (8): 1353–1360.

142 Laufer, R. and Kleinrock, L. (2016). The capacity of wireless CSMA/CA networks. *IEEE/ACM Transactions on Networking* 24 (3): 1518–1532.

143 Lázaro, F. and Stefanovic, C. (2017). Finite-length analysis of frameless ALOHA with multi-user detection. *IEEE Communications Letters* 21 (4): 769–772.

144 Le Boudec, J.-Y. and Thiran, P. (2001). *Network Calculus: A Theory of Deterministic Queuing Systems for the Internet*. Springer, LNCS.

145 Le Boudec, J.-Y. (2010). *Performance Evaluation of Computer Communication Systems*. Lausanne, Switzerland: EPFL Press.

146 Lee, J., Lee, H., Yi, Y. et al. (2016). Making 802.11 DCF near- optimal: design, implementation, and evaluation. *IEEE/ACM Transactions on Networking* 24 (3): 1745–1758.

147 Liew, S.C., Kai, C.H., Leung, H.C., and Wong, P. (2010). Back-of-the-envelope computation of throughput distributions in CSMA wireless networks. *IEEE Transactions on Mobile Computing* 9 (9): 1319–1331.

148 Little, J.D.C. (1961). A proof of the queueing formula $L = \lambda W$. *Operations Research* 9: 383–387.

149 Liu, L. and Shi, D.-H. (1996). Busy period in $GI^X/G/1$. *Journal of Applied Probability* 33 (3): 815–829.

150 Liu, J., Yi, Y., Proutiere, A. et al. (2010). Towards utility–optimal random access without message passing. *Wireless Communications and Mobile Computing* 10 (1): 115–128.

151 Liu, K. and Lee, J.Y. (2016). On improving TCP performance over mobile data networks. *IEEE Transactions on Mobile Computing* 15 (10): 2522–2536.

152 LoRa Alliance. (2016). LoRaWAN™ Specification. Version: V1.0.2.

153 Luchak, G. (1958). The continuous-time solution of the equations of the single-channel queue with a general class of service-time distribution by the method of generating functions. *Journal of the Royal Statistical Society. Series B (Methodological)* 20 (1): 176–181.

154 Madueño, G.C., Nielsen, J.J., Kim, D.M. et al. (2016). Assessment of LTE wireless access for monitoring of energy distribution in the smart grid. *IEEE Journal on Selected Areas in Communications* 34 (3): 675–688.

155 Maity, M., Raman, B., and Vutukuru, M. (2017). TCP download performance in dense WiFi scenarios: analysis and solution. *IEEE Transactions on Mobile Computing* 16 (1): 213–227.

156 Magistretti, E., Gurewitz, O., and Knightly, E.W. (2014). 802.11ec: Collision avoidance without control messages. *IEEE/ACM Transactions on Networking* 22 (6): 1845–1858.

157 Maguluri, S.T., Srikant, R., and Ying, L. (2014). Heavy traffic optimal resource allocation algorithms for cloud computing clusters. *Performance Evaluation* 81: 20–39.

158 Markov, A.A. (1907). Extension of the limit theorems of probability theory to a sum of variables connected in a chain. The notes of the Imperial Academy of Sciences of St. Petersburg. *VIII Series, Physio-Mathematical College* 22 (9).

159 Mathis, M., Mahdavi, J., Floyd, S., and Romanow, A. (October 1996). TCP selective acknowledgment options, RFC 2018.

160 Meyn, S.P. (2009). *Control Techniques for Complex Networks*. New York, NY: Cambridge University Press.

161 Miorandi, D. and Altman, E. (2006). Connectivity in one-dimensional ad hoc networks: a queueing theoretical approach. *Wireless Networks (Springer)* 12 (5): 573–587.

162 Mitzenmacher, M. (2001). The power of two choices in randomized load balancing. *IEEE Transactions on Parallel and Distributed Systems* 12 (10): 1094–1104.

163 Miyazawa, K., Sasaki, K., Oda, N., and Yamaguchi, S. (2018). Cyclic performance fluctuation of TCP BBR. In: *IEEE 42nd Annual Computer Software and Applications Conference (COMPSAC)*, vol. 1, 811–812. IEEE.

164 Mitra, D. (1992). Asymptotically optimal design of congestion control for high speed data networks. *IEEE Transactions on Communications* 40 (2): 301–311.

165 Morgan, Y.L. (2010). Notes on DSRC & WAVE standards suite: its architecture, design, and characteristics. *IEEE Communications Surveys & Tutorials* 12 (4): 504–518.

166 Morsalin, S., Mahmud, K., and Town, G.E. (2018). Scalability of vehicular M2M communications in a 4G cellular network. *IEEE Transactions on Intelligent Transportation Systems* 19 (10): 3113–3120.

167 Neuts, M.F. (1981). *Matrix-Geometric Solutions in Stochastic Models: An Algorithmic Approach*. Baltimore, MD/London: The John Hopkins University Press.

168 Neuts, M.F. (1989). *Structured Stochastic Matrices of M/G/1 Type and Their Applications*. Boca Raton, FL: CRC Press.

169 Newman, M. (2010). *Networks: An Introduction*. New York, NY: Oxford University Press.

170 Nichols, K. and Jacobson, V. (2012). Controlling queue delay. *Queue* 10 (5): 20–34.

171 Paolini, E., Stefanovic, C., Liva, G., and Popovski, P. (2015). Coded random access: applying codes on graphs to design random access protocols. *IEEE Communications Magazine* (June) 53 (6): 144–150.

172 Papoulis, A. and Unnikrishna, S. (2002). *Probability, Random Variables and Stochastic Processes*, 4e. New York: McGraw-Hill.

173 Parekh, A.K. and Gallager, R.G. (1993). A generalized processor sharing approach to flow control in integrated services networks: the single-node case. *IEEE/ACM Transactions on Networking* 1 (3): 344–357.

174 Parekh, A.K. and Gallager, R.G. (1994). A generalized processor sharing approach to flow control: the multiple node case. *IEEE/ACM Transactions on Networking* 2 (2): 137–150.

175 Polese, M., Centenaro, M., Zanella, A., and Zorzi, M. (2016). M2M massive access in LTE: RACH performance evaluation in a Smart City scenario. In: *Proceedings of the 2016 IEEE International Conference on Communications (ICC)*, 1–6. Kuala Lumpur: IEEE.

176 Polese, M., Chiariotti, F., Bonetto, E., Rigotto, F., Zanella, A., and Zorzi, M. (2019). A Survey on Recent Advances in Transport Layer Protocols. IEEE Communications Surveys & Tutorials 21 (4): 3584–3608.

177 Ramaiyan, V., Kumar, A., and Altman, E. (2008). Fixed point analysis of single cell IEEE 802.11e WLANs: uniqueness and multistability. *IEEE Transactions on Networking* 16 (5): 1080–1093.

178 Ramakrishnan, K., Floyd, S., and Black, D. (September 2001). The addition of Explicit Congestion Notification (ECN) to IP. IETF, RFC 3168.

179 Reiser, M. and Kobayashi, H. (1975). Queueing networks with multiple closed chains: theory and computational algorithms. *IBM Journal of Research and Development* 19 (3): 283–294.

180 Ren, F. and Lin, C. (2011). Modeling and improving TCP performance over cellular link with variable bandwidth. *IEEE Transactions on Mobile Computing* 10 (8): 1057–1070.

181 Rhee, I., Xu, L., Ha, S., et al. (February 2018). CUBIC for fast long-distance networks, RFC 8312.

182 Roberts, J.W. (1994). Virtual spacing for flexible traffic control. *International Journal of Communication Systems* 7: 307–318.

183 Roberts, J. W. (1981). A service system with heterogeneous user require-ments – Application to multi-services telecommunications systems. *Performance of Data Communications Systems and Their Application*, G. Pujolle, Ed., The Netherlands: North-Holland, 1981, pp. 423–431.

184 Ross, S.M. (1996). *Stochastic Processes*, 2e. New York, NY: Wiley.

185 Rybko, A.N. and Stolyar, A.L. (1992). Ergodicity of stochastic processes describing the operation of open queueing networks. *Problems of Information Transmission (Problemy Peredachi Informatsii)* 28 (3): 199–220.

186 Rubino, G. and Sericola, B. (2014). *Markov Chains and Dependability Theory*. UK: Cambridge University Press.

187 Sardar, B. and Saha, D. (2006). A survey of TCP enhancements for last-hop wireless networks. *IEEE Communications Surveys & Tutorials* 8 (3): 20–34.

188 Serfozo, R. (1999). *Introduction to Stochastic Networks*. New York, NY: Springer-Verlag.

189 Sharma, G., Ganesh, A.J., and Key, P.B. (2006). Performance analysis of contention based medium access control protocols. In: *Proceedings of IEEE INFOCOM 2006* (23–28 April 2006). Barcelona (Spain): IEEE.

190 Shreedhar, M. and Varghese, G. (1996). Efficient fair queuing using deficit round-robin. *IEEE/ACM Transactions on Networking* 4 (3): 375–385.

191 Sim, M.S., Chung, M., Kim, D. et al. (2017). Nonlinear self-interference cancellation for full-duplex radios: from link-level and system-level performance perspectives. *IEEE Communications Magazine* 55 (9): 158–167.

192 Sommer, C. and Dressler, F. (2011). Using the right two-ray model? A measurement-based evaluation of PHY models in VANETs. 17th ACM International Conference on Mobile Computing and Networking (MobiCom). Poster Session, Las Vegas, NV (September 2011).

193 Srikant, R. and Ying, L. (2014). *Communication Networks: An Optimization, Control, and Stochastic Networks Perspective*. Cambridge University Press.

194 Stefanovic, C., Paolini, E., and Liva, G. (2018). Asymptotic performance of coded slotted ALOHA with multipacket reception. *IEEE Communications Letters* 22 (1): 105–108.

195 Stevens, W.R. and Fall, K.R. (2012). *TCP-IP Illustrated—Volume 1: The protocols*, 2e. Addison-Wesley.

196 Stewart, W.J. (2009). *Probability, Markov Chains, Queues, and Simulation: The Mathematical Basis of Performance Modeling*. Princeton, NJ: Princeton University Press.

197 Stidham, S. Jr., (1974). A last word on $L = \lambda W$. *Operations Research* 22.

198 Takács, L. (1963). Delay distributions for one line with Poisson input, general holding times, and various orders of service. *The Bell System Technical Journal* 42 (2): 487–503.

199 Takács, L. (1969). On Erlang's formula. *Annals of Mathematical Statistics* 40 (1): 71–78.

200 Takagi, H. (1986). *Analysis of Polling Systems*. Cambridge, MA: MIT Press.

201 Tello-Oquendo, L., Leyva-Mayorga, I., Pla, V. et al. (2018). Performance analysis and optimal access class barring parameter configuration in LTE-A networks with massive M2M traffic. *IEEE Transactions on Vehicular Technology* 67 (4): 3505–3520.

202 Toledo, A.L., Vercauteren, T., and Wang, X. (2006). Adaptive optimization of IEEE 802.11 DCF based on Bayesian estimation of the number of competing terminals. *IEEE Transactions on Mobile Computing* 5 (9): 1283–1296.

203 Vvedenskaya, N.D., Dobrushin, R.L.'v., and Karpelevich, F.I. (1996). Queueing system with selection of the shortest of two queues: an asymptotic approach. *Problemy Peredachi Informatsii* 32 (1): 20–34.

204 Whitt, W. (1983). The queueing network analyzer. *The Bell System Technical Journal* 62 (9): 2779–2815.

205 Whitt, W. (1984). Heavy traffic approximations for service systems with blocking. *AT&T Bell Laboratories Technical Journal* 63 (5): 708–723.

206 Whitt, W. (2002). *Stochastic Process Limits*. New York, NY: Springer-Verlag.

207 Wierman, A. (2007). Scheduling for today's computer systems: bridging theory and practice. PhD Dissertation. School of Computer Science, Carnegie Mellon University, Pittsburgh, PA.

208 Wilkinson, R.I. (1956). Theories for toll traffic engineering in the U.S.A. *Bell System Technical Journal* 35: 421–514.

209 Wolff, R.W. (1989). *Stochastic Modeling and the Theory of Queues*. Englewood Cliffs, NJ: Prentice Hall.

210 Zhan, W. and Dai, L. (2018). Massive random access of machine-to-machine communications in LTE networks: modeling and throughput optimization. *IEEE Transactions on Wireless Communications* 17 (4): 2771–2785.

211 Zhang, J., Ren, F., and Lin, C. (2013). Survey on transport control in data center networks. *IEEE Network* 27 (4): 22–26.

212 Zhang, M., Polese, M., Mezzavilla, M. et al. (2019). Will TCP work in mmWave 5G cellular networks? *IEEE Communications Magazine* 57 (1): 65–71.

213 Zhong, Z., Hamchaoui, I., Khatoun, R., and Serhrouchni, A. (2018). Performance evaluation of CQIC and TCP BBR in mobile network. In 21st Conference on Innovation in Clouds, Internet and Networks and Workshops (ICIN) (February 2018).

214 Zhou, X., Wu, F., Tan, J. et al. (2017). Designing low-complexity heavy-traffic delay-optimal load balancing schemes: theory to algorithms. *Proceedings of the ACM on Measurement and Analysis of Computing Systems* 1 (2), Article 39.

215 Zhu, X., Pan, R., Ramalho, M. et al. (2019). NADA: A Unified Congestion Control Scheme for Real-Time Media. IETF Internet Draft, draft-ietf-rmcatnada-10.

Index

Network Traffic Engineering: Stochastic Models and Applications, First Edition. Andrea Baiocchi.
© 2020 John Wiley & Sons, Inc. Published 2020 by John Wiley & Sons, Inc.

Printed and bound by CPI Group (UK) Ltd, Croydon, CR0 4YY

27/10/2024

14580269-0004